Telecommunication System Engineering

WILEY SERIES IN TELECOMMUNICATIONS AND SIGNAL PROCESSING

John G. Proakis, Editor
Northeastern University

Worldwide Telecommunications Guide for the Business Manager
Walter H. Vignault

Expert System Applications to Telecommunications
Jay Liebowitz

Business Earth Stations for Telecommunications
Walter L. Morgan and Denis Rouffet

Introduction to Communications Engineering, 2nd Edition
Robert M. Gagliardi

Satellite Communications: The First Quarter Century of Service
David W. E. Rees

Synchronization in Digital Communications, Volume 1
Heinrich Meyr and Gerd Ascheid

Digital Telephony, 2nd Edition
John Bellamy

Elements of Information Theory
Thomas M. Cover and Joy A. Thomas

Telecommunication Transmission Handbook, 3rd Edition
Roger L. Freeman

Digital Signal Estimation
Robert J. Mammone, Editor

Telecommunication Circuit Design
Patrick D. van der Puije

Meteor Burst Communications: Theory and Practice
Donald L. Schilling, Editor

Mobile Communications Design Fundamentals, 2nd Edition
William C. Y. Lee

Fundamentals of Telecommunication Networks
Tarek N. Saadawi, Mostafa Ammar, with Ahmed El Hakeem

Optical Communications, 2nd Edition
Robert M. Gagliardi and Sherman Karp

Wireless Information Networks
Kaveh Pahlavan and Allen H. Levesque

Practical Data Communications
Roger L. Freeman

Active Noise Control Systems: Algorithms and DSP Implementations
Sen M. Kuo and Dennis R. Morgan

Telecommunication System Engineering, 3rd Edition
Roger L. Freeman

Telecommunication System Engineering

Third Edition

Roger L. Freeman

A Wiley-Interscience Publication
JOHN WILEY & SONS, INC.
New York • Chichester • Brisbane • Toronto • Singapore

Marc Hammond
Somersworth NH

The texts extracted from the ITU material have been reproduced with the prior authorization of the Union as the copyright holder. The sole responsibility for selecting extracts for reproduction lies with the author alone and can in no way be attributed to the ITU. The complete volumes of the ITU material from which the texts reproduced here are extracted, can be obtained from

International Telecommunication Union
General Secretariat—Sales and Marketing Service
Place de Nations
CH-1211 GENEVA 20 (Switzerland)
Telephone: +41 22 730 51 11   Telex: 421 000 itu ch
Telegram: ITU GENEVE   Fax: +41 22 730 51 94
X.400: S=Sales; P=itu; A=Arcom; C=ch Internet: Sales@itu.ch

As referenced, some figures and tables throughout this text are copyrighted by The Institute of Electrical and Electronics Engineers, Inc. As referenced, some figures and tables from Chapters 12 and 14 are reprinted with permission from IEEE Stds. 802.2 © 1989; 802.3 © 1992; 802.4 © 1990; 802.5 © 1989, 802.6 © 1990, and 802.7 © 1989. The IEEE disclaims any responsibility or liability resulting from the placement and use in this publication. All figures are reprinted with permission.

As referenced, figures and tables in Chapters 12, 14, and 15 are reproduced with permission from and copyrighted by the American National Standards Institute. Copies of the standards referenced can be purchased from the American National Standards Institute, 11 West 42nd Street, New York, NY 10036.

As referenced, some figures and tables are copyrighted by Bellcore and are reprinted with permission.

This text is printed on acid-free paper.

Library of Congress Cataloging-in-Publication Data:
Freeman, Roger L.
   Telecommunication system engineering / Roger L. Freeman. — 3rd
ed.
      p.   cm. — (Wiley series in telecommunications and signal
processing)
   Includes index.
   ISBN 0-471-13302-7 (cloth : alk. paper)
   1. Telecommunication systems — Design and construction.
2. Telephone systems — Design and construction.   I. Title.
II. Series.
TK5103.F68   1996
621.382—dc20   96-10943
                                                   CIP

Printed in the United States of America
10 9 8 7 6 5 4 3

**To our daughters,
Rosalind and Cristina**

CONTENTS

Chapter 1   Some Basics in Conventional Telephony   **1**

1   Background and Concept   1
    1.1   Telecommunication Networks   1
2   The Simple Telephone Connection   2
3   Sources and Sinks   5
4   Telephone Networks: Introductory Terminology   6
5   Essentials of Traffic Engineering   7
    5.1   Introduction and Terminology   7
    5.2   Measurement of Telephone Traffic   9
    5.3   Blockage, Lost Calls, and Grade of Service   11
    5.4   Availability   12
    5.5   "Handling" of Lost Calls   13
    5.6   Infinite and Finite Traffic Sources   14
    5.7   Probability-Distribution Curves   14
    5.8   Smooth, Rough, and Random Traffic   16
6   Erlang and Poisson Traffic Formulas   17
    6.1   Alternative Traffic Formula Conventions   20
    6.2   Computer Programs for Traffic Calculations   31
7   Waiting Systems (Queueing)   37
    7.1   Server-pool Traffic   37
8   Dimensioning and Efficiency   39
    8.1   Alternative Routing   40
    8.2   Efficiency Versus Circuit Group Size   42
9   Bases of Network Configurations   42
    9.1   Introductory Concepts   42
    9.2   Hierarchical Networks   45
    9.3   Hierarchical Network Structures   46
    9.4   Rules for Conventional Hierarchical Networks   47
    9.5   Homing Arrangements and Interconnecting Network in
          North America   49
10 Routing Methods   51

11   Variations in Traffic Flow   52
12   Both-Way Circuits   52
13   Quality of Service   53

Chapter 2   Local Networks   **59**

1   Introduction   59
2   Subscriber Loop Design   61
     2.1   General   61
     2.2   Subscriber Loop Length Limits   62
     2.3   Quality of a Telephone Speech Connection   63
     2.4   Subscriber Loop Design Techniques   71
3   Other Limiting Conditions   77
4   Shape of a Serving Area   78
5   Exchange Location   80
6   Other Loop Design Techniques   84
     6.1   Minigauge Techniques   84
     6.2   New Loop Design Plans for US Regional Bell Operating
           Companies   86
7   Design of Local Area Trunks (Junctions)   88
8   Voice-Frequency Repeaters   89
9   Tandem Routing   90
10   Dimensioning of Trunks   93
11   Community of Interest   93

Chapter 3   Conventional Analog Switching in Telephony   **99**

1   Introduction   99
     1.1   Approach   99
     1.2   Switching in the Telephone Network   99
     1.3   Some Historical Background   100
     1.4   Two-Wire and Four-Wire Switches   101
2   Numbering: One Basis of Switching   102
3   Concentration   103
4   Basic Switching Functions   104
5   Some Introductory Switching Concepts   106
6   Types of Electromechanical Switches   108
7   Multiples and Links   109
8   Definitions: Degeneration, Availability, and Grading   110
     8.1   Degeneration   110
     8.2   Availability   110
     8.3   Grading   111

9	Conventional Electromechanical Switches	111
	9.1 Step-by-Step or Strowger Switch	111
	9.2 Crossbar Switch	113
10	System Control	114
	10.1 Introduction	114
	10.2 Progressive Control	114
	10.3 Common Control (Hard-Wired)	118
11	Stored-Program Control	124
	11.1 Introduction	124
	11.2 Basic Functions of Stored-Program Control	125
	11.3 Additional Functions in a Stored-Program-Control Exchange	125
	11.4 A Typical Stored-Program-Control Exchange: European Design	128
	11.5 Evolutionary Stored-Program-Control and Distributed Processing	129
12	Concentrators and Satellites	133
13	Call Charging: European Versus North American Approaches	134
14	Transmission Factors in Analog Switching	135
	14.1 Introduction	135
	14.2 Basic Considerations	135
	14.3 Zero Test-Level Point	136
	14.4 Transmission Specifications	136
15	Numbering Concepts for Telephony	137
	15.1 Introduction	137
	15.2 Definitions	137
	15.3 Factors Affecting Numbering	140
	15.4 In-Dialing	143
16	Telephone Traffic Measurement	144
17	Dial Service Observation	146

Chapter 4   Signaling for Analog Telephone Networks   **151**

1	Introduction	151
2	Supervisory Signaling	152
	2.1 E and M Signaling	153
3	ac Signaling	155
	3.1 General	155
	3.2 Low-Frequency ac Signaling Systems	155
	3.3 In-Band Signaling	156
	3.4 Out-of-Band Signaling	157
4	Address Signaling: Introduction	159
	4.1 Two-Frequency Pulse Signaling	159
	4.2 Multifrequency Signaling	161

5 Compelled Signaling   165
6 Link-by-Link Versus End-to-End Signaling   168
7 The Effects of Numbering on Signaling   170
8 Associated and Disassociated Channel Signaling   170
9 Signaling in the Subscriber Loop   172
   9.1 Background and Purpose   172
10 Metallic Trunk Signaling   173
   10.1 Basic Loop Signaling   173
   10.2 Reverse-Battery Signaling   174

Chapter 5   Introduction to Transmission for Telephony   181

1 General   181
2 The Three Basic Impairments to Voice Channel Transmission   182
   2.1 Attenuation Distortion   182
   2.2 Phase Distortion   183
   2.3 Noise   184
   2.4 Level   187
   2.5 Signal-to-Noise Ratio   188
3 Two-Wire and Four-Wire Transmission   189
   3.1 Two-Wire Transmission   189
   3.2 Four-Wire Transmission   189
   3.3 Operation of a Hybrid   190
4 Multiplexing   193
   4.1 General   193
   4.2 Frequency Division Multiplex   193
   4.3 Mixing   194
   4.4 CCITT Modulation Plan   195
   4.5 Loading of Multichannel Frequency-Division-Multiplex
        Systems   202
   4.6 Pilot Tones   206
   4.7 Noise and Noise Calculations in FDM Carrier Systems   206
5 Shaping of a Voice Channel and Its Meaning in Noise Units   208

Chapter 6   Long-Distance Networks   215

1 General   215
2 The Design Problem   216
3 Link Limitation   217
4 International Network   219
5 Exchange Location   220
6 Network Design Procedures   222

7 Traffic Routing in the National Network 227
  7.1 New Routing Techniques 227
  7.2 Logic of Routing 228
  7.3 Call Control Procedures 230
  7.4 Applications 231
  7.5 AT&T's Dynamic Nonhierarchical Routing (DNHR) 234

8 Network Design and Configurations from a Bellcore Perspective 237
  8.1 Introduction 237
  8.2 Basic Routing Fundamentals 239
  8.3 Application of Alternate Routing 242
  8.4 Fundamentals of Dynamic Routing Technique 242

9 Transmission Factors in Long-Distance Telephony 244
  9.1 Introduction 244
  9.2 Echo 245
  9.3 Singing 245
  9.4 Causes of Echo and Singing 245
  9.5 Transmission Design to Control Echo and Singing 247
  9.6 Introduction to Transmission–Loss Engineering 248
  9.7 Via Net Loss 253
  9.8 Loss Plan for the Evolving Digital Networks (U.S.) 259

Chapter 7 The Design of Long-Distance Links **265**

1 Introduction 265

2 The Bearer 265

3 Introduction to Radio Transmission 266

4 Design Essentials for Line-of-Sight Microwave Systems 267
  4.1 Introduction 267
  4.2 Setting Performance Requirements 269
  4.3 Site Selection and Preparation of a Path Profile 269
  4.4 Path Analysis or Link Budget 275
  4.5 Running a Path/Site Survey 286
  4.6 System Test Prior to Cutover 286
  4.7 Fades, Fading, and Fade Margins 287
  4.8 Diversity and Hot-Standby Operation 289
  4.9 LOS Microwave Repeaters 292
  4.10 Frequency Planning and Frequency Assignment 292

5 Satellite Communications 293
  5.1 Introduction 293
  5.2 Application 293
  5.3 Definition 294
  5.4 The Satellite 294
  5.5 Three Basic Technical Problems 295

5.6   Frequency Bands: Desirable and Available   296
5.7   Multiple Access of a Satellite   297
5.8   Earth Station Link Engineering   303
5.9   Digital Communications by Satellite   311
5.10 Very Small Aperture Terminal (VSAT) Networks   311

6 Fiber-Optics Communication Links   317
6.1   Application   317
6.2   Introduction to Optical Fiber as a Transmission Medium 318
6.3   Types of Optical Fiber   321
6.4   Splices and Connectors   322
6.5   Light Sources   323
6.6   Light Detectors   325
6.7   Optical-Fiber Amplifiers   327
6.8   Fiber-Optic Link Design   329

Chapter 8   Digital Transmission Systems   339

1 Digital Versus Analog Transmission   339

2 Basis of Pulse-Code Modulation   340

3 Development of a Pulse-Code Modulation Signal   341
3.1   Sampling   341
3.2   Quantization   342
3.3   Coding   347

4 Pulse-Code Modulation System Operation   354

5 Practical Applications   355
5.1   General   355
5.2   Practical System Block Diagram   357
5.3   The Line Code   360
5.4   Signal-to-Gaussian-Noise Ratio on Pulse-Code Modulation
      Repeater Lines   360
5.5   Regenerative Repeaters   361
5.6   PCM System Enhancements   363

6 Higher-Order PCM Multiplex Systems   367
6.1   Introduction   367
6.2   Stuffing and Justification   367
6.3   North American Higher-Level Multiplex   368
6.4   The European E1 Digital Hierarchy   369

7 Long-Distance PCM Transmission   371
7.1   Transmission Limitations   371
7.2   Jitter   373
7.3   Distortion   374
7.4   Thermal Noise   374
7.5   Crosstalk   374
7.6   Echo   375

8   Digital Loop Carrier (DLC)   375
9   SONET and SDH   375
  9.1   Introduction   375
  9.2   SONET   376
  9.3   Synchronous Digital Hierarchy (SDH)   383
10 Delta Modulation   391
  10.1   Introduction   391
  10.2   Continuous Variable-Slope Delta Modulation   392
11 Summary of Advantages and Disadvantages of Digital
    Transmission   396

Chapter 9   Digital Switching and Networks   403
1   Introduction   403
2   Advantages and Issues of PCM Switching   404
3   Approaches to PCM Switching   405
  3.1   General   405
  3.2   Time Switch   405
  3.3   Space Switch   408
  3.4   Time–Space–Time Switch   409
  3.5   Space–Time–Space Switch   411
  3.6   TST Compared to STS   411
4   Review of Some Digital Switching Concepts   412
  4.1   Early Ideas   412
  4.2   Higher-Level Multiplex Structures Internal to a Digital
       Switch   413
  4.3   An Overview of the AT&T 5ESS   414
  4.4   An Overview of the NTI DMS-100 with Supernode/ENET   417
5   The Digital Network   425
  5.1   Introduction   425
  5.2   Digital Extension to the Subscriber   427
  5.3   Change of Profile of Services   430
  5.4   Digital Transmission Network Models: CCITT   431
  5.5   Digital Network Synchronization   431
  5.6   Digital Network Performance Requirements   441
  5.7 A-Law Conversion to μ-Law; Digital Loss   452

Chapter 10   Introduction to Data Communications   459
1   Overview   459
2   The Bit   460
3   Removing Ambiguity—Binary Convention   460
4   Coding   461

4.1	Introduction to Binary Coding Techniques	461
4.2	Hexadecimal Representation and the BCD Code	463
4.3	Some Specific Binary Codes for Information Interchange	464
5	**Errors in Data Transmission**	464
5.1	Introduction	464
5.2	Throughput	468
5.3	The Nature of Errors	468
5.4	Error Detection and Error Correction	469
5.5	Error Correction with Feedback Channel	472
6	**The dc Nature of Data Transmission**	473
6.1	Loops	473
6.2	Neutral and Polar dc Transmission Systems	473
7	**Binary Transmission and the Concept of Time**	475
7.1	Introduction	475
7.2	Asynchronous and Synchronous Transmission	476
7.3	Timing	479
7.4	Distortion	481
7.5	Bits, Bauds and Symbols	483
7.6	Digital Data Waveforms	483
8	**Data Interface—The Physical Layer**	485
9	**Digital Transmission on an Analog Channel**	489
9.1	Introduction	489
9.2	Modulation–Demodulation Schemes	489
9.3	Critical Parameters	491
9.4	Channel Capacity	499
9.5	Some Modem Selection Considerations	501
9.6	Equalization	503
9.7	Practical Modem Applications	504
9.8	Data Transmission on the Digital Network	509

Chapter 11 Data Networks and Their Operation		**517**
1	**Introduction**	517
1.1	Applications	519
2	**Initial Design Considerations**	520
2.1	General	520
2.2	Data Terminals and Workstations	521
3	**Network Topologies and Configurations**	521
4	**Three Approaches to Data Switching**	526
5	**Circuit Optimization**	527
5.1	Throughput	530
5.2	Some Cost-Effective Options to Meet "Throughput" Requirements	530

6   Data Network Operation   532
    6.1   Introduction   532
    6.2   Protocols   533
    6.3   X.25: A Packet-Switched Network Access Standard   548

7   TCP/IP and Related Protocols   557
    7.1   Background and Scope   557
    7.2   TCP/IP and Data Link Layers   559
    7.3   The IP Routing Function   560
    7.4   The Transmission Control Protocol (TCP)   569

8   IBM System Network Architecture (SNA)   577
    8.1   Background   577
    8.2   The SNA Layered Architecture   579
    8.3   Advanced Peer-to-Peer Networking (APPN)   579
    8.4   Architectural Components of an SNA Network   582
    8.5   Nodes with Both Hierarchical and Peer-Oriented Function 587
    8.6   Links and Transmission Groups   588
    8.7   SNA Network Configurations   589
    8.8   The Transport Network and Network Accessible Units   591
    8.9   SNA Logical Units (LUs)   594
    8.10 Some SNA Data Formats   596

Chapter 12   Local Area Networks   **605**

1   Definition and Applications   605

2   LAN Topologies   606

3   The Two Broad Categories of LAN Transmission Techniques   608
    3.1   Broadband Transmission Considerations   610
    3.2   Fiber-Optic LANs   612

4   Overview of IEEE/ANSI LAN Protocols   613
    4.1   General   613
    4.2   How LAN Protocols Relate to OSI   615
    4.3   Logical Link Control (LLC)   616

5   LAN Access Protocols   622
    5.1   Introduction   622
    5.2   Background: Contention and Polling   622
    5.3   CSMA and CSMA/CD Access Techniques   623
    5.4   Token Bus   632
    5.5   Token Ring   638
    5.6   Fiber Distributed Data Interface   642
    5.7   LAN Performance   655
    5.8   LAN Internetworking via Spanning Devices   657

Chapter 13 Integrated Services Digital Networks   **665**

1 Background and Goals of ISDN   665
2 ISDN Structures   667
  2.1 ISDN User Channels   667
  2.2 Basic and Primary User Interfaces   668
3 User Access and Interface   669
  3.1 General   669
4 ISDN Protocols and Protocol Issues   671
5 ISDN Networks   673
6 ISDN Protocol Structures   674
  6.1 ISDN and OSI   674
  6.2 Layer 1 Interface, Basic Rate   677
  6.3 Layer 1 Interface, Primary Rate   685
7 Overview of the Layer 2 Interface: Link Access Procedure for
  the D-Channel   689
  7.1 Layer 2 Frame Structure for Peer-to-Peer Communication   693
  7.2 LAPD Primitives   697
8 Overview of Layer 3   697
  8.1 Layer 3 Specification   700
9 ISDN Packet-Mode Review   705
  9.1 Introduction   705
  9.2 Case A: Configuration When Accessing PSPDN Services   705
  9.3 Case B: Configuration for the ISDN Virtual Circuit Service   707
  9.4 Service Aspects   708

Chapter 14 Emerging Broadband Data Technologies   **713**

1 Introduction   713
2 How Can the Network Be Speeded Up?—Frame Relay   713
  2.1 Background and Rationale   713
  2.2 The Genesis of Frame Relay   715
  2.3 Introduction to Frame Relay   717
  2.4 The Frame Structure   718
  2.5 DL-CORE Parameters (as Defined by ANSI)   724
  2.6 Procedures   724
  2.7 Traffic and Billing on Frame Relay   724
  2.8 Congestion Control, a Discussion   726
  2.9 Policing a Frame Relay Network   728
  2.10 Quality of Service Parameters   731
3 Distributed Queue Dual Bus (DQDB)   733
  3.1 Introduction and Purpose   733
  3.2 An Overview of DQDB   734
  3.3 Access to the DQDB Network   735

3.4 DQDB Layer Services 743
3.5 DQDB Layer Protocol Data Unit (PDU) Formats 747
4 Switched Multimegabit Data Service (SMDS) 751
4.1 Introduction 751
4.2 Relationship to DQDB 752
4.3 Overview of SMDS 753
4.4 Subscriber–Network Interface (SNI) 757
4.5 SMDS Interface Protocol (SIP) 757
4.6 Access Classes and Enforcement 765
4.7 SMDS in an Operating Environment 767

Chapter 15   The Asynchronous Transfer Mode and
                Broadband ISDN   **773**

1 Where Are We Going? 773
2 Introduction to ATM 774
3 User–Network Interface (UNI) Configuration and Architecture 776
4 The ATM Cell—Key to Operation 779
4.1 ATM Cell Structure 779
4.2 Idle Cells 785
5 Cell Delineation and Scrambling 785
5.1 Delineation and Scrambling Objectives 785
5.2 Cell Delineation Algorithm 786
6 ATM Layering and B-ISDN 787
6.1 Functions of Individual ATM/B-ISDN Layers 788
7 Services: Connection-Oriented and Connectionless 799
7.1 Functional Architecture 799
7.2 CLNAP Protocol Data Unit (PDU) and Encoding 801
8 Some Aspects of a B-ISDN/ATM Network 804
8.1 ATM Routing and Switching 804
9 Signaling Requirements 805
9.1 Setup and Release of VCCs 805
9.2 Signaling Virtual Channels 807
9.3 Meta-Signaling 807
10 Quality of Service (QoS) 808
10.1 ATM Service Quality Review 808
10.2 QoS Parameter Descriptions 809
11 Traffic Control and Congestion Control 811
11.1 Generic Functions 811
11.2 Events, Actions, Time Scales, and Response 812
11.3 Quality of Service, Network Performance, and Cell Loss
     Priority 812
11.4 Traffic Descriptors and Parameters 813
11.5 User–Network Traffic Contract 814

12 Transporting ATM Cells   817
     12.1 In the DS3 Frame   817
     12.2 DS1 Mapping   818
     12.3 E1 Mapping   819
     12.4 Mapping ATM Cells into SDH   820
     12.5 Mapping ATM Cells into SONET   822

Chapter 16   CCITT Signaling System No. 7   **827**

  1 Introduction   827

  2 Overview of SS No. 7 Architecture   828

  3 SS No. 7 Relationship to OSI   829

  4 Signaling System Structure   831
     4.1 Signaling Network Management   833

  5 Signaling Data-Link Layer (Layer 1)   835

  6 The Signaling Link (Layer 2)   836
     6.1 Basic Signal Unit Format   839

  7 Signaling Network Functions and Messages (Layer 3)   842
     7.1 Introduction   842
     7.2 Signaling Message-Handling Functions   842
     7.3 Signaling Network Management   846

  8 Signaling Network Structure   847
     8.1 Introduction   847
     8.2 International and National Signaling Networks   847

  9 Signaling Performance—Message Transfer Part   848
     9.1 Basic Performance Parameters   848
     9.2 Traffic Characteristics   849
     9.3 Transmission Parameters   849
     9.4 Signaling Link Delays over Terrestrial and Satellite Links   849

 10 Numbering Plan for International Signaling Point Codes   850

 11 Hypothetical Signaling Reference Connections   851

 12 Signaling Connection Control Part (SCCP)   853
     12.1 Introduction   853
     12.2 Services Provided by the SCCP   853
     12.3 Peer-to-Peer Communication   854
     12.4 Primitives and Parameters   855
     12.5 Connection-Oriented Function: Temporary Signaling
            Connections   856
     12.6 Structure of the SCCP   857

 13 User Parts   858
     13.1 Introduction   858
     13.2 Telephone User Part (TUP)   860
     13.3 ISDN User Part (ISUP)   863

Chapter 17   Cellular/Mobile Radio and PCN/PCS   **883**

1   Introduction   883
  1.1   Background   883
  1.2   Scope and Objective   883

2   Some Basic Concepts of Cellular Radio   884

3   Radio Propagation in the Mobile Environment   888
  3.1   The Propagation Problem   888
  3.2   Several Propagation Models   889
  3.3   Microcell Prediction According to Lee   892

4   Impairments—Fading in the Mobile Environment   895
  4.1   Introduction   895
  4.2   Classification of Fading   896
  4.3   Diversity—A Technique to Mitigate the Effects of Fading
        and Dispersion   897
  4.4   Cellular Radio Path Calculations   901

5   The Cellular Radio Bandwidth Dilemma   901
  5.1   Background and Objectives   901
  5.2   Bit Rate Reduction of the Digital Voice Channel   902

6   Network Access Techniques   905
  6.1   Introduction   905
  6.2   Frequency-Division Multiple Access (FDMA)   905
  6.3   Time-Division Multiple Access (TDMA)   906
  6.4   Code-Division Multiple Access (CDMA)   910

7   Frequency Reuse   915

8   Paging Systems   918
  8.1   What Are Paging Systems?   918
  8.2   Radio-Frequency Bands for Pagers   918
  8.3   Radio Propagation into Buildings   918
  8.4   Techniques Available for Multiple Transmitter Zones   919
  8.5   Paging Receivers   919
  8.6   System Capacity   920
  8.7   Codes and Formats for Paging Systems   920
  8.8   Considerations for Selecting Codes and Formats   921

9   Personal Communication Systems   921
  9.1   Defining Personal Communication   921
  9.2   Narrow-band Microcell Propagation at PCS Distances   922

10 Mobile Satellite Communications   926
  10.1   Background and Scope   926
  10.2   Overview of Satellite Mobile Services   926
  10.3   System Trends   928
  10.4   IRIDIUM   928
  10.5   Subscriber Unit Segment   934

Chapter 18   Network Management   **941**

1   What is Network Management?   941

2   The Bigger Picture   942

3   Traditional Breakout by Tasks   942
    3.1   Fault Management   942
    3.2   Configuration Management   943
    3.3   Performance Management   943
    3.4   Security Management   943
    3.5   Accounting Management   943

4   Survivability—Where Network Management Really Pays   944
    4.1   Survivability Enhancement—Rapid Troubleshooting   944

5   System Depth—A Network Management Problem   947
    5.1   Aids in Network Management Provisioning   948
    5.2   Communications Channels for the Network Management
          System   951

6   Network Management from a PSTN Perspective   952
    6.1   Objectives and Functions   952
    6.2   Network Traffic Management Center   953
    6.3   Network Traffic Management Principles   953
    6.4   Network Traffic Management Functions   954
    6.5   Network Traffic Management Controls   957

7   Network Management Systems   961
    7.1   What Are Network Management Systems?   961
    7.2   An Introduction to Network Management Protocols   961
    7.3   Remote Monitoring (RMON)   967
    7.4   SNMP Version 2   968
    7.5   Common Management Information Protocol (CMIP)   970

8   Network Management in ATM   971
    8.1   Interim Local Management Interface (ILMI) Functions   974
    8.2   ILMI Service Interface   975

Index   **979**

PREFACE

Dramatic changes have taken place in the telecommunications industry since the publication of the second edition of *Telecommunication System Engineering* in 1989. Microwave radio has taken a backseat as fiber-optic bearer systems found universal application. Digital formats compatible with the high bit rate of fiber-optic communications are now being implemented. These are SONET and Synchronous Digital Hierarchy (SDH). Digital switching too has taken a great leap forward with the introduction of very high capacity time switches.

In the world of data communications a new term, the *enterprise network*, was introduced. Local area networks have kept pace with these developments with the introduction of new devices and higher speeds. Their interconnectivity over long distances has brought about a revolutionary technique called frame relay. In answer to this need for greater speed and capacity, Bellcore has developed switched multimegabit data service (SMDS), which is just starting operation.

Cellular radio has spawned Personal Communication Service (PCS), which promises to completely change our personal lives by its "universal tether." In the future we will be in constant touch with the world no matter where we may be, with voice, data, facsimile, and possibly even video capability. The weight of these PCS devices will be measurable in ounces (grams) rather than in pounds (kilos).

One of the most promising developments of all is the asynchronous tranfer mode (ATM), which truly integrates voice, data, and image communication with a common format, the cell.

I have incorporated these and other developments in this edition emphasizing the system aspects of each. I have also maintained an overview of what some may call dated technology. Included in this overview are analog network signaling and space division switching. There are two valid reasons for this: first, to lay the tutorial groundwork for digital network signaling and switching, and second, analog technology continues to be widely used in many parts of the world. At the same time, I have increased the emphasis on digital transmission, switching, and networks. I have also increased the length and breadth of my coverage of data communications by adding TCP/IP and SNA protocols.

Some technologists claim there are now two separate worlds in the field of communications: telecommunications meaning voice communications and data communications. I do not hold with that argument, and thus I give equal space to data communications as a vital discipline under the telecommunications umbrella. The two technologies are inextricably intertwined.

In the preparation of this third edition, as with the previous editions and with my other books, I have relied heavily on national and international standards. Certainly the largest contributor is the International Telecommunication Union (ITU) in Geneva, Switzerland. This organization has undergone a major reorganization which officially took place on January 1, 1993. Prior to that date its two subsidiary organizations were the CCITT (International Consult[ta]tive Committee on Telephone and Telegraph) and CCIR (International Consult[at]ive Committee on Radio). After that date, CCITT became ITU Telecommunications Standardization Sector (ITU-T), which I have abbreviated as ITU-T. CCIR after that date became the ITU Radiocommunication Sector, which I have abbreviated as ITU-R.

The goal of this edition continues to be that of the popular second edition, namely to impart practical knowledge of telecommunications to a wide readership. Specifically, it is a treatise on telecommunication system engineering, and it is meant for people who need a technical grounding in the subject. I have tried to reduce both the amount and complexity of the mathematics prerequisite to that of a first-year university level. I would also expect the reader to be able to handle the physics of basic electricity, have some knowledge of modulation, and understand such terms as bandwidth.

I am indebted to a large group of people who have graciously provided me help and advice and have afforded me input material. John Lawlor, independent consultant in Sharon, MA, critiqued the first chapter and provided reference material. Dr. Ron Brown, independent consultant based in Melrose, MA and his wife, Annette, were there when I needed them. Bill Otasky of Beverly Farms, MA supplied input material and encouragement. Fran Drake, telecommunication seminar program director at the University of Wisconsin, suggested constructive additions and changes. Fran always feeds me material, especially on cellular radio. The telecommunication seminars I give in Madison are always excellent sounding boards on how to present a difficult technical subject to a broad-based audience.

Dr. Ernst Munter and Paul Roddick of Nortel gave me invaluable assistance in the digital switching arena and reviewed the digital switching chapter. Lisa Lepore, AT&T consultant liaison, opened doors for me on AT&T's 5ESS digital switch. Likewise, I am indebted to Joseph Golden of the Telesector Resources Group of Nynex for material on CCITT Signaling System No. 7. Dr. Don Schilling, professor emeritus of City College (New York), supported me on this edition as he did on the second edition.

My wife, Paquita, has shown her usual patience and forebearing during the preparation and production phases of this third edition.

ROGER L. FREEMAN

Scottsdale, Arizona

Telecommunication System Engineering

1

SOME BASICS IN CONVENTIONAL TELEPHONY

1 BACKGROUND AND CONCEPT

Telecommunications deals with the service of providing electrical communications at a distance. The service is supported by an industry that depends on a large body of increasingly specialized scientists and engineers. The service may be private or open to public correspondence (i.e., access). The most cogent example is the telephone company, when based on private enterprise, or telephone administration, when government owned.

1.1 Telecommunication Networks

The public switched telecommunication network (PSTN) is immense. There are literally hundreds of national networks interconnected to form a gigantic international network. There are about 1,000,000,000 telephones connected to this worldwide network. Hundreds of thousands of computers also use the network in various overlaid subnets such as the popular Internet. This same PSTN serves as a vehicle for video conferencing and for point-to-point facsimile connectivity on dial-up connections. An adjunct to the network is cellular service. In the United States alone there are over 25 million cellular subscribers, and that number may double in less than three years.

The total investment in the network is in the hundreds of thousands of millions of U.S. dollars for the total world. This is one important reason that it is slow to adopt new technology. The network ownership must amortize its very large investments.

There are other networks. One type that is showing great growth is the enterprise network. This is a corporate network, but most always implies a local area network (LAN), optimized for data communications. An enterprise network may consist of multiple LANs. They may be interconnected over a distance using some form of the PSTN. Frame relay is one possibility.

Another network is the private network, which partially or wholly may be an enterprise network. It may just be a private form of the PSTN, providing many of the services of the PSTN but dedicated to one corporation. However, we find more and more that such single corporation networks lease out their excess capacity to other corporations, possibly even to the PSTN. In some situations, such private networks are set up simply to save money. In other situations, especially where the PSTN provides substandard service, such private networks exist to provide just service or acceptable service. In some cases, if the private network didn't exist, there would be no service at all. Some private networks are simple overlays on the PSTN, where a corporation just leases transmission capacity, providing their own switching. There are many possible variations here.

Some argue that telecommunications with all its possible facets is the world's largest business. We don't take sides on this issue. What we do wish to do is to impart to the reader a technical knowledge of telecommunication networks from a system viewpoint. By "system" we mean how one discipline can interact with another to reach a certain end goal. If we do it right, that interaction can work for us; if not, it may work against obtaining our goal.

Therefore, a primary concern of this book is to describe the development of the PSTN, why it is built as it is, and how it is evolving. We also intend to show how it will expand and carry other than voice communication, how special services are evolving as based originally on the existing telephone network, and where certain split-offs will occur at a later date. Today, the bulk of the telecommunication industry is dedicated to the telephone network.

Telecommunication engineering has traditionally been broken down into two basic segments: transmission and switching. This division is most apparent in telephony. Transmission deals with delivering a quality electrical signal from point X to point Y. Let us say that switching connects X to Y, rather than to Z. Until some years ago, transmission and switching were two very separate and distinct disciplines. Today that distinction is disappearing. As we proceed, we deal with both disciplines and show in later chapters how the two are melding together.

2 THE SIMPLE TELEPHONE CONNECTION

The common telephone as we know it today is a device connected to the outside world by a pair of wires. It consists of a handset and its cradle with a signaling device, consisting of either a dial or push buttons. The handset is made up of two electroacoustic transducers, the earpiece or receiver and the mouthpiece or transmitter. There is also a sidetone circuit that allows some of the transmitted energy to be fed back to the receiver.

The transmitter or mouthpiece converts acoustic energy into electric energy by means of a carbon granule transmitter. The transmitter requires a direct-current (dc) potential, usually on the order of 3–5 V, across its electrodes. We call this the *talk battery*, and in modern telephone systems it is supplied over the line (central battery) from the switching center and has been standardized at −48 V dc. Current from the battery flows through the carbon granules or grains when the telephone is lifted from its cradle or goes "off hook."* When sound impinges on the diaphragm of the transmitter, variations of air pressure are transferred to the carbon, and the resistance of the electrical path through the carbon changes in proportion to the pressure. A pulsating direct current results.

The typical receiver consists of a diaphragm of magnetic material, often soft iron alloy, placed in a steady magnetic field supplied by a permanent magnet, and a varying magnetic field caused by voice currents flowing through the voice coils. Such voice currents are alternating (ac) in nature and originate at the far-end telephone transmitter. These currents cause the magnetic field of the receiver to alternately increase and decrease, making the diaphragm move and respond to the variations. Thus an acoustic pressure wave is set up, more or less exactly reproducing the original sound wave from the distant telephone transmitter. The telephone receiver, as a converter of electrical energy to acoustic energy, has a comparatively low efficiency, on the order of 2–3%.

Sidetone is the sound of the talker's voice heard in his (or her) own receiver. Sidetone level must be controlled. When the level is high, the natural human reaction is for the talker to lower his or her voice. Thus by regulating sidetone, talker levels can be regulated. If too much sidetone is fed back to the receiver, the output level of the transmitter is reduced as a result of the talker lowering his or her voice, thereby reducing the level (voice volume) at the distant receiver and deteriorating performance.

To develop our discussion, let us connect two telephone handsets by a pair of wires, and at middistance between the handsets a battery is connected to provide that all-important talk battery. Such a connection is shown diagrammatically in Figure 1.1. Distance D is the overall separation of the two handsets and is the sum of distances d_1 and d_2; d_1 and d_2 are the distances from each handset to the central battery supply. The exercise is to extend the distance D to determine limiting factors given a fixed battery voltage, say, 48 V dc. We find that there are two limiting factors to the extension of the wire pair between the handsets. These are the IR drop, limiting the voltage across the handset transmitter, and the attenuation. For 19-gauge wire, the limiting distance is about 30 km, depending on the efficiency of the handsets. If the limiting characteristic is attenuation and we desire to extend the pair farther, amplifiers could be used in the line. If the battery voltage is limiting,

---

*The opposite action of "off hook" is "on hook"—that is, placing the telephone back in its cradle, thereby terminating a connection.

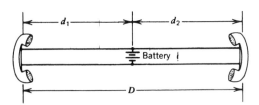

Figure 1.1. A simple telephone connection.

then the battery voltage could be increased. With the telephone system depicted in Figure 1.1, only two people can communicate. As soon as we add a third person, some difficulties begin to arise. The simplest approach would be to provide each person with two handsets. Thus party A would have one set to talk to B, another to talk to C, and so forth. Or the sets could be hooked up in parallel. Now suppose A wants to talk to C and doesn't wish to bother B. Then A must have some method of selectively alerting C. As stations are added to the system, the alerting problem becomes quite complex. Of course, the proper name for this selection and alerting is *signaling*. If we allow that the pair of wires through which current flows is a loop, we are dealing with loops. Let us also call the holder of a telephone station a *subscriber*. The loops connecting them are subscriber loops.

Let us now look at an eight-subscriber system, each subscriber connected directly to every other subscriber. This is shown in Figure 1.2. When we connect each and every station with every other one in the system, this is called a *mesh* connection, or sometimes full mesh. Without the use of amplifiers and with 19-gauge copper wire size, the limiting distance is 30 km. Thus any connecting segment of the octagon may be no greater than 30 km. The only way we can justify a mesh connection of subscribers economically is when each and every subscriber wishes to communicate with every other

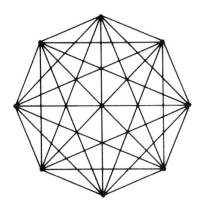

Figure 1.2. An 8-point mesh connection.

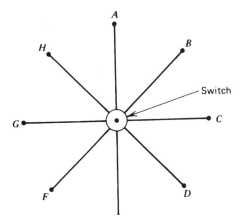

Figure 1.3. Subscribers connected in a star arrangement.

subscriber in the network for virtually the entire day (full period). As we know, however, most telephone subscribers do not use their telephones on a full-time basis. The telephone is used at what appear to be random intervals throughout the day. Furthermore, the ordinary subscriber or telephone user will normally talk to only one other subscriber at a time. He/she will not need to talk to all other subscribers simultaneously.

If more subscribers are added and the network is extended beyond about 30 km, it is obvious that transmission costs will spiral, because that is what we are dealing with exclusively here—transmission. We are connecting each and every subscriber together with wire transmission means, requiring many amplifiers and talk batteries. Thus it would seem wiser to share these facilities in some way and cut down on the transmission costs. We now discuss this when switch and switching enter the picture. Let us define a *switch* as a device that connects inlets to outlets. The inlet may be a calling subscriber line, and the outlet may be the line of a called subscriber. The techniques of switching and the switch as a concept are widely discussed later in this text. Switching devices and how they work are covered in Chapters 3 and 9. Consider Figure 1.3, which shows our subscribers connected in a *star* network with a switch at the center. All the switch really does in this case is to reduce the transmission cost outlay. Actually, this switch reduces the number of links between subscribers, which really is a form of concentration. Later in our discussion it becomes evident that switching is used to concentrate traffic, thus reducing the cost of transmission facilities.

3   SOURCES AND SINKS*

Traffic is a term that quantifies usage. A subscriber *uses* the telephone when he/she wishes to talk to somebody. We can make the same statement for a

*The traffic engineer may wish to use the terminology "origins and destinations."

telex (teleprinter service) subscriber or a data-service subscriber. But let us stay with the telephone.

A network is a means of connecting subscribers. We have seen two simple network configurations, the mesh and star connections, in Figures 1.2 and 1.3. When talking about networks, we often talk of sources and sinks. A call is initiated at a traffic source and received at a traffic sink. Nodal points or nodes in a network are the switches.

4 TELEPHONE NETWORKS: INTRODUCTORY TERMINOLOGY

From our discussion we can say that a telephone network can be regarded as a systematic development of interconnecting transmission media arranged so that one telephone user can talk to any other within that network. The evolving layout of the network is primarily a function of economics. For example, subscribers share common transmission facilities; switches permit this sharing by concentration.

Consider a very simplified example. Two towns are separated by, say, 20 mi, and each town has 100 telephone subscribers. Logically, most of the telephone activity (the traffic) will be among the subscribers of the first town and among those of the second town. There will be some traffic, but considerably less, from one town to the other. In this example let each town have its own switch. With the fairly low traffic volume from one town to the other, perhaps only six lines would be required to interconnect the switch of the first town to that of the second. If no more than six people want to talk simultaneously between the two towns, a number as low as six can be selected. Economics has mandated that we install the minimum number of connecting telephone lines from the first town to the second to serve the calling needs between the two towns. The telephone lines connecting one telephone switch or exchange with another are called *trunks* in North America and *junctions* in Europe. The telephone lines connecting a subscriber to the switch or exchange that serves the subscriber are called *lines, subscriber lines*, or *loops*. Concentration is a line-to-trunk ratio. In the simple case above it was 100 lines to six trunks (or junctions), or about a 16:1 ratio.

A telephone subscriber looking into the network is served by a *local exchange*. This means that the subscriber's telephone line is connected to the network via the local exchange or central office, in North American parlance. A local exchange has a serving area, which is the geographical area in which the exchange is located: all subscribers in that area are served by that exchange.

The term *local area*, as opposed to *toll area*, is that geographical area containing a number of local exchanges and inside which any subscriber can call any other subscriber without incurring tolls (extra charges for a call).

Toll calls and long-distance calls are synonymous. For instance, a local call in North America, where telephones have detailed billing, shows up on the bill as a time-metered call or is covered by a flat monthly rate. Toll calls in North America appear as separate detailed entries on the telephone bill. This is not so in most European countries and in those countries following European practice. In these countries there is no detailed billing on direct-distance-dialed (subscriber-trunk-dialed) calls. All such subscriber-dialed calls, even international ones, are just metered, and the subscriber pays for the meter steps used per billing period, which is often one or two months. In European practice a long-distance call, a toll call if you will, is one involving the dialing of additional digits (e.g., more than six or seven digits).

Let us call a network a *grouping of interworking telephone exchanges.* As the discussion proceeds, the differences between local networks and national networks are shown. Two other types of network are also discussed. These are specialized versions of a local network and are the rural network (rural area) and metropolitan network (metropolitan area). (Also consult Refs. 9, 16, 17, and 18.)

5 ESSENTIALS OF TRAFFIC ENGINEERING

5.1 Introduction and Terminology

As we have already mentioned, telephone exchanges are connected by trunks or junctions. The number of trunks connecting exchange X with exchange Y is the number of voice pairs or their equivalent used in the connection. One of the most important steps in telecommunication engineering practice is to determine the number of trunks required on a route or connection between exchanges. We could say we are *dimensioning* the route. To dimension a route correctly, we must have some idea of its usage—that is, how many people will wish to talk at once over the route. The usage of a transmission route or a switch brings us into the realm of traffic engineering, and the usage may be defined by two parameters: (1) *calling rate*, or the number of times a route or traffic path is used per unit period; or, more properly defined, "the call intensity per traffic path during the busy hour";* and (2) *holding time*, or "the duration of occupancy of a traffic path by a call,"* or sometimes, "the average duration of occupancy of one or more paths by calls."* A *traffic path* is "a channel, time slot, frequency band, line, trunk, switch, or circuit over which individual communications pass in sequence."* *Carried traffic* is the volume of traffic actually carried by a switch, and *offered traffic* is the volume of traffic offered to a switch.

To dimension a traffic path or size a telephone exchange, we must know the traffic intensity representative of the normal busy season. There are

---

Reference Data for Radio Engineers [1], p. 31–8.

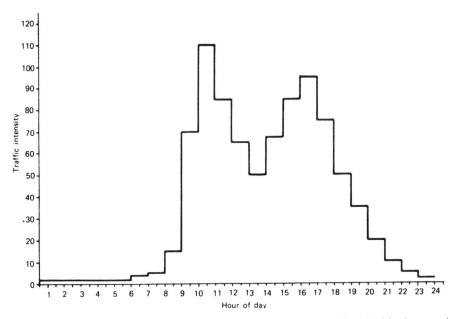

Figure 1.4. Bar chart of traffic intensity over a typical working day (U.S., mixed business and residential).

weekly and daily variations in traffic within the busy season. Traffic is very random in nature. However, there is a certain consistency we can look for. For one thing, there usually is more traffic on Mondays and Fridays and a lower volume on Wednesdays. A certain consistency can also be found in the normal workday hourly variation. Across the typical day the variation is such that a 1-h period shows greater usage than any other. From the hour with least traffic to the hour of greatest traffic, the variation can exceed 100 : 1. Figure 1.4 shows a typical hour-by-hour traffic variation for a serving switch in the United States.* It can be seen that the busiest period, the *busy hour* (BH), is between 10 A.M. and 11 A.M.. From one workday to the next, originating BH calls can vary as much as 25%. To these fairly "regular" variations, there are also unpredictable peaks caused by stock market or money market activity, weather, natural disaster, international events, sporting events, and so on. Normal system growth must also be taken into account. Nevertheless, suitable forecasts of BH traffic can be made. However, before proceeding, consider the five most common definitions of BH:

Busy Hour Definitions (CCITT Rec. E.600)

1. *Busy Hour.* The busy hour refers to the traffic volume or number of call attempts, and is that continuous 1-h period lying wholly in the time

---

*The busy hour will vary from country to country because of cultural differences.

interval concerned for which this quantity (i.e., traffic volume or call attempts) is greatest.

2. *Peak Busy Hour*. The busy hour each day; it usually is not the same over a number of days.

3. *Time Consistent Busy Hour*. The 1-h period starting at the same time each day for which the average traffic volume or call-attempt count of the exchange or resource group concerned is greatest over the days under consideration.

From *Engineering and Operations in the Bell System*, 2nd ed. [23]

4. The engineering period (where the grade of service criteria is applied) is defined as the Busy Season Busy Hour (BSBH), which is the busiest clock hour of the busiest weeks of the year.

5. The Average Busy Season Busy Hour (ABSBH) is used for trunk groups and always has a grade of service criterion applied. For example, for the ABSBH load, a call requiring a circuit in a trunk group should encounter "all trunks busy" (ATB) no more than 1% of the time.

Reference 23 goes on to state that peak loads are of more concern than average loads when engineering switching equipment and engineering periods other than the ABSBH are defined. Examples of these are the highest BSBH and the average of the ten highest BSBHs. Sometimes the engineering period is the weekly peak hour (which may not even be the BSBH).

When dimensioning telephone exchanges and transmission routes, we shall be working with BH traffic levels and care must be used in the definition of busy hour.

5.2 Measurement of Telephone Traffic

If we define *telephone traffic* as the aggregate of telephone calls over a group of circuits or trunks with regard to the duration of calls as well as their number [2], we can say that traffic flow (A)

$$A = C \times T$$

where C designates the number of calls originated during a period of 1 h and T is the average holding time, usually given in hours. A is a dimensionless unit because we are multiplying calls/hour by hour/call.

Suppose that the average holding time is 2.5 min and the calling rate in the BH for a particular day is 237. The traffic flow (A) would then be 237×2.5, or 592.5 call-minutes (Cm) or 592.5/60, or about 9.87 call-hours (Ch).

Ramses Mina [2] states that a distinction should be made between the terms "traffic density" and "traffic intensity." The former represents the number of simultaneous calls at a given moment, while the latter represents the average traffic density during a 1-h period. The quantity of traffic used in the calculation for dimensioning of switches is the traffic intensity.

The preferred unit of traffic intensity is the erlang, named after the Danish mathematician A. K. Erlang [5]. The erlang is a dimensionless unit. One erlang represents a circuit occupied for 1 h. Considering a group of circuits, traffic intensity in erlangs is the number of call-seconds per second or the number of call-hours per hour. If we knew that a group of 10 circuits had a call intensity of 5 erlangs, we would expect half of the circuits to be busy at the time of measurement.

In the United States the term "unit call" (UC), or its synonymous term, "hundred call-second," abbreviated CCS,* generally is used. These terms express the sum of the number of busy circuits, provided that the busy trunks were observed once every 100 s (36 observations in 1 h) [2].

There are other traffic units. For instance: call-hour (Ch)—1 Ch is the quantity represented by one or more calls having an aggregate duration of 1 h; call-second (Cs)—1 Cs is the quantity represented by one or more calls having an aggregate duration of 1 s; traffic unit (TU), a unit of traffic intensity. One TU is the average intensity in one or more traffic paths carrying an aggregate traffic of 1 Ch in 1 h (the busy hour unless otherwise specified). 1 TU = 1 E (erlang) (numerically). The *equated busy hour call* (EBHC) is a European unit of traffic intensity. 1 EBHC is the average intensity in one or more traffic paths occupied in the BH by one 2-min call or an aggregate duration of 2 min. Thus we can relate our terms as follows:

$$1 \text{ erlang} = 30 \text{ EBHC} = 36 \text{ ccs} = 60 \text{ Cm}$$

assuming a 1-hour time-unit interval.

Traffic measurements used for long-term network planning are usually based on the traffic in the busy hour (BH), which is usually determined based on observations and studies.

The traditional traffic measurements on trunks during a measurement interval are:

- Peg count†—calls offered
- Usage—traffic (CCS or erlangs) carried
- Overflow—call encountering all trunks busy

*The first letter C in CCS stands for the Roman numeral 100.

†A term taken from telephony in the older days where manual switching was prevalent. A peg board was installed by the telephone operator to keep count of offered calls. The present definition is taken from Ref. 23. "A count of all calls offered to a trunk group, usually measured for one hour. As applied to units of switching systems with common control, *peg count*, or *carried peg count*, means the number of calls actually handled."

From these measurements, the blocking probability and mean traffic load carried by the trunk group can be calculated.

Extensive traffic measurements are made on switching systems because of their numerous traffic sensitive components. Usual measurements for a component such as a service circuit include calls carried peg count and usage. The typical holding time for a common-control element in a switch is considerably shorter than that for a trunk, and short sampling intervals (e.g., 10 s) or continuous monitoring are used to measure usage.

Traffic measurements for short-term network management purposes are usually concerned with detecting network congestion. Calls offered peg count and overflow count can be used to calculate attempts per circuit per hour (ACH) and connections per circuit per hour (CCH), with these measurements being calculated over very short time periods (e.g., 10-min intervals).

Under normal circumstances, ACH and CCH are approximately equal. Examples of abnormal conditions are:

- ACH high, CCH normal—heavy demand, excessive blockage, normal holding times for connected calls indicating that most calls switched are completed, heavy traffic but low congestion [25].
- ACH high, CCH high—heavy traffic, short trunk holding times indicate uncompleted call attempts being switched, congestion [25]. (Consult Ref. 24.)

5.3 Blockage, Lost Calls, and Grade of Service

Assume that an isolated telephone exchange serves 5000 subscribers and that no more than 10% of the subscribers wish service simultaneously. Therefore, the exchange is dimensioned with sufficient equipment to complete 500 simultaneous connections. Each connection would be, of course, between any two of the 5000 subscribers. Now let subscriber 501 attempt to originate a call. He/she cannot because all the connecting equipment is busy, even though the line he wishes to reach may be idle. This call from subscriber 501 is termed a *lost call* or *blocked call*. He has met blockage. The probability of meeting blockage is an important parameter in traffic engineering of telecommunication systems. If congestion conditions are to be met in a telephone system, we can expect that those conditions will usually be met during the BH. A switch is engineered (dimensioned) to handle the BH load. But how well? We could, indeed, far overdimension the switch such that it could handle any sort of traffic peaks. However, that is uneconomical. So with a well-designed switch, during the busiest of BHs we may expect some moments of congestion such that additional call attempts will meet blockage. *Grade of service* expresses the probability of meeting blockage during the BH and is expressed by the letter p. A typical grade of service is $p = 0.01$. This means that an average of one call in 100 will be blocked or "lost" during the

BH. Grade of service, a term in the Erlang formula, is more accurately defined as the *probability of blockage*. It is important to remember that lost calls (blocked calls) refer to calls that fail at *first* trial. We discuss attempts (at dialing) later, that is, the way blocked calls are handled.

We exemplify grade of service by the following problem. If we know that there are 354 seizures (lines connected for service) and 6 blocked calls (lost calls) during the BH, what is the grade of service?

$$\text{Grade of service} = \frac{\text{Number of lost calls}}{\text{Total number of offered calls}}$$
$$= \frac{6}{354 + 6} = \frac{6}{360} \tag{1.1}$$

or

$$p = 0.017$$

The average grade of service for a network may be obtained by adding the grade of service contributed by each constituent switch, switching network, or trunk group. The *Reference Data for Radio Engineers* [1, Section 31] states that the grade of service provided by a particular group of trunks or circuits of specified size and carrying a specified traffic intensity is the probability that a call offered to the group will find available trunks already occupied on first attempt. That probability depends on a number of factors, the most important of which are (1) the distribution in time and duration of offered traffic (e.g., random or periodic arrival and constant or exponentially distributed holding time), (2) the number of traffic sources [limited or high (infinite)], (3) the availability of trunks in a group to traffic sources (full or restricted availability), and (4) the manner in which lost calls are "handled."

Several new concepts are suggested in these four factors. These must be explained before continuing.

5.4   Availability

Switches were previously discussed as devices with lines and trunks, but better terms for describing a switch are "inlets" and "outlets." When a switch has full availability, each inlet has access to any outlet. When not all the free outlets in a switching system can be reached by inlets, the switching system is referred to as one with "limited availability." Examples of switches with limited and full availability are shown in Figures 1.5A and 1.5B.

Of course, full availability switching is more desirable than limited availability but is more expensive for larger switches. Thus full availability switching is generally found only in small switching configurations and in many new digital switches (see Chapter 9). *Grading* is one method of improving the traffic-handling capacities of switching configurations with limited availability. Grading is a scheme for interconnecting switching subgroups to make the switching load more uniform.

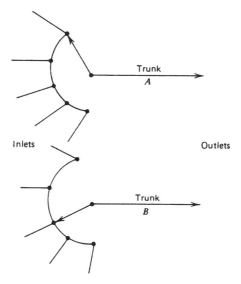

Figure 1.5A. An example of a switch with limited availability.

5.5 "Handling" of Lost Calls

In conventional telephone traffic theory, three methods are considered for the handling or dispensing of lost calls: (1) lost calls held (LCH), (2) lost calls cleared (LCC), and (3) lost calls delayed (LCD). The LCH concept assumes that the telephone user will immediately reattempt the call on receipt of a congestion signal and will continue to redial. The user hopes to seize connection equipment or a trunk as soon as switching equipment becomes available for the call to be handled. It is the assumption in the LCH concept that lost calls are held or waiting at the user's telephone. This concept further assumes that such lost calls extend the average holding time theoretically, and in this case the average holding time is zero, and all the time is waiting time. The principal traffic formula used in North America is based on the LCH concept.

The LCC concept, which is used primarily in Europe or those countries accepting European practice, assumes that the user will hang up and wait some time interval before reattempting if the user hears the congestion signal on the first attempt. Such calls, it is assumed, disappear from the system. A reattempt (after the delay) is considered as initiating a new call. The Erlang formula is based on this criterion.

The LCD concept assumes that the user is automatically put in queue (a waiting line or pool). For example, this is done when the operator is dialed. It is also done on most modern computer-controlled switching systems, generally referred to under the blanket term *stored program control* (SPC). The LCD category may be broken down into three subcategories, depending

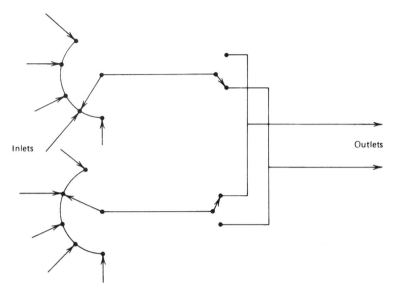

Figure 1.5B. An example of a switch with full availability.

on how the queue or pool of waiting calls is handled. The waiting calls may be handled last in first out, first in line first served, or at random.

5.6 Infinite and Finite Traffic Sources

We can assume that traffic sources are infinite or finite. For the infinite-traffic-sources case the probability of call arrival is constant and does not depend on the state of occupancy of the system. It also implies an infinite number of call arrivals, each with an infinitely small holding time. An example of finite traffic sources is when the number of sources offering traffic to a group of trunks or circuits is comparatively small in comparison to the number of circuits. We can also say that with a finite number of sources the arrival rate is proportional to the number of sources that are not already engaged in sending a call.

5.7 Probability-Distribution Curves

Telephone-call originations in any particular area are random in nature. We find that originating calls or call arrivals at an exchange closely fit a family of probability-distribution curves following a Poisson distribution. The Poisson distribution is fundamental to traffic theory.

Most of the common probability-distribution curves are two-parameter curves; that is, they may be described by two parameters, mean and variance. The mean is a point on the probability-distribution curve where an equal number of events occur to the right of the point as to the left of the point. "Mean" is synonymous with "average." We define mean as the x-coordinate of the center of the area under the probability-density curve for the population. The lowercase Greek letter mu (μ) is the traditional indication of the mean; $\bar{x}$ is also used.

The second parameter used to describe a distribution curve is the dispersion, which tells us how the values or population are dispersed about the center or mean of the curve. There are several measures of dispersion. One is the familiar *standard deviation*, where "the standard deviation s of a sample of n observations $x_1, x_2, \ldots, x_n$ is

$$s = \sqrt{\frac{1}{n-1} \sum_{i=1}^{n} (x_i - \bar{x})^2} \tag{1.2}$$

The *variance* V of the sample values is the square of s. The parameters for dispersion s and s^2, the standard deviation and variance, respectively, are usually denoted σ and σ^2 and give us an idea of the squatness of a distribution curve. Mean and standard deviation of a normal distribution curve are shown in Figure 1.6, where we can see that σ^2 is another measure of dispersion, the variance, or essentially the average of the squares of the distances from mean aside from the factor $n/(n-1)$.

We have introduced two distribution functions describing the probability of distribution, often called the *distribution* of x or just $f(x)$. Both functions are used in traffic engineering. But before proceeding, the variance-to-mean ratio (VMR) must also be introduced. Sometimes VMR is called the *coefficient of overdispersion*. The formula for VMR is

$$\alpha = \frac{\sigma^2}{\mu} \tag{1.3}$$

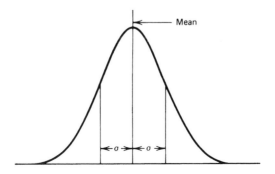

Figure 1.6. A normal distribution curve showing the mean and the standard deviation, σ.

5.8 Smooth, Rough, and Random Traffic

Traffic probability distributions can be divided into three distinct categories: (1) smooth, (2) rough, and (3) random. Each may be defined by α, the VMR. For smooth traffic, α is less than 1. For rough traffic, α is greater than 1. When α is equal to 1, the traffic distribution is called *random*. The Poisson distribution function is an example of random traffic where the VMR $= 1$. Rough traffic tends to be peakier than random or smooth traffic. For a given grade of service, more circuits are required for rough traffic because of the greater spread of the distribution curve (greater dispersion).

Smooth traffic behaves like random traffic that has been filtered. The filter is the local exchange. The local exchange looking out at its subscribers sees call arrivals as random traffic, assuming that the exchange has not been overdimensioned. The smooth traffic is the traffic on the local exchange outlets. The filtering or limiting of the peakiness is done by call blockage during the BH. Of course, the blocked traffic may actually overflow to alternative routes. Smooth traffic is characterized by a positive binomial distribution function, perhaps better known to traffic people as the *Bernoulli distribution*. An example of the Bernoulli distribution is as follows [6]. If we assume subscribers make calls independently of each other and that each has a probability p of being engaged in conversation, then if n subscribers are examined, the probability that x of them will be engaged is

$$B(x) = C_x^n p^x (1 - p)^{n-x}; \qquad 0 < x < n$$

$$\text{Its mean} = np \qquad (1.4)$$

$$\text{Its variance} = np(1 - p)$$

where the symbol C_x^n means the number of ways that x entities can be taken n at a time. Smooth traffic is assumed in dealing with small groups of subscribers; the number 200 is often used as the breakpoint [6]. That is, groups of subscribers are considered small when the subscribers number less than 200. And as mentioned, smooth traffic is also used with carried traffic. In this case the rough or random traffic would be the offered traffic.

Let's consider the binomial distribution for rough traffic. This is characterized by a negative index. Therefore, if the distribution parameters are k and q, where k is a positive number representing a hypothetical number of traffic sources and q represents the occupancy per source and may vary between 0 and 1, then

$$R'(x, k, q) = \binom{x + k - 1}{k - 1} q^x (1 - q)^k \qquad (1.5)$$

where R' is the probability of finding x calls in progress for the parameters k and q [2]. Rough traffic is used in dimensioning toll trunks with alternative routing. The symbol B (Bernoulli) is used by traffic engineers for smooth traffic and R for rough traffic. Although P may designate probability, in traffic engineering it designates Poissonian, and hence we have "P" tables such as those in Table 1.2.

The Bernoulli formula is

$$B'(x, s, h) = C_s^x h^x (1 - h)^{s-x} \qquad (1.6)$$

where C_s^x indicates the number of combinations of s things taken x at a time, h is the probability of finding the first line of an exchange busy, $1 - h$ is the probability of finding the first line idle, and s is the number of subscribers. The probability of finding two lines busy is h^2, the probability of finding s lines busy is h^s, and so on. We are interested in finding the probability of x of the s subscribers with busy lines.

The Poisson probability function can be derived from the binomial distribution, assuming that the number of subscribers s is very large and the calling rate per line h is low* such that the product $sh = m$ remains constant and letting s increase to infinity in the limit

$$P(x) = \frac{m^x}{x!} e^{-m} \qquad (1.7)$$

where

$$x = 0, 1, 2, \ldots$$

For most of our future discussion, we consider call-holding times to have a negative exponential distribution in the form

$$P = e^{-t/h} \qquad (1.8)$$

where t/h is the average holding time and in this case P is the probability of a call lasting longer than t, some arbitrary time interval.

Figure 1.7 compares smooth, random, and rough traffic probability distributions.

6   ERLANG AND POISSON TRAFFIC FORMULAS

When dimensioning a route, we want to find the number of circuits that serve the route. There are several formulas at our disposal to determine that number of circuits based on the BH traffic load. In Section 5.3 four factors were discussed that will help us to determine which traffic formula to use given a particular set of circumstances. These factors primarily dealt with (1) call arrivals and holding-time distribution, (2) number of traffic sources, (3) availability, and (4) handling of lost calls.

---

*For example, less than 50 millierlangs (mE).

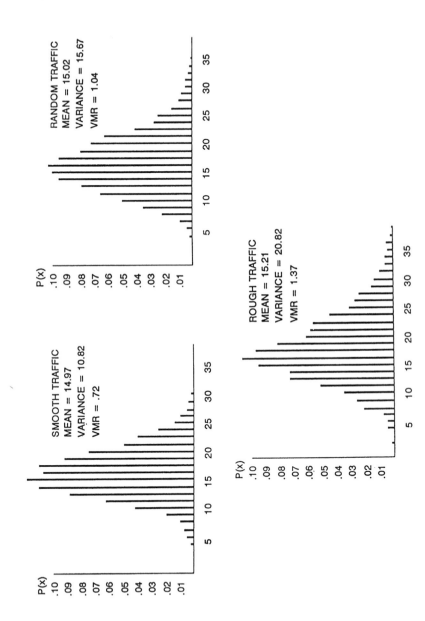

Figure 1.7. Traffic probability distributions: smooth, random, and rough traffic. Courtesy of John Lawlor Associates, Sharon, Massachusetts. (Ref. 25)

The Erlang B loss formula is probably the most common one used today outside the United States. Loss here means the probability of blockage at the switch due to congestion or to "all trunks busy" (ATB). This is expressed as grade of service E_B or the probability of finding x channels busy. The other two factors in the Erlang B formula are the mean of the *offered* traffic and the number of trunks or servicing channels available. Thus

$$E_B = \frac{A^n/n!}{1 + A + A^2/2! + \cdots + A^n/n!} \tag{1.9}$$

where n is the number of trunks or servicing channels, A is the mean of the offered traffic, and E_B is the grade of service using the Erlang B formula. This formula assumes that

- Traffic originates from an infinite number of sources.
- Lost calls are cleared assuming a zero holding time.
- The number of trunks or servicing channels is limited.
- Full availability exists.

At this point in our discussion of traffic we suggest that the reader learn to differentiate between time congestion and call congestion when dealing with grade of service. *Time congestion*, of course, refers to the decimal fraction of an hour during which all trunks are busy simultaneously. *Call congestion*, on the other hand, refers to the number of calls that fail at first attempt, which we term *lost calls*. Keep in mind that the Erlang B formula deals with offered traffic, which differs from carried traffic by the number of lost calls.

Table 1.1 is based on the Erlang B formula and gives trunk-dimensioning information for some specific grades of service, from 0.001 to 0.05 and from 1 to 150 trunks. Table 1.1 uses traffic-intensity units UC (unit call) and TU (traffic unit), where TU is in erlangs assuming BH and UC is in CCS (100 call-seconds); 1 erlang = 36 CCS (based on a 1-h time interval). To exemplify the use of Table 1.1, suppose that a route carried 16.68 erlangs of traffic with a desired grade of service of 0.001; then 30 trunks would be required. If the grade of service were reduced to 0.05, the 30 trunks could carry 24.80 erlangs of traffic. When sizing a route for trunks or an exchange, we often come up with a fractional number of servicing channels or trunks. In this case we would opt for the next highest integer because we cannot install a fraction of a trunk. For instance, if calculations show that a trunk route should have 31.4 trunks, it would be designed for 32 trunks.

The Erlang B formula, based on lost calls cleared, has been standardized by the CCITT (CCITT Rec. Q.87) and has been generally accepted outside the United States. In the United States the Poisson formula [2] is favored. This formula is often called the *Molina formula*. It is based on the LCH concept. Table 1.2 provides trunking sizes for various grades of service deriving from the P formula; such tables are sometimes called "P" tables (Poisson) and assume full availability. We must remember that the Poisson equation also assumes that traffic originates from a large (infinite) number of independent subscribers or sources (random traffic input), with a limited number of trunks or servicing channels and LCH.

It is not as straightforward as it may seem when comparing grades of service between Poisson and Erlang B formulas (or tables). The grade of service $p = 0.01$ for the Erlang B formula is equivalent to a grade of service of 0.005 when applying the Poisson (Molina) formula. Given these grades of service, assuming LCC with the Erlang B formula permits up to several tenths of erlangs less of traffic when dimensioning up to 22 trunks, where the two approaches equate (e.g., where each formula allows 12.6 erlangs over the 22 trunks). Above 22 trunks the Erlang B formula permits the trunks to carry somewhat more traffic and at 100 trunks, 2.7 erlangs more than for the Poisson formula under the LCH assumption.

6.1   Alternative Traffic Formula Conventions

Some readers may be more comfortable using traffic formulas with a different convention and notation. The Erlang B and Poisson formulas were derived from Ref. 2. The formulas and notation used in this subsection have been taken from Ref. 20. The following is the notation used in the formulas given below:

> A = The expected traffic density, expressed in busy hour erlangs.
> P = The probability that calls will be lost (or delayed) because of insufficient channels.
> n = The number of channels in the group of channels.
> s = The number of sources in the group of sources.
> p = The probability that a single source will be busy at an instant of observation. This is equal to A/s.
> x = A variable representing a number of busy sources or busy channels.
> e = The Naperian logarithmic base, which is the constant $2.71828+$.
> $\binom{m}{n}$ = The combination of m things taken n at a time.
> $\sum_{X=m}^{n}$ = The summation of all values obtained when each integer or whole number value, from m to n inclusive, is substituted for the value x in the expression following the symbol.
> ∞ = The conventional symbol for infinity.

The Poisson formula has the following assumptions: (1) infinite sources, (2) equal traffic density per source, and (3) lost calls held (LCH). The formula is

$$P = e^{-A} \sum_{x=n}^{\infty} \frac{A^x}{x!} \tag{1.10}$$

TABLE 1.1 Trunk-Loading Capacity, Based on Erlang B Formula, Full Availability

Trunks	Grade of Service 1 in 1000		Grade of Service 1 in 500		Grade of Service 1 in 200		Grade of Service 1 in 100		Grade of Service 1 in 50		Grade of Service 1 in 20	
	UC	TU	UC	TU	UC	TU	UC	TU	UC	TU	UC	TU
1	0.04	0.001	0.07	0.002	0.2	0.005	0.4	0.01	0.7	0.02	1.8	0.05
2	1.8	0.05	2.5	0.07	4	0.11	5.4	0.15	7.9	0.22	14	0.38
3	6.8	0.19	9	0.25	13	0.35	17	0.46	22	0.60	32	0.90
4	16	0.44	19	0.53	25	0.70	31	0.87	39	1.09	55	1.52
5	27	0.76	32	0.90	41	1.13	49	1.36	60	1.66	80	2.22
6	41	1.15	48	1.33	58	1.62	69	1.91	82	2.28	107	2.96
7	57	1.58	65	1.80	78	2.16	90	2.50	106	2.94	135	3.74
8	74	2.05	83	2.31	98	2.73	113	3.13	131	3.63	163	4.54
9	92	2.56	103	2.85	120	3.33	136	3.78	156	4.34	193	5.37
10	111	3.09	123	3.43	143	3.96	161	4.46	183	5.08	224	6.22
11	131	3.65	145	4.02	166	4.61	186	5.16	210	5.84	255	7.08
12	152	4.23	167	4.64	190	5.28	212	5.88	238	6.62	286	7.95
13	174	4.83	190	5.27	215	5.96	238	6.61	267	7.41	318	8.83
14	196	5.45	213	5.92	240	6.66	265	7.35	295	8.20	350	9.73
15	219	6.08	237	6.58	266	7.38	292	8.11	324	9.01	383	10.63
16	242	6.72	261	7.26	292	8.10	319	8.87	354	9.83	415	11.54
17	266	7.38	286	7.95	318	8.83	347	9.65	384	10.66	449	12.46
18	290	8.05	311	8.64	345	9.58	376	10.44	414	11.49	482	13.38
19	314	8.72	337	9.35	372	10.33	404	11.23	444	12.33	515	14.31
20	339	9.41	363	10.07	399	11.09	433	12.03	474	13.18	549	15.25
21	364	10.11	388	10.79	427	11.86	462	12.84	505	14.04	583	16.19
22	389	10.81	415	11.53	455	12.63	491	13.65	536	14.90	617	17.13
23	415	11.52	442	12.27	483	13.42	521	14.47	567	15.76	651	18.08
24	441	12.24	468	13.01	511	14.20	550	15.29	599	16.63	685	19.03

(continued)

TABLE 1.1 (continued)

Trunks	Grade of Service 1 in 1000		Grade of Service 1 in 500		Grade of Service 1 in 200		Grade of Service 1 in 100		Grade of Service 1 in 50		Grade of Service 1 in 20	
	UC	TU	UC	TU	UC	TU	UC	TU	UC	TU	UC	TU
25	467	12.97	495	13.76	540	15.00	580	16.12	630	17.50	720	19.99
26	493	13.70	523	14.52	569	15.80	611	16.96	662	18.38	754	20.94
27	520	14.44	550	15.28	598	16.60	641	17.80	693	19.26	788	21.90
28	546	15.18	578	16.05	627	17.41	671	18.64	725	20.15	823	22.87
29	573	15.93	606	16.83	656	18.22	702	19.49	757	21.04	858	23.83
30	600	16.68	634	17.61	685	19.03	732	20.34	789	21.93	893	24.80
31	628	17.44	662	18.39	715	19.85	763	21.19	822	22.83	928	25.77
32	655	18.20	690	19.18	744	20.68	794	22.05	854	23.73	963	26.75
33	683	18.97	719	19.97	774	21.51	825	22.91	887	24.63	998	27.72
34	711	19.74	747	20.76	804	22.34	856	23.77	919	25.53	1033	28.70
35	739	20.52	776	21.56	834	23.17	887	24.64	951	26.43	1068	29.68
36	767	21.30	805	22.36	864	24.01	918	25.51	984	27.34	1104	30.66
37	795	22.03	834	23.17	895	24.85	950	26.38	1017	28.25	1139	31.64
38	823	22.86	863	23.97	925	25.69	981	27.25	1050	29.17	1175	32.63
39	851	23.65	892	24.78	955	26.53	1013	28.13	1083	30.08	1210	33.61
40	880	24.44	922	25.60	986	27.38	1044	29.01	1116	31.00	1246	34.60
41	909	25.24	951	26.42	1016	28.23	1076	29.89	1149	31.92	1281	35.59
42	937	26.04	981	27.24	1047	29.08	1108	30.77	1182	32.84	1317	36.58
43	966	26.84	1010	28.06	1078	29.94	1140	31.66	1215	33.76	1353	37.57
44	995	27.64	1040	28.88	1109	30.80	1171	32.54	1248	34.68	1388	38.56
45	1024	28.45	1070	29.71	1140	31.66	1203	33.43	1282	35.61	1424	39.55
46	1053	29.26	1099	30.54	1171	32.52	1236	34.32	1315	36.53	1459	40.54
47	1083	30.07	1129	31.37	1202	33.38	1268	35.21	1349	37.46	1495	41.54
48	1111	30.88	1159	32.20	1233	34.25	1300	36.11	1382	38.39	1531	42.54
49	1141	31.69	1189	33.04	1264	35.11	1332	37.00	1415	39.32	1567	43.54

Trunks	Grade of Service 1 in 1000		Grade of Service 1 in 500		Grade of Service 1 in 200		Grade of Service 1 in 100		Grade of Service 1 in 50		Grade of Service 1 in 20	
	UC	TU	UC	TU	UC	TU	UC	TU	UC	TU	UC	TU
50	1170	32.51	1220	33.88	1295	35.98	1364	37.90	1449	40.25	1603	44.53
51	1200	33.33	1250	34.72	1327	36.85	1397	38.80				
52	1229	34.15	1280	35.56	1358	37.72	1429	39.70				
53	1259	34.98	1310	36.40	1390	38.60	1462	40.60				
54	1289	35.80	1341	37.25	1421	39.47	1494	41.50				
55	1319	36.63	1371	38.09	1453	40.35	1527	42.41				
56	1349	37.46	1402	38.94	1484	41.23	1559	43.31				
57	1378	38.29	1432	39.79	1516	42.11	1592	44.22				
58	1408	39.12	1463	40.64	1548	42.99	1625	45.13				
59	1439	39.96	1494	41.50	1579	43.87	1657	46.04				
60	1468	40.79	1525	42.35	1611	44.76	1690	46.95				
61	1499	41.63	1556	43.21	1643	45.64	1723	47.86				
62	1529	42.47	1587	44.07	1675	46.53	1756	48.77				
63	1559	43.31	1617	44.93	1707	47.42	1789	49.69				
64	1590	44.16	1648	45.79	1739	48.31	1822	50.60				
65	1620	45.00	1679	46.65	1771	49.20	1855	51.52				
66	1650	45.84	1710	47.51	1803	50.09	1888	52.44				
67	1681	46.69	1742	48.38	1835	50.98	1921	53.35				
68	1711	47.54	1773	49.24	1867	51.87	1954	54.27				
69	1742	48.39	1804	50.11	1900	52.77	1987	55.19				
70	1773	49.24	1835	50.98	1932	53.66	2020	56.11				
71	1803	50.09	1867	51.85	1964	54.56	2053	57.03				
72	1834	50.94	1898	52.72	1996	55.45	2087	57.96				
73	1865	51.80	1929	53.59	2029	56.35	2120	58.88				
74	1895	52.65	1960	54.46	2061	57.25	2153	59.80				
75	1926	53.51	1992	55.34	2093	58.15	2186	60.73				

(continued)

TABLE 1.1 (continued)

Trunks	Grade of Service 1 in 1000		Grade of Service 1 in 500		Grade of Service 1 in 200		Grade of Service 1 in 100		Grade of Service 1 in 50		Grade of Service 1 in 20	
	UC	TU	UC	TU	UC	TU	UC	TU	UC	TU	UC	TU
76	1957	54.37	2024	56.21	2126	59.05	2219	61.65				
77	1988	55.23	2055	57.09	2159	59.96	2253	62.58				
78	2019	56.09	2087	57.96	2191	60.86	2286	63.51				
79	2050	56.95	2118	58.84	2223	61.76	2319	64.43				
80	2081	57.81	2150	59.72	2256	62.67	2353	65.36				
81	2112	58.67	2182	60.60	2289	63.57	2386	66.29				
82	2143	59.54	2213	61.48	2321	64.48	2420	67.22				
83	2174	60.40	2245	62.36	2354	65.38	2453	68.15				
84	2206	61.27	2277	63.24	2386	66.29	2487	69.08				
85	2237	62.14	2308	64.13	2419	67.20	2521	70.02				
86	2268	63.00	2340	65.01	2452	68.11	2554	70.95				
87	2299	63.87	2372	65.90	2485	69.02	2588	71.88				
88	2330	64.74	2404	66.78	2517	69.93	2621	72.81				
89	2362	65.61	2436	67.67	2550	70.84	2655	73.75				
90	2393	66.48	2468	68.56	2583	71.76	2688	74.68				
91	2425	67.36	2500	69.44	2616	72.67	2722	75.62				
92	2456	68.23	2532	70.33	2650	73.58	2756	76.56				
93	2488	69.10	2564	71.22	2682	74.49	2790	77.49				
94	2519	69.98	2596	72.11	2715	75.41	2823	78.43				
95	2551	70.85	2628	73.00	2748	76.32	2857	79.37				
96	2582	71.73	2660	73.90	2781	77.24	2891	80.31				
97	2614	72.61	2692	74.79	2814	78.16	2925	81.24				
98	2645	73.48	2724	75.68	2847	79.07	2958	82.18				
99	2677	74.36	2757	76.57	2880	79.99	2992	83.12				
100	2709	75.24	2789	77.47	2913	80.91	3026	84.06				

Trunks	Grade of Service 1 in 1000		Grade of Service 1 in 500		Grade of Service 1 in 200		Grade of Service 1 in 100		Grade of Service 1 in 50		Grade of Service 1 in 20	
	UC	TU	UC	TU	UC	TU	UC	TU	UC	TU	UC	TU
101	2740	76.12	2821	78.36	2946	81.83	3060	85.00				
102	2772	77.00	2853	79.26	2979	82.75	3094	85.95				
103	2804	77.88	2886	80.16	3012	83.67	3128	86.89				
104	2836	78.77	2918	81.05	3045	84.59	3162	87.83				
105	2867	79.65	2950	81.95	3078	85.51	3196	88.77				
106	2899	80.53	2983	82.85	3111	86.43	3230	89.72				
107	2931	81.42	3015	83.75	3145	87.35	3264	90.66				
108	2963	82.30	3047	84.65	3178	88.27	3298	91.60				
109	2995	83.19	3080	85.55	3211	89.20	3332	92.55				
110	3027	84.07	3112	86.45	3244	90.12	3366	93.49				
111	3059	84.96	3145	87.35	3277	91.04	3400	94.44				
112	3091	85.85	3177	88.25	3311	91.97	3434	95.38				
113	3122	86.73	3209	89.15	3344	92.89	3468	96.33				
114	3154	87.62	3242	90.06	3378	93.82	3502	97.28				
115	3186	88.51	3275	90.96	3411	94.74	3536	98.22				
116	3218	89.40	3307	91.86	3444	95.67	3570	99.17				
117	3250	90.29	3340	92.77	3478	96.60	3604	100.12				
118	3282	91.18	3372	93.67	3511	97.53	3639	101.07				
119	3315	92.07	3405	94.58	3544	98.45	3673	102.02				
120	3347	92.96	3437	95.48	3578	99.38	3707	102.96				
121	3379	93.86	3470	96.39	3611	100.31	3741	103.91				
122	3411	94.75	3503	97.30	3645	101.24	3775	104.86				
123	3443	95.64	3535	98.20	3678	102.17	3809	105.81				
124	3475	96.54	3568	99.11	3712	103.10	3843	106.76				
125	3507	97.43	3601	100.02	3745	104.03	3878	107.71				

(continued)

TABLE 1.1 *(continued)*

Trunks	Grade of Service 1 in 1000		Grade of Service 1 in 500		Grade of Service 1 in 200		Grade of Service 1 in 100		Grade of Service 1 in 50		Grade of Service 1 in 20	
	UC	TU	UC	TU	UC	TU	UC	TU	UC	TU	UC	TU
126	3540	98.33	3633	100.93	3779	104.96	3912	108.66				
127	3572	99.22	3666	101.84	3812	105.89	3946	109.62				
128	3604	100.12	3699	102.75	3846	106.82	3981	110.57				
129	3636	101.01	3732	103.66	3879	107.75	4015	111.52				
130	3669	101.91	3765	104.57	3912	108.68	4049	112.47				
131	3701	102.81	3797	105.48	3946	109.62	4083	113.42				
132	3733	103.70	3830	106.39	3980	110.55	4118	114.38				
133	3766	104.60	3863	107.30	4013	111.48	4152	115.33				
134	3798	105.50	3896	108.22	4047	112.42	4186	116.28				
135	3830	106.40	3929	109.13	4081	113.35	4221	117.24				
136	3863	107.30	3961	110.04	4114	114.28	4255	118.19				
137	3895	108.20	3994	110.95	4148	115.22	4289	119.14				
138	3928	109.10	4027	111.87	4181	116.15	4324	120.10				
139	3960	110.00	4060	112.78	4215	117.09	4358	121.05				
140	3992	110.90	4093	113.70	4249	118.02	4392	122.01				
141	4025	111.81	4126	114.61	4283	118.96	4427	122.96				
142	4058	112.71	4159	115.53	4316	119.90	4461	123.92				
143	4090	113.61	4192	116.44	4350	120.83	4496	124.88				
144	4122	114.51	4225	117.36	4384	121.77	4530	125.83				
145	4155	115.42	4258	118.28	4418	122.71	4564	126.79				
146	4188	116.32	4291	119.19	4451	123.64	4599	127.74				
147	4220	117.22	4324	120.11	4485	124.58	4633	128.70				
148	4253	118.13	4357	121.03	4519	125.52	4668	129.66				
149	4285	119.03	4390	121.95	4552	126.46	4702	130.62				
150	4318	119.94	4423	122.86	4586	127.40	4737	131.58				

Source: Courtesy of GTE Automatic Electric Company (Bulletin No. 485).

TABLE 1.2 Trunk-Loading Capacity, Based on Poisson Formula, Full Availability

Trunks	Grade of Service 1 in 1000		Grade of Service 1 in 100		Grade of Service 1 in 50		Grade of Service 1 in 20		Grade of Service 1 in 10	
	UC	TU	UC	TU	UC	TU	UC	TU	UC	TU
1	0.1	0.003	0.4	0.01	0.7	0.02	1.9	0.05	3.8	0.10
2	1.6	0.05	5.4	0.15	7.9	0.20	12.9	0.35	19.1	0.55
3	6.9	0.20	16	0.45	20	0.55	29.4	0.80	39.6	1.10
4	15	0.40	30	0.85	37	1.05	49	1.35	63	1.75
5	27	0.75	46	1.30	56	1.55	71	1.95	88	2.45
6	40	1.10	64	1.80	76	2.10	94	2.60	113	3.15
7	55	1.55	84	2.35	97	2.70	118	3.25	140	3.90
8	71	1.95	105	2.90	119	3.30	143	3.95	168	4.65
9	88	2.45	126	3.50	142	3.95	169	4.70	195	5.40
10	107	2.95	149	4.15	166	4.60	195	5.40	224	6.20
11	126	3.50	172	4.80	191	5.30	222	6.15	253	7.05
12	145	4.05	195	5.40	216	6.00	249	6.90	282	7.85
13	166	4.60	220	6.10	241	6.70	277	7.70	311	8.65
14	187	5.20	244	6.80	267	7.40	305	8.45	341	9.45
15	208	5.80	269	7.45	293	8.15	333	9.25	370	10.30
16	231	6.40	294	8.15	320	8.90	362	10.05	401	11.15
17	253	7.05	320	8.90	347	9.65	390	10.85	431	11.95
18	276	7.65	346	9.60	374	10.40	419	11.65	462	12.85
19	299	8.30	373	10.35	401	11.15	448	12.45	492	13.65
20	323	8.95	399	11.10	429	11.90	477	13.25	523	14.55
21	346	9.60	426	11.85	458	12.70	507	14.10	554	15.40
22	370	10.30	453	12.60	486	13.50	536	14.90	585	16.25
23	395	10.95	480	13.35	514	14.30	566	15.70	616	17.10
24	419	11.65	507	14.10	542	15.05	596	16.55	647	17.95

TABLE 1.2 *(continued)*

Trunks	Grade of Service 1 in 1000		Grade of Service 1 in 100		Grade of Service 1 in 50		Grade of Service 1 in 20		Grade of Service 1 in 10	
	UC	TU	UC	TU	UC	TU	UC	TU	UC	TU
25	444	12.35	535	14.85	572	15.90	626	17.40	678	18.85
26	469	13.05	562	15.60	599	16.65	656	18.20	710	19.70
27	495	13.75	590	16.40	627	17.40	686	19.05	741	20.60
28	520	14.45	618	17.15	656	18.20	717	19.90	773	21.45
29	545	15.15	647	17.95	685	19.05	747	20.75	805	22.35
30	571	15.85	675	18.75	715	19.85	778	21.60	836	23.20
31	597	16.60	703	19.55	744	20.65	809	22.45	868	24.10
32	624	17.35	732	20.35	773	21.45	840	23.35	900	25.00
33	650	18.05	760	21.10	803	22.30	871	24.20	932	25.90
34	676	18.80	789	21.90	832	23.10	902	25.05	964	26.80
35	703	19.55	818	22.70	862	23.95	933	25.90	996	27.65
36	729	20.25	847	23.55	892	24.80	964	26.80	1028	28.55
37	756	21.00	876	24.35	922	25.60	995	27.65	1060	29.45
38	783	21.75	905	25.15	951	26.40	1026	28.50	1092	30.35
39	810	22.50	935	25.95	982	27.30	1057	29.35	1125	31.25
40	837	23.25	964	26.80	1012	28.10	1088	30.20	1157	32.14
41	865	24.05	993	27.60	1042	28.95	1120	31.10	1190	33.05
42	892	24.80	1023	28.40	1072	29.80	1151	31.95	1222	33.95
43	919	25.55	1052	29.20	1103	30.65	1183	32.85	1255	34.85
44	947	26.30	1082	30.05	1133	31.45	1214	33.70	1287	35.75
45	975	27.10	1112	30.90	1164	32.35	1246	34.60	1320	36.65
46	1003	27.85	1142	31.70	1194	33.15	1277	35.45	1352	37.55
47	1030	28.60	1171	32.55	1225	34.05	1309	36.35	1385	38.45
48	1058	29.40	1201	33.35	1255	34.85	1340	37.20	1417	39.35
49	1086	30.15	1231	34.20	1286	35.70	1372	38.10	1450	40.30
50	1115	30.95	1261	35.05	1317	36.60	1403	38.95	1482	41.15

TABLE 1.2 *(continued)*

Trunks	Grade of Service 1 in 1000		Grade of Service 1 in 100		Grade of Service 1 in 50		Grade of Service 1 in 20		Grade of Service 1 in 10	
	UC	TU	UC	TU	UC	TU	UC	TU	UC	TU
51	1143	31.75	1291	35.85	1349	37.45				
52	1171	32.55	1322	36.70	1380	38.35				
53	1200	33.35	1352	37.55	1410	39.15				
54	1228	34.10	1382	38.40	1441	40.05				
55	1256	34.90	1412	39.20	1472	40.90				
56	1285	35.70	1443	40.10	1503	41.75				
57	1313	36.45	1473	40.90	1534	42.60				
58	1342	37.30	1504	41.80	1565	43.45				
59	1371	38.10	1534	42.60	1596	44.35				
60	1400	38.90	1565	43.45	1627	45.20				
61	1428	39.65	1595	44.30	1659	46.10				
62	1457	40.45	1626	45.15	1690	46.95				
63	1486	41.30	1657	46.05	1722	47.85				
64	1516	42.10	1687	46.85	1752	48.65				
65	1544	42.90	1718	47.70	1784	49.55				
66	1574	43.70	1749	48.60	1816	50.45				
67	1603	44.55	1780	49.45	1847	51.30				
68	1632	45.35	1811	50.30	1878	52.15				
69	1661	46.15	1842	51.15	1910	53.05				
70	1691	46.95	1873	52.05	1941	53.90				
71	1720	47.80	1904	52.90	1973	54.80				
72	1750	48.60	1935	53.75	2004	55.65				
73	1779	49.40	1966	54.60	2036	56.55				
74	1809	50.25	1997	55.45	2067	57.40				

(continued)

29

TABLE 1.2 *(continued)*

Trunks	Grade of Service 1 in 1000		Grade of Service 1 in 100		Grade of Service 1 in 50		Grade of Service 1 in 20		Grade of Service 1 in 10	
	UC	TU	UC	TU	UC	TU	UC	TU	UC	TU
75	1838	51.05	2028	56.35	2099	58.30				
76	1868	51.90	2059	57.20	2130	59.15				
77	1898	52.70	2091	58.10	2162	60.05				
78	1927	53.55	2122	58.95	2194	60.95				
79	1957	54.35	2153	59.80	2226	61.85				
80	1986	55.15	2184	60.65	2258	62.70				
81	2016	56.00	2215	61.55	2290	63.60				
82	2046	56.85	2247	62.40	2321	64.45				
83	2076	57.65	2278	63.30	2354	65.40				
84	2106	58.50	2310	64.15	2386	66.30				
85	2136	59.35	2341	65.05	2418	67.15				
86	2166	60.15	2373	65.90	2451	68.10				
87	2196	61.00	2404	66.80	2483	68.95				
88	2226	61.85	2436	67.65	2515	69.85				
89	2256	62.65	2467	68.55	2547	70.75				
90	2286	63.50	2499	69.40	2579	71.65				
91	2317	64.35	2530	70.30	2611	72.55				
92	2346	65.15	2562	71.15	2643	73.40				
93	2377	66.05	2594	72.05	2674	74.30				
94	2407	66.85	2625	72.90	2706	75.15				
95	2437	67.70	2657	73.80	2739	76.10				
96	2468	68.55	2689	74.70	2771	76.95				
97	2498	69.40	2721	75.60	2803	77.85				
98	2528	70.20	2752	76.45	2836	78.80				
99	2559	71.10	2784	77.35	2868	79.65				
100	2589	71.90	2816	78.20	2900	80.55				

The Erlang B formula assumes (1) infinite sources, (2) equal traffic density per source, and (3) lost calls to cleared (LCC). The formula is

$$P = \frac{\dfrac{A^n}{n!}}{\displaystyle\sum_{x=0}^{n} \dfrac{A^x}{x!}} \tag{1.11}$$

The Erlang C formula assumes (1) infinite sources, (2) lost calls delayed (LCD), (3) exponential holding times, and (4) calls served in order of arrival. The formula is

$$P = \frac{\dfrac{A^n}{n!} \cdot \dfrac{n}{n-A}}{\displaystyle\sum_{x=0}^{n-1} \dfrac{A^x}{x!} + \dfrac{A^n}{n!}\dfrac{n}{n-A}} \tag{1.12}$$

The binomial formula assumes (1) finite sources, (2) equal traffic density per source, and (3) lost calls held (LCH). The formula is

$$P = \left(\frac{s-A}{s}\right)^{s-1} \sum_{x=n}^{s-1} \binom{s-1}{x}\left(\frac{A}{s-A}\right)^x \tag{1.13}$$

(Also consult Ref. 19.)

6.2 Computer Programs for Traffic Calculations

6.2.1 Erlang B Computer Program

Relatively simple programs can be written in BASIC computer language to solve the Erlang B equation shown below:

$$P = \frac{A^N/N!}{1 + A + A^2/2! + A^3/3! + \cdots + A^N/N!}$$

Following is a sample program. It is written using GWBASIC, Version 3.11.

```
10   PRINT "ERLANG B, TRAFFIC ON EACH TRUNK."
20   INPUT "OFFERED TRAFFIC (ERLANG) = ",A
30   LET B = 0: LET H = 0
40   INPUT "MINIMUM GRADE OF SERVICE (DEFAULT = 1) = ",H
50   IF H = 0 THEN LET H = 1
60   INPUT "OBJECTIVE GRADE OF SERVICE (DEFAULT = 0.001) = ",B
70   IF B = 0 THEN LET B = .001
80   LET N = 0: LET T = 1: LET T1 = 1: LET 0 = A: LET Z = 15
90   PRINT "TRUNKS        OFFERED        CARRIED CUMULATIVE P = G/S"
```

```
100   PRINT
"  - - - - - - - - - - - - - - - - - - - - - - - - - - - - - - - - - - - - - - - - - - - - - - - - - - ''
110   IF T1 > 9.999999E + 20 THEN 230
120   LET N = N + 1: LET T = T * A/N: LET T1 = T1 + T: LET P = T/T1
130   LET L = A * P: LET S = A - L: LET C = 0 - L
140   IF P > H THEN LET Z = N + 14: GOTO 110
150   PRINT USING "###                ";N,
160   PRINT USING "###.####           ";0,
170   PRINT USING "    .####           ";C,
180   PRINT USING "###.####           ";S,
190   PRINT USING ".####";P
200   LET 0 = L: IF P < B THEN 240
210   IF N < Z THEN 110
220   STOP: LET Z = Z + 15: PRINT "OFFERED TRAFFIC (ERLANG) IS ",A: GOTO 90
230   LET T = T * 1E - 37: LET T1 = T1 * 1E - 37: GOTO 120
240   END
```

For a given traffic load, this program shows the load offered to each trunk, the load carried by each trunk, the cumulative load carried by the trunk group, and the probability of blocking for a group with "N" trunks. This assumes that the trunks are used in sequential order. The results are presented on a trunk-by-trunk basis until the probability of blocking reaches some minimum value [$Pr(Blocking) = 0.001$ in this version]. Both the minimum and maximum probability values can be adjusted by entering input values when starting the program.

The variable "Z" is used to stop the calculation routine every "Z" lines so that the data will fill one screen at a time. Calculation will continue when the instruction "CONT" is entered (or by pressing the "F5" key in some computers).

In order to avoid overflow in the computer registers when very large numbers are encountered, the least significant digits are dropped for large traffic loads (see Instruction 110 and Instruction 230). The variables affected by this adjustment (T and T1) are used only to determine the probability ratio, and the error introduced is negligible. As the traffic load A becomes very large, the error increases.

Following is a sample program calculation for an offered traffic load of 10 erlangs:

ERLANG B, TRAFFIC ON EACH TRUNK.
OFFERED TRAFFIC (ERLANG) = 10
MINIMUM GRADE OF SERVICE (DEFAULT = 1) =
OBJECTIVE GRADE OF SERVICE (DEFAULT = 0.001) =

TRUNKS	OFFERED	CARRIED	CUMULATIVE	P = G/S
1	10.0000	0.9091	0.9091	0.9091
2	9.0909	0.8942	1.8033	0.8197
3	8.1967	0.8761	2.6794	0.7321
4	7.3206	0.8540	3.5334	0.6467
5	6.4666	0.8271	4.3605	0.5640

TRUNKS	OFFERED	CARRIED	CUMULATIVE	P = G/S
6	5.6395	0.7944	5.1549	0.4845
7	4.8451	0.7547	5.9096	0.4090
8	4.0904	0.7072	6.6168	0.3383
9	3.3832	0.6511	7.2679	0.2732
10	2.7321	0.5863	7.8542	0.2146
11	2.1458	0.5135	8.3677	0.1632
12	1.6323	0.4349	8.8026	0.1197
13	1.1974	0.3540	9.1566	0.0843
14	0.8434	0.2752	9.4318	0.0568
15	0.5682	0.2032	9.6350	0.0365
16	0.3650	0.1420	9.7770	0.0223
17	0.2230	0.0935	9.8705	0.0129
18	0.1295	0.0581	9.9286	0.0071
19	0.0714	0.0340	9.9625	0.0037
20	0.0375	0.0188	9.9813	0.0019
21	0.0187	0.0098	9.9911	0.0009

6.2.2 Poisson Computer Program

Relatively simple programs can be written in BASIC computer language to solve the Poisson equation shown below:

$$P = \frac{A^N/N!}{1 + A + A^2/2! + A^3/3! + \cdots} = \frac{A^N/N!}{e^A} = \frac{A^N}{N!}e^{-A}$$

Following is a sample program which can be used for traffic loads up to 86 erlangs. (Loads greater than 86 erlangs may cause register overflow in some computers.) It is written using GWBASIC, Version 3.11.

```
10   PRINT "POISSON (86 ERLANG MAXIMUM)"
20   PRINT "P = PROBABILITY OF N TRUNKS BUSY"
30   PRINT "P1 = PROBABILITY OF BLOCKING"
40   INPUT "OFFERED TRAFFIC IN ERLANGS = ",A: LET E = EXP(A)
50   LET N = 0: LET T = 1: LET T1 = 1: LET T2 = 0: LET Z = 15
60   PRINT "TRUNKS PR (N TRUNKS BUSY) PR (BLOCKING)"
70   PRINT "- - - - - - - - - - - - - - - - - - - - - - - - - - -"
80   LET P1 = 1 - T2: LET T2 = T1/E: LET P = T/E
90   IF P<.00001 THEN LET Z = N + 14: GOTO 170
100  PRINT USING "###     ";N,
110  PRINT USING "        #.####      ";P,P1
120  IF N<Z THEN 180
130  STOP: LET Z = Z + 15
140  PRINT "OFFERED TRAFFIC IN ERLANGS IS ";A
150  PRINT "TRUNKS PR (N TRUNKS BUSY) PR (BLOCKING)"
160  PRINT "- - - - - - - - - - - - - - - - - - - - - - - - - - -"
170  IF N<Z THEN 120: LET Z = Z + 15
180  LET N = N + 1: LET T = T * A/N: LET T1 = T1 + T
190  IF P1>.0001 THEN 80
200  END
```

This program calculates both (a) the probability that exactly "N" trunks are busy and (b) the probability of blocking (the probability that "N" or more service requests are received). The results are presented on a trunk-by-trunk basis until the probability of blocking reaches some minimum value [P1 = Pr(Blocking) = 0.0001 in this version]. Both the minimum and maximum probabilities can be adjusted by changing the values of P (Instruction 90) and P1 (Instruction 190).

The variable "Z" is used to stop the calculation routine every "Z" lines so that the data will fill one screen at a time. Calculation will continue when the instruction "CONT" is entered (or by pressing the "F5" key in some computers).

Following is a sample program calculation for an offered traffic load of 10 erlangs:

```
POISSON (86 ERLANG MAXIMUM)
P = PROBABILITY OF N TRUNKS BUSY
P1 = PROBABILITY OF BLOCKING
OFFERED TRAFFIC IN ERLANGS = 10
```

TRUNKS	PR (N TRUNKS BUSY)	PR (BLOCKING)
0	0.0000	1.0000
1	0.0005	1.0000
2	0.0023	0.9995
3	0.0076	0.9972
4	0.0189	0.9897
5	0.0378	0.9707
6	0.0631	0.9329
7	0.0901	0.8699
8	0.1126	0.7798
9	0.1251	0.6672
10	0.1251	0.5421
11	0.1137	0.4170
12	0.0948	0.3032
13	0.0729	0.2084
14	0.0521	0.1355
15	0.0347	0.0835
16	0.0217	0.0487
17	0.0128	0.0270
18	0.0071	0.0143
19	0.0037	0.0072
20	0.0019	0.0035
21	0.0009	0.0016
22	0.0004	0.0007
23	0.0002	0.0003
24	0.0001	0.0001
25	0.0000	0.0000

6.2.3 Erlang C Computer Program

Relatively simple programs can be written in BASIC computer language to solve the Erlang C equation shown below:

$$P = \frac{(A^N/N!)(N/(N-A))}{1 + A + A^2/2! + A^3/3! + \cdots + (A^N/N!)(N/(N-A))}$$

Following is a sample program. It is written using GWBASIC, Version 3.11.

```
10   PRINT "ERLANG C CALCULATES DELAY"
20   INPUT "OFFERED TRAFFIC (ERLANGS) = ",A
30   PRINT "TRUNKS      P        D1        D2        Q1        Q2";
40   PRINT "    P8      P4      P2      P1      PP"
50   PRINT
"- - - - - - - - - - - - - - - - - - - - - - - - - - - - - - - - - - -."
60   LET N = 1: LET T = 1: LET T1 = 1
70   IF T1>1E + 21 THEN 200
80   IF N<=A THEN 190
90   IF N = A + 1 THEN LET Z = A + 15
100  LET T2 = T * (A / N) * (N / (N - A)): LET P = T2 / (T1 + T2)
110  LET D2 = 1 / (N - A): LET D1 = P * D2: LET Q2 = A * D2: LET Q1 = P * Q2
120  LET PO = P / EXP(2 / D2)
130  PRINT USING "###         ";N,
140  PRINT USING ".####       ";P,
150  PRINT USING "###.##   ";D1,D2,Q1,Q2,
160  PRINT USING ".##         ";P0
170  IF N > Z THEN 210
180  IF P < .02 THEN 250
190  LET T = T * A / N: LET T1 = T1 + T: LET N = N + 1: GOTO 70
200  LET T = T * 1E - 37: LET T1 = T1 * 1E - 37: GOTO 80
210  STOP: LET Z = Z + 15
220  PRINT "OFFERED TRAFFIC (ERLANGS) = ",A
230  PRINT "TRUNKS      P     D1     D2     Q1     Q2     PP"
240  PRINT " - - - - - - - - - - - - - - - - - - - - - - - - - ": GOTO 180
250  END
```

For a given traffic load, this program shows the probability of delay for a group with "N" trunks, with call delays expressed in units of average holding time:

P   probability that a call will be delayed
D1   average delay on all calls, including those not delayed
D2   average delay on calls that are delayed
Q1   average calls in queue
Q2   average calls in queue when all servers are busy
PP   probability of delay exceeding two holding time intervals

The program calculates delay and queue characteristics by using the following relationships derived from the Erlang C formula:

The average delay on all calls, D1, including those not delayed, is given by

$$D1 = P(> 0)(h/(N - A)), \quad \text{where h is the average holding time}$$

The average delay, D2, on calls delayed is given by

$$D2 = h/(N - A), \quad \text{where h is the average holding time}$$

The average calls in queue, Q2, when all servers are busy is given by

$$Q2 = Ah/(N - A) = AD2$$

The average calls in queue, Q1, is given by

$$Q1 = P(> 0)(Ah/(N - A)) = P(> 0)Q2$$

The probability of delay exceeding some multiple of the average holding time, PP (PP > 2 ∗ HT is used in the program), is given by

$$PP = P(> 0)/e^{(XHT/D2)}, \quad \text{where X is the number of HT intervals}$$

This assumes that the trunks are used in sequential order. The results are presented on a trunk-by-trunk basis beginning with "A + 1" trunks to ensure that there are sufficient trunks available to carry the busy hour traffic after some delay has occurred. This process continues until the probability of delay is small enough that delays are relatively short [Pr(Delay) = .02 in this version]. The minimum delay probability can be adjusted as desired.

The variable "Z" is used to stop the calculation routine every "Z" lines so that the data will fill one screen at a time. Calculation will continue when the instruction "CONT" is entered (or by pressing the "F5" key in some computers).

In order to avoid overflow in the computer registers when very large numbers are encountered, the least significant digits are dropped for large traffic loads (see Instruction 70 and Instruction 200). The variables affected by this adjustment (T and T1) are used only to determine the probability ratio, and the error introduced is negligible. As the traffic load A becomes very large, the error increases.

Following is a sample program calculation for an offered traffic load of 10 erlangs:

```
ERLANG C CALCULATES DELAY
OFFERED TRAFFIC (ERLANGS) = 10
```

TRUNKS	P	D1	D2	Q1	Q2	PP
11	0.6821	0.68	1.00	6.82	10.00	0.09
12	0.4494	0.22	0.50	2.25	5.00	0.01
13	0.2853	0.10	0.33	0.95	3.33	0.00
14	0.1741	0.04	0.25	0.44	2.50	0.00
15	0.1020	0.02	0.20	0.20	2.00	0.00

TRUNKS	P	D1	D2	Q1	Q2	PP
16	0.0573	0.01	0.17	0.10	1.67	0.00
17	0.0309	0.00	0.14	0.04	1.43	0.00
18	0.0159	0.00	0.13	0.02	1.25	0.00

The information in Section 6.2 was graciously provided by John Lawlor Associates, Sharon, MA.

7   WAITING SYSTEMS (QUEUEING)

A short discussion follows regarding traffic in queueing systems. Queueing or waiting systems, when dealing with traffic, are based on the third assumption, namely, lost calls delayed (LCD). Of course, a queue in this case is a pool of callers waiting to be served by a switch. The term *serving time* is the time a call takes to be served from the moment of arrival in the queue to the moment of being served by the switch. For traffic calculations in most telecommunication queueing systems, the mathematics is based on the assumption that call arrivals are random and Poissonian. The traffic engineer is given the parameters of offered traffic, the size of the queue, and a specified grade of service and will determine the number of serving circuits or trunks required.

The method by which a waiting call is selected to be served from the pool of waiting calls is called *queue discipline*. The most common discipline is the first-come, first-served discipline, where the call waiting longest in the queue is served first. This can turn out to be costly because of the equipment required to keep order in the queue. Another type is random selection, where the time a call has waited is disregarded and those waiting are selected in random order. There is also the last-come, first-served discipline and bulk service discipline, where batches of waiting calls are admitted, and there are also priority service disciplines, which can be preemptive and nonpreemptive. In queueing systems the grade of service may be defined as the probability of delay. This is expressed as $P(t)$, the probability that a call is not being immediately served and has to wait a period of time greater than t. The average delay on all calls is another parameter that can be used to express grade of service, and the length of queue is another.

The probability of delay, the most common index of grade of service for waiting systems when dealing with full availability and a Poissonian call arrival process, is calculated by using the Erlang C formula, which assumes an infinitely long queue length. Syski [3] provides a good guide to Erlang C and other, more general waiting systems. (Also consult Refs. 13, 14, and 15.)

7.1   Server-Pool Traffic

Server pools are groups of traffic resources, such as signaling registers and operator positions, that are used on a shared basis. Service requests that cannot be satisfied immediately are placed in a queue and served on a first-in, first-out (FIFO) basis. Server-pool traffic is directly related to offered traffic, server-holding time, and call-attempt factor, and inversely related to call-holding time. This is expressed in equation 1.14.

$$A_S = \frac{A_T \cdot T_S \cdot C}{T_C} \qquad (1.14)$$

where A_S = server-pool traffic in erlangs
A_T = total traffic served in erlangs
T_S = mean server-holding time in hours
T_C = mean call-holding time in hours
C = call-attempt factor (dimensionless)

Total traffic served refers to the total offered traffic that requires the services of the specific server pool for some portion of the call. For example, a dual-tone multifrequency (DTMF) receiver pool is dimensioned to serve only the DTMF tone-dialing portion of total switch traffic generated by DTMF signaling sources.

Table 1.3 gives representative server-holding times for typical signaling registers as a function of the number of digits received or sent.

The mean server-holding time is the arithmetic average of all server-holding times for the specific server pool. Equation 1.15 can be used to calculate mean server-holding time for calls with different holding-time characteristics.

$$T_S = a \cdot T_1 + b \cdot T_2 + \cdots + k \cdot T_n \qquad (1.15)$$

where T_S = mean server-holding time in hours
$T_1, T_2, \ldots, T_n$ = individual server-holding times in hours
$a, b, \ldots, k$ = fractions of total traffic served $(a + b + \cdots + k = 1)$

Consider the following example. Determine the mean DTMF receiver-holding time for a switch where subscribers dial local calls using a 7-digit

TABLE 1.3 Typical Signaling Register Holding Times in Seconds

	Number of Digits Received or Sent				
Signaling Register	1	4	7	10	11
Local dial-pulse (DP) receiver	3.7	8.3	12.8	17.6	19.1
Local DTMF receiver	2.3	5.2	8.1	11.0	12.0
Incoming MF receiver	1.0	1.4	1.8	2.2	2.3
Outgoing MF sender	1.5	1.9	2.3	2.8	3.0

number and long-distance calls using an 11-digit number. Assume that 70% of the calls are local calls, the remainder are long-distance calls, and the typical signaling register holding times found in Table 1.3 are applicable.

$$T_s = (0.7)(8.1 \text{ s}) + (0.3)(12.0 \text{ s}) = 9.27 \text{ s}$$

Call-attempt factors are dimensionless numbers that adjust offered traffic intensity to compensate for call attempts that do not result in completed calls. Therefore, call-attempt factors are inversely proportional to the fraction of completed calls as defined in equation 1.16.

$$C = 1/k \tag{1.16}$$

where $C =$ call-attempt factor (dimensionless)
$k =$ fraction of calls completed (decimal fraction)

Here is another example. Table 1.4 gives representative subscriber call-attempt dispositions based on empirical data taken from the large North American PSTN data base. Determine the call-attempt factors for these data, where 70.7% of the calls were completed ($k = 0.707$).

$$C = \frac{1}{k} = \frac{1}{0.707} = 1.414$$

Section 7.1 is based on Section 1.2.2 of *Traffic System Design Handbook* by James R. Boucher, IEEE Press, 1992 [26].

8 DIMENSIONING AND EFFICIENCY

By definition, if we were to dimension a route or estimate the required number of servicing channels, where the number of trunks (or servicing channels) just equaled the erlang load, we would attain 100% efficiency. All trunks would be busy with calls all the time or at least for the entire BH. This would not even allow several moments for a trunk to be idle while the switch

TABLE 1.4 Typical Call-Attempt Dispositions

Call-Attempt Disposition	Percentage
Call was completed	70.7
Called subscriber did not answer	12.7
Called subscriber line was busy	10.1
Call abandoned without system response	2.6
Equipment blockage or failure	1.9
Customer dialing error	1.6
Called directory number changed or disconnected	0.4

decided the next call to service. In practice, if we engineered our trunks, trunk routes, or switches this way, there would be many unhappy subscribers.

On the other hand, we do, indeed, want to size our routes (and switches) to have a high efficiency and still keep our customers relatively happy. The goal of our previous exercises in traffic engineering was just that. The grade of service is one measure of subscriber satisfaction. As an example, let us assume that between cities X and Y there are 100 trunks on the interconnecting telephone route. The tariffs, from which the telephone company derives revenue, are a function of the erlangs of carried traffic. Suppose we allow a dollar per erlang-hour. The very upper limit of service on the route is 100 erlangs. If the route carried 100 erlangs of traffic per day, the maximum return on investment would be $2400 a day for that trunk route and the portion of the switches and local plant involved with these calls. As we well know, many of the telephone company's subscribers would be unhappy because they would have to wait excessively to get calls through from X to Y. How, then, do we optimize a trunk route (or serving circuits) and keep the customers as happy as possible?

Remember from Table 1.1, with an excellent grade of service of 0.001, that we relate grade of service to subscriber satisfaction and that the 100 circuits could carry up to 75.24 erlangs during the BH. Assuming the route did carry 75.24 erlangs for the BH, it would earn $75.24 for that hour and something far less than $2400 per day. If the grade of service were reduced to 0.01, 100 trunks would bring in $84.06 for the busy hour. Note the improvement in revenue at the cost of reducing grade of service. Another approach to saving money is to hold the erlang load constant and decrease the number of trunks and switch facilities accordingly as the grade of service is reduced. For instance, 70 erlangs of traffic at $p = 0.001$ requires 96 trunks and at $p = 0.01$, only 86 trunks.

8.1 Alternative Routing

One method of improving efficiency is to use alternative routing (called *alternate routing* in North America). Suppose that we have three serving areas, X, Y, and Z, served by three switches, X, Y, and Z, as shown in Figure 1.8.

Let the grade of service be 0.005 (1 in 200 in Table 1.1).We found that it would require 67 trunks to carry 50 erlangs of traffic during the BH to meet that grade of service between X and Y. Suppose that we reduced the number of trunks between X and Y, still keeping the BH traffic intensity at 50 erlangs. We would thereby increase the efficiency on the X–Y route at the cost of reducing the grade of service. With a modification of the switch at X, we could route the traffic bound for Y that met congestion on the X–Y route via Z. Then Z would route this traffic on the Z–Y link. Essentially, this is alternative routing in its simplest form. Congestion probably would only occur during very short peaky periods in the BH, and chances are that these

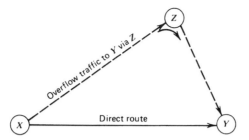

Figure 1.8. Simplified diagram of the alternative routing concept (solid line represents direct route, dashed line represents alternative route carrying the overflow from X to Y).

peaks would not occur simultaneously with peaks in traffic intensity on the Z–Y route. Furthermore, the added load on the X–Z–Y route would be very small. Some idea of traffic peakiness that would overflow onto the secondary route $(X + Z + Y)$ is shown in Figure 1.9.

One of the most accepted methods of dimensioning switches and trunks using alternative routing is the equivalent random group (ERG) method developed by Wilkinson [11]. The Wilkinson method uses the mean M and the variance V. Here the *overflow traffic* is the "lost" traffic in the Erlang B calculations, which were discussed earlier. Let M be the mean value of that overflow and A be the random traffic offered to a group of n circuits (trunks). Then

$$V = M\left(1 - M + \frac{A}{1 + n + M - A}\right) \tag{1.17}$$

When the overflow traffic from several sources is combined and offered to a single second (or third, fourth, etc.) choice of a group of circuits, both the mean and the variance of the combined traffic are the arithmetical sums of the means and variances of the contributors.

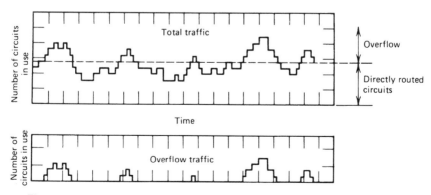

Figure 1.9. Traffic peakiness, the peaks representing overflow onto alternative routes.

The basic problem in alternative routing is to optimize circuit group efficiency (e.g., to dimension a route with an optimum number of trunks). Thus we are to find what circuit quantities result in minimum cost for a given grade of service, or to find the optimum number of circuits (trunks) to assign to a direct route allowing the remainder to overflow on alternative choices. There are two approaches to the optimization. The first method is to solve the problem by successive approximations, and this lends itself well to the application of the computer [12]. Then there are the manual approaches, two of which are suggested in the annex to CCITT Rec. Q.88 (1976). Alternate (alternative) routing is further discussed in Chapter 6.

8.2 Efficiency Versus Circuit Group Size

In the present context a *circuit group* refers to a group of circuits performing a specific function. For instance, all the trunks (circuits) routed from X to Y in Figure 1.8 make up a circuit group, irrespective of size. This circuit group should not be confused with the "group" used in transmission-engineering carrier systems.

If we assume full loading, it can be stated that efficiency improves with circuit group size. From Table 1.1, given $p = 0.01$, 5 erlangs of traffic requires a group with 11 trunks, more than a $2:1$ ratio of trunks to erlangs, and 20 erlangs requires 30 trunks, a $3:2$ ratio. Note how the efficiency has improved. One hundred and twenty trunks will carry 100 erlangs, or 6 trunks for every 5 erlangs for a group of this size. Figure 1.10 shows how efficiency improves with group size.

9 BASES OF NETWORK CONFIGURATIONS

9.1 Introductory Concepts

A network in telecommunications may be defined as a method of connecting exchanges so that any one subscriber in the network can communicate with

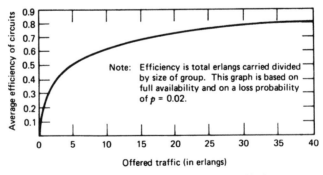

Figure 1.10. Group efficiency increases with size.

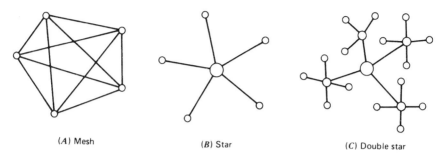

(A) Mesh   (B) Star   (C) Double star

Figure 1.11. Examples of star, double-star, and mesh configurations.

any other subscriber. For this introductory discussion, let us assume that subscribers access the network by a nearby local exchange. Thus the problem is essentially how to connect exchanges efficiently. There are three basic methods of connection in conventional telephony: (1) mesh, (2) star, and (3) double and higher-order star (see Section 2 of this chapter). The mesh connection is one in which each and every exchange is connected by trunks (or junctions) to each and every other exchange as shown in Figure 1.11A. A star connection utilizes an intervening exchange, called a *tandem exchange*, such that each and every exchange is interconnected via a *single* tandem exchange. An example of a star connection is shown in Figure 1.11B. A double-star configuration is one where sets of pure star subnetworks are connected via higher-order tandem exchanges, as shown in Figure 1.11C. This trend can be carried still further, as we see later on, when hierarchical networks are discussed.

As a general rule we can say that mesh connections are used when there are comparatively high traffic levels between exchanges, such as in metropolitan networks. On the other hand, a star network may be applied when traffic levels are comparatively low.

Another factor that leads to star and multiple-star network configurations is network complexity in the trunking outlets (and inlets) of a switch in a full mesh. For instance, an area with 20 exchanges would require 380 traffic groups (or links), and an area with 100 exchanges would require 9900 traffic groups. This assumes what are called *one-way groups*. A one-way group is best defined considering the connection between two exchanges, A and B. Traffic originating at A bound for B is carried in one group and the traffic originating at B bound for A is carried in another group, as shown in the following diagram:

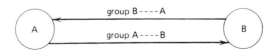

Thus, in practice, most networks are compromises between mesh and star configurations. For instance, outlying suburban exchanges may be connected to a nearby major exchange in the central metropolitan area. This exchange may serve nearby subscribers and be connected in mesh to other large exchanges in the city proper. Another example is the city's long-distance exchange, which is a tandem exchange looking into the national long-distance network, whereas the major exchanges in the city are connected to it in mesh. An example of a real-life compromise among mesh, star, and multiple-star configurations is shown in Figure 1.12.

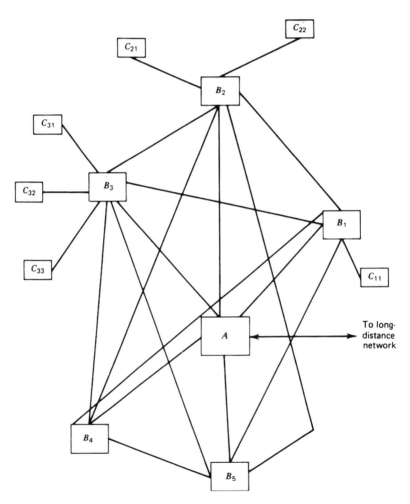

Figure 1.12. A typical telephone network serving a small city as an example of a compromise between mesh and star configuration. *A* is the highest level in this simple hierarchy. *A* might house the "point of presence" (POP) in the U.S. network. *B* is a local exchange. *C* may be a satellite exchange or a concentrator. Consult Ref. 24.

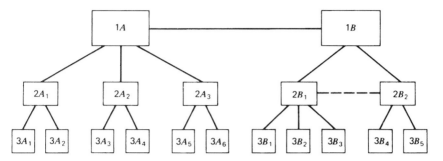

Figure 1.13. A higher-order star network.

9.2 Hierarchical Networks

To bring order out of this confusion, hierarchical networks evolved. That is, a systematic network was developed that reduces the trunk group outlets (and inlets) of a switch to some reasonable amount, permits the handling of high traffic intensities on certain routes where necessary, and allows for overflow and a means of restoral in certain circumstances. Consider Figure 1.13, which is a simplified example of a higher-order star network. The term "order" here is significant and leads to the discussion of hierarchical networks.

A hierarchical network has levels giving orders of importance of the exchanges making up the network, and certain restrictions are placed on traffic flow. For instance, in Figure 1.13 there are three levels or ranks of exchange. The smallest boxes in the diagram are the lowest-ranked exchanges, which have been marked with a "3" to indicate the third level or rank. Note the restrictions (or rules) of traffic flow. As the figure is drawn, traffic from $3A_1$ bound for $3A_2$ would have to flow through exchange $2A_1$. Likewise, traffic from exchange $2A_2$ to $2A_3$ would have to flow through exchange $1A$. Carrying the concept somewhat further, traffic from any A exchange to any B exchange would necessarily have to be routed through exchange $1A$.

The next consideration is the high-usage route. For instance, if we found that there were high traffic intensities between $2B_1$ and $2B_2$, trunks and switch gear might well be saved by establishing a high-usage route between the two (shown by dashed line). Thus we might call the high-usage route a *highly traveled shortcut*. Of course, high-usage routes could be established between any pair of exchanges in the network if traffic intensities and distances involved proved this strategy economical. When high-usage routes are established, traffic between the exchanges involved will first be offered to the high-usage route and overflow would take place through hierarchical structure, or as shown in our Figure 1.13, up to the next level and down. If routing is through the highest level in the hierarchy, we call this route the

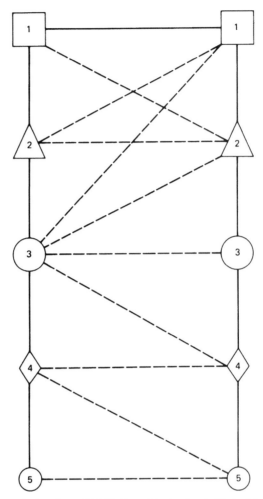

Figure 1.14. The North American (AT & T) hierarchical network (dashed lines show high-usage trunks). Note how the two highest ranks are connected in mesh.

final route. Figure 1.13 shows traffic routed *between* exchanges $2B_1$ and $2B_2$ via exchange $1B$ being routed on the final route.

9.3   Hierarchical Network Structures

Since the inception of national and international long-distance telephone service, networks have evolved into a hierarchical structure. Figure 1.13 shows an elementary hierarchical network structure. Today the trend is away from a hierarchical structure as we discuss in Section 9.6.

Certain points in terminology should be clarified. In Europe especially and also in North America, we should distinguish "transit exchanges" and "tandem exchanges." Although both perform the same function, the switching of trunks, a tandem exchange serves the local area or the lowest levels of a hierarchy. A transit exchange switches trunks in the toll or long-distance area. The term "via exchange" is also used in North America for a transit exchange.

In older CCITT documents we should expect to see the term "CT," meaning "central transit" in French or "central tránsito" in Spanish. In English the term is simply *transit exchange*. In these older CCITT documents, a number was often placed after CT, to indicate the level of the hierarchy. For instance, CT1 was the highest-order transit exchange in the now defunct CCITT routing scheme, CT2 was the next-to-highest order, and CT3 was the third order from the top.

CCITT also uses the British term *junction*, and we use it in this text where appropriate. This is a trunk serving the local area. In CCITT terminology, trunks serve as high-order connections. Primary centers are collecting centers (exchanges) for traffic to interconnect the toll or long-distance network. The term *center* may be related to "central," meaning a switching node or exchange, usually of higher order in the hierarchy.

Figure 1.14 presents the AT&T hierarchical "routing pattern." The highest order or rank in the hierarchy is the class 1 center and the lowest, a class 5 office ("office" is taken from the North American term "central office," meaning switch). It should be noted, according to hierarchical rules, that a high-usage (HU) trunk group may be established between any two switching centers regardless of location or rank, whenever the traffic volume justifies it. The table below illustrates the AT&T nomenclature used.

Order	Name	Comments
Class 1	Regional center	Top of the hierarchy
Class 2	Sectional center	
Class 3	Primary center	
Class 4	Toll center	Now "point of presence" (POP) where LEC meets IXC
Class 5	End office	In the local exchange carrier (LEC) area

9.4 Rules for Conventional Hierarchical Networks

A backbone structure to a hierarchical network is noted in Figure 1.14: from left to right or from right to left, the outside vertical lines are connected by the top horizontal line, which we refer to as "up, across, and down," as shown in the diagram that follows.

The CCITT terms these routes "theoretical final routes." For our argument, a final route is a route from which no overflow is permitted. A hierarchical network is characterized by a full set of final routes from source to sink. Any other routes are supplementary to the pure hierarchy, regardless of whether overflow is permitted on them.

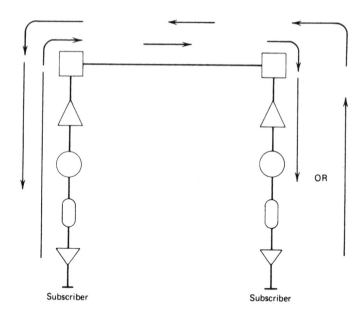

A hierarchical system of routing leads to simplified switch design. A common expression used when discussing hierarchical routing and star connections is that lower-rank exchanges "home" on higher-rank exchanges. If a call is destined for an exchange of lower rank in its chain, the call proceeds down the chain. Likewise, if a call is destined for another exchange outside the chain, it proceeds up the chain. Or when such high-usage routes exist, a call may be routed on a route additional or supplementary to the pure hierarchy, proceeding to the distant transit center and then descending to the destination. Of course, at the highest level in a pure hierarchy the call crosses from one chain over to the other. In hierarchical networks only the order of each switch in the hierarchy and those additional links (high-usage routes) that provide access need be known. In such networks administration is simplified, and storage or routing information is reduced when compared to the full-mesh type of network, for example.

The CCITT (Rec. Q.13) suggests the far-to-near criterion, whereby the first-choice route in advancing a call is to advance the call as far as possible from its origin using the backbone route to measure distances. The second choice is the next best and so forth. The additional routes (diagonal dashed lines in Figure 1.14) result in decreasing the number of links traversed, thus minimizing the total links traversed in a call connection and improving transmission and signaling characteristics and, on long international connections, keeping the total number of links required to 12 or below. See Chapter 6, Section 3.

One weakness in conventional hierarchical structured networks is circuit security. When we say "security" in this context, we mean that the loss of one (or several) links due to fire, explosion, natural disaster, cutting, or sabotage will not cause the full breakdown of communication in the network. Rather, we mean that communication can be maintained, perhaps with reduced capability and increased blockage, but nevertheless maintained. Now consider a hierarchical network. Higher in the network the nodes of each rank become fewer and fewer. For instance, take the United States and Canada. There are only 10 nodes of class 1 in the United States and 2 in Canada. Even with these 12 nodes completely interconnected in mesh, as they actually are in practice, the loss of one or more nodes or links at this level in the hierarchy might seriously jeopardize communications. Thus the current tendency is to reduce the number of levels or ranks in the hierarchy, thus increasing the number of higher-level nodal points. Improved circuit security is the result when they are fully interconnected (mesh), offering more combinations of alternative route configurations. The desirability of this trend is obvious. It becomes highly feasible as the network becomes modernized, replacing electromechanical switching with computer-controlled digital switching. We see future large national networks with only three levels of hierarchy.

9.5 Homing Arrangements and Interconnecting Network in North America

We spoke of "homing" of a dependent switching node to the next-higher rank nodal point in purely hierarchical networks. The use of HU routes so widely used in practice makes the rule really valid only on the backbone final routes. For example, consider the North American network where class 5, the lowest-ranking switch, may be served directly from any higher-ranking location. Possible homing arrangements for each class (rank) of switching node are shown in Table 1.5.

Final trunk groups in this network are engineered for a lost-call probability of 0.01. As we have seen, direct high-usage trunk groups are provided between switching nodal points of any rank (class) when the volume

TABLE 1.5 Homing Arrangements in North America

Rank and Class	May Home at Switches of Classes
End office, 5	Class 4, 3, 2, or 1
Toll center, 4	Class 3, 2, or 1
Primary center, 3	Class 2 or 1
Sectional center, 2	Class 1
Regional center, 1	All regional centers mutually interconnected

of traffic and economics warrant these groups and where automatic alternative routing features are available. The rule is that high-usage trunk groups carry most but not all of the offered traffic in the BH. The proportion of the offered traffic carried on a direct high-usage trunk group ordinarily is determined, in part, by the relative costs of the direct route and the alternative route. The economic factors to be considered are the additional switching costs on the alternative routes. In some instances, high-usage trunk groups may be designed on a "no overflow" basis with the 0.01 lost-call objective. Homing arrangements are not changed in this case, and these trunks are called "full groups." Such full groups effectively truncate or limit the hierarchical chain of final routes for the traffic offered them.

9.6   The Trend Away from a Hierarchical Structure

There is a decided trend away from hierarchical routing and network structure. However, there will always be some form of hierarchical structure into the foreseeable future.

There are several reasons for this trend in our opinion, and one interplays with the other. The first deals with transmission while the second deals with switching.

Since about 1965, transmission techniques have taken leaps forward. Satellite communications allowed direct routes some one-third the way around the world. This was followed by fiber-optics transmission links with nearly infinite bandwidths and excellent performance properties. We discuss these transmission techniques in Chapter 7.

In the switching domain, the stored program control (SPC) switch had the computer brains to make nearly real-time decisions for routing. The advent of CCITT Signaling System No. 7 (Chapter 16) and improvements in network management (Chapter 18) made it possible for optimum routing based on real-time availability of route capacity. The complex hierarchy is becoming obsolete.

Nearly all reference to a routing hierarchy disappeared from CCITT in the 1988 Plenary Session (Melbourne) documents. International connectivity is by means of direct/high-usage routes. In fact, CCITT Rec. E.172 (Geneva 10/92) states that "In the ISDN era, it is suggested that the network structure should be non-hierarchical, ..."

In the United States a minimum of a two-layer hierarchy will evolve. This is due to the Bell System divestiture court decision where the lower level is the "local exchange carrier (LEC)" and the higher level consists of the interexchange carriers (IXCs). This also may change. There is a possibility that the LECs will be allowed by the courts to enter the long-distance carrier business.

AT&T is implementing a new routing scheme called *dynamic nonhierarchical routing* (DNHR) and has shown some 15% improvement in routing performance with DNHR. A brief description of DNHR is given in Chapter 6.

10 ROUTING METHODS

There are generally three methods of routing calls from source to sink through one, several, or many intermediate switching nodes. As we have seen, there may be many possible patterns through which a given call can traverse. The problem is to *decide* how the call should proceed through the many possible path combinations in the network. The three methods are (1) right-through routing, (2) own-exchange routing, and (3) computer-controlled routing (with common-channel signaling). In right-through routing the originating exchange determines the route from source to sink. Alternative routing is not allowed at intermediate switching points. However, the initial outgoing circuit group may be arranged so that one or more alternative routes are presented. Because of its inherent limitations in alternative routing and the requirement that a change in network configuration or the addition of new exchanges entail alteration in each existing complex switch (i.e., switches with translators), right-through routing is limited almost exclusively to the local area.

Own-exchange routing allows for changes in routing as the call proceeds to its destination. This routing system is particularly suited to networks with alternative routing and changes in routing patterns in response to changes in load configuration. Another advantage in own-exchange routing is that when new exchanges are added or the network is modified, minimal switch modifications are required in the network. One disadvantage is the possibility of establishing a closed routing loop where a call may be routed such that it is eventually routed back to its originating exchange or other exchange through which it has already been routed in attempting to reach its destination. However, a hierarchical routing system ensures that such loops cannot be generated. If routing loops are established in an operating network, there can be disastrous consequences, as the reader can appreciate.

Conventional telephone networks have signaling information for a particular call carried on the same path (pair of wires or their equivalent) that carries the speech, often called the *conversation path*. Signaling, as we discuss later in considerable detail, is the generation and transmission of information that sets up a desired call and routes it through the network to its destination. New and more modern computer-controlled networks often use a separate path to carry the required signaling information. In this case the computer in the originating exchange or originating long-distance exchange can "optimally" route the call through the network on a separate signaling path. The originating computer would have a "map in memory" of the network with updated details of network conditions such as traffic load at the various nodes and trunks and outages. The necessary adaptive information is broadcast on the separate path that connects the various computers in the network. This is computer-controlled routing. Such routing is termed "routing with common-channel signaling" and with adaptive network management signals. DNHR, described in Chapter 6, is such a routing scheme. (Consult Ref. 8.)

11 VARIATIONS IN TRAFFIC FLOW

In networks covering large geographic expanses and even in cases of certain local networks, there may be a variation in time of day of the BH or in the direction of traffic flow. In the United States, business traffic peaks during several hours before and after the noon lunch period on weekdays, and social calls peak in early evening. Traffic flow tends to be from suburban living areas to urban centers in the morning, and the reverse in the evening.

In national networks covering several time zones where the differences in local time may be appreciable, long-distance traffic tends to be concentrated in a few hours common to BH peaks at both ends. In such cases it is possible to direct traffic so that peaks of traffic in one area fall into valleys of traffic in another (noncoincident busy hour). The network design can be made more economical if configured to take advantage of these phenomena, particularly in the design and configuration of direct routes versus overflow.

12 BOTH-WAY CIRCUITS

We defined one-way circuits in Section 9.1. Here traffic from A to B is assigned to one group of circuits, and traffic from B to A is assigned to another separate group. In both-way (or two-way) operation a circuit group may be engineered to carry traffic in both directions. The individual circuits in the group may be used in either direction, depending on which exchange seizes the circuit first.

In engineering networks it is most economical to have a combination of one-way and both-way circuits on longer routes. Signaling and control arrangements on both-way circuits are substantially more expensive. However, when dimensioning a system for a given traffic intensity, fewer circuits are needed in both-way operation, with notable savings on low-intensity routes (i.e., below about 10 erlangs in each direction). For long circuits, both-way operation has obvious advantages when dealing with a noncoincident BH. During overload conditions, both-way operation is also advantageous because the direction of traffic flow in these conditions is usually unequal.

The major detriment to two-way operation, besides its increased signaling cost, is the possibility of double seizure. This occurs when both ends seize a circuit at the same time. There is a period of time when double seizure can occur in a two-way circuit; this extends from the moment the circuit is seized to send a call and the moment when it becomes blocked at the other end. Signaling arrangements can help to circumvent this problem. Likewise, switching arrangements can be made such that double seizure can occur only on the last free circuit of a group. This can be done by arranging in turn the sequence of scanning circuits so that the sequence on one end of a two-way circuit is reversed from that of the other end. Of course, great care must be taken on circuits having long propagation times, such as satellite and long undersea cable circuits. By extending the time between initial seizure and blockage at the other end, these circuits are the most susceptible, just because a blocking signal takes that much longer to reach the other end.

13 QUALITY OF SERVICE

Quality of service appears at the outset to be an intangible concept. However, it is very tangible for a telephone subscriber unhappy with his or her service. The concept of service quality must be mentioned early in any all-encompassing text on telecommunications systems. System engineers should never once lose sight of the concept, no matter what segment of the system they may be responsible for. Quality of service also means *how happy* the telephone company (or other common carrier) is keeping the customer. For instance, we might find that about half the time a customer dials, the call goes awry or the caller cannot get a dial tone or cannot hear what is being said by the party at the other end. All these have an impact on quality of service. So we begin to find that quality of service is an important factor in many areas of the telecommunications business and means different things to different people. In the old days of telegraphy, a rough measure of how well the system was working was the number of service messages received at a switching center. In modern telephony we now talk about service observing (see Chapter 3, Section 17).

The transmission engineer calls quality of service "customer satisfaction," which is commonly measured by how well the customer can hear the calling party. It is called *reference equivalent*,* which is measured in decibels (dB). In our discussion of traffic, lost calls certainly constitute one measure of service quality, and if measured in decimal quantity, one target figure would be $p = 0.01$. Other items to be listed under service quality are:

- Delay before receiving dial tone ("dial-tone delay").
- Post dial(ing) delay (time from completion of dialing a number to first ring of telephone called).
- Availability of service tones (busy tone, telephone out of order, ATB, etc.).

*Or "corrected reference equivalent," or "loudness rating."

- Correctness of billing.
- Reasonable cost to customer of service.
- Responsiveness to servicing requests.
- Responsiveness and courtesy of operators.
- Time to installation of new telephone, and, by some, the additional services offered by the telephone company [22].

One way or another each item, depending on service quality goal, will have an impact on the design of the system.

Furthermore, each item on the list can be quantified—usually statistically, such as reference equivalent, or in time, such as time taken to install a telephone. In some countries this can be measured in years. Good reading can be found in CCITT Recs. Q.60, Q.60 bis, and Q.61.

REVIEW QUESTIONS

1. Give the standard telephone battery voltage with respect to ground.

2. What is *on-hook* and *off-hook*? When a subscriber subset (the telephone) goes "off-hook," what occurs at the serving switch? List two items.

3. Suppose that the sidetone level of a telephone is increased. What is the natural reaction of the subscriber?

4. A subscriber pair, with a fixed battery voltage, is extended. As we extend the loop further, two limiting performance factors come into play. Name them.

5. Define a mesh connection. Draw a star arrangement.

6. In the context of the argument presented in the chapter, what is the principal purpose of a local switch?

7. What are the two basic parameters that define "traffic"?

8. Distinguish offered traffic from carried traffic.

9. Give one valid definition of the *busy hour*.

10. On a particular traffic relation the calling rate is 461 and the average call duration is 1.5 min during the BH. What is the traffic intensity in CCS, in erlangs?

11. Define *grade of service*.

12. A particular exchange has been dimensioned to handle 1000 calls during the busy hour. On a certain day during the BH 1100 calls are offered. What is the resulting grade of service?

13. Distinguish a full availability switch from a limited availability switch.

14. In traffic theory there are three ways lost calls are handled. What are they?

15. Call arrivals at a large switch can be characterized by what type of mathematical distribution? Such arrivals are_____in nature.

16. Based on the Erlang B formula and given a BH requirement for a grade of service of 0.005 and a BH traffic intensity of 25 erlangs on a certain traffic relation, how many trunks are required?

17. Carry out the same exercise as in question 16 but use the Poisson tables to determine the number of trunks required.

18. Give at least two queueing disciplines.

19. As the grade of service is improved, what is the effect on trunk efficiency?

20. What is the basic purpose of alternative routing? What does it improve?

21. How does circuit group size (number of trunks) affect efficiency for a fixed grade of service?

22. Give the three basic methods of connecting exchanges. (These are the three basic network types.)

23. At what erlang value on a certain traffic relation does it pay to use tandem routing? Is this a maximum or a minimum value?

24. Differentiate between one-way and both-way circuits.

25. What is the drawback of one-way circuits? of both-way circuits?

26. Hierarchical networks are used universally in national and international telephone networks. Differentiate between high-usage (HU) connectivities and final route.

27. Distinguish a tandem exchange from a transit exchange.

28. Define the term *homing* on hierarchical networks.

29. What advantage is there in reducing the number of hierarchical levels?

30. What is the recommended lost call probability for final trunk groups?

31. Name the three basic routing methods.

32. How can we take advantage of the noncoincident busy hour in a large national network?

33. Name at least five items that can be listed under *quality of service*.

REFERENCES

1. International Telephone and Telegraph Corporation, *Reference Data for Radio Engineers*, 5th ed., Howard W. Sams, Indianapolis, 1968.

2. R. R. Mina, "The Theory and Reality of Teletraffic Engineering," *Telephony*, a series of articles (April 1971).

3. R. Syski, *Introduction to Congestion Theory in Telephone Systems*, Oliver and Boyd, Edinburgh, 1960.

4. G. Dietrich et al., *Teletraffic Engineering Manual*, Standard Electric Lorenz, Stuttgart, Germany, 1971.

5. E. Brockmeyer et al., "The Life and Works of A. K. Erlang," *Acta Polytechnica Scandinavia*, The Danish Academy of Technical Sciences, Copenhagen, 1960.

6. *A Course in Telephone Traffic Engineering*, Australian Post Office, Planning Branch, 1967.

7. Arne Jensen, *Moe's Principle*, The Copenhagen Telephone Company, Copenhagen, Denmark, 1950.

8. *Networks*, Laboratorios ITT de Standard Eléctrica SA, Madrid, 1973 (limited circulation).

9. *Local Telephone Networks*, The International Telecommunications Union, Geneva, 1968.

10. *Electrical Communication System Engineering Traffic,* U.S. Department of the Army, TM-11-486-2, August 1956.

11. R. I. Wilkinson, "Theories for Toll Traffic Engineering in the USA," *BSTJ*, **35** (March 1956).

12. *Optimization of Telephone Trunking Networks with Alternate Routing,* ITT Laboratories of Standard Eléctrica (Spain), Madrid, 1974 (limited circulation).

13. J. Riordan, *Stochastic Service Systems*, John Wiley & Sons, New York, 1962.

14. L. Kleinrock, *Queueing Systems*, Vols. 1 and 2, John Wiley & Sons, New York, 1975.

15. T. L. Saaty, *Elements of Queueing Theory with Applications*, McGraw-Hill, New York, 1961.

16. J. E. Flood, *Telecommunications Networks,* IEE Telecommunications Series 1, Peter Peregrinus, London, 1975.

17. *Notes on The Network—1980*, American Telephone and Telegraph Company, New York, 1980.

18. *National Telephone Networks for the Automatic Service*, International Telecommunications Union–CCITT, Geneva, 1964.

19. D. Bear, *Principles of Telecommunication Traffic Engineering*, IEE Telecommunications Series 2, Peter Peregrinus, London, 1976.

20. R. L. Freeman, *Reference Manual for Telecommunications Engineering*, 2nd ed., John Wiley & Sons, New York, 1994.

21. *CCITT Blue Books*, IXth Plenary Assembly, Melbourne, 1988.

22. "Telecommunications Quality," *IEEE Communications Magazine* (October 1988, entire issue).

23. *Engineering and Operations in the Bell System*, 2nd ed., AT&T Bell Laboratories, Murray Hill, NJ, 1983.

24. *BOC Notes on the LEC Networks—1994*, issue 2, SR-TSV-002275, Bellcore, Piscataway, NJ, April 1994.

25. Private Communications—John Lawlor/John Lawlor Associates, Sharon, MA, December 1994.

26. J. R. Boucher, *Traffic System Design Handbook*, IEEE Press, New York, 1992.

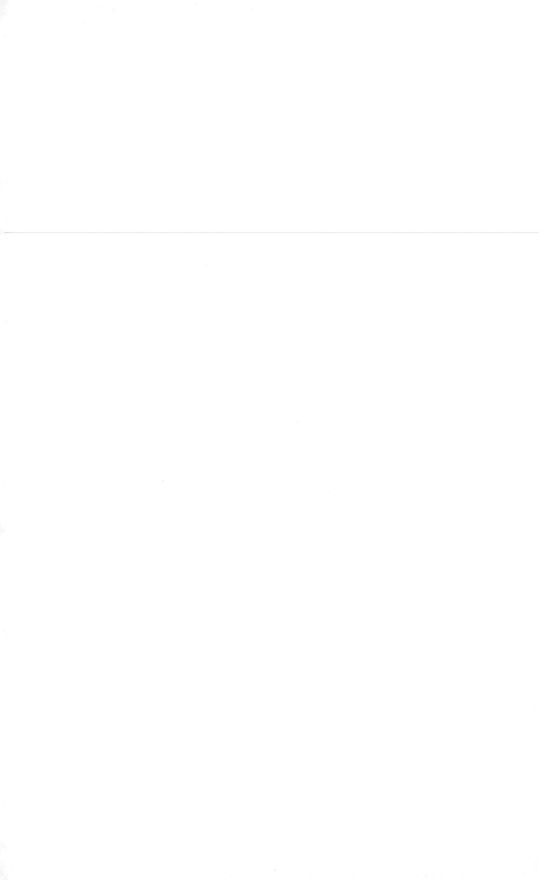

LOCAL NETWORKS

1 INTRODUCTION

The importance of local network design, whether standing on its own merit or part of an overall national network, cannot be overstressed. In comparison to the long-distance sector, the local sector is not the big income producer per capita invested, but there would be no national network without it. Telephone companies or administrations invest, on the average, more than 50% in their local areas. In the larger, more developed countries the investment in local plant may reach 70% of total plant investment.

The local area, as distinguished from the long-distance or national network, was discussed in Section 4 of Chapter 1. In this chapter we are more precise in defining the local area itself. Let us concede that the local area includes the subscriber plant, local exchanges, and the trunk plant interconnecting these exchanges as well as those trunks connecting a local area to the next level of network hierarchy, or the point of presence (POP)* (USA) or primary center (CCITT).

To further emphasize the importance of the local area, consider Table 2.1, which was taken from Ref. 1 (CCITT). Figure 2.1 is a simplified diagram of a local network with five local exchanges and illustrates the makeup of a typical small local area.

The design of such a network (Figure 2.1) involves a number of limiting factors, the most important of which is economic. Investment and its return are not treated in this text. However, our goal is to build the most economical network assuming an established quality of service. Considering both quality of service and economy, certain restraints will have to be placed on the design. For example, we will want to know:

*POP is a point of interface with interexchange carriers (e.g., AT&T, MCI, etc.).

TABLE 2.1 Average Percentage of Investments in Public Telephone Equipment

Item	Average for 16 Countries (%)
Subscriber plant	13
Outside plant for local networks	27
Exchanges	27
Long-distance trunks	23
Buildings and land	10

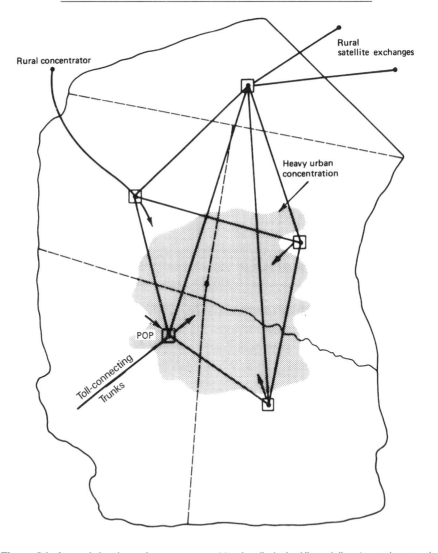

Figure 2.1. A sample local area (arrows represent trunk pull; dashed lines delineate serving areas).

- Geographic extension of the local area of interest.
- Number of inhabitants and existing telephone density.
- Calling habits.
- Percentage of business telephones.
- Location of existing telephone exchanges and extension of their serving areas.
- Trunking scheme.
- Present signaling and transmission characteristics.

Each of these criteria or limiting factors is treated separately, and interexchange signaling and switching per se are dealt with later in separate chapters. Let us also assume that each exchange in the sample will be capable of serving up to 10,000 subscribers. Also assume that all telephones in the area have seven-digit numbers, the last four of which are the subscriber number of the respective serving area of each exchange. The reasoning behind these assumptions becomes apparent in later chapters.

A further assumption is that all subscribers are connected to their respective serving exchanges by wire pairs, resulting in some limiting subscriber loop length. This leads to the first constraining factor dealing with transmission and signaling characteristics. In general terms, the subscriber should be able to hear the distant calling party reasonably well (transmission) and to "signal" that party's serving switch. These items are treated at length in the following section.

2 SUBSCRIBER LOOP DESIGN

2.1 General

The pair of wires connecting the subscriber to the local serving switch has been defined as the *subscriber loop*. It is a dc loop in that it is a wire pair supplying a metallic path for the following:

- Talk battery for the telephone transmitter (Chapter 1, Section 2).
- An ac ringing voltage for the bell on the telephone instrument supplied from a special ringing source voltage.
- Current to flow through the loop when the telephone instrument is taken out of its cradle ("off hook"), telling the serving switch that it requires "access," thus causing a line seizure at that switch.
- The telephone dial that, when operated, makes and breaks the direct current on the closed loop, which indicates to the switching equipment the number of the distant telephone with which communication is desired; alternatively, a touch-tone pad with digit buttons. Unique pairs

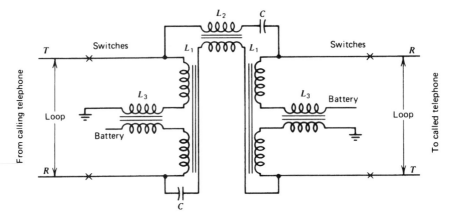

Figure 2.2. Battery feed circuit [22]. Note: Battery and ground are fed through inductors *L*2 and *L*.1 through switch to loops. Copyright © 1961 by Bell Telephone Laboratories, Inc.

of audio tones, representing digits 1–0 are transmitted to the serving exchange switching equipment.

The typical subscriber loop is powered by means of a battery feed circuit at the switch. Such a circuit is shown in Figure 2.2. Telephone battery source voltage has been fairly well standardized at −48 V dc.

2.2 Subscriber Loop Length Limits

The two basic criteria that are considered when designing subscriber loops and that limit their length are attenuation limits and signaling limits. Attenuation in this case refers to loop loss at reference frequency measured in decibels (or nepers). Reference frequency is 1000 Hz in North America and 800 Hz in Europe and many other parts of the world. As a telephone loop is extended in length, its loss at reference frequency increases. It follows that at some point as the loop is extended, level will be attenuated such that the subscriber cannot hear *sufficiently well*.

Likewise, as a loop is extended in length while the battery (supply) voltage is kept constant, the effectiveness of signaling is ultimately lost. This limit is a function of the *IR* drop of the line. We know that *R*, the resistance, increases as length increases. With today's modern telephone sets, the first feature to suffer is usually "signaling," particularly that area of loop signaling called "supervision." In this case it is a signal sent to the switching equipment requesting "seizure" of a switch circuit and at the same time indicating to callers that the line is busy. "Off hook" is the term more commonly used to describe this signal condition. When a telephone is taken "off hook" (i.e., out of its cradle), the telephone loop is closed and current flows, closing a relay at the switch. If current flow is insufficient, the relay will not close or will close and open intermittently ("chatter") such that the line seizure cannot be effected.

Signaling limits are a function for the conductivity of the loop conductor and its diameter or gauge. For this introductory discussion, we can consider that the attenuation limits are controlled by the same parameters. Consider a copper conductor. The larger the conductor, the greater the ability to conduct current and thus the longer the loop conductors may be for signaling purposes. Because copper is expensive, we may not make the conductor as large as we would wish and extend subscriber cable loops over long distances. This is an economic constraint. Before we go further, we describe a method of measuring what a subscriber considers as "hearing sufficiently well."

2.3 Quality of a Telephone Speech Connection

2.3.1 Reference Equivalent

2.3.1.1 Definition. "Hearing sufficiently well" on a telephone connection is a subjective matter under the blanket heading of "customer satisfaction." Various methods have been derived over the years to rate telephone connections regarding customer (subscriber) satisfaction. Regarding the received telephone signal, subscriber satisfaction is affected by the level (signal power), the signal-to-noise ratio, and the response or attenuation frequency characteristic. A common rating system internationally in use today for grading of customer satisfaction is the *reference equivalent* system. This system considers only the first criterion (viz., level). It must be emphasized that subscriber satisfaction is subjective. To measure satisfaction, the world regulative body for telecommunications, the International Telecommunications Union (ITU), devised a system of rating the level sufficient to "satisfy" using the familiar decibel as the unit of measurement. The reference equivalent is broken down into two basic parts. The first is a subjective value in decibel rating of a particular type of subset. The second part is simply the losses (measured at 800 Hz) end-to-end of the intervening network. To determine the reference equivalent of a particular circuit, we add algebraically the decibel value assigned to the subset to the losses of the connecting circuit. Let us look at how the reference equivalent system was developed, keeping in mind again that it is a subjective measurement dealing with the likes and dislikes of the "average" human being. A standard for reference equivalent was determined in Europe by a team of qualified personnel in a laboratory. A telephone connection, intended to be the most efficient telephone system known, was established in the laboratory. The original reference system or unique master reference consisted of the following:

- A solid-black telephone transmitter.
- Bell telephone receiver.
- Interconnecting these, a "zero-decibel-loss" subscriber loop.

• Connecting the loop, a manual central battery, 22-V dc telephone exchange (switch).

To avoid ambiguity of language, the test team used a test language that consisted of logatoms. A "logatom" is a one-syllable word consisting of a consonant, a vowel, and another consonant.

More accurate measurement methods have been developed since. A more modern reference system is now available at the ITU laboratory in Geneva, Switzerland, called the NOSFER. From this master reference, field test standards are available to telephone companies, administrations, and industry to establish the reference equivalents of telephone subsets in use. These field test sets are calibrated for equivalence with the NOSFER. The NOSFER is made up of a standard telephone transmitter, a receiver, and a network. The reference equivalent of a subscriber's subset, together with the associated subscriber line and feeding bridge, is a quantity obtained by balancing the loadness of receiver speech signals and is expressed relative to the whole or to a corresponding part of the NOSFER (or field) reference system.

2.3.1.2 Application. Most telephone companies or administrations consider that a standard telephone subset* is used. The objective is to measure the capabilities of these subsets regarding loudness. Thus type tests are run on the subsets against calibrated field standards. As mentioned earlier, these may be done on the set alone or on the set plus a fixed length of subscriber loop and feed bridge of known characteristics. The tests are subjective and are carried out in a laboratory. The microphone or transmitter and the earpiece or receiver are each rated separately and are called the *transmit reference equivalent* (TRE) and the *receive reference equivalent* (RRE), respectively. The unit of measurement is the decibel, and negative values indicate that the reference equivalent is better than the laboratory standard (see CCITT Rec. P.72).

In telephone systems the *overall reference equivalent* (ORE) is the more common measurement. Simply, this is the sum of the TRE, the RRE, and the losses of the intervening network. Now consider the simplified telephone network shown in the following diagram:

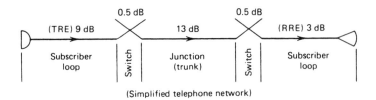

(Simplified telephone network)

---

*In the United States, the standard is the 500/2500 subset.

The reference equivalent for this circuit is 26 dB, including a 0.5-dB loss for each switch. As defined previously, a junction or trunk is a circuit connecting two switches (exchanges). These may or may not be adjacent. The circuit shown in the preceding diagram may be called a *small transmission plan*. For this discussion, we can define a transmission plan as a method of assigning losses end to end on a telephone circuit. Further on in this text we discuss why all telephone circuits, at least all conventional analog circuits, must be lossy. The reference equivalent is a handy device for rating such a plan regarding subscriber satisfaction (see CCITT Recs. P.42 and P.72).

For more in-depth studies of transmission quality (subscriber satisfaction), the reader should review the AEN (articulation reference equivalent) method described in CCITT Recs. P.12, P.12A, and P.13.

When studying transmission plans or developing them, we usually consider that all sections of a circuit are symmetrical. Let us examine the one shown in the preceding diagram. On each end of a circuit we have a subscriber loop. Thus the same loss is assigned to each loop in the plan, which may not be the case at all in real life. From the local exchange to the first long-distance exchange. variously termed *junctions* or *toll-connecting trunks*, a loss is assigned that is identical at each end, and so forth.

To maintain the symmetry regarding reference equivalent of telephone subsets, we use the term $(T + R)/2$. As we see from the preceding diagram, the TRE and the RRE of the subset have different values. We get the $(T + R)/2$ by summing the TRE and RRE and dividing by 2. This is done to arrive at the desired symmetry. Table 2.2 gives reference equivalent data on a number of standard subscriber subsets used in various parts of the world.

It is stated in CCITT Rec. G.121 that the reference equivalent from the subscriber set to an international connection should not exceed 20.8 dB (TRE) and to the subscriber set at the other end from the same point of reference (RRE), 12.2 dB (the intervening losses are already included in these figures). By adding 12.2 dB and 20.8 dB, we find 33 dB to be the ORE recommended as a maximum* for an international connection. Table 2.3 should be of interest in this regard. Supplementing the table, it should also be noted that when the overall reference equivalent drops to about 6 dB, subscribers begin to complain that calls are too loud.

The reader may ask why international connections are being discussed in a chapter on local networks. The answer is simple and the concept is very important. All calls originate and terminate in a local area. If we are to follow CCITT Rec. G.121, no calls (or very few) should exceed an ORE of 33 dB. In fact, a national transmission plan should reflect the obvious results of Table 2.3 in that we could improve subscriber satisfaction by reducing the ORE from the 33-dB level. And more than half of the 33 dB can be attributed to the local area(s), source, and sink. The limitation, the constraint on local area design, is obvious; the total transmission loss should be under

---

*This is the recommendation for 97% of the connections made in a country of average size.

TABLE 2.2   Reference Equivalents for Subscriber Sets in Various Countries

Country	Sending (dB)	Receiving (dB)
With limiting subscriber lines and exchange feeding bridges		
Australia	14.0[a]	6.0[a]
Austria	11.0	2.6
France	11.0	7.0
Norway	12.0	7.0
Germany	11.0	2.0
Hungary	12.0	3.0
The Netherlands	17.0	4.0
United Kingdom	12.0	1.0
South Africa	9.0	1.0
Sweden	13.0	5.0
Japan	7.0	1.0
New Zealand	11.0	0.0
Spain	12.0	2.0
Finland	9.5	0.9
With no subscriber lines		
Italy	2.0	−5
Norway	3.0	−3
Sweden	3.0	−3
Japan	2.0	−1
United States (loop length 1000 ft, 83 Ω)	5.0	−1[b]

[a]Minimum acceptable performance.
[b]Freeman [8].

Source: CCITT, Local Telephone Networks, ITU, Geneva, July 1968; and National Telephone Networks for the Automatic Service (Ref. 1).

TABLE 2.3   British Post Office Survey of Subscribers for Percentage of Unsatisfactory Calls

Overall Reference Equivalent (dB)	Unsatisfactory Calls (%)
40	33.6
36	18.9
32	9.7
28	4.2
24	1.7
20	0.67
16	0.228

9 dB (approximately half of 18 dB), with no more than 6 or 7 dB (depending on the transmission plan) for subscriber loop loss. A transmission plan assigns losses to the various segments of a telephone network to meet an ORE goal among other factors.

TABLE 2.4 Subscriber Opinion Survey: Corrected Reference Equivalent

Planning Value of the Overall Corrected Reference Equivalent (dB)	Representative Opinion Results[a]	
	Percent "Good plus Excellent"	Percent "Poor plus Bad"
5–15	>90	<1
20	80	2
25	65	5
30	40	15

[a]Based on a composite opinion model.

Source: CCITT Rec. P.11, Table 1/P.11, page 14, Vol. V, Red Book. Copyright ITU (Ref. 6).

2.3.2 Corrected Reference Equivalent. Reference equivalent was introduced in Section 2.3.1 to familiarize the reader with a concept of transmission quality. Reference equivalents express the loudness loss of telephone connections and have been typically stated in terms of the planning value of the ORE of a complete connection. Because difficulties have been encountered in the use of reference equivalents, the planning value of the ORE has been replaced by the *corrected reference equivalent* (CRE) (CCITT Rec. G.111) around 1980. This has required some adjustment in the planning values of loudness loss of complete or partial connections.

CCITT Recs. G.121 and P.11 recommended the following objectives for maximum values for sending and receiving CREs:

- For the sending system between a subscriber and the first international circuit, the CRE should not exceed 25 dB.
- For the receiving system between the same two points, the CRE should not exceed 14 dB.

For large countries with five national circuits forming part of the four-wire chain, the CRE values are then 26 dB and 15 dB for sending CRE and receiving CRE, respectively. Table 2.4 gives customer opinion results for various CRE values in decibels (from CCITT Rec. P.11). Table 2.5 gives approximate equivalents between the older reference equivalent and the present corrected reference equivalent.

2.3.3 Loudness Rating

2.3.3.1 Introduction. An essential purpose of a telephone connection is to provide a transmission path for speech between a talker's mouth and the ear of the far-end listener. The loudness of the received speech signal depends on acoustic pressure provided by the talker and the **loudness loss** of the acoustic-to-acoustic path from the input to a telephone microphone at one end of the

TABLE 2.5 Corresponding Values of Reference Equivalent and Corrected Reference Equivalent for Typical Connections

Connection		Previously Recommended RE(q)	Presently Recommended CRE(y)
	Min	6	5[a]
Optimum range for a connection	Optimum	9	7[a]–11
(Rec. G.111 §3.2)	Max	18	16
Traffic weighted mean values			
Long-term objectives			
Connection (Rec. G.111 §3.2)	Min	13	13
	Max	18	16
National system send (Rec. G.121 §1)	Min	10	11.5
	Max	13	13
National system receive (Rec. G.121 §1)	Min	2.5	2.5
	Max	4.5	4
Short-term objectives			
Connection (Rec. G.111 §3.2)	Max	23	25.5
National system send (Rec. G.121 §1)	Max	16	19
National system receive (Rec. G.121 §1)	Max	6.5	7.5
Maximum values for national system	Send	21	25
(Rec. G.121 §2.1) of an average-sized country	Receive	12	14
Minimum for the national sending system			
(Rec. G.121 §3)		6	7

[a]These values apply for conditions free from echo; customers may prefer slightly larger values if some echo is present.

Source: CCITT Rec. P.11, Table 2a/P.11, page 3, ITU-T Rec. P.11, Helsinki 3/93. Copyright ITU (Ref. 24).

connection to the output of a telephone receiver at the other end of the connection. The effectiveness of speech communication over telephone connections and customer satisfaction depend, to a large extent, on the loudness loss which is provided. As the loudness loss is increased from a preferred range, the listening effort is increased and customer satisfaction decreases. At still higher values of loudness loss, the intelligibility decreases and it takes longer to convey a given quantity of information. On the other hand, if too little loudness loss is provided, customer satisfaction is decreased because the received speech is too loud.

Over the years, various methods have been used by transmission engineers to measure and express the loudness loss of telephone connections. We described reference equivalent in Section 2.3.1 and it is covered in CCITT Recs. P.42 and P.72.

Because difficulties were encountered in the use of reference equivalents, the planning value of the overall reference equivalent was replaced by the corrected reference equivalent (CRE), which we discussed in Section 2.3.2. It is also covered in CCITT Rec. G.111.

TABLE 2.6 Overall Loudness Rating Opinion Results

Overall Loudness Rating (dB)	Representative Opinion Results[a]	
	Percent "Good plus Excellent"	Percent "Poor plus Bad"
5–15	>90	<1
20	80	4
25	65	10
30	45	20

[a]Based on opinion relationship derived from the transmission quality index (see Annex A, ITU-T Rec. P.11).
Source: ITU-T Rec. P.11, Table 1/P.11, page 2, Helsinki 3/93.

CCITT Recs. P.76, P.78, and P.79 provide information on subjective and objective methods for the determination of **loudness ratings** (LRs) which are now recommended by the ITU-T organization. These methods are expected to eliminate the need for subjective determinations of loudness loss in terms of the corrected reference equivalent. The currently recommended values of loudness loss in terms of loudness ratings are give in ITU-T Recs. G.111 and G.121 [25, 26].

2.3.3.2 Customer Opinion. Customer opinion, as a function of loudness loss, can vary with the test group and the particular test design. Table 2.6 gives opinion results for various values of overall loudness rating (OLR) in decibels (dB). These are based on representative laboratory conversation test results for telephone connections in which other characteristics such as circuit noise have little contribution to impairment.

2.3.3.3 Recommended Values of Loudness Rating. Table 2.7 provides further information on selected values of loudness rating which have been recommended or are under study by the ITU-T organization.*

2.3.3.4 Determination of Loudness Rating. The designation of loudness ratings (LRs) in an international connection is given in Figure 2.3. Telephone sensitivity must be measured, and this can be carried out using ITU-T Recs. P.66, P.76, P.78, and P.79. Telephone sensitivity includes both microphone output and earpiece sensitivity. Overall loudness rating (OLR) can be calculated using the following formula:

$$OLR = SLR + CLR + RLR$$

*Previously called CCITT.

TABLE 2.7   LR Values (dB) Cited in ITU-T Recs. G.111 and G.121

	SLR[a]	CLR[a]	RLR[a]	OLR[a]
Traffic weighted mean values				
Long term	$7-9^b$	$0-0.5^e$	$1-3^{b,f}$	$8-12^{e-g}$
Short term	$7-15^b$	$0-0.5^e$	$1-6^{b,f}$	$8-21^{e-g}$
Maximum values for an average-				
sized country	16.5^c		13^c	
Minimum value	-1.5^d			

[a]As in Figure 2.3.
[b]Clause 1/G.121, ITU-T Rec. G.121.
[c]Subclause 2.1/G.121, ITU-T Rec. G.121.
[d]Clause 3/G.121, ITU-T Rec. G.121.
[e]When the international chain is digital, CLR = 0. If the international chain consists of one analog circuit, CLR = 0.5 and then OLR is increased by 0.5 dB. (If the attenuation distortion with frequency of this circuit is pronounced, the CLR may increase by another 0.2 dB. See A.4.2/G.111, ITU-T Rec. G.111.)
[f]See also the remarks made in 3.2/G.111, ITU-T Rec. G.111.
[g]Subclause 3.2/G.111, ITU-T Rec. G.111.

Source: ITU-T Rec. P.11, Table 2b/P.11, page 4, Helsinki 3/93 (Ref. 24).

The overall loudness rating is defined as the loudness loss between the speaking subscriber's mouth and the listening subscriber's ear via a connection.

The send loudness rating (SLR) is defined as the loudness loss between the speaking subscriber's mouth and an electric interface in the network. (The loudness loss here is defined as the weighted decibel average of driving sound pressure to measured voltage.)

The receive loudness rating (RLR) is the loudness loss between an electric interface in the network and the listening subscriber's ear. (The loudness loss in this case is defined as the weighted decibel average of driving electromotive force to measured sound pressure.)

Circuit loudness rating (CLR) is the loudness loss between two electrical interfaces in a connection or circuit, each interface terminated by its nominal

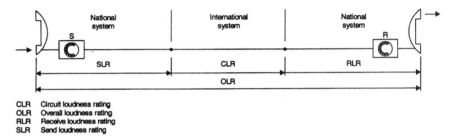

CLR   Circuit loudness rating
OLR   Overall loudness rating
RLR   Receive loudness rating
SLR   Send loudness rating

Figure 2.3. Designation of LRs in an international connection. From ITU-T Rec. P.11, Figure 1/P.11, page 4, Helsinki 3/93 (Ref. 24).

impedance, which may be complex. (The loudness loss here is approximately equivalent to the weighted decibel average of the composite electric loss.)

These definitions are taken from ITU-T Rec. G.111, Helsinki 3/93 (Ref. 25).

2.4 Subscriber Loop Design Techniques

2.4.1 *Introduction.* Consider the following drawing of a simplified subscriber loop:

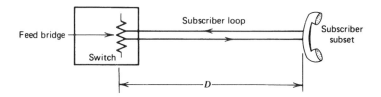

Distance D, the loop length, is most important. We know from this diagram that D must be limited because of attenuation of the voice signal. Likewise, there is a limit to D due to dc resistance, so signaling the local switch can be effected.

The attenuation limit would be taken from the national transmission plan, and for our discussion 6 dB is assigned as the limit (referenced to 800 Hz). For the loop resistance limit we must look to the switch. For instance, many conventional crossbar switches will accept up to 1300 Ω.* From this figure we subtract 300 Ω, the nominal resistance for the telephone subset in series with the loop, leaving a 1000-Ω limit for the wire pair if we disregard the feed bridge resistance. Therefore, in the paragraphs that follow, the figures 6 dB (attenuation limit for loop)† and 1000 Ω (resistance limit) are used.

2.4.2 *Calculating the Resistance Limit.* To calculate the dc loop resistance for copper conductors, the following formula is applicable:

$$R_{dc} = \frac{0.1095}{d^2} \qquad (2.1)$$

where R_{dc} is the loop resistance in ohms per mile (statute) and d is the diameter of the conductor (in inches).

If we wish a 10-mi loop and allow 100 Ω per mile of loop (for the stated 1000-Ω limit), what diameter of copper wire would be needed?

---

*Many semielectronic switches will accept 1800-Ω loops and, with special line equipment, 2400 Ω.
†In the United States this value may be as high as 9 dB.

TABLE 2.8 Loss and Resistance per 1000 ft pf Subscriber Cable[a]

Cable Gauge	Loss/1000 ft (dB)	Ω/1000 ft of Loop
26	0.51	83.5
24	0.41	51.9
22	0.32	32.4
19	0.21	16.1

[a]Cable is low-capacitance type (i.e., ≤0.075 nF/mi).

$$100 = \frac{0.1095}{d^2}$$

$$d^2 = \frac{0.1095}{100}$$

$$d = 0.033 \text{ in. or } 0.76 \text{ mm (round off to } 0.80 \text{ mm)}$$

Using Table 2.8, we can compute maximum loop lengths for 1000-Ω signaling resistance. As an example, for a 26-gauge loop, we have

$$\frac{1000}{83.5} = 11.97 \text{ or } 11,970 \text{ ft}$$

This, then, is the signaling limit, and not the loss (attenuation) limit, or what some call the "transmission limit," referred to in Section 2.4.3. As the reader has certainly inferred by now, resistance design is a method of designing subscriber loops using resistance limits as a basis or limiting parameter.

2.4.3 Calculating the Loss Limit. Attenuation or loop loss is the basis of transmission design of subscriber loops. The attenuation of a wire pair varies with frequency, resistance, inductance, capacitance, and leakage conductance. Also, resistance of the line will depend on temperature. For open-wire lines, attenuation may vary by ±12% between winter and summer conditions. For buried cable, which we are more concerned with in this context, variations due to temperature are much less.

Table 2.8 gives losses of some common subscriber cable per 1000 ft. If we are limited to 6-dB (loss) on a subscriber loop, then by simple division we can derive the maximum loop length permissible for transmission design considerations for the wire gauges shown.

$$26 \quad \frac{6}{0.51} = 11.7 \text{ kft}$$

$$24 \quad \frac{6}{0.41} = 14.6 \text{ kft}$$

TABLE 2.9   Code for Load-Coil Spacing

Code Letter	Spacing (ft)	Spacing (m)
A	700	213.5
B	3000	915.0
C	929	283.3
D	4500	1372.5
E	5575	1700.4
F	2787	850.0
H	6000	1830.0
X	680	207.4
Y	2130	649.6

$$22 \quad \frac{6}{0.32} = 19.0 \text{ kft}$$

$$19 \quad \frac{6}{0.21} = 28.5 \text{ kft}$$

2.4.4 Loading. In many situations it is desirable to extend subscriber loop lengths beyond the limits described in Sections 2.2 and 2.4. Common methods to attain longer loops without exceeding loss limits are to increase conductor diameter, use amplifiers and/or range extenders,* and use the inductive loading.

Inductive loading tends to reduce transmission loss on subscriber loops and other types of voice pair at the expense of good attenuation-frequency response beyond 3000 Hz. Loading a particular voice-pair loop consists of inserting series inductances (loading coils) into the loop at fixed intervals. Adding load coils tends to decrease the velocity of propagation and increase impedance. Loaded cables are coded according to the spacing of the load coils. The standard code for load coils regarding spacing is shown in Table 2.9.

Loaded cables typically are designated 19-H-44, 24-B-88, and so forth. The first number indicates the wire gauge, with the letter taken from Table 2.9 and indicative of the spacing, and the third item is the inductance of the coil in millihenries (mH). For instance, 19-H-66 is a cable commonly used for long-distance operation in Europe. Thus the cable has 19-gauge voice pairs loaded at 1830-m intervals with coils of 66-mH inductance. The most commonly used spacings are B, D, and H.

Table 2.10 will be useful for calculation of attenuation of loaded loops for a given length. For example, for 19-H-88 cable (last entry in the table), the

---

*A range extender is a device that increases battery voltage on a loop, which extends its signaling range. It may also contain an amplifier, thereby extending transmission loss limits as well.

TABLE 2.10   Some Properties of Cable Conductors

Diameter (mm)	AWG No.	Mutual Capacitance (nF/km)	Type of Loading	Loop Resistance (Ω/km)	Attenuation at 1000 Hz (dB/km)
0.32	28	40	None	433	2.03
		50	None		2.27
0.40		40	None	277	1.62
		50	H-66		1.42
		50	H-88		1.24
0.405	26	40	None	270	1.61
		50	None		1.79
		40	H-66	273	1.25
		50	H-66		1.39
		40	H-88	274	1.09
		50	H-88		1.21
0.50		40	None	177	1.30
		50	H-66	180	0.92
		50	H-88	181	0.80
0.511	24	40	None	170	1.27
		50	None		1.42
		40	H-66	173	0.79
		50	H-66		0.88
		40	H-88	174	0.69
		50	H-88		0.77
0.60		40	None	123	1.08
		50	None		1.21
		40	H-66	126	0.58
		50	H-88	127	0.56
0.644	22	40	None	107	1.01
		50	None		1.12
		40	H-66	110	0.50
		50	H-66		0.56
		40	H-88	111	0.44
0.70		40	None	90	0.92
		50	H-66		0.48
		40	H-88	94	0.37
0.80		40	None	69	0.81
		50	H-66	72	0.38
		40	H-88	73	0.29
0.90		40	None	55	0.72
0.91	19	40	None	53	0.71
		50	None		0.79
		40	H-44	55	0.31
		50	H-66	56	0.29
		50	H-88	57	0.26

Source: ITT, Telecommunication Planning Documents—*Outside Plant* (Ref. 7). Reprinted with permission of ALCATEL, Stuttgart, Germany.

attenuation per kilometer is 0.26 dB (0.42 dB/statute mile). Thus for our 6-dB loop loss limit, we have 6/0.26, limiting the loop to 23 km in length (14.3 statute miles).

In the design of a subscriber loop, budget 400 Ω resistance for the telephone subset. The minimum loop current is 20 mA. Budget about 5.5 Ω per load coil and 8.5 Ω for each miniload coil.

2.4.5 Summary of Limiting Conditions: Transmission and Signaling. We have been made aware that the size of an exchange serving area is limited by factors of economy involving signaling and transmission. Signaling limitations are a function of the type of exchange and the diameter of the subscriber pairs and their conductivity, whereas transmission is influenced by pair characteristics. Both limiting factors can be extended, but that extension costs money, particularly when there may be many thousands of pairs involved. The decision boils down to the following:

1. If the pairs to be extended are few, they should be extended.
2. If the pairs to be extended are many, it probably is worthwhile to set up a new exchange area or a satellite exchange or to use an outside plant module in the area.

These economies are linked to the cost of copper. The current tendency is to reduce the wire gauge wherever possible or even resort to the use of aluminum as the pair conductor.

2.4.6 Subscriber Loop Impedances. For a conventional two-wire switch, the characteristic impedance is 900 Ω. This is called a *compromise impedance*, and it is the impedance looking into the line circuit.

Since the late 1970s with the advent of digital transmission, a lot of work was done to improve impedance matching and characterization of the subscriber loop looking into the switch. A family of subscriber loop impedances is shown in Table 2.11.

Most equipment to be attached to a two-wire* loop is considered to have a 600-Ω impedance (Table 2.11d) Also, termination impedances at each end of a four-wire circuit are considered to be 600 Ω resistive. Both the 600-Ω and 900-Ω values are conventions and are compromises. Another value is 735 Ω (Table 2.11e). Some test instruments use this value, which is still another compromise, between 600 and 900 Ω.

Consider Figure 2.4. It shows how characteristic impedance varies with frequency. The example in the figure is a 26-gauge nonloaded wire pair. Reference frequency in North America is 1000 Hz, whereas in Europe and in countries under European hegemony it is 800 Hz. The figure shows that at about 1000 Hz the characteristic impedance is indeed 900 Ω.

---

*Two-wire and four-wire transmission is described in Chapter 5.

TABLE 2.11   Standard Termination Impedances

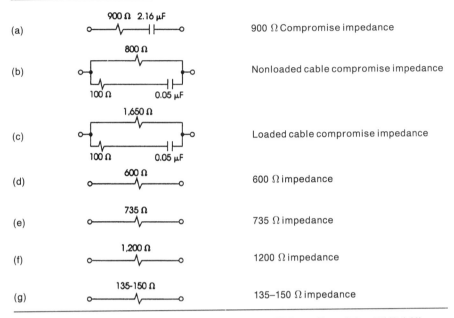

(a)	900 Ω 2.16 μF	900 Ω Compromise impedance
(b)	800 Ω / 100 Ω / 0.05 μF	Nonloaded cable compromise impedance
(c)	1,650 Ω / 100 Ω / 0.05 μF	Loaded cable compromise impedance
(d)	600 Ω	600 Ω impedance
(e)	735 Ω	735 Ω impedance
(f)	1,200 Ω	1200 Ω impedance
(g)	135-150 Ω	135–150 Ω impedance

Source: Subscriber Loop Signaling and Transmission Handbook, IEEE Press, Figure 5-6, p. 100 (Ref. 23).

Trunk plant often uses heavier-gauge cable, say 22 AWG (American Wire Gauge), where the characteristic impedance (Z_0) approaches 600 Ω at 1000 Hz (Table 2.11d). Loaded subscriber loops have a Z_0 approaching

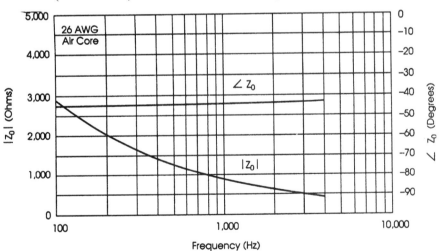

Figure 2.4. Characteristic impedance, Z_0, varies with frequency. From *Subscriber Loop Signaling and Transmission Handbook*, IEEE Press, Figure 5-2D, page 88 (Ref. 23).

TABLE 2.12   Average Occupation Time During the Busy Hour per Subscriber Line

Subscriber Type	BH Traffic Intensity (erlangs)
Residence	0.01–0.04
Business	0.03–0.06
PABX	0.1–0.6
Coin box	0.07

1200 Ω (Table 2.11f). Some high data-rate modems designed to connect to a subscriber loop may have a Z_0 in the range of 135–150 Ω (Table 2.11g). These lower values were selected because the impedance of the loop may go as low as 100 Ω at higher frequencies.

3   OTHER LIMITING CONDITIONS

The size of an exchange area obviously will depend greatly on subscriber (or potential subscriber) density and distribution. The subscriber traffic is another factor to be considered. The CCITT offers the values in Table 2.12 for subscriber line traffic intensity.

Exchange sizes are often in units of 10,000 lines. Although the number of subscribers initially connected should be considerably smaller than when an exchange is installed, 10,000 is the number of subscribers that may be connected when an exchange reaches "exhaust," where it is filled and no more subscribers can be connected.

Ten thousand is not a magic number, but it is a convenient one. It lends itself to crossbar (switch) unit size and is a mean unit for subscriber densities in suburban areas and mid-sized towns in fairly well developed countries. More important, though, is its significance in telephone numbering (the assignment of telephone numbers). Consider a seven-digit number. Now break that down into a three-number group and a four-number group. The first three digits, that is, the first three dialed, identify the local exchange. The last four identify the individual subscriber and is called the subscriber number. Note the breakdown in the following sample:

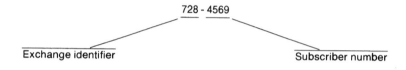

For the subscriber number there are 10,000 number combination possibilities, from 0000 to 9999. Of course there are up to 1000 possibilities for the exchange identifier. Numbering is discussed at length in Chapter 3 as a consideration under switching.

The foregoing discussion does not preclude exchanges larger than 10,000 lines. But we still deal in units of 10,000 lines, at least in conventional telephony. The term "wire center" is often used to denote a single location housing one or more 10,000 line exchanges. Some wire centers house up to 100,000 lines with a specific local serving area. Wire centers with an ultimate capacity of up to 140,000 lines can be economically justified under certain circumstances. Subscriber density, of course, is the key. Nevertheless, many exchanges will have extended loops requiring some sort of special conditioning, such as larger-gauge wire pairs, loading, range extenders, amplifiers, and the application of carrier techniques (Chapter 5). Leaving aside rural areas, 5–25% of an exchange's subscriber loops may well require such conditioning or may be called "long loops."

4 SHAPE OF A SERVING AREA

The shape of a serving area has considerable effect on optimum exchange size. If a serving area has sharply angular contours, the exchange size may have to be reduced to avoid excessively long loops (e.g., revert to the use of more exchanges in a given local geographical area of coverage).

There is an optimum trade-off between exchange size, and we mean here the economies of large exchanges (centralization), and the high cost of long subscriber loops. An equation that can assist in determining the trade-off is as follows, which is based on uniform subscriber density, a circular serving area A of radius r such that

$$C = \frac{A}{\pi r^3}(a + bd\pi r^3) + Ad(f + gL) \tag{2.2}$$

where C is the total cost of exchanges, which decreases when r increases and to which is added the cost of subscriber loops, which increases with r, and L is the average loop length, which may be related as $L = (2r/3)$, the straight-line distance. To determine the minimum cost of C with respect to r, the equation is differentiated and the result is set equal to zero. Thus

$$\frac{2Aa}{\pi r^3} = \frac{2Adg}{3} \tag{2.3}$$

and this may be fairly well approximated by

$$r = \left(\frac{a}{dg}\right)^{1/3} \tag{2.4}$$

The cost of exchange equipment is $a + bn$, where n is the number of lines, and the cost of subscriber loops is equal to $(f + gL)$ where, as we stated, L is the average loop length given a uniform density of subscribers, d; and a, b, f, and g are constants.

Since r varies as the cube root, its value does not change greatly for wide ranges of values of d. One flaw is that loops are seldom straight-line distances, and this can be compensated for by increasing g in the ratio (average loop length). This theory is simplified by making exchange areas into circles.

If an entire local area is to be covered, fully circular exchange serving areas are impractical. Either the circles will overlap or uncovered spaces will result, neither of which is desirable. There are then two possibilities: square serving areas or hexagonal serving areas. Of the two, a hexagon more nearly approaches a circle. The size of the hexagon can vary with density with a goal of 10,000 lines per exchange as the ultimate capacity. Again, a serving area could have a wire center of 100,000 lines or more, particularly in heavily populated metropolitan areas.

Besides the hexagon, full coverage of local areas may only be accomplished using serving areas of equal triangles or squares. This assumes, of course, that the local area was *ideally* divided into identical geometric figures and would apply only under the hypothetical situation of nearly equal telephone density throughout. A typical hexagon subdivision is shown in the diagram that follows.

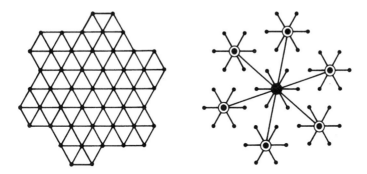

The routing problem then arises. How should the serving areas with their respective local exchanges be interconnected? From our previous discussion we know that two extremes are offered, mesh and star. We are also probably aware that as the number of exchanges involved increases, full mesh becomes very complicated and is not cost-effective. Certainly it is not as cost-effective as a simple hierarchical network of two or three levels permitting high-usage (HU) connections between selected nodes. For instance, given the hexagon formation in the preceding diagram, a full-mesh or two-level star network can be derived, as shown in the diagram that follows.

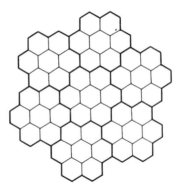

From the routing pattern in the preceding diagram it should be noted that fan-outs of 6 and 8 allow symmetry, whereas fan-outs of 5 and 7 lead to inequalities. A fan-out of 4 is usually too small for economic routing.

It will be appreciated that some of the foregoing assumptions are rarely found in any real telecommunications environment. Uniform telephone density was one real assumption, and the implication of uniform traffic flow was another. Serving areas are not uniform geometric figures, exchanges seldom may be placed at serving area centers, and routing will end up as a mix of star and mesh. Small local areas serving 5–15 exchanges or even more may well be fully mesh connected. Some cities are connected in full mesh with over 50 exchanges. But as telephone growth continues, tandem routing will become the economic alternative [5, 11].

5 EXCHANGE LOCATION

A fairly simple, straightforward method for determination of the theoretical optimum exchange location is described in Ref. 1, Chapter 6. Basically, the method determines the center of subscriber density in much the same way the center of gravity would be calculated. In fact, other publications call it the center-of-gravity method.

Using a map to scale, a defined area is divided into small squares of 100–500 m on a side. One guide for determination of side length would be to use a standard length of the side of a standard city block in the serving area of interest. The next step is to write in the total number of subscribers in each of the blocks. This total is the sum of three figures: (1) existing subscribers, (2) waiting list, and (3) forecast of subscribers for 15 or 20 years into the future. It follows that the squares used for this calculation should coincide with the squares used in the local forecast. The third step is to trace two lines over the subscriber area. One is a horizontal line that has approximately the same number of total subscribers above the line as below. The second is a vertical line where the number of subscribers to the left of the line is the same as that to the right. The point of intersection of these two lines is the theoretical optimum center or exchange location. A sample of this method is shown in Figure 2.5, where S_1 is the sum of subscribers across a single line and C_s is the cumulative sum.

Now that the ideal location is known, where will the real optimum location be? This will depend considerably on secondary parameters such as availability of buildings and land; existing and potential cable or feeder runs; the so-called trunk pull; and layout of streets, roads, and highways. "Trunk pull" refers to the tendency to place a new exchange near the one or several other exchanges with which it will be interconnected by trunks (junctions). Of course, this situation occurs on the fringes of urban areas where a new exchange location will tend to be placed nearer to the more populated area, thereby tending to shorten trunk routes. This is illustrated in Figure 2.1 and discussed in more detail below.

The preceding discussion assumed a bounded exchange area; in other words, the exchange area boundaries were known. Assume now that exchange locations are known and that the boundaries are to be determined. What follows is also valid for redistributing an entire area and cutting it up into serving areas. A great deal of this chapter has dealt with subscriber loop length limits. Thus an outer boundary will be the signaling limits of loops as

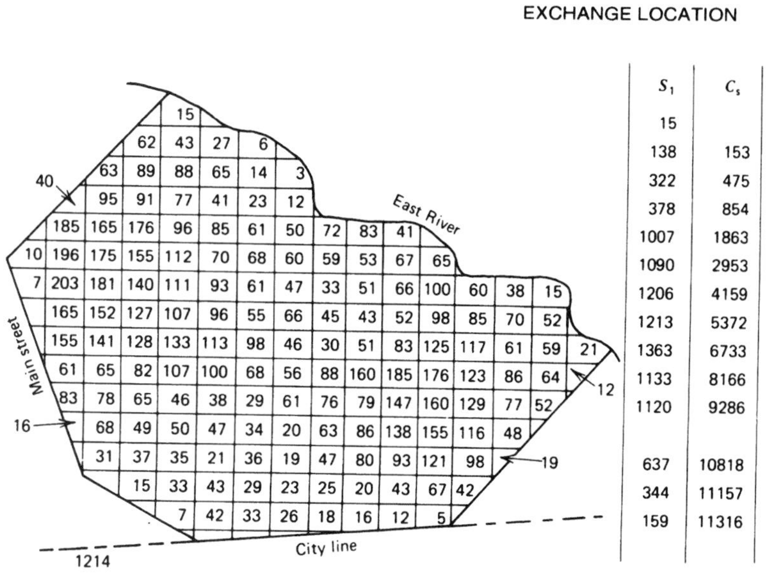

EXCHANGE LOCATION

Figure 2.5. Sample wire centering exercise using "center-of-gravity" method [9].

TABLE 2.13   Resistance Limits for Several Types of Exchange

Exchange Type	Resistance Limit (Ω)
No. 1 step-by-step (USA)	1300
No. 1 crossbar (USA)	1300
No. 5 crossbar (USA)	1520
ESS (USA)	2000
Pentaconta (ITT) (crossbar)	1250
Rotary (ITT) (Europe)	1200
Metaconta (local) (ITT)	2000
Pentaconta 2000 (ITT)	1250
DMS-100	1900 (4000)
Other digital local exchanges	2000–4000 (Ref. 21)

described previously. The optimum cost–benefit trade-off is found when all or nearly all loops in a serving area remain nonconditioned and of small diameter, say, 26 gauge. We note that with H-66 loading the outer boundary will be just under 5 km in this case. It also would be desirable to have a hexagonal area if possible. In practice, however, natural boundaries may well be the most likely real boundaries of a serving area. "Main street," "East River," and "City line" in Figure 2.5 illustrate this point. In fact, these boundaries may set the limits such that they may be considerably greater or less than the maximum signaling (supervisory) limits suggested earlier. Of course, there are two types of serving areas where the argument does not hold. These are rural areas and densely populated urban areas. For the rural areas we can imagine very large serving areas and, for urban areas of dense population, considerably smaller serving areas than those set out with maximum supervisory signaling limits.

To determine boundaries of serving areas when dealing with an exchange that is already installed and a new exchange, we could use the so-called ratio technique. Again, we use signaling (supervisory) limits as the basis. As we are aware, these limits are basically determined by the type of exchange and copper wire gauge utilized for subscriber loops. Table 2.13 gives resistance limits for several of the more common telephone exchanges found in practice.

The ratio method is as follows. Given an existing exchange A and a new exchange B that will be established on a cable route from A, assume that A is a step-by-step exchange and that B is a Pentaconta exchange. The distance from A to B along a cable route can be computed by equating distance to the sum of the resistances of exchanges A and B. Use 26-gauge wire in this case, and take the resistance from Table 2.13. From Table 2.8 using H-66 loading, resistance can be equated to 273 Ω/km. Sum the resistance limits of exchanges $(A + B)$:

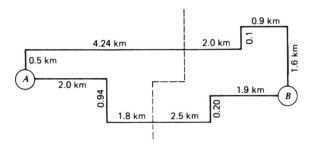

Figure 2.6. Determining serving area boundaries with ratio method [17].

$$A + B = 1250 + 1300 = 2550 \, \Omega$$

Then divide by 273:

$$2550/273 \simeq 9.34 \, \text{km (distance } A \text{ to } B)$$

With no other factors influencing the decision, the boundary would be established along the feeder (cable) route (distance from A):

$$D_A = 1300 \times 9.34/(1250 + 1300)$$

or $\simeq 4.74$ km from A. This exercise is shown diagrammatically in Figure 2.6.

Continue the exercise and examine the exchange serving area at A. Assume that the area is a square with A at its center. Allowing for *non*-crow-fly feeder routes, we can assume that the square has 8 km on a side, as shown in the following diagram:

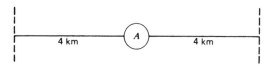

or 64 km^2. If the largest exchange wanted had 100,000 subscribers and the smallest has 10,000 at the end of the forecast period (15–20 years), then subscriber density would be 10,000/64 to 100,000/64 or 156 subscribers per km^2 to 1560 subscribers per km^2.

When serving areas exceed 100,000 lines at the end of a forecast period in densely populated urban areas, breaking up of the areas into smaller ones with exchanges of smaller capacity should be considered. One problem with large conventional exchanges is size; another is cable entry and mainframe size. The latter is more evident with new SPC* [electronic switching system (ESS)] exchanges in that cable entry and mainframe are much larger than the exchange itself. However, the really major problem is reliability or survivability. If a very large exchange is knocked out as a result of fire,

---

*SPC stands for stored program control (i.e., switch control is computer-based).

explosion, sabotage, or other disaster, it is much more catastrophic than the loss of a smaller single exchange.

Another consideration in exchange location is what is called "trunk pull" discussed briefly above. This is a secondary factor in exchange placement and refers to the tendency, in certain circumstances, to shift a proposed exchange location to shorten trunks (junctions). Trunk pull (see Fig. 2.1) becomes a significant factor only where population fringe areas around urban centers and trunks extend toward the city center and the exchanges in question may well be connected in full mesh in their respective local area. Some possible saving may be accrued by shifting the exchange more toward the center of population, thereby shortening trunks at the expense of lengthening subscriber loops in the direction of more sparse population, even with the implication of conditioning on well over 10% of the loops. The best way to determine if a shift is worthwhile is to carry out an economic study comparing costs in PWAC (present worth of annual charges) [18, 19], comparing the proposed exchange as located by the center-of-gravity method to the cost of the shifted exchange. Basic factors in the comparison are in the cost of subscriber plant, trunk plant, plant construction, and cost of land (or buildings). The various methods of handling the sparsely populated sections of the fringe serving area in question should also be considered. For instance, we might find that in those areas, much like rural areas, there may be a tendency for people to bunch up in little villages or other small centers of population. This situation is particularly true in Europe. In this case the system designer may find considerable savings in telephone-plant costs by resorting to the use of satellite exchanges or concentrators. Trunk connections between satellite or concentrator and main exchange may be made by using carrier techniques. Concentrator and satellite exchanges are discussed in Chapter 3 and carrier techniques in Chapter 5.

A typical fringe-area situation is shown in Figure 2.7. The "bunching" and the other possibility, "thinning out," are shown in the figure. "Thinning out" is just the population density per unit area decreases as we proceed from an urban center to the countryside. A topological line of population density of 10 inhabitants per square kilometer (26 inhabitants per square mile) is a fair guideline for separation of the rural part of the fringe area from the urban–suburban part. Of course, in the latter part we would have to resort to widespread conditioning, to the use of concentrator–satellites, or to subscriber carrier or subscriber pulse-code modulation (PCM) (Chapter 8).

6   OTHER LOOP DESIGN TECHNIQUES

6.1   Minigauge Techniques

Fine-gauge and "minigauge" techniques essentially are refinements of the unigauge concept. In each case the principal object is to reduce the amount of

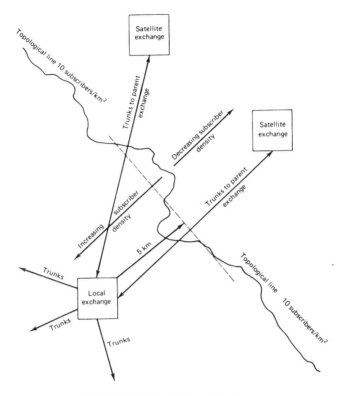

Figure 2.7. Fringe-area considerations.

copper in the subscriber plant. Obviously, one method is to use still smaller gauge pairs on shorter loops. For instance, the use of gauges as small as 32 is being considered. Another approach is to use aluminum as the conductor. When aluminum is used, a handy rule of thumb to follow is that the ohmic and attenuation losses of aluminum may be equated to copper in that aluminum wire should always be the next "standard gauge" larger than its copper counterpart if copper were used. Some of the more common gauges are compared in the table that follows.

Copper	Aluminum
19	17
22	20
24	22
26	24

Aluminum has some drawbacks as well. The major ones are summed up as follows:

- It should not be used on the first 500 yd (m) of cable where the cable has a larger diameter (and here we mean more loops before branching).
- It is more difficult to splice than copper.
- It is more brittle.
- Because the equivalent conductor is larger than its copper counterpart, an equivalent aluminum cable with the same conductivity-loss characteristics will have a smaller pair count in the same sheath.

6.2 New Loop Design Plans for US Regional Bell Operating Companies*

Before 1980 US Bell operating companies designed subscriber loops in accordance with one of three loop design plans: (1) resistance design (RD), 96%; (2) unigauge design (UG), 1%; (3) long route design (LRD), 3%. (Percentages show approximate percentage of total plant for each plan.) The objective, of course, is to design subscriber loops on a global basis rather than on an individual basis, which is extremely costly. If the system engineers fully comply with these design rules, then no loop will exceed switch signaling limitations (resistance) and there will be an adequate distribution of transmission losses.

A 1980 survey [20] studied loss design in a population of loops having an average length greater than the mean length of the general population. It was found that approximately 4% of the measured losses were greater than 8.5 dB, and 2% exceeded 10 dB. With the design rules in effect at the time, losses exceeding 8.5 dB were expected. Under the new, revised resistance design (RRD) rules, no properly designed loop should display a loss greater than 8.5 dB at 1000-Hz reference frequency.

6.2.1 Previous Rules

6.2.1.1 Resistance Design (RD)

Resistance: 0–1300 Ω (includes only resistance of cable and load coils).

Load coils: H-88 beyond 18 kft [not including bridged tap (BT)].

End sections (ES) and bridged tap (BT): Nonloaded $BT = 6$ kft (max); loaded $ES + BT = 15$ kft (max), 3 kft (min), 12 kft recommended.

Transmission limitations: None.

Cable gauges: Any combination of 19–26 gauge.

6.2.1.2 Long Route Design (LRD) (Rural Areas)

Loop resistance: 1301–3000 Ω (including only resistance of cable and load coils).

---

*The source of Section 6.2 is Ref. 20, Section 7.

Load coils: Full H-88.

End sections (ES) and bridged tap (BT): ES + BT = 12 kft.

Transmission limitations: For loop resistances > 1600 Ω, loop gain required.

Cable gauges: Any combination of 19–26 gauge.

6.2.2 Present Loop Design Rules

6.2.2.1 Revised Resistance Design (RRD)

Loop resistance: 0–18 kft, 1300 Ω (max) (includes resistance of cable and load coils only); 18–24 kft, 1500 Ω (max) (includes resistance of cable and load coils only); >24 kft, use digital loop carrier (DLC).

Load coils: Full H-88 for loops longer than 18 kft.

End sections (ES) and bridged tap (BT): Nonloaded total cable + BT = 18 kft, maximum BT = 6 kft. Loaded ES + BT = 3–12 kft.

Transmission limitations: None.

Cable gauges: Two gauge combinations preferred (22-, 24-, 26-gauge).

6.2.2.2 Concentration Range Extension with Gain (CREG)

Loop resistance: 0–2800 Ω (includes only resistance of cable and load coils).

Load coils: Full H-88 >15 kft.

End sections (ES) and bridged tap (BT): Nonloaded cable + BT = 15 kft, maximum BT = 6 kft. Loaded ES + BT = 3–12 kft.

Transmission limitations: Gain range extension required for loop resistance > 1500 Ω.

Cable gauges: Two gauge combinations preferred (22-, 24-, 26-gauge).

6.2.2.3 Modified Long Route Design (MLRD)

Loop resistance: 1501–2800 Ω.

Load coils: Full H-88.

End sections (ES) and bridged tap (BT) (max): ES + BT = 3–12 kft.

Transmission limitations: Range extension with gain required for loop resistances greater than 1500 Ω.

Cable gauges: Two gauge combinations preferred (22-, 24-, 26-gauge).

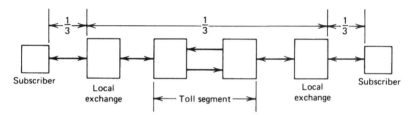

Figure 2.8. One conceptual thumb rule for network loss assignment.

7   DESIGN OF LOCAL AREA TRUNKS (JUNCTIONS)

Exchanges in a common local area are interconnected by trunks, called junctions in the United Kingdom. Depending on length and certain other economic factors, these trunks use voice-frequency (VF) transmission over wire pairs formed up in cables. In view of the relatively small number of such trunk circuits in comparison to the total number of subscriber lines* in the area, it is generally economical to minimize attenuation in this portion of the network.

One approach used by some telephone companies or administrations is to allot $\frac{1}{3}$ of the total end-to-end reference equivalent to each subscriber's loop and $\frac{1}{3}$ to the trunk network. Figure 2.8 illustrates this concept. For instance, if the transmission plan called for a 24-dB ORE, then $\frac{1}{3}$ of 24 dB, or 8 dB, would be assigned to the trunk plant. Of this we may assign 4 dB to the four-wire portion of the long-distance (toll) network, leaving 4 dB for local VF trunks or 2 dB at each end. The example has been highly simplified, of course. For the toll-connecting trunks (e.g., those trunks connecting the local network to the toll network), if a good return loss† cannot be maintained on all or nearly all connections, the losses on two-wire toll connecting trunks may have to be increased to reduce possibilities of echo and singing. Sometimes the range of loss for these two-wire circuits must be extended to 5 or 6 dB. It is just these circuits into which the four-wire toll network looks directly. Two-wire and four-wire circuits, echo and singing, are discussed in Chapter 5. Thus it can be seen that the approach to the design of VF trunks varies considerably from that used for subscriber loop design. Although we must ensure that signaling limits are not exceeded, the transmission limits will almost always be exceeded well before the signaling limit. The tendency to use larger-diameter cable on long routes is also evident. If loading is to be used, the first load coil is installed at distance $D/2$, where D is the normal separation distance between load points. Take the case of H loading, for instance. The distance between load points is 1830 m (Table 2.9), but the first

---

*Because of the inherent concentration in local switches, approximately 1 trunk is allotted from 8 to 25 subscribers, depending on design.

†Return loss is discussed in Chapter 5.

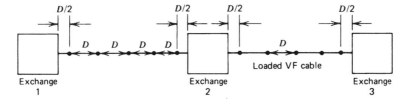

Figure 2.9. Loading of VF trunks (junctions).

load coil from the exchange is placed at $D/2$, or 915 m from the exchange. Then, if an exchange is bypassed, a fully loaded section exists. This concept is illustrated in Figure 2.9.

Now consider this example. A loaded 500-pair VF trunk cable extends across town. A new switching center is to be installed along the route where 50 pairs are to be dropped and 50 pairs inserted. It would be desirable to establish the new switch midway between load points. At the switch 450 circuits will bypass the new switching center. Using this $D/2$ technique, these circuits will need no conditioning; they will be fully loaded sections (i.e., $D/2 + D/2 = 1D$, a fully loaded section). Meanwhile, the 50 circuits entering from each direction are terminated for switching and need conditioning, so each electrically resembles a fully loaded section. However, the physical distance from the switch out to the first load point is $D/2$ or, in the case of H loading, 915 m. To make the load distance electrically equivalent to 1830 m, line build-out (LBO) is used. This is done simply by adding capacitance to the line. Suppose the location of the new switching center was such that it was not midway between load points but some other fractional distance. For the section consisting of the shorter distance, LBO is used. For the other longer run, often a half-load coil is installed at the switching center, and the LBO is added to trim up the remaining electrical distance.

By this time the reader has gathered what conditioning is in this context. For loaded trunks and loops, LBO is the action taken to change the electrical characteristics of the line by adding capacitance or inductance. It may be done at line midpoint where there is access, such as at a load-coil location. However, it is more commonly done at the switching center because of accessibility of cable pairs at the mainframe associated with the center.

8   VOICE-FREQUENCY REPEATERS

In telephone terminology, *voice-frequency repeaters* imply the use of *uni-directional* amplifiers at voice frequency on VF trunks. On a two-wire trunk we must resort to four-wire transmission techniques at the repeater. Two-wire and four-wire transmission are discussed in Chapter 5. Thus on a two-wire trunk two amplifiers must be used on each pair with a hybrid in and a hybrid out. The hybrid converts the two-wire circuit to a four-wire circuit and vice

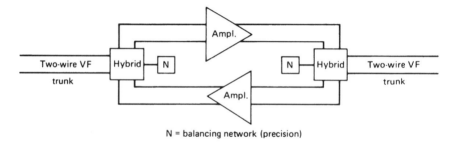

N = balancing network (precision)

Figure 2.10. Simplified block diagram of a VF repeater.

versa. A simplified block diagram of a VF repeater is shown in Figure 2.10. The gain of a VF repeater can be run up as high as 20 or 25 dB, and originally they were used on 50-mi 19-gauge loaded cable in the long-distance (toll) plant. Today they are seldom found on long-distance circuits but do have application on local trunk circuits, where the gain requirements are considerably less. Trunks using VF repeaters have the repeater's gain adjusted to the equivalent loss of the circuit minus the 4-dB loss to provide the necessary singing margin (see Chapter 5). In practice, a repeater is installed at each end of the trunk circuit to simplify maintenance and power feeding. Gains may be as high as 6–8 dB. It can be appreciated that the application of VF repeaters would be on trunks where the losses were excessive, beyond those called for in the appropriate transmission plan, such as on long loops or long VF trunks.

Another repeater more commonly used on two-wire trunks is the negative impedance repeater. This repeater can provide a gain as high as 12 dB, but 7 or 8 dB is more common in actual practice. The negative impedance repeater requires an LBO at each port and is a true two-way, two-wire repeater. The repeater action is based on regenerative feedback of two amplifiers. The advantage of negative impedance repeaters is that they are transparent to dc signaling. On the other hand, VF repeaters require a composite arrangement to pass the dc signaling. This consists of a transformer bypass.

9 TANDEM ROUTING

The local-area trunking scheme evolutionally has been mesh connection of exchanges, and in many areas of the world it remains full mesh. We said initially that mesh connection is desirable and viable for heavy traffic flows. As traffic flows reduce, going from one situation to another, the use of tandem routing in the local area becomes an interesting, economical alternative.

Furthermore, it can be shown that a local trunk network can be optimized, under certain circumstances, with a mix of tandem, high-usage (overflow to alternative routes), and direct connection (mesh). We often refer to these three possibilities as THD. The system designer wishes to determine, on a particular trunk circuit, if it should be "tandem," "high-usage," or "direct" (T, H, or D). This determination is based on incremental cost of the trunk (junction), making the total network costs as low as possible. Such incremental cost can be stated:

$$B = c + (bl) \tag{2.5}$$

where c is the cost of switching equipment per circuit, b is the incremental costs for trunks per mile or kilometer, and l is the length of the trunk (or junction) circuit.

To carry out a THD decision for a particular trunk route, the input data required are the *offered* traffic between the local exchanges in question and the grade of service. We can now say that the THD decision is to a greater extent determined by the offered traffic A between the exchanges and the cost ratio ε, where

$$\varepsilon = \frac{B}{B_1 + B_3} \tag{2.6}$$

and where B is the cost for the direct route, with B_1 from exchange 1 to the proposed tandem and B_2 from the proposed tandem exchange to exchange 2, for incremental costs between direct and tandem routing. Of course, before starting such an exercise, the provisional tandem points must be known. Figure 2.11, which was taken from Ref. 4, may be helpful as a decision guide.

To approximate the number of high-usage (HU) trunk circuits required, the following formula may be used

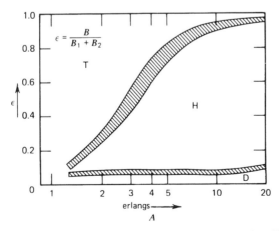

Figure 2.11. THD diagram. Courtesy of the International Telecommunications Union–CCITT [4].

$$F(n, A) = AE(n, A) - E(n + 1, A) = \varepsilon\psi \tag{2.7}$$

where E is the grade of service, n is the number of high-usage circuits, A is the offered traffic between exchanges, $F(n, A)$ is the "improvement" function [i.e., increase in traffic carried on high-usage trunk group on increasing the number of these trunk circuits from n to $(n + 1)$], ε is the cost ratio (as previously), and ψ is the efficiency of incremental trunks [marginal utilization (of magnitude 0.6–0.8)]. However, a still better approximation will result if the following formula is used:

$$F(n, A) = \varepsilon[1 - 0.3(1 - \varepsilon^2)] \tag{2.8}$$

The following exercise emphasizes several practical points on the application of tandem routing in the local area. Assume that there are 120,000 subscribers in a certain local area. If we allow a BH calling rate of 0.5 erlangs per subscriber (see Table 2.12) and assume full-mesh connection, we can assemble Table 2.14. On the basis of Table 2.14 it would not be reasonable to expect any appreciable local area trunk economy through tandem working with less than 30–40 exchanges in the area. Of course, Table 2.14 assumes unity of community of interest with the more distant exchanges so that a few attractive tandem routings might exist in the 20-to-30-exchange range. Areas smaller than those in our sample will naturally reduce the number of exchanges at which tandem working is viable, and a larger area and higher calling rate will increase this number. The results of any study to determine feasibility of using tandem exchange is very sensitive to the number of exchanges in the local area and considerably less sensitive to the size of the area and calling rate. Hence we can see that the economy of tandem routing is least in areas of dense subscriber population, where exchanges can be placed without exceeding subscriber loop length limits and where the trunks will be short. Relatively sparsely populated areas are more favorable and are

TABLE 2.14   Traffic Table: Full-Mesh Connection (120,000 Subscribers)

Number of Exchanges	Average Size of Exchange	Originated Traffic per Exchange (erlangs)	Average Traffic to Each Distant Exchange (erlangs)
10	12000	600	60
12	10000	500	42
15	8000	400	27
20	6000	300	15
30	4000	200	6.7
40	3000	150	3.8

likely to show relatively low community-of-interest factors between distant points in the area. (1, 4, 5, 13).

10 DIMENSIONING OF TRUNKS

A primary effort for the system engineer in the design of the local trunk network is the dimensioning of the trunks of that network. Here we simply wish to establish the economic optimum number of trunk circuits between exchanges X and Y. If we are given the traffic (in erlangs) between the two exchanges and the grade of service, we can assign the number of trunk circuits between X and Y. As discussed in Chapter 1, it is assumed that all traffic values are BH values. Once given this input information, the erlang value (the number of trunk circuits) can be derived from Table 1.1 (no overflow assumed).

It can be appreciated by the reader that the trunk traffic intensities used in the design of trunk routes should allow for growth—that is, the increase of traffic from the present with the passage of time. This increase is attributable to several factors: (1) the increase in the number of telephones in the area that generate traffic, (2) the probable increase in telephone usage, and (3) the possibility of a change in character of the area in question, such as from rural to suburban or from residential to commercial. Thus the designer must use properly forecast future traffic values. These values in practice are for the forecast period 5–8 years in the future. The art of arriving at these figures is called *forecasting*. Present traffic values should always be available as a base or point of departure.

Suppose we use a sample local area with five exchanges: *A, B, C, D,* and *E.* For an 8-year forecast period we could possibly come up with the traffic matrix like that shown in Table 2.15. Thus from the traffic matrix in the direction exchange *C* to exchange *B* there is a traffic intensity of 11 erlangs. Applying the 11 erlangs to Table 1.1, we see that 23 circuits would be required. For the distance *B* to *C,* the traffic intensity is 13 erlangs, and again from Table 1.1, 26 circuits would be required. These circuit figures suppose a grade of service of $p = 0.001$. For a grade of service of $p = 0.01$, 19 and 22 circuits, respectively, would be required.

The preceding discussion of routing implied an on-paper routing that would probably vary in the practice of actual cable lays and facility drawings to something quite different. In simplified terms, these differences are shown in Fig. 2.12.

11 COMMUNITY OF INTEREST

We have referred to the community-of-interest concept in passing. This is a method used as an aid to estimate calling rate and traffic distribution for a

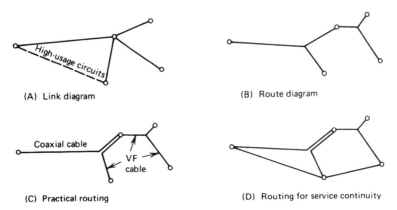

(A) Link diagram   (B) Route diagram

(C) Practical routing   (D) Routing for service continuity

Figure 2.12. Routing diagrams and practical routing (VF stands for voice frequency): (A) link diagram; (B) route diagram; (C) practical routing; (D) routing for service continuity [5].

new exchange and its connecting trunks. The community-of-interest factor K can be defined as follows:

$$\text{Traffic } A - B = K \times \frac{\text{Traffic originating at } A \times \text{Traffic originating at } B}{\text{Total originating traffic in area}}$$

$$(2.9)$$

If all subscribers are equally likely to call all others, the proportion results:

$$\frac{\text{Traffic originating at } B}{\text{Traffic originating in area}} = \frac{\text{Traffic terminating at } B}{\text{Traffic terminating in area}} \qquad (2.10)$$

This is the expected proportion of all A traffic that is directed to B. In this case $K = 1$, corresponding to equal community of interest between all subscribers. The condition of $K > 1$ or $K < 1$ indicates a greater or lesser interest than average between exchanges A and B. The factor K is affected by the type of area, whether residential or business, as well as the distances between exchanges. For instance, in metropolitan areas K may range from 4 for calls originating and terminating on the same exchange to 0.25 for calls on opposite sides of the city or metropolitan area.

The K factor is a useful reference for the installation of a new exchange where the serving area of the new exchange will be cut out from serving areas of other exchanges. The community-of-interest factors may then be taken from traffic data from the old exchanges, and the values averaged and then applied to the new exchange. The same principle may be followed for exchange extensions in a multiexchange area [5].

TABLE 2.15 Sample Traffic Matrix[a]

From/to	A	B	C	D	E	Toll	Total Orig.	Lines Working	Traffic per Line
				Traffic (erlangs)					
A	21	20	65	2.5	2.5	1.5	112.5	9200	0.122
B	22	80	13	6.0	5.0	4.0	130	26000	0.005
C	5	11	7	2.5	2.5	1.0	29	7500	0.004
D	2	7	1	0.3	0.2	0.5	11	3000	0.004
E	2	5	2	0.2	0.3	0.5	10	2800	0.004
Toll	2	3	1	0.5	0.5	—	7	—	—
Total	53	126	89	12	11	7.5	298.5	48500	0.006
Total-line	0.006	0.005	0.012	0.004	0.004	0.00015	—	—	—

[a]For 8-year forecast period.

Example: Traffic A to A is traffic originated at A and terminating at A.

REVIEW QUESTIONS

1. Local networks can be defined in a number of ways. Give the definition of a local network that is provided in the text.

2. Considering both quality of service and economy, give at least five constraints or planning factors that go into the design of a local telephone network.

3. What are the two limiting performance factors in the design of a subscriber loop that constrain length?

4. Reference equivalent is a measure of transmission quality. What singular parameter does it measure?

5. What are the three elements (measured in decibels) that, when summed, constitute the value of ORE?

6. What important parameter contained in a national transmission plan is vital in calculating ORE?

7. In reviewing the quality of service for speech telephony, the earliest measure of quality was *reference equivalent*; this was followed by an improved measure of customer satisfaction with a telephone connection which was called *corrected reference equivalent* (CRE). Yet, as time went on, around 1990 CCTTT came up with still a third measure of quality of a telephone connection. What was it called?

8. If we assume that a modern local exchange is designed for a maximum of 2000 Ω of loop resistance, not including the resistance of the end instrument (i.e., the telephone subset), what is the maximum loop length of 26-gauge copper wire that can be used? Assume 83 Ω per 1000 ft of loop.

9. The National Transmission Plan allows a 7-dB maximum subscriber loop loss. Assume a 0.51-dB loss per 1000 ft of loop for a 26-gauge wire pair. What is the maximum subscriber loop length considering only loss?

10. What is the maximum subscriber loop length permissible using inputs from questions 7 and 8?

11. The exchange involved in question 9 must serve some rural subscribers who are over the maximum loop length. Name at least four expedients that can be used to serve these subscribers. Differentiate whether these expedients have an impact on resistance limit or transmission limit.

12. What is a reasonable range of values for subscriber loop impedance?

13. A subscriber connected to a local exchange commonly (not always) is identified by a seven-digit number made up of a three-digit prefix

followed by a four-digit number. What does the three-digit prefix identify? What is the four-digit sequence called?

14. How many different individual subscribers can be served by a four-digit number assuming no blocked numbers?

15. Using only subscriber density, describe a method of idealized exchange placement.

16. Both resistance design and revised resistance design are, as the names imply, based solely on loop resistance. Explain why loss seems to be of no concern as a limiting factor.

17. Describe LBO and how it is used on loaded wire-pair trunks.

18. If D is the separation of trunk load coils, why is the first load coil outward from an exchange at $D/2$?

19. Name two types of VF repeaters.

20. What is the philosophy behind tandem routing? Why use it in the first place?

21. In network design, what is the THD decision? What are the three input parameters necessary for this decision?

22. Define community of interest.

23. Describe a traffic matrix and what it is used for.

REFERENCES

1. *Local Telephone Networks*, International Telecommunications Union, Geneva, 1968.
2. International Telephone and Telegraph Corporation, *Reference Data for Radio Engineers*, 6th ed., Howard W. Sams, Indianapolis, 1976.
3. *Transmission Planning Aspects of Speech Service in Digital Public Land Mobile Networks*, ITU-T Rec. G.173, Helsinki, March 1993.
4. *National Networks for the Automatic Service*, International Telecommunications Union, Geneva, 1968.
5. *Networks*, Telecommunication Planning Documents, ITT Laboratories (Spain), Madrid, 1973.
6. *CCITT Red Books, Malaga-Torremolinos*, 1984, in particular Vols. III and V.
7. *Outside Plant*, Telecommunication Planning Documents, ITT Laboratories (Spain), Madrid, 1973.
8. Roger L. Freeman, *Telecommunication Transmission Handbook*, 3rd ed., John Wiley & Sons, New York, 1991.
9. Y. Rapp, "Algunos Puntos de Vista Económicos para el Planeamiento a Largo Plazo de la Red Telefónica," L. M. Ericsson, Stockholm, 1964.

10. *Placement of Exchanges in Urban Areas—Computer Program*, ITT Laboratories (Spain), Madrid, 1974.

11. J. C. Emerson, *Local Area Planning*, Telecommunications Planning Symposium [ITT Laboratories (Spain)], Boksburg, South Africa, 1972.

12. L. Alvarez Mazo and P. H. Williams, *Influence of Different Factors on the Optimum Size of Local Exchanges* [ITT Laboratories (Spain)], Boksburg, South Africa, 1972.

13. J. C. Emerson, *Factors Affecting the Use of Tandem Exchanges in the Local Area* [ITT Laboratories (Spain)], Boksburg, South Africa, 1972.

14. IEEE ComSoc, *The International Symposium on Subscriber Loops and Services*, Atlanta, Ga., 1977.

15. IEEE ComSoc, *Second International Symposium of Subscriber Loops and Services*, London, 1976.

16. J. E. Flood, *Telecommunications Networks*, IEE Telecommunications Series 1, Peter Peregrinus, London, 1975.

17. Y. Rapp, *Planning of Exchange Locations and Boundaries in Multi-Exchange Networks*, Vol. 18, Ericsson Technology, 1962, p. 94.

18. *Telecommunications Planning*, ITT Laboratories (Spain), Madrid, 1974.

19. O. Smidt, *Engineering Economics*, Telephony Publishing Co., Chicago, 1970.

20. *BOC Notes on the LEC Networks—1994*, Issue 2, SR-TSV-002275, Bellcore, Piscataway, NJ, April 1994.

21. *Electronic Switching: Digital Central Offices of the World*, Amos Joel, ed., IEEE Press, New York, 1982.

22. *Transmission Systems for Communications*, 5th ed., Bell Telephone Laboratories, Holmdel, NJ, 1982.

23. W. D. Reeve, *Subscriber Loop Signaling and Transmission Handbook Analog*, IEEE Press, New York, 1992.

24. *Effect of Transmission Impairments*, ITU-T Rec. P.11, ITU Helsinki, March 1993.

25. *Loudness Ratings (LRs) in an International Connection*, ITU-T Rec. G.111, ITU Helsinki, March 1993.

26. *Loudness Ratings (LRs) of National Systems*, ITU-T Rec. G.121, ITU Helsinki, March 1993.

CONVENTIONAL ANALOG SWITCHING IN TELEPHONY

1 INTRODUCTION

1.1 Approach

Today switching in the public switched telecommunications network (PSTN) is becoming entirely digital. Looking five years into the future, we estimate that PSTN switching will be entirely digital in North America as well as in many other areas in the world. However, this chapter provides a description of analog switching for this third edition for several important reasons:

1. Most of the concepts covered here hold for digital switching as well.
2. Some analog switching is still used and pockets of analog switching will continue to be used in the PSTN in various parts of the world.
3. The chapter provides an excellent historical perspective for the evolvement of the telecommunications network.

Such factors as functional description, desirable features, and operational requirements, as well as those basic switching concepts, are introduced here and will be carried on in the digital switching chapter.

1.2 Switching in the Telephone Network

A network of telephones consists of pathways connecting switching nodes so that each telephone in the network can connect with any other telephone for which the network provides service. Today there are hundreds of millions of telephones in the world, and nearly each and every one can communicate with any other one. Chapter 1 discussed the two basic technologies in the engineering of a telephone network, transmission and switching. Transmission allows any two subscribers in the network to be heard satisfactorily. Switching permits the network to be built economically by concentration of transmission facilities. These facilities are the pathways (trunks) connecting the switching nodes.

Switching establishes a path between two specified terminals, which we call *subscribers* in *telephony*. The term *subscriber* implies a public telephone network. There is, however, no reason why these same system criteria cannot be used on private or quasipublic networks. Likewise, there is no reason why that network cannot be used to carry information other than speech telephony. In fact, in later chapters these "other" applications are discussed, as well as modifications in design and features specific for special needs.

A switch sets up a communication path on demand and takes it down when the path is no longer needed. It performs logical operations to establish the path and automatically charges the subscriber for usage. A commercial switching system satisfies, in broad terms, the following user requirements:

1. Each user has need for the capability of communicating with any other user.

2. The speed of connection is not critical, but the connection time should be relatively small compared to holding time or conversation time.

3. The grade of service, or the probability of completion of a call, is also not critical but should be high. Minimum acceptable percentage of completed calls during the BH may average as low as 95%, although the general grade of service goal for the system should be 99%* (equivalent to $p = 0.01$).

4. The user expects and assumes conversation privacy but usually does not specifically request it, nor, except in special cases, can it be guaranteed.

5. The primary mode of communication for most users will be voice (or the voice channel).

6. The system must be available to the user at any time the user may wish to use it [2, 3, 4].

1.3   Some Historical Background

The earliest telephones required no switching because they were installed on a dedicated-line basis. It quickly became evident that such a one-on-one arrangement was inefficient at best to meet growing economic and social needs. A network began to evolve in which telephones were connected to a central location where an operator could connect and disconnect telephone lines from any pair of subscribers to the service whose lines terminated in

---

*See CCITT Rec. Q.95; $p = 0.01$ per link on an international connection.

that location. In North America the location was called an *office* or *central office*. We can also see where the word *exchange* came from, because the operator changed or exchanged wire pairs.

Reference 22 describes the earliest "switching" as brass strips (mid-1870s) that looked like door hinges. These "switches" could interconnect eight or ten subscribers and were adapted from those used in the telegraph business.

The plug-and-jack technique with a flexible wire cord was invented by a Western Electric engineer in Chicago in 1878. An operator could interconnect 50 or more subscribers from one position. Such a technique continued to be used for some 50 years and still is used today on some older manual PBXs.

The next major development occurred in a curious way in Kansas City. An undertaker, Almon B. Strowger, had his business compromised by the local telephone operator. Some say that she had a boyfriend who was a competing undertaker. Apparently she listened in on telephone calls from prospective customers and tipped off her boyfriend, who then scooped up the corpse business that ordinarily would have been Strowger's. Instead of complaining, Strowger set about inventing an automatic switchboard that would eliminate the intervention of the operator, thus ensuring his fair share of the undertaking business. The first Strowger switch went into commercial operation in 1892. This became the step-by-step switch, called the *Strowger switch* in the United Kingdom. Some step-by-step switches are still in operation today.

There followed the fully mechanical panel switch in the 1920s, the last of which was finally taken out of service in 1982. The No. 1 crossbar design was proposed in 1913, but the first crossbar switch did not go into operation until 1938. Crossbar switch installation peaked about 1983.

The first transistorized electronic logic switch was introduced for field trials in 1960 (No. 1 ESS) and the first digital stored-program control (SPC) switch in 1976 (4ESS). The second-generation TDM-SPC switch, 5ESS, went into operation in 1982 [22].

1.4 Two-Wire and Four-Wire Switches

Two-wire and four-wire transmission is discussed in Chapter 5, Section 3. Essentially, two-wire transmission is where both the transmit and receive paths are carried on the same wire pair or other single medium. Four-wire transmission is where the transmit and receive paths are separate and a wire pair is assigned to each path. All multiplex equipment is four-wire.

We also have two-wire and four-wire switches. Two-wire switches are generally found in the local area. Four-wire switches are used in conjunction with multiplexing equipment, and they are generally found in the tandem and interexchange carrier (toll) area. All digital switches (Chapter 9) are four-wire. When switches are used in conjunction with multiplex equipment, they may be six-wire switches: two wires for transmit, two wires for receive, and two wires for supervisory signaling.

2 NUMBERING: ONE BASIS OF SWITCHING

A telephone subscriber looking into a telecommunication network sees a repeatedly branching tree of links. At each branch point there are multiple choices. Assume that a calling subscriber wishes to contact one particular distant subscriber. To reach that distant subscriber, a connection is built up utilizing one choice at each branch point. Of course, some choices lead to the desired end point, and others lead away from it. Alternative paths are also presented. A call is directed through this maze, which we call a *telephone network*, by a telephone number. It is this number that activates the switch or switches at the "maze" branch point(s).

Actually, a telephone number performs two important functions: (1) it routes the call, and (2) it activates the necessary equipment for proper call charging. Each telephone subscriber is assigned a distinct number, which is cross-referenced in the telephone directory with the subscriber's name and address; in the local serving exchange (switch), this number is associated with a distinct subscriber line.

If a subscriber wishes to make a telephone call, she lifts her receiver "off hook" (i.e., takes the handset out of its cradle) and awaits a dial tone that indicates readiness of her serving switch to receive instructions. These "instructions" are the number that the subscriber dials (or the buttons that she punches) giving the switch certain information necessary to (a) route the call to the distant subscriber with whom she wishes to communicate and (b) set up the call-charging equipment.

A subscriber number is the number to be dialed or called to reach a subscriber in the same local (serving) area. Remember that our definition of a local "serving area" is the area served by a single switch (exchange). The thinking that follows ties that switch capacity in total lines to the number of digits in the telephone number.

If we had a switch with a capacity of 100 lines,
it could serve up to 100 subscribers and we could assign telephone numbers 00 through 99.

If we had a switch with a capacity of 1000 lines,
it could serve up to 1000 subscribers and we could assign telephone numbers 000 through 999.

If we had a switch with a capacity of 10,000 lines,
it could serve up to 10,000 subscribers and we could assign telephone numbers 0000 through 9999.

Thus the critical points occur where the number of subscribers reaches numbers such as 100, 1000, and 10,000.

In most present switching systems there is a top limit to the number of subscribers that can be served by one switching unit. Increase beyond this number is either impossible or uneconomical. A given switch unit is usually most economical when operating with the number of subscribers near the maximum of its design. However, it is necessary for practical purposes to hold some spare capacity in reserve. As we proceed in the discussion of switching, we consider exchanges with seven-digit subscriber numbers, such as

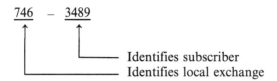

The subscriber is identified by the last four digits, permitting up to 10,000 subscribers, 0000 through 9999, allowing for no blocked numbers, such as

$$746 - 0000$$

The three-digit calling area has a capacity of 999 local exchanges, again allowing for no blocked numbers, such as

000
911 (emergency number – USA)

Section 15 of this chapter presents a more detailed discussion of numbering.

3   CONCENTRATION

One key to switching and network design is concentration. A *local* switching exchange concentrates traffic. This concept is often depicted as shown in the following diagram:

Let us dwell on the term *concentration* a bit more. Concentration reduces the number of switching paths or links *within* the exchange and the number of trunks connecting the local exchange to other exchanges. A switch also performs the function of *expansion* to provide all subscribers served by the exchange with access to incoming trunks and local switching paths.

Consider trunks in a long-distance network. A small number of trunks is inefficient not only in terms of loading (see Chapter 1, Section 8.2) but also in terms of economy. Cost is amortized on a per circuit basis; 100 trunks in a link or traffic relation are much more economical than 10 on the same link.* Tandem exchanges concentrate trunks in the local area for traffic relations (links) from sources of low traffic intensity, particularly below 20 erlangs, improving trunk efficiency.

4   BASIC SWITCHING FUNCTIONS

In a local exchange means are provided to connect each subscriber line to any other in the same exchange. In addition, any incoming trunk can connect to any subscriber line and any subscriber to any outgoing trunk. The switching functions are remotely controlled by the calling subscriber, whether he is a local or long-distance subscriber. These remote instructions are transmitted to the exchange by "off hook," "on hook," and dial information. There are eight basic functions of a conventional switch or exchange:

1. Interconnection
2. Control
3. Alerting
4. Attending
5. Information receiving
6. Information transmitting
7. Busy testing
8. Supervisory

Consider a typical manual switching center (Figure 3.1) where the eight basic functions are carried on for each call. The important interconnecting function is illustrated by the jacks appearing in front of the operator, subscriber-line jacks, and jacks for incoming and outgoing trunks. The interconnection is made by double-ended connecting cords, connecting subscriber to subscriber or subscriber to trunk. The cords available are always less than half the number of jacks appearing on the board, because one interconnecting cord occupies two jacks (by definition). Concentration takes place at this point on a manual exchange. Distribution is also carried out because any cord may be used to complete a connection to any of the terminating jacks. The operator is alerted by a lamp when there is an incoming call requiring connection. This is the attending–alerting function. The operator then assumes the control function, determining an idle connecting cord and plugging into the incoming

---

*Assuming an efficient traffic loading in each case.

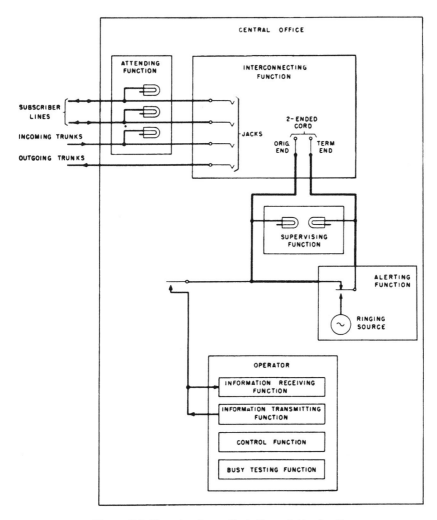

Figure 3.1. Manual exchange illustrating switching functions.

jacks. She then determines call destination, continuing her control function by plugging the cord into the terminating jack of the called subscriber or proper trunk to terminate her portion of control of the incoming call. Of course, before plugging into the terminating jack, she carries out a busy-test function to determine that the called line–trunk is not busy. To alert the called subscriber that there is a call, she uses the manual ring-down by connecting the called line to a ringing current source, as shown in Figure 3.1. Other signaling means are usually used for trunk signaling if the incoming call is destined for another exchange. On such a call the operator performs the information function orally or by dialing the call information to the next exchange.

The supervision function is performed by lamps to show when a call is completed and the cord taken down. The operator performs numerous control functions to set up a call, such as selecting a cord, plugging it into the originating jack of the calling line, connecting her headset to determine calling information, selecting (and busy testing) the called subscriber jack, and then plugging the other end of the cord into the proper terminating jack and alerting the called subscriber by ring-down. Concentration is the ratio of the field of incoming jacks to cord positions. Expansion is the number of cord positions to outgoing (terminating) jacks. The terminating jacks and originating jacks can be interchangeable. The called subscriber at another moment in time may become a calling subscriber. On the other hand, incoming and outgoing trunks may be separated. In this case they would be one-way circuits. If not separated, they would be both-way circuits, accepting both incoming and outgoing traffic [2, 3, 4].

5 SOME INTRODUCTORY SWITCHING CONCEPTS

All telephone switches have, as a minimum, three functional elements: concentration, distribution, and expansion. Concentration (and expansion) was briefly introduced in Chapter 1, Section 2 to explain the basic rationale of switching. Viewing a switch another way, we can say that it has originating line appearances and terminating line appearances. These are shown in the simplified conceptual drawing in Figure 3.2. Figure 3.2 shows the three different call possibilities of a typical local exchange (switch):

1. A call originated by a subscriber who is served by the exchange and bound for a subscriber who is served by the same exchange (route *A–B–C–D–E*).

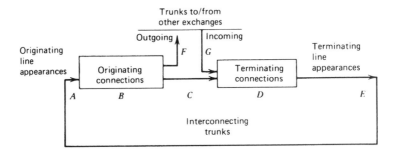

Figure 3.2. Originating and terminating line appearances.

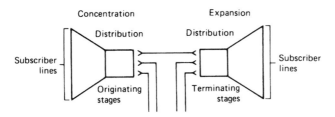

Figure 3.3. The concept of distribution.

2. A call originated by a subscriber who is served by the exchange and bound for a subscriber who is served by another exchange (route *A–B–F*).

3. A call originated by a subscriber who is served by another exchange and bound for a subscriber served by the exchange in question (route *G–D–E*).

Call concentration takes place in *B* and call expansion at *D*. Figure 3.3 is simply a redrawing of Figure 3.2 to show the concept of distribution. The distribution stage in switching serves to connect by switching the concentration stage to the expansion stage.

The symbols used in switching diagrams are as those in the following diagram, where concentration is shown on the left and expansion is shown on the right.

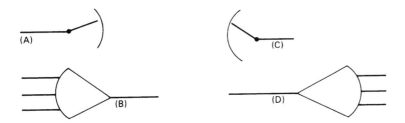

The number of inputs to a concentration stage is determined by the number of subscribers connected to the exchange. Likewise, the number of outputs of the expansion stage is equal to the number of connected subscribers whom the exchange serves. The outputs of the concentration stage are less than the inputs. These outputs are called *trunks* and are formed in groups; thus we refer to *trunk groups*. The sizing or dimensioning of the number of trunks per group is a major task of the systems engineer. The number is determined by the erlangs of traffic originated by the subscribers and the calling rate (see Chapter 1, Sections 6 and 8).

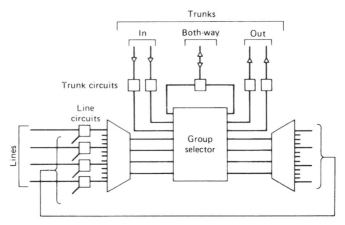

Figure 3.4. The group selector concept where both lines and trunks can be switched by the same matrix.

A group selector is used in distribution switching to switch one trunk (between concentration and expansion stages) to another and is often found not only in switches that switch subscriber lines but also where trunk switching is required. Such a requirement may be found in small cities where a switch may carry out a dual function, both subscriber switching and trunk switching from concentrators or other local switches. A group selector alone is a tandem switch. Figure 3.4 illustrates the group selector principle [1].

6   TYPES OF ELECTROMECHANICAL SWITCHES

There are basically three types of electromechanical switches. Sometimes these types are broken down into gross-motion switches and fine-motion switches. All are relay (or equivalent) operated. In the gross-motion category the oldest (patented in 1891) is the Strowger switch or step-by-step switch (S×S). The second type of gross-motion switch is the rotary switch, such as the ITT 7D. Both switches are still used. The switches are operated electromechanically by ratchets such as a stepping relay. Conveniently, the relays (S×S) are in steps of 10 to fit our decimal-number system. They were termed *gross-motion switches* because of the space traversed between terminals.

Fine-motion switches are typified by the crossbar switch. This is a coordinate switch or matrix. A speech-path connection proceeding through a switch is made by cross-points. Similar matrices can be constructed of reed relays or solid-state cross-points. Such matrices may be presented by a block diagram, such as that shown in Figure 3.5 [5].

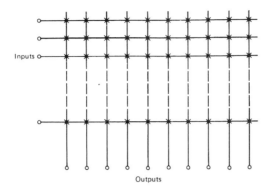

Figure 3.5. Typical diagram of a cross-point matrix.

7   MULTIPLES AND LINKS

A multiple multiplies. It is a method of obtaining several outputs from one input. Thus access is extended (see Figure 3.6A). Or the reverse may be true, where multiple inputs gain access to one output (see Figure 3.6B). Links provide connection for a multiple of switch inputs from one stage to a multiple of switch outputs in another stage (see Figure 3.7) [5].

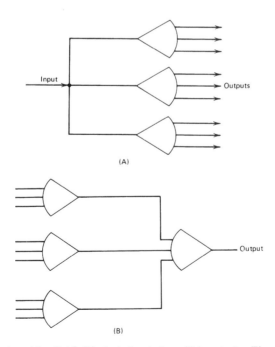

Figure 3.6. Examples of "multiple": (A) single inputs to multiple outputs; (B) multiple inputs to a single output.

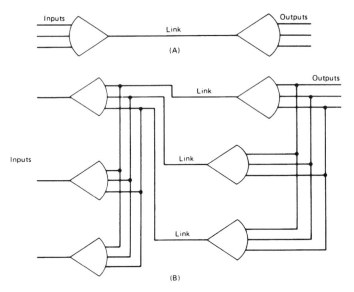

Figure 3.7. Examples of links.

8   DEFINITIONS: DEGENERATION, AVAILABILITY, AND GRADING

8.1   Degeneration

Degeneration can be expressed by the following ratio:

$$\text{Degeneration on a link} = \frac{\text{Variance of offered traffic}}{\text{Mean of offered traffic}} \qquad (3.1)$$

Degeneration is a measure of the extent to which the traffic on a given link varies from pure random traffic. For pure random traffic, degeneration (the preceding ratio) equals 1. For overflow traffic, the variance is equal to or, in the majority of cases, greater than the mean. The more degenerate the traffic, the heavier the demand during peak periods and the greater the number of transmission facilities required.

8.2   Availability

At a switching array, availability describes the number of outlets that a free inlet is able to reach and test for free or busy condition:

1. *Full availability.* Every free inlet is at all times able to test every outlet.
2. *Limited availability.* The absence of full availability. The availability at a switching array can be assigned a value, namely, the number of outlets available to each inlet (see Section 8.3).

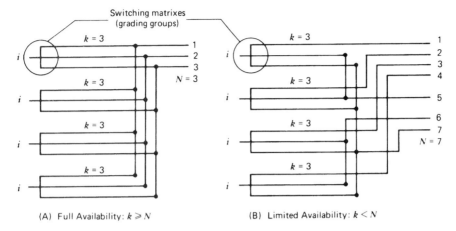

Figure 3.8. Full and limited availability. i = number of inlets per switching matrix (grading group), m = number of switching matrices (grading groups) (4, in A and B), N = number of outlets (three in A; seven in B), k = availability [number of outlets per switching matrix (grading group)]. (*Note*: This is a simplified illustration. In a typical switching array, k would equal 10 or even 20.) [5]

8.3 Grading

At a switching array with limited availability the inlets are arranged into groups, called *grading groups*. All the inlets in a grading group always have access to the same outlets. Grading is a method of assigning outlets to grading groups in such a way that they assist each other in handling the traffic (see Figure 3.8).

9 CONVENTIONAL ELECTROMECHANICAL SWITCHES

9.1 Step-by-Step or Strowger Switch

A step-by-step switch is conveniently based on a stepping relay with 10 levels. In its simplest form, which uses direct progressive control, dial pulses from the subscriber's telephone activate the switch, with each pulse stepping the switch one level. If the subscriber dials a 3, three pulses are generated by the subscriber subset and transmitted to the switch. The switch then steps to level 3. We can then imagine the second digit dialed passing to the second stepping relay bank, and the third, to the third bank. A dialed number, say 375, may be stepped through three sets of banks of 10, as shown in the following diagram:

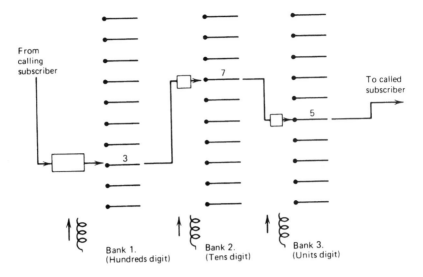

To save space and reduce postdial delay, the step-by-step switch evolved into a two-motion switch, and two banks became one. For the first digit dialed, the switch steps vertically, as shown in the preceding diagram. For the second digit, the same bank steps horizontally. Thus a bank covers 10×10 or 100 digits, and two banks in series can cover 10,000 (0 through 9999).

The line-finder technique is used in more modern $S \times S$ (post-1928) switches. Line finders are more simple switches, with several available per group of incoming subscriber lines. On an incoming call, when a subscriber goes "off hook," a line finder automatically seeks the line desiring service and extends the connection to a line selector. The line finder provides the first stage of concentration. The line finder, once connected, supplies dial tone to the calling subscriber. A 1000-line $S \times S$ switch is illustrated in Figure 3.9.

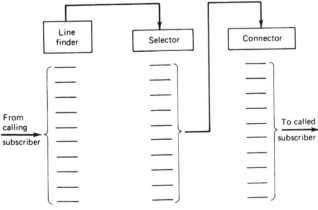

Figure 3.9. A step-by-step switch showing line finders, selectors, and connectors serving 1000 lines.

The connector in Figure 3.9 is called the *final selector* in Europe. To extend the switch in Figure 3.9 to 10,000 lines, a second selector stage is added between the selector and the connector stages. The connector stage shown in Figure 3.9 may also be called the *final connecting stage*. Normally, the last two digits dialed by the calling subscriber control the connector, which typically has access to 100 lines. The connector is more complex than the preceding stages. It must busy-test the called line and if busy, return a "busy-back" (the audible "line busy" signal) to the calling subscriber. If the called line is idle, it must apply ringing current to the called line. It will supply talk battery to both calling and called subscriber once the called subscriber goes "off hook." It provides supervision, holding the talk path in operation until one or both conversing parties go "on hook."

9.2 Crossbar Switch

Crossbar switching dates back to 1938 and reached a peak of installed lines in 1983. Its life has been extended by using stored-program control (SPC) rather than hard-wired control in the more conventional crossbar configuration. The crossbar is actually a matrix switch used to establish the speech path. An electrical contact is made by actuating a horizontal and a vertical relay. Consider the switch in Figure 3.10. To make contact at point B_4 on the matrix, horizontal relay B and vertical relay 4 must close to establish connection. Such a closing is usually momentary but sufficient to cause "latching." Two forms of latching are found in conventional crossbar practice, mechanical and electrical. The latch keeps the speech-path connection until an "on-hook" condition results, freeing the horizontal and vertical relays to establish other connections, whereas connection B_4 in Figure 3.10 has been "busied out."

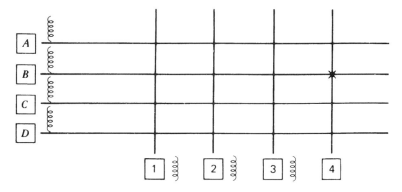

Figure 3.10. The crossbar concept.

Private industry has fielded numerous types of crossbar configuration. Early crossbar switches were made up of basic 10×10 switching matrices or blocks. Northern Electric (Canada) has a 10×20 (20 vertical) matrix in a miniswitch configuration. The ITT Pentaconta 1000 has 22 vertical and 14 horizontal bars, one of which is used as a changeover bar to provide for switching 52 outlets. The output of the basic switch block matrix for Pentaconta has 50 lines. Modern local crossbar switches, such as Pentaconta 1000 or the ATT No. 5 crossbar, handle up to 10,000 subscribers in a basic switch. In the case of Pentaconta 1000, this can be extended to 20,000. A basic 10,000-line Pentaconta has 22 primary selection switch blocks.

10 SYSTEM CONTROL

10.1 Introduction

The basic function of the control system in a switch is to establish an appropriate path through the switch matrix. Thus the control system must know the calling and called ports on the matrix and be able to find a free path between them. There are two methods of establishing a path, progressive control and common control.

10.2 Progressive Control

As the term indicates, "progressive control" implies that a speech path is set up progressively through the switch stage by stage. At each stage the selecting action chooses a group of identical paths that lead toward the ultimate call destination in the switch. Hunting action selects one of a group of competing circuits as the call progresses. The control system has no foreknowledge of conditions ahead in the next stage or step in the call setup. The call could run into blocking or a busy line and must wait through this extensive setup process before returning a "busy-back" or congestion signal.

If the control system functions directly as a result of the subscriber dial pulses, it is called *direct progressive control*. This was the case in our previous discussion of step-by-step switching. In more modern progressive control systems a register is interposed (buffers) between the subscriber dial digits. The register accepts dial digits, interprets them, and then controls switch functions. This is called *register progressive control*. With direct progressive control the called telephone number is directly associated with a specific path through the selection tree. Because the number is decimal-based, progression through the switching tree is carried out in branches of 10, as shown in the diagram that follows.

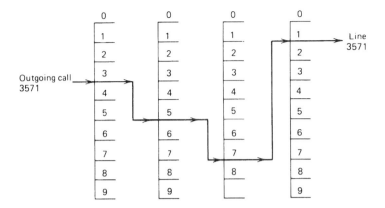

In direct progressive control the switching timing is in direct sequence and in synchronism with the subscriber-dialed digits. The principal advantage of direct progressive control is economy. Control circuitry is minimized, and such switches were used in small-community dial offices (exchanges). A functional block diagram of a direct-control switch is shown in Figure 3.11.

Disadvantages are that direct progressive control switches require a certain amount of overbuild to provide a satisfactory grade of service because of the stage-by-stage hunt-and-choose requirement without foreknowledge that a certain path is busy. The result is a lower efficiency than that in other systems on interconnecting paths. It can, however, be more tolerant of overloads concomitant with a poorer grade of service. Of course, direct progressive control requires subscriber lines to terminate directly on switch connector stages in strict correspondence with the subscribers' numbers. There is also an inherent inflexibility of trunk assignments that must conform directly with the exchange identifying digits.

Register progressive control eliminates many of the disadvantages of direct progressive control. It buffers incoming dial information, thus providing number translation. On step-by-step installations, register progressive control consists of an A-unit hunter and a director. The A-unit hunter connects the incoming line to a free A-digit selector. This unit steps the selector to the group level corresponding to the first dialed digit and then rotates to a free director. The remaining two digits of the exchange code (see Section 15 of this chapter) and the four digits of the subscriber number are registered in the director. The first two remaining digits route the call to (own) exchange or proper distant exchange. The director translates digits two and three to a code that is more versatile for switch operation and stores the last four digits of the number dialed. The translated code is then transmitted (second and third digits of called numbers) to the first code selector, and the call is extended to the proper exchange. The remainder of the digits (i.e., the subscriber number) are then transmitted as is for the indicated exchange to act on. Once these operations are completed, the A-digit hunter, the selector,

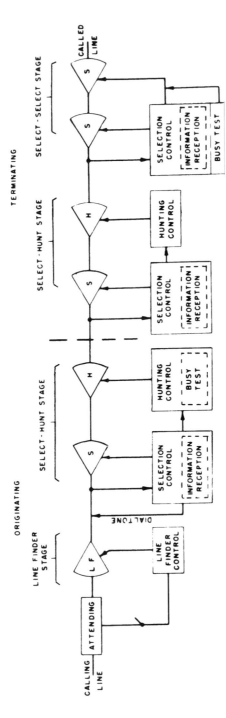

Figure 3.11. Simplified diagram of control circuits for direct progressive control. Copyright © AT&T, 1961. Reprinted with permission.

and the director are freed to act on another incoming call. The setup of the call proceeds as in direct progressive control.

It will be noted that a seven-digit telephone number was used in the preceding example. The call setup proceeds as follows:

1. Seven incoming dialed digits.
2. A unit selector hunts/finds free director.
3. The director then:
   a. Acts on the first digit.
   b. Translates second and third digits.
   c. Transmits second and third digits.
   d. Registers fourth through seventh digits.
   e. Transmits digits four through seven (untranslated).
4. Second and third translated digits are acted on by the first selector for own exchange or indicated distant exchange.
5. Fourth through seventh digits are fed to the second selector at own exchange or the call is routed to a distant exchange.
6. A-digit switch and director is released for the next incoming call.

Register progressive control was introduced in 1923 in the United Kingdom with director system. This first attempt at number translation lacked the capability of alternative routing when congestion occurred. Call routing remained inflexible.

As switching evolved, some of these deficiencies were eliminated or, at least, alleviated by adding further translation or by associating register translation with the later stages of switches as well as the first stages.

To reduce postdial delay,* a switch may be made to start a call setup once the first three dialed digits are received. Initial translations and trunk selection can take place during the period in which the calling subscriber is dialing the last four digits.

As automatic telephone service grew, switching engineers were confronted with the problem of numerous types of switches in a common area. Some would be older progressive control, others would be register progressive control, and still others would be common control. Even greater variance occurred in internal switch codes, the digital codes used in a switch internally to set up and route a call.

Switching design engineers have since resorted to the technique of the control register with information transmitting function added. The register translator (control register) carries out three consecutive steps in this case: (1) information reception, (2) internal control signaling, and (3) information

---

*Postdial delay is the time interval between the last digit dialed by a subscriber and the first indication to that subscriber that there is ringing on the distant telephone. Postdial delay is the primary measure of performance of a signaling system.

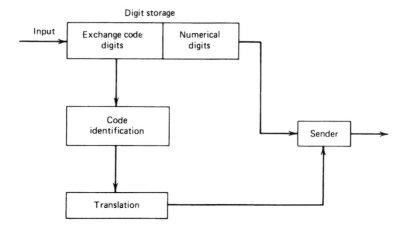

Figure 3.12. Interexchange control register.

transmission. The translator in this case determines from the exchange code (the first three dialed digits) the outgoing trunk group *and* the type of signals required by the group and the amount of information to be transmitted. This flexibility in design and operation of multiexchange areas was vastly improved. The interexchange control register concept is shown in Figure 3.12. Outgoing trunk-control registers facilitate the use of tandem exchange operation. Also, incoming trunk calls to an exchange so equipped can be handled directly from a distant exchange if they are similarly equipped.

Interexchange control registers enabled alternative routing. When first-attempt routing is blocked, this same technique can be used for second-attempt routing through an exchange by setting up a speech path with switching components different from the first setup that failed. The sender in Figure 3.12 generates and transmits signaling information to distant exchanges.

10.3 Common Control (Hard-Wired)

10.3.1 General. "Common control" is an ambiguous term. Any control circuitry in a switch that is used for more than one switching device may be termed "common control." We try to distinguish common control from the progressive control referred to in the previous section, yet common control is used in the register translators of progressive control.

For purposes of this discussion, common control is defined as providing a means of control of the interconnecting switch network, first identifying the input and output terminals of the network that are free and then establishing a path between them. This implies a busy-test of a speech path before establishment of the path. Common control may cover the entire switch or separate control for the originating and terminating halves. Common-control systems employ markers, which are discussed subsequently. Such marker systems are most applicable to grid switching networks (e.g., networks internal to the switch).

10.3.2 Grid Networks. A grid in switching networks may be defined as a combination of two or more stages of switch blocks connected together to accommodate numbers of input and output circuits as needed to meet the speech-path switching requirements of a given exchange design. The grid network differs from others previously discussed because it can be considered bidirectional.

This means that the input could also be considered an output and vice versa. Thus parts of the network may be used for both originating and terminating purposes. Concentration and expansion can be combined, and thus a subscriber line has only one appearance on the network. A basic two-stage primary–secondary grid network is shown in Figure 3.13, and an elemental three-stage grid is shown in Figure 3.14. Within each block in a grid network any input can connect to any output. If the blocks in Figure 3.13 had 10 inputs each, with 10 blocks in each stage and fulfilling the basic requirement for at least 1 access from each primary group block to each secondary group, then at least 100 links would be required.

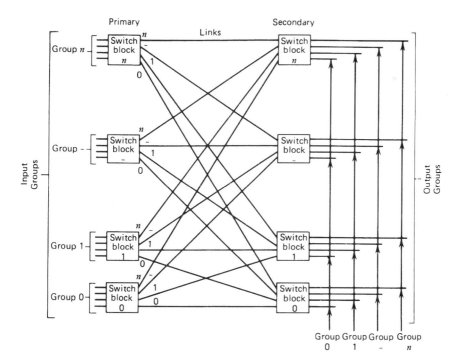

Figure 3.13. Basic two-stage grid network. Copyright © AT&T, 1963 [5]. Reprinted with permission.

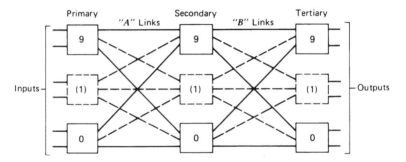

Figure 3.14. An elemental three-stage grid. Copyright © AT&T, 1963 [5]. Reprinted with permission.

The fundamental difference between progressive and grid networks is in the number of paths a specific call looks ahead to inside the network on call setup. The progressive network is a hunt–select network, where the "tree" principle holds (see Section 10.2). As a call progresses through the network, the path is built up stage by stage. At each stage it looks at a choice of 10 branches ahead of it. Only when looking at the grid network as a whole is there a choice of 10 paths. On a stage-by-stage basis in the grid network, however, there is only a choice of one path for a given call at any intermediate stage. As we see later on, this path has been selected and busy-tested before actually being set up.

Conventional crossbar switches for large switching centers are usually four-stage grid networks. Such a network is shown in Figure 3.15. It is actually made up of two stages of primary and secondary grids. The two stages are connected by what are called *junctors*. Note that only one junctor as a minimum is required per secondary switch of the input grid to connect to each input grid of the primary switch of the output grid. One junctor matches any pair of originating and terminating links. As can be seen in Figure 3.15, it is particularly important that traffic balance be carefully maintained on the grid inputs and outputs since the junctors from each input grid are divided equally among all grids. However, the busy-test and trunk-hunting functions of common-control systems provide for many different path setups to service a particular call.

10.3.3 Common-Control Principles. The "marker" sets aside common-control systems as we define them from other more generalized common-control systems discussed previously. On grid circuits the characteristics are such that with specified input and output terminals, various sets of linkage paths exist that can provide connection between the terminals. It is the marker, with terminal points identified, that locates a path, busy-tests it, and finally sets up a particular channel through the switch grid network.

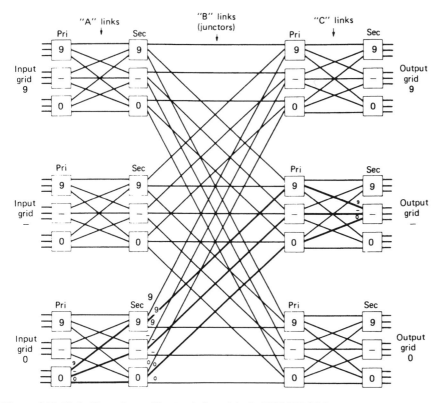

Figure 3.15. Typical four-stage grid network. Copyright © AT&T, 1963 [5]. Reprinted with permission.

A marker always works with one or more registers. A marker is a rapidly operating device serving many calls per minute. It cannot wait on the comparatively slow input information supplied by an incoming line or trunk. Such information, whether dial pulses, touch tones, or interregister signaling information, is stored in the register and released to the marker on demand. The register may receive the entire dialed number and store it before dumping to the marker or may take only the exchange code or area code plus exchange code (see Section 15 of this chapter) which is sufficient to identify a trunk group. The register will also identify the location of the call input or set up a control path to the input for the marker. Figure 3.16 illustrates the basic functions of a marker.

Whereas the register provides the dialed exchange number to the marker, the translator provides the marker information on access to the proper trunk group. Of course, the translator may also provide other information, such as the type of signaling required on that trunk group. The reader must bear in mind that modern common-control switches, particularly grid-type switches, use a control code that differs from the numerical dialed code. The dialed

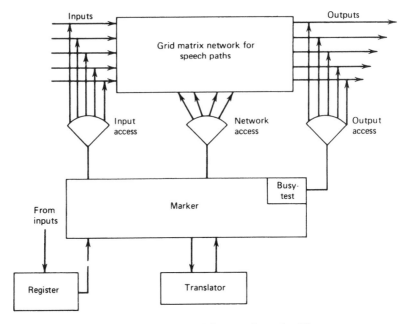

Figure 3.16. Functional diagram of a marker [5].

code is the directory number (DN) of the called subscriber. The equipment number (EN) is the number the equipment uses and is often a series of five one- or two-digit numbers to indicate location on the grid matrix of the speech-path network. The equipment codes are arbitrary and require changes from time to time to accommodate changes in number assignments. The changes are done on a patch field associated with the translator. One equipment number code is the "two out of five" code, which has ten possibilities to correspond to our decimal base number system and is used on the subscriber dial (or push buttons). As the number implies, combinations of five elements are taken two at a time. This code is shown in Table 3.1.

The number combinations are additive, corresponding to the decimal equivalent except the 4–7 combination that adds to 11 (not to 0). For

TABLE 3.1   Two Out of Five Code

Digit	Two Out of Five 0–1–2–4–7	Digit	Two Out of Five 0–1–2–4–7
1	0–1	6	2–4
2	0–2	7	0–7
3	1–2	8	1–7
4	0–4	9	2–7
5	1–4	0	4–7

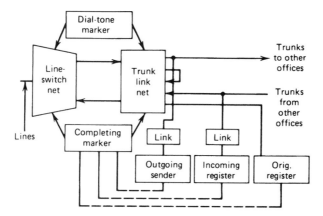

Figure 3.17. American Telephone and Telegraph No. 5 crossbar; a typical marker application. Copyright © AT&T, 1963 [5]. Reprinted with permission.

transmission through the exchange, the numbers are represented by two out of five possible audio frequencies.

On conventional exchanges, one marker cannot serve all incoming calls, particularly during the busy hour (BH). Good marker holding time per call is of the order of half a second. Call attempts may be much greater than two per second, so several markers are required. Each marker must have access from any input grid to any output grid, and thus we can see where markers might compete. This possibility of "double seizure" can cause blockage. Therefore only one control circuit is allowed into a specific grid at any one time.

To reduce marker holding time, fast action switches are required. This is one reason why common (marker) control is more applicable to fine-motion switches such as the crossbar switch. The control circuits themselves must also be fast-acting. Older switches thus used all-relay devices and some vacuum tubes. Newer switches used solid-state control circuitry, which was more reliable and even faster acting.

Figure 3.17 is a simplified functional diagram of the ATT (North American) No. 5 crossbar system. The originating and terminating networks are combined. Two different types of marker are used, the dial-tone marker and the completing marker. A typical No. 5 crossbar has a maximum of four dial-tone markers and eight call-completing markers. The dial-tone marker sets up connections between the calling subscriber line and an originating register. Call-completing markers carry out the remainder of the control functions, such as trunk selection, identification of calling and called line terminals, channel busy-test and selection, junctor group and pattern control, route advance, sender link control, trunk charge control when applicable, overall timing, automatic message accounting (AMA) information, and trouble recorder control.

The marker is also used for exchange troubleshooting and fault location. It is the most convenient check of the switching network because it samples the entire network at frequent intervals. Many markers provide trouble-recording circuits. Trouble on one call setup attempt may have no effect on another. The marker provides a trouble output, returns to service, and reattempts the call. Reattempts by a marker are usually limited to two so that the marker can return to service other calls.

If a marker finds an "all trunks busy" (ATB) condition on a tandem or transit call, it can ask for instructions to reroute the call. It then handles the call as a first attempt. If again an ATB condition is found and a second alternative route is available, it can make a call attempt on the second rerouting. When all possible routes are busy, the marker returns an ATB signal to the calling subscriber.

11 STORED-PROGRAM CONTROL

11.1 Introduction

Stored-program control (SPC) is a broad term designating switches where common control is carried out to a greater extent or entirely by computerware. Computerware can be a full-scale computer, minimicrocomputer, microprocessors, or other electronic logic circuits. Control functions may be entirely carried out by a central computer in one extreme for centralized processing or partially or wholly by distributed processing utilizing microprocessors. Software may be hard-wired or programmable. Telephone switches are logical candidates for digital computers. A switch is digital in nature,* as it works with discrete values. Most of the control circuitry, such as the marker, works in a binary mode.

The conventional crossbar marker requires about half a second to service a call. Up to 40 expensive markers are required on a large exchange. Strapping points on the marker are available to laboriously reconfigure the exchange for subscriber change, new subscribers, changes in traffic patterns, reconfiguration of existing trunks or their interface, and so on.

Replacing register markers with programmable logic—a computer, if you will—permits one device to carry out the work of 40. A simple input sequence on the keyboard of the computer workstation replaces strapping procedures. System faults are printed out as they occur, and circuit status may be printed out periodically. Due to the speed of the computer, postdial delay is reduced and so on. Computer-controlled exchanges permit numerous new service offerings, such as conference calls, abbreviated dialing, "camp on busy," call forwarding, and incoming-call signal to a busy line [7, 9].

---

*This important concept is the basis for the argument set forth in Chapter 9 regarding the rationale for digital switching.

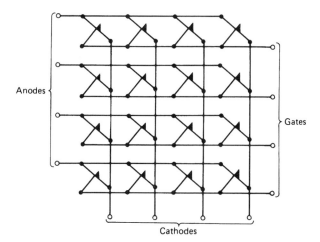

Figure 3.18. An SCR cross-point matrix.

11.2 Basic Functions of Stored-Program Control

There are four basic functional elements of an SPC switching system: (1) switching matrix, (2) call store (memory), (3) program store (memory), and (4) central processor.

The switching matrix may be made up of electromechanical cross-points, such as in the crossbar switch, reed, correed, or ferreed cross-points or switching semiconductor diodes, often SCR (silicon-controlled rectifier). An SCR matrix is shown in Figure 3.18.

The call store is often referred to as the "scratch pad" memory. This is a temporary storage of incoming call information ready for use, on command from the central processor. It also contains availability and status information of lines, trunks, and service circuits, and internal switch-circuit conditions. Circuit status information is brought to the memory by a method of scanning. All speech circuits are scanned for a busy/idle condition.

The program store provides the basic instructions to the controller (central processor). In many installations translation information is held in this store, such as DN to EN translation and trunk signaling information.

A simplified functional diagram of a basic (full) SPC system is shown in Figure 3.19.

11.3 Additional Functions in a Stored-Program-Control Exchange

Figure 3.20 is a conceptual block diagram of a typical North American SPC exchange. As we can see, an SPC exchange can be broken down into three functional levels: (1) line and trunk switching network, (2) input–output equipment, and (3) common-control equipment. Figure 3.20 is an

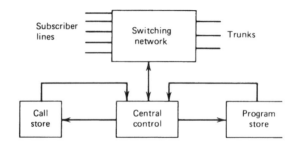

Figure 3.19. Simplified functional diagram of an SPC exchange.

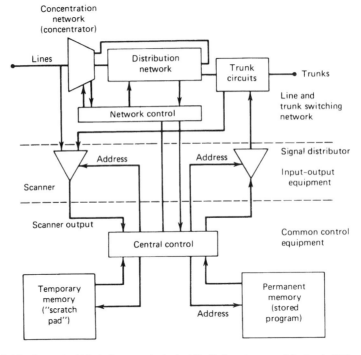

Figure 3.20. Conceptual block diagram of a typical North American semielectronic SPC exchange.

expansion of Figure 3.19, showing, in addition, scanner circuitry and signal distribution.

The control network executes the orders given by the central control processor (computer). These orders usually consist of instructions such as "connect" or "release," along with location information on where to carry out the action in the switching network. In the ATT ESS system the network control circuits are classified in three major functional categories:

1. Selectors, which set up and release a connection on receipt of location information.
2. Identifiers, which determine the location (called an "address") of a network terminal on one side that is to be connected to a known terminal (address) on the other side of the network.
3. Enablers, which are circuits sequentially enabling junctors to be connected together.

The input–output equipment consists of a line scanner and a signal distributor. Both circuits operate under control of the central control processor. The scanner and distributor carry out the "time-sharing" concept of SPC. The term "time sharing" is the basis where one (or several) computers can control literally thousands of circuits, with each circuit being served serially. The concept of "holding time," important in SPC, is the time taken to serve a circuit, and "delay" is the time each circuit must wait to be served. A computer read–write cycle in a typical SPC is 2–5 µs, with a scanning rate of 2–5 µs per terminal, line scanning during digit reception every 10 µs, and 100-ms supervisory scan. The scanner is an input circuit used for sampling the states (idle or busy) of subscriber lines, trunks, and switch test points to permit monitoring the operation of the system. The signal distributor, on the other hand, is an output circuit directing output signals to various points in the system. In the ATT ESS the signal distributor is primarily used on trunk circuits for supervisory and signaling actions.

Common-control equipment, as mentioned previously, is made up of the central processor, call store, and program store. The three units can be considered as making up the control computer, which is capable of transmitting orders to the system as well as detecting signals from the system. The SPC systems with centralized control that we have discussed above have a human interface (I/O = input–output) with the central controller. This, in many instances, is a teleprinter (i.e., keyboard send, printer receive) or keyboard with a visual display unit. The installation adds many advantages and conveniences not found in more conventional switching installations. Several of these are:

- Rerouting and reallocation of trunks
- Traffic statistics
- Renumbering of lines, subscriber move
- Changes in subscriber class
- Exchange status

- Fault finding
- Charge records

All these functions can be carried out via the I/O equipment connected to the central processor. In the basic ATT ESS exchanges there are two teleprinter accesses [7, 9].

11.4   A Typical Stored-Program-Control Exchange: European Design

The ITT Metaconta is a joint French–Spanish development installed worldwide. Its switching network can handle up to 32,000 lines. The control section can handle up to about 60,000 BH call attempts, depending on the facilities offered and the type and number of different signaling systems to be handled.

The central processor is made up of dual ITT 1600 16-bit binary parallel minicomputers under a load-sharing principle. The memory is of the random access type with a word length of 17 bits (16 bits + 1 parity bit) and a maximum capacity of $2^{16} = 65,536$ words, consisting of 8 blocks of 8192 words. The memory is a ferrite core type with an access time of 0.35 µs and a cycle time of 0.85 µs.

A Metaconta L exchange uses two classes of operational programs, consisting of a resident package and an on-demand package. The *resident package* has four types of computer programs.

1. Call-processing programs, which control all the call treatment as well as charging and statistics recording.

2. Man–machine communication programs, which handle communications between the system and operating personnel and also permit the loading of on-demand programs.

3. On-line test and defensive programs, which observe abnormal behavior of the system and either restore normal conditions or make appropriate decisions to disconnect faulty devices.

4. Start-up and recovery programs, which handle the system status evolution. In this group are programs that permit taking over call processing when one machine fails.

The on-demand package contains programs that have to be called up by the operating personnel through teleprinter messages. The on-demand programs carry out such functions as traffic observation, charging information dump, call tracing, and routine tests. Service features and facilities offered by Metaconta L, listed in Tables 3.2 and 3.3, are broken down into two areas,

TABLE 3.2   Service Features and Facilities: Subscriber Related

Line Classes	Subscriber Facilities
One-party lines	Abbreviated numbering
With dial or push-button-type subscriber sets	One or two digits
With or without special charging categories	For individual lines, or groups of lines,
Two-party lines	or for all lines
With or without privacy, separate ringing,	Transfer of terminating calls
separate charging, revertive call facility	Conversation hold and transfer
Multiparty lines (up to 10 main stations)	Calling party's ring-back
Without privacy	Toll-call offering
With selective or semiselective ringing	Hot line
PABXs	Automatic wake-up
Unlimited number of trunks	Doctor-on-duty service
Uni- or bidirectional trunks	Do-not-disturb service
With or without in-dialing	Absentee service
Coin-box lines	Immediate time and charge information
Local traffic	Conference calls
Toll traffic	Centrex facilities (optional)
Special applications	Voice mail
Restricted lines	Call-waiting
To own exchange	Call-forwarding
To urban, regional, national, or toll areas	
To some specified routes	
Priority lines	
Toll essential	
Essential	
Priority during emergency, overload situations, etc.	

Source: Ref. 9.

subscriber related and administration related. Figure 3.21 is a functional block diagram of the Metaconta L system.

11.5   Evolutionary Stored-Program Control and Distributed Processing

11.5.1 Introduction. The presently installed switching plant represents an extremely large investment yet to be fully amortized. This plant is essentially made up of step-by-step, rotary, and crossbar equipment that is quickly being replaced by digital switches. The peak in crossbar installations measured in lines served was not reached until well into the 1980s. In 1973 SPC served less than 10% of ATT subscribers. One current approach is to modify existing common-control exchanges with some limited SPC techniques, and another is to install conventional crossbar with SPC control, either partial or full. The ITT Pentaconta 2000 is an example. Still another approach is to resort to distributed processing rather than use a comparatively large computer as described in Sections 11.3 and 11.4.

TABLE 3.3   Service Features and Facilities: Administration Related

---

Administration Facilities

Interoffice signaling
  Direct-current signaling codes (step-by-step, rotary 7A and 7D, R6 with register or direct control, North American dc codes)
  Alternating-current pulse signaling codes
  MF signaling codes for register-controlled exchanges (MFCR 2 code, MF Socotel, North American MF codes)
  Direct data transmission over common signaling link between processor-controlled exchanges of the time or space division types

Charging
  Control of charge indicators at subscriber premises
  Metering on a single fee or a multifee basis
  Free number service

Numbering
  Full flexibility for equipment number—directory number translation
  For local calls, national toll calls, international toll calls
  Private automatic branch exchanges with direct inward dialing

Routing
  Prefix translation for outgoing or transit calls
  Alternative or overflow routing on route busy or congestion condition
  Resignaling on route busy or congestion condition
  Called side release control

Maintenance
  Plug-in boards
  Automatic fault detection and identification means by diagnostic programs

Operation
  Generalized use of teleprinters; workstations
  Possibility of a remote centralized maintenance and operation center

---

11.5.2 Modification of Existing Plant. Consider conventional electromechanical switching with common control (in its widest meaning). Processing by relays is distributed in registers, markers, translators, directors, and finders. Whereas implementation of new services and features available in full SPC systems is costly and difficult or even impossible because of the performance limitations of conventional electromechanical design, it is feasible or even desirable to substitute, on a limited basis, microprocessor control to replace hard-wired relay-based control elements such as registers and translators. In essence, software replaces much of the hardware in the control systems in electromechanical switches. Many administrations and telephone companies have found this to be a cost-effective alternative to full SPC implementation. The resulting modified equipment will provide many of the new services of SPC as well as cost savings in maintenance, administration, and operation of existing installations. The basic design of crossbar and $S \times S$ features distributed control and goes against the main concept of full SPC, which features completely centralized control.

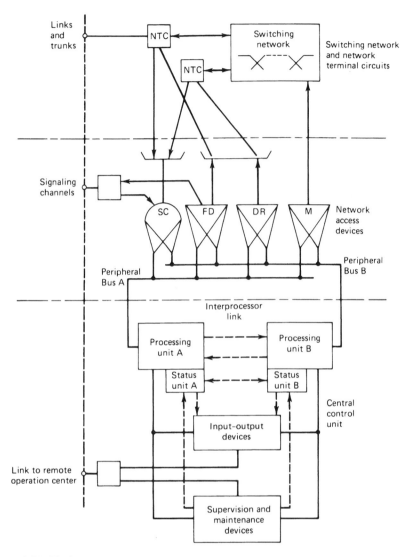

Figure 3.21. Block schematic of Metaconta L showing switching network and network terminal circuits, network access devices, and the central control unit (SC = scanner, FD = fast driver, DR = slow driver, M = marker, NTC = network terminal circuit). Courtesy of the International Telephone and Telegraph Corporation [8]. Reprinted with permission of ALCATEL, Stuttgart, Germany.

There are other incompatibilities as well, one of which is environmental. Relay circuitry can tolerate wide ranges of temperature and humidity; solid state cannot. Relay circuits use comparatively high voltage (48 V) and current (mA) ranges. Solid state, particularly integrated circuits (ICs), use low voltage and low current. "Low" in this case means <5 V and several microamperes per circuit. Electromechanical systems are essentially immune to electromagnetic interference (EMI) and in many cases are good generators of EMI. Solid-state circuitry, on the other hand, is particularly susceptible to EMI. Replacement of electromechanical control circuits with solid-state programmable units just cannot be done at random. Interfacing costs are a major consideration.

Survivability is another argument against full SPC with centralized control. There is a certain amount of redundancy in the control circuitry of a conventional crossbar switch. Calls can be processed several at a time, each with its own marker, registers, and translator. There is "graceful degradation" in service when one control unit fails. However, let the singular central control fail in a full SPC switch, and catastrophic failure results. This is one important reason why the Metaconta L has two central processors.

To render the system cost-effective, the design engineer must select the point of interface between the conventional electromechanical switch and the SPC modules to be installed. One consideration for the interface point would be the concentration links of the register–receiver–sender–translator group.

A major area that is a prime candidate for replacement of hardware by a software-oriented device [processor(s)] is in the routing and control sector of an electromechanical switch. It is here where most changes are required in a switch during its lifetime. These are the administrative and functional changes of subscriber data, routing, and signaling. By implementing SPC, many of the advantages are achieved by carrying out changes in the stored programs rather than the laborious strapping required on the electromechanical units. There is also the advantage of substitution of one for many. One control processor can replace 64 or more conventional registers, although it is desirable to use two processors for the additional reliability provided. One processor in this case is used for backup with automatic switchover on failure of the operational processor. As a switch grows (i.e., as more subscribers are added) during its lifetime, additional processors may be added, as needed, to share the load.

One approach to centralization is to have the several control processors be served by a common memory for the switch as a whole. This simplifies the problem of programming new information or modifying existing programs. It also simplifies the I/O such that one keyboard teleprinter can serve the entire switch.

Administration and maintenance procedures are often carried out by a separate processor in a modified electromechanical switch. Connection is made to the routing and control processor bus as well as to electromechanical elements. A separate I/O is also provided [14–17].

12   CONCENTRATORS AND SATELLITES

Concentrators or "line concentrators" consolidate subscriber loops, are remotely operated, and are the concentration (and expansion) portion of a

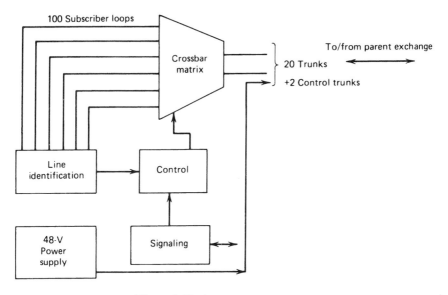

Figure 3.22. A crossbar concentrator.

switch placed at a remote location. This is the conventional meaning of a concentrator. The concentrator may use $S \times S$ or crossbar facilities for the concentration matrix. Control may be by conventional relay or by a hard-wired processor. A typical crossbar type is shown in Figure 3.22, where 100 subscriber loops are consolidated to 20 trunks plus 2 trunks for control from a nearby exchange. Concentrators are used in sparsely populated areas that require a long trunk connection to the nearest exchange. In effect, the concentrators extend the serving area of an exchange. However, all calls originating in the concentrator serving area must be serviced by the parent exchange. Conventional concentrators can serve up to several hundred subscribers.

Concentration can also be effectively carried out using carrier techniques, either pulse-code modulation (PCM) (Chapter 8) or frequency-division multiplex (FDM) (Chapter 5).

A satellite switch originates and terminates calls from a parent exchange. It differs from a concentrator in that local calls (i.e., calls originating and terminating inside the same satellite serving area) are served by the satellite and do not have to traverse the parent exchange. A block of numbers is assigned to the satellite serving area and is usually part of the basic number block assigned to the parent exchange. Because of its numbering arrangement, the satellite can discriminate between local calls and calls to be handled by the parent exchange. The satellite can be regarded as a component of the parent exchange that has been dislocated and moved to a distant site. This is still another method of extending the serving area of a main exchange to reduce the cost of serving small groups of subscribers, which, under ordinary circumstances, could only be served by excessively long loops to the parent exchange. Satellites range in size from 300 to 2000 lines. Concentrators are usually more cost-effective than satellites when serving 300 lines or fewer.

13 CALL CHARGING: EUROPEAN VERSUS NORTH AMERICAN APPROACHES

In Europe and in countries following European practice, telephone-call charging is simple and straightforward. Each subscriber line is equipped with a meter with a stepping motor at the local exchange. Calls are metered on a time basis. The number of pulses per second actuating the meter are derived from the exchange code or the area code of the dialed number. A local call to a neighbor (same exchange) may be 1 pulse (one step) per minute, a call to a nearby city 3 pulses, or a call to a distant city 10 pulses per minute. International direct-dial calls require checking the digits of the country code that will then set the meter to pulse at an even higher rate. All completed calls are charged; thus the metering circuitry must also sense call supervision to respond to call completion—that is, when the called subscriber goes "off hook" (which starts the meter pulses)—and also to respond to call termination (i.e., when either subscriber goes "on hook") to stop meter pulses. Many administrations and telephone companies refer to this as the *Karlson method* of pulse metering. Some form of number translation is required, in any event, to convert dialed number information to key the metering circuit for the proper number of pulses per unit period.

Pulse metering or "flat rate" billing requires no further record keeping other than periodic meter readings. Charging equipment cost is minimal, and administrative expenses are nearly inconsequential. Detailed billing on toll calls is used in North America (and often flat rate on local calls). Detailed billing is comparatively expensive to install and requires considerable administrative upkeep. Such billing information is recorded by determining first the calling subscriber number and then the called number. Also, of course, there is the requirement of timing call duration.

Automatic message accounting (AMA) is a term used in North American practice. When the accounting is done locally, it is called *local automatic message accounting* (LAMA), and when it is carried out at a central location, it is called *centralized automatic message accounting* (CAMA). Sometimes operator intervention is required, especially when automatic number identification (ANI) is not available or for special lines such as hotels. With automatic identified outward dialing (AIOD), independent data links are sometimes used, or trunk outpulsing may be another alternative.

Only some of the complexities have been described in North American billing practice. One overriding reason why detailed billing is provided on long-distance service is customer satisfaction. In European practice it is difficult to verify a telephone bill. A customer can complain that it is too high but can do little to prove it. Ordinarily subscriber-dialed long-distance service is lumped with local service, and a bulk amount for the number of meter steps incurred for the intervening billing period is shown on the bill. With detailed billing, individual subscriber-dialed calls can be verified and the bulk local charge is separate. In this case a subscriber has verification for a complaint if it is a valid one.

14 TRANSMISSION FACTORS IN ANALOG SWITCHING

14.I Introduction

Transmission in telephony deals with the delivery of a "quality" signal to the far-end user. If the signal is speech, it should suffer minimum distortion as well as meet certain requirements regarding level and signal-to-noise ratio. If the signal is data, it should meet a certain minimal error rate. A switch placed in a transmission path is one more element that will affect signal quality as it traverses the network. On many calls, whether speech, data, or facsimile, there could be as many as 13 switches in tandem. Thus the extent to which a switch distorts transmission must be specified.

14.2 Basic Considerations

Normally a switch sets up a speech path. Space-division switching, the type of switching discussed in this chapter, provides a metallic path made up of wire and relay or semiconductor cross-points. Thus we would expect an analog space-division switch to act on the signal passing through it in the following ways:

1. It will attenuate the signal; this is called *insertion loss*.
2. It will add noise to the signal, thus deteriorating signal-to-noise ratio.
3. As longitudinal balance is improved, it will tend to decrease added noise.
4. It will be a source of impulse noise, especially with gross-motion switches.
5. It will tend to distort the attenuation response.
6. It will tend to deteriorate envelope delay characteristics.
7. It will have a return loss characteristic when looking into the switch from either direction. This may affect echo characteristics.
8. It will be a source of crosstalk.
9. It will be a source of absolute delay.
10. It will be a source of harmonic distortion.

14.3   Zero Test-Level Point

Signal level in telephony (see Chapter 5, Section 2.3) is a most important consideration. In a telephone network we often deal with relative levels measured in dBr, which can be related to the familiar dBm as follows:

$$dBm = dBm0 + dBr \qquad (3.2)$$

where dBm0 is an absolute unit of power in dBm referred to as the 0 TLP.

The 0 TLP can be located anywhere in the network. North American practice places it at the outgoing switch of the local switching network, whereas CCITT Rec. G.122 [11] sets levels outgoing from the first international switch in a country (the CT; see Chapter 6) at -3.5 dBr and entering the same switch from the international network at -4.0 dBr. The CCITT then leaves it to the national telephone administration to set the 0 TLP, provided that it meets the relative levels just specified [19].

System gain is usually achieved and levels adjusted at the inputs and outputs of carrier equipment (FDM, see Chapter 5; PCM, Chapters 9 and 10) to meet the requirements of a national transmission plan.

14.4   Transmission Specifications

Table 3.4 lists transmission characteristics for four-wire switches. Loss dispersion deals with the variation of loss through a switch. The loss of a four-wire switch (Table 3.4) is 0.5 dB. On some paths through the switch the loss may only be 0.4 dB, while on others it may be 0.6 dB. Therefore the loss dispersion is 0.2 dB. This factor is of particular importance when dealing with network stability. If loss dispersion is too great, singing may occur, particularly when circuits operate near their singing margin. Return loss at local exchanges (two-wire) should be at least 11 dB (North American practice), with a standard deviation of 3 dB in the band 500–2500 Hz; in the band 250–3200 Hz, return loss should be 6 dB with a standard deviation of 2 dB. Return loss measurements on a local exchange are made against a 600-Ω or 900-Ω resistor in series with a 2.14-microfarad (μF) capacitor [12].

Longitudinal balance refers to balance to ground for both sides of the circuit. Open telephone circuit measurements from one of the two wire leads to ground can display a 100-V peak, and it is not unusual to find several milliamperes of current through a 1000-Ω terminating resistor to ground. The objective is to balance the two sides to ground, producing no net voltage between the two sides. When there is imbalance, the noise produced is often referred to as *metallic noise*. The method of measurement of longitudinal balance must be stated [19].

TABLE 3.4 Transmission Characteristics for Four-Wire Switches

Item	1 [(CCITT) Q.45][c]	2 (Ref. 18)[c]
Loss	0.5 dB	0.5 dB
Loss, dispersion[a]	<0.8 dB	<0.2 dB
Attenuation/frequency response	300–400 Hz: −0.2/+0.5 dB	−0.1/+0.2 dB
	400–2400 Hz: −0.2/+0.3 dB	−0.1/+0.2 dB
	2400–3400 Hz: −0.2/+0.5 dB	−0.1/+0.3 dB
Impulse noise	5 in 5 min above −35 dBm0	5 in 5 min, 12 dB above floor of random noise[d]
Noise		
Weighted	200 pWp	25 pWp
Unweighted	100,000 pW0	3000 pW
Unbalance against ground	300–600 Hz: 40 dB	300–3000 Hz: 55 dB
	600–3400 Hz: 46 dB	3000–3400 Hz: 53 dB
Crosstalk		
Between go and return paths	60 dB	65 dB ⎫ in the band
Between any two paths	70 dB	80 dB ⎭ 200–3200 Hz
Harmonic distortion		50 dB down with a −10 dBm signal for second harmonic, 60 dB down for third harmonic
		200 Hz: 15 dB
Impedance variation with	300–600 Hz: 15 dB	300 Hz: 18 dB
frequency[b]	600–3400 Hz: 20 dB	500–2500 Hz: 20 dB
		3000 Hz: 18 dB
		3400 Hz: 15 dB
Group delay (600–3000 Hz)	100 μs	

[a]Dispersion loss is the variation in loss from calls with the highest loss to those with the lowest loss. This important parameter affects circuit stability.
[b]Expressed as return loss.
[c]Reference frequency, where required, is 800 Hz for column 1 and 1000 Hz for column 2.
[d]Taken from standard measurement techniques for impulse noise.

15 NUMBERING CONCEPTS FOR TELEPHONY

15.1 Introduction

Numbering was introduced in Section 2 as a basic element of switching in telephony. This section discusses, in greater detail, numbering as a factor in the design of a telephone network.

15.2 Definitions

There are four elements to an international telephone number. CCITT Rec. E.163 recommends that not more than 12 digits make up an international number. These 12 digits exclude the international prefix, which is that combination of digits used by a calling subscriber to a subscriber in another country to obtain access to the automatic outgoing international equipment; thus we have 12 digits maximum made up of 4 elements. For example, dialing from Madrid to a specific subscriber in Brussels requires only 10 digits (inside the 12-digit maximum).

07	32	2	4561234
International prefix	Country code	National significant number	Subscriber number
		Trunk code (area code)	

Thus the international number is

32 2 456 1234

According to CCITT international usage (Recs. E.160, E.161, and E.162), we define the following terms.

Numbering Area (Local Numbering Area). This is the area in which any two subscribers use the same dialing procedure to reach another subscriber in the telephone network. Subscribers belonging to the same numbering area may call one another simply by dialing the subscriber number. If they belong to different numbering areas, they must dial the trunk prefix plus the trunk code in front of the subscriber number.

Subscriber Number. This is the number to be dialed or called to reach a subscriber in the same local network or numbering area.

Trunk Prefix (Toll-Access Code). This is a digit or combination of digits to be dialed by a calling subscriber making a call to a subscriber in his own country but outside his own numbering area. The trunk prefix provides access to the automatic outgoing trunk equipment.

Trunk Code (Area Code). This is a digit or combination of digits (not including the trunk prefix) characterizing the called numbering area within a country.

Country Code. This is the combination of one, two, or three digits characterizing the called country.

Local Code. This is a digit or combination of digits for obtaining access to an adjacent numbering area or to an individual exchange (or exchanges) in that area. The national significant number is not used in this situation.

From Madrid, dialing a subscriber in Copenhagen requires 9 digits, Brussels 10, near London (Croydon) 10, Harlow (England) outskirts 11,

Harlow center 10, and New York City 11 (not including the international prefix). This raises the concepts of uniform and nonuniform numbering as well as some ambiguity.

The CCITT defines *uniform numbering* as a numbering scheme in which the length of the subscriber numbers is uniform inside a given numbering area. It defines *nonuniform numbering* as a scheme in which the subscriber numbers vary within a given numbering area. With uniform numbering, each subscriber, by using the same number of digits, can be reached inside a numbering area and from one numbering area to another inside national boundaries. Theoretically, this is true for North America (north of the Rio Grande River), where a subscriber can dial seven digits and always seven digits for that subscriber to reach any other subscriber inside the calling area. Ten digits is required to reach any subscriber outside the numbering area. Note that in North America the trunk code is called the *area code*. This arrangement is shown in the following formula as it was used at the end of 1973, before the introduction of "interchangeable codes" discussed briefly further in this section.

$$
\begin{array}{cc}
 & \text{Telephone number} \\
\text{Area code} & \text{(subscriber number)} \\
0 & \\
N\!-\!X* & NNX\!-\!XXXX \\
1 &
\end{array}
$$

where X is any number from 0 to 9, N is any number from 2 to 9, and 0/1 is the number 0 or 1.

Following a plan set up as in this formula, ATT (the major North American telephone administration prior to 1983) provided for an initial arrangement of 152 area codes, each with a capacity of 540 exchange codes. Remember that there are 10 digits involved in dialing between areas including all the 50 United States, Canada, Puerto Rico, and parts of Mexico.

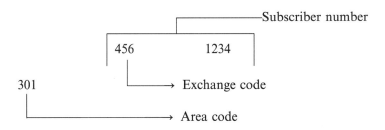

Subsequently, the 540 exchange codes were expanded to 640 when ANC (all number calling) was introduced. This simply means that the use of letters was eliminated and digit combinations 55, 57, 95, and 97 could be utilized where

*Excludes the combination $N11$.

names and resulting letters could not be structured from such combinations, and the code group $NN0$ was also added. This provided needed number relief. This relief was not sufficient to meet telephone growth on the continent, and further code relief had to be provided. This was carried out by realigning code areas, introducing interchangeable codes with code areas where required, and splitting existing code areas and introducing new area codes [23]. The 1995 interchangeable NPA codes provide 792 NPA codes (code areas) with its revised NXX format.

It should be appreciated that a switch must be able to distinguish between receipt of calls bound both out of and within the calling area. The introduction of "interchangeable" codes precludes the ability of a switch to determine whether a seven-digit number (bound for a subscriber inside the same calling area) or a 10-digit number (for a call bound for a subscriber in another calling area) can be expected. This was previously based on the presence of a "0" or "1" in the second digit position. Now the switch must use either of two different methods to distinguish seven-digit numbers: the "timing method," where the exchange is designed so that the equipment waits for 3–5 s after receiving seven digits to distinguish between seven- and ten-digit calls (if one or more digits are received in the waiting period, the switch then expects a ten-digit call), or the "prefix method," which utilizes the presence of either a "1" or "0" prefix that identifies the call as having a ten-digit format. This, of course, is the addition of an extra digit.

The prefix method described here for ATT is the "trunk prefix" defined previously. Such a trunk prefix is used in Spain, for example. In the automatic service, Spain uses either six or seven digits for the subscriber number. Numbering areas with high population density and high telephone growth, such as Madrid, Barcelona, and Valencia, have seven digits. For numbering areas of low telephone density, six digits are used. However, when dialing between numbering areas, the subscriber always dials nine digits. This is also referred to as a *uniform system*. Inside numbering areas the subscriber number is made up of six or seven digits. Area codes (trunk codes) consist of one or two digits. The ninth digit is used for toll access. To dial Madrid, $91 + 7$ digits is used, and to dial Huelva (a small province in southwestern Spain), $955 + 6$ digits is used. The United Kingdom presents an example of nonuniform numbering where subscriber numbers in the same numbering area may be five, six, or seven digits. Dialing from one area to another often involves different procedures.

15.3 Factors Affecting Numbering

15.3.1 General. In telecommunications network design there are numerous trade-offs between economy and operability. Operability covers a large realm, one aspect of which is the human interface. The subscriber must use his/her telephone, and its use should be easy to understand and apply. Uniform numbering and number length notably improve operability.

Number assignment should leave as large a reserve of numbers as possible for growth. The simplest method to accomplish this tends toward longer numbers and nonuniform numbers. Another goal is to reduce switching costs. One way is to reduce number analysis, that is, the number of digits to be analyzed by a switch for proper routing and charging. We find that an economical analysis becomes increasingly necessary as the network becomes more complex with more and more direct routes.

15.3.2 Routing. Consider the scheme for number analysis in routing shown in Figure 3.23. Only one-digit analysis is required to route any call from exchange X to any station in the network. The first digit selects the required outlet. However, if a direct route was established between X and Z_1, two-digit analysis would be required. If a direct route was established to Z_2, three-digit analysis would be necessary. Note that some freedom of number assignment has been lost because all numbers beginning with digit 2 home on X, 3 on Y, and 4 on Z. With the loss of such freedom, we have gained simplification of switches. We could add more digit analysis capability at each exchange and gain more freedom of digit assignment.

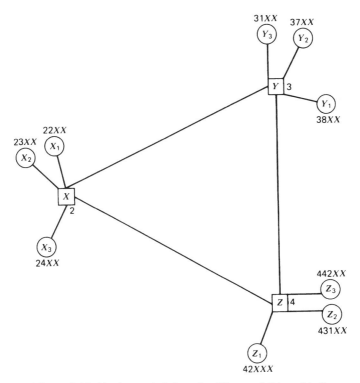

Figure 3.23. Number analysis in routing (X = any digit from 0 to 9).

This brings up one additional point, geographical significance and the definition of numbering areas. When dealing with trunk codes, "geographical significance" means that neighboring call areas are assigned digits beginning with the same number. In Spain there is geographical significance; in the United States there is not. In the northwestern corner of Spain, all trunk codes begin with 8. Aside from Barcelona, all codes in northeastern Spain begin with 7. In Andalusia all begin with 5, and around Valencia the codes begin with 6; Alicante is 65 and Murcia is 68. In the United States, New York City is 212 and 718, just to the north is 914, across the Hudson River is 301, and Long Island is 516.

However, both countries try to use political administrative boundaries to coincide with call-area boundaries. In the United States in many cases we have a whole state or a grouping of counties, but never crossing state boundaries. In Spain we are dealing with provinces, with each province bearing its own call-area number (although some provinces share the same number). Nevertheless, boundaries do coincide with political demarcations. This eases subscriber understanding and simplifies tariffing procedures.

15.3.3 Tariffs (Charging). The billing of telephone calls is automated in the automatic service. There are two basic methods for charging, bulk billing and detailed billing (see Section 13 of this chapter). Detailed billing is essentially a North American practice. Detailed billing includes in the subscriber's bill a listing of each toll call made, number called, date and time, charge time, and individual charge entry for each toll call. Such a form of billing requires extensive number analysis, which is usually carried out by data-processing equipment in centralized locations.

When charges are determined typically by accumulated meter steps, billing is defined as bulk billing. The subscriber is periodically presented with a bill indicating the total number of steps incurred for the period at a certain money rate per step. Bulk billing is much more economical to install and administer. It does lead to much greater subscriber misunderstanding and, in the long run, to dissatisfaction.

With bulk billing, stepping meters are part of the switching equipment. The switching equipment determines the tariff and call duration and actuates the meters accordingly. For bulk billing the relationship between numbering and billing is much closer than in the case of detailed billing. Ideally, tariff zones should coincide with call areas (routing areas). Care should be taken in the compatibility of charging for tariff zones and the numerical series used in numbering. It follows that the larger the tariff zone, the less numerical analysis required.

15.3.4 Size of Numbering Area and Number Length. Size relates not only to geographical dimension but also to the number of telephones encompassed in the area. Areas of very large geographical size present problems in network and switch design. Principally, large areas show a low level of community of interest between towns. In this case, switching requires more digit analysis, and tariff problems may be complicated. If the area has a large telephone population, a larger number of digits may be required, with implied longer dialing times and more postdial delay for analyses. However, large areas are more efficient for number assignment.

With smaller numbering areas, short-length subscriber numbers may be used for those with a higher community of interest. Smaller digit storage would be required for intra-area calls with shorter holding times for switch control units. Smaller areas do offer less flexibility, particularly in the future when uneven growth takes place and forecasts are in error. In the mid-1970s, North America faced a situation requiring reconfiguration and the addition of new areas and also the use of an extra digit or access code.

One report [10] recommends, as a goal, that a numbering area be no greater than 70,000 km^2 nor have less than 100,000 subscribers at the end of a numbering-plan validity period. Subscribers should be able to dial a shorter number when calling other subscribers inside their own numbering area. All other subscribers would be reached by dialing the national number. This is done by adding a simple trunk code in front of the subscriber number. The use of a trunk prefix has advantages in switch design. This should be a single digit, preferably 0. Zero is recommended because few, if any, subscriber numbers start with zero. Thus when a switch receives the initial digit as zero, it is prepared to receive the longer toll number for interarea dialing, whereas if it receives any other digit, it is prepared for receipt of a subscriber number for intra-area dialing. In the United Sates, the initial digit 1 is often used to indicate a toll call.

For dialed international calls, the CCITT recommends number length no greater than 12 digits. A little research will show that few international calls require 12 digits. Number length, of course, deals with number of subscribers, code blocks reserved, immediate and future spare capacity, call-area size, and trunk prefix assignment. Uniform and nonuniform numbering are also factors. Other factors are 40-year forecast accuracy, subscriber habits, routing and translation facility availability, switching system capabilities, and existing numbering scheme.

15.4   In-Dialing

Numbers, particularly in uniform numbering systems, must be set aside for PABXs with in-dialing capability. This is particularly true when the PABX numbering scheme is built into the national scheme for in-dialing. For example, suppose that a PABX main number is 543-7000, with an extension to that PABX of 678. To dial the PABX extension directly from the outside, we would dial 543-7678, and the 543-7*XXX* block would be lost for use except for 999 extension possibilities for that PABX. If it were a small installation, the numbering block loss could be reduced by 543-7600 (PABX main number), extension 78—543-7678. The PABX would be limited to 99 (or 98) direct in-dialed extensions. If the PABX added digits onto the end of the main number, there would be no impact on national or area numbering. For example:

Main PABX number:	543-7000
PABX extension number:	678
Extension dialed directly from outside:	543-7000678

Here we face the CCITT limitation of 12 digits for international dialing; 10 of the 12 have been used. We must also consider digit storage in switches handling such calls. Of course, the realm of nonuniform numbering has been entered. A third method suggested is shown in the following example:

Main PABX number:	53-87654
PABX extension number:	789
Extension dialed directly from outside:	503-8789

Here the first and third digits locate the local exchange on which the PABX is a subscriber. That local exchange must then analyze the fourth, fifth, and possibly the sixth digits. Disadvantages are that translation is always required and that there is some loss to the numbering system.

16   TELEPHONE TRAFFIC MEASUREMENT

Statistics on telephone traffic are mandatory both at the individual exchange level and for the network as a whole. Traffic measurements provide the required statistics when carried out in accordance with a well-organized plan. The statistics include traffic intensity (erlangs or ccs) and its distribution by type of subscriber and service on each route and circuit group and its variation daily, weekly, and seasonally. In the process of traffic measurement, congestion is indicated, if present, and switch and network efficiency can be calculated. The most common measurer of "efficiency" is grade of service.

Congestion and its causes tell the systems engineer whether an exchange is overdimensioned, underdimensioned, or of improper traffic balance. Traffic statistics provide a concrete base or starting point from which to forecast growth. They give past traffic evolution of subscriber-generated traffic growth by type of subscriber and class of service. They also provide a prediction of the evolution of local traffic between exchanges of national long-distance (toll) traffic and international traffic. Traffic measurements (plus forecasts) supply the data necessary to dimension new exchanges and for the extension of existing ones. They are especially important when a new exchange is to replace an existing exchange or when a new exchange will replace several existing exchanges.

Traffic measurement involves a number of parameters, including seizures (call attempts), completed calls, traffic intensity (involving holding times), and congestion. The term "seizures" indicates the number of times a switching unit or groups are seized without taking into account holding time. Seizures may be equated to call attempts and give an expression of how much exchange control equipment is being worked. The number of completed calls is of interest in the operation and administration of an exchange. Of real interest are statistics on uncompleted calls that are not attributed to busy lines or lack of answer or that can be attributed to the specified grade of service. However, at a particular exchange a completed call means only that the switch in question has carried out its function, which does not necessarily imply that a connection has been established between two subscribers. The intensity of traffic or traffic volume, a most important parameter, directly provides a measure of usage of a circuit. It is especially useful in those switching units involved directly in the speech network. On the other hand, it is not directly indicative of grade of service. Approximate grade of service should be taken from the traffic tables used to dimension the exchange by using measured traffic intensity as an input. The approximation is most accurate when traffic intensity approaches dimensioned intensity.

Congestion involves three characteristics: "all circuits busy," overflow, and dial-tone delay. "All circuits busy" is an indication of the number of times, and eventually the duration, where all units of a switching group are handling traffic simultaneously. Thus it represents an index of the real grade of service. Its use is particularly effective for those switching groups that are operating near maximum capacity. For an overdimensioned exchange, the "all circuits busy" index is useless and tells us nothing. Overflow is an index of the number of call attempts that have not been able to proceed due to congestion. For networks with alternative routing, overflow tells us the number of offered calls not handled by a specific switching equipment group. Dial-tone delay is directly indicative of overall grade of service that an exchange provides to its subscribers, particularly in the preselection switching stages. Dial-tone delay is normally expressed in the time required in getting dial tone compared to a fixed time, usually 3 s, as a percentage of total calls.

Traffic measurements should be made through the busy hour (BH), which can be determined by reading amperage of exchange battery over the estimated period of occurrence, say, 9:30 A.M. to 12:30 P.M., every 10 min (or by peg count usage devices, microprocessors on lines, equipment or trunk circuits). These measurements should be done daily for at least three weeks. Traffic measurements should be carried out for the work week, one week per month for an entire year. The means of measurement depends on available equipment and exchange type. This may range from simple observation to fully automatic traffic-measuring equipment with recorders on magnetic or paper tape. The engineer may have to resort to the use of electromechanical counters and/or electromechanical or electronic hunting devices with which many exchanges, particularly those with common control, are equipped [21].

17 DIAL-SERVICE OBSERVATION

Dial-service observation provides an index or measurement of the telephone service provided to the customer (see Chapter 1, Section 13). It gives an administration or telephone company a sample measure of maintenance required (or how well it is being carried out), load balance, and adequacy of installed equipment. The number of customer-originated calls sampled per day is of the order of 1–1.5% of the total calls originated; 200 observations per exchange per day is a minimum [20].

Traditionally, service observation has been done manually, requiring the presence of an observer. Service observation positions have some forms of automation, such as automatic recording of a calling number and/or the mark sensing or keypunching to be employed for computer-processing input. Tape recorders may be placed at exchanges to automatically record calls on selected lines. The tapes are then periodically sent to the service observation desk. Tapes usually have a "time hack" to record the time and the duration of calls. A service observation desk usually serves many exchanges or an entire local area. Tandem exchanges are good candidates because they concentrate the service function. The results of service observation are intended to represent average service; thus care should be taken to ensure that subscribers selected represent the average customer. The following is one list of data to be collected per line observed:

- Total call attempts (local and long-distance)
- Completed calls
- Ineffective calls due to calling subscriber: incomplete call, late dialing, unavailable number dialed, call abandoned prematurely
- Ineffective calls due to called subscriber: busy, no answer
- Ineffective calls due to equipment: wrong number, congestion, no ringing, busy tone, no answer, interruptions
- Transmission quality (level, distortion, noise, etc.)
- Wrong number dialed by subscriber
- Dial-tone delay
- Postdial delay

Results of service observation are computerized, and summary tables are published monthly or quarterly. Individual tables are often made for each exchange, relating it to the remainder of the network. Quality of route can be determined since observations identify the called local exchange. A similar set of observations may be made for the long-distance (toll) service exclusively.

There is a tendency to fully automate service observation without any human intervention. The major weakness to this approach is the lack of subjective observation of factors such as type of noise and its effect on a call, subscriber behavior to certain stimuli, and so forth.

REVIEW QUESTIONS

1. List at least four user requirements with regard to switching.

2. Why do we use switching in the first place? Explain the rationale of switching versus transmission.

3. Give two basic functions of a telephone number.

4. What is a *serving area*?

5. Without digit blocking, how many distinct entities can be served by a three-digit number? a four-digit number?

6. What are the two basic functions of a local switch?

7. In a local area a certain traffic relation is 12 erlangs. What would be the most economical form of routing: direct or tandem?

8. There are eight basic functions that a conventional generic switch carries out. Name at least six.

9. What does the term *supervision* mean in switching?

10. In the concentration stage of a local switch, are there more or less inputs than outputs?

11. In the expansion stage of a local switch, the number of outputs equals the number of _____.

12. Define a trunk group using the term *traffic relation*.

13. There are two basic categories of electromechanical space-division switches. Name them and give an example switch type for each.

14. What does *degeneration* tell us about a particular switch?

15. Differentiate between *limited availability* and *full availability*.

16. Define *grading*.

17. What is the function of latching on a crossbar switching matrix?

18. Differentiate between direct progressive control and register progressive control.

19. Based on the definition in this text, what major advance does *common control* have over a progressive control system?

20. In a crossbar switch, what unit carries out the common-control function?

21. What is the function of a *junctor* on crossbar switches?

22. Why would a marker need more than one register?

23. Why is it advantageous to use *equipment number* (EN) rather than *dial number* (DN) in a switch?

24. SPC simply means that a switch is _____-controlled.

25. Give at least three major advantages of SPC over hard-wired control switches.

26. What is the meaning of *holding time* and *delay* with regard to SPC systems?

27. Give at least two advantages and one disadvantage of using centralized control (versus distributed control).

28. What is the one major difference between a *concentrator* and a *satellite*?

29. Describe the differences between European and North American charging methods. Bring in the acronyms *CAMA* and *LAMA*.

30. Give at least five transmission impairments encountered in an analog switch.

31. Give the CCITT and North American definitions of 0 TLP.

32. Discuss the importance of longitudinal balance on a wire-pair trunk.

33. Give the two different characteristic impedances that may be encountered on local exchange VF switch ports.

34. Differentiate between uniform and nonuniform numbering.

35. With switch complexity in mind, discuss the importance of numbering on routing.

36. Describe at least two methods of measuring telephone traffic through a switch. Describe one indirect method that is fairly accurate.

37. Give at least three traffic parameters that are measured at a switch that give valuable insight to planners, system design engineers, and telephone company/administration managers.

38. In what period during a weekday is traffic measurement most important?

39. Give at least five parameters to be included in dial-service observation.

REFERENCES

1. *Switching Systems*, American Telephone and Telegraph Company, New York, 1961.
2. M. Hobbs, *Modern Communication Switching Systems*, Tab Books, Blue Ridge Summit, PA, 1974.
3. T. H. Flowers, *Introduction to Exchange Systems*, John Wiley & Sons, New York, 1976.
4. A. F. Joel, "What Is Telecommunication Circuit Switching?" *Proc. IEEE*, **65** (9) (September 1977).
5. *Fundamental Principles of Switching Circuits and Systems*, American Telephone and Telegraph Company, New York, 1963.
6. International Telephone and Telegraph Corporation, *Reference Data for Radio Engineers*, 6th ed., Howard W. Sams, Indianapolis, 1976.
7. J. G. Pearce, *Electronic Switching*, Telephony Publishing Company, Chicago, 1968.
8. J. P. Dartois, "Metaconta L Medium Size Local Exchanges," *Electr. Commun.*, **48** (3) (1973).
9. J. G. Pearce, "The New Possibilities of Telephone Switching," *Proc. IEEE*, **65** (9) (September 1977).
10. "Numbering," Telecommunication Planning, ITT Laboratories (Spain), Madrid, 1973.
11. CCITT, Q Recommendations (Rec. Q), Redbooks, VIII Plenary Assembly, Malaga-Torremolinos, Spain, October 1984.
12. L. F. Goeller, *Design Background for Telephone Switching*, Lees ABC of the Telephone, Geneva, Ill., 1977.
13. *Notes on the Network 1980*, American Telephone and Telegraph Company, New York, 1980.
14. B. E. Briley and W. N. Toy, "Telecommunication Processors," *Proc. IEEE*, **65** (9) (September 1977).
15. T. H. Flowers, "Processors and Processing in Telephone Exchanges," *Proc. IEEE*, **119** (3) (March 1972).
16. E. Aro, "Stored Program Control-Assisted Electromechanical Switching—An Overview," *Proc. IEEE*, **65** (9) (September 1977).
17. P. J. Hiner, "TXE4 and the New Technology," *Telecommunications* (January 1977).
18. USITA Symposium, April 1970, Open Questions 18–37.
19. Roger L. Freeman, *Telecommunication Transmission Handbook*, 3rd ed., John Wiley & Sons, New York, 1991.

20. L. Alvarez Mazo and M. Poza Martinez, "Dial Service Observation," Telecommunication Planning Symposium, STC (SA) Boksburg, South Africa, June 1972.

21. L. Alvarez Mazo, "Network Traffic Measurements," Telecommunication Planning Symposium, STC (SA) Boksburg, South Africa, June 1972.

22. R. Deese, "A History of Switching," *Telecommunications* (February 1984).

23. *Notes on the BOC Intra-LATA Networks—1994*, SR-TSV-002275, Bellcore, Livingston, NJ, April 1994. Bell Communications Research Technical Reference.

4

SIGNALING FOR ANALOG
TELEPHONE NETWORKS

1 INTRODUCTION

In a switched telephone network, signaling conveys the intelligence needed for one subscriber to interconnect with any other in that network. Signaling tells the switch that a subscriber desires service and then gives the local switch the data necessary to identify the required distant subscriber and hence to route the call properly. It also provides supervision of the call along its path. Signaling also gives the subscriber certain status information, such as dial tone, busy tone (busy back), and ringing. Metering pulses for call charging may also be considered a form of signaling.

There are several classifications of signaling:

1. General.
   a. Subscriber signaling.
   b. Interswitch signaling.
2. Functional.
   a. Audible–visual (call progress and alerting).
   b. Supervisory.
   c. Address signaling.

Figure 4.1 shows a more detailed breakdown of these functions.

It should he appreciated that on many telephone calls, more than one switch is involved in call routing. Therefore switches must interchange information among switches in fully automatic service. Address information is provided between modern switching machines by interregister signaling, and the supervisory function is provided by line signaling. The audible–visual category of signaling functions inform the calling subscriber regarding call

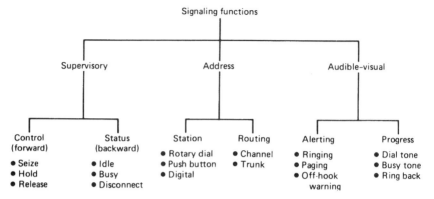

Figure 4.1. A functional breakdown of signaling.

progress, as shown in Figure 4.1. The *alerting* function informs the called subscriber of a call waiting or an extended "off-hook" condition of his or her handset. Signaling information can be conveyed by a number of means from subscriber to switch or between (and among) switches. Signaling information can be transmitted by means such as

- Duration of pulses (pulse duration bears a specific meaning)
- Combination of pulses
- Frequency of signal
- Combination of frequencies
- Presence or absence of a signal
- Binary code
- For dc systems, the direction or level of transmitted current

From Ref. 1.

2 SUPERVISORY SIGNALING

Supervisory signaling provides information on line or circuit condition and indicates whether a circuit is in use or is idle. It informs the switch and interconnecting trunk circuits whether a calling party is "off hook," or "on hook," or a called party is "off hook," or "on hook." The terms "on hook" and "off hook" were derived from the position of the receiver of an old-fashioned telephone set in relation to the mounting (in this case a hook) provided for it. If a subscriber subset is on hook, the conductor (subscriber loop) between the subscriber and his local exchange is open and no current is flowing. For the reverse or off-hook condition, there is a dc shunt across the line, and current is flowing. These terms have also been found convenient for designating the two signaling conditions of a trunk (junction). Usually, if a trunk is not in use, the on-hook condition is indicated toward both ends. Seizure of the trunk at the calling end initiates an off-hook signal transmitted toward the called end.

The reader must appreciate that supervisory information–status must be maintained end to end on every telephone call. It is necessary to know when a calling subscriber lifts his/her telephone off hook, thereby requesting service. It is equally important that we know when the called subscriber answers (i.e., lifts her telephone off hook), because that is when we may start metering the call to establish charges. It is also important to know when the called and calling subscribers return their telephones to the on-hook condition. Charges stop, and the intervening trunks comprising the talk path as well as the switching points are then rendered idle for use by another pair of subscribers. During the period of occupancy of a talk path end to end, we must know that this particular path is busy (is occupied) so that no other call attempt can seize it.

Dialing of a subscriber line is merely interruption of the subscriber loop's off-hook condition, often called "make and break." The "make" is a current flow condition (or off hook), and the "break" is the no-current condition (or on hook). How do we know the difference between supervisory and dialing? Primarily by duration—the on-hook interval of a dial pulse is relatively short and is distinguishable from an on-hook disconnect signal (subscriber hangs up), which is transmitted in the same direction for a longer duration. Thus the switch is sensitized to duration to distinguish between supervisory and dialing of a subscriber loop. Figure 4.2 is a simplified diagram of a subscriber loop showing its functional signaling elements.

2.1 E and M Signaling

Probably the most common form of trunk supervision is E and M signaling, particularly with FDM multiplex equipment (Chapter 5). Yet it only becomes true E and M signaling where the trunk interfaces with the switch (see Figure 4.3). E-lead and M-lead signaling systems are semantically derived from historical designation of signaling leads on circuit drawings covering these systems. Historically, the E and M signaling interface provides two leads between the switch and what we may call *trunk-signaling equipment* (signaling interface). One lead is called the "E-lead," which carries signals *to* the switching equipment. Such signal directions are shown in Figure 4.3, where we see that signals from switch *A* and switch *B* leave *A* on the M-lead and are delivered to *B* on the E-lead. Likewise, from *B* to *A*, supervisory information leaves *B* on the M-lead and is delivered to *A* on the E-lead.

For conventional E and M signaling (referring to electromechanical exchanges), the following supervisory conditions are valid:

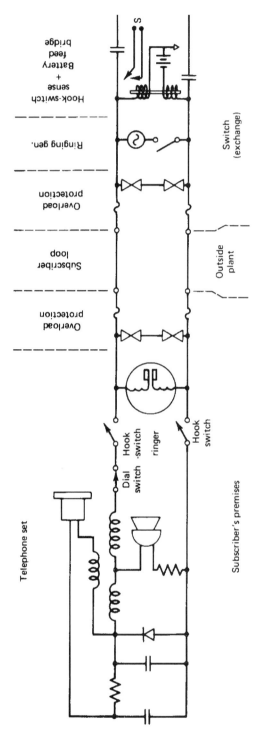

Figure 4.2. Signaling with a conventional telephone subset. Note functions of hook switch, dial, and ringer. From [12].

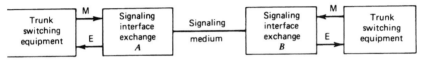

Figure 4.3. E and M signaling.

Direction		Condition at A		Condition at B	
Signal A to B	Signal B to A	M-Lead	E-Lead	M-Lead	E-Lead
On hook	On hook	Ground	Open	Ground	Open
Off hook	On hook	Battery	Open	Ground	Ground
On hook	Off hook	Ground	Ground	Battery	Open
Off hook	Off hook	Battery	Ground	Battery	Ground

Source: Ref. 8.

3   AC SIGNALING

3.1   General

Up to this point we have reviewed the most employed means of supervisory trunk signaling (or line signaling). Direct-current signaling, such as reverse-battery signaling, has notable limits on distance because it cannot be applied directly to carrier systems (Chapters 5 and 8) and is limited on metallic pairs due to the IR drop of the lines involved. Direct-current trunk signaling is addressed in Section 10.

There are many ways to extend these limits, but from a cost-effectiveness standpoint there is a limit that we cannot afford to exceed. On trunks exceeding dc capabilities, some form of ac signaling will be used. Traditionally, ac signaling systems are divided into three categories: low-frequency, in-band, and out-band (out-of-band) systems. Each of these can derive the four E and M signaling states.

3.2   Low-Frequency ac Signaling Systems

An ac signaling system operating below the limits of the conventional voice channel (i.e., <300 Hz) are termed *low frequency*. Low-frequency signaling systems are one-frequency systems, typically 50 Hz, 80 Hz, 135 Hz, or 200 Hz. It is impossible to operate such systems over carrier-derived channels (see Chapter 5) because of the excessive distortion and band limitation introduced. Thus low-frequency signaling is limited to metallic-pair transmission systems. Even on these systems, cumulative distortion limits circuit length. A maximum of two repeaters may be used, and, depending on the type of circuit (open wire, aerial cable, or buried cable) and wire gauge, a rough rule of thumb is a distance limit of 80–100 km.

3.3   In-Band Signaling

In-band signaling refers to signaling systems using an audio tone, or tones inside the conventional voice channel, to convey signaling information. In-band signaling is broken down into three categories: (1) one frequency (SF or single frequency), (2) two frequency (2VF), and (3) multifrequency (MF). As the term implies, in-band signaling is where signaling is carried out directly in the voice channel. As the reader is aware, the conventional voice channel as defined by the CCITT occupies the band of frequencies from 300 Hz to 3400 Hz. Single-frequency and two-frequency signaling systems utilize the 2000- to 3000-Hz portion, where less speech energy is concentrated.

3.3.1 Single-Frequency Signaling. Single-frequency signaling is used almost exclusively for supervision. In some locations it is used still for interregister signaling, but the practice is diminishing in favor of more versatile methods such as MF signaling. The most commonly used frequency is 2600 Hz, particularly in North America. On two-wire trunks 2600 Hz is used in one direction and 2400 Hz in the other. A diagram showing application of SF signaling on a four-wire trunk is shown in Figure 4.4.

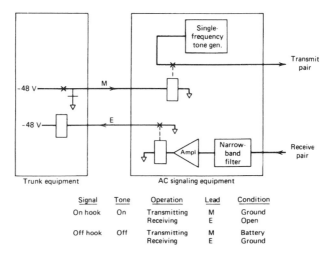

Figure 4.4. Functional block diagram of a single-frequency signaling circuit. (*Note*: Wire pairs, "receive" and "transmit," derive from carrier-equipment "receive" and "transmit" channels.) [2, 6].

3.3.2 *Two-Frequency Signaling.* Two-frequency signaling is used for both

supervision (line signaling) and address signaling. We often associate SF and 2VF supervisory signaling systems with carrier (FDM) operation. Of course, when we discuss such types of line signaling (supervision), we know that the term "idle" refers to the on-hook condition while "busy" refers to the off-hook condition. Thus, for such types of line signaling that are governed by audio tones of which SF and 2VF are typical, we have the conditions of "tone on when idle" and "tone on when busy." The discussion holds equally well for in-band and out-of-band signaling methods. However, for in-band signaling, supervision is by necessity tone-on idle; otherwise subscribers would have an annoying 2600-Hz tone on throughout the call.

A major problem with in-band signaling is the possibility of "talk-down," which refers to the premature activation or deactivation of supervisory equipment by an inadvertent sequence of voice tones through the normal use of the channel. Such tones could simulate the SF tone, forcing a channel dropout (i.e., the supervisory equipment would return the channel to the idle state). Chances of simulating a 2VF tone set are much less likely. To avoid the possibility of talk-down on SF circuits, a time-delay circuit or slot filters to bypass signaling tones may be used. Such filters do offer some degradation to speech unless they are switched out during conversation. They must be switched out if the circuit is going to be used for data transmission [7].

It becomes apparent why some administrations and telephone companies have turned to the use of 2VF supervision, or out-of-band signaling for that matter. For example, a typical 2VF line signaling arrangement is the CCITT No. 5 code, where f_1 (one of the two VF frequencies) is 2400 Hz and f_2 is 2600 Hz. 2VF signaling is also used widely for address signaling (see Section 4.1 of this chapter) [4].

3.4 Out-of-Band Signaling

With out-of-band signaling, supervisory information is transmitted out of band (i.e., above 3400 Hz). In all cases it is a single-frequency system. Some out-of-band systems use "tone on when idle," indicating the on-hook condition, whereas others use "tone off." The advantage of out-of-band signaling is that either system, tone on or tone off, may be used when idle. Talk-down cannot occur because all supervisory information is passed out of band, away from the speech-information portion of the channel.

The preferred CCITT out-of-band frequency is 3825 Hz, whereas 3700 Hz is commonly used in the United States. It also must be kept in mind that out-of-band signaling is used exclusively on carrier systems, not on wire trunks. On the wire side, inside an exchange, its application is E and M signaling. In other words, out-of-band signaling is one method of extending E and M signaling over a carrier system.

In the short run, out-of-band signaling is attractive in terms of both economy and design. One drawback is that when channel patching is

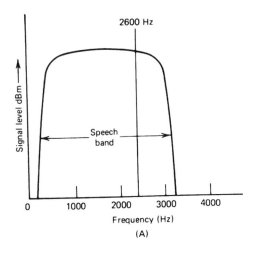

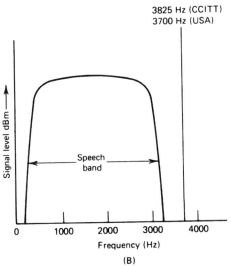

Figure 4.5. Single-frequency signaling: (A) in-band; (B) out-of-band.

required, signaling leads have to be patched as well. In the long run, the signaling equipment required may indeed make out-of-band signaling even more costly because of the extra supervisory signaling equipment and signaling lead extensions required at each end and at each time that the carrier (FDM) equipment demodulates to voice. The major advantage of out-of-band signaling is that continuous supervision is provided, whether tone on or tone off, during the entire telephone conversation. In-band SF signaling and out-of-band signaling are illustrated in Figure 4.5. An example of out-of-band signaling is the CCITT R-2 System (CCITT Rec. Q.351) (see Table 4.1) [9, 11].

TABLE 4.1   R-2 Line Signaling (3825 Hz)

	Direction	
Circuit State	Forward (Go)	Backward (Return)
Idle	Tone on	Tone on
Seized	Tone off	Tone on
Answered	Tone off	Tone off
Clear back	Tone off	Tone on
Release	Tone on	Tone on or off
Blocked	Tone on	Tone off

Source: Ref. 11.

4   ADDRESS SIGNALING: INTRODUCTION

Address signaling originates as dialed digits (or activated push buttons) from a calling subscriber, whose local switch accepts these digits and, using that information, directs the telephone call to the desired distant subscriber. If more than one switch is involved in the call setup, signaling is required between switches (both address and supervisory). Address signaling between switches in conventional systems is called *interregister signaling.*

The paragraphs that follow discuss various more popular standard ac signaling techniques such as 2VF, MF pulse, and MF tone. Although interregister signaling is stressed where appropriate, some supervisory techniques are also reviewed. Common-channel signaling is discussed in Chapter 16 where we describe CCITT No. 7 signaling system.

4.1   Two-Frequency Pulse Signaling

Two-frequency signaling is commonly used as an interregister mode of signaling employing the speech band for the transmission of information. It may also be used for line signaling.* There are various methods of using two-voice frequencies to transmit signaling information. For example, CCITT No. 4 uses 2040 Hz and 2400 Hz to represent binary 0 and 1, respectively. It uses a four-element code, permitting 16 different coded characters, as shown in Table 4.2. With the CCITT No. 4 code both interregister and line signaling utilize the 2VF technique. Interregister signaling in this case is a pulse-type signaling, and line signaling utilizes the combination of the two frequencies and the duration of signal to convey the necessary supervisory information, as shown in Table 4.3.

*Supervision (interswitch).

TABLE 4.2   CCITT Signal Code System No. 4

Digit	Number	Combination Elements			
		1	2	3	4
1	1	y	y	y	x
2	2	y	y	x	y
3	3	y	y	x	x
4	4	y	x	y	y
5	5	y	x	y	x
6	6	y	x	x	y
7	7	y	x	x	x
8	8	x	y	y	y
9	9	x	y	y	x
0	10	x	y	x	y
Call operator code 11	11	x	y	x	x
Call operator code 12	12	x	x	y	y
Spare code (see CCITT Rec. Q.104)	13	x	x	y	x
Incoming half-echo suppressor required	14	x	x	x	y
End of pulsing	15	x	x	x	x
Spare code	16	y	y	y	y

Sending duration of binary elements 35 ± 7 ms. Sending duration of blank elements between binary elements 35 ± 7 ms. Element x is 2040 Hz; element y is 2400 Hz.

Source: Ref. 10.

TABLE 4.3   CCITT No. 4 Line Signaling

Forward signals	
Terminal seizing	*Px*
Transit seizing	*Py*
Numerical signals	As in Table 4.2
Clear forward	*Pxx*
Forward transfer	*Pyy*
Backward signals	
Proceed to send	
Terminal	*x*
International transit	*y*
Number received	*p*
Busy flash	*pX*
Answer	*pY*
Clear back	*Px*
Release guard	*Pyy*
Blocking	*Px* (congestion)
Unblocking	*Pyy*

Source: Ref. 10.

As we mentioned, line signaling with the CCITT No. 4 code is based on signal duration as well as frequency. The line-signaling format uses both tone frequencies, 2040 Hz and 2400 Hz. Each line signal consists of an initial *prefix* (*P*) signal followed by a control signal element, called a *suffix*. The *P* signal consists of both frequencies (2VF), and the suffix signal consists of one frequency, where *x* is 2040 Hz and *y* is 2400 Hz (see Table 4.3). Now consider the durations of the following signal elements used for line signaling:

$$P = 150 \pm 30 \text{ ms}$$
$$x, y = 100 \pm 20 \text{ ms (each)}$$
$$xxyy = 350 \pm 70 \text{ ms (each)}$$

This set of values refers to transmitted signal duration—that is, as transmitted by the signaling sender.

Let us see how these signals are used in the line (interswitch supervision) signaling (see Table 4.3).

The supervisory functions "clear forward" and "release guard" do the reverse of "call setup." They take the call down or disconnect, readying the circuit for the next user. "Clear back" is another example.

4.2 Multifrequency Signaling

Multifrequency signaling is in wide use today for interregister signaling. It is an in-band method utilizing five or six tone frequencies, two at a time. Multifrequency signaling works equally well over metallic and carrier (FDM) systems. Four commonly used MF signaling systems follow with a short discussion for each [5].

4.2.1 SOCOTEL. SOCOTEL is an interregister signaling system used principally in France, areas of French influence, and Spain with some modifications. The frequency pairs and their digit equivalents are shown in Table 4.4. Line signaling used with SOCOTEL may be dc, 50 Hz, or 2000 Hz. The same frequencies are used in both directions.

4.2.2 Multifrequency Signaling in North America: The R-1 Code. The MF signaling system principally used in the United States and Canada is recognized by the CCITT as the R-1 code. It is a two-out-of-five frequency-pulse system. Additional signals for control functions are provided by combinations using a sixth frequency. Table 4.5 shows digits and other applications and their corresponding frequency combinations as well as a brief explanation of "other applications."

TABLE 4.4 Basic SOCOTEL MF Signaling Code

Tone Frequencies (Hz)[a]	Digit
700 + 900	1
700 + 1100	2
900 + 1100	3
700 + 1300	4
900 + 1300	5
1100 + 1300	6
700 + 1500	7
900 + 1500	8
1100 + 1500	9
1300 + 1500	0

[a]The 1700-Hz frequency is also used for signaling system check and when more code groups are required by a telephone company or national administration. When 1700 Hz is used for coding, 1900 Hz is used for checking the system; 1700 Hz and/or 1900 Hz may also be used for control purposes.

Source: Ref. 3.

TABLE 4.5 The R-1 Code[a] (North American MF)

Digit	Frequency Pair (Hz)	
1	700 + 900	
2	700 + 1100	
3	900 + 1100	
4	700 + 1300	
5	900 + 1300	
6	1100 + 1300	
7	700 + 1500	
8	900 + 1500	
9	1100 + 1500	
10 (0)	1300 + 1500	

Use	Frequency Pair	Explanation
KP	1100 + 1700	Preparatory for digits
ST	1500 + 1700	End-of-pulsing sequence
STP	900 + 1700 ⎫	
ST2P	1300 + 1100 ⎬	Used with TSPS (traffic service position system)
ST3P	700 + 1700 ⎭	
Coin collect	700 + 1100	Coin control
Coin return	1100 + 1700	Coin control
Ring-back	700 + 1700	Coin control
Code 11	700 + 1700	Inward operator (CCITT No. 5)
Code 12	900 + 1700	Delay operator
KP1	1100 + 1700	Terminal call
KP2	1300 + 1700	Transit call

[a]Pulsing of digits is at the rate of about seven digits per second with an interdigital period of 68 ± 7 ms. For intercontinental dialing for CCITT No. 5 code compatibility, the R-1 rate is increased to 10 digits per second. The KP pulse duration is 100 ms [3, 11].

Source: Ref. 10.

TABLE 4.6   CCITT No. 5 Code[a] **Showing Variations with R-1 Code**

Signal	Frequencies (Hz)	Remarks
KP1	1100 + 1700	Terminal traffic
KP2	1300 + 1700	Transit traffic
1	700 + 900	
2	700 + 1100	
3–0	Same as Table 4.5	
ST	1500 + 1700	
Code 11	700 + 1700	Code 11 operator
Code 12	900 + 1700	Code 12 operator

[a]Line signaling for CCITT No. 5 code is 2VF, with f_1 2400 Hz and f_2 2600 Hz. Line-signaling conditions are shown in Table 4.7.

Source: Ref. 3.

TABLE 4.7   CCITT No. 5 Line-Signaling Code

Signal	Direction	Frequency	Sending Duration	Recognition Time (ms)
Seizing	$\longrightarrow$	f_1	Continuous	40 ± 10
Proceed to send	$\longleftarrow$	f_2	Continuous	40 ± 10
Busy flash	$\longleftarrow$	f_2	Continuous	125 ± 25
Acknowledgment	$\longrightarrow$	f_1	Continuous	125 ± 25
Answer	$\longleftarrow$	f_1	Continuous	125 ± 25
Acknowledgment	$\longrightarrow$	f_1	Continuous	125 ± 25
Clear back	$\longleftarrow$	f_2	Continuous	125 ± 25
Acknowledgment	$\longrightarrow$	f_1	Continuous	125 ± 25
Forward transfer	$\longrightarrow$	f_2	850 ± 200 ms	125 ± 25
Clear forward	$\longrightarrow$	$f_1 + f_2$	Continuous	125 ± 25
Release guard	$\longleftarrow$	$f_1 + f_2$	Continuous	125 ± 25

Source: Ref. 10.

4.2.3 CCITT No. 5 Signaling Code. Interregister signaling with the CCITT No. 5 code is very similar in makeup to the North American R-1 code. Variations with R-1 are shown in Table 4.6. The CCITT No. 5 line-signaling code is shown in Table 4.7.

4.2.4 *The R-2 Code.* The R-2 code is listed by CCITT (Rec. Q.361) [11] as a European regional signaling code. Taking full advantage of combinations of two-out-of-six tone frequencies, 15 frequency-pair possibilities are available. This number is doubled in each direction by having meaning

TABLE 4.8 European R-2 System

	Frequencies (Hz)						
	1380	1500	1620	1740	1860	1980	Forward Direction I/II
Index No. for Groups I/II and A/B	1140	1020	900	780	660	540	Backward Direction A/B
1	x	x					
2	x		x				
3		x	x				
4	x			x			
5		x		x			
6			x	x			
7	x				x		
8		x			x		
9			x		x		
10				x	x		
11	x					x	
12		x				x	
13			x			x	
14				x		x	
15					x	x	

Source: Ref. 11.

groups I and II in the forward direction and groups A and B in the backward direction (see Table 4.8).

Groups I and A are said to be of primary meaning, and groups II and B are said to be of secondary meaning. The change from primary to secondary meaning is commanded by the backward signal A-3 or A-5. Secondary meanings can be changed back to primary meanings only when the original change from primary to secondary was made by the use of the A-5 signal. Referring to Table 4.8, the 10 digits to be sent in the forward direction in the R-2 system are in group I and are index numbers 1 through 10 in the table. The index 15 signal (group A) indicates "congestion in an international exchange or at its output." This is a typical backward information signal giving circuit status information. Group B consists of nearly all "backward information" and in particular deals with subscriber status.

The R-2 line-signaling system has two versions: the one used on analog networks is discussed here; the other, on digital (PCM) networks, is briefly covered in Chapter 8. The analog version is an out-of-band tone-on-when-idle system. Table 4.9 shows the line conditions in each direction, forward and backward. Note that the code takes advantage of a signal sequence that has six characteristic operating conditions.

Let us consider several of these conditions.

TABLE 4.9   Line Conditions for R-2 Code

Operating Condition of the Circuit	Signaling Conditions	
	Forward	Backward
1. Idle	Tone on	Tone on
2. Seized	Tone off	Tone on
3. Answered	Tone off	Tone off
4. Clear back	Tone off	Tone on
5. Release	Tone on	Tone on or off
6. Blocked	Tone on	Tone off

Source: Ref. 11.

Seized. The outgoing exchange (call-originating exchange) removes the tone in the forward direction. If seizure is immediately followed by release, removal of the tone must be maintained for at least 100 ms to ensure that it is recognized at the incoming end.

Answered. The incoming end removes the tone in the backward direction. When another link of the connection using tone-on-when-idle continuous signaling precedes the outgoing exchange, the "tone-off" condition must be established on the link as soon as it is recognized in this exchange.

Clear Back. The incoming end restores the tone in the backward direction. When another link of the connection using tone-on-when-idle continuous signaling precedes the outgoing exchange, the "tone-off" condition must be established on this link as soon as it is recognized in this exchange.

Clear Forward. The outgoing end restores the tone in the forward direction.

Blocked. At the outgoing exchange the circuit stays blocked as long as the tone remains off in the backward direction.

4.2.5 Subscriber Tones and Push-Button Codes (North America).
Subscriber subsets in many places in the world are either dial or push button. The push-button type is more versatile, and more rapid dialing can be accomplished by a subscriber. Table 4.10 compares digital dialed, dial pulses (breaks), and multifrequency (MF) push-button tones. Table 4.11 shows the audible tones commonly used in North America. Functionally, these are the call-progress tones presented to the subscriber.

5   COMPELLED SIGNALING

In many of the signaling systems discussed thus far, signal element duration is an important parameter. For instance, in a call setup an initiating exchange sends a 100-ms seizure signal. Once this signal is received at the distant end,

TABLE 4.10   North American Push-Button Codes

Digit	Dial Pulse (Breaks)	Multifrequency Push-Button Tones
0	10	941,1336 Hz
1	1	697,1209 Hz
2	2	697,1336 Hz
3	3	697,1474 Hz
4	4	770,1209 Hz
5	5	770,1336 Hz
6	6	770,1477 Hz
7	7	852,1209 Hz
8	8	852,1336 Hz
9	9	852,1477 Hz

Source: Ref. 2.

TABLE 4.11   Audible Tones Commonly Used in North America

Tone	Frequencies (Hz)	Cadence
Dial	$350 + 440$	Continuous
Busy (station)	$480 + 620$	0.5 s on, 0.5 s off
Busy (network congestion)	$480 + 620$	0.2 s on, 0.3 s off
Ring return	$440 + 480$	2 s on, 4 s off
Off-hook alert	Multifrequency howl	1 s on, 1 s off
Recording warning	1400	0.5 s on, 15 s off
Call waiting	440	0.3 s on, 9.7 s off

Source: Ref. 2.

the distant exchange sends a "proceed to send" signal back to the originating exchange; in the case of the R-1 system, this signal is 140 ms or more in duration. Then, on receipt of "proceed to send" the initiating exchange spills all digits forward. In the case of R-1, each digit is an MF pulse of 68-ms duration with 68 ms between each pulse. After the last address digit an ST (end-of-pulsing) signal is sent. In the case of R-1 the incoming (far-end) switch register knows the number of digits to expect. Thus there is an explicit acknowledgment that the call setup has proceeded satisfactorily. Thus R-1 is a good example of noncompelled signaling.

A fully compelled signaling system is one in which each signal continues to be sent until an acknowledgment is received. Thus signal duration is not significant and bears no meaning. The R-2 and SOCOTEL are examples of fully compelled signaling systems. Figure 4.6 shows a fully compelled signaling sequence. Note the small overlap of signals, causing the acknowledging (reverse) signal to start after a fixed time on receipt of the forward

Outgoing register

Incoming register

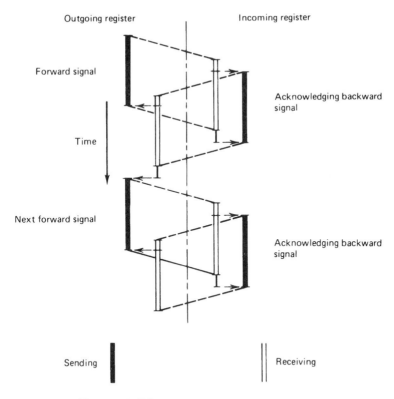

Forward signal

Acknowledging backward
signal

Time

Next forward signal

Acknowledging backward
signal

Sending

Receiving

Figure 4.6. Fully compelled signaling procedure.

signal. This is because of the minimum time required for recognition of the incoming signal. After the initial forward signal, further forward signals are delayed for that short recognition time (see Figure 4.6). Recognition time is normally less than 80 ms.

Fully compelled signaling is advantageous in that signaling receivers do not have to measure duration of each signal, thus making signaling equipment simpler and more economical. Fully compelled signaling adapts automatically to the velocity of propagation, to long circuits, to short circuits, to metallic pairs, or to carrier and is designed to withstand short interruptions in the transmission path. The principal drawback of compelled signaling is its inherent lower speed, thus requiring more time for setup. Setup time over space-satellite circuits with compelled signaling is appreci-able and may force the system engineer to seek a compromise signaling system.

There is also a partially compelled type of signaling, where signal duration is fixed in both forward and backward directions according to system specifications; or the forward signal is of indefinite duration and the backward signal is of fixed duration. The forward signal ceases once the backward signal has been received correctly. The CCITT No. 4 is a variation of partially compelled signaling [3].

6   LINK-BY-LINK VERSUS END-TO-END SIGNALING

An important factor to be considered in switching system design that directly affects both signaling and customer satisfaction is postdialing delay. This is the amount of time it takes after the calling subscriber completes dialing until ring-back is received. Ring-back is a backward signal to the calling subscriber telling her that her dialed number is ringing. Postdialing delay must be made as short as possible.

Another important consideration is register occupancy time for call setup as the setup proceeds from originating exchange to terminating exchange. Call-setup equipment, that equipment used to establish a speech path through a switch and to select the proper outgoing trunk, is expensive. By reducing register occupancy per call, we may be able to reduce the number of registers (and markers) per switch, thus saving money.

Link-by-link and end-to-end signaling each affect register occupancy and postdialing delay, each differently. Of course, we are considering calls involving one or more tandem exchanges in a call setup, because this situation usually occurs on long-distance or toll calls. Link-by-link signaling may be defined as a signaling system where *all* interregister address information must be transferred to the subsequent exchange in the call-setup routing. Once this information is received at this exchange, the preceding exchange control unit (register) releases. This same operation is carried on from the originating exchange through each tandem (transit) exchange to the terminating exchange of the call. The R-1 system is an example of link-by-link signaling.

End-to-end signaling abbreviates the process such that tandem (transit) exchanges receive only the minimum information necessary to route the call. For instance, the last four digits of a seven-digit telephone number need be exchanged only between the originating exchange (e.g., the calling sub-scriber's local exchange or the first toll exchange in the call setup) and the terminating exchange in the call setup. With this type of signaling, fewer digits are required to be sent (and acknowledged) for the overall call-setup sequence. Thus the signaling process may be carried out much more rapidly, decreasing postdialing delay. Intervening exchanges on the call route work much less, handling only the digits necessary to pass the call to the next exchange in the sequence.

The key to end-to-end signaling is the concept of "leading register." This is the register (control unit) in the originating exchange that controls the call routing until a speech path is set up to the terminating exchange before releasing to prepare for another call setup. For example, consider a call from subscriber *X* to subscriber *Y*.

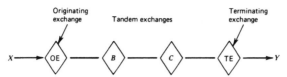

The telephone number of subscriber *Y* is 345-6789. The sequence of events is as follows using end-to-end signaling:

- A register at exchange OE receives and stores the dialed number 345-6789 from subscriber *X*.
- Exchange OE analyzes the number and then seizes a trunk (junction) to exchange *B*. It then receives a "proceed-to-send" signal indicating that the register at *B* is ready to receive routing information (digits).
- Exchange OE then sends digits 34, which are the minimum necessary to effect correct transit.
- Exchange *B* analyzes the digits 34 and then seizes a trunk to exchange *C*. Exchanges OE and *C* are now in direct contact and exchange *B*'s register releases.
- Exchange OE receives the "proceed-to-send" signal from exchange *C* and then sends digits 45, those required to effect proper transit at *C*.
- Exchange *C* analyzes digits 45 and then seizes a trunk to exchange TE. Direct communication is then established between the leading register for this call at OE and the register at TE being used on this call setup. The register at *C* then releases.
- Exchange OE receives the "proceed-to-send" signal from exchange TE, to which it sends digits 5678, the subscriber number.
- Exchange TE selects the correct subscriber line and returns to *A* ring-back, line busy, out of order, or other information after which all registers are released.

Thus we see that a signaling path is opened between the leading register and the terminating exchange. To accomplish this, each exchange in the route must "know" its local routing arrangements and request from the leading register those digits it needs to route the call further along its proper course.

Again, the need for backward information becomes evident, and backward signaling capabilities must be nearly as rich as forward signaling capabilities when such a system is implemented.

R-1 is a system inherently requiring little backward information (inter-register). The little information that is needed, such as "proceed to send," is sent via line signaling. The R-2 system has major backward information requirements, and backward information and even congestion and busy signals are sent back by interregister signals [3].

7   THE EFFECTS OF NUMBERING ON SIGNALING

Numbering, the assignment and use of telephone numbers, affects signaling as well as switching. It is the number or the translated number, as we found out in Chapter 3, that routes the call. There is "uniform" numbering and "nonuniform" numbering. How does each affect signaling? Uniform numbering can simplify a signaling system. Most uniform systems in the nontoll or local-area case are based on seven digits, although some are based on six. The last four digits identify the subscriber. The first three digits (or the first two in the case of a six-digit system) identify the exchange. Thus the local exchange or transit exchanges know when all digits are received. There are two advantages to this sort of scheme:

1. The switch can proceed with the call once all digits are received because it "knows" when the last digit (either the sixth or seventh) has been received.
2. "Knowing" the number of digits to expect provides inherent error control and makes "time out"* simpler.

For nonuniform numbering, particularly on direct distance dialing in the international service, switches require considerably more intelligence built in. It is the initial digit or digits that will tell how many digits are to follow, at least in theory.

However, in local or national systems with nonuniform numbering, the originating register has no way of knowing whether it has received the last digit, with the exception of receiving the maximum total used in the national system. With nonuniform numbering, an incompletely dialed call can cause a useless call setup across a network up to the terminating exchange, and the call setup is released only after time out has run its course. It is evident that with nonuniform numbering systems, national (and international) networks are better suited to signaling systems operating end to end with good features of backward information, such as the R-2 system [3].

8   ASSOCIATED AND DISASSOCIATED CHANNEL SIGNALING

Here we introduce a new concept: diassociated channel signaling. Up to now, we have only considered associated channel signaling. Figure 4.7 is meant to

---

*"Time out" is the resetting of call-setup equipment and return of dial tone to subscriber as a result of incomplete signaling procedure, subset left off hook, and so forth.

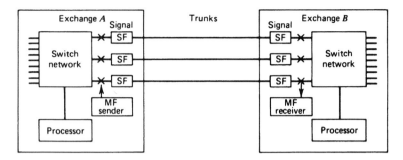

Associated Channel Signaling

(Conventional SF-MF)

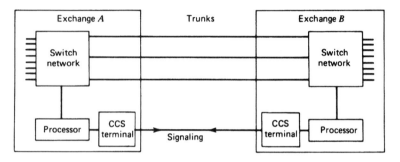

Separate Channel Signaling

Figure 4.7. Conventional analog associated channel signaling (upper drawing) versus separate channel signaling (lower drawing). *Note*: Signaling on upper drawing accompanies voice paths; signaling on the lower drawing is conveyed on a separate circuit (or time slot). CCS = common channel signaling such as CCITT Signaling System No. 7.

show the difference. The upper drawing in Figure 4.7 shows conventional signaling on an analog circuit. This fits all of the signaling techniques discussed so far. The signaling goes right along with the channel it is associated with, on the same medium. The lower drawing in Figure 4.7 shows "separate channel signaling." Now this signaling may or may not go on that same medium or path. Most often with this type of signaling, that separate channel handles all signaling (especially supervisory signaling) for a group of channels. Typically, the European PCM system called E1 (Chapter 8) uses this type of signaling. One separate digital channel covers all supervisory signaling for 30 traffic channels. If it travels on the same medium and path as its associated traffic channels, it is still associated channel signaling.

We can carry this one step further. That separate channel can follow a different path using, perhaps, different media. CCITT Signaling System No. 7 is always separate channel, but can be associated or disassociated. Figure 4.8 shows separate channel signaling, but associated. Figure 4.9 shows fully disassociated channel signaling.

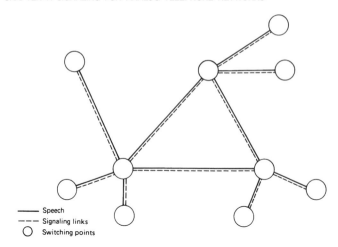

Figure 4.8. Associated channel signaling. As shown, the signaling channel is separate, but associated. With conventional analog signaling, it would be a single solid line, where the signaling is embedded with its associated traffic.

9 SIGNALING IN THE SUBSCRIBER LOOP

9.1 Background and Purpose

In Chapter 2 we described loop start signaling, although we didn't call it that. When a subscriber takes a telephone off hook (i.e., out of its cradle), there is a switch closure in the subset (see Figure 4.2 and Section 2 of this chapter), current flows in the loop alerting the serving exchange that service is desired

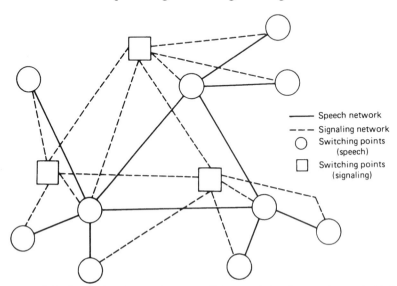

Figure 4.9. Fully disassociated channel signaling. This type of signaling may be used with CCITT Signaling System No. 7, described in Chapter 16.

on that telephone. As a result, dial tone is returned to the subscriber. This is basic supervisory signaling on the subscriber loop.

A problem can arise from this form of signaling. It is called *glare*. Glare is the result of attempting to seize a particular subscriber loop from each direction. In this case it would be an outgoing call and an incoming call nearly simultaneously. There is a much greater probability of glare with a PABX than with an individual subscriber.

Ground-start signaling is the preferred signaling system when lines terminate in a switching system such as a PABX. It operates as follows. When a call is from the local serving switch to the PABX, the local switch immediately grounds the conductor tip to seize the line. With some several seconds delay, ringing voltage is applied to the line (where required). The PABX immediately detects the grounded tip conductor and will not allow an outgoing call from the PABX to use this circuit, thus avoiding glare.

In a similar fashion, if a call originates at the PABX and is outgoing to the local serving switch, the PABX grounds the ring conductor to seize the line. The serving switch recognizes this condition and prevents other calls from attempting to terminate on the circuit. The switch then grounds the tip conductor and then returns dial tone after it connects a digit receiver. There can be a rare situation when double seizure occurs, causing glare. Usually one or the other end of the circuit is programmed to back down and allow the other call to proceed. A ground start interface is shown in Figure 4.10.

Terminology in signaling often refers back to manual switchboards or, specifically, to the plug used with these boards and its corresponding jack as shown in Figure 4.11. Thus we have tip (T), ring (R), and sleeve (S). Often, only the tip and ring are used, and the sleeve is grounded and has no real function.

10 METALLIC TRUNK SIGNALING

10.1 Basic Loop Signaling

Many trunks serving the local area are metallic-pair trunks. They are dc loops much like the subscriber loop. Many still use dial pulses for address signaling as well as some sort of supervisory signaling. Figure 4.12 illustrates a typical metallic trunk circuit with subscriber loops on each end. The example in the figure uses a step-by-step (SXS) switch, which is rather dated; however, the concepts illustrated are still very valid.

Of course these loop-signaling circuits provide two signaling states: one when the circuit is opened and the other when the circuit is closed. A third signaling state is obtained by reversing the direction or changing the magnitude of the current in the circuit. Combinations of (1) open/close, (2) polarity reversal, and (3) high/low current are used for distinguishing signals intended for one direction of signaling (e.g., dial-pulse signals) from those

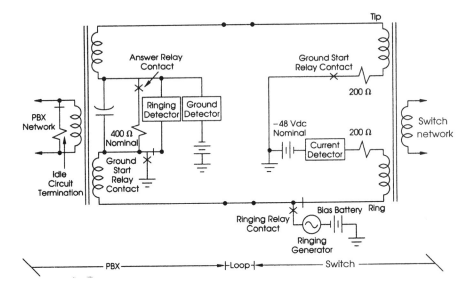

Figure 4.10. Ground start interface block diagram. From Figure 2-7, Ref. 14. Reprinted with permission.

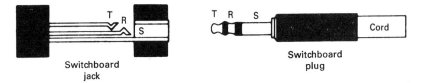

Figure 4.11. Switchboard plug with corresponding jack (R, S, and T are ring, sleeve, and tip, respectively).

intended for the opposite direction (e.g., answer signals). We describe the most popular loop method below, namely, reverse-battery signaling.

We illustrate loop-signaling methods with electromechanical components. These methods are also applicable to SPC (stored-program control) switching systems. SPC switching systems can connect to loop signaling on physical facilities, to analog or digital carrier systems, or can work directly with a bit stream from a digital carrier (e.g., SLC-96, Chapter 8).

10.2 Reverse-Battery Signaling

Reverse-battery signaling employs basic methods (1) and (2) above, and takes its name from the fact that battery and ground are reversed on the tip and ring to change the signal toward the calling end from on hook to off hook. Figure 4.13 shows a typical application of reverse-battery signaling in a common-control switch.

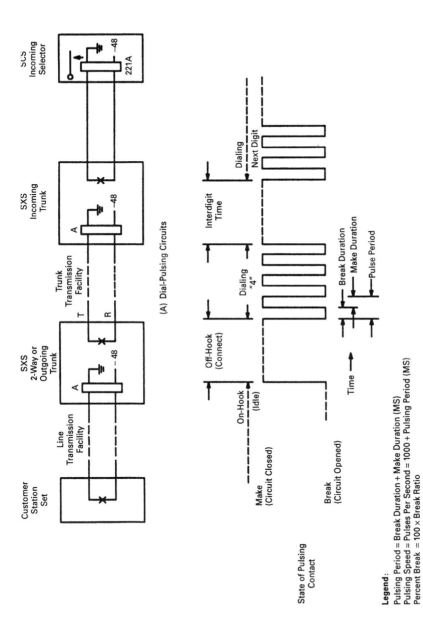

Figure 4.12. Dial-pulse signaling. From Bellcore, Figure 6-19, page 6-85, Ref. 2.

Legend:
Pulsing Period = Break Duration + Make Duration (MS)
Pulsing Speed = Pulses Per Second = 1000 ÷ Pulsing Period (MS)
Percent Break = 100 × Break Ratio

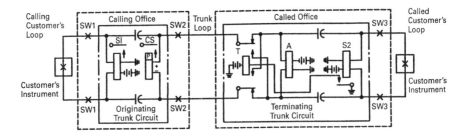

Figure 4.13. Reverse-battery signaling. From Bellcore, Figure 6-27, page 6-102, Ref. 2. Reprinted with permission.

In the idle or on-hook condition, all relays are unoperated and the switch (SW) contacts are open. Upon seizure of the outgoing trunk by the calling switch (exchange) (trunk group selection based on the switch or exchange code dialed by the calling subscriber), the following occurs:

- SW1 and SW2 contacts close, thereby closing loop to called office (exchange) and causing the A relay to operate.
- Operation of the A relay signals off-hook (connect) indication to the called switch (exchange).
- Upon completion of pulsing between switches, SW3 contacts close and the called subscriber is alerted. When the called subscriber answers, the S2 relay is operated.
- Operation of the S2 relay operates the T relay, which reverses the voltage polarity on the loop to the calling end.
- The voltage polarity causes the CS relay to operate, transmitting an off-hook (answer) signal to the calling end.

When the calling subscriber hangs up, disconnect timing starts (between 150 and 400 ms). After the timing is completed, SW1 and SW2 contacts are released in the calling switch. This opens the loop to the A relay in the called switch and releases the calling subscriber. The disconnect timing (150–400 ms) is started in the called switch as soon as the A relay releases. When the disconnect timing is completed, the following occurs:

- If the called subscriber has returned to on hook, SW3 contacts release. The called subscriber is now free to place another call.
- If the called subscriber is still off hook, disconnect timing is started in the called switch. On the completion of the timing interval, SW3 contacts open. The called subscriber is then returned to dial tone. If the circuit is seized again from the calling switch during the disconnect timing, the disconnect timing is terminated and the called subscriber is returned to

dial tone. The new call will be completed without interference from the previous call.

When the called subscriber hangs up, the CS relay in the calling switch releases. Then the following occurs:

- If the calling subscriber has also hung up, disconnection takes place as described above.
- If the calling subscriber is still off hook, disconnect timing is started. On the completion of the disconnect timing, SW1 and SW2 contacts are opened. This returns the calling subscriber to dial tone and releases the A relay in the called switch. The calling subscriber is free to place a new call at this time. After the disconnect timing, the SW3 contacts are released, which releases the called subscriber. The called subscriber can place a new call at this time.

Section 10 is based on information taken from Ref. 2.

REVIEW QUESTIONS

1. Give the three generic signaling functions, and explain the purpose of each.

2. Differentiate between line signaling and interregister signaling.

3. There are seven ways to transmit signaling information, one of which is frequency. Name five others.

4. How does a switch know whether a particular talk path is busy or idle?

5. A most common form of line signaling is E and M signaling. Describe how it works in three sentences or less.

6. Compare in-band and out-of-band supervisory signaling regarding tone-on idle/busy, advantages, and disadvantages.

7. What is the most common form of in-band supervisory signaling in North America (analog)?

8. What is the standard out-of-band signaling frequency for North America?

9. Give the principal advantage of 2VF over SF supervisory signaling.

10. Compare CCITT No. 5, R-1, and R-2 supervisory signaling systems.

11. List three types of interregister signaling using MF.

12. Clearly distinguish compelled and noncompelled signaling. Give advantages and disadvantages of each.

13. On connections involving communication satellites, what would be more attractive regarding postdial delay, compelled or noncompelled signaling?

14. Describe and compare link-by-link and end-to-end signaling. Associate R-1 and R-2 with each.

15. Describe at least four types of backward information.

16. Describe the effects of uniform and nonuniform numbering on signaling/switching. Why is uniform numbering more advantageous to signaling/switching?

17. Distinguish associated channel signaling from separate channel signaling.

18. What is "disassociated channel" signaling? Give some examples.

19. What is the meaning of *glare*?

20. Describe how loop signaling works from the subscriber subset.

21. Ground-start signaling is commonly used on analog circuits connecting a PABX with the local serving exchange. How does it nearly eliminate glare?

22. Reverse-battery signaling is commonly used on two-wire_____. It is useful on these two-wire circuits because it can_____.

REFERENCES

1. C. A. Dahlbom and C. Breen, "Signaling Systems for Control of Telephone Switching," *BSTJ*, **39** (November 1960).

2. *BOC Notes on the LEC Networks—1990*, SR-TSV-002275, Issue 1, Bellcore, Piscataway, NJ, March 1991.

3. "Signalling," from Telecommunication Planning Documents, ITT Laboratories (Spain), Madrid, November 1974.

4. *National Networks for the Automatic Service*," CCITT-ITU, Geneva, 1964.

5. M. Den Hertog, "Inter-register Multifrequency Signaling for Telephone Switching in Europe," *Electrical Communications*, **38**(1) (1972).

6. *Reference Data for Engineers: Radio, Electronics, Computers and Communications*, 7th ed., Howard W. Sams, Indianapolis, 1985.

7. R. L. Freeman, *Telecommunication Transmission Handbook*, 3rd ed., John Wiley & Sons, New York, 1991.

8. "Lenkurt Demodulator"—*World-Wide E&M Signaling* (July–August 1977); *A Glossary of Signaling Terms*, April 1974, GTE Lenkurt, San Carlos, CA.

9. *General Recommendations on Telephone Switching and Signaling*, CCITT Recommendations, Fascicle VI.1, IXth Plenary Assembly, Melbourne, 1988.

10. *Specifications of Signalling Systems 4 and 5*, CCITT Recommendations, Fascicle VI.2, IXth Plenary Assembly, Melbourne, 1988.

11. *Specifications of Signalling Systems R1 and R2,"* CCITT Recommendations, Fascicle VI.4, IXth Plenary Assembly, Melbourne, 1988.

12. *Access Area Switching and Signaling: Concepts, Issues and Alternatives*, NTIA, Boulder, Colorado, May 1978.

13. C. A. Dahlbom, "Signaling Systems and Technology," *Proc. IEEE*, **65**, 1349–1353 (1977).

14. W. D. Reeve, *Subscriber Loop Signaling and Transmission Handbook*, *Analog*, IEEE Press, IEEE, New York, 1992.

5

INTRODUCTION TO TRANSMISSION
FOR TELEPHONY

1 GENERAL

The basic building block for transmission is the telephone channel or voice channel. "Voice channel" implies spectral occupancy, whether the voice path is over wire, radio, coaxial cable, or over a fiber-optic system. If a pair of wires of a simple subscriber loop is extended without loading, we can expect to see the spectral content from the average talker with frequencies as low as 20 Hz and as high as 20 kHz if the transducer of the telephone set was at all efficient across this band. Our ear, at least in younger people, is sensitive to frequencies from about 30 Hz to as high as 20 kHz. However, the primary content of a voice signal (energy plus emotion) will occupy a much narrower band of frequencies (approximately 100–4000 Hz). Considering these and other factors, we say that the nominal voice channel occupies the band from 0 to 4 kHz. CCITT defines the voice channel as the band of frequencies between 300 and 3400 Hz. Bell Laboratories [4] states that "the optimum tradeoff between economics and quality of transmission occurs when the telephone speech signal is band-limited to the range from about 200 to 3200 Hz."

There are three basic impairments we must deal with regarding the voice channel.

- Attenuation distortion (frequency response)
- Phase distortion
- Noise

Two additional impairments are echo and singing. We will deal with these two later.

Level is another important parameter. Level must be controlled in a telecommunications network. It most surely can impact quality of service.

2 THE THREE BASIC IMPAIRMENTS TO VOICE CHANNEL TRANSMISSION

2.1 Attenuation Distortion

A signal transmitted over a voice channel suffers various forms of distortion. That is, the output signal from the channel is distorted in some manner such that it is not an exact replica of the input. One form of distortion is called *attenuation distortion* and is the result of imperfect amplitude–frequency response. Attenuation distortion can be avoided if all frequencies within the passband are subjected to exactly the same loss (or gain). Whatever the transmission medium, however, some frequencies are attenuated more than others. For example, on loaded wire-pair systems, higher frequencies are attenuated more than lower ones. On carrier equipment (see Section 4 of this chapter), band-pass filters are used on channel units, where, by definition, attenuation increases as the band edges are approached. Figure 5.1 is a good example of the attenuation characteristics of a voice channel operating over carrier equipment.

Attenuation distortion across the voice channel is measured against a reference frequency. The CCITT specifies 800 Hz as a reference, which is universally used in Europe, Africa, and parts of Hispanic America, whereas 1000 Hz is the common reference frequency in North America. Let us look at some ways attenuation distortion may be stated. For example, one European requirement may state that between 600 Hz and 2800 Hz the level will vary no more than -1 to $+2$ dB, where the plus sign means more loss and the minus sign means less loss. Thus if a signal at -10 dBm is placed at the input

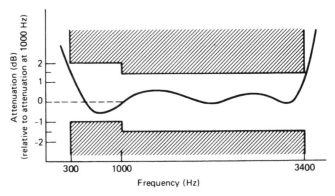

Figure 5.1. Typical attenuation–frequency response (attenuation distortion) for a voice channel. Hatched areas show specified limits.

of the channel, we would expect −10 dBm at the output at 800 Hz (if there were no overall loss or gain), but at other frequencies we could expect a variation between −1 and +2 dB. For instance, we might measure the level at the output at 2500 Hz at −11.9 dBm and at 1100 Hz at −9 dBm.

2.2 Phase Distortion

A voice channel may be regarded as a band-pass filter. A signal takes a finite time to pass through a telecommunication network. This time is a function of the velocity of propagation, which varies with the media involved.

The velocity of propagation also tends to vary with frequency because of the electrical characteristics associated with the network. Considering the voice channel, therefore, the velocity of propagation tends to increase toward band center and decrease toward band edge. This is illustrated in Figure 5.2, which shows relative delay across the voice channel.

The finite time it takes a signal to pass through the total extension of a voice channel or any network is called *delay*. Absolute delay is the delay a signal experiences while passing through the channel end-to-end at a reference frequency. But we see that the propagation time is different for different frequencies with the wavefront of one frequency arriving before the wavefront of another in the passband. A modulated signal will not be distorted on passing through the channel if the phase shift changes uniformly with frequency, whereas if the phase shift is nonlinear with respect to frequency, the output signal is distorted compared to the input.

In essence we are dealing with phase linearity of a circuit. If the phase–frequency relationship over a passband is not linear, distortion will occur in

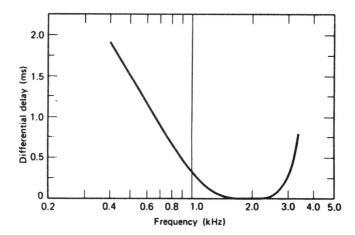

Figure 5.2. Typical differential delay across a voice channel. Frequency-division-multiplex (FDM) equipment back-to-back.

the transmitted signal. This phase distortion is often measured by a parameter called *envelope delay distortion* (EDD). Mathematically, envelope delay is the derivative of the phase shift with respect to frequency. The maximum difference in the derivative over any frequency interval is called envelope delay distortion. Therefore EDD is always a difference between the envelope delay at one frequency and that at another frequency of interest in the passband. Note that envelope delay is often defined the same as group delay—that is, the ratio of change, with angular frequency, of the phase shift between two points in a network [2].

2.2.1 Notes on Phase Distortion.

Absolute delay is minimum around 1700 and 1800 Hz in the voice channel. This is shown in Figure 5.2. The figure also shows that around 1700 or 1800 Hz, envelope delay distortion is flattest. It is for this reason that so many data modems use 1700 or 1800 Hz for the characteristic tone frequency which is modulated by the data.

This brings up the next point. Phase distortion (or EDD) has little effect on speech communications over the telecommunication network. However, for data transmission, phase distortion is the greatest bottleneck for data rate (i.e., number of bits per second that the channel can support). It has probably more effect on limiting data rate than any other parameter [3].

2.3 Noise

2.3.1 General.

Noise, in its broadest definition, consists of any undesired signal in a communication circuit. The subject of noise and noise reduction is probably the most important single consideration in transmission engineering. It is the major limiting factor in system performance. For the discussion in this text, noise is broken down into four categories:

1. Thermal noise
2. Intermodulation noise
3. Crosstalk
4. Impulse noise

2.3.2 Thermal Noise.

Thermal noise occurs in all transmission media and all communication equipment, including passive devices. It arises from random electron motion and is characterized by a uniform distribution of energy over the frequency spectrum with a Gaussian distribution of levels.

Every equipment element and the transmission medium proper contribute thermal noise to a communication system if the temperature of that element or medium is above absolute zero. Thermal noise is the factor that sets the lower limit of sensitivity of a receiving system and is often expressed as a temperature, usually given in units referred to absolute zero. These units are kelvins.

Thermal noise is a general expression referring to noise based on thermal agitations. The term "white noise" refers to the average uniform spectral distribution of noise energy with respect to frequency. Thermal noise is directly proportional to bandwidth and temperature. The amount of thermal noise to be found in 1 Hz of bandwidth in an actual device is

$$P_n = kT \text{ (W/Hz)} \tag{5.1}$$

where k is Boltzmann's constant, equal to 1.3803×10^{-23} J/K, and T is the absolute temperature (K) of the circuit (device). At room temperature, $T = 17°C$ or 290 K; thus

$$P_n = 4.00 \times 10^{-21} \text{ W/Hz of bandwidth}$$
$$= -204 \text{ dBW/Hz of bandwidth}$$
$$= -174 \text{ dBm/Hz of bandwidth}$$

For a band-limited system (i.e., a system with a specific bandwidth), $P_n = kTB$ (W), where B refers to the so-called noise bandwidth in hertz. Thus at 0 K we obtain $P_n = -228.6$ dBW/Hz of bandwidth; for a system with a noise bandwidth measured in hertz (B) and whose noise temperature is T we obtain:

$$P_n = -228.6 \text{ dBW} + 10 \log T + 10 \log B \tag{5.2}$$

2.3.3 Intermodulation Noise. Intermodulation (IM) noise is the result of the presence of intermodulation products. If two signals with frequencies F_1 and F_2 are passed through a nonlinear device or medium, the result will contain IM products that are spurious frequency energy components. These components may be present either inside and/or outside the band of interest for a particular device. IM products may be produced from harmonics of the desired signals in question, either as products between harmonics or as one of the signals and the harmonic of the other(s) or between both signals themselves. The products result when two (or more) signals beat together or "mix." Look at the mixing possibilities when passing F_1 and F_2 through a nonlinear device. The coefficients indicate the first, second, or third harmonics.

- Second-order products $F_1 \pm F_2$
- Third-order products $2F_1 \pm F_2$; $2F_2 \pm F_1$
- Fourth-order products $2F_1 \pm 2F_2$; $3F_1 \pm F_2 \ldots$

Devices passing multiple signals simultaneously, such as multichannel radio equipment, develop intermodulation products that are so varied that they resemble white noise.

Intermodulation noise may result from a number of causes:

- Improper level setting. If the level of input to a device is too high, the device is driven into its nonlinear operating region (overdrive).
- Improper alignment causing a device to function nonlinearly.
- Nonlinear envelope delay.
- Device malfunction.

To summarize, intermodulation noise results from either a nonlinearity or a malfunction that has the effect of nonlinearity. The cause of intermodulation noise is different from that of thermal noise. However, its detrimental effects and physical nature can be identical with those of thermal noise, particularly in multichannel systems carrying complex signals.

2.3.4 Crosstalk

Crosstalk refers to unwanted coupling between signal paths. There are essentially three causes of crosstalk: (1) electrical coupling between transmission media, such as between wire pairs on a voice-frequency (VF) cable system, (2) poor control of frequency response (i.e., defective filters or poor filter design), and the nonlinear performance in analog (FDM) multiplex systems. Excessive level may exacerbate crosstalk.

There are two types of crosstalk:

1. *Intelligible*, where at least four words are intelligible to the listener from extraneous conversation(s) in a 7-s period.
2. *Unintelligible*: crosstalk resulting from any other form of disturbing effects of one channel on another.

Intelligible crosstalk presents the greatest impairment because of its distraction to the listener. Distraction is considered to be caused either by fear of loss of privacy or primarily by the user of the primary line consciously or unconsciously trying to understand what is being said on the secondary or interfering circuits; this would be true for any interference that is syllabic in nature.

Received crosstalk varies with the volume of the disturbing talker, the loss from the disturbing talker to the point of crosstalk, the coupling loss between the two circuits under consideration, and the loss from the point of crosstalk to the listener. The most important of these factors for this discussion is the coupling loss between the two circuits under consideration. Talker levels have been discussed elsewhere in this text. Also, we must not lose sight of the fact that the effects of crosstalk are subjective, and other factors have to be considered when crosstalk impairments are to be measured. Among these factors are the type of people who use the channel, the acuity of listeners, traffic patterns, and operating practices [3].

2.3.5 Impulse Noise. Impulse noise is noncontinuous, consisting of irregular pulses or noise spikes of short duration and of relatively high amplitude. These spikes are often called "hits," and each spike has a broad spectral content.* Impulse noise degrades voice telephony usually only marginally, if at all. However, it may seriously degrade error performance on data or other digital circuits. We will discuss impulse noise further when we discuss data transmission in Chapter 10.

2.4 Level

Level is an important parameter in the telecommunications network, particularly in the analog network or in the analog portion of the network. In the context of this book, when we use the word *level*, we mean *signal magnitude*. Level could be comparative: The output of an amplifier is 20 dB higher than the input. But more commonly, we mean absolute level, and in telephony it is measured in dBm (decibels referenced to 1 milliwatt), and milliamperes; and in radio systems we are more apt to use dBW (decibels referenced to 1 watt). The voltage example is video (i.e., television). It uses the unit dBmV, meaning decibels referenced to 1 millivolt.

In the telecommunication network, if levels are too high, amplifiers become overloaded, resulting in increases in intermodulation noise or increased crosstalk. If levels are too low, customer satisfaction may suffer.

System levels are used for engineering a communication system. These are usually taken from a level chart or reference system drawing made by a planning group or as a part of an engineered job. On the chart a 0 TLP (zero test-level point) is established. A test-level point is a location in a circuit or system at which a specified test-tone level is expected during alignment. A 0 TLP is a point at which the test-tone level should be 0 dBm.

From the 0 TLP other points may be shown using the unit dBr (decibel reference). A minus sign shows that the level is so many decibels below reference and a positive sign, above. The unit dBm0 is an absolute unit of power in dBm referred to the 0 TLP. The dBm can be related to dBr and dBm0 by the following formula:

$$dBm = dBm0 + dBr \qquad (5.3)$$

For instance, a value of -32 dBm at a -22-dBr point corresponds to a reference level of -10 dBm0. A -10-dBm0 signal introduced at the 0-dBr point (0 TLP) has an absolute signal level of -10 dBm [6].

2.4.1 Typical Levels. Earlier measurements of speech level used the unit of measure VU, standing for volume unit. For a 1000-Hz sinusoid signal, 0 VU = 0 dBm. When a VU meter is used to measure the level of a voice

---

*Impulse noise "smears" a broad frequency bandwidth.

signal, it is difficult to equate VU and dBm. One of the problems, of course, is that speech transmission is characterized by spurts of signal. We can relate VU to dBm in the following formula:

$$\text{Average power of a telephone talker} \approx \text{VU} - 1.4 \ (\text{dBm})$$

There are two new measurements of voice signals. These are *equivalent peak level* (EPL) and *long-term conversational level*. The unit of measurement for both is the dBm0 (dBm referenced to the zero test-level point or 0 TLP).

EPL can be described approximately as the 95% point on the cumulative probability distribution of instantaneous talker power. Periods of silence are excluded from the distribution.

In the long-distance network, EPLs can be characterized by a mean of −11.1 dBm0 with a standard deviation of 4.7 dB. Average power for calls in the long-distance network can be characterized by a mean of −25.5 dBm0 and a standard deviation of 5.3 dB. For the local area, the corresponding quantities are a mean EPL of −11.8 dBm0 with a standard deviation of 4.7 dB and a mean average power of −26.5 dBm0 with a standard deviation of 5.4 dB. The distributions of EPL and average power are approximately log-normal (i.e., normal when the independent variable is expressed in dB units) [4].

2.5 Signal-to-Noise Ratio

When dealing with transmission engineering, signal-to-noise ratio is perhaps more frequently used than any other criterion when designing a telecommunication system. Signal-to-noise ratio expresses in decibels the amount by which a signal level exceeds the noise within a specified bandwidth.

As we review several types of material to be transmitted, each will require a minimum signal-to-noise ratio to satisfy the customer or to make the receiving instrument function within certain specified criteria. We might require the following signal-to-noise ratios (S/N) with the corresponding end instruments:

- Voice: 40 dB ⎫ based on customer satisfaction.
- Video: 45 dB ⎭
- Data: ~15 dB, based on a specified error rate and modulation type.

In Figure 5.3* a 1000-Hz signal has a signal-to-noise ratio of 10 dB. The level of the noise is +5 dBm and the signal, +15 dBm. Thus

---

*Assume a nominal 4-kHz bandwidth in the example.

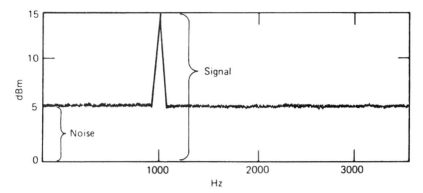

Figure 5.3. Signal-to-noise ratio.

$$\left(\frac{S}{N}\right)_{\text{dB}} = \text{level}_{(\text{signal in dBm})} - \text{level}_{(\text{noise in dBm})} \qquad (5.4)$$

Signal-to-noise ratio in the PSTN really has limited use in characterizing speech transmission because of the "spurtiness" of the human voice. We can appreciate that individual talker signal power can fluctuate widely so that S/N ratio is far from constant from telephone call to telephone call. In lieu of actual voice, we use a test tone to measure level and S/N, which has a constant amplitude and, of course, no silent intervals [4].

3   TWO-WIRE AND FOUR-WIRE TRANSMISSION

3.1   Two-Wire Transmission

A telephone conversation inherently requires transmission in both directions. When both directions are carried on the same pair of wires, it is called *two-wire transmission*. The telephones in our homes and offices are connected to a local switching center (exchange) by means of two-wire circuits. A more proper definition for transmitting and switching purposes is that when oppositely directed portions of a single telephone conversation occur over the same electrical transmission channel or path, we call this *two-wire operation*.

3.2   Four-Wire Transmission

Carrier and radio systems require that oppositely directed portions of a single conversation occur over separate transmission channels or paths (or use mutually exclusive time periods). Thus we have two wires for the transmit path and two wires for the receive path, or a total of four wires for a full-duplex (two-way) telephone conversation. For almost all operational

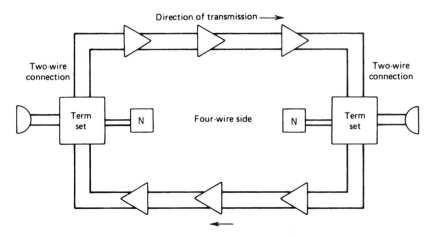

Figure 5.4. A typical long-distance (toll) telephone connection.

telephone systems, the end instrument (i.e., the telephone subset) is connected to its intervening network on a two-wire basis.

Nearly all long-distance (toll) telephone connections traverse four-wire links. From the near-end user the connection to the long-distance network is two-wire or via a two-wire link. Likewise, the far-end user is also connected to the long-distance (toll) network via a two-wire link. Such a long-distance connection is shown in Figure 5.4. Schematically, the four-wire inter-connection is shown as if it were a single-channel wire-line with amplifiers; however, it would more likely be a multichannel carrier on cable and/or multiplex on radio. However, the amplifiers in Figure 5.4 serve to convey the ideas that this section considers. As shown in Figure 5.4, conversion from two-wire to four-wire operation is carried out by a terminating set, more commonly referred to in the industry as a *term set*, which contains a four-port balanced transformer (a hybrid) or, less commonly, a resistive net-work.

3.3 Operation of a Hybrid

A hybrid, in terms of telephony (at voice frequency), is a transformer. For a simplified description, a hybrid may be viewed as a power splitter with four sets of wire-pair connections. A functional block diagram of a hybrid device is shown in Figure 5.5. Two of the wire-pair connections belong to the four-wire path, which consists of a transmit pair and a receive pair. The third pair is the connection to the two-wire link that is eventually connected to the subscriber subset via one or more switches. The last wire pair of the four connects the hybrid to a resistance–capacitance balancing network, which

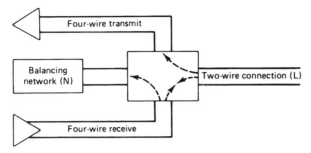

Figure 5.5. Operation of a hybrid transformer.

electrically balances the hybrid with the two-wire connection to the subscriber subset over the frequency range of the balancing network. An artificial line may be used for this purpose.

Signal energy entering from the two-wire subset connection divides equally, half of it dissipating in the impedance of the four-wire side receive path and the other half going to the four-wire side transmit path, as shown in Figure 5.5. Here the *ideal* situation is that no energy is to be dissipated by the balancing network (i.e., there is a perfect balance). The balancing network is supposed to display the characteristic impedance of the two-wire line (subscriber connection) to the hybrid. Signal energy entering from the four-wire side receive path is also split in half in the ideal situation where there is perfect balance. Half of the energy is dissipated by the balancing network (N) and half at the two-wire port (L) (see Figure 5.5).

The reader notes that in the description of the hybrid, in every case, ideally half of the signal energy entering the hybrid is used to advantage and half is dissipated or wasted. Also keep in mind that any passive device inserted in a circuit, such as a hybrid, has an insertion loss. As a rule of thumb, we say that the insertion loss of a hybrid is 0.5 dB. Thus there are two losses here that the reader must not lose sight of:

Hybrid insertion loss	0.5 dB
Hybrid dissipation loss	3.0 dB (half of the power)
	3.5 dB (total)

As far as this section is concerned, any signal passing through a hybrid suffers a 3.5-dB loss. This is a good design number for gross engineering practice. However, some hybrids used on short subscriber connections purposely have higher losses, as do special resistance-type hybrids.

In Figure 5.5 consider the balancing network (N) and the two-wire side of the hybrid (L). In all probability (L), the two-wire side, will connect to a subscriber through at least one switch. Thus the two-wire port on the hybrid could look into at least 10,000 possible subscriber connections—some short loops, some long loops, and other loops in poor condition. Because of fixed conditions on the four-wire side, we can pretty much depend on holding a good impedance match. Our concern under these conditions is impedance match with the two-wire side. That is the match between the compromise network (N) and the two-wire side (L).

We measure the capability of impedance match by *return loss*. In this particular case we call it *balance return loss*:

$$\text{Balance return loss}_{dB} = 20 \log_{10} \frac{Z_L + Z_N}{Z_L - Z_N}$$

Let's say, for argument's sake, that we have a perfect match. In other words the impedance of the two-wire subscriber loop side on this particular call was exactly 900 Ω and the balancing network (N) was 900 Ω. Substitute these numbers in the formula above and we get

$$\text{Balance return loss}_{dB} = 20 \log \frac{900 + 900}{900 - 900}$$

Examine the denominator. It is zero. Any number divided by zero is infinity. Thus we have an infinitely high return loss. And this happens when we have a perfect match, an ideal condition. Of course it is seldom realized in real life.

In real life we find that the balance return loss for a large population of hybrids connected in service and serving a large population of two-wire users has a median more of the order of 11 dB with a standard deviation of 3 dB (Ref. 4). This is valid for North America. For some other areas of the world, balance return loss median may be lower with a larger standard deviation.

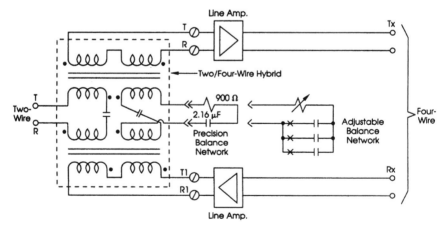

Figure 5.6. Block diagram of two-wire-to-four-wire conversion using a hybrid. From Figure 5-9, page 104, Ref. 13. Reprinted with permission of IEEE Press.

When return loss becomes low (i.e., there is a poor impedance match), there is a reflection of the speech signal. This is echo. We define the cause of echo as any impedance mismatch in the network. Most commonly this mismatch occurs at the hybrid. Echo that is excessive becomes singing. This is caused by high positive feedback on intervening amplifiers. Singing is a highly undesirable impairment. We discuss the control of echo and singing in Chapter 6. Figure 5.6 is a block diagram of a hybrid circuit.

4 MULTIPLEXING

4.1 General

Reference 1 defines multiplexing as the combining of two or more signals into a single wave from which the signals can be individually recovered. In our case the signals are voice channels and we can literally combine more than 1000 such channels for transmission over a medium. Of course the medium has to be able to accommodate the required bandwidth. On a wire pair we might combine 24, 30, or 48 channels. On LOS microwave we can commonly carry 1800 analog channels or hundreds of digital channels. On fiber optics we can carry thousands of digital channels.

There are essentially two generic ways we can multiplex voice channels:

1. In the frequency domain using frequency division multiplex (FDM)
2. In the time domain using time division multiplex (TDM).

We provide a brief review of FDM in this chapter. TDM, using pulse-code modulation (PCM), is discussed in Chapter 8.

4.2 Frequency Division Multiplex

4.2.1 General. Frequency division multiplex (FDM) is a method of allotting a unique band of frequencies in a comparatively wide band-frequency spectrum of the transmission medium to each communication channel on a continuous time basis. The communication channel may be a voice channel 4 kHz wide, a 120-Hz telegraph channel (for example), a 15-kHz broadcast channel, a 48-kHz data channel, or a 4.2-MHz television channel.

Before launching into multiplexing, keep in mind that all multiplex systems work on a four-wire basis. The transmit and receive paths are separate (see Figure 5.7). For the discussion that follows, it is assumed that the reader has some background on how an SSB (single sideband) signal is developed and how a carrier is suppressed in SSBSC systems.

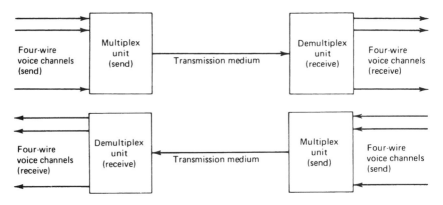

Figure 5.7. Simplified block diagram of a frequency division multiplex link.

4.3 Mixing

The heterodyning or mixing of signals of frequencies A and B is shown in the following diagram. What frequencies may be found at the output of the mixer?

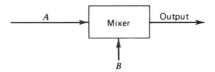

Both of the original signals will be present, as well as signals representing their sum and difference in the frequency domain. Thus signals of frequency A, B, $A + B$, and $A - B$ will be present at the output of the mixer.* Such a mixing process is repeated many times in FDM equipment.

Let us look at the boundaries of the nominal 4-kHz voice channel. These are 300 Hz and 3400 Hz. Let us further consider these frequencies as simple tones of 300 Hz and 3400 Hz. Now consider the mixer below and examine the possibilities at its output. First, the output may be summed, such that

20,000 Hz	20,000 Hz
+ 300 Hz	+3,400 Hz
20,300 Hz	23,400 Hz

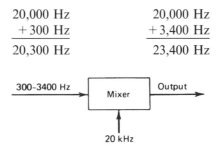

*Of course, if the mixer is a balanced mixer, only signals $A + B$ and $A - B$ will be present.

A simple low-pass filter could filter out all frequencies below 20,300 Hz. Now imagine that instead of two frequencies we have a continuous spectrum of frequencies between 300 Hz and 3400 Hz (i.e., we have the CCITT voice channel). We represent the spectrum as a triangle:

As a result of the mixing process (translation), we have another triangle:

When we take the sum, as we did previously, and filter out all other frequencies, we say we have selected the upper sideband. Thus we have a triangle facing to the right, termed an *upright* or *erect* sideband.

We can also take the difference, such that

$$
\begin{array}{ll}
\quad 20{,}000 \text{ Hz} & \quad 20{,}000 \text{ Hz} \\
\underline{- \ \ 300 \text{ Hz}} & \underline{-3{,}400 \text{ Hz}} \\
\quad 19{,}700 \text{ Hz} & \quad 16{,}600 \text{ Hz}
\end{array}
$$

and we see that in the translation (mixing process) we have had an inversion of frequencies. The higher frequencies of the voice channel become the lower frequencies of the translated spectrum, and the reverse occurs when the difference is taken. We represent this by a right triangle facing the other direction (left):

16,600   19,700 Hz

This is called an *inverted sideband*. To review, when we take the sum, we get an erect sideband. When we take the difference in the mixing process, frequencies invert and we have an inverted sideband represented by a triangle facing left.

4.4 CCITT Modulation Plan

4.4.1 Introduction. A modulation plan sets forth the development of a band of frequencies called the *line frequency* (i.e., ready for transmission on the line or other transmission medium). The modulation plan usually is a diagram showing the necessary mixing, local oscillator insertion frequencies, and the sidebands selected by means of the triangles described previously, in a step-by-step process from voice-channel input to line-frequency output. The CCITT has recommended a standardized modulation plan with a common terminology allowing large telephone networks, in both national and multinational systems, to interconnect. In the paragraphs that follow, the reader is advised to be careful with terminology.

4.4.2 Formation of Standard CCITT Group. The standard *group* as defined by CCITT occupies the frequency band 60–108 kHz and contains 12 voice channels. Each voice channel is the nominal 4-kHz voice channel occupying the 300- to 3400-Hz spectrum. The group is formed by mixing each of the 12 voice channels with a particular carrier frequency associated with the channel. Lower sidebands are then selected. Figure 5.8 shows the basic (and preferred) approach to the formation of the standard CCITT group. It should be noted that the 60- to 108-kHz band voice channel 1 occupies the highest frequency segment by convention, between 104 kHz and 108 kHz. The layout of the standard group is shown in Figure 5.9. For more information refer to CCITT Rec. G.232.

Single sideband suppressed carrier (SSBSC) modulation techniques are used in all cases. The CCITT recommends that the carrier leak be down to at least -26 dBm0, referred to the zero test level point (0 TLP) (see Section 2.4 of this chapter).

4.4.3 Formation of Standard CCITT Supergroup. A supergroup contains five standard CCITT groups, equivalent to 60 voice channels. The standard supergroup occupies the frequency band 312–552 kHz. Each group making up the supergroup is translated in frequency to the supergroup band by mixing with the proper carrier frequency. The carrier frequencies are 420 kHz for group 1, 468 kHz for group 2, 516 kHz for group 3, 564 kHz for group 4, and 612 kHz for group 5. In the mixing process the difference is taken (lower sidebands are selected). This translation process is shown in Figure 5.10.

4.4.4 Formation of Standard CCITT Basic Mastergroup and Supermastergroup. The basic mastergroup contains five supergroups (300 voice channels) and occupies the spectrum 812–2044 kHz. It is formed by translating the five standard supergroups, each occupying the 312- to 552-kHz band, by a process similar to that used to form the supergroup from five standard CCITT groups. This process is shown in Figure 5.11.

The basic supermastergroup contains three mastergroups and occupies the band 8516–12,388 kHz. The formation of the supermastergroup is shown in Figure 5.12.

4.4.5 Line Frequency. The band of frequencies that the multiplex applies to the line, whether the line is a radiolink, coaxial cable, wire pair, or open

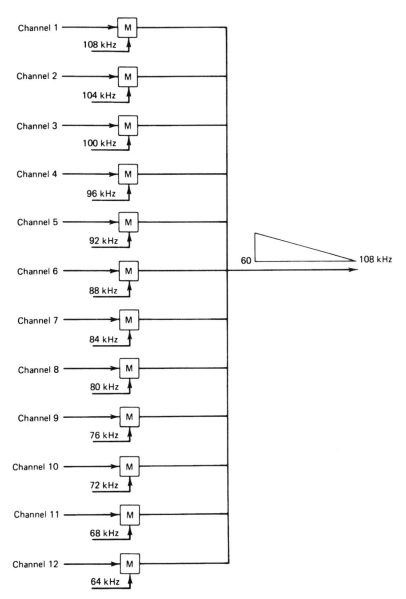

Figure 5.8. Formation of standard CCITT group.

wire, is called the *line frequency*. Another expression is HF (high frequency), which is ambiguous and often confused with HF radio transmission.

The line frequency in any one particular case may be just the direct application of a group or supergroup to the line. However, a final translation stage more commonly occurs, particularly in high-density systems. One of the

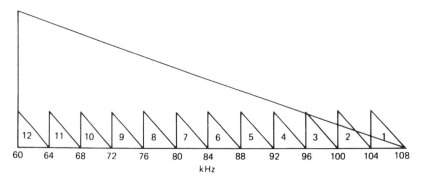

Figure 5.9. Layout of standard CCITT group.

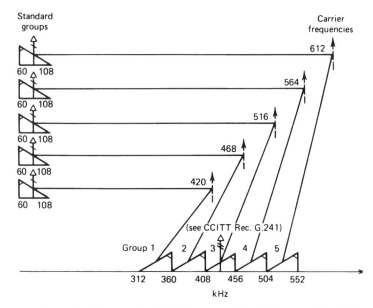

Figure 5.10. Formation of the standard CCITT supergroup. Vertical arrows with solid lines are level-regulating pilot tones; arrows with dashed lines are translation carrier frequencies. Courtesy of the International Telecommunication Union–CCITT [6].

recognized CCITT configurations is shown in Figure 5.13, namely, the standard CCITT 15-supergroup assembly.

4.4.6 North American L-Carrier. "L-Carrier" is the generic name given by the Bell System of North America to their long-haul SSB carrier system. Its development of the basic group and supergroup assemblies is essentially the same as that of the CCITT described previously. There is a variance in levels and pilot tone frequencies. The basic mastergroup differs, however. It consists of 600 VF channels (i.e., 10 standard supergroups). The L600

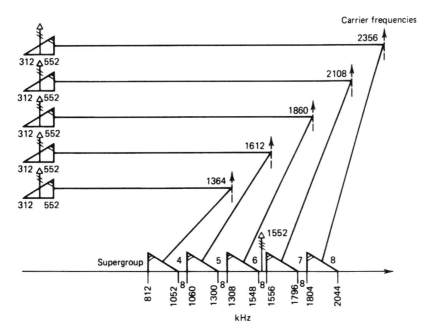

Figure 5.11. Formation of standard CCITT mastergroup. Courtesy of the International Telecommunication Union—CCITT [6].

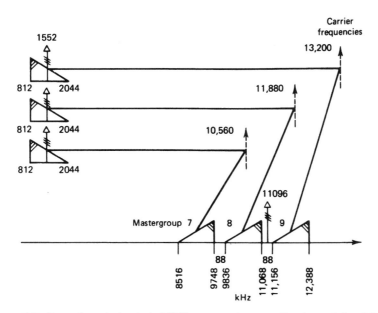

Figure 5.12. Formation of standard CCITT supermastergroup. Courtesy of the International Telecommunication Union—CCITT [6].

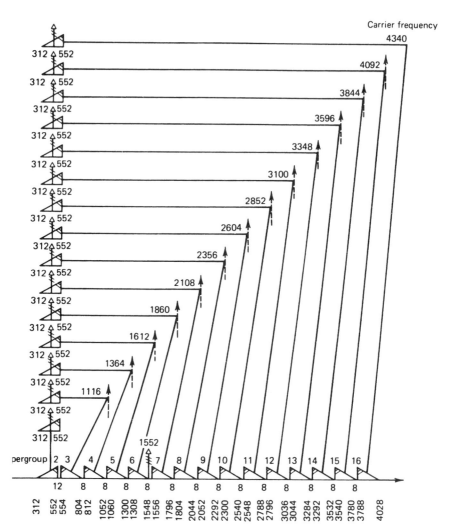

Figure 5.13. Makeup of basic CCITT 15-supergroup assembly. Courtesy of the International Telecommunication Union–CCITT [6].

configuration occupies the band 60–2788 kHz, and the U600 configuration occupies the band 564–3084 kHz. The relevant mastergroup assemblies are shown in Figure 5.14. The Bell System (ATT) identifies specific long-haul line-frequency configurations by adding a simple number after the letter "L." For example, the L3 carrier, which is used on coaxial cable and one type of ATT microwave (TH), has three mastergroups (see Figure 5.14) plus one supergroup, comprising 1800 VF channels on line occupying the band 312–8284 kHz. Table 5.1 compares L-Carrier with CCITT carrier (FDM) standards. Subsequent paragraphs discuss the parameters and their meanings [12].

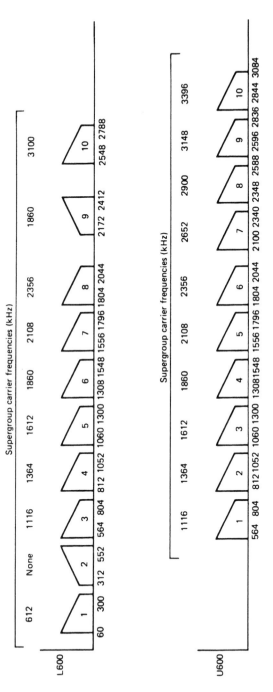

Figure 5.14. L-Carrier mastergroup assemblies (all frequencies in kilohertz). Copyright © Bell Telephone Laboratories, 1964 [12].

TABLE 5.1   L-Carrier and CCITT Comparison Table

Item	ATT L-Carrier	CCITT
Level		
Group		
Transmit	−42 dBm	−37 dBm
Receive	−5 dBm	−8 dBm
Supergroup		
Transmit	−25 dBm	−35 dBm
Receive	−25 dBm	−30 dBm
Impedance		
Group	130 Ω balanced	75 Ω unbalanced
Supergroup	75 Ω unbalanced	75 Ω unbalanced
VF channel	200–3300 Hz	300–3400 Hz
Response	+1.0 to −1.0 dB	+0.9 to −3.5 dB
Channel carrier		
Levels	0 dBm	Not specified
Impedances	130 Ω balanced	Not specified
Signaling	2600 Hz in-band	3825 Hz out-of-band
Group pilot		
Frequencies	92 or 104.08 kHz	84.08 kHz
Relative levels	−20 dBm0	−20 dBm0
Supergroup carrier		
Levels	+19.0 dBm per mod or demod	Not specified
Impedances (ohm)	75 unbalanced	Not specified
Supergroup pilot frequency	315.92 kHz	411.92 kHz
Relative supergroup pilot levels	−20 dBm0	−20 dBm0
Frequency synchronization	yes, 64 kHz	Not specified
Line pilot frequency	64 kHz	308, 12,435 kHz
Relative line pilot level	−14 dBm0	−10 dBm0
Regulation		
Group	Yes	Yes
Supergroup	Yes	Yes

4.5   Loading of Multichannel Frequency-Division-Multiplex Systems

4.5.1 Introduction. Most of the FDM (carrier) equipment in use today carries speech traffic, which is sometimes misnamed "message traffic" in North America. In this context we refer to full-duplex conversations by telephone between two "talkers." However, the reader should remember that there is a marked increase in the use of these same intervening talker facilities for the transmission of data and facsimile.

For this discussion, the problem essentially concerns human speech and how multiple users may load a carrier system. If we load a carrier system too heavily—if the input levels are too high—IM noise and crosstalk would become intolerable, eventually leading to the breakdown of the system. If we do not load the system sufficiently, the signal-to-noise ratio will suffer. The problem is fairly complex because speech amplitude varies:

1. With talker volume.

2. At a syllabic rate.

3. At an audio rate.

4. With varying circuit losses as different loops and trunks are switched into the same channel bank voice-channel input.

4.5.2 Loading. For the loading of multichannel FDM systems, CCITT Rec. G.223 [6] recommends the following:

> It will be assumed for the calculation of intermodulation below the overload point that the multiplex signal during the busy hour can be represented by a uniform spectrum [of] random noise signal, the mean absolute power level of which, at a zero relative level point (in dBm0), is given by

$$P_{av} = -15 + 10 \log N* \qquad [5.5]$$

> when N is equal to or greater than 240 and

$$P_{av} = -1 + 4 \log N \qquad [5.6]$$

> when N is equal to or greater than 12 and less than 240 [where N is the number of voice channels (all logs to the base 10)].

4.5.3 Single-Channel Loading. Many telephone administrations have attempted to standardize on -16 dBm0 for single-channel speech input to multichannel FDM equipment. With this input, peaks in speech level may reach -3 dBm0. Tests indicate that such peaks will not be exceeded more than 1% of the time. However, the conventional value of average power per voice channel allowed by the CCITT is -15 dBm0 (see CCITT Rec. G.223). This assumes a standard deviation of 5.8 dB and the traditional activity factor of 0.25 (see Section 4.5.5). Average talker level is assumed to be at -11.5 VU (volume units). We must turn to the use of standard deviation because we are dealing with talker levels that vary with each talker and thus with the mean or average.

4.5.4 Loading with Constant-Amplitude Signals. Speech on multi-channel systems has a low duty cycle or activity factor. We exemplify a simplified derivation of the traditional activity factor of 0.25%. Certain other types of signals transmitted over multichannel equipment have an activity factor of 1. This means that they are transmitted continuously, or continuously over fixed time frames. They are also usually characterized by constant amplitude. The following are some examples of these typesof signal:

---

*For North American practice, use $-16 + 10 \log N$.

- Telegraph tone or tones.
- Signaling tone or tones.
- Pilot tones.
- Data signals [particularly PSK (phase shift keying) and FSK (frequency shift keying) modulation].
- Facsimile (in digital or FM mode).

To provide a safety factor on overload, the tendency would be to reduce level. Here again, if we reduce level too much, the signal-to-noise ratio and hence the error rate will suffer.

Using the loading formulas in Section 4.5.2 may permit the use of several data or telegraph channels without serious consequence. But if any wide use is to be made of the system for data-telegraph-facsimile, then other loading criteria must be used. For typical constant-amplitude signals, traditional (CCITT) transmit levels as seen at the input of the channel modulator of FDM carrier equipment are as follows:

- Data: -13 dBm0.
- Signaling [single-frequency supervision, tone on when idle (see Chapter 4)]: -20 dBm0.
- Composite telegraph: -8.7 dBm0.

For one FDM system now on the market with 75% speech loading and 25% data loading with more than 240 channels (total), the manufacturer recommends the following:

$$P_{\text{rms}} = -11 + 10 \log N \qquad (5.7)$$

(units in dBm0) using -5 dBm0 per channel for data-input levels and -8 dBm0 for the composite telegraph signal level. However, the manufacturer hastens to mention that all VF channels may be loaded with data signals (or telegraph) at -8 dBm0 level per channel. But for the -5-dBm0 signal level only two channels per group may be assigned to this level, with the remainder speech, or the group must be "deloaded," which means that a certain number of channels must remain idle.

4.5.5 Activity Factor. Speech is characterized by large variations in amplitude, ranging from 30 dB to 50 dB. We often use the VU (volume unit) to measure speech levels. The VU can be equated to the dBm for a simple sine wave in the VF range across 600 Ω. For a complex signal such as our speech signal, the average power measured in dBm of a typical single talker is

$$P_{\text{dBm}} = V_{\text{VU}} - 1.4 \qquad (5.8)$$

or that a 0-VU talker has an average power of -1.4 dBm. Empirically, the peak power is about 18.6 dB higher than average power for a typical talker. The peakiness of speech level means that carrier (FDM) equipment must be operated at a low average power to withstand voice peaks to avoid overloading and distortion. These can be related to an activity factor T_a, which is defined as that proportion of the time that the rectified speech envelope exceeds some threshold. If the threshold is about 20 dB below the average power, the activity dependence on threshold is fairly weak. The preceding equation for average talker power can now be rewritten in relation to the activity factor as follows:

$$P_{dBm} = V_{VU} + 10 \log T_a \qquad (5.9)$$

If $T_a = 0.725$, the results will be the same for the equation relating VU to dBm.

Now consider adding a second talker at a different frequency segment on the same equipment, but independent of the first talker. Of course, we are describing here the operation of FDM equipment discussed previously. With the second talker added, the system average power will increase 3 dB as expected. If we have N talkers, each on a different frequency segment, the average power developed will be

$$P_{dBm} = V_{VU} - 1.4 + 10 \log N \qquad (5.10)$$

where P_{dBm} is the power developed across the frequency band occupied by all talkers.

Empirically, it has been found that the peakiness or peak factor of many talkers over a multichannel analog system reaches the characteristics of random noise peaks when the number of talkers, N, exceeds 64. When $N = 2$, the peaking factor is 18 dB; when $N = 10$, it is 16 dB; when $N = 50$, it is 14 dB; and so forth.

An activity factor—a term analogous to a "duty cycle" of 1—which we have been using in the preceding argument is quite unrealistic. If it is 1, it means that somebody is talking on each channel all the time. The traditional figure for activity factor accepted by the CCITT and used in North American practice is 0.25. Follow the argument in the next paragraph on how to reach this lower figure.

The multichannel FDM equipment cannot be designed for N callers and no more. If this were true, a new call would have to be initiated every time a call terminated, or calls would have to be turned away because of congestion. Thus the equipment must be "overdimensioned" for BH service. For this dimensioning problem, we drop the activity factor from 1 to 0.70. Other causes will reduce this figure even more. For instance, circuits are essentially inactive during call setup as well as during pauses for thinking during a conversation. The 0.70 figure now must be dropped to 0.50. This latter figure is divided in half because of the talk–listen condition. If we disregard cases of "double-talking," it is obvious that while one end is talking on a full-duplex telephone circuit, the other is listening. Thus a circuit (in one direction) is idle half the time during the "listen" period. The resulting activity factor is 0.25. (Also consult Ref. 7.)

4.6 Pilot Tones

In FDM equipment pilot tones are used primarily for level regulation and secondarily for fault alarms. The nature of speech, particularly in varying amplitude, makes it a poor prospect as a reference for level control. Ideally, simple single-sinusoid constant-amplitude signals with 100% duty cycles provide simple control information for level-regulating equipment. Frequency division multiplex level regulators operate very much in the same manner as automatic-gain control circuits on radio systems, except that their dynamic range is considerably smaller. Modern FDM carrier systems initiate a level-regulating pilot tone on each group at the transmit end of the system. Individual level-regulating pilots are also inserted on all supergroups and mastergroups. The intent is to regulate system level within ±0.5 dB.

Pilots are assigned frequencies that are part of the transmitted spectrum yet do not interfere with voice-channel operation. They usually are assigned a frequency appearing in the guard band between voice channels or are residual carriers (i.e., partially suppressed carriers). The CCITT recommends one of the following as group-regulation pilots:

84.080 kHz (at a level of −20 dBm0)
84.140 kHz (at a level of −25 dBm0)

The Defense Communications Agency of the U.S. Department of Defense recommends 104.08 kHz for group regulation and alarm. For CCITT group pilots, the maximum level of interference permissible in the voice channel is −73 dBm0p. CCITT pilot filters have a bandwidth at the 3-dB points of 50 Hz (see CCITT Rec. G.241 for further information on other CCITT pilot frequencies and levels). The operating range of level control equipment activated by pilot tones is usually about ±4 or ±5 dB. If the incoming level of a pilot tone in the multiplex receive equipment drops outside the level-regulating range, an alarm will be indicated (if such an alarm is included in the system design); CCITT Rec. G.241 suggests such an alarm when the incoming level varies 4 dB up or down from the nominal.

4.7 Noise and Noise Calculations in FDM Carrier Systems

4.7.1 General. Carrier equipment is the principal contributor of noise on coaxial cable systems and other metallic transmission media. On radiolinks (line-of-sight microwave) it makes up about 25% of the total noise on the overall system link. The traditional approach is to consider noise with respect to a hypothetical reference circuit. Several methods are possible, depending on the application. These include the CCITT method, which is based on a 2500-km hypothetical reference circuit, and the U.S. Department of Defense method used in specifying communication systems. Such military systems are based on a 6000-nautical-mile reference circuit with 1000-mi links and 333-mi sections.

4.7.2 CCITT Approach. CCITT Rec. G.222 [6] states the following:

The mean psophometric* power, which corresponds to the noise produced by all modulating (multiplex) equipment ..., shall not exceed 2500 pW† at a zero relative level point. This value of power refers to the whole of the noise due to various causes (thermal, intermodulation, crosstalk, power supplies, etc.). Its allocation between various equipments can be to a certain extent left to the discretion of design engineers. However, to ensure a measurement agreement in the allocation chosen by different administrations, the following values are given as a guide to the target values:

for 1 pair of channel modulators	200 pW
for 1 pair of group modulators	80 pW
for 1 pair of supergroup modulators	60 pW

The following values are recommended on a provisional basis:

for 1 pair of mastergroup modulators	60 pW
for 1 pair of supermastergroup modulators	60 pW
for 1 pair of 15-supergroup assembly modulators	60 pW

Experience has shown that these target figures can often be improved considerably. The CCITT has purposely loosened the noise value allotted to channel modulators so that other modulation schemes from voice channel to group can be used rather than the direct modulation approach that we have described here. For instance, one solid-state FDM equipment now on the market, when operated with CCITT loading, has the following characteristics:

1 pair of channel modulators	31 pWp‡
1 pair of group modulators	50 pWp
1 pair of supergroup modulators	50 pWp

Using another solid-state equipment and increasing the loading to 75% voice, 17% telegraph tones, and 8% data, the following noise information is applicable:

---

*See Section 5 of this chapter.

†1 pW = 1×10^{-9} mW.

‡Picowatts psophometrically weighted. See Section 5 of this chapter.

1 pair of channel modulators	322 pWp
1 pair of group modulators	100 pWp
1 pair of supergroup modulators	63 pWp

If this equipment is used on a real circuit with heavier loading, the sum for noise for channel modulators, group modulators, and supergroup modulator pairs is 485 pWp. Thus by simple arithmetic we see that such a system would be permitted to demodulate to voice only five times over a 2500-km route if we were to adhere to CCITT recommendations (i.e., $5 \times 485 = 2425$ pWp). This leads to the use of through-group and through-supergroup techniques discussed in Section 4.7. Figure 5.15 shows a typical application of this same equipment using CCITT loading.

5 SHAPING OF A VOICE CHANNEL AND ITS MEANING IN NOISE UNITS

The attenuation distortion or frequency response of a voice channel is termed "flat" because when it is tested from input to output, the response from, say, 300 Hz to 3400 Hz may vary by only several decibels. Of course, from the input of a voice channel modulator of an FDM carrier link to the output of the companion demodulator, flat response, which is indeed what we have, should vary no more than perhaps ±0.5 dB.

Now connect a telephone handset transmitter (with appropriate talk battery) to the input of the voice channel modulator and handset receiver to the output of the companion voice channel demodulator and include the acuity of the "average" human ear. Now we see a "shaping" effect. The handset and the ear acuity "shape" the channel when the audio level is compared at various frequencies at the input of the acoustical–electrical transducer (handset transmitter) to the audio output from the electrical–acoustical transducer (the handset receiver).*

The frequency response measurement uses a reference frequency located at the point of minimum attenuation in the voice channel. In North America the reference frequency is 1000 Hz, whereas in Europe and many other locations in the world it is 800 Hz. All CCITT recommendations dealing with the voice channel use 800 Hz as reference.

Considering now that shaping effect we just mentioned combining the mouth, handset, and ear, we see that for speech transmission, certain frequencies in the voice channel are attenuated more than others. When

---

*The acoustical properties of the typical human ear must also be taken into account because the design of a weighting network, using a particular handset, is based on the annoyance factor of a certain single-frequency tone in the voice channel compared to the reference frequency.

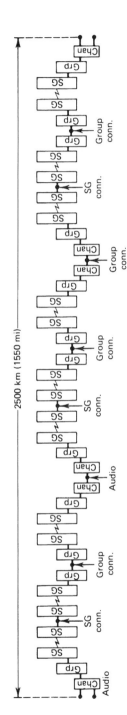

2500 km (1550 mi)

Figure 5.15. Typical CCITT reference-system noise calculations. These calculations are for typical equipment using CCITT loading. Courtesy of GTE Lenkurt, Inc., San Carlos, California.

Multiplex

		pWp0	dBrnC0
3 complete sets of channel bank equipment (XMT and REC)	at 110 pW	330	26
6 complete sets of group bank equipment	at 45 pW	270	25
9 complete sets of supergroup bank equipment	at 58 pW	522	28
3 sets of group connector equipment	at 5 pW	15	12.6
3 sets of supergroup connector equipment	at 16 pW48	17.6	
Total 46A3 Carrier Noise		1185	31.5

Microwave

		pWp0	dBrnC0
9 complete radio sections of 6 hops each			
78A3 radio (600 ch.) at 500 pWp0/section		4500	37.3
75A3 radio (960 ch.) at 400 pWp0/section		3600	36.4

Overall System Noise

Using 78A3 radio (600 channels)	(4500 + 1185)	5685
Using 75A3 radio (960 channels)	(3600 + 1185)	4785

Note: Conversion from pWp0 to dBrnC0 is $dBrnC0 = (10 \log pWp0) + 0.8$.
Note: Calculations are for a typical equipment with CCITT loading.
Courtesy GTE Lenkurt Incorporated, San Carlos, California.

transmission systems are tested, *weighting networks* are used to simulate these effects. There are two types of weighting networks now in use:

- C-message used in North America
- Psophometric used in the rest of the world, and is starting to be accepted in North America

Let's look at this weighting from another viewpoint. Admit that noise we hear is annoying. Let the noise be a single-sinusoid tone from an audio signal generator. Pass this tone through the microphone on a standard handset and have someone listen on the earpiece. Adjust the signal generator to 1000 Hz, the reference frequency, and adjust the level to a point where the listener is annoyed. Read the level. Let's say the level is 0 dBm. Now set the signal generator to 300 Hz and increase the level until the point of same annoyance is reached. It will be noted that we had to increase the level to about + 20 dBm for equal annoyance (at 1000 Hz). We can adjust the generator to 3000 Hz, and it will require about + 3 dBm for equal annoyance. We can draw a curve for frequencies inside the voice channel which we could call an *annoyance curve*. Such a curve is shown in Figure 5.16. It is the C-message weighting curve. A similar curve can be drawn for psophometric weighting. It, too, is shown in Figure 5.16.

Figure 5.17 shows the C-message curve with the noise advantage. This is the hatched area between 300 and 3400 Hz. If we are to use a circuit for

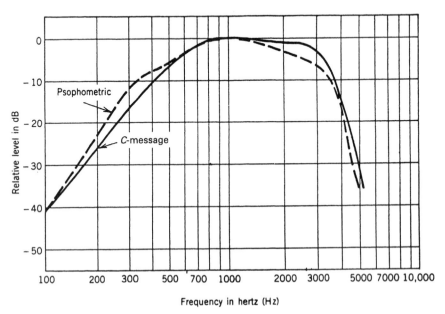

Figure 5.16. Line weighting curves for telephone channel noise.

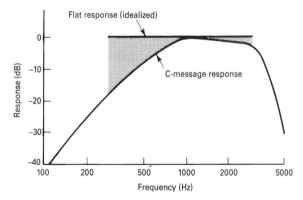

Figure 5.17. C-message weighting curve showing how we achieve about a 2.0-dB noise advantage with speech transmission over a voice channel. The hatched area shows the noise advantage.

speech, the noise in the hatched area is not counted against use, but is instead a noise advantage, about 2.5 dB of advantage for psophometric weighting and about 2.0 dB for C-message weighting.

The noise measurement unit with C-message weighting is dBrnC, and when referenced to the 0-TLP, it is dBrnC0. dBrn might be called decibels reference noise. For psophometric weighting, the most common unit for noise measurement is the pWp and it is a linear unit. pWp stands for picowatts psophometrically weighted. $pW = 1 \times 10^{-12}$ watts or 1×10^{-9} mW. If we wish to use a dB unit, in telephony we'd opt for dBmp or dBm psophometrically weighted. With psophometric weighting, the reference frequency is 800 Hz; with C-message it is 1000 Hz.

If we are to use this same circuit (voice channel) for data transmission, the noise advantage goes away, and we must consider the channel as flat as shown in Figure 5.17.

The reader must be certain to differentiate between a weighted voice channel and a flat channel. In the case of a weighted channel, noise measurements are weighted because the eventual user of the channel is the human mouth–ear combination along with the associated handset as the transducer of electrical energy to speech or audio energy. The flat channel, on the other hand, has a comparatively flat frequency response across the voice channel, and we generally define the voice channel as occupying the spectrum from 300 to 3400 Hz. Data, facsimile and video* transmission systems utilize "flat" channels, that is, where the amplitude–frequency response varies little between 300 and 3400 Hz—say, of the order of ±0.5 dB or better. Of course, on long networks with numerous points of modulation–demodulation, the amplitude–frequency response can vary considerably more, but we still refer to the voice channel as being flat.

*Under certain conditions, video signal-to-noise ratios are weighted [i.e., TASO (Television Allocations Studies Organization) weightings].

To convert from flat channel noise measurements in pW or dBm, the following excerpt from CCITT Rec. G.223 may be useful:

If uniform-spectrum random noise is measured in a 3.1 kHz band with flat attenuation frequency characteristic, the noise level must be reduced 2.5 dB to obtain the psophometric power level. For another bandwidth, B, the weighting factor will be equal to:

$$2.5 + 10 \log \left(\frac{B}{3.1} \right) (\text{dB})$$

When $B = 4$ kHz, for example, this formula gives a weighting factor of 3.6 dB.

The following relationships will be useful for converting from one noise unit to another:

$$-90 \text{ dBm} = -2 \text{ dBrnC and thus } -92 \text{ dBm} = 0 \text{ dBrnC}$$
$$(\text{white noise loading})$$
$$-92.5 \text{ dBmp} = -90 \text{ dBm (flat, white noise)}$$
$$1 \text{ pWp} = -90 \text{ dBmp}$$
$$\text{value in dBm} = 10 \log (\text{value in pWp} \times 10^{-9}) + 2.5 \text{ dB}$$
$$\text{dBrnC} = 10 (\log \text{pWp} \times 10^{-9}) - 0.5 \text{ dB} + 90 \text{ dB}$$
$$\text{value in pW} \times 0.56 = \text{value in pWp}$$
$$\text{value in pWp}/0.56 = \text{value in pW}$$

(Material based on Ref. 5.)

In telecommunications and in other areas of electronics, as a general rule a passband is defined between 3-dB points. Remember that if power is dropped 3 dB, it is cut in half. The voice channel is the one exception to this rule. The voice channel in this context uses 10-dB points or where the power drops to 1/10th.

REVIEW QUESTIONS

1. There are four basic transmission parameters for a telephone voice channel, three of which are impairments. Name and define each.

2. Define the specified CCITT voice channel.

3. What are the two reference frequencies for the voice channel? One is European and accepted by CCITT and the other is North American.

4. The reference test-tone input to a telephone channel is −16 dBm and the output is +7 dBm. What range of values in dBm can be expected at the output if the attenuation distortion is −1 + 2.5 dB from 600 Hz to 2600 Hz?

5. What causes phase distortion?

6. What test-tone level should one expect at 0 TLP? A value of −10 dBm at the −12-dBr point corresponds to what dBm0 value?

7. Name the four basic categories of noise.

8. What thermal noise level would one expect with a receiver with 1000 K noise temperature and a bandwidth of 1 kHz?

9. Two frequencies of 1000 Hz and 1200 Hz appear at the input of a nonlinear device. Give values of third-order products.

10. Give at least two causes of intermodulation products.

11. Impulse noise generally does not affect speech telephony. What can it affect seriously?

12. The signal level coming out of a receiver is +7 dBm and the thermal noise level is −35 dBm. What is the signal-to-noise ratio in this case?

13. Define two-wire and four-wire transmission. Where would each be applied?

14. What are the two loss components across a model hybrid? In each case explain what causes the loss.

15. Channel 1 of the standard CCITT FDM group has an injection local oscillator carrier frequency of 108 kHz. What band does channel 1 occupy (3-dB points)?

16. What frequency band does the CCITT group occupy? How many VF channels does it accommodate?

17. What is the standard activity factor for FDM telephony?

18. Considering question 17, what is the problem when voice channels carry conventional data signals on an FDM system? Discuss.

19. When an FDM system is loaded too heavily (i.e., levels are too high), what will result? It can be disastrous.

20. What are the primary and secondary functions of pilot tones on an FDM system?

21. What are the standard North American and CCITT test-tone inputs on an FDM VF channel?

22. What signal level might we expect in the subscriber plant? In the long-distance plant? Give the answer in dBm.

23. What is the problem measuring signal-to-noise ratio for a speech signal? How might we overcome that problem?

24. C-message weighted noise uses what noise measurement units?

25. Express 1 pWp in watts and milliwatts. Use the powers of 10.

26. What is the power level in dBm of 200 pWp? Give the same level in dBmp.

REFERENCES

1. *The New IEEE Standard Dictionary of Electrical and Electronic Terms*, 5th ed., IEEE Press, New York, 1992.

2. International Telephone and Telegraph Corporation, *Reference Data for Radio Engineers*, 6th ed., Howard W. Sams, Indianapolis, 1976.

3. *The Lenkurt Demodulator*, Lenkurt Electric Co., San Carlos, CA, December 1964, June 1965, and September 1965.

4. *Transmission Systems for Communications*, 5th ed., Bell Telephone Laboratories, 1982.

5. H. H. Smith, *Noise Transmission Level Terms in American and International Practice*, ITT Communication Systems, Paramus, NJ, 1964.

6. *CCITT G. Recommendations*, Fascicles III.1 and III.2, IXth Plenary Assembly, Melbourne, 1988.

7. B. D. Holbrook and J. T. Dixon, "Load Rating Theory for Multichannel Amplifiers," *BSTJ*, 624–644. (October 1939).

8. *BOC Notes on the LEC Networks—1990*, SR-TSV-002275, Issue 1, Bellcore, Piscataway, NJ, March 1991.

9. W. Oliver, *White Noise Loading of Multichannel Communication Systems*, Marconi Instruments, St. Albans, Herts, UK, 1976.

10. R. L. Freeman, *Telecommunication Transmission Handbook*, 3rd ed., John Wiley & Sons, New York, 1991.

11. W. D. Reeve, *Subscriber Loop Signaling and Transmission Handbook—Analog*, IEEE Press, New York, 1992.

12. *Telecommunication Transmission Engineering*, Vol. 1, 2nd ed., AT&T, New York, 1977.

13. *IEEE Standard Methodologies for Specifying Voicegrade Channel Transmission Parameters and Evaluating Connection Transmission Performance for Speech Telephony*, IEEE Std 823-1989, IEEE, New York, 1989.

6

LONG-DISTANCE NETWORKS

1 GENERAL

The design of a long-distance network involves basically three considerations: (1) routing scheme given inlet and outlet points and their traffic intensities, (2) switching scheme and associated signaling, and (3) transmission plan. In the design each criterion will interact with the others. In addition, the system designer must specify type of traffic, lost-call criteria or grade of service, a survivability criterion, forecast growth, and quality of service. The trade-off of all these factors with "economy" is probably the most vital part of initial planning and downstream system design.

Consider transcontinental communications in the United States. Service is now available for people in New York to talk to people in San Francisco. From the history of this service, we have some idea of how many people wish to talk, how often, and for how long. These factors are embodied in traffic intensity and calling rate. There are also other cities on the West Coast to be served and other cities on the East Coast. In addition, there are existing traffic nodes at intermediate points such as Chicago and St. Louis. An obvious approach would be to concentrate all traffic into one transcontinental route with drops and inserts at intermediate points.

Again, we must point out that switching enhances the transmission facilities. From an economic point of view, it would be desirable to make transmission facilities (carrier, radio, and cable systems) adaptive to traffic load. These facilities taken alone are inflexible. The property of adaptivity, even when the transmission potential for it has been predesigned through redundancy, cannot be exercised except through the mechanism of switching in some form. It is switching that makes transmission adaptive.

The following requirements for switching ameliorate the weaknesses of transmission systems: Concentrate light, discretely offered traffic from a multiplicity of sources and thus enhance the utilization factor of transmission trunks; select and make connections to a statistically described distribution of destinations per source; and restore connections interrupted by internal or external disturbances, thus improving reliabilities (and survivability) from the levels on the order of 90% to 99% to levels on the order of 99% to 99.9% or better. Switching cannot carry out this task alone. Constraints have to be iterated or fed back to the transmission systems, even to the local area. The transmission system must not excessively degrade the signal to be transported; it must meet a reliability constraint expressed in MTBF (mean time between failures) and availability and must have an alternative route scheme in case of facility loss, whether switching node or trunk route. This latter may be termed *survivability* and is only partially related to overflow (e.g., alternative routing).

The single transcontinental main traffic route in the United States suggested earlier has the drawback of being highly vulnerable. Its level of survivability is poor. At least one other route would be required. Then why not route that one south to pick up drops and inserts? Reducing the concentration in the one route would result in a savings. Capital, of course, would be required for the second route. We could examine third and fourth routes to improve reliability–survivability and reduce long feeders for concentration at the expense of less centralization. In fact, with overflow, one to the other, dimensioning can be reduced without reduction of overall grade of service.

2 THE DESIGN PROBLEM

The same factors enter into long-distance network design as were discussed for the local area in Chapter 2. The first step is exchange placement. Here we follow North American practice and call an exchange in the long-distance network a "toll exchange."* Rather than base the placement decision on subscriber density and their calling rates, the basic criterion is economy, the most cost-effective optimum. Toll-center placement is discussed in Section 5 of this chapter.

Having selected toll-center locations, the design procedure is to construct the familiar traffic matrix, where cost ratio studies are carried out to determine whether routing will be direct or tandem. The direct routes are called high-usage (HU) routes. The tendency is to use tandem (or "transit" exchanges) working and direct (HU) routes with overflow. The economic decision arises to balance switching against transmission (considering our arguments in Section 1). Compare local versus long-distance networks:

---

*In European and CCITT terminology it is a transit exchange, and the toll network is often called the *transit network*.

	Switching Cost per Circuit	Transmission Cost per Circuit	Favored Network
Local network	Relatively high	Low	Mesh
Toll network	Relatively low	High	Star

In the past, for the long-distance network we could nearly always assume a hierarchical structure with three, four, or even five levels in the hierarchy. Ideally, the highest levels would be connected in mesh for survivability [1].

Our thinking has changed. We are moving away from the hierarchical concept (albeit slowly) to one using more direct routes. Vestiges of a hierarchy remain in the United States for the general network structure. However, we must also take into account structural considerations with the breakup of the Bell System and the formation of the Regional Bell Operating Companies (RBOCs). The RBOCs with their local exchange carriers (LECs) formed the lower level of this artificial hierarchy, and the interexchange (long-distance) carriers (IXCs) formed an upper level.

In local network design, particularly in metropolitan areas, we could often assume a mesh connection. There might be an exceptional case where tandem working would prove to be economical, where traffic flows were 20 erlangs or less. Because of inherent (comparative) low traffic flows in the long-distance network, a star topology can be assumed at the outset, and we would then proceed to determine cases where direct links may be justified with or without alternative routing.

3 LINK LIMITATION

CCITT Rec. Q.40, para. 3 [2], states the following:

For reasons of transmission quality and the efficient operation of signalling, it is desirable to limit as much as possible the number of circuits connected in tandem.

The apportionment between national and international circuits in such a chain may vary.

The maximum number of circuits to be used for an international call is 12 with up to a maximum of four of the circuits being international.

In exceptional cases and for a low number of calls, the total number of circuits may be 14, but even in this case the maximum number of international circuits is four.

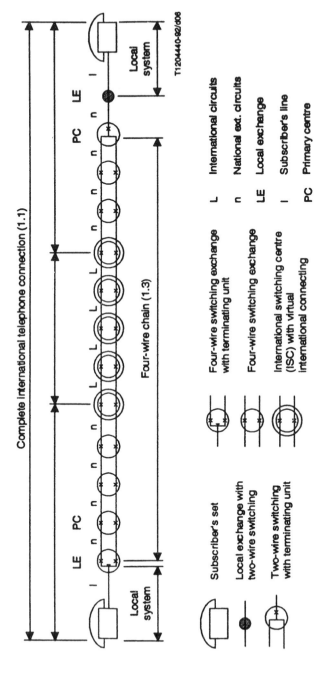

Figure 6.1. An international connection to illustrate the nomenclature adopted and the maximum number of links in tandem for an international connection (CCITT Rec. G.101). Fig. 6/G.101. Courtesy of CCITT-ITU. [2]

If 12 circuits in tandem are the absolute maximum and we subtract four for the international portion, eight are left for the national portions. We then allow four maximum for each national network. This limit is crucial in the national network design. The concept is illustrated in Figure 6.1.

4 INTERNATIONAL NETWORK

Before 1980 the CCITT routing plan was based on a network with a hierarchical structure with descending levels called CT1, CT2, CT3, and CTX. Since 1980 CCITT has made a radical change in its international routing plan. The new plan might be called a "free routing structure." It assumes that national administrations (telephone companies) will maintain national hierarch¦cal networks. Obviously the change was brought about by the long reach of satellite communications with which international high-usage (HU) trunks can terminate practically anywhere in the territory of a national administration.

The CCITT International Telephone Routing Plan is contained in CCITT Rec. E.171 [3] and is reviewed below.

In practice, the large majority of international telephone traffic is routed on direct circuits (i.e., no intermediate switching point) between international switching centers (ISCs). It should be noted that it is the rules governing routing of connections consisting of a number of circuits in tandem that this recommendation primarily addresses. These connections have an importance in the network because:

- they are used as alternate routes to carry overflow traffic in busy periods to increase network efficiency
- they can provide a degree of service protection in the event of failure of other routes
- they can facilitate network management when associated with ISCs having temporary alternative routing capabilities.

This plan replaces the previous one established in 1964. Rec. E.171 continues under "Principles":

The Plan preserves the freedom of administrations: (a) to route their originating traffic directly or via any transit administration they choose; (b) to offer transit capabilities to as wide a range of destinations as possible in accordance with the guidelines it provides . . . .

The governing features of this plan are:
(a) it is *not* hierarchical,
(b) administrations are free to offer whatever transit capabilities they wish, providing they conform to the Recommendation,

(c) direct traffic should be routed over final (fully provided) or high usage circuit groups,

(d) no more than 4 international circuits in tandem between the originating and terminating ISCs,

(e) advantage should be taken of the non-coincidence of international traffic by the use of alternative routings and provide route diversity (Rec. E.523),

(f) the routing of transit switched traffic should be planned to avoid possibility of circular routings,

(g) when a circuit group has both terrestrial and satellite circuits, the choice of routing should be governed by:

—the guidance given in (CCITT) Rec. G.114 (e.g., no more than 400 ms one-way propagation time),

—the number of satellite circuits likely to be utilized in the overall connection,

—the circuit which provides the better transmission quality and overall service quality,

(h) the inclusion of two or more satellite circuits in the same connection should be avoided in all but exceptional cases. Regarding (h), reference should be made to Annex A of Recs. E.171 and Q.14 [3 and 2].

5 EXCHANGE LOCATION

The design of the toll network is closely related to the layout of toll areas; thus the system designer would start by placing a toll exchange in each toll area, probably in or near a large city in the area. Tariffs (tolls) for long-distance calls are usually based on the crow-fly distance traversed by the call. For instance, there may be fixed charges for calls over distances 0–20 km, 20–50 km, 50–100 km, 100–200 km, 200–300 km, and so on. It is expensive to measure route distance for every call. Hence a country or telephone serving area (in the macro sense) is usually divided into tariff areas and charged the same amount for calls between two different areas, no matter what the exact origin and destination in each of the areas. Tariff areas should not be made too small, because this would result in numerous areas with costly charging and billing equipment. When tariff areas are too large, tariffs may tend to be inequitable to subscribers. An area 50 km in diameter may be used as a guide for desirable size of tariff area. Areas will be considerably larger in sparsely populated regions. After tentatively placing a toll exchange in each toll area, the system designer should then examine adjacent pairs of toll exchanges to determine whether one exchange could serve both areas.

The next step is to examine assignments of toll exchanges regarding numbering. In Chapters 3 and 4 we saw how numbering routes a call and inputs call accounting equipment or call metering. Numbering may entail consideration of more than one toll exchange in geographically large tariff areas. Another consideration is maximum size of a toll exchange. Depending on expected long-distance calling rate and holding times, we might suppose 0.003 erlangs* per subscriber line; thus a 4000-line toll exchange could serve just under a million subscribers. The exchange capacity should be dimensioned to the forecast long-distance traffic load 10 years after installation. If the system undergoes a 15% expansion in long-distance traffic volume per year, it will grow to four times its present size in 10 years. Exchange location in the toll area is not very sensitive to traffic; it is more important to make maximum use of existing plant.

Hierarchy is another essential aspect. One important criterion is establishing the number of hierarchical levels in a national network. The tendency today is to reduce the number of hierarchical levels. Factors leading to more than two levels are:

- Geographical size
- Telephone density, usually per 100 inhabitants
- Toll traffic trends
- Political factors†

The trend toward greater use of direct HU routes may also force the use of less hierarchical levels. Once the number of hierarchical levels has been established, the number of fan outs must be considered to establish the number of long-distance exchanges in the network. Fan outs of six to eight are desirable. Thus with a two-stage hierarchy with a fan out of six and then eight, there would be 48 of the lowest-level long-distance (toll) exchanges. A three-level hierarchy, using the same rules, would have $48 \times 8 = 384$, a formidable number.

As can be seen, there are many choices open to the system engineer to establish the route-plan hierarchy. For example, if there are 24 long-distance exchanges in an area, the network will initially be a star connection, either three-stage, two-stage with low (initial) fan out, or two-stage with high initial fan out. Here "fan out" refers to the highest level and works downward. Figure 6.2 illustrates these principles. Figure 6.2A shows an area with 26 exchanges and part of a larger area with perhaps 100 or more exchanges, with the principal city in the upper left-hand corner. Three choices of fan outs are shown. Figure 6.2B is a three-level hierarchy with a four-to-five fan out at each stage. For a two-level hierarchy, two possibilities are suggested. Figure 6.2C has low initial fan out and 6.2D has a high one. The choice between 6.2C and 6.2D may depend on traffic intensity between nodes or availability of routes. For national networks, the fan out in Figure 6.2D may

---

*There is a marked tendency of toll traffic growth per subscriber in highly developed countries.
†Such as the divestiture of the Bell System.

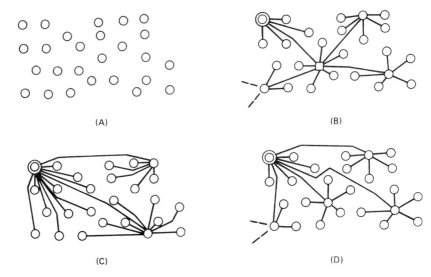

Figure 6.2. Choice of fan outs: (A) basic pattern; (B) three-stage interconnection; (C) two-stage interconnection with low initial fan out; (D) two-stage interconnection with high initial fan out.

be most economical because traffic is brought to a common point more quickly, leaving the individual branches to be least traffic efficient [4]. (Also consult Refs. 13, 14, and 15).

6   NETWORK DESIGN PROCEDURES

The attempt to attain a final design of an optimum national network is a major "cut-and-try" process. It lends itself well to computer techniques such as a program developed by ITT, "Optimization of Telephone Trunking Networks with Alternate Routing." A second program complements the series, namely, "Optimization of Telephone Networks with Hierarchical Structure" [10, 11].

In every case the design must first take into account the existing network. Major changes in that network require large expenditure. The network also represents an existing investment that should be amortized over time. Elements of the system, such as switches, are of varying age; some switches remain in service for up to 40 years, and others have been recently installed. Removal of a switch with only several years of service would not be economical. These switches also may have specific signaling characteristics* and are interconnected by a trunk network.

To simplify the design process, visualize a group of local areas. That is, the geographical and demographic area of interest in which a national network is

---

*Universal implementation of ITU-T Signaling System No. 7 is apparent. It provides a uniform signaling regime across the entire toll network.

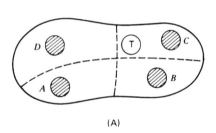
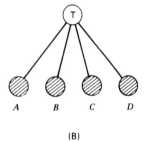

(A)   (B)

Figure 6.3. Areas and exchange relationships.

to be designed is made up of contiguous local areas (Chapter 2). There are now three bases to work from:

1. There are existing local areas, each of which has a toll exchange.
2. There is one or more ISCs placed at the top of the network hierarchy.
3. There will be no more than four links in tandem (Section 3) on any connection to reach an ISC.

Point 1 may be redefined as a toll area made up of a grouping of local areas probably coinciding with a numbering (plan) area. This is illustrated in a very simplified manner in Figure 6.3, where T, in European (CCITT) terminology, is a primary center, or a class 4 exchange in North American terminology. Center T, of course, is a tandem exchange with a fan out of four; these are four local exchanges, *A*, *B*, *C*, and *D* homing on T. The entire national geographic area will be made up of small segments, as shown in Figure 6.3, and each may be represented by a single exchange such as T.

The next step is to examine traffic flows to and from (originating and terminating at) each T. This information is organized and tabulated on a traffic matrix. A simplified example is shown in Table 6.1. Care must be taken in the preparation and subsequent use of such a table. The convention used here is that values are read *from* the exchange in the left-hand column *to* the exchange in the top row. For example, traffic from exchange 1 to exchange 5 is 23 erlangs, and traffic from exchange 5 to exchange 1 is 25 erlangs. It is often useful to set up a companion matrix of distances between exchange pairs. The matrix (Table 6.1) immediately offers candidates for high-usage routes. Nonetheless, this step is carried out after a basic hierarchical structure is established.

We recommend that a hierarchical structure be established at the outset, being fully aware that the structure may be modified or even done away with entirely in the future as dynamic routing disciplines are incorporated (see Section 7). At the top of a country's hierarchy is (are) the international switching center(s). The next level down, as a minimum, would be the long-distance network, thence down to a local network consisting of local serving

TABLE 6.1 Toll traffic Matrix (Sample) (in Erlangs)

From Exchange	To Exchange									
	1	2	3	4	5	6	7	8	9	10
1		57	39	73	23	60	17	21	23	5
2	62		19	30	18	26	25	2	9	6
3	42	18		28	17	31	19	8	10	12
4	70	31	23		6	7	5	8	4	3
5	25	19	32	5		22	19	31	13	50
6	62	23	19	8	20		30	27	19	27
7	21	30	17	40	16	32		15	16	17
8	21	5	12	3	25	19	17		18	29
9	25	10	9	1	16	22	18	19		19
10	7	8	7	2	47	25	13	30	17	

exchanges and tandem exchanges. The long-distance network might be divided into two layers.

Figure 1.14 shows the AT&T North American hierarchical network. It is a five-level hierarchy. North America (i.e., United States and Canada) covers a very large geographical area. Medium-sized countries may require fewer hierarchical levels, possibly four or three, and small countries may require two levels.

The outline of another five-level hierarchy is shown in Figure 6.4 with HU routes. Note that the lowest level is not included in the figure, that of the local exchange. HU routes ameliorate the problems of excessive links in tandem on the great majority of completed calls, thereby meeting the intent of CCITT Rec. Q.40.

Suppose, for example, that a country had four major population centers and could be divided into four areas around each center. Each of the four major population centers would have a tertiary center assigned, one of which would be the ISC. Each tertiary center would have one or several secondary centers homing to it, and a number of primary centers would home to the secondary centers. This procedure is illustrated in Figure 6.5 and is represented systematically in Figure 6.6, thus establishing a hierarchy and setting out the final routes. In this case, one of the tertiary exchanges would be an ISC. We define a final route as a route from which no traffic can overflow to an alternative route. It is a route that connects an exchange immediately above or below it in the network hierarchy and there is also connection of the two exchanges at the top level of the network. Final routes are said to make up the "backbone" of a network. Calls that are offered to the backbone but cannot be completed are lost calls.

A *high-usage route* is defined as any route that is not a final route; it may connect exchanges at a level of the network hierarchy *other than* the top level, such as between T_1 and T_2 in Figure 6.6. It may also be a route between exchanges on different hierarchical levels when the lower-level exchange does

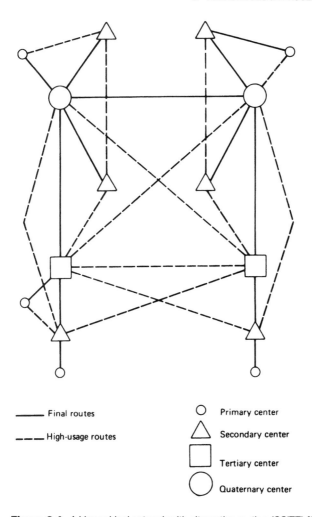

Figure 6.4. A hierarchical network with alternative routing (CCITT) [2].

not home on the higher level. A *direct route* is a special type of high-usage route connecting exchanges of the lowest rank in the hierarchy. Figure 6.7 illustrates these two definitions. High-usage routes are between exchanges 1 and 2 and 3 and 2. The direct route is also between exchanges 1 and 2, with exchanges 1 and 2 the lowest level in the hierarchy.

Before final dimensioning can be carried out, a grade of service p must be established, usually at no greater than 1% per link on the final route during the busy hour (BH). If the maximum number of links in an international call is established at 12 (in some cases 14), the very worst grade of service would be $12 \times 1\%$; however, on most calls the overall grade of service would be significantly better. These would be figures for direct-dialed international calls

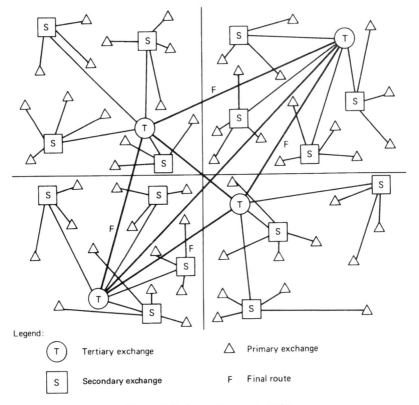

Legend:

(T) Tertiary exchange   △ Primary exchange

[S] Secondary exchange   F Final route

Figure 6.5. A sample network design.

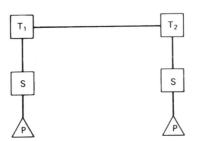

Figure 6.6. Hierarchical representation showing final routes.

(see CCITT Recs. Q.13 and Q.95 bis). If there are three links in tandem on final choice routes, there would be up to 3% grade of service under worst conditions; with four links it would be 4%. These latter figures are for national connections. The use of HU connections reduces tandem operation and tends to improve overall grade of service.

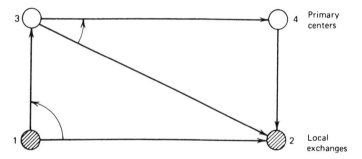

Figure 6.7. Hierarchical network segment.

The next step is to lay out high-usage routes. As mentioned earlier, this is done with the aid of the traffic matrix. One method of dimensioning with overflow (alternative routing) was discussed in Section 8.1 of Chapter 1, and the discussion continued in Section 9 of that chapter. Reference should also be made to Section 10 of Chapter 2. Trunks are costly. The exercise is to optimize the number of trunks and maintain a given grade of service. The methodologies given in Chapters 1 and 2 will help carry out this exercise. However, an increasing number of network designers are using computers to carry out this function. Two such computer programs are referenced at the beginning of this subsection. Traffic intensities used in the traffic matrices should be those taken from a 10-year forecast. (Also consult Ref. 8.)

7 TRAFFIC ROUTING IN THE NATIONAL NETWORK

7.1 New Routing Techniques

7.1.1 *Objective of Routing.* The objective of routing is to establish a successful connection between any two exchanges in the network. The function of traffic routing is the selection of a particular circuit group, for a given call attempt or traffic stream, at an exchange in the network. The choice of a circuit group may be affected by information on the availability of downstream elements of the network on a quasi-real-time basis.

7.1.2 *Network Topology.* A network comprises a number of nodes (i.e., switching centers) interconnected by circuit groups. There may be several direct circuit groups between a pair of nodes and these may be one-way or both-way (two-way). A simplified illustration of this idea is shown in Figure 6.8.

Remember that a direct route consists of one or more circuit groups connecting adjacent nodes. We define an indirect route as a series of circuit

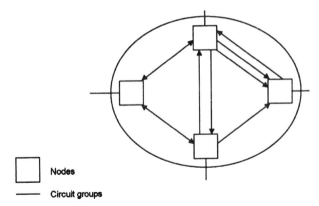

☐ **Nodes**

—— **Circuit groups**

Figure 6.8. A simplified network with circuit groups connecting pairs of nodes with one-way and both-way working.

groups connecting two nodes providing an end-to-end connection via other nodes.

7.1.2.1 Network Architecture. With a national network under many circumstances we would adopt a hierarchy of switching centers (e.g., local area, trunk, regional trunk, and international), with each level of the hierarchy performing different functions. As we mentioned above, there is no hierarchy for international switching centers (ISCs). Here telecommunication companies and administrations are free to determine the most suitable utilization of their individual ISCs. (See Section 4 and CCITT Rec. E.171. (Also consult Ref. 22.)

7.2   Logic of Routing

7.2.1 Routing Structure. Conceptually, hierarchical routing need not be directly related to a concept of a hierarchy of switching centers as we described above. A routing structure is hierarchical if, for all traffic streams, all calls offered to a given route, at a specific node, overflow to the same set of routes irrespective of the routes already tested. The routes in the set will always be tested in the same sequence, although some routes may not be available for certain types of calls. The last choice route is final (i.e., the final route) in the sense that no traffic streams using this route may overflow further.

A routing structure is nonhierarchical if it violates the above-mentioned definition (e.g., mutual overflow between circuit groups originating at the same exchange). An example of hierarchical routing in a nonhierarchical network of exchanges is shown in Figure 6.9.

7.2.2 Routing Scheme. A routing scheme defines how a set of routes is made available for calls between a pair of nodes. The "scheme" may be *fixed*

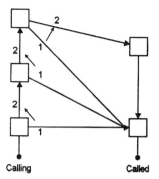

Calling   Called

Note – All nodes are of equal status.

Figure 6.9. Hierarchical routing in a nonhierarchical network of exchanges.

or *dynamic*. For a fixed scheme, the set of routes in the routing pattern is always the same. In the case of a dynamic scheme, the set of routes in the routing pattern varies.

7.2.2.1 Fixed Routing Scheme. Here routing patterns in the network may be fixed, in that changes to the route choices for a given type of call attempt require manual intervention. Changes then represent a "permanent change" to the routing scheme (e.g., the introduction of new routes requires a change to a fixed routing scheme).

7.2.2.2 Dynamic Routing Scheme. Routing schemes may also incorporate frequent automatic variations. Such changes may be time-dependent, state-dependent, and/or event-dependent. The updating of routing patterns may take place periodically or aperiodically, predetermined, depending on the state of the network or depending on whether calls succeed or fail in the setup of a route.

Time-Dependent Routing. Here routing patterns are altered at fixed times during the day or week to allow for changing traffic demands. It is important to note that these changes are preplanned and are implemented consistently over a long time period.

State-Dependent Routing. In this case, routing patterns are varied automatically according to the state of the network. These are called *adaptive routing schemes*. To support such a routing scheme, information is collected about the status of the network. For example, each toll exchange may compile records of successful calls or outgoing trunk occupancies. This information may then be distributed through the network to other exchanges or passed to a centralized database. Based on this network status information, routing decisions are made either in each exchange or at a central processor serving all exchanges. The concept is shown in Figure 6.10.

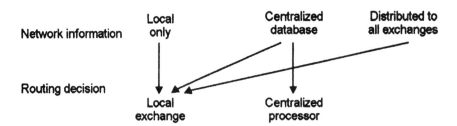

Figure 6.10. Adaptive or state-dependent routing, network information versus routing decisions.

Event-Dependent Routing. Here routing patterns are updated locally on the basis of whether calls succeed or fail on a given route choice. Each exchange has a list of choices, and the updating favors those choices which succeed and discourages those which suffer congestion.

7.2.3 Route Selection. Route selection is the action to actually select a definite route for a specific call. The selection may be *sequential* or *nonsequential.*

In the case of sequential selection, routes in a set are always tested in sequence and the first available route is selected. For the nonsequential case, the routes are tested in no specific order.

The decision to select a route can be based on the state of the outgoing circuit group or the states of series of circuit groups in the route. In either case, it can also be based on the incoming path of entry, class of service, or type of call to be routed. One example of the above is *selective trunk reservation.*

7.3 Call Control Procedures

Call control procedures define the entire set of interactive signals necessary to establish, maintain, and release a connection between exchanges. Two such call control procedures are *progressive call control* and *originating call control.*

7.3.1 Progressive Call Control. This type of call control uses link-by-link signaling (see Chapter 4) to pass supervisory controls sequentially from one exchange to the next. Progressive call control can be either irreversible or reversible. In the irreversible case, call control is always passed downstream toward the destination exchange. Call control is reversible when it can be passed backwards (maximum of one node) using automatic rerouting or crankback actions.

7.3.2 Originating Call Control. Here the originating exchange maintains control of the call setup until a connection between the originating and terminating exchanges has been completed.

7.4 Applications

7.4.1 Automatic Alternative Routing. One type of progressive (irreversible) routing is automatic alternative routing (AAR). When an exchange has the option of using more than one route to the next exchange, an alternative routing scheme can be employed. Two main types of AAR are available:

(1) When there is a choice of direct-circuit groups between two exchanges.
(2) When there is a choice of direct and indirect routes between the two exchanges.

Alternative routing takes place when all appropriate circuits in a group are busy. Several circuit groups may be tested sequentially. The test order is fixed or time-dependent.

7.4.2 Automatic Rerouting (Crankback). Automatic rerouting (ARR) is a routing facility enabling connection of call attempts encountering congestion during the initial call setup phase. Thus, if a signal indicating congestion is received from exchange B, subsequent to the seizure of an outgoing trunk from exchange A, the call can be rerouted at A. This concept is shown in Figure 6.11.

ARR performance can be improved through the use of different signals to indicate congestion, S1 and S2 (see Figure 6.11):

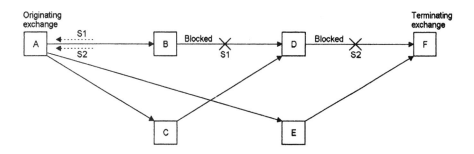

Note – Blocking from B to D activates signal S1 to A. Blocking from D to F activates signal S2 to A.

Figure 6.11. The automatic rerouting (ARR) or crankback concept. From CCITT Rec. E.170, Figure 4/E.170, page 5, Geneva, 10/92 [5].

- S1 indicates that congestion has occurred on outgoing trunks from exchange B;
- S2 indicates that congestion has occurred further downstream—for example, on outgoing trunks from D.

The action to be taken at exchange A upon receiving S1 or S2 may be either to block the call or to reroute it.

In the example shown in Figure 6.11, a call from A to D is routed via C because the circuit group B–D is congested (S1-indicator) and a call from A to F is routed via E because circuit group D–F is congested (S2-indicator).

One positive consequence of this alternative is to increase the signaling load and number of call setup operations resulting from the use of these signals. If such an increase is unacceptable, it may be advisable to restrict the number of reroutings or limit the signaling capability to fewer exchanges. Of course, care must be taken to avoid circular routings ("ring-around-the-rosy"), which return the call to the point at which blocking previously occurred during call setup.

7.4.3 Load Sharing. All routing schemes should result in the sharing of traffic load between network elements. Routing schemes can, however, be developed to ensure that call attempts are offered to route choices according to a preplanned distribution. Figure 6.12 illustrates this application of load sharing, which can be made available as a software function of SPC exchanges. The system works by distributing the call attempts to a particular destination in a fixed ratio between the specified routing patterns.

7.4.4 Dynamic Routing

7.4.4.1 Example of State-Dependent Routing. A centralized routing processor is employed to select optimum routing patterns on the basis of actual occupancy level of circuit groups and exchanges in the network which are monitored on a periodic basis (e.g., 10 s). Figure 6.13 illustrates this concept. In addition, qualitative traffic parameters may also be taken into consideration in the determination of the optimal routing pattern.

This routing technique inherently incorporates fundamental principles of network manage-ment in determining routing patterns. These principles include:

- Avoiding occupied circuit groups.
- Not using overloaded exchanges for transit.
- In overload circumstances, restriction of routing direct connections.

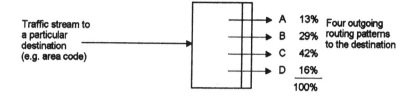

Note – Each outgoing routing pattern (A, B, C, D) may include alternative routing options.

Figure 6.12.. An example of preplanned distribution of load sharing.

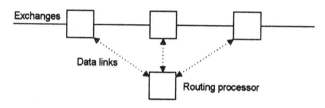

Figure 6.13. State-dependent routing example with centralized processor.

7.4.4.2 Example of Time-Dependent Routing. For each originating and terminating exchange pair, a particular route pattern is planned depending on the time of day and day of the week. This is shown in Figure 6.14. A weekday, for example, can be divided into different time periods, with each time period resulting in different route patterns being defined to route traffic streams between the same pair of exchanges.

This type of routing takes advantage of idle circuit capacity in other possible routes between the originating and terminating exchanges which may exist due to noncoincident busy hours. Crankback may be utilized to identify downstream blocking on the second link of each two-link alternative path.

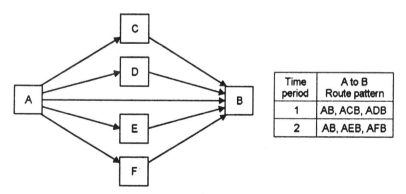

Time period	A to B Route pattern
1	AB, ACB, ADB
2	AB, AEB, AFB

Figure 6.14. Example of time-dependent routing. From CCITT Rec. E.170, Figure 7/E.170, page 6, Geneva, 10/92 [5].

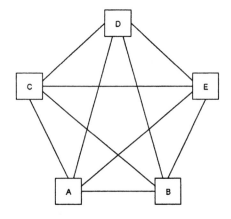

	A to B route pattern	
Choice	Current	After call failure
1	AB	AB
2	AEB	ACB

Figure 6.15. Event-dependent routing in a mesh network.

7.4.4.3 Example of Event-Dependent Routing. In a fully connected (mesh) net-work, calls between each originating and terminating exchange pair try the direct route with a two-link alternative path selected dynamically. While calls are successfully routed on a two-link path, that alternative is retained. Other-wise, a new two-link alternative path is selected. This updating, for example, could be random or weighted by the success of previous calls.

This type of routing scheme routes traffic away from congested links by retaining routing choices where calls are successful. It is simple, adapts quickly to changing traffic patterns, and requires only local information. Such a scheme is illustrated in Figure 6.15.

Sections 7.1 through 7.4 are based on CCITT Rec. E.170 [5]. (Also consult Ref. 9.)

7.5 AT&T's Dynamic Nonhierarchical Routing (DNHR)

Figure 6.16 is a model of a network structure that incorporates DNHR into the AT&T intercity network. DNHR is made possible through SPC switches and the incorporation of CCITT No. 7 signaling in the network. As shown in Figure 6.16, a DNHR network has only one level of tandem switching, and local exchange carrier local serving switches and customer equipment are directly connected or "home" to the higher-level tandem switches in the DNHR environment, as they were in the hierarchy. The local exchange

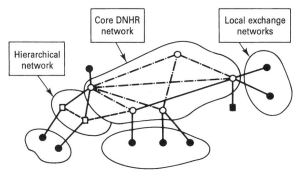

○ DNHR tandem switch
□ Hierarchical tandem switch
● Local exchange carrier end office
■ Customer premises equipment

Figure 6.16. DNHR network configuration. Reprinted with permission of IEEE Press [6].

carrier (LEC) end offices and customer premises equipment switch customer telephone lines to each other or to the interexchange carrier (IXC) network.

The DNHR portion of the AT&T intercity network consists of 4ESS digital switches interconnected by the CCITT Signaling System No. 7 common channel signaling network. Dynamic routing rules are used only among DNHR switches, and conventional hierarchy routing rules are used among all other pairs of switches. There are a number of smaller switching systems employing only hierarchical rules. Many of these smaller systems, designated as "hierarchical tandem switches" in Figure 6.16, home directly on DNHR switches. From the viewpoint of the hierarchical switches, the DNHR switches appear as a large network of the highest-level hierarchical centers, and hierarchical routing patterns are therefore defined by this hybrid structure.

The dynamic routing method, called two-link DNHR with crankback,* is shown in Figure 6.17. The example is for intercity DNHR, but a similar routing technique can also be applied to metropolitan DNHR or international DNHR. The DNHR strategy capitalizes on two factors:

(1) Selection of minimum-cost paths between originating and terminating switches;

(2) Design of optimal, time-varying routing patterns to achieve minimum cost trunking by taking advantage of noncoincident network busy periods.

The dynamic, or time-varying, nature of the routing scheme is achieved by introducing several route choices. The DNHR routes consist of different

---

*Crankback—the return of circuit control to an originating exchange for other alternative routing possibilities.

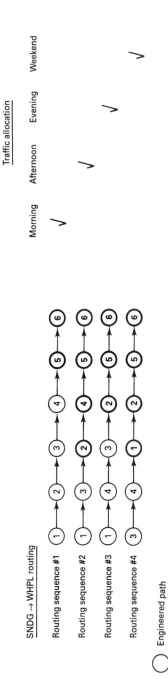

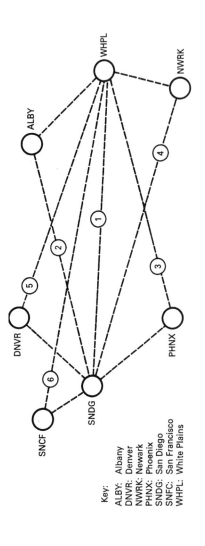

Figure 6.17. Two-Link DNHR with crankback. Reprinted with permission of IEEE Press [6].

sequences of paths, and each path has one or, at most, two links or trunk groups in tandem. In Figure 6.17, the originating switch at San Diego (SNDG) retains control over a dynamically routed call until it is either completed to its destination at White Plains (WHPL) or blocked. A call overflowing the second leg of a two-link connection—for example, the ALBY-WHPL link of the SNDG-ALBY-WHPL path—is returned to the originating switch (SNDG) for possible further alternate routing. Control is returned by sending a common channel signaling crankback signal from the via switch (ALBY) to the originating switch.

Each of the four routing sequences illustrated in Figure 6.17 uses a subset of the six paths in a different order. Each routing sequence results in a different allocation of link flows; for example, the first-choice paths normally carry the maximum flow of a switch-to-switch load, but all satisfy a switch-to-switch blocking requirement.

As shown in the figure, the paths used in various time periods need not be the same. Allocating traffic to the optimum route choice during each time period leads to design benefits due to the noncoincidence of loads. Because many intercity traffic demands change with time in a reasonably predictable manner, the routing also changes with time to achieve maximum trunk utilization and minimum network cost. Ten DNHR time periods are used to divide the hours of an average business day into contiguous routing intervals, and five time periods are used to divide the hours of the weekend.

The only decisions necessary in real time involve network conditions that become known in real time, such as actual (not forecast) loads, failures, and overload. The real-time traffic-sensitive (or adaptive) aspect of the routing strategy, also shown in Figure 6.17, involves the use of additional real-time paths for possible completion of calls that overflow the preplanned routing sequences.

AT&T's experience with DNHR shows that with over 100 DNHR switches in the national network, cost savings of approximately 14–16 percentage points of the hierarchical network design cost are achieved. AT&T also reports drastic reduction in blockage on national holidays such as Christmas when excessive traffic demands are placed upon the network.

This section is based on the article by Dr. Gerald R. Ash, "Design and Control of Networks with Dynamic Nonhierarchical Routing," *IEEE Communications*, October 1990 [6].

8   NETWORK DESIGN AND CONFIGURATIONS FROM A BELLCORE PERSPECTIVE

8.1   Introduction

The PSTN in the United States has two elements: the local exchange carriers (LECs) providing local access and the interexchange carriers (IXCs) providing long-distance (toll) service.

A LATA is a "local access and transport area." LATAs serve the following two basic purposes:

1. They provide a method of delineating the area within which BOCs* and independent telecommunication companies may offer services.
2. They provided a basis for determining how the assets of the former Bell System were to be divided between the BOCs and AT&T at divestiture.

The court decision for divestiture, called "the Modification of Final Judgement" (MFJ) in its Appendix B, requires each BOC to offer equal access through BOC end offices (local exchanges) in a LATA to all interexchange carriers (IXCs). All carriers must be provided services that are equal in type, quality, and price to those provided to AT&T.

8.1.1 Definition of Point of Presence (POP). A point of presence (POP) is a location within a LATA that has been designated by an IXC for the connection of its facilities with those of a LEC. Typically, a POP will be at a building that houses an IXC's switching system or facility node, and it must be located within the LATA that the IXC serves. An IXC may have more than one POP within a LATA, and a POP may be for public and private, switched and nonswitched services. One location may serve as a POP for several service types.†

The IXC is required to designate at each POP a physical point of termination (POT), consistent with the technical and operational characteristics specified by the LEC. The POT provides a clear demarcation between the LEC's exchange-access functions and the IXC's interLATA functions, and it enables the LEC to meet its tariff obligations. This POT generally will be a distribution frame or other item of equipment at which the LEC's access lines terminate and where cross-connection, testing, and service verification can occur. The IXC can also have a multiplex (TDM or FDM) POT interface, which requires a special testing procedure.

8.1.2 Clarification. We do not wish the reader to confuse a LATA with the local area as defined in Chapter 2. Often a LATA covers a very large geographical coverage area such that many trunks and the encumbent tandem switches give more of an appearance of a toll area. Thus, there is no reason why the methods and means described in this section and recommended by Bellcore cannot be used for a toll (long-distance) network strategy and implementation. The material in this section is based on Bellcore's document *BOC Notes on the LEC Networks—1994* [18] and offers some

---

*BOC stands for Bell Operating Companies, more properly referred to as Regional Bell Operating Companies or RBOCs (formed after divestiture of the Bell System).
†It should be noted that a POP can serve multiple IXCs.

unique network design and routing implementations, although tailored for the special architecture of the U.S. network, not found at this time any-where else in the world. This does not mean, however, that some of these approaches do not have a more universal application.

8.2 Basic Routing Fundamentals

Alternate routing is advantageous because it provides the opportunity to minimize the cost per unit of carried traffic. With alternate routing, the load is allocated to high-usage and final routes in the most economical manner. Alternate routing also permits the meshing of traffic streams that have differing busy hours or seasons.

Figure 6.18 illustrates a one-way, high-usage (HU) trunk group from local exchange A to local exchange B, with an alternate (final) route via a tandem (exchange). In general, the direct or HU route is shorter and less expensive than the alternate-route path. However, because each leg of the alternate route is used by other calls, a number of traffic items can be combined for improved efficiency on that route. The basic design problem, then, is to minimize the cost of carrying the offered load (that is, to determine how much of the offered load should be carried on the direct route and how much should be overflowed to the alternate route).

Figure 6.19 shows the relationships involved. The graph shows, as a func-tion of the number of trunks in the HU trunk group, the cost of the direct route, the cost of the alternate route, and total cost for serving the given offered load. (See Chapter 2 where we discuss tandem routing in the local area. There is some parallelism in the approach.)

The HU trunk group cost, of course, increases in direct proportion to the number of HU trunks. If there are no HU trunks, all of the offered traffic must be carried on the alternate route, so the incremental alternate-route cost decreases. This cost decreases very rapidly as the first trunks are added to the

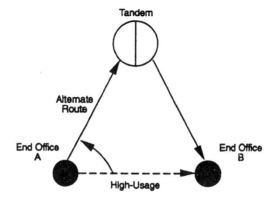

Figure 6.18. Alternate routing arrangement.

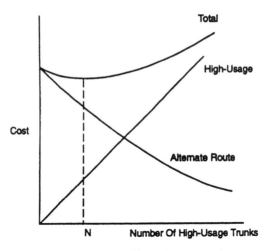

Figure 6.19. Relationships involved in an alternate routing arrangement.

HU trunk group. This is due to the efficiency of each of these trunks, which relieves a substantial amount of load from the alternate route. As more HU trunks are added, each successive HU trunk carries less traffic, while each alternate-route trunk continues to carry a significant amount of traffic. Eventually it becomes undesirable to add any more HU trunks. The point at which this threshold occurs is where the total cost (the sum of the two curves) is minimized. This point is designated as N in Figure 6.19.

A method commonly used to determine N is called *economic hundred call seconds* (ECCS) engineering. This method determines the maximum number of HU trunks for which the cost per CCS carried on the "last" trunk of the HU trunk group is less than or equal to the cost per CCS on an additional alternate-route trunk.

This relationship can be expressed by the following equation and is the basis of ECCS engineering:

$$CALT/CHU = 28/ECCS$$

where

$$CALT = \text{cost of a path on the alternate route}$$
$$CHU = \text{cost of a trunk on the HU route}$$
$$28 = \text{capacity in CCS added to the alternate route by the}$$
$$\text{addition of an incremental trunk (path)}$$

The equation is solved for the ECCS—the load to be carried by the "last" or least-efficient trunk in the HU trunk group. Given the ECCS and the offered load, standard trunking tables (as found in Chapter 1) can be entered to determine the number of trunks required and the estimated amount of overflow. This is the largest number of trunks for which the load carried on the last trunk is not less than the ECCS.

Because the equation is solved for the ECCS, the other elements of the equation must be known. The left side of the equation (CALT/CHU) is the cost ratio, or the relationship of the cost of a path on the alternate route to the cost of a trunk on a direct route. Cost ratios used for alternate-route engineering are always greater than unity.

The "28" shown in the equation is the incremental capacity of the alternate route (the capacity that would be added to the alternate route by the addition of one path). This value is usually assumed to be a constant of 28 CCS, thereby permitting calculation of the ECCS as a function of a single variable, the cost ratio.

Thus it can be seen that with low cost ratios, the ECCS will be high and fewer high-usage trunks will be provided. Conversely, a low ECCS would result from a high cost ratio and a greater number of HU trunks will be provided. Simply, the more expensive the alternative route relative to the HU trunk group, the less traffic that will be overflowed to it.

The total cost curve is rather flat near the minimum. As a result, errors in ECCS that might result from minor cost ratio or incremental CCS errors will not have a significant impact on network cost.

The number of HU trunks to be provided in a group depends not only on the ECCS and offered load, but on the variability of the offered load as well. This variability can be either within the hour (usually peakedness) or can be day to day. Such variability can be the result of traffic patterns as in the case of day-to-day variations, or it can be system induced as is usually the case with peakedness. Where such variability is present, equivalent random-engineering techniques are required for the HU groups, and Neal–Wilkinson capacity tables are used to size grade-of-service engineered final trunk groups.

Traffic volumes reach peaks during certain hours. Trunks are usually provided to care for average time-consistent busy-hour loads in the busy season of the year. Where only one outlet (trunk group) is available, trunks must be provided for the group busy-hour load. If two routes (a direct and an alternate route) are available, however, the busy hours on each of the two routes will frequently be different. Where this is the case, trunks only need to be provided on the direct route to care for that portion of busy-hour offered load that cannot be carried on idle trunks in the alternate route. The alternate route may be sized for a different busy hour and this is not fully loaded in the busy hour of the direct route.

Often there are two or more potential first-choice alternate routes for an HU group. The selection of alternate routes is based on a routing discipline if overall cost differences are not significant, or the choice can be based on the economics of each individual case (that is, selection of the least expensive alternate route). In general, the overall network economics are not highly sensitive to variation in alternate-route costs.

8.3 Application of Alternate Routing

The principle of alternate routing is basic to the design of the network and is used extensively to provide economic and service advantages (see also Section 7.4.1). Modern switching equipment automatically seeks alternate routes. Calls can be offered in succession to a series of alternate routes via one or more tandems. At each switching center, all of the HU trunk groups to which a call can be offered are kept very busy with a portion of the traffic overflowing to another route. The final trunk groups are fewer in number and have low blocking so that the engineered level of service is good. The overall chance of completing a call is improved by the fact that it can be offered to more than one trunk group. Switching equipment operates rapidly, and the change in speed of service between the selection of direct and alternate routes is not significant.

In an emergency situation of limited impact and extent, such as a localized equipment failure, the ability to use an alternate route adds another measure of protection to service. However, if there is a heavy surge of traffic over an entire area (for example, during a major disaster such as a hurricane), there is little margin to absorb such surges in load, and service may not be immediately available as it would be with an only-route-type network.

In addition to the final trunk groups that connect end offices (local serving switches) to their home tandem, high-usage trunk groups are provided from end offices to other end offices and tandems when justified by economics and traffic volumes. The traffic items that should be considered when evaluating traffic volumes for the purpose of proving-in new HU groups are subject to the rules of the network hierarchy.

8.4 Fundamentals of Dynamic Routing Technique

A dynamic routing technique (DRT) is a traffic routing method in which one or more central controllers determine near-real-time routes for a switched network, based on the state of network congestion measured as trunk group busy/idle status and switch congestion. The choice of traffic routes in a hierarchical network is static (i.e., preplanned and fixed) over time, other than the manual or automated controls performed by network traffic management (NTM) in response to localized or general network overload.

Using DRT, network traffic can be more efficiently distributed over the network trunk groups and switches than traffic routed on the hierarchical network. Based on near-real-time network traffic congestion, DRT selects routes that provide lower blocking than today's fixed routes can attain. Consequently, DRT can reduce the level of demand servicing and react more easily to relieve problems associated with forecast errors. Also, if a switch or fiber link fails, networks using DRT can carry significantly more traffic than networks using hierarchical routing.

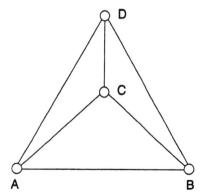

Figure 6.20. Example of nonhierarchical routing using the DR-T routing scheme.

DRT differs from current routing operations in at least two important areas:

1. The frequency of traffic data collection
2. Application of this data to the selection of routes

This dynamic routing algorithm, commonly called DR-T, can use traffic data collected about every T minutes to change routes after each data update, where T is fixed from near zero to about five minutes.

Routing using the DR-T scheme is illustrated in Figure 6.20, where any node can originate traffic and also act as a via node (a transit/tandem exchange). The choice of routes can change in the next T-minute period.

In nonhierarchical routing, switches could originate and terminate traffic and also be used for via traffic (transit or tandem traffic).

At the beginning of the update interval of length T (e.g., $T = 5$ minutes), the ordered routes for first-offered traffic from switches A to B determined by the DR-T algorithm could be:

1. A—B
2. A—C—B

The ordered routes from switches C to B could be

1. C—D—B
2. C—B

See Section 7.5 for dynamic routing from an AT&T perspective. Section 8 is based on Section 4 of *BOC Notes on the LEC Networks—1994*, SR-TSV-002275, Issue 2, April 1994 [18].

9 TRANSMISSION FACTORS IN LONG-DISTANCE TELEPHONY

9.1 Introduction

Long-distance analog communication systems require some method to overcome losses. As a wire-pair telephone circuit is extended, there is some point where loss accumulates so as to attenuate signals to such a degree that the far-end subscriber is dissatisfied. The subscriber cannot hear the near-end talker sufficiently well. Extending the wire connection still further, the signal level can drop below the noise level. For good received level, a 40-dB signal-to-noise ratio is desirable (see Chapter 5, Section 2.5). To overcome the loss, amplifiers can be added; in fact, amplifiers are installed on many wire-pair trunks. Early North American transcontinental circuits were on open-wire lines using amplifiers quite widely spaced. However, as BH demand increased to thousands of circuits, the limited capacity of such an approach was not cost-effective.

System designers turned to wideband radio and coaxial cable systems where each bearer or pipe* carried hundreds (and now thousands) of simultaneous telephone conversations. Carrier (frequency division) multiplex techniques made this possible (see Chapter 5). Frequency division multiplex (FDM) requires separation of transmit and receive voice paths. In other words, the circuit must convert from two-wire to four-wire transmission. This is normally carried out by a hybrid transformer, or resistive hybrid. Figure 6.21 is a simplified block diagram of a telephone circuit with transformation from two-wire to four-wire operation at one end and conversion back to two-wire operation at the other end. This concept was introduced in Chapter 5, Section 3.

*On a pair of coaxial cables, a pair of fiber-optic light guides, or a pair of radio-frequency carriers, one coming and one going.

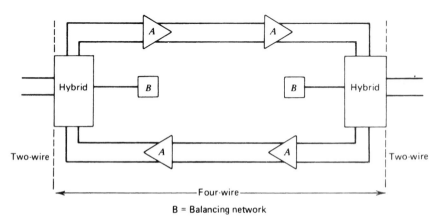

B = Balancing network

Figure 6.21. Simplified schematic of two-wire/four-wire operation.

The two factors that must be considered that greatly affect transmission design are "echo" and "singing." These are related, as we see in Sections 9.2 and 9.3.

9.2 Echo

As the name implies, echo in telephone systems is the return of a talker's voice. To be an impairment, the returned voice must suffer some noticeable delay. Thus we can say that echo is a reflection of the voice. Analogously, it may be considered as that part of the voice energy that bounces off obstacles in a telephone connection. These obstacles are impedance irregularities, more properly called *impedance mismatches*. Echo is a major annoyance to the telephone user. It affects the talker more than the listener. Two factors determine the degree of annoyance of echo: its loudness and the length of its delay.

9.3 Singing

Singing is the result of sustained oscillations due to positive feedback in telephone amplifiers or amplifying circuits. Circuits that sing are unusable and promptly overload multichannel carrier equipment (FDM; see Chapter 5).

Singing may be regarded as echo that is completely out of control. This can occur at the frequency at which the circuit is resonant. Under such conditions the circuit losses at the singing frequency are so low that oscillation will continue, even after cessation of its original impulse.

9.4 Causes of Echo and Singing

Echo and singing can generally be attributed to the mismatch between the balancing network of the hybrid and its two-wire connection associated with the subscriber loop. It is at this point that the major impedance mismatch usually occurs and an echo path exists. To understand the cause of the mismatch, remember that we always have at least one two-wire switch between the hybrid and the subscriber. Ideally, the hybrid-balancing network must match each subscriber line to which it may be switched. Obviously, the impedances of the four-wire trunks (lines) may be kept fairly uniform. However, the two-wire subscriber lines may vary over a wide range. The subscriber loop may be long or short, may or may not have inductive loading, and may or may not be carrier derived. The hybrid imbalance causes signal reflection or signal "return." The better the match, the more the return signal is attenuated. The amount that the return signal (or reflected signal) is attenuated is called the *return loss*, and it is expressed in decibels. The reader should remember that any four-wire circuit may be switched to hundreds or even thousands of different subscribers; if not, it would be a simple matter to match the four-wire circuit to its single subscriber through the hybrid. This is why the hybrid to which we refer has a compromise balancing network rather than a precision network. A compromise network is usually adjusted for a compromise in the expected range of impedance (Z) encountered on the two-wire side.

Let us now consider the problem of match. If the impedance match is between the balancing network (N) and the two-wire line (L) (see Figures 5.4 and 5.5), then

$$\text{Return loss dB} = 20 \log_{10} \frac{Z_N + Z_L}{Z_N - Z_L} \tag{6.1}$$

If the network perfectly balances the line, then $Z_N = Z_L$, and the return loss would be infinite.

Return loss may also be expressed in terms of reflection coefficient, or

$$\text{Return loss dB} = 20 \log_{10} \frac{1}{\text{Reflection coefficient}} \tag{6.2}$$

where the reflection coefficient is equal to the reflected signal/incident signal.

We use the term "balance return loss" (see CCITT Rec. G.122) and classify it as two types:

1. Balance return loss from the point of view of echo.* This is the return loss across the band of frequencies from 300 Hz to 3400 Hz.
2. Balance return loss from the point of view of stability. This is the return loss between 0 and 4000 Hz.

The band of frequencies most important in terms of echo for the voice channel is that from 300 Hz to 3400 Hz. A good value for echo return loss for toll telephone plant is 11 dB, with values on some connections dropping to as low as 6 dB. For further information, the reader should consult CCITT Recs. G.122 and G.131 [7].

Echo and singing may be controlled by:

- Improved return loss at the term set (hybrid).
- Adding loss on the four-wire side (or on the two-wire side).
- Reducing the gain of the individual four-wire amplifiers.

The annoyance of echo to a subscriber is also a function of its delay. Delay is a function of the velocity of propagation of the intervening transmission facility. A telephone signal requires considerably more time to traverse

---

*Called *echo return loss* (ERL) in via net loss (VNL) (North American practice; Section 9.7) but uses a weighted distribution of level.

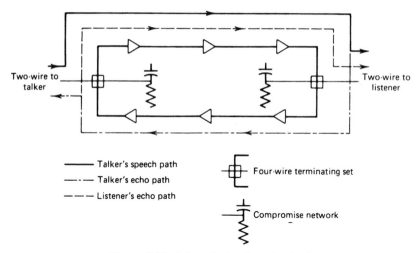

Figure 6.22. Echo paths in a four-wire circuit.

100 km of a voice-pair cable facility, particularly if it has inductive loading, than it requires to traverse 100 km of radio facility (as low as 22,000 km/s for a loaded cable facility and 240,000 km/s for a carrier facility). Delay is measured in one-way or round-trip propagation time measured in milliseconds. The CCITT recommends that if the mean round-trip propagation time exceeds 50 ms for a particular circuit, an echo suppressor or echo canceler should be used. Practice in North America uses 45 ms as a dividing line. In other words, where echo delay is less than that stated previously here, echo can be controlled by adding loss.

An echo suppressor is an electronic device inserted in a four-wire circuit that effectively blocks passage of reflected signal energy. The device is voice operated with a sufficiently fast reaction time to "reverse" the direction of transmission, depending on which subscriber is talking at the moment. The blocking of reflected energy is carried out by simply inserting a high loss in the return four-wire path. Figure 6.22 shows the echo path on a four-wire circuit. An echo canceler generates an echo-canceling signal [12].

9.5 Transmission Design to Control Echo and Singing

As stated previously, echo is an annoyance to the subscriber. Figure 6.23 relates echo path delay to echo path loss. The curve in Figure 6.23 traces a group of points at which the average subscriber will tolerate echo as a function of its delay. Remember that the longer the return signal is delayed, the more annoying it is to the telephone talker (i.e., the more the echo signal must be attenuated). For instance, if the echo path delay on a particular circuit is 20 ms, an 11-dB loss must be inserted to make echo tolerable to the talker. The careful reader will note that the 11 dB designed into the circuit will increase the end-to-end reference equivalent by that amount, which is

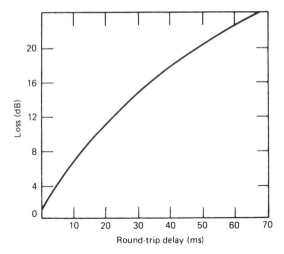

Figure 6.23. Talker echo tolerance for average telephone users.

quite undesirable. The effect of loss design on reference equivalents and the trade-offs available are discussed in the paragraphs that follow.

If singing is to be controlled, all four-wire paths must have some loss. Once they go into a gain condition, and we refer here to overall circuit gain, positive feedback will result and the amplifiers will begin to oscillate or "sing." North American practice calls for a 4-dB loss on all four-wire circuits to ensure against singing. The CCITT recommends a minimum loss for a national network of 10 dB (CCITT Rec. G.122, p. 3) [19].

Almost all four-wire circuits have some form of amplifier and level control. Such amplifiers are often embodied in the channel banks of the carrier (FDM) equipment.

9.6   Introduction to Transmission-Loss Engineering

One major aspect of transmission system design for a telephone network is to establish a transmission-loss plan. Such a plan, when implemented, is formulated to accomplish three goals:

1. Control singing (stability).
2. Keep echo levels within limits tolerable to the subscriber.
3. Provide an acceptable overall reference equivalent (loudness rating) to the subscriber.

For North America the via net loss (VNL) concept embodies the transmission plan idea (VNL is covered in Section 9.7).

From our preceding discussions we have much of the basic background necessary to develop a transmission-loss plan. We know the following:

1. A certain minimum loss must be maintained in four-wire circuits to ensure against singing.

2. Up to a certain limit of round-trip delay, echo is controlled by loss.

3. It is desirable to limit these losses as much as possible to improve reference equivalent (loudness rating).

National transmission plans vary considerably. Obviously, length of circuit is important, as well as the velocity of propagation of the transmission media. Two approaches are available in the preparation of a loss plan, variable-loss plan (i.e., VNL, as used in North America) and fixed-loss plan (i.e., as used in Europe). A national transmission-loss plan for a small country (i.e., small in geographic area) such as Belgium could be quite simple. Assume that a 4-dB loss is inserted in all four-wire circuits to prevent singing. Consult Figure 6.23, where 4 dB allows for 4 ms of round-trip delay. If we assume carrier transmission for the entire length of the connection and use 105,000 mi/s for the velocity of propagation, we can satisfy Belgium's echo problem. The velocity of propagation used comes out to 105 mi/ms (169 km/ms). By simple arithmetic, we see that a 4-dB loss on all four-wire circuits will make echo tolerable for all circuits extending 210 mi (338 km) (i.e., 2×105). This is an application of the fixed-loss type of transmission plan. In the case of small countries or telephone companies operating over a small geographic extension, the minimum loss inserted to control singing controls echo as well for the entire country.

Let us try another example. Assume that all four-wire connections have a 7-dB loss. Figure 6.23 indicates that 7 dB permits an 11-ms round-trip delay. Assume that the velocity of propagation is 105,000 mi/s. Remember that we are dealing with round-trip delay. The talker's voice reaches the far-end hybrid and is then reflected back. This means that the signal traverses the system twice, as shown in Figure 6.24. Thus 7 dB of loss for the given velocity of propagation allows about 578 mi of extension or, for all intents and purposes, the distance between subscribers and will satisfy the loss requirements for a country with a maximum extension of 578 mi (925 km).

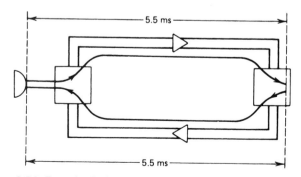

Figure 6.24. Example of echo round-trip delay (5.5 + 5.5 = 11-ms round-trip delay).

It has become evident by now that we cannot continue increasing losses indefinitely to compensate for echo on longer circuits. Most telephone companies and administrations have set a 45- or 50-ms round-trip delay criterion, which sets a top figure above which echo suppressors are to be used. One major goal of the transmission-loss plan is to improve overall loudness rating or to apportion more loss to the subscriber plant so that subscriber loops can be longer or to allow the use of less copper (i.e., smaller-diameter conductors). The question arises as to what measures can be taken to reduce losses and still keep echo within tolerable limits. One obvious target is to improve return losses at the hybrids. If all hybrid return losses are improved, the echo tolerance curve shifts; this is because improved return losses reduce the intensity of the echo returned to the talker. Thus the talker is less annoyed by the echo effect.

One way of improving return loss is to make all two-wire lines out of the hybrid look alike—that is, have the same impedance. The switch at the other end of the hybrid (i.e., on the two-wire side) connects two-wire loops of varying length, thus causing the resulting impedances to vary greatly. One approach is to extend four-wire transmission to the local office such that each hybrid can be better balanced. This is being carried out with success in Japan. The U.S. Department of Defense has its Autovon (automatic voice network), in which every subscriber line is operated on a four-wire basis. Two-wire subscribers connect through the system via PABXs (private automatic branch exchanges).

Let us return to standard telephone networks using two-wire switches in the subscriber area; suppose that balance return loss could be improved to 27 dB. Thus minimum loss to ensure against singing could be reduced to 0.4 dB. Now suppose that we distribute this loss across four four-wire circuits in tandem. Thus each four-wire circuit would be assigned a 0.1-dB loss. If we have gain in the network, singing will result. The safety factor between loss and gain is 0.4 dB. The loss in each circuit or link is maintained by amplifiers. It is difficult to adjust the gain of an amplifier to 0.1 dB, much less keep it there over long periods, even with good automatic regulation. *Stability* or *gain stability* is the term used to describe how well a circuit can maintain a desired level. Of course, in this case we refer to a test-tone level. In the preceding example it would take only one amplifier to shift 0.4 dB, two to shift in the positive direction 0.2 dB, and so forth. The importance of stability, then, becomes evident.

The stability of a telephone connection depends on three criteria:

1. The variation of transmission level with time.
2. The attenuation–frequency characteristics of the links in tandem.
3. The distribution of balance return loss.

Each criterion becomes magnified when circuits are switched in tandem. To handle the problem properly, we must talk about statistical methods and

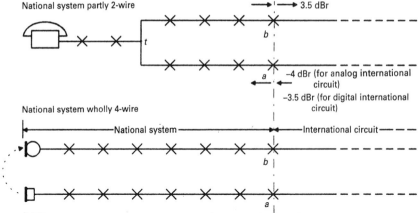

NOTE: *a, b* are virtual analog switching points of the international circuit.

Figure 6.25. National system representation and the definition of virtual analog switching points *a–t–b* and points *a–b*. From ITU-T Rec. G.122, Figure 1/G.122, page 1, Helsinki, 3/93 [19].

standard distributions. In the case of criteria 1 and 2, we refer to the tandem nature of the four-wire circuits. Criterion 3 refers to switching subscriber loops–hybrid combinations that will give a poorer return loss than will the 11 dB stated earlier. Return losses on some connections can drop to 3 dB or less.

Stability is discussed in CCITT Recs. G.122 [19], G.131, and G.151C [7]. In essence, the loss through points *a–t–b* in Figure 6.25 should have a value not less than $(10 + N)$ dB for established connections, where N is the number of four-wire circuits in the national chain. Thus the minimum loss is stated (CCITT Rec. G.122), and Rec. G.131 is quoted in part as follows:

> The standard deviation of transmission loss among international circuits routed in groups equipped with automatic regulation is 1 dB . . . . This accords with . . . the tests . . . [which] indicate that this target is being approached in that 1.1 dB was the standard deviation of the recorded data.

> It is also evident that those national networks which can exhibit no better stability balance return loss than 3 dB, 1.5 dB standard deviation, are unlikely to seriously jeopardize the stability of international connections as far as oscillation is concerned. However, the near-singing [rain-barrel effect] distortion and echo effects that may result give no grounds for complacency in this matter.

Stability requirements in regard to North American practice are embodied in the VNL concept discussed in Section 9.7.

9.6.1 CCITT Loss plan

9.6.1.1 Introduction. CCITT (now the ITU-T organization) concern is with international connections. However, for an international connection there must be two national connectivities: one from the originating country of the connection and one for the terminating country. In all probability the cause of echo will be in one or both national extensions of an international connectivity; and the control of echo will take place on the national extensions. For this very practical reason, CCITT makes certain echo/singing control recommendations for national networks. Our following discussion briefly outlines the CCITT approach to control echo and singing on national networks. The information provided is based on ITU-T Rec. G.122, Helsinki, 1993 [19].

9.6.1.2 Rationale for Transmission Loss. Figure 6.25 shows a national system representation with the virtual analog switching points *a* and *b*. CCITT states in G.122 that the transmission loss introduced between *a* and *b* by the national system, referred to as the loss (*a–b*), is important from three points of view:

(a) It contributes to the margin that the international connection has against oscillation during the setting-up and clearing-down of the connection. A minimum loss over the band 0–4 kHz is the characteristic value.

(b) It contributes to the margin of stability during a communication. Again, a minimum loss in the band 0–4 kHz is the characteristic value, but in this case the subscribers' apparatus (telephone, modem, etc.) are assumed to be connected and in the operating condition.

(c) It contributes to the control of echoes and, in respect of the subjective effect of talker echo, a weighted sum of the loss (*a–b*) over the band 300–3400 Hz is the characteristic value.

Rec. G.122 gives three facets to loss design, which are outlined below.

9.6.1.3 Loss (a–b) to Avoid Instability During Setup, Clear-Down, and Changes in a Connection. To ensure adequate stability of international connections, the distribution (taken over many actual calls) of the loss (*a–b*) during the worst situation should be such that the risk of a loss (*a–b*) of 0 dB or less does not exceed 6 in 1000 calls when using the calculation given below. The requirement should be observed at any frequency in the band 0–4 kHz.

To ensure adequate stability under this paragraph, the following simultaneous conditions must be met on the national network:

1. The sum of the nominal transmission losses in both directions of transmission *a–b* and *t–b* measured between the two-wire input of the terminating set *t*, and one or other of the virtual switching points on the international circuit (see Figure 6.25), *a* or *b*, should not be less than (4 + *n*) dB, where *n* is the number of analog or mixed analog–digital four-wire circuits in the national chain.

2. The stability balance return loss at the terminating set t should have a value not less than 2 dB for the terminal conditions encountered during normal operation.

3. The standard deviation of variations of transmission loss of a circuit should not exceed 1 dB.

9.6.1.4 Unweighted Loss (a–b) on Established Connections. The objective is that the risk of the loss $(a–b)$ reaching low values at any frequency in the range 0–4 kHz should be as small as practicable. This requires restrictions on the distribution of values of stability loss $(a–b)$ for the population of actual international calls established over the national system. Such a distribution can be characterized by a mean value and a standard deviation.

The objective will be obtained by a national system sharing a mean value of at least $(10 + n)$ dB together with a standard deviation not larger than $(6.25 + 4n)^{1/2}$ dB in the band 0–4 kHz, where n is the number of analog or mixed analog–digital four-wire circuits in the national chain.

The distribution of stability loss $(a–b)$ recommended above could, for example, be attained if, in addition to meeting the conditions stated in Section 9.6.1.3, the mean value of the stability balance return loss at the terminating set is not less than 6 dB and the standard deviation not larger than 2.5 dB.

9.6.1.5 Echo Loss (a–b) on Established Connections. In order to minimize the effects of echo on international connections, it is recommended that the distribution of echo loss $(a–b)$ for the population of actual international calls established over the national system should have a mean value of not less than $15 + n$ dB with a standard deviation not exceeding $(9 + 4n)^{1/2}$, where n is the number of analog or mixed analog–digital four-wire circuits in the national chain.

9.7 Via Net Loss

Via net loss (VNL) is a concept or method of transmission planning that permits a relatively close approach to an overall zero transmission loss in the telephone network (lowest practicably attainable) and maintains singing and echo within specified limits. The two criteria that follow are basic to VNL design:

1. Customer–customer talker echo should be satisfactorily low on more than 99% of all telephone connections that encounter the maximum delay likely to be experienced.

2. The total amount of overall loss is distributed throughout the trunk segments of the connection by allocation of loss to the echo characteristics of each segment.

One important concept in the development of the discussion on VNL is that of echo return loss (ERL) (see Section 9.4). For this discussion, we consider ERL as a single-valued weighted figure of return losses in the frequency band 500–2500 Hz. Echo return loss differs from return loss in that it takes into account a weighted distribution of level versus frequency to simulate the non-linear characteristics of the transmitter and receiver of the telephone instrument. By using ERL measurements, it is possible to arrive at a basic design factor for the development of the VNL formula. This design factor states that the average return loss at class 5 offices (local exchanges) is 11 dB, with a standard deviation of 3 dB. Considering a standard distribution curve and the one, two, or three σ points on the curve, we could thus expect practically all measurements of ERL to fall between 2 dB and 20 dB at class 5 offices (local exchanges). Via net loss also considers that reflection occurs at the far end in relation to the talker where the four-wire trunks connect to the two-wire circuits (i.e., at the far-end hybrid).

The next concept in the development of the VNL discussion is overall connection loss (OCL), which is the value of one-way trunk loss between two end (local) offices (not subscribers). Consider that

$$\text{Echo path loss} = 2 \times \text{Trunk loss (one-way)} + \text{Return loss (hybrid)}$$

where all units are in decibels. Now let us consider the average tolerance for a particular echo path loss. Average echo tolerance is taken from the curve in Figure 6.23. Therefore

$$\text{OCL} = \frac{\text{Average echo tolerance (loss)} - \text{Return loss}}{2} \tag{6.3}$$

where all units are in decibels. Return loss in this case is the average echo return loss that must be maintained at the distant local exchange—the 11 dB given earlier.

An important variability factor not considered in the formula is trunk stability, which determines how close assigned levels are maintained on a trunk. Via net loss practice dictates trunk stability to be maintained with a normal distribution of levels and a standard deviation of 1 dB in each direction. For a round-trip echo path the deviation is taken as 2 dB. This variability applies to each trunk in a tandem connection. If there are three trunks in tandem, this deviation must be applied to each.

The reader will recall that the service requirement in VNL practice is satisfactory echo performance for 99% of all connections. This may be considered a cumulative distribution, or 2.33 standard deviations, summing from negative infinity toward the positive direction. The OCL formula may now be rewritten as follows:

$$\text{OCL} = \frac{\text{Average echo tolerance} - \text{Average return loss} + 2.33D}{2} \tag{6.4}$$

where D is the composite standard deviation of all functions and all units are in decibels. The derivation of D, the composite standard deviation of all functions, is as follows:

$$D = \sqrt{D_t^2 + D_{rl}^2 + ND_l^2} \qquad (6.5)$$

where D_t is the standard deviation of distribution of echo tolerance among a large group of observers, given as 2.5 dB; D_{rl} is the standard deviation of distribution of return loss, given as 3 dB; D_l is the standard deviation of distribution of the variability of trunk loss for a round-trip echo path, given as 2 dB; and N is the number of trunks switched in tandem to form a connection class 5 office to class 5 office. Now consider several trunks in tandem; it can be calculated that at just about any given echo path delay, the OCL increases approximately 0.4 dB for each trunk added. With this simplification, once we have the OCL for one trunk, all that is needed to compute the OCL for additional trunks is to add 0.4 dB times the number of trunks added in tandem. This loss may be regarded as an additional constant needed to compensate for variations in trunk loss in the VNL formula.

Figure 6.26 relates echo path delay (round-trip) to overall connection loss (OCL for one trunk, then for a second trunk in tandem, and for four and six trunks in tandem). Although the straight-line curve has been simplified, the approximation is sufficient for engineering VNL circuits. Note that the straight-line curve in Figure 6.26 cuts the Y axis at 4.4 dB, where round-trip delay is 0. This 4.4 dB is based on two conditions, namely, that all trunks have a minimum of 4 dB to control singing and that there is 0.4 dB protection against negative variation of trunk loss. Another important point to be defined on the linear curve in Figure 6.26 is a round-trip delay of 45 ms, which corresponds to an OCL of 9.3 dB. Empirically, it has been determined that echo suppressors must be used for delays greater than 45 ms. From this same linear curve the following formula for OCL may be derived:

$$\text{OCL} = (0.102)(\text{Path delay in ms})$$
$$+ (0.4 \text{ dB})(\text{Number of trunks in tandem}) + 4 \text{ dB} \qquad (6.6)$$

Usually the 4 dB* as shown in the preceding OCL equation is applied to the extremity of each trunk network, namely, to the toll-connecting trunks, 2 dB to each.

Overall connection loss deals with the losses of an entire network consisting of trunks in tandem, whereas VNL deals with the losses assigned to one trunk. The VNL formula follows from the OCL formula. The key here is the round-trip delay on the trunk in question. The delay time for a

---

*This value in equation 6.6 has been changed to 5 dB or 2.5 dB at each end applied to the toll-connecting trunks. The additional 0.5 dB added to each end accommodates nominal switch loss at the extreme ends of the circuit.

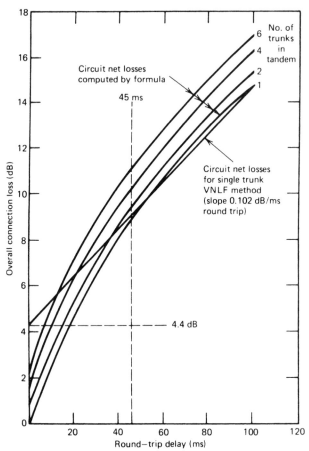

Figure 6.26. Approximate relationship between round-trip echo delay and overall connection loss (OCL) [16].

transmission facility employing only one particular medium is equal to the reciprocal of the velocity of propagation of the medium multiplied by the length of the trunk. To obtain round-trip time, this figure must be multiplied by 2; thus

$$\text{VNL} = 0.102 \times 2 \times \frac{1}{\text{Velocity of propagation}}$$
$$\times \text{(One-way length of the trunk)} + 0.4 \text{ dB} \qquad (6.7)$$

Another term is often introduced to simplify the equation, the via net loss factor (VNLF):

$$\text{VNL} = \text{VNLF} \times \text{(One-way length of trunk in miles)} + 0.4 \text{ dB} \qquad (6.8)$$

$$\text{VNLF} = \frac{2 \times 0.102}{\text{Velocity of propagation of the medium}} \text{ (dB/mi)} \qquad (6.9)$$

The velocity of propagation of the medium used here must be modified by such things as delays caused by repeaters, intermediate modulation points, and facility terminals.

Via net loss factors for loaded two-wire facilities are 0.03 dB/mi, with H-88 loading on 19-gauge wire and increased to 0.04 dB/mi on B-88 and H-44 facilities. On four-wire carrier and radio facilities, the factor improves to 0.0015 dB/mi. For connections with round-trip delay times in excess of 45 ms, the standard VNL approach must be modified. As mentioned previously, these circuits use echo suppressors that automatically switch about 50 dB into the echo return path, and the switch actuates when speech is received in the "return" path, thus switching the pad into the "go" path.

Via net loss practice in North America treats long delay circuits with up to a maximum of 45 ms of delay in the following manner (see Figure 6.27). Here the total round-trip delay is arbitrarily split into two parts for connections involving regional intertoll trunks. If the regional intertoll delay exceeds 22 ms, echo suppressors are used. If the figure is 22 ms or less, echo is controlled by VNL design. Thus allow 22 ms for the maximum delay for the regional intertoll segments to the connection. This leaves 23-ms maximum delay for the other segments $(45-22)$. Now apply the VNL formula for a delay of 22 ms. Thus

$$\text{VNL} = 0.102 \times 22 + 0.4 = 2.6 \text{ B}$$

This loss is equivalent to the maximum length of an intertoll trunk without an echo suppressor. What is that length?

$$\begin{aligned}
\text{Length (one-way)} &= \frac{\text{VNL} - 0.4}{\text{VNLF}} \text{ mi} \\
&= \frac{0.102 \times 22 + 0.4 - 0.4}{0.0015} = 1498 \text{ mi*} \qquad (6.10)
\end{aligned}$$

In summary, in VNL design we have three types of loss that may be assigned to a trunk:

Type	Loss
Toll-connecting trunk	VNL + 2.5 dB
Intertoll trunk (no echo suppressor)	VNL
Intertoll trunk (with echo suppressor)	0 dB

From Ref. 16 (also consult Ref. 20).

*The VNLF value in this equation indicates carrier and/or radio for the whole trunk.

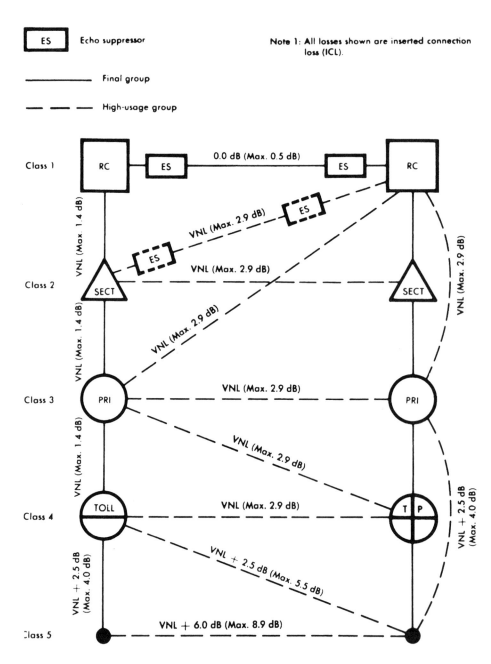

Figure 6.27. Trunk losses with VNL design. Losses include 0.4-dB design loss allowed for maintenance. Copyright © American Telephone and Telegraph Co., 1977 [23].

9.8   Loss Plan for the Evolving Digital Networks (US)

For digital connections terminated in analog access lines, the required loss values are dependent on the connection architecture:

- For interLATA or interconnecting network connections, the requirement is 6 dB.

- For intraLATA connections involving different LECs (local exchange carriers), 6 dB is the preferred value although 3 dB may apply to connections not involving a tandem switch.

- For intraLATA connections involving the same LEC, the guidelines are:

$$0\text{--}6 \text{ dB (typically 0 dB, 3 dB, or 6 dB)}$$

The choice of network loss value depends on performance considerations, administrative simplicity, and current network design. (From Bellcore *BOC Notes on the LEC Networks—1994* [18].)

ANSI adds the following guidelines for intraLATA, same exchange carrier operation.

- For intra-DEO/RSU* connections between metallic access lines, 0–6 dB is recommended for DEOs or RSUs with line lengths up to 12 kilofeet, and 0 dB for DEOs or RSUs with line lengths above 12 kilofeet. The selection of appropriate loss values are administered by the exchange carrier and is not required to be an automatic function of the switch.

- For connections involving an RSU/RDT† or multiple RSUs/RDTs, two considerations apply:

  1. The length of the connected metallic loops.

  2. The value of the round-trip delay between the host and remote.

When considering the length of the connected metallic loops, the desired value of network loss is the same as that for intra-DEO/RSU connections between metallic access lines. The following guidelines can be applied for the selection of the network loss based on the round-trip delay between the host and remote:

(a) 0 dB can be used up to 3.2 ms round-trip delay (approximately 80 miles).

(b) 3 dB can be used up to 8 ms round-trip delay (approximately 360 miles).

---

*DEO is digital end office; RSU is remote switching unit.

†RDT is a remote digital terminal.

(c) 6 dB can be used up to 12 ms round-trip delay (approximately 600 miles).

Network loss, when added to the connection, is added only in the receive path.

To facilitate the use of the same network for both analog and digital services, it is desirable to insert loss, where required, as near the end-user terminal as possible. However, for practical reasons, it may be necessary to administer network loss that is dependent upon the type of connection, at the point of switching nearest the end-user terminal. Loss values that are not dependent upon the type of connection can be inserted at the D/A conversion point, which may be at the DEO, RSU, RDT, IDLC (integrated digital loop carrier), or ONU (optical network unit), or at the last point of switching. (From ANSI T1.508–1992 [22].) (Consult also Refs 18 and 23.)

REVIEW QUESTIONS

1. What are the three basic underlying considerations in the design of a long-distance (toll) network?

2. What is the fallacy of providing just one high-capacity trunk across the United States to serve all major population centers by means of tributaries off that main trunk?

3. How can the utilization factor of trunks be improved?

4. For long-distance (toll) switching centers, what is the principal factor involved in the placement of such exchanges (differing from local exchange placement substantially)—economy, subscriber density, altitude, or what?

5. How are the highest levels of a national hierarchical network connected, and why is this approach used?

6. On a long-distance (toll) connection, why must the number of links in tandem be limited?

7. Describe in one short sentence the current approach to structure of the international network as recommended by CCITT.

8. What type of routing is used on the majority of international connections?

9. Why do we limit the number of satellites used on an international full-duplex speech telephone connection?

10. Name three principal factors used in deciding how many and where toll (long-distance) exchanges will be located in a given geographic area.

11. Discuss the impact of fan outs on the number of hierarchical levels in a national network.

12. Name the three principal bases required at the outset for the design of a toll (long-distance) network.

13. In the design of a long-distance (toll) network, once the hierarchical levels have been established, what is assembled next?

14. Define a final route.

15. Define high-usage (HU) route and direct route.

16. A grade of service no greater than _____% per link is recommended on a final route.

17. When assembling a traffic matrix, at the ends of what forecast period are the traffic intensities valid for?

18. Define a network hierarchy from the point of view of a routing structure.

19. There are two generic types of routing schemes. What are they?

20. Name the three different types of dynamic routing and explain each type in one sentence.

21. Compare progressive call control procedures with originating call control.

22. What is *crankback*?

23. Give an example of state-dependent routing.

24. What two factors does AT&T's DNHR routing depend on?

25. One advantage of alternate (alternative) routing is economy. What is the second advantage?

26. Dynamic routing technique (DRT) differs from current routing operations in two important areas. What are these two areas?

27. What is the principal cause of echo in the telephone network?

28. What causes singing in the telephone network?

29. Differentiate balance return loss and echo return loss.

30. What is the return loss of a two-wire terminal of a hybrid where the subscriber loop has a 300-Ω impedance and the balancing network is set for 900 Ω?

31. How can we control echo? Two answers are required.

32. The stability of a telephone connection depends on three factors. Give just two of these factors.

33. Based on the CCITT (ITU-T) loss plan for established connections, where a particular connectivity traverses three links, what loss will be inserted to control echo/singing?

34. On the new loss plan for North America for the digital network, how much loss is inserted at each for a circuit 500 mi (800 km) in length?

REFERENCES

1. *National Networks for the Automatic Service*, ITU, Geneva, 1964.

2. *CCITT Recommendations*, Blue Books, Fascicle VI.1, "General Recommendations on Telephone Switching and Signaling," Recs. Q.1–Q.118 bis (Study Group XI), IXth Plenary Assembly, Melbourne, 1988.

3. *International Routing Plan*, CCITT Rec. E.171, Blue Books, Vol. II, Fascicle II.2, IXth Plenary Assembly, Melbourne, 1988.

4. *Telecommunication Planning*, ITT Laboratories Madrid (Spain), 1973 (in particular, Section 2, "Networks").

5. *Traffic Routing*, CCITT Rec. E.170, ITU Geneva, October 1992.

6. G. R. Ash, "Design and Control of Networks with Dynamic Nonhierarchical Routing," *IEEE Communications Magazine*, IEEE, New York, October 1990.

7. *CCITT Recommendations*, Vol. III, Fascicle III.1, Blue Books, IXth Plenary Assembly, Melbourne, 1988.

8. J. E. Flood, *Telecommunication Networks*, IEE Series, London, 1974.

9. *General Network Planning*, CCITT, ITU Geneva, 1983.

10. "Optimization of Telephone Networks with Hierarchical Structure," (computer program), ITT Laboratories Madrid, Spain, 1973.

11. "Optimization of Telephone Trunking Networks with Alternate Routing" (computer program), ITT Laboratories Madrid, Spain, 1973.

12. R. L. Freeman, *Telecommunication Transmission Handbook*, 3rd ed., John Wiley & Sons, New York, 1991.

13. F. T. Andrews and R. W. Hatch, "National Telephone Network Planning in the AT&T," *IEEE Commun. Tech. J.* (June 1971).

14. Ramses R. Mina, *Introduction to Teletraffic Engineering,"* Telephony Publishing Corporation, Chicago, 1974.

15. *Theory of Telephone Traffic: Tables and Diagrams*, Siemens, Berlin–Munich, Part 1, 1971.

16. M. A. Clement, "Transmission," reprint from *Telephony* (magazine), Telephony Publishing Corporation, Chicago, 1969.

17. *Transmission Systems for Communications*, 5th ed., Bell Telephone Laboratories, Holmdel, NJ, 1982.

18. *BOC Notes on the LEC Networks—1994*, Issue 2, SR-TSV-002275, Bellcore, Piscataway, NJ, 1994.

19. *Influence of National Systems on Stability and Talker Echo in International Connections*, ITU-T Rec. G.122, Helsinki, March 1993.

20. *Engineering and Operations in the Bell System*, 2nd ed., Bell Telephone Laboratories, Holmdel, NJ, 1984.

21. *ISDN Routing Plan*, CCITT Rec. E.172, ITU, Geneva, October 1992.

22. "Network Performance—Loss Plan for Evolving Digital Networks," ANSI T1.508–1992, ANSI, New York, July 1992.

23. *Telecommunication Transmission Engineering*, 2nd ed., Vols. 1 and 2, AT&T, New York, 1977.

7

THE DESIGN OF LONG-DISTANCE LINKS

1 INTRODUCTION

In Chapter 6 we proposed a methodology for the design of a long-distance network. The network may be defined as a group of switching nodes inter-connected by links. We may refer to a link as a transmission highway be-tween switches carrying one or more traffic relations. The link could appear as that in the following diagram, where switches A, B, and C are connected to switches X, Y, and Z over a link as shown. The discussion that follows introduces the essentials of transmission design of such links.

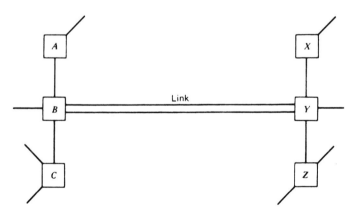

2 THE BEARER

British telecommunication engineers are fond of the term "bearer," which is quite descriptive. The bearer is what carries the information signal(s). It could be a pair of wires or two pairs on a four-wire basis, a radio carrier in each direction, a coaxial cable, or a fiber-optics cable. The wire pair could be open wire lines, aerial cable, or buried cable. In the text that follows it is assumed that the bearer will be transporting some sort of multiplex configuration, probably in a digital format as discussed in Chapter 8.

Modern long-distance links use either radio as the medium of choice or fiber-optic cable. The decision on which one to use is driven by economics more than any other factor. However, capacity can well be another deciding factor. If a requirement for a certain trunk group connectivity exceeds 10,000 equivalent digital voice channels at the end of a forecast period, optical fiber would certainly be the medium of choice at the outset. We do not configure a link for today's traffic requirements. We size the link for probable forecast requirements, or at least engineer a link for that future expected expansion.

Coaxial cable has purposefully been left out of the discussion. It remains, of course, as a transmission medium such as a radio-frequency transmission line for radio systems, and it is very much a contender in cable television systems. Optical fiber is so far superior as transport for telecommunication digital configurations that coaxial cable must be removed from contention. Coaxial cable requires many more active repeaters per unit length than fiber optics. Jitter, a major transmission impairment on digital systems, builds up as a function of the number of repeaters in tandem. Another reason to favor optical fiber is that it needs no equalization, whereas coaxial cable needs equalizers even for modest digital configurations.

3   INTRODUCTION TO RADIO TRANSMISSION

Wire, cable, and fiber are well-behaved transmission media, and they display little variability in performance. The radio medium, on the other hand, displays notable variability in performance. The radio-frequency spectrum is shared with others and requires licensing. Metallic and fiber media need not be shared and do not require licensing (but often require right-of-way).

A major factor in the selection process is information bandwidth. Fiber optics seems to have nearly an infinite bandwidth. Radio systems have very limited information bandwidths. It is for this reason that radio-frequency bands 2 GHz and above are used for PSTN and private network applications. In fact the U.S. Federal Communications Commission requires that users in the 2-GHz band must have systems supporting 96 digital voice channels where bandwidths are still modest. In the 4- and 6-GHz bands, available bandwidths are 500 MHz allocated in 20- and 30-MHz segments for each radio-frequency carrier.

One might ask why use radio in the first place if it has so many drawbacks. Often, it turns out to be less expensive compared to fiber-optic cable. But there are other factors such as:

- No requirement for right-of-way
- Less vulnerable to vandalism
- Not susceptible to "accidental" cutting of the link
- Often more suited to crossing rough terrain
- Often more practical in heavily urbanized areas
- As a backup to fiber-optic cable links.

Fiber-optic cable systems provide strong competition with line-of-sight (LOS) microwave, but LOS microwave does have a place and a good market.

Satellite communications is an extension of line-of-sight microwave. It is also feeling the "pinch" of competition from fiber-optic systems. It has two drawbacks. First, of course, is limited information bandwidth. The second is excessive delay when the popular geostationary satellite systems are utilized. It also shares frequency bands with LOS microwave.

One application showing explosive growth is very small aperture terminal (VSAT) systems. It is very specialized and has great promise for certain enterprise networks, and there are literally thousands of these networks now in operation.

Another application that will start to show explosive deployment in the next several years will be forms of mobile/cellular and personal communications such as Motorola's Iridium system. This system nearly eliminates the notorious delay problem by using low earth orbit (LEO) satellites.

4 DESIGN ESSENTIALS FOR LINE-OF-SIGHT MICROWAVE SYSTEMS

4.1 Introduction

Line-of-sight (LOS) microwave provides broadband bearer connectivity over a link or series of links in tandem. We can take advantage of this "line-of-sight" phenomenon at frequencies from 150 MHz and upwards into the millimeter spectrum. Each link can be up to 30 miles (46 km) long or more depending on terrain topology. Some links extend over 100 miles (160 km). A series of LOS links is shown in Figure 7.1. Perhaps the key term here is line-of-sight. It implies that the antenna of the radiolink on one end has to be able to "see" the antenna on the other end. This may not necessarily be true, but it does give some idea of the problem.

Let us suppose *smooth earth*. This means earth with no mountains or ridges, buildings, or sloping ground of any sort. Here our LOS distance is limited by the horizon. Given an LOS microwave antenna height h_{ft} or h'_m above ground surface, the distance d_{mi} or d_{km} to the horizon just where the ray beam will graze the rounded earth surface horizon can be calculated using one of the formulas given below.

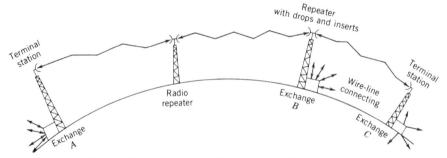

Figure 7.1. A sketch of an LOS microwave radio relay system.

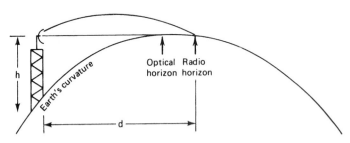

Figure 7.2. Radio and optical horizon (smooth earth)

To optical horizon: $(k = 1)$

$$d = \sqrt{\frac{3h}{2}} \tag{7.1A}$$

and the radio horizon: $(k = 4/3)$

$$d = \sqrt{2h} \tag{7.1B}$$

$$d' = 2.9(2h')^{1/2} \tag{7.1C}$$

Equation 7.1A gives the distance to the optical horizon (i.e., the ray travels a straight line) in miles (d) and feet (h), equation 7.1B gives the distance to the radio horizon in miles and h is in feet, and equation 7.1C gives the distance to the radio horizon in kilometers and h is in meters. The concept of optical and radio horizon is shown in Figure 7.2.

The distance to the radio horizon varies with the index of refraction of the intervening space. Some designers say it is 4/3 the distance to the radio horizon, the microwave ray beam being bent toward the earth. However, this generalization may be overly optimistic under certain circumstances.

The design of a microwave LOS link involves five basic steps:

1. Setting performance requirements.
2. Site selection and preparation of a path profile to determine antenna tower heights.

3. Carrying out a path analysis, also called a *link budget*.

4. Running a path/site survey.

5. Test of the system prior to cutover to traffic.

In the following paragraphs we review each of the five steps.

4.2 Setting Performance Requirements

Often a microwave link is part of an extensive system of multiple links in tandem. Thus we must first set system requirements based on the output of the far-end receiver of the several or many links. If the system were analog, the specification would be given for noise in a voice channel; if it were video, a signal-to-noise ratio specification would be provided. In the case we will emphasize here, it will be a bit error rate on a digital bit stream. This is an electrical signal of "1s" and "0s," probably in one of the serial formats described in Chapter 8.

The specification should be based on an existing standard. If the link being designed is part of the North American PSTN, we would use *BOC Notes on the LEC Networks–1994* [1]. Otherwise, we might turn to the CCITT, now called the ITU-T organization; or to CCIR, now called the ITU-R organization. For example, CCITT Recs. G.821 and G.826 and CCIR Rec. 594-3 [2–4] are excellent sources. Rec. 594-3 states that the bit error ratio should not exceed 1×10^{-6} during more than 0.4% of any month and 1×10^{-3} during more than 0.054% of any month. These values are taken at the end of a hypothetical reference digital path. Often the BER (bit error rate or ratio) is set for 1×10^{-9} during unfaded conditions for a single link in a large system to assure conformance to the CCITT and CCIR Recommendations. The specification should be set for a certain time availability (e.g., 99.9% of a month) to be in conformance with the time distributions given in these recommendations (e.g., 0.054% of a month).

4.3 Site Selection and Preparation of a Path Profile

4.3.1 Site Selection

In this step we will select operational sites where we will install and operate radio equipment. After site selection, we will prepare a path profile of each link to determine the heights of radio towers to achieve "line of sight." Sites are selected using large topographical maps. If we are dealing with a long system crossing a distance of hundreds of miles or kilometers, we should minimize the number of sites involved. There will be two terminal sites, where the system begins and ends. Along the way, repeater sites will be required. At some repeater sites, we may have need to drop and insert traffic. Other sites will just be repeaters. This concept is shown in Figure 7.3. The figure shows the drops and inserts of traffic at telephone exchanges. These drop and insert

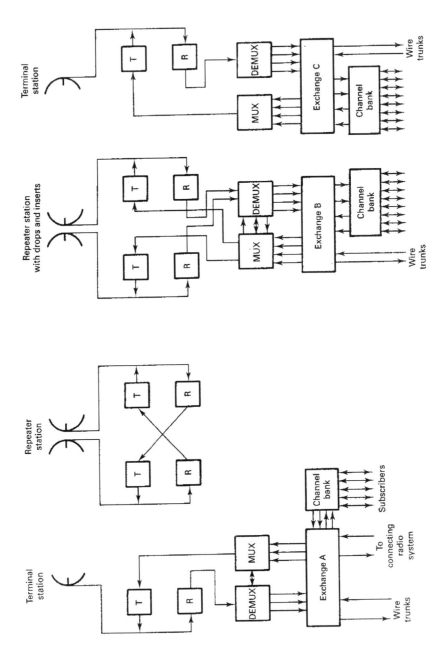

Figure 7.3. Simplified functional block diagram of the LOS microwave system shown in Figure 7.1.

points may just as well be buildings or other facilities in a private/corporate network. There is considerable iteration between site selection and path profile preparation to optimize the route.

In essence, the sites selected for drops and inserts will be points of traffic concentration. There are several trade-offs to be considered:

1. Bringing traffic in by wire or cable rather than adding drop and insert (add–drop) capabilities at relay points.
2. Siting based on propagation advantages (or constraints) only versus colocation with exchange (or corporate facility) (saving money for land and buildings).
3. Method of feeding (feeders)*: by light-route radio, fiber-optic cable, and wire-pair cable.

In gross system design, exchange location (or corporate facility location), particularly with tandem/transit exchanges, must be considered in light of probable radio and cable routes. Another consideration is electromagnetic compatibility (EMC). Midcity repeater-relay or terminal sites have the following advantages:

- Colocation with a local or toll exchange.
- Use of tall buildings as natural towers.

And they have the following disadvantages:

- Wave reflections (multipath) off buildings.
- Electromagnetic compatibility (EMC) problems, particularly from other nearby emitters and industrial emission.
- Low-grade labor market.

Sitings in the country have fewer EMC problems and usually a better labor force (for operators, other operational personnel, and technicians), and right-of-way for cable is easier.

We are now led to propagation constraints. Terminal sites will be in or near heavily populated areas and preferably colocated with a toll exchange. The tops of modern large office buildings, if properly selected, are natural towers. Relay sites are heavily influenced by intermediate terrain. Accessible hilltops or mountain tops are good prospective locations. Draw a line along the path of the desired route. Sites would zigzag along the line with optical or "radio" separation distances. If tower costs were $300 per foot ($900 per meter), 300-ft or 100-m towers might be the height limit for economic

---

*Here the word *feeders* refers to feeding the mainline trunk radio system. Feeders could be called *spurs*.

reasons. If hilltop or mountaintop sites are well selected, towers that high may never have to be considered. High towers are the rule over flat country. The higher the tower, the longer the line-of-sight distance. Thus, on a given link, fewer repeaters would be required if towers could be higher. Hence there is a trade-off between tower height and number of repeaters.

4.3.2 Calculation of Tower Heights

Assume now that sites along a microwave radio relay route have been carefully selected. The next step in engineering is the determination of tower heights. The objective is to keep the tower height as low as possible and still maintain effective communication. The towers must just be high enough for the radio beam to surmount obstacles in the path. As the discussion proceeds, the term "high *enough*" is carefully defined. What obstacles might there be in the path? To name some, there are terrain such as mountains, ridges, hills, and earth curvature—which is highest at midpath—and buildings, towers, grain elevators, and so on.

All obstacles along the path must be scaled on graph paper in an exercise called *path profiling*. Good topographical maps are required of the region. Ideally, such maps should be 1:24,000, although 1:62,500 maps are acceptable. A straight line is drawn between the sites in question and then on linear graph paper scaled to 1 in. for 2 mi on the horizontal or 1 cm for 1 or 2 km. Vertical scales depend on the rate of change of elevation along the path. An ideal scaling is 100 ft/in. or 1 cm for 10 m and over hilly country, 1 in. equivalent to 200 ft or 1 cm equivalent to 20 m. In mountainous country the vertical scale may have to be as much as 1 in. equivalent to 1000 ft or 1 cm equivalent to 100 m. Each obstacle encountered must be identified with a letter or number on the horizontal scale. The next step is to establish a point directly on top of each obstacle, giving altitude above mean sea level. The bottom of the chart need not be mean sea level; it may be mean sea level plus so many meters. Once the reference altitude has been established, we must give several additional clearances. If the obstacle is terrain with vegetation, especially trees, a clearance for trees and growth must be established. If no other values are available, use 40 ft (12 m) and 10 ft (3 m), respectively.

To the altitude or height of each obstacle must be added "earth bulge," the number of feet or meters an obstacle is raised higher in elevation (into the path) as a result of earth curvature (EC) or "earth bulge." The amount of earth bulge at any point in the path may be calculated by the formula(s):

$$h = 0.677d_1 d_2 \qquad (h \text{ in feet; } d \text{ in miles}) \qquad (7.2\text{A})$$

$$h = 0.078d_1 d_2 \qquad (h \text{ in meters; } d \text{ in km}) \qquad (7.2\text{B})$$

where d_1 is the distance from the near end of the hop to the obstacle in question and d_2 is the distance from the far end of the hop to the obstacle in question. Equation 7.2 is for a ray beam that is a straight line (i.e., no bending). Atmospheric refraction may cause the beam to be bent either toward or away from the earth. This bending effect is handled by adding the factor K to equation 7.2, where

$$K = \frac{\text{Effective earth radius}}{\text{True earth radius}}$$

such that

$$h_{\text{ft}} = \frac{0.667 d_1 d_2}{K} \qquad (d \text{ in miles}) \qquad (7.3A)$$

$$h_{\text{m}} = \frac{0.078 d_1 d_2}{K} \qquad (d \text{ in km}) \qquad (7.3B)$$

If the factor K is greater than 1, the ray beam is bent toward the earth and the radio horizon is greater than the optical horizon. If K is less than 1, the radio horizon is less than the optical horizon. For general system planning purposes, $K = \frac{4}{3}$ may be used. However, for specific path engineering, K must be selected with care. The value of h or earth curvature corrected for K from equation 7.3 must be added to obstacle height in the path-profile exercise for each obstacle.

Still another factor must be added to obstacle height, namely, Fresnel zone clearance. This factor derives from the electromagnetic wave theory that a wavefront, which our ray beam is, has expanding properties as it travels through space. These expanding properties result in reflections and phase transitions as the wave passes over an obstacle. The outcome is an increase or a decrease in received signal level. The amount of additional clearance over obstacles that must be allowed to avoid problems of the Fresnel phenomenon (diffraction) is expressed in Fresnel zones. The first Fresnel zone radius may be calculated from the following formula:

$$R_{\text{ft}} = 72.1 \sqrt{\frac{d_1 d_2}{FD}} \qquad (7.4A)$$

where F is the frequency in gigahertz, d_1 is the distance from transmit antenna to obstacle (statute miles), d_2 is the distance from path obstacle to receive antenna (statute miles), and $D = d_1 + d_2$. For metric units:

$$R_{\text{m}} = 17.3 \sqrt{\frac{d_1 d_2}{FD}} \qquad (7.4B)$$

where F is the frequency in gigahertz and d_1, d_2, and D are the same as in equation 7.4A, but d and D are in kilometers and R in meters.

Previously, a clearance of 0.6 Fresnel zone (0.6 is the value of R in equation 7.4) was considered sufficient. A new rule of thumb is evolving,

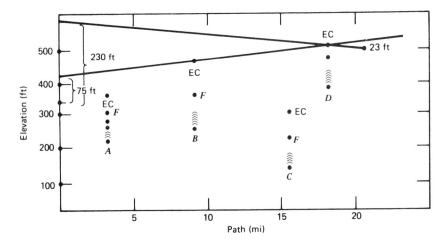

	Obstacle	d_1	d_2	Basic Height (ft)	F (Fresnel) (ft)	EC (ft)	T and G (ft)	Adjusted Total Height (ft)
Tree conditions: 40 + 10 ft growth (T and G)	A	3.5	19.0	220	30	49	50	349
Frequency band: 6 GHz	B	10	12.5	270	41	91.7	50	452.7
Midpath Fresnel (0.6)	C	17	5.5	160	36	68.6	50	314.6
= 42 ft	D	20	2.5	390	25.2	36.6	50	501.8

Figure 7.4. Practice path profile (x in miles, y in feet; assume that $K=0.9$). Note: EC = earth curvature of each bulge.

namely, when $K = \frac{2}{3}$, at least 0.3 Fresnel zone clearance is required, and 1.0 Fresnel zone clearance must be allowed when $K = \frac{4}{3}$. At points near the ends of a path, Fresnel zone clearances should be at least 6 m or 20 ft [5].

The three basic increment factors that must be added to obstacle heights are now available: vegetation height and its growth, earth bulge corrected for K factor, and Fresnel zone clearance. These are marked as indicated previously on our path-profile chart. A straight line is drawn from right to left, just clearing the obstacle points as corrected for the three factors. Another line is then drawn from left to right. A sample profile is shown in Figure 7.4. Some balance is desirable so at one extreme we have a very tall tower and at the other extreme we have a little stubby tower. This is true but for one exception: when a reflection point exists at an inconvenient spot along the path.

4.3.3 Reflection Point. Possible reflection points may be obtained from the profile. The objective is to adjust tower heights such that the reflection

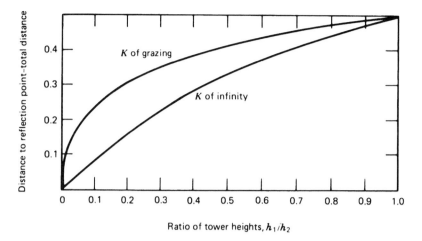

Figure 7.5. Calculation of reflection points.

point is adjusted to fall on land area where the reflected energy will be broken up and scattered. Bodies of water and other smooth surfaces cause reflections that are undesirable. Figure 7.5 can facilitate calculations for the adjustment of the reflection point. It uses a ratio of tower heights, h_1/h_2, and the shorter tower height is always h_1. The reflection area lies between a K factor of grazing ($K = 1$) and a K factor of infinity. The distance expressed is always from h_1, the shorter tower. The reflection point can be moved by adjusting the ratio h_1/h_2.

For a path that is highly reflective for much of its length, space-diversity operation may minimize the effects of multipath reception.

4.4 Path Analysis or Link Budget

4.4.1 Introduction

The path analysis (or link budget) is carried out to dimension the link. What is meant here is to establish operating parameters such as transmitter power output, parabolic antenna aperture (diameter), and receiver noise figure, among others. The link is assumed to be digital. Digital formats are described in Chapter 8. The type of modulation and modulation rate (number of transitions per second) are also important parameters.

Table 7.1 shows basic parameters in two columns. The first we call "normal" and would be the most economic; the second column is titled "special," giving improved performance parameters, but at an increased price.

Diversity reception is another option that may wish to be considered. It entails greater expense. The options in Table 7.1 and diversity reception will be addressed further on.

TABLE 7.1 LOS Microwave Basic Equipment Parameters

Parameter	Normal	Special	Comments
Transmitter power	1 W	10 W	500 mW common above 10 GHz
Receiver noise figure	8–12 dB	Down to 1.2 dB	Use of LNA (low noise amplifier)
Antenna	Parabolic 2–12 ft	Same	Antennas over 12' not recommended
Modulation	64–128 QAM	Up to 512 QAM	Based on bandwidth/bit rate constraints

4.4.2 Approach

We can directly relate the desired performance to the receive signal level (RSL) at the first active stage of the far-end receiver and the receiver's noise characteristics. A reference RSL is established.

Next, we calculate the free-space loss between the transmit antenna and receive antenna. This is a function of distance and frequency (i.e., the microwave transmitter operational frequency). We then calculate the EIRP (effective isotropically radiated power) at the transmit antenna. The EIRP is the sum of the transmitter power output, minus transmission line losses plus the antenna gain, all in decibel units.

When we add the EIRP to the free-space loss (in dB), the result is the isotropic* receive level (IRL). When we add the receive antenna gain to the IRL and subtract the receive transmission line losses, we get the receive signal level (RSL).

This relationship of path gains and losses is shown in Figure 7.6.

Path Loss. For all intents and purposes, path loss up to about 10 GHz can be considered as only "free-space loss." To introduce the reader to the problem, consider an isotropic antenna—that is, an antenna that radiates uniformly in all directions. If the isotropic radiator is fed by a transmitted power P_t, it radiates $P_t/4\pi d^2$ (W/m^2) at a distance d, and if a radiator has a gain G_t, the power flow is enhanced by the factor G_t. Finally, the power intercepted by an antenna of effective cross section A (related to the gain by $G_r = 4\pi A/\lambda^2$) is $P_t G_t G_r (\lambda/4\pi d)^2$. The term $(\lambda/4\pi d)^2$ is known as the free-space loss and represents the steady decrease of power flow (in W/m^2) as the wave propagates. From this we can derive the more common formula of free-space path loss, which reduces to

$$L = 96.6 + 20 \log_{10} F + 20 \log_{10} D \qquad (7.5A)$$

*An isotropic antenna is an antenna that is uniformly omnidirectional with 0 dB gain. It is an imaginary reference antenna. The isotropic receive level is the power level we would expect to achieve at that point using an isotropic antenna.

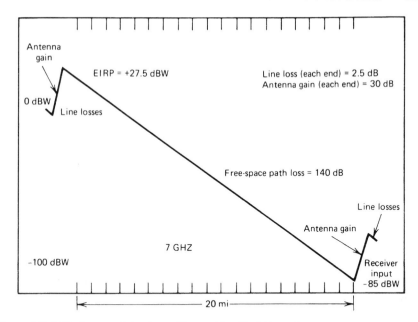

Figure 7.6. LOS microwave link gains and losses (simplified). Transmitter output is 1 watt or 0 dBW.

where L is the free-space attenuation between isotropic antennas in dB, F is the frequency in GHz, and D is the path distance in statute miles. In the metric system we obtain

$$L_{dB} = 92.4 + 20 \log_{10} F_{GHz} + 20 \log_{10} D_{km} \qquad (7.5B)$$

Consider the problem from a different aspect. It requires 22 dB to launch a wave to just 1 wavelength (1λ) distant from an antenna. Thus for an antenna emitting $+10$ dBW, we could expect the signal one wavelength away to be 22 dB down, or -12 dBW. Whenever we double the distance, we incur an additional 6 dB of loss. Hence at 2λ from the $+10$-dBW radiator, we would find -18 dBW; at 4λ, -24 dBW; 8λ, -30 dBW; and so on. Now suppose that we have an emitter where $F = 1$ GHz. What is the path loss at 1 statute mile?

$$L = 96.6 + 20 \log_{10} 1 + 20 \log_{10} 1 = 96.6 \text{ dB}$$

From rough calculations, the 6-dB relationship is worthwhile and also gives insight in that if we have a 20-mi path and shorten or lengthen it by a mile, our signal level will be affected little.

Calculation of EIRP. Effective isotropically radiated power is calculated by adding decibel units: the transmitter power output (in dBm or dBW), the

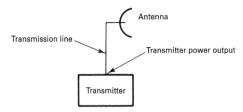

EIRP = Trans. output (dBW) – Trans. line loss (dB) + Ant. gain (dB) (7.6)

Figure 7.7. Elements in the calculation of EIRP.

transmission line losses in dB (a negative value because it is a loss), and the antenna gain in dBi.* Figure 7.7 shows this graphically.

Example
If a microwave transmitter has a 1-watt (0-dBW) power output, the waveguide loss is 3 dB, and the antenna gain is 34 dBi, what is the EIRP in dBW?

$$\text{EIRP}_{dBW} = 0 \text{ dBW} - 3 \text{ dB} + 34 \text{ dBi}$$
$$= +31 \text{ dBW}$$

Calculation of Isotropic Receive Level (IRL). The IRL is the RF power level impinging on the receive antenna. It would be the power we would measure at the base of an isotropic receive antenna. The calculation is shown graphically in Figure 7.8.

Calculation of Receive Signal Level (RSL). The receive signal level (RSL) is the power level entering the first active stage of the receiver:

$$\text{RSL}_{dBW} = \text{IRL}_{dBW} + \text{Rec. ant. gain (dB)} - \text{Rec. trans. line losses (dB)} \quad (7.8)$$

(*Note*: Power levels can be in dBm as well, but we must be consistent.)

Example
Suppose the isotropic receive level (IRL) was −121 dBW, the receive antenna gain was 31 dB, and the line losses were 5.6 dB. What would the RSL be?

$$\text{RSL} = -121 \text{ dBW} + 31 \text{ dB} - 5.6 \text{ dB}$$
$$= -95.6 \text{ dBW}$$

Calculation of Receiver Noise Level. The thermal noise level of a receiver is a function of the receiver noise figure and its bandwidth. For analog radio

---

*dBi = decibels referenced to an isotropic (antenna).

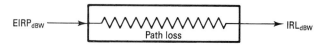

$$IRL_{dBW} = EIRP_{dBW} - Path\ loss_{dB}$$
below 10-GHz Path loss = Free-space loss (FSL) (7.7)

Figure 7.8. Calculation of isotropic receive level.

systems, receiver thermal noise level is calculated using the bandwidth of the intermediate frequency (IF). For digital systems, the noise level of interest is in only 1 Hz of bandwidth using the notation N_0, the noise level in a 1-Hz bandwidth.

The noise that a device self-generates is given by its noise figure (dB) or a noise temperature value. Any device, even passive devices, above absolute zero generates thermal noise. We know the thermal noise power level in a 1-Hz bandwidth of a perfect receiver operating at absolute zero. It is:

$$P_n = -228.6\ dBW/Hz$$

where P_n is the noise power level. Many will recognize this as Boltzmann's constant expressed in dBW.

We can calculate the thermal noise level of a perfect receiver operating at room temperature using the following formula:

$$P_n = -228.6\ dBW/Hz + 10\ log\ 290\ (K)$$
$$P_n = -204\ dBW/Hz$$
$$(7.9)$$

The value, 290 kelvins, is room temperature, or about 17°C or 68°F.

Noise figure simply tells us how much noise has been added to a signal while passing through a device in question. Noise figure (dB) is the difference in signal-to-noise ratio between the input to the device and the output of that same device.

We can convert noise figure to noise temperature in kelvins with the following formula:

$$NF_{dB} = 10\ log(1 + T_e/290)$$
$$(7.10)$$

where T_e is the effective noise temperature of a device. Suppose the noise figure of a device is 3 dB. What is the noise temperature? 290 K.

$$3\ dB = 10\ log(1 + T_e/290)$$
$$0.3 = log(1 + T_e/290)$$
$$1.995 = 1 + T_e/290$$

We round 1.995 to 2; thus

$$2 - 1 = T_e/290$$
$$T_e = 290 \text{ K}$$

The thermal noise power level of a device operating at room temperature is:

$$P_n = -204 \text{ dBW/Hz} + \text{NF}_{dB} + 10 \log \text{BW}_{Hz} \qquad (7.11)$$

where BW is the bandwidth of the device in Hz.

Example

A microwave receiver has a noise figure of 8 dB and its bandwidth is 10 MHz. What is the thermal noise level (sometimes called the thermal noise threshold)?

$$P_n = -204 \text{ dBW/Hz} + 8 \text{ dB} + 10 \log(10 \times 10^6)$$
$$= -204 \text{ dBW/Hz} + 8 \text{ dB} + 70 \text{ dB}$$
$$= -126 \text{ dBW}$$

If the receiver in the above example was operating in a digital regime, we'd want to calculate N_0.

$$N_0 = -204 \text{ dBW/Hz} + \text{NF}_{dB}$$
$$= -196 \text{ dBW/Hz}$$

Calculation of E_b/N_0 in Digital Radio Systems. Many readers are familiar with signal-to-noise ratio (S/N). It was introduced in Chapter 5. In digital systems we use E_b/N_0, meaning energy per bit per noise spectral density ratio. We can relate E_b/N_0 to bit error rate (BER) given the modulation type in question.

We defined N_0 above. E_b is the energy per bit. Suppose the RSL was 1 watt and we were receiving 1000 bits per second. How much energy is imparted to 1 bit? It is 1 mW. We simply divided 1 watt by 1000 bits per second. In radio work it is easier to do the division logarithmically because we work with decibels. E_b can be stated as follows:

$$E_b = \text{RSL} - 10 \log(\text{bit rate}) \qquad (7.12)$$

Example

A certain radio system receives 1.544 Mbps and the RSL is -108 dBW. What is the energy per bit (E_b)?

$$E_b = -108 \text{ dBW} - 10 \log(1.544 \times 10^6)$$
$$= -108 \text{ dBW} - 61.88 \text{ dB}$$
$$= -169.88 \text{ dBW}$$

We can now develop a formula for E_b/N_0:

$$E_b/N_0 = \text{RSL}_{\text{dBW}} - 10 \log(\text{bit rate}) - (-204 \text{ dBW} + \text{NF}_{\text{dB}}) \qquad (7.13)$$

Simplifying we obtain

$$E_b/N_0 = \text{RSL}_{\text{dBW}} - 10 \log(\text{bit rate}) + 204 \text{ dBW} - \text{NF}_{\text{dB}} \qquad (7.14)$$

Some Notes on E_b/N_0 and Its Use
E_b/N_0, for a given BER, will be different for different types of modulation (e.g., FSK, PSK, QAM, etc.).

When working with E_b, we divide RSL by the bit rate, not the symbol rate nor the baud rate.

There is a theoretical E_b/N_0 and a practical E_b/N_0. The practical is always a greater value than the theoretical, greater by the *modulation implementation loss* in dB.

Figure 7.9 is an example where we can relate BER to E_b/N_0. There are two curves in the figure: The first is from the left if for BPSK/QPSK (binary phase shift keying/quadrature phase shift keying), and the second is for 8-ary PSK (an 8-level PSK modulation). The values are for coherent detection. Coherent detection means that the receiver has a phase reference as a basis to make its binary decisions.

Digital Modulation of LOS Microwave Radios. Digital systems, typically standard PCM as discussed in Chapter 8, are notoriously wasteful of bandwidth compared to their analog counterparts. For example, the analog voice channel is nominally 4 kHz, whereas the digital voice channel, assuming one bit per hertz of bandwidth, is 64 kHz. This is a 16-to-1 difference in required bandwidth. Thus various national regulatory authorities, such as the U.S. FCC, require that digital systems be bandwidth conservative. One term that is used is *bit packing*. This means packing more bits into a hertz of bandwidth. Roughly, the FCC requires about 4.5 bits per hertz of bandwidth. To meet these requirements, digital LOS microwave utilizes some form of quadrature amplitude modulation (QAM), and as a minimum at the 64-QAM level, or often 128-QAM or 256-QAM. 64-QAM has a theoretical bit packing capability of 6 bits per Hz, 128-QAM at 7 bits/Hz, and 256-QAM at 8 bits/Hz. Figure 7.10 compares bit error rate performance versus E_b/N_0 for various QAM schemes [12]. (Also consult Ref. 18.)

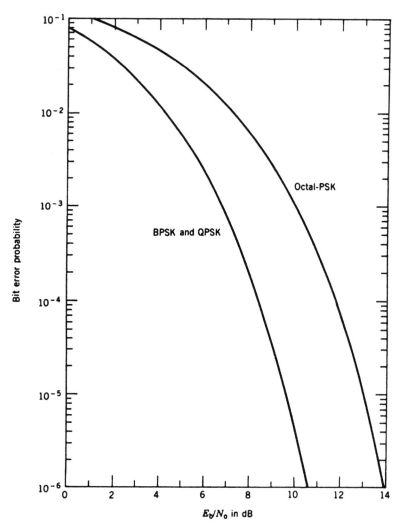

Figure 7.9. Bit error probability (BER) versus E_b/N_0 performance of coherent BPSK/QPSK and 8-ary PSK (octal PSK).

To sum up this section on digital LOS microwave, we will work an example problem. A digital link operates in the 7-GHz band with a link 37 km long. The bit rate is 155 Mbps and the modulation is 64-QAM. The specified BER for the link is 1×10^{-7} and the modulation implementation loss is 2 dB. The receiver noise figure is 8 dB. The antennas have 35-dB gain at each end, and transmission line losses are 1.8 dB at each end. What link margin can be expected?

First turn to Figure 7.10 and derive the required E_b/N_0. This is 19.5 dB; add to this the modulation implementation loss of 2 dB and the result is that the required value for E_b/N_0 is 21.5 dB.

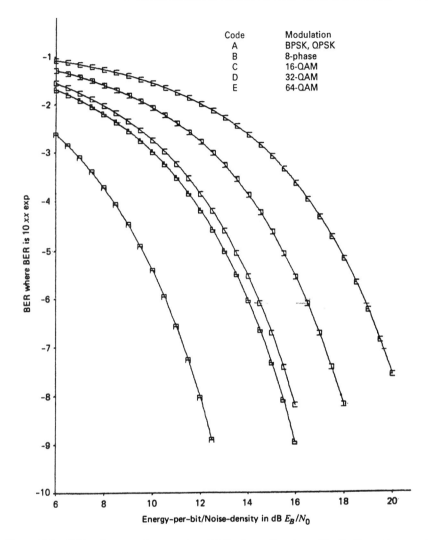

Figure 7.10. BER performance for several modulation types. Courtesy of the Raytheon Company [5].

The next step is to calculate a candidate RSL value. We know that E_b must be 21.5 dB above N_0. We can calculate N_0 because we have the receiver noise figure.

$$N_0 = -204 \text{ dBW} + 8 \text{ dB}$$
$$= -196 \text{ dBW} \quad \text{and}$$
$$E_b = -196 \text{ dBW} + 21.15 \text{ dB}$$
$$= -174.5 \text{ dBW}$$

Thus RSL, in this case, is $10 \log(1.544 \times 10^6)$ greater than E_b.

$$\text{RSL}_{\text{dBW}} = E_b + 10 \log(1.544 \times 10^6) \quad \text{(from equation 7.8)}$$
$$= -174.5 \text{ dBW} + 61.88 \text{ dB}$$
$$= -112.61 \text{ dBW}$$

We will hold this minimum RSL value for future reference, and now turn to the transmit side of the link. Assume the transmitter has a 1-watt output or 0 dBW. Calculate EIRP in dBW.

$$\text{EIRP}_{\text{dBW}} = 0 \text{ dBW} - 1.8 \text{ dB} + 35 \text{ dB} \quad \text{(equation 7.6)}$$
$$= +33.2 \text{ dBW}$$

Calculate the free-space loss (path loss):

$$\text{FSL}_{\text{dB}} = 92.4 + 20 \log 37 + 20 \log 7 \quad \text{(equation 7.5B)}$$
$$= 92.4 + 31.36 + 16.90$$
$$= 140.66 \text{ dB}$$

Calculate the IRL:

$$\text{IRL}_{\text{dBW}} = +33.2 \text{ dBW} - 140.66 \text{ dB} \quad \text{(equation 7.7)}$$
$$= -107.46 \text{ dBW}$$

Calculate RSL:

$$\text{RSL} = -107.46 \text{ dBW} + 35 \text{ dB} - 1.8 \text{ dB} \quad \text{(from equation 7.8)}$$
$$= -74.26 \text{ dBW}$$

Calculate the margin:

$$\text{Margin} = -74.26 \text{ dBW} - (-112.61 \text{ dBW})$$
$$= 38.35 \text{ dB}$$

Often we are faced with the problem of "what antenna gain will provide the margin or provide the gain necessary to meet performance objectives?"

Parabolic Antenna Gain. At a given frequency the gain of a parabolic antenna is a function of its effective area and may be expressed by the formula

$$G = 10 \log_{10}(4\pi A\eta/\lambda^2) \tag{7.15}$$

where G is the gain in decibels relative to an isotropic antenna, A is the area of antenna aperture, η is the aperture efficiency, and λ is the wavelength at the operating frequency. Commercially available parabolic antennas with a conventional horn feed at their focus usually display a 55% efficiency or somewhat better. With such an efficiency, gain (G, in decibels) is then

$$G = 20 \log_{10} D + 20 \log_{10} F + 7.5 \qquad (7.16)$$

where F is the frequency in gigahertz and D is the parabolic diameter in feet. In metric units, we have

$$G = 20 \log_{10} D + 20 \log_{10} F + 17.8 \qquad (7.17)$$

where D is measured in meters and F in gigahertz.

What size antenna would be required in the preceding example? Let $G = 35$ dB and $F = 6$ GHz.

$$35 \text{ dB} = 20 \log_{10} D + 20 \log_{10} 7 + 7.5$$
$$20 \log D = 35 - 20 \times 0.8451 - 7.5$$
$$= [10.598/20]$$
$$= 3.38 \text{ feet}$$

Parabolic dish antennas, with waveguide (horn) feeds (see Figure 7.11), are probably the most economic antennas for radiolinks operating from 3 GHz

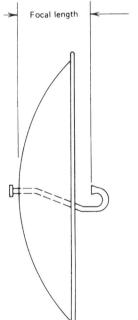

Focal length

Figure 7.11. Typical parabolic antenna with front feed.

upward. From 50 MHz to about 3 GHz, coaxial feeds are used, and often the antennas are Yagi's. Coaxial cable transmission lines deliver the RF energy from/to transmitter/receiver to the antenna in this range. Above 3 GHz, coaxial cable becomes too lossy and waveguide is more practical.

Other types of antennas may also be used, such as the "cornucopia," horn, and spiral. Besides cost and gain, other features are front-to-back ratio, side lobes, and efficiency. For instance the "cornucopia," called such because it looks like "the horn of plenty," has efficiencies in excess of 60% and improved side-lobe discrimination but is more costly.

4.5 Running a Path/Site Survey

This can turn out to be the most important step in the design of an LOS microwave link (or hop). We have found through experience that mountains move (i.e., map error), buildings grow, grain elevators appear where none were before, east of Madrid a whole high-rise community goes up, and so forth.

Another point from experience: If someone says "line-of-sight" conditions exist on a certain path, **don't believe it!** Line of sight must be precisely defined. We reiterate that for each obstacle in the LOS microwave path, earth curvature with proper K-factor must be added to obstacle height, 0.6 of the first Fresnel zone must be added on top of that,* and then 50 ft for trees and 10 ft more for growth must be added if in a vegetated area (to avoid foliage loss penalties).

Much of the survey is to verify findings and conclusions of the path profile. Of course each site must be visited to determine the location of the radio equipment shelter, the location of the tower, whether site improvement is required, the nearest prime power lines, and site access, among other items to be investigated.

Site/path survey personnel must personally inspect the sites in question, walking/driving the path or flying the path in a helicopter, or a combination thereof. The use of GPS† receivers are helpful to verify geographical positions along the path, including altitudes.

4.6 System Test Prior to Cutover

A series of tests should be carried out to verify if the link (or system) meets the performance requirements established in Section 4.2. There are two important tests for a digital link or system. The first is the measurement of receive signal level (RSL). The second test is the bit error rate test (BERT).

---

*Often it is advisable to add 10 ft (or 3 m) of safety factor on top of the 0.6 first Fresnel zone clearance to avoid any diffraction loss penalties.
†GPS stands for geographical positioning system, a satellite navigation system that is extremely accurate.

The tests ideally should be made over time. Here we mean to run the BERT continuously for several hours or a day to capture the effects of fading.

4.7 Fades, Fading, and Fade Margins

In Section 4.4.2 we showed how path loss (free-space loss) can be calculated. This was a fixed loss which can be simulated in the laboratory with an attenuator. On very short radio paths below about 10 GHz, the signal level impinging on the distant end receiving antenna, assuming full LOS conditions, can be calculated to less than 1 dB. If the transmitter continues to give the same output, the receive signal level will remain uniformly the same over long periods of time, for years. As the path is extended, the measured RSL will vary around a median. The signal level may remain at that median for minutes or hours, and then suddenly drop and then return to the median again. In other periods and/or on other links, this level variation can be continuous for periods of time. Drops in level can be as much as 30 dB or more. This phenomenon is called *fading*. The system and link design must take fading into account when sizing or dimensioning the system/link.

As the RSL drops in level, so does the E_b/N_0. As the E_b/N_0 decreases, there is a deterioration in error performance; the BER degrades. Fades vary in depth, duration, and frequency (i.e., number of fade events per unit of time). We cannot eliminate the fades, but we can mitigate their effects. The primary tool we have is to overbuild each link by increasing the margin.

Link margin is the number of dB we have as a surplus in the link design. We could design an LOS microwave link so we just achieve the RSL at the distant receiver to satisfy the E_b/N_0 (and BER) requirements using free-space loss as the only factor in link attenuation (besides transmission line loss). Unfortunately we will only meet our specified requirements about 50% of the time. So we must add margin to compensate for the fading.

We have to determine what percentage of the time the link meets BER performance requirements. We call this *time availability*.* If a link meets its performance requirements 99% of the time, then it does not meet performance requirements 1% of the time. We call this latter factor *unavailability*.

To improve time availability, we must increase the link margin, often called the *fade margin*. How many additional dB are necessary? There are several approaches to the calculation of a required fade margin. One of the simplest and most straightforward approaches is to assume that the fading follows a Rayleigh distribution, often considered worst-case fading. If we

---

*Other texts call this "reliability." The use of this term should be deprecated because it is ambiguous and confusing. In our opinion, reliability should relate to equipment failure rate, not propagation performance.

base our premise on a Rayleigh distribution, then the following fade margins
can be used:

Time Availability (%)	Required Fade Margin (dB)
90	8
99	18
99.9	28
99.99	38
99.999	48

More often than not, LOS microwave systems consist of multiple hops. Here
our primary interest is the time availability at the far-end receiver in the
system after the signal has progressed across all of the hops. From this time
availability value we will want to assign an availability value for each hop or
link.

Suppose a system has nine hops and the system time availability specified
is 99.95%, and we want to calculate the time availability per hop or link.
The first step is to calculate the system time unavailability. This is simply
$1.0000 - 0.9995 = 0.0005$. We now divide this value by 9 (i.e., there are nine
hops or links):

$$0.0005/9 = 0.0000555$$

Now we convert this value to time availability:

$$\text{Per-hop time availability} = 1.0000000 - 0.0000555$$
$$= 0.99994 \text{ or } 99.994\%$$

The most common cause of fading is multipath conditions. Refer to Figure
7.12. As the term implies, signal energy follows multiple paths from the transmit antenna to the receive antenna. Two additional paths, besides the main
ray beam, are shown in Figure 7.12. Most of the time the delayed signal

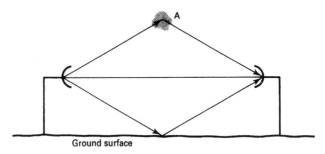

Ground surface

A = layers of different refractive index

Figure 7.12. Multipath is the most common cause of fading.

energy (from the reflected/refracted paths) will be out of phase with the principal ray beam which causes fading. In digital systems, there is the additional impairment of dispersion caused by multipath. Of course, the delay energy arrives later, spilling into the next bit or binary symbol position, increasing the probability that that bit decision will be in error.

Probably the most economic way to overbuild a link is to increase the antenna aperture. Every time we double the aperture (i.e., in this case, doubling the diameter of the parabolic dish), we increase the gain by 6 dB (see equations 7.16 and 7.17). We recommend that apertures for LOS microwave antennas not exceed 12 ft (3.7 m). Not only does the cost of the antenna get notably greater as aperture increases over 8 ft (2.5 m), but the equivalent sail area of the dish starts to have an impact on system design. Wind pressure on large dishes increases tower twist and sway, resulting in movement out of the capture area of the ray beam at the receive antenna. This forces us to stiffen the tower, which could dramatically increase system cost. Also, as antenna aperture increases, gain increases and beamwidth decreases.

Other measures we can take to overbuild a link are:

- Insert a low-noise amplifier (LNA) in front of the receiver–mixer. Improvement: 6–12 dB.

- Use an HPA (high-power amplifier). Usually a traveling-wave tube (TWT) amplifier; 10 watts output. Improvement: 10 dB.

- Implement FEC (forward error correction). Improvement: 1–5 dB. Involves adding a printed circuit board at each end. It will affect link bandwidth. See Ref. 17 for description of FEC.

- Implement some form of diversity. Space diversity is preferable in many countries. Can be a fairly expensive measure. Improvement: 5–20 dB or more. Diversity is described below.

It should be appreciated that fading varies with path length, frequency, climate, and terrain. The rougher the terrain, the more reflections are broken up. Flat terrain, and especially paths over water, tends to increase the incidence of fading. For example, in dry, windy, mountainous areas the multipath fading phenomenon may be nonexistent. In hot, humid coastal regions a very high incidence of fading may be expected.

4.8   Diversity and Hot-Standby Operation

Diversity reception means the simultaneous reception of the same radio signal over two or more paths. Each "path" is handled by a separate receiver chain and then combined by predetection or postdetection combiners in the radio equipment so that effects of fading are mitigated. The separate diversity paths

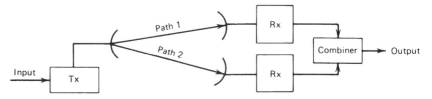

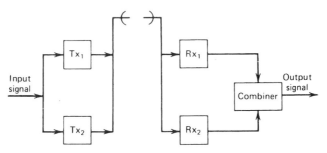

Figure 7.13. A space-diversity configuration.

Figure 7.14. A frequency-diversity configuration.

can be based on space, frequency, and/or time diversity. The simplest form of diversity is space diversity. Such a configuration is shown in Figure 7.13.

The two diversity paths in space diversity are derived at the receiver end from two separate receivers with a combined output. Each receiver is connected to its own antenna, separated vertically on the same tower. The separation distance should be at least 70 wavelengths and preferably 100 wavelengths. In theory, fading will not occur on both paths simultaneously.

Frequency diversity is more complex and more costly than space diversity. It has advantages as well as disadvantages. Frequency diversity requires two transmitters at the near end of the link. The transmitters are modulated simultaneously by the same signal but transmit on different frequencies. Frequency separation must be at least 2%, but 5% is preferable. Figure 7.14 is an example of a frequency-diversity configuration. The two diversity paths are derived in the frequency domain. When a fade occurs on one frequency, it will probably not occur on the other frequency. The more one frequency is separated from the other, the less chance there is that fades will occur simultaneously on each path.

Frequency diversity is more expensive, but there is greater assurance of path reliability. It provides full and simple equipment redundancy and has the great operational advantage of two complete end-to-end electrical paths. In this case, failure of one transmitter or one receiver will not interrupt service, and a transmitter and/or a receiver can be taken out of service for maintenance. The primary disadvantage of frequency diversity is that it doubles the amount of frequency spectrum required in this day and age when spectrum is at a premium. In many cases it is prohibited by national licensing authorities. For example, the U.S. Federal Communications Commission

(FCC) does not permit frequency diversity for industrial users. It also should be appreciated that it will be difficult to get the desired frequency spacing.

The full equipment redundancy aspect is very attractive to the system designer. Another approach to achieve diversity improvement in propagation plus reliability improvement by fully redundant equipment is to resort to the "hot-standby" technique. On the receive end of the path, a space-diversity configuration is used. On the transmit end a second transmitter is installed as in Figure 7.14, but the second transmitter is on "hot standby." This means that the second transmitter is on but its signal is not radiated by the antenna. On a one-for-one basis the second transmitter is on the same frequency as the first transmitter. On failure of transmitter 1, transmitter 2 is switched on automatically.

One-for-N hot standby is utilized on large radiolink systems employing several radio carriers, where the cost for duplicate equipment for each channel may be prohibitive. In this case, one full set of spare equipment in the "on" condition serves to replace one of several operational channels, and the spare equipment is assigned its own frequency. On the receive side there is just one extra receiver. Relay cut-over must be provided, and no space-diversity improvement is afforded. Likewise, there is no paralleling of inputs on the transmit side; thus the switching from the operational pair to the standby pair is much more complex. On such multi-RF channel arrangements it is customary to assign some sort of priority arrangement. Often the priority channel enjoys the advantage of frequency diversity, whereas the other RF channels do not. On failure of one of the other channels, the diversity improvement is lost on the priority channel, with the diversity pair switched to carry the traffic on the failed pair. In another arrangement the standby channel carries low-priority traffic and does not operate in a frequency-diversity arrangement, while providing protection for perhaps two or three other channels carrying the higher-priority traffic. On failure of a high-priority channel, the lower-priority channel drops its traffic, replacing one of the RF channels carrying the more important traffic flow. Once this occurs, the remaining channels operate without standby equipment protection.

Diversity Improvement. Propagation reliability improvement can be exemplified as follows. If a 30-mi path required a 51-dB fade margin to achieve a 99.999% reliability on 6.7 GHz without diversity, with space diversity on the same path, only a 33-dB fade margin would be required for the same propagation reliability, namely, 99.999% (Vigants, *IEEE Trans. Commun.*, December 1968 and Ref. 6). For frequency diversity in the nondiversity condition, assuming Rayleigh fading, a 30-dB fade margin would display something better than a 99.9% path reliability. But under the same circumstances with frequency diversity, with only a 1% frequency separation, propagation reliability on the same path would be improved to 99.995% [6].

4.9   LOS Microwave Repeaters

Digital LOS microwave repeaters completely demodulate the incoming signal to baseband (i.e., to the raw electrical signal of "1s" and "0s."). This full demodulation causes regeneration of the "1s" and "0s." The outgoing signal is "squared up" and retimed. Regeneration is explained in Chapter 8. The regenerated baseband signal is remodulated, upconverted, and retransmitted on a different frequency. RF and IF repeaters are not recommended for digital microwave systems because such repeaters do not have the vital regeneration stage.

4.10   Frequency Planning and Frequency Assignment

4.10.1   General. To derive optimum performance from an LOS microwave system, the design engineer must set out a frequency-usage plan that may or may not have to be approved by the national regulatory organization.

The problem has many aspects. First, the useful RF spectrum is limited from above dc to about 150 GHz. The upper limit is technology-restricted. To some extent it is also propagation-restricted. The frequency ranges for this discussion cover the bands in Table 7.2. Those frequencies above 10 GHz could be called rainfall-restricted, because at about 10 GHz is where excess attenuation due to rainfall can become an important design factor.

Then there is the problem of congestion. Around urban and built-up areas, frequency assignments below 10 GHz are hard to obtain from national regulatory authorities. If we plan properly for excess rainfall attenuation, nearly equal performance is available at those higher frequencies.

4.10.2   Radio-Frequency Interference (RFI). There are three facets to RFI in this context. (1) Own microwave can interfere with other LOS microwave and satellite communication earth stations nearby, (2) nearby LOS microwave and satellite communication facilities can interfere with own microwave, and (3) own microwave can interfere with itself. To avoid self-interference (No. 3), it is advisable to use frequency plans of CCIR (ITU-R organization) as set forth in the RF Series (Fixed Service). Advantage is taken of proper frequency separation, transmit and receive, and polarization isolation. CCIR also provides methods for interference analysis (coordination

TABLE 7.2   Some LOS Microwave Frequency Bands

2110–2130 MHz	18,920–19,160 MHz
3700–4200 MHz	19,260–19,700 MHz
5925–6425 MHz	21,200–23,600 MHz
6525–6875 MHz	27,500–29,500 MHz
10,700–11,700 MHz	31,000–31,300 MHz
17,700–18,820 MHz	38,600–40,000 MHz

contour), also in the RF Series. Another alternative is specialist companies that provide a service of electromagnetic compatibility (EMC) analysis.

5 SATELLITE COMMUNICATIONS

5.1 Introduction

Satellite communications is an extension of LOS microwave technology covered in Section 4. The satellite must be within LOS of each participating earth terminal. We are more concerned about noise in satellite communication links than we were with LOS microwave. In most cases, received signals will be of a lower level. On satellite systems operating below 10 GHz, very little link margin is required; there is essentially no fading, as experienced in LOS microwave.

5.2 Application

Satellite communication is another method of extending the digital network (Chapters 8 and 9). These digital trunks may be used as any other digital trunks for telephony, data, facsimile, and video. Satellite links may prove optimum for a variety of applications, including the following:

1. On international high-usage trunks country to country.
2. On national trunks, between switching nodes that are fairly well separated in distance [i.e., > 200 miles (320 km)] in highly developed countries. Again, the tendency is to use satellite links for direct high-usage connectivity. It may serve as an adjunct to LOS microwave and fiber optics.
3. In areas under development where satellite links replace HF radio and a high growth is expected to be eventually supplemented by radiolink and fiber optic cable.
4. In sparsely populated, highly rural, "out-back" areas where it may be the only form of communication. Northern Canada and Alaska are good examples.
5. On final routes for overflow on a demand-assignment basis. Route length again is a major consideration.
6. In many cases, on international connections reducing such connections to one link.
7. On private and industrial networks including VSAT* networks.
8. On specialized common carriers.
9. On thin-line communications and tracking systems.

*VSAT = very small aperture terminal.

5.3   Definition

A number of world bodies, including the CCIR and the U.S. Federal Communications Commission (FCC), have now accepted the term "earth station" as a radio facility located on the earth's surface that communicates with satellites. A "terrestrial station" is a radio facility on the earth's surface that communicates with other similar facilities on the earth's surface. Section 4 of this chapter dealt with one form of terrestrial stations. The term "earth station" as used today has come more to mean a radio station operating with other stations on the earth via an orbiting satellite relay.

Nearly all commercial communication satellites are geostationary. Such satellites orbit the earth in a 24-hr period. Thus they appear stationary over a particular geographic location on earth. For a 24-hr synchronous orbit the altitude of a geostationary satellite is 22,300 statute miles or 35,900 km above the earth's equator.

5.4   The Satellite

Most of the presently employed communication satellites are RF repeaters. A typical RF repeater used in a communication satellite is shown in Figure 7.15. The tendency today is to call these types of satellite "bent pipe" satellites as opposed to processing satellites. A processing satellite, as a minimum, regenerates the received digital signal. It may decode and recode a digital bit stream. It also may have some bulk switching capability, switching to crosslinks connecting to other satellites. Theoretically, as mentioned earlier, three such satellites placed correctly in equatorial geostationary orbit could provide communication from one earth station to any other located anywhere on the earth surface (see Figure 7.16). However, high latitude service is marginal and nil north of 80°N and south of 80°S.

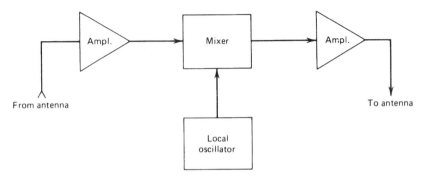

Figure 7.15. Simplified functional block diagram of one transponder of a typical communication satellite.

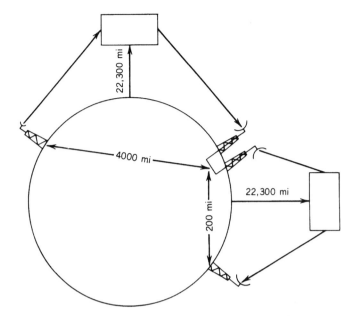

Figure 7.16. Distance involved in satellite communications.

5.5 Three Basic Technical Problems

As the reader can appreciate, satellite communication is nothing more than radiolink (microwave LOS) communication using one or two RF repeaters located at great distances from the terminal earth stations, as shown in Figure 7.16. Because of the distance involved, consider the slant range from earth antenna to satellite to be the same as the satellite altitude. This would be true if the antenna were pointing at zenith to the satellite. Distance increases as the pointing angle to the satellite decreases (elevation angles).

We thus are dealing with very long distances. The time required to traverse these distances—namely, earth station to satellite to another earth station—is on the order of 250 ms. Round-trip delay will be 2 × 250 or 500 ms. These propagation times are much greater than those encountered on conventional terrestrial systems. So one major problem is propagation time and resulting echo on telephone circuits. It influences certain data circuits in delay to reply for block or packet transmission systems and requires careful selection of telephone signaling systems, or call-setup time may become excessive.

Naturally, there are far greater losses. For LOS microwave we encounter free-space losses possibly as high as 145 dB. In the case of a satellite with a range of 22,300 mi operating on 4.2 GHz, the free-space loss is 196 dB and at 6 GHz, 199 dB. At 14 GHz the loss is about 207 dB. This presents no insurmountable problem from earth to satellite, where comparatively high power transmitters and very high gain antennas may be used. On the contrary, from satellite to earth the link is power-limited for two reasons:

(1) in bands shared with terrestrial services such as the popular 4-GHz band to ensure noninterference with those services and (2) in the satellite itself, which can derive power only from solar cells. It takes a great number of solar cells to produce the RF power necessary; thus the down-link, from satellite to earth, is critical, and received signal levels will be much lower than on comparative radiolinks, as low as -150 dBW. A third problem is crowding. The equatorial orbit is filling with geostationary satellites. Radio-frequency interference from one satellite system to another is increasing. This is particularly true for systems employing smaller antennas at earth stations with their inherent wider beamwidths. It all boils down to a frequency congestion of emitters.

It should be noted that by the year 2000, we can expect to see several low earth-orbit satellite systems in operation. These satellites typically orbit some 500 km above the earth.

5.6   Frequency Bands: Desirable and Available

The most desirable frequency bands for commercial satellite communication are in the spectrum 1000–10,000 MHz. These bands are:

3700–4200 MHz (satellite-to-earth or down-link)
5925–6425 MHz (earth-to-satellite or up-link)
7250–7750 MHz* (down-link)
7900–8400 MHz* (up-link)

These bands are preferred by design engineers for the following primary reasons:

- Less atmospheric absorption than higher frequencies.
- Rainfall loss not a concern.
- Less noise, both galactic and man-made.
- A well-developed technology.
- Less free-space loss compared to the higher frequencies.

There are two factors contraindicating application of these bands and pushing for the use of higher frequencies:

- The bands are shared with terrestrial services.
- There is orbital crowding (discussed earlier).

Higher-frequency bands for commercial satellite service are:

*These two bands are intended mainly for military application.

10.95–11.2 GHz (down-link)

11.45–12.2 GHz (down-link)

14.0-14.5 GHz (up-link)

17.7-20.2 GHz (downlink)

27.5-30.0 GHz (up-link)

Above 10 GHz rainfall attenuation and scattering and other moisture and gaseous absorption must be taken into account. The satellite link must meet a BER of 1×10^{-6} at least 99.9% of the time. One solution is a space-diversity scheme where we can be fairly well assured that one of the two antenna installations will not be seriously affected by the heavy rainfall cell affecting the other installation. Antenna separations of 4–10 km are being employed. Another advantage with the higher frequencies is that requirements for down-link interference are less; thus satellites may radiate more power. This is often carried out on the satellite using spot-beam antennas rather than general-coverage antennas.

5.7 Multiple Access of a Satellite

Multiple access is defined as the ability of a number of earth stations to interconnect their respective communication links through a common satellite. Satellite access is classified (1) by assignment, whether quasi-permanent or temporary, namely, (a) preassigned multiple access or (b) demand-assigned multiple access (DAMA), and (2) according to whether the assignment is in the frequency domain or the time domain, namely, (a) frequency-division multiple access (FDMA) or (b) time-division multiple access (TDMA). On comparatively heavy routes (≥ 10 erlangs), preassigned multiple access may become economical. Other factors, of course, must be considered, such as whether the earth station is "INTELSAT" standard as well as the space-segment charge that is levied for use of the satellite. In telephone terminology, "preassigned" means dedicated circuits. Demand-assigned multiple access is useful for low-traffic multipoint routes where it becomes interesting from an economic standpoint. Also, an earth station may resort to DAMA as a remedy to overflow for its FDMA circuits.

5.7.1 *Frequency-Division Multiple Access.* Historically, FDMA has the highest usage and application of the various access techniques. The several RF bands available (from Section 5.6) have a 500-MHz bandwidth. A satellite contains a number of transponders, each of which covers a frequency segment of the 500-MHz bandwidth. One method of segmenting the 500-MHz is by utilizing 12 transponders, each with a 36-MHz bandwidth. Sophisticated satellites such as INTELSAT V segment the 500-MHz available with transponders up to 77-MHz bandwidth at 4–6 GHz and at 11–14 GHz have one transponder with a 241-MHz bandwidth.

TABLE 7.3   INTELSAT V, VA, and VI Global Beam: Voice-Channel Capacity Versus Bandwidth Assignments (Partial Listing) FM/FDM-FDMA

Carrier capacity (number of voice channels)	24.0	60.0	96.0	132.0	252.0	432.0	792.0
Top baseband frequency (kHZ)	108.0	252.0	408.0	552.0	1052.0	796.0	4028.0
Allocated satellite bandwidth (MHz)	1.25	2.5	5.0	5.0	7.5	10.0	25.0
Occupied bandwidth (MHz)	1.1	2.1	3.3	4.3	6.3	8.3	21.6

Source: Ref. 7.

With FDMA operation, each earth station is assigned a segment or a portion of a segment. For a nominal 36-MHz transponder, 14 earth stations may access in an FDMA format, each with 24 voice channels in a standard CCITT modulation plan (Chapter 5). The INTELSAT VI assignments for a 36-MHz transponder are shown in Table 7.3, where it can be seen that when larger channel groups are used, fewer earth stations can access the same transponder.

Consider the following hypothetical example of how FDMA works. Frequency translation is via a 2225-MHz local oscillator in the satellite, and the difference mode is used in the mixer; for example:

Satellite Receive Frequency (MHz)		Mixer Frequency (MHz)		Satellite Transmit Frequency (MHz)
5925	−	2225	=	3700
6425	−	2225	=	4200

If a transponder had the 6262- to 6298-MHz segment, then

6262	−	2225	=	4037
6298	−	2225	=	4073

In the segment, assign three RF carriers from three locations in common view of an Atlantic satellite. Each carrier will be frequency modulated with two FDM supergroups, as in Table 7.3. Subgroup A, the remaining 12 channels of the FDM 132 VF channel total, are spare and are disregarded. Let us assign Spain (Buitrago) the transponder subsegment 6262–6267 MHz; Etam, West Virginia (USA), 6272–6277 MHz; and Longoville, Chile, 6282–

6287 MHz. These are the up-link frequencies. When converted in the satellite, they are as follows:

	Up-link (MHz)	Down-link (MHz)
Buitrago	6262–6267	4037–4042
Etam	6272–6277	4047–4052
Longoville	6282–6287	4057–4062

The United States would transmit one carrier at Etam to communicate with both Buitrago and Longoville. The carrier contains two supergroups, as shown in the following diagram.

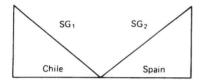

The United States would receive two carriers at Etam for the receive end, one from Buitrago and one from Longoville, shown in the following diagram, and would pick off supergroup 1 (as in our example) from the Chile carrier and supergroup 2 from the Spanish carrier. A similar exercise can be carried out for the Buitrago and Longoville situations. Note that one transmitter can access many locations, but to receive multiple locations, a receiver chain is required for each distant location to be received. If a single up-link carrier occupied an entire transponder (Table 7.3), it could serve (36/1.25) or 28 down-link separate locations, provided that each location required no more than 24 VF channels.

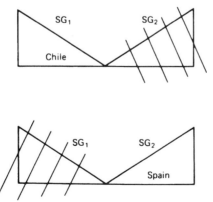

5.7.2 Time-Division Multiple Access.   Time-division multiple access (TDMA) operates in the time domain, and may only be used for digital network connectivity. Use of the satellite transponder is on a time-sharing basis. Individual time slots are assigned to earth stations in a sequential order. Each earth station has full and exclusive use of the transponder bandwidth during its time-assigned segment. Depending on the bandwidth of the transponder, bit rates of 10–100 Mbps (megabits per second) are used.

With TDMA operation, earth stations use digital modulation and transmit with bursts of information. The duration of a burst lasts for the time period of the slot assigned. Timing synchronization is a major problem.

A frame, in digital format, may be defined as a repeating cycle of events. It occurs in a time period containing a single digital burst from each accessing earth station and the guard periods or guard times between each burst. A sample frame is shown in Figure 7.17 for earth stations 1, 2, and 3, to N. Typical frame periods are 750 μs for INTELSAT and 250 μs for the Canadian Telesat.

The reader will appreciate that timing is crucial to effective TDMA operation. The greater N becomes (i.e., the more stations operating in the frame period), the more clock timing affects the system. The secret lies in the "carrier and clock (timing) recovery pattern" as shown in Figure 7.17. One way to ensure that all stations synchronize to a master clock is to place a

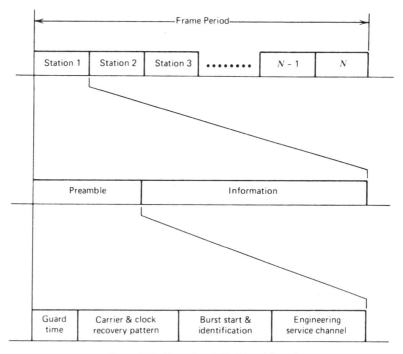

Figure 7.17. Example of TDMA burst format.

sync burst as the first element in the format frame. The INTELSAT does just this. The burst carries 44 bits, starting with 30 bits carrier and bit timing recovery, 10 bits for the "unique word," and 4 bits for the station identification code.

Why use TDMA in the first place? It lies in a major detraction of FDMA. Satellites use traveling-wave tubes (TWTs) in their transmitter final amplifiers. A TWT has the undesirable property of nonlinearity in its input–output characteristics when operated at full power. When there is more than one carrier accessing the transponder simultaneously, high levels of inter-modulation (IM) products are produced, thus increasing noise and crosstalk. When a transponder is operated at full power output, such noise can be excessive and intolerable. Thus input must be backed off (i.e., level reduced) by ≥ 3 dB. This, of course, reduces the EIRP and results in reduced efficiency and reduced information capacity. Consequently, each earth station's up-link power must be carefully coordinated to ensure proper loading of the satellite. The complexity of the problem increases when a large number of earth stations access a transponder, each with varying traffic loads.

On the other hand, TDMA allows the transponder's TWT to operate at full power because only one earth-station carrier is providing input to the satellite transponder at any one instant.

To summarize, consider the following advantages and disadvantages of FDMA and TDMA. The major advantages of FDMA are as follows:

- No network timing is required.
- Channel assignment is simple and straightforward.

The major disadvantages of FDMA are as follows:

- Up-link power levels must be closely coordinated to obtain efficient use of transponder RF output power.
- Intermodulation difficulties require power back-off as the number of RF carriers increases with inherent loss of efficiency.

The major advantages of TDMA are as follows:

- There is no power sharing and IM product problems do not occur.
- The system is flexible with respect to user differences in up-link EIRP and data rates.
- Accesses can be reconfigured for traffic load in almost real time.

The major disadvantages of TDMA are as follows:

- Accurate network timing is required.

- There is some loss of throughput due to guard times and preambles.
- Large buffer storage may be required if frame lengths are long.

5.7.3 Demand-Assignment Multiple Access. The demand-assignment multiple access (DAMA) method has single voice channels allocated to an earth station on demand. A pool of idle channels is available, and assignments from the pool are made on request. When a call has been completed on the channel, the channel is returned to the idle pool for reassignment.

The DAMA method is analogous to a telephone switch. When a subscriber goes off hook, a line is seized; on dialing, a connection is made; and when the call is completed and there is an on-hook condition, the voice path through the switch is returned to "idle" and is ready for use by another subscriber. There are three methods available for handling DAMA in a satellite system:

- Polling.
- Random access–central control.
- Random access–distributed control.

The polling method is fairly self-explanatory. A master station "polls" all other stations in the system sequentially. When a positive reply is received, a channel is assigned accordingly. As the number of stations increases, the polling interval becomes longer and the system tends to become unwieldy. With the random access–central control method, status of channels is coordinated by a central control computer, which is usually located at a "master" earth station. Call requests (call attempts in switching) are passed to the central processor via digital order wire (digitally over the radio service channel), and a channel is assigned if available. Once the call is completed and the subscriber goes on hook, the speech path is taken down and the channel used is returned to the demand-access pool. According to the system design there are various methods to handle blocked calls ["all trunks busy" (ATB)], such as queuing and other repeat attempts.

The distributed control–random access method utilizes a processor control at each earth station in the system. All earth stations in the network monitor the status of all channels where channel status is continuously updated via a digital order-wire circuit. When an idle circuit is seized, all users are informed and the circuit is removed from the pool. Similar information is transmitted to all users when the circuit returns to the idle condition. The same problems arise regarding blockage (ATB) as in the central control system. Distributed control is more costly, particularly in large systems with many users. It is attractive in the international environment, as it eliminates the "politics" of a master station.

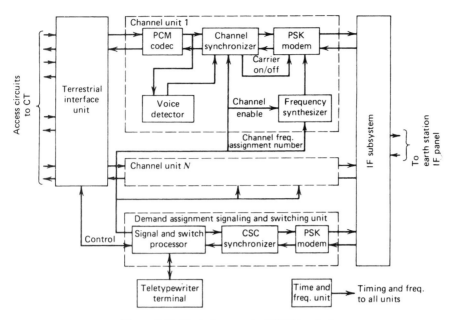

Figure 7.18. Block diagram of a SPADE terminal.

Many systems use a mix of preassigned channels on an FDMA basis and DAMA channels. The DAMA concept uses only transponder "space" when in use, and all DAMA stations in the system can *directly* access each other. On low-usage routes and at earth stations with a low traffic volume, DAMA is attractive, as proven on low space-segment-usage costs. DAMA employs the technique of single channel per carrier (SCPC), whereas TDMA and FDMA systems are multichannel. SCPC may use FM or phase-shift keying (PSK) modulation. Frequency modulation systems operate on an analog basis, usually with preemphasis, threshold extension of the carrier, and syllabic commanding. With PSK systems the voice is digitized in a PCM or delta-modulation format. Modulation rates for PCM are 64 kbps per channel, and with delta modulation they are 32 or 40 kbps (see Chapter 8).

A typical distributed control DAMA system (SPADE) used by INTELSAT is shown in Figure 7.18. This system uses PCM on the voice circuit with four-phase modulation. A typical SCPC FM system is illustrated in Figure 7.19.

5.8   Earth Station Link Engineering

5.8.1   Introduction. Up to this point we have discussed basic earth communication problems such as access and coverage. This section reviews some of the link engineering problems associated with earth station system engineering. The approach used to introduce the reader to essential path engineering expands on the basic principles previously discussed (Section 4) in this

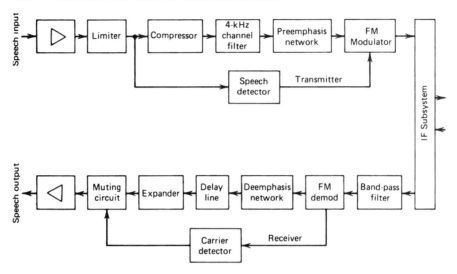

Figure 7.19. Block diagram of a typical SCPC FM station unit.

chapter on radiolinks. As we saw in Section 5.3, an earth station is a distant RF repeater. By international agreement the repeater's EIRP is limited because nearly all bands are shared by terrestrial services. The limit for satellites transmitting in the 4-GHz band is -142 dBW/m^2 (flux density) in a 4-kHz bandwidth and -140 dBW/m^2 for the 11-GHz band.

5.8.2   *Earth Station Receiving System Figure of Merit, G/T.*

The figure of merit of an earth station receiving system, G/T, has been introduced into the technology to describe the capability of an earth station or a satellite to "*receive*" a signal. It is also a convenient tool in the link budget analysis. A link budget is used by the system engineer to size components of earth stations and satellites, such as RF output power, antenna gain and directivity, and receiver front-end characteristics.

G/T can be written as a mathematical identity:

$$G/T = G_{\mathrm{dB}} - 10 \log T_{\mathrm{sys}} \qquad (7.18)$$

where G is the net antenna gain up to an arbitrary reference point or reference plane in the down-link receive chain (for an earth station). Conventionally, in commercial practice the reference plane is taken at the input of the low-noise amplifier (LNA). Thus G is simply the gross gain of the antenna minus all losses up to the LNA. These losses include feed loss, waveguide loss, bandpass filter loss, and, where applicable, directional coupler loss, waveguide switch insertion loss, radome loss, and transition losses.

T_{sys} is the effective noise temperature of the receiving system and

$$T_{\mathrm{sys}} = T_{\mathrm{ant}} + T_{\mathrm{recvr}} \qquad (7.19)$$

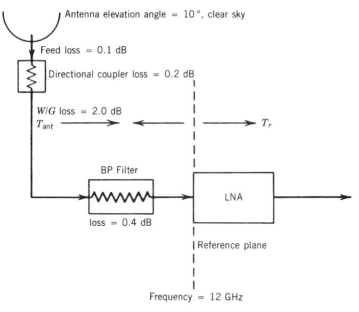

Figure 7.20. Example of an earth station receiving system.

T_{ant} or the antenna noise temperature includes all noise-generating components up to the reference plane. The components include sky noise (T_{sky}) plus the thermal noise generated by ohmic losses created by all devices inserted into the system up to the reference plane, including the radome. A typical earth station receiving system is shown in Figure 7.20 for an 11-GHz down-link. For a 4-GHz down-link the minimum elevation angle would be 5°. The elevation angle is that angle measured from the horizon (0°) to the antenna main beam when pointed at the satellite. Antenna noise (T_{ant}) is calculated by the following formula:

$$T_{\text{ant}} = \frac{(l_a - 1)290 + T_{\text{sky}}}{l_a} \qquad (7.20)$$

where l_a is the numeric equivalent of the sum of the ohmic losses up to the reference plane. l_a is calculated by

$$l_a = \log_{10}^{-1}\frac{L_a}{10} \qquad (7.21)$$

where L_a is the sum of the losses in decibels.

Sky noise varies with frequency and elevation angle. Typical values of sky noise are (from CCIR Rep. 720) [8]:

Frequency (GHz)	Elevation Angle	Sky Noise (K)
4.0	5°	25
7.5	5°	35
11.7	10°	30
20.0	10°	110

An earth station operating at 12 GHz with a 10° elevation angle will typically display an antenna noise temperature of 150 K [9].

Using values given in Figure 7.20 and T_{sky} of 30 K, we can calculate T_{ant} for a typical earth station where the down-link operating frequency is 12 GHz.

	Sum the losses:	Feed	0.1 dB
		Directional coupler	0.2 dB
		Waveguide	2.0
		Bandpass filter	0.4 dB
		Total	2.7 dB $= L_a$

Using equation 7.21 we obtain

$$l_a = \log_{10}^{-1}(2.7/10)$$
$$= 1.86$$

Now use the value for l_a and T_{sky} in equation 7.20:

$$T_{ant} = [(1.86 - 1)290 + 30]/1.86$$
$$= 150 \text{ K}$$

The noise figure for a typical commercial LNA or down-converter is 4 dB at 12 GHz. Convert the 4-dB value to equivalent noise temperature. Noise figure can be related to noise temperature (T_e) by the following formula:

$$NF_{dB} = 10 \log(1 + T_e/290) \tag{7.22}$$

For the sample problem, $T_e = T_{recvr}$:

$$4 \text{ dB} = 10 \log(1 + T_e/290)$$
$$0.4 = \log(1 + T_e/290)$$

Take the antilog of 0.4, and

$$2.51 = 1 + T_{\mathrm{e}}/290$$
$$T_{\mathrm{e}} = 438 \text{ K} = T_{\mathrm{recvr}}$$
$$T_{\mathrm{sys}} = T_{\mathrm{ant}} + T_{\mathrm{recvr}}$$
$$= 150 + 438$$
$$= 588 \text{ K}$$

Suppose that the antenna in Figure 7.20 had a 47-dB gross gain. What, then, is the G/T of the receiving system at a $10°$ elevation angle? Calculate net antenna gain.

$$G_{\mathrm{net}} = 47 - 2.7$$
$$= 44.3 \text{ dB}$$
$$G/T = 44.3 - 10 \log T_{\mathrm{sys}}$$
$$= 44.3 - 10 \log 588$$
$$= 44.3 - 27.7$$
$$= +16.7 \text{ dB/K}$$

5.8.3 Station Margin. One major consideration in the design of LOS microwave systems is the fade margin, which is the additional signal level added in the system calculations to allow for fading. This value is often on the order of 20–50 dB. In other words, the receive signal level system in question was overbuilt to provide 20–50 dB above threshold to overcome most fading conditions or to ensure that noise would not exceed a certain norm for a fixed time frame.

As we saw in Section 4, fading is caused by anomalies in the intervening medium between stations or by the reflected signal, thus causing interference to the direct ray signal. There would be no fading phenomenon on a radio signal being transmitted through a vacuum well above the earth's surface. Thus satellite earth station signals are subject to fade only during the time they traverse the atmosphere. For this case, most fades, if any, may be attributed to rainfall or very low elevation angle refractive anomalies.

Margin or station margin is an additional design advantage that compensates for deteriorated propagation conditions or fading. The margin designed into an LOS system is large and is achieved by increasing antenna size, improving the receiver noise figure, or increasing transmitter output power. The station margin of a satellite earth station in comparison is small, on the order of 4–6 dB. Typical rainfall attenuation exceeding 0.01% of a year may be from 1 dB to 2 dB (in the 4-GHz band) without a radome on the antenna and when the antenna is at $5°$ elevation angle. As the antenna elevation increases to zenith, the rain attenuation notably decreases because the signal passes through less atmosphere. The addition of a radome could increase the attenuation to 6 dB or greater during precipitation. Receive station margin for an earth station, those extra decibels on the down-link, is sometimes achieved by use of threshold-extension demodulation techniques for FM systems and forward error correction coding on digital systems. Up-link margin is provided by using larger transmitters and by increasing power output when necessary. A G/T ratio in excess of the minimum required for clear sky conditions at the 5° elevation angle will also provide margin but may prove expensive to provide.

5.8.4 Typical Down-Link Power Budget.

A link budget is a tabular method of calculating space communication system parameters. The approach is similar to that used on line-of-sight microwave links (radiolinks) (see Section 4.4). We start with the EIRP of the satellite for the down-link or the EIRP of the earth station for the up-link. The bottom line is C/N_0 and link margin (in decibels). C/N_0 is the carrier-to-noise ratio in 1 Hz of bandwidth at the input of the LNA. (*Note:* RSL or receive signal level and C are synonymous.) Expressed as an equation:

$$\frac{C}{N_0} = \text{EIRP} - \text{FSL}_{\text{dB}} - (\text{other losses}) + G/T_{\text{dB/K}} - k \qquad (7.23)$$

where FSL* is the free-space loss to the satellite for the frequency of interest and k is Boltzmann's constant expressed in decibel-watts. "Other losses" may include (where applicable):

- Polarization loss (0.5 dB).
- Pointing losses, terminal and satellite (0.5 dB each).
- Off-contour loss (depends on satellite antenna characteristics).
- Gaseous absorption loss (varies with frequency, altitude, and elevation angle).
- Excess attenuation due to rainfall (for systems operating above 10 GHz).

The loss values in parentheses are conservative estimates and should be used only if no definitive information is available.

The off-contour loss refers to spacecraft antennas that provide a spot or zone beam with a footprint on a specific geographical coverage area. There are usually two contours, one for G/T (up-link) and the other for EIRP (down-link). Remember that these contours are looking from the satellite down to the earth's surface. Naturally, an off-contour loss would be invoked only for earth stations located outside of the contour line. This must be distinguished from satellite pointing loss, which is a loss value to take into account that satellite pointing is not perfect. The contour lines are drawn as if the satellite pointing were "perfect."

---

*Remember that geostationary satellite range varies with elevation angle and is minimum at zenith.

Gaseous absorption loss (or atmospheric absorption) varies with frequency, elevation angle, and altitude of the earth station. As one would expect, the higher the altitude, the less dense the air and thus the less loss. Gaseous absorption losses vary with frequency and inversely with elevation angle. Often, for systems operating below 10 GHz, such losses are neglected. Reference 17 suggests a 1-dB loss at 7.25 GHz for elevation angles under 10° and for 4 GHz, 0.5 dB below 8° elevation angle.

Example of a Link Budget

Assume the following: a 4-GHz down-link, 5° elevation angle, EIRP is $+30$ dBW; satellite range is 25,573 statute miles (sm), and the terminal G/T is $+20.0$ dB/K. Calculate the down-link C/N_0.

First calculate the free-space loss. Use equation 7.5:

$$\begin{aligned} L_{dB} &= 96.6 + 20 \log F_{GHz} + 20 \log D_{sm} \\ &= 96.6 + 20 \log 4.0 + 20 \log 25,573 \\ &= 96.6 + 12.04 + 88.16 \\ &= 196.8 \text{ dB} \end{aligned}$$

Example Link Budget: Down-Link

EIRP of satellite	$+30$ dBW
Free-space loss	-196.8 dB
Satellite pointing loss	-0.5 dB
Off-contour loss	0.0 dB
Excess attenuation rainfall	0.0 dB
Gaseous absorption loss	-0.5 dB
Polarization loss	-0.5 dB
Terminal pointing loss	-0.5 dB
Isotropic receive level	-168.8 dBW
Terminal G/T	$+20.0$ dB/K
Sum	-148.8 dBW
Boltzmann's constant (dBW)	$-(-228.6$ dBW)
C/N_0	79.8 dB

On repeatered satellite systems, sometimes called "bent-pipe satellite systems" (those that we are dealing with here), the link budget is carried out only as far as C/N_0, as we did above. It is calculated for the up-link and for the down-link separately. We then calculate an equivalent C/N_0 for the system (i.e., up-link and down-link combined). Use the following formula to carry out this calculation:

$$\left(\frac{C}{N_0}\right)_{(s)} = \frac{1}{1/(C/N_0)_{(u)} + 1/(C/N_0)_{(d)}} \tag{7.24}$$

Example

Suppose that an up-link has a C/N_0 of 82.2 dB and its companion down-link has a C/N_0 of 79.8 dB. Calculate the C/N_0 for the system $(C/N_0)_s$. First calculate the equivalent numeric value (NV) for each C/N_0 value.

$$\text{NV}(1) = \log^{-1}(79.8/10) = 95.5 \times 10^6$$
$$\text{NV}(2) = \log^{-1}(82.2/10) = 166 \times 10^6$$
$$C/N_0 = 1/[(10^{-6}/95.5) + (10^{-6}/166)]$$
$$= 1/(0.016 \times 10^{-6}) = 62.5 \times 10^6 = 77.96 \text{ dB}$$

This is the carrier-to-noise ratio in 1 Hz of bandwidth. To derive C/N for a particular RF bandwidth, use the following formula:

$$C/N = C/N_0 - 10 \log BW_{\text{Hz}} \tag{7.25}$$

Suppose the example system had a 1.2-MHz bandwidth with the C/N_0 of 77.96 dB. What is the C/N?

$$C/N = 77.96 \text{ dB} - 10 \log(1.2 \times 10^6)$$
$$= 77.96 - 60.79$$
$$= 17.17 \text{ dB}$$

5.8.5 Up-Link Considerations. A typical specification for INTELSAT states that the EIRP per voice channel must be $+61$ dBW (example); thus to determine the EIRP for a specific number of voice channels to be transmitted on a carrier, we take the required output per voice channel in dBW (the above) and add logarithmically $10 \log N$, where N is the number of voice channels to be transmitted.

For example, consider the case for an up-link transmitting 60 voice channels; thus

$$+61 \text{ dBW} + 10 \log 60 = 61 + 17.78 = +78.78 \text{ dBW}$$

If the nominal 50-ft (15-m) antenna has a gain of 57 dB (at 6 GHz) and losses typically of 3 dB, the transmitter output power, P_t, required is

$$\text{EIRP}_{\text{dBW}} = P_t + G_{\text{ant}} - \text{line losses}_{\text{dB}} \tag{7.26}$$

where P_t is the output power of the transmitter (in decibel-watts) and G_{ant} is the antenna gain (in decibels) (up-link). Then in the example we have

$$+78.78 \text{ dBW} = P_t + 53 - 3$$
$$P_t = +24.78 \text{ dBW}$$
$$= 300.1 \text{ W}$$

5.9 Digital Communications by Satellite

There are three methods of handling digital communications by satellite: TDMA, FDMA, and over a VSAT network. TDMA was covered in Section 5.7.2, and VSAT networks are discussed in Section 5.10. Digital access by FDMA is used in a similar fashion as an analog FDM/FM configuration. Several users may share a common transponder and the same backoff rules hold; in fact they are even more important when using a digital format because the IM products generated can notably degrade error performance.

As we have mentioned previously, satellite communications is down-link-limited because down-link EIRP level is strictly restricted. Still we want to receive sufficient power to meet the error performance objectives. One way to achieve such a goal is to FEC-code the links where lower E_b/N_0 ratios will still meet error objectives. Thus INTELSAT requires coding on their digital accesses. Typical INTELSAT digital link parameters are given in Table 7.4. This is for the Intermediate Data Rate (IDR) Digital Carrier System. All IDR carriers are required to use $R^* = 3/4$ FEC convolutional coding. Reference 17 provides a description of various FEC channel coding schemes. INTELSAT recommends using standard information rates specified by CCITT. (See Table 7.5.)

The occupied satellite bandwidth unit for IDR carriers is approximately equal to 0.6 times the transmission rate. The transmission rate is defined as the coded symbol rate. To provide guardbands between adjacent carriers on the same transponder, the nominal satellite bandwidth unit is 0.7 times the transmission rate.

IDR carriers are designed to provide a service in accordance with CCIR Recs. 522, 614, and 579. To achieve these requirements, the system is designed to provide a nominal BER of 1×10^{-7} under clear sky conditions. Under degraded sky conditions (typically with rainfall), a worst-case BER of 1×10^{-3} for all but 0.04% of the year is provided. Table 7.5 gives other important IDR carrier parameters.

5.10 Very Small Aperture Terminal (VSAT) Networks

5.10.1 Rationale. VSATs are defined by their antenna aperture (diameter), which can vary from 0.5 meters (1.6 ft) to 2 meters (6.5 ft). A VSAT network consists of one comparatively large hub earth terminal and remote VSAT

---

*R is the coding rate. $R =$ information bit rate/coded symbol rate. When $R = 3/4$ and the information rate is 1.544 Mbps, the coded rate is 4/3 that value or 2,058,666 symbols per second.

TABLE 7.4 QPSK Characteristics and Transmission Parameters for IDR Carriers

Parameter	Requirement
1. Information rate (IR)	64 kbit/s to 44.736 Mbit/s
2. Overhead data rate for carriers with IR $\geq$ 1.544 Mbit/s	96 kbit/s
3. Forward error correction encoding	Rate 3/4 convolutional encoding/Viterbi decoding
4. Energy dispersal (scrambling)	As per ITU-R5.524-4
5. Modulation	Four-phase Coherent PSK
6. Ambiguity resolution	Combination of differential encoding (180°) and FEC (90°)
7. Clock recovery	Clock timing must be recovered from the received data stream
8. Minimum carrier bandwidth (allocated)	0.7 R Hz or [0.933 (IR + Overhead)]
9. Noise bandwidth (and occupied bandwidth)	0.6 R Hz or [0.8 (IR + Overhead)]
10. E_b/N_0 at BER (Rate 3/4 FEC)	10^{-3} 10^{-7} 10^{-8}
a. Modems back-to-back	5.3 dB 8.3 dB 8.8 dB
b. Through satellite channel	5.7 dB 8.7 dB 9.2 dB
11. C/T at nominal operating point	$-219.9 + 10 \log_{10}$(IR + OH), dBW/K
12. C/N in noise bandwidth at nominal operating point (BER $\leq 10^{-7}$)	9.7 dB
13. Nominal bit error rate at operating point	1×10^{-7}
14. C/T at threshold (BER = 1×10^{-3})	$-222.9 + 10 \log_{10}$(IR + OH), dBW/K
15. C/N in noise bandwidth at threshold (BER = 1×10^{-3})	6.7 dB
16. Threshold bit error rate	1×10^{-3}

Source: IESS-308, Rev. 7, Ref. 10. Courtesy of INTELSAT.
Notes:
(1) IR is the information rate in bits per second.
(2) R is the transmission rate in bits per second and equals (IR + OH) times 4/3 for carriers employing Rate 3/4 FEC.
(3) The allocated bandwidth will be equal to 0.7 times the transmission rate, rounded up to the next highest odd integer multiple of 22.5 kHz increment (for information rates less than or equal to 10 Mbit/s) or 125 kHz increment (for information rates greater than 10 Mbit/s).
(4) Rate 3/4 FEC is mandatory for all IDR carriers.
(5) OH = overhead.

terminals. Some networks in the United States have more than 2000 remote VSAT terminals (a large drugstore chain). Many such networks exist.

There are three underlying reasons for the use of VSAT networks:

1. An economic alternative to establish a data network, particularly if traffic flow is to/from a central facility, usually a corporate head-quarters to/from outlying remotes.

TABLE 7.5 Transmission Parameters for IDR Carriers with Rate 3/4 Coding

Information Rate (bit/s)	Overhead Rate (kbit/s)	Data Rate, (IR + OH) (bit/s)	Transmission Rate (bit/s)	Occupied Bandwidth (Hz)	Allocated Bandwidth (Hz)	C/T (dBW/K)	C/N_0 (dB — Hz)	C/N (dB)
64 k	0	64 k	85.33 k	51.2 k	67.5 k	−171.8	56.8	9.7
192 k	0	192 k	256.00 k	153.6 k	202.5 k	−167.1	61.5	9.7
384 k	0	384 k	512.00 k	307.2 k	382.5 k	−164.1	64.5	9.7
1.544 M	96	1.640 M	2.187 M	1.31 M	1552.5 k	−157.8	70.8	9.7
2.048 M	96	2.144 M	2.859 M	1.72 M	2002.5 k	−156.6	72.0	9.7
6.312 M	96	6.408 M	8.544 M	5.13 M	6007.5 k	−151.8	76.8	9.7
8.448 M	96	8.544 M	11.392 M	6.84 M	7987.5 k	−150.6	78.0	9.7
32.064 M	96	32.160 M	42.880 M	25.73 M	30125.0 k	−144.8	83.8	9.7
34.368 M	96	34.464 M	45.952 M	27.57 M	32250.0 k	−144.5	84.1	9.7
44.736 M	96	44.832 M	59.776 M	35.87 M	41875.0 k	−143.4	85.2	9.7

Source: IESS-308, Rev. 7, Ref. 10. Courtesy of INTELSAT.

Notes:

(1) The above table illustrates parameters for recommended carrier sizes. However, any other information rate between 64 kbit/s and 44.736 Mbit/s can be used.

(2) C/T, C/N_0, and C/N values have been calculated for a 10^{-7} BER and assume the use of Rate 3/4 FEC.

(3) For carrier information rates of 10 Mbit/s and below, carrier frequency spacings will be odd multiples of 22.5 kHz. For greater rates, they will be on multiples of 125 kHz.

(4) Rate 3/4 FEC is mandatory for all IDR carriers.

2. To bypass telephone companies with a completely private network.

3. To provide quality telecommunication connectivity where other means are substandard or nonexistent.

Regarding reason 3, the author is aware of one emerging nation where 124 bank branches had no electrical communication whatsoever with the headquarters institution in the capital city.

5.10.2 Characteristics of a VSAT Network. On conventional VSAT networks, the hub is designed to compensate for the VSAT handicap (i.e., its small size). For example, a hub antenna aperture is 5–11 meters (16–50 ft) [11]. High-power amplifiers (HPAs) run from 100 to 600 watts output power. Low-noise amplifiers, typically at 12 GHz, display (a) noise figures from 0.5 to 1.0 dB and (b) low-noise downconverters in the range of 1.5-dB noise figure. Hub G/T values range from $+29$ to $+34$ dB/K.

VSAT terminals have transmitter output powers ranging from 1 to 50 watts, depending on service characteristics. Receiver noise performance using a low-noise downconverter is about 1.5 dB, otherwise 1 dB with an LNA. G/T values for 12.5-GHz down-links are between $+14$ and $+22$ dB/K, depending greatly on antenna aperture. The idea is to make a VSAT terminal as inexpensive as possible.

Figure 7.21 shows the hub/VSAT concept of a star network with the hub at the center.

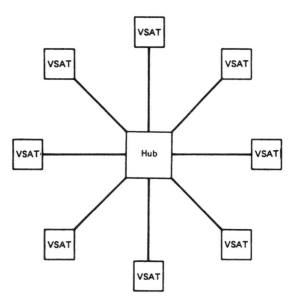

Figure 7.21. VSAT network topology. Note the star configuration and that the outlying VSAT remotes can number in the thousands.

5.10.3 Access Techniques

Inbound and Outbound. Inbound refers to traffic from VSAT(s) to hub, and *outbound* refers to traffic from hub to VSAT(s). The outbound link is commonly a time-division multiplex (TDM) serial bit stream, often 56 kbps, and some high-capacity systems reach 1.544 or 2.048 Mbps. The inbound links can take on any one of a number of flavors, typically 9600 bps.

More frequently VSAT systems support interactive data transactions, which are very short in duration. Thus, we can expect bursty operation from a remote VSAT terminal. One application is to deliver, in near real time, point-of-sale (POS) information, forwarding it to headquarters where the VSAT hub is located. Efficiency of bandwidth use is not a primary motivating factor in system design. Thus, for the interactive VSAT data network environment, low delay, simplicity of implementation, and robust operation are generally of greater importance than the bandwidth efficiency achieved.

Message access on any shared system can be of three types: fixed assigned, contention (random access), or reservation (controlled access). There are hybrid schemes between contention and reservation.

In the fixed assigned multiple access, VSAT protocols are SCPC*/FDMA, CDMA (a spread spectrum technique), and TDMA. All three are comparatively inefficient in the bursty environment with hundreds or thousands of potential users.

With the contention/random access category, there is the famous ALOHA protocol and selective reject (SREJ) ALOHA. Both have to have some form of collision resolution. In ALOHA, stations transmit new messages on the channel as they are generated. Collision resolution is achieved simply by retransmitting colliding packets with random delay. ALOHA has low efficiency of channel use but this is offset by low access delay, the ability to handle variable-length packets, robust operation, and minimal equipment complexity.

SREJ ALOHA employs subpacketization of messages in conjunction with a selective reject strategy and is able to achieve notably higher random access throughout than pure ALOHA. Here data are formed into a contiguous sequence of independently detectable fixed-length subpackets, each with its own header and acquisition preamble. The idea here is that most collisions in an asynchronous channel result in partial overlap of contending packets, so that only the smaller subpackets encounter conflict and must be retransmitted. SREJ ALOHA is an efficient access protocol.

Slotted ALOHA is based on the principle of reducing the vulnerability of a packet by constraining transmission to fixed-length packets which begin and end in TDMA-like slot boundaries. Thus maximum throughput is

---

*SCPC stands for single channel per carrier.

increased and the excellent delay characteristics of pure ALOHA are maintained.

Another variant is the tree collision resolution algorithm (CRA) random access system, which uses fixed-length packets. Here packets involved in collision participate in a systematic partitioning procedure for collision resolution, during which time new messages are not allowed to access the channel. The tree-type protocols have excellent capacity capabilities but can suffer deadlocks due to incorrect channel observation.

The third group of protocols is the reservation/controlled access type. These are based on demand assignment multiple access (DAMA) schemes and are useful variable length data frames. As in our previous discussion in Section 5.7.3, DAMA is a two-step process. There was a call-setup stage before the actual call went into operation. In this case, there is phase 1, where short reservation packets are transmitted from the VSAT requesting service and giving information regarding the station's demand. The second phase is the actual passing of data.

Reservation can be made by either a contention process or a fixed assigned process, both described above. One major advantage of a DAMA protocol is that data messages can be scheduled in a conflict-free manner. Well-designed DAMA systems have advantages with variable-length message traffic with relatively high overall channel throughput. However, this higher throughput is accompanied by a relatively large minimum latency (>0.500 ms) delay due to the reservation mechanism.

A subset of DAMA is DAMA with TDMA reservations. In this case, channel frames are made up into request and packet transmission intervals. Slots are available in each frame for VSATs to request data slot allocation. Of course, the allocation assignment is returned to the VSAT in the outbound TDM bit stream. Due to satellite propagation delays and TDMA frame structure, such protocols are characterized by a relatively high latency delay and are limited regarding the maximum number of accesses.

Another form of DAMA is DAMA with slotted ALOHA reservations. This access protocol can be used to support large VSAT populations because it uses contention access rather than fixed assigned TDMA-type access. DAMA with slotted ALOHA access provides good overall performance and can handle mixed interactive file-transfer traffic [11].

5.10.4 VSAT Transponder Operation. A total VSAT system may occupy no more than 1 MHz of transponder space. Other, larger VSAT systems may require more transponder bandwidth. A typical 1-MHz frequency assignment may be apportioned as shown in Figure 7.22.

TDMA inbound carriers to the hub, as shown in Figure 7.22, can be configured for one user per carrier (more similar to an FDMA configuration) or multiple users per carrier using one of several TDMA access techniques, or a DAMA discipline. There are many subnetworking possibilities as well. For example, each TDMA carrier may be assigned to a family of users. Forward

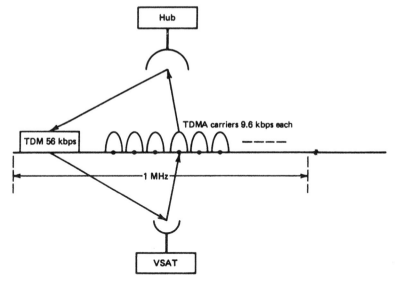

Figure 7.22. VSAT system operation with a 1-MHz allocation on a satellite transponder [12].

error correction (FEC) is commonly used and the employment of ARQ* is almost universal.

Carrier modulation techniques for VSAT systems is commonly BPSK, often favored over QPSK because it is more robust.

6   FIBER-OPTICS COMMUNICATION LINKS

6.1   Application

Fiber optics as a transmission medium has a comparatively unlimited bandwidth. It has excellent attenuation properties, as low as 0.25 dB/km. A major advantage fiber has when compared to coaxial cable is that no equalization is necessary. Also, repeater separation is on the order of 10–100 times that of coaxial cable for equal transmission bandwidths. Other advantages are:

- Electromagnetic immunity.
- Ground loop elimination.
- Security.
- Small size and lightweight.
- Expansion capabilities requiring change out of electronics only, in most cases.
- No licensing required.

*ARQ stands for automatic repeat request. See Chapter 11 for a discussion of ARQ.

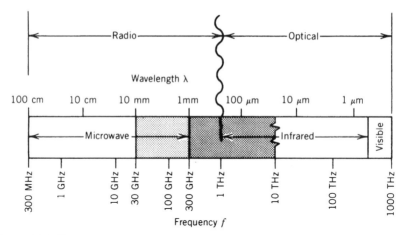

Figure 7.23. Frequency spectrum above 300 MHz. The usable wavelengths are just above and below 1 μm.

Fiber has analog transmission application, particularly for video/TV. However, for this discussion we will be considering only digital applications, principally as a PCM highway or "bearer."

Fiber-optics transmission is used for links under 1 ft in length all the way up to and including transoceanic undersea cable. In fact, all transoceanic cables presently being installed and planned for the future are based on fiber optics.

Fiber-optics technology was developed by physicists, and, following the convention of optics, wavelength rather than frequency is used to denote the position of light emission in the electromagnetic spectrum. The fiber optics of today uses three wavelength bands: around 800 nm, 1300 nm, and 1600 nm or near-visible infrared. This is shown in Figure 7.23.

This section includes an overview of how a fiber-optics link works, including types of fiber and a discussion of sources, detectors, and connectors and splices.

6.2   Introduction to Optical Fiber as a Transmission Medium

Optical fiber consists of a core and a cladding as shown in Figure 7.24. At present the most efficient core material is silica SiO_2.

The practical propagation of light through an optical fiber may best be explained using ray theory and Snell's law. Simply stated, we can say that when light passes from a medium of higher refractive index (n_1) into a medium of lower refractive index (n_2), the refractive ray is bent away from the normal. For instance, a ray traveling in water and passing into an air region is bent away from the normal to the interface between the two regions. As the angle of incidence becomes more oblique, the refracted ray is bent more until finally the refracted energy emerges at an angle of 90° with respect

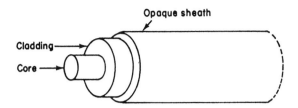

Figure 7.24. Structure of optical fiber consisting of a central core and a peripheral transparent cladding surrounded by protective packaging.

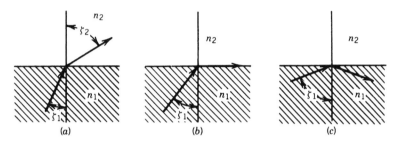

Figure 7.25. Ray paths for several angles of incidence ($n_1 > n_2$).

to the normal and just grazes the surface. Figure 7.25 shows the various incidence angles. Figure 7.25b illustrates what is called the *critical angle*, where the refracted ray just grazes the surface. Figure 7.25c is an example of total internal reflection. This occurs when the angle of incidence exceeds the critical angle. A glass fiber, for the effective transmission of light, requires total internal reflection.

Another property of the fiber for a given wavelength λ is the normalized frequency V; then

$$V = \frac{2\pi a}{\lambda} \sqrt{n_1^2 - n_2^2} \tag{7.27}$$

where a is the core radius, n_1 is the index of refraction of the core, and n_2 is the index of refraction of the cladding. n_2 of unclad fiber equals 1 (air). In equation 7.27 the term $\sqrt{n_1^2 - n_2^2}$ is called the *numerical aperture* (NA).

In essence the numerical aperture is used to describe the light-gathering ability of fiber. In fact, the amount of optical power accepted by a fiber varies as the square of the numerical aperture. It is also interesting to note that the numerical aperture is independent of any physical dimension of the fiber [12].

As shown in Figure 7.26, there are three basic elements in an optical-fiber transmission system: the optical source, the fiber link, and the optical detector. Regarding the fiber link itself, there are two basic impairments that can limit the length of such a link without resorting to repeaters or that can

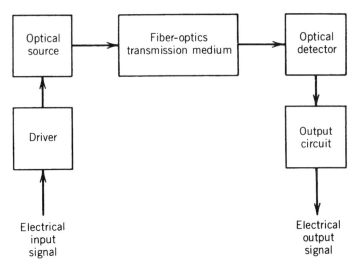

Figure 7.26. Typical fiber-optic communication link.

limit the distance between repeaters. These impairments are loss, usually expressed in decibels per kilometer, and dispersion, usually expressed as bandwidth per unit length, such as megahertz per kilometer. A particular fiber-optic link may be *power-limited* or *dispersion-limited.*

Dispersion, manifesting itself in intersymbol interference at the receive end, can be brought about by several factors. There is material dispersion, modal dispersion, and chromatic dispersion. Material dispersion can manifest itself when the emission spectral line is very broad, such as with a light-emitting diode (LED) optical source. Certain frequencies inside the emission line travel faster than others, causing some transmitted energy from a pulse to arrive later than other energy. This causes intersymbol interference. Modal dispersion occurs when several different modes are launched. Some have more reflections inside the fiber than other modes, thus, again, causing some energy from the higher-order modes to be delayed compared to lower-order modes.

One way of limiting the number of modes N that a fiber can support is by applying equation 7.27. The modes propagated can be reduced by reducing V, the normalized frequency, and by keeping the ratio n_1/n_2 as small as practical, often 1.01 or less. V can be reduced by reducing a, the radius of the core. When the radius is about 4 μm, $V = 2.405$ and only the HE_{11} mode will propagate. This very thin glass fiber is called *monomode fiber*, and using this type of fiber with some of the longer light wavelengths puts us well on the way to reducing or nearly eliminating dispersion [13].

We should only be concerned with chromatic dispersion on monomode fiber carrying a transmission data rate greater than 1 Gbps. We will handle chromatic dispersion later on.

6.3 Types of Optical Fiber

There are three categories of optical fiber as distinguished by their modal and physical properties:

- Step index (multimode)
- Graded index (multimode)
- Single mode (also called monomode)

Step-index fiber is characterized by an abrupt change in refractive index, and graded index is characterized by a continuous and smooth change in refractive index (i.e., from n_1 to n_2). Figure 7.27 shows the fiber construction and refractive index profile for step-index fiber (7.27a) and graded-index fiber (7.27b). Both step-index and graded-index fibers are characterized as "multimode" because more than one mode can propagate. Graded index has a superior bandwidth–distance product compared to that of step index. In other words it can carry a higher bit rate further than step index. It is also more expensive.

Single-mode fiber is designed such that only one mode can propagate. To do this, $V \leq 2.405$. Such fiber exhibits no modal dispersion at all (theoretically). Typically we might encounter a fiber with indices of refraction of $n_1 = 1.48$ and $n_2 = 1.46$. If the optical source wavelength is 1.2 μm, for single-mode operation we'd find the core radius to be about 4 μm [15].

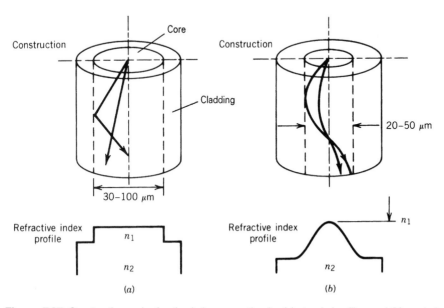

Figure 7.27. Construction and refractive index properties for (a) step-index fiber and (b) graded-index fiber.

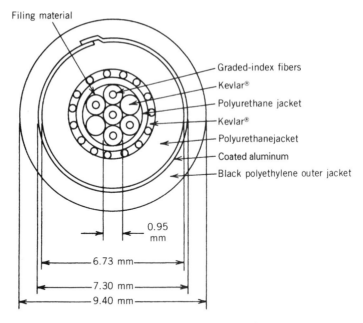

Filing material

Graded-index fibers
Kevlar®
Polyurethane jacket
Kevlar®
Polyurethanejacket
Coated aluminum
Black polyethylene outer jacket

0.95 mm
6.73 mm
7.30 mm
9.40 mm

Figure 7.28. Direct-burial optical fiber cable.

There are two additional factors that the fiber-optics communication system designer must take into account, namely, minimum bending radius and fiber strength. Radiation losses at fiber waveguide bends are usually quite small and may be neglected in system design unless the bending radius is smaller than that specified by the fiber manufacturer. Minimum bending radii vary from about 2 cm to 10 cm, depending on the cable characteristics, or, as a rule of thumb, about 10 times the cable diameter. Fiber cable strength is also specified by the manufacturer. For example, one manufacturer for a specific cable type specifies a maximum pulling tension of 1780 N (400 lb) at 20°C, a maximum permissible compression load of 655 N/cm (375 lb/in.) flat plate, and a maximum permissible impact force of 280 N/cm (160 lb in.).

Figure 7.28 shows a typical five-fiber cable for direct burial.

6.4 Splices and Connectors

Optical fiber cable is commonly available in 1-km sections; it is also available in longer sections, in some types up to 10 km or more. In any case there must be some way of connecting the fiber to the source and to the detector as well as connecting the reels of cable together, whether in 1 km or more lengths, as required. There are two methods of connection, namely, splicing or using connectors. The objective in either case is to transfer as much light as possible through the coupling. A good splice couples more light than the best connectors.

A good splice can have an insertion loss as low as 0.09 dB, whereas the best connector loss can be as low as 0.3 dB.

An optical fiber splice requires highly accurate alignment and an excellent end finish to the fibers. There are three causes of loss at a splice:

1. Lateral displacement of fiber axes.

2. Fiber end separation.

3. Angular misalignment.

Splice loss also varies directly with the numerical aperture of the fiber in question.

There are two types of splice now available, the mechanical splice and the fusion splice. With a mechanical splice an optical matching substance is used to reduce splicing losses. The matching substance must have a refractive index close to the index of the fiber core. A cement with similar properties is also used, serving the dual purpose of refractive index matching and fiber bonding. The fusion splice, also called a *hot splice*, is where the fibers are fused together. The fibers to be spliced are butted together and heated with a flame or electric arc until softening and fusion occur.

Splices require special splicing equipment and trained technicians. Thus it can be seen that splices are generally hard to handle in a field environment such as a cable manhole. Connectors are much more amenable to field connecting. However, connectors are lossier and can be expensive. Repeated mating of a connector may also be a problem, particularly if dirt or dust deposits occur in the area where the fiber mating takes place.

However, it should be pointed out that splicing equipment is becoming more economic, more foolproof, and more user-friendly. Technician training is also becoming less of a burden.

Connectors are nearly universally used at the source and at the detector to connect the main fiber to these units. This makes easier change-out of the detector and source when they fail or have degraded operation.

6.5 Light Sources

A light source, perhaps more properly called a *photon source*, has the fundamental function in a fiber-optics communication system to convert efficiently electrical energy (current) into optical energy (light) in a manner that permits the light output to be effectively launched into the optical fiber. The light signal so generated must also accurately track the input electrical signal so that noise and distortion are minimized.

The two most widely used light sources for fiber-optics communication systems are the light-emitting diode (LED) and the semiconductor laser, sometimes called a *laser diode* (LD). LEDs and LDs are fabricated from the same basic semiconductor compounds and have similar heterojunction

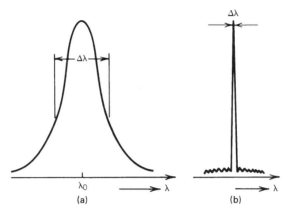

Figure 7.29. Spectral distribution of the emission from (a) an LED and (b) a semiconductor laser (LD). λ is the optical wavelength.

structures. They do differ in the way they emit light and in their performance characteristics.

An LED is a forward-biased p–n junction that emits light through spontaneous emission, a phenomenon referred to as *electroluminescence*. LDs emit light through stimulated emission. LEDs are less efficient than LDs but are considerably more economical. They also have a longer operational life. The emitted light of an LED is incoherent with a relatively wide spectral line width (from 30 to 60 nm) and a relatively large angular spread, about 100°. On the other hand, a semiconductor laser emits a comparatively narrow line width (from <2 to 4 nm). Figure 7.29a shows the spectral line for an LED, and Figure 7.29b shows the spectral line for a semiconductor laser.

With present technology the LED is capable of launching about 100 µW (−10 dBm) or less of optical power into the core of a fiber with a numerical aperture of 0.2 or better. A semiconductor laser with the same input power can couple up to 7 mW (+8.5 dBm) into the same cable. The coupling efficiency of an LED is on the order of 2%, whereas the coupling efficiency of an LD (semiconductor laser) is better than 50%.

Methods of coupling a source into an optical fiber vary, as do coupling efficiencies. To avoid ambiguous specifications on source output powers, such powers should be stated at the *pigtail*. A pigtail is a short piece of optical fiber coupled to the source at the factory and, as such, is an integral part of the source. Of course, the pigtail should be the same type of fiber as that specified for the link.

Component lifetimes for LEDs are on the order of 100,000 hr (MTBF) with up to a million hours reported in the literature. Many manufacturers guarantee a semiconductor laser for 20,000 hr and more. About 150,000 hours can be expected from semiconductor lasers after stressing and culling of unstable units. Such semiconductor lasers are used in TAT8/9 undersea cables connecting North America to Europe.

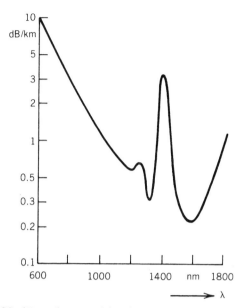

Figure 7.30. Attenuation per unit length versus wavelength of glass fiber [12].

It should be noted that the life expectancy of a semiconductor laser is reduced when it is overdriven to derive more coupled power, such as more than 7 mW. Often these devices are operated at 1 mW (0 dBm). LEDs also have a drawback. Output power tends to drop off with age.

The semiconductor laser is a temperature-dependent device. Its threshold current increases nonlinearly with temperature. Rather than attempt to control the device's temperature, a negative feedback circuit is used whereby a portion of the emitted light is sampled, detected, and fed back to control the drive current. Such circuits are similar to the familiar AGC circuits used on radio receivers.

Fiber optic communication systems operate in the nominal wavelength regions of 820 nm, 1330 nm, and 1550 nm. If we examine the attenuation versus wavelength curve in Figure 7.30, we see that for the 820-nm region the lowest attenuation we can expect is 3 dB/km. As wavelength increases, going to the right on the curve, we see another valley around 1330 nm, where the loss per unit length drops to 0.5 dB/km, and there is still another valley at 1550 nm, where the loss is only 0.25 dB/km or a little less. There is mature technology available for all three wavelengths.

6.6   Light Detectors

The most commonly used detectors (receivers) for fiber-optics communication systems are photodiodes, either PIN or APD. The terminology *PIN* derives from the semiconductor construction of the device where an intrinsic (I) material is used between the p–n junction of the diode.

A photodiode can be considered a photon counter. The photon energy E is a function of frequency and is given by

$$E = h\nu \tag{7.28}$$

where h is Planck's constant (W/s^2) and ν is the frequency in hertz. E is measured in watt-seconds or kilowatt-hours.

The receiver power in the optical domain can be measured by counting, in quantum steps, the number of photons received by a detector per second. The power in watts may be derived by multiplying this count by the photon energy, as given in equation 7.28.

The efficiency of the optical-to-electrical power conversion is defined by a photodiode's *quantum efficiency*, which is the average number of electrons released by each incident photon. A highly efficient photodiode would have a quantum efficiency of 1, and decreasing from 1 indicates progressively poorer efficiencies. The quantum efficiency, in general, varies with wavelength and temperature.

For the fiber-optic communication system engineer, *responsivity* is a most important parameter when dealing with photodiode detectors. Responsivity is expressed in amperes per watt or volts per watt and is sometimes called *sensitivity*. Responsivity is the ratio of the root mean square (rms) value of the output current or voltage of a photodetector to the rms value of the incident optical power. In other words, responsivity is a measure of the amount of electrical power we can expect at the output of a photodiode, given a certain incident light power signal input. For a photodiode the responsivity R is related to the wavelength λ of the light flux and to the quantum efficiency η, the fraction of the incident photons that produce a hole-electron pair. Thus

$$R = \frac{\eta\lambda}{1234} \, (A/W) \tag{7.29}$$

with λ measured in nanometers.

The avalanche photodiode (APD) is a gain device displaying gains on the order of 15–20 dB. The PIN diode is not a gain device. Table 7.6 summarizes detector sensitivities with the standard BER of 1×10^{-9} for some common bit rates. Noise equivalent power (NEP) is often used as the figure of merit of a photodiode. NEP is defined as the rms value of optical power required to produce a unit signal-to-noise ratio (i.e., signal-to-noise ratio = 1) at the output of a light-detecting device. NEPs vary for specific diode detectors between 1×10^{-13} W/Hz$^{1/2}$ and 1×10^{-14} W/Hz$^{1/2}$.

Of the two types of photodiodes discussed here, the PIN is more economical and requires less complex circuitry than does its APD counterpart. The PIN diode has peak responsivity from about 800 nm to 900 nm for silicon devices. These responsivities range from 300 μA/mW to 600 μA/mW.

TABLE 7.6 Summary of Experimental Receiver Diode Sensitivities (Average Received Optical Power, P, in dBm; BER = 1 $\times$ 10^{-9})

Bit Rate (Mbps)	InGaAs PIN		InGaAs SAM/SAGM[a] APD		Ge APD		InGaAs Photoconductor	
	1.3 μm	1.55 μm	1.3 μm	1.55 μm	1.3 μm	1.55 μm	1.3 μm	1.55 μm
34	−52.5				−46	−55.8		
45	−49.9		−51.7		−51.9			
100					−40.5			
140	−46				−45.2	−49.3		
274	−43		−45	−38.7	−36			
320	−43.5							
420			−43	−41.5				
450			−42.5		−39.5	−40.5		
565					−33			
650	−36							
1000			−38	−37.5	−28		−34.4	
1200	−33.2	−36.5						
1800			−31.3		−30.1			
2000				−36.6		−31		−28.8
4000				−32.6				

[a]SAGM = separated absorption, grating, and multiplication regions. SAM = separated absorption and multiplication regions.

Source: Ref. 16.

The overall response time for the PIN diode is good for about 90% of the transient but sluggish for the remaining 10%, which is a "tail." The power response of the tail portion of a pulse may limit the net bit rate on digital systems.

The PIN detector does not display gain, whereas the APD does. The response time of the APD is far better than that of the PIN diode, but the APD displays certain temperature instabilities where responsivity can change significantly with temperature. Compensation for temperature is usually required in APD detectors and is often accomplished by a feedback control of bias voltage. It should be noted that bias voltages for APDs are much higher than for PIN diodes, and some APDs require bias voltages as high as 200 V. Both the temperature problem and the high-voltage bias supply complicate repeater design.

6.7 Optical Fiber Amplifiers

Optical amplifiers amplify incident light through stimulated emission, the same mechanism as used with lasers. These amplifiers are the same as lasers

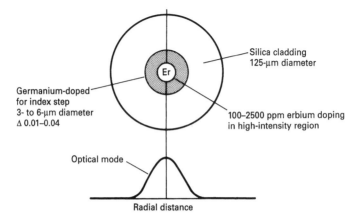

Figure 7.31. Erbium-doped fiber core geometry. From Ref. 14. Courtesy of Hewlett-Packard.

Two-Stage Design

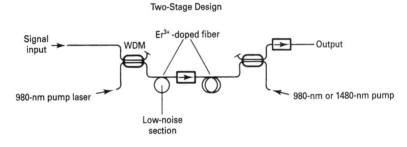

High gain 20–50 dB

Figure 7.32. EFDA block diagram. From Ref. 14. Courtesy of Hewlett-Packard.

without feedback. Optical gain is achieved when the amplifier is pumped either electrically or optically to realize population inversion.

There are semiconductor laser amplifiers, Raman amplifiers, Brillouin amplifiers, and erbium-doped fiber amplifiers (EDFAs). Certainly the EDFA shows the widest acceptance. One reason is that they operate near the 1.55-μm wavelength region where fiber loss is at a minimum. Reference 15 states that it is possible to achieve high amplifier gains in the range of 30–40 dB with only a few milliwatts of pump power when EDFAs are pumped by using 0.980-μm or 1.480-μm semiconductor lasers.

Figure 7.31 shows erbium-doped fiber core geometry and Figure 7.32 illustrates a typical block diagram of a modern, low-noise EDFA.

In Figure 7.32, optical pumping is provided by fiber pigtailed semiconductor lasers with typically 100 mW of power. Low-loss wavelength division multiplexers efficiently combine pump and signal powers and can also be used to provide a pump power bypass around the internal isolator. The EDFA has an input stage that is codirectionally pumped and an output stage counterdirectionally pumped. Such multistage EDFA designs have

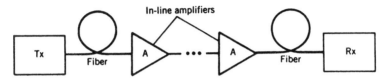

Figure 7.33. Applications of EFDAs.

simultaneously achieved a low noise figure of 3.1 dB and a high gain of 54 dB [14, 15].

The loops of fiber should be noted in Figure 7.32. These are lengths of fiber with the dopant. The length of erbium-doped fiber required for a particular amplifier application depends on the available pump power, doping concentration, the design topology, and gain and noise requirements. Generally, low-noise amplifiers will have shorter active fiber lengths (e.g., 6–8 meters at 300 ppm erbium doping where the optimum gain length may be 10 meters). For a constant pump power, reducing the erbium-doped fiber length increases the fractional population of erbium ions in the metastable state. This reduces the amount of spontaneous emission generated by the amplifier. Since the noise figure and gain characteristics vary slowly about the optimum values, great precision is not generally required in the active fiber length.

Erbium-doped fiber amplifiers are often installed directly after a semiconductor laser source (transmitter) and/or directly before the PIN or APD receiver at the distant end. Figure 7.33 illustrates this concept. This can extend the length of a fiber-optic link without repeaters about 100–200 km, or it can extend the distance between repeaters by a similar amount.

6.8 Fiber-Optic Link Design

The design of a fiber-optics communication link involves several steps. Certainly the first consideration is to determine the feasibility of such a transmission system for a desired application. There are two aspects of this decision: economic and technical. Can we get equal or better performance for less money using some other transmission medium such as wire pair, coaxial cable, LOS microwave, and so on?

Fiber-optics communication links have wide application. Analog applications for cable television (CATV) trunks are showing particularly rapid growth. Fiber is also used for low-level signal transmission in radio systems such as for long runs of IF, and even for RF. However, in this text we stress digital applications, some of which are listed below:

- On-premises data bus
- LANs (e.g., fiber distributed data interface)

- High-level PCM or CVSD* configurations; SONET and SDH
- Radar data links
- Conventional data links where bit/symbol rates exceed 19.2 kbps
- Digital video including cable television

It seems that the present trend of cost erosion will continue for fiber cable and components. Fiber-optic repeaters are considerably more expensive than their PCM metallic counterparts. The powering of the repeaters can be more involved, particularly if power is to be taken from the cable itself. This means that the cable must have a metallic element to supply power to downstream repeaters, thereby losing a fiber-optic advantage. Metal in the cable, particularly for supplying power, can be a conductive path for ground loops. Another approach is to supply power locally to repeaters with a floating battery backup.

A key advantage to fiber over metallic cable is *less repeaters* per unit length. In Chapter 8 we will show that repeaters in tandem are the principal cause of jitter, a major impairment to a digital system such as PCM. Reducing the number of repeaters reduces jitter accordingly. In fact, fiber-optic systems require a small fraction of the number of repeaters compared to PCM for the same unit length, either on wire pair or coaxial cable.

6.8.1   Design Approach. The first step in designing a fiber-optic communication system is to establish the basic system parameters. Among these we would wish to know:

- Signal to be transmitted (e.g., PCM, CATV analog; bit rate and format)
- System length, fiber portion end-to-end
- Growth requirements (additional circuits, increased bit rates)
- Tolerable signal impairment level stated as signal-to-noise ratio or BER at the output of the terminal-end detector

The link BER should be established at the common ad-hoc standard of 1×10^{-9}.

Throughout the design procedure, when working with trade-offs, the designer must establish whether he/she is working in the power-limited domain or dispersion-limited domain. For instance, with the lower bit rates such as DS1 (T1) 1.544 Mbps or E1 2.048 Mbps, we would expect to be in the power-limited domain under all circumstances.

Some years ago, we would start concerning ourselves about the dispersion-limited domain at bit rates of about 45 Mbps (DS3) and above. Now it is a

---

*CVSD stands for continuous variable slope delta (modulation), a form of digital modulation where the coding is 1 bit at a time. Very popular with the armed forces.

whole different matter. The majority of installations today use single-mode (monomode) fiber at the longer wavelengths. Single-mode fiber eliminates the possibility of modal dispersion. When we use a semiconductor laser with single-mode fiber, we eliminate material dispersion. At about 500 Mbps, chromatic dispersion may become a concern.

However, there is still another saving factor or two. If we have operation at the zero-dispersion wavelength, about 1300 nm, dispersion may really not be a concern until about the 1-Gbps rate. Of course, at 1300 nm we have lost the use of the really low loss band at around 1550 nm. There is an answer to that, too. Use a fiber where the minimum dispersion window has been shifted to the 1550-nm region. Such fiber, of course, is more costly, but the cost may be worth it. It is another trade-off.

The designer must select the most economic alternatives among the following factors:

- Fiber parameters: single mode or multimode; if multimode, step index or graded index; number of fibers, cable makeup, strength.
- Transmission wavelength: 820 nm, 1330 nm, or 1550 nm.
- Source type: LED or semiconductor laser; there are subsets to each source type.
- Detector type: PIN or APD.
- Use of EDFA (amplifiers).
- Repeaters, if required, and how they will be powered.
- Modulation will probably be intensity modulation (IM), but the electrical waveform entering the source is important; possibly consider Manchester coding.

There is the splice and connector trade-off as well as the type of splice and type of connector. Permanently installed systems would opt for splices because of lower insertion losses. Temporarily installed systems, such as used by the military in the tactical environment, may prefer connectors because of ease and speed of mating and demating.

6.8.2 Loss Design. As a first step, assume that the system is power-limited. Probably a majority of systems being installed today can stay in the power-limited regime if monomode fiber is used with semiconductor lasers. When designing systems for bit rates in excess of 600 or 1000 Mbps, consider using semiconductor lasers with very narrow line widths and zero dispersion window shifted monomode fiber.

For a system operating at these high bit rates, even with the attribute of monomode fiber, chromatic dispersion can become a problem, particularly at the desirable 1550-nm band. Chromatic dispersion is really a form of material dispersion described earlier. It is the sum of two effects: "material dispersion" and waveguide dispersion. As one would expect, with material dispersion, different wavelengths travel at different velocities of propagation. This is true even with the narrow line width of semiconductor lasers. *Waveguide dispersion* is a result of lightwaves traveling through single-mode fibers that extend into the cladding. Its effect is more pronounced at the longer wavelengths because there is more penetration of the cladding and the "effective" refractive index is reduced. This causes another wavelength dependence on the velocity of light through the fiber, and therefore another form of dispersion. Thus the use of semiconductor lasers with very narrow line widths (e.g., <0.5 nm) helps mitigate chromatic dispersion [15].

Link margin is another factor for trade-off. We set this dB value aside in reserve for the following contingencies:

• Cable reel loss variability

• Future added splices (due to cable repair) and their insertion loss.

• Component degradation over the life of the system. This is particularly pronounced for LED output.

CCITT recommends 3 dB for link margin; others [14] recommend 6 dB. Ideally, for system reliability, a large margin is desirable. To optimize system first cost, as low a value as possible would be desired.

The system designer develops a power budget, similar in many respects to the path analysis or link budget of LOS microwave and satellite communication link design. However, there is little variability in a fiber-optic link budget; for example, there is no fading.

For a first-cut design, there are two source types, LED and semiconductor laser. Expect a power output of an LED in the range of −10 dBm; and for the semiconductor laser budget 0 dBm, although up to nearly +10 dBm is possible. There are two types of detectors, PIN and APD. For long links with high bit rates, the APD may become the choice. We would expect that the longer wavelengths would be used, but 820/850-nm links are still being installed. We must not forget reliability in our equation for choices. For lower bit rates and shorter links, we would give LEDs a hard look. They are cheaper and are much more reliable (MTBF).

Let's try a link design for 140 Mbps and see how far it can be extended without the use of repeaters. We will use a PIN detector. Turn to Table 7.6 and find the threshold in dBm for a link BER of 1×10^{-9}. This is −46 dBm at 1.3 μm. Here is the starting point for the power budget.

The source is a semiconductor laser with 0-dBm output. However, we must work up an output penalty due to what is called the *extinction ratio*. With a semiconductor laser, using intensity modulation, the 1s and 0s are not distinguished by true on and off conditions. For the "0" condition, the laser semiconductor still displays a very low power output (i.e., it is not completely turned off). The power emitted during the "0" bits is due to spontaneous emission. We call this "off" state power P_0, which depends on the bias

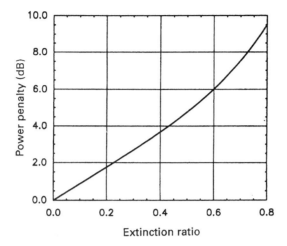

Figure 7.34. Extinction ratio (r_{ex}) versus power penalty (dB). From Ref. 15. Reprinted with permission.

current I_b and the threshold current I_{th}. If $I_b < I_{th}$, the power emitted during the "0" bits is due to spontaneous emission, and generally $P_0 \ll P_1$, where P_1 is the on-state power corresponding to the "1" bit. By contrast, P_0 can be a significant fraction of P_1 if the laser is biased close to or above threshold. The extinction ratio is defined as:

$$r_{ex} = P_0/P_1 \tag{7.30}$$

P_0 and P_1 are expressed in mW. Figure 7.34 relates r_{ex}, the extinction ratio, to power penalty in dB. For semiconductor lasers biased below threshold, r_{ex} is typically below 0.05 and the resulting penalty is <0.4 dB [15].

Return now to the example power budget. The power out for a "1" is 1 mW and the related power out for the symbol 0 is 0.1 mW. The wavelength is 1550 nm (1.550 μm). The monomode fiber is in 1-km reels. Fusion splicing is employed. Calculate the maximum distance to the far-end terminal or to the first repeater. One EDFA is used with a 40-dB gain.

Calculate the power penalty for the r_{ex}, which is 0.1. From Figure 7.34, the power penalty is 0.7 dB. The output of the source is 0 dBm with a power penalty of 0.7 dB, leaving a net output of −0.3 dBm. The receiver threshold is −46 dBm, giving 45.3 dB to be allocated to the link budget. Add to this value the gain of the EDFA, 40 dB, providing a final value of 85.3 dB to be allocated to the following:

- Fiber at 0.25 dB/km
- Two connectors at 0.5 dB each or 1 dB total
- Splices every kilometer; allow 0.25 dB per splice
- A margin of 4 dB

If we subtract the 1 dB for connectors and the 4 dB for margin from the 85.3 dB, we are left with 80.3 dB. Add the splice loss and kilometer loss for 1 km, and the result is 0.5 dB. Divide this value into 80.3 dB, and the maximum length is 160 km between a terminal and first repeater or between repeaters. Of course, there is one less splice than 1-km lengths, plus the 0.3 dB left over from the 80.3 dB. This can be called *additional margin*. It is advisable to set these calculations up in tabular form as follows:

Connector loss (@ 0.5 dB/conn), 2 connectors	1 dB
Margin	4 dB
Splice loss, 0.25 dB/splice, 159 splices	39.75 dB
Fiber loss, 160-km fiber @ 0.25 dB/km	40.0 dB
Total	84.75 dB
Additional margin	0.55 dB

For an analysis of dispersion and system bandwidth, consult Ref. 12.

REVIEW QUESTIONS

1. What transmission medium should be considered as a principal candidate for a digital waveform that would carry more than 10,000 full-duplex telephone circuits at the end of the forecast period?

2. What are the advantages of using the RF bands from 2 GHz to 10 GHz for trunk telephony? Name at least two.

3. Frequencies above 10 GHz have an additional major impairment that must be taken into account in radiolink design. What is the cause of this impairment?

4. Discuss the problem of delay in speech telephone circuits traversing a satellite. Can you think of any problems in telephone signaling or in data transmission?

5. Give three of the five basic procedure steps in the design of line-of-sight radiolinks.

6. Name three basic planning considerations in siting radiolinks.

7. Name two advantages and at least two disadvantages to a radiolink terminal sited in the middle of an urban area.

8. Why do we limit the heights of radio towers serving radiolink (line-of-sight) systems?

9. Describe how earth bulge varies from one end of a radiolink hop to the other. Where is earth bulge maximum?

10. When a K factor of $\frac{4}{3}$ is used, does the microwave ray beam bend toward or away from the earth?

11. In the path profile, what are the three basic increment factors that must be added to obstacle height?

12. Name at least five of the basic factors that a radiolink design engineer must deal with when carrying out the path calculation (analysis) stage of the design effort.

13. Calculate the free-space loss in decibels of a radiolink hop 24 statute miles long operating at 4100 MHz.

14. Express Boltzmann's constant in decibel-watts.

15. If the signal-to-noise ratio into a device is 31 dB and the signal-to-noise ratio at the device output is 28 dB, what is the device noise figure?

16. A receiving system operates at room temperature; its noise figure is 11 dB, and its bandwidth is 2000 kHz. What is the thermal noise threshold of the system?

17. A receive signal level is -120 dBW and the bit rate is 2.048 Mbps. What is E_b?

18. The receiver on a digital LOS microwave link has a noise figure of 7 dB. What is N_0 for this receiver?

19. What is modulation implementation loss? Refer it to E_b/N_0.

20. Why is the selection of modulation type so important on digital LOS microwave?

21. The gain of a parabolic dish antenna is a function of which three variables?

22. What is the standard efficiency (%) used with LOS microwave parabolic dish antennas?

23. Calculate the EIRP of an LOS microwave transmitting system where the transmitter has 1.0-W output, 125 ft of waveguide with a loss of 0.015 dB/ft, and an antenna gain of 35 dB.

24. What is the most common type of fading encountered on an LOS microwave link? Discuss its causes and techniques for mitigation.

25. On a 20-mi (32-km) LOS microwave hop, the desired path propagation reliability is 99.995%. Based on a Rayleigh fading criterion, what fade margin (in dB) is required?

26. Why should "line of sight" be precisely defined?

27. A picowatt equals how many watts?

28. There are two types of diversity which may be employed on line-of-sight microwave links. What are they? Discuss pros and cons of each.

29. Why is frequency diversity not advisable to use?

30. An LOS microwave link does not meet error rate specifications. What steps can be taken, in ascending order of cost, to assure the system meets specifications.

31. What are the two ways of configurating hot-standby equipment?

32. Satellite communications is an extension of line-of-sight microwave; thus a satellite must be in _____ ___ _____ of an earth station.

33. For speech telephony circuits, why would we want to limit satellite links to just one in such a connectivity?

34. How does range (free-space loss) vary with elevation angle for a satellite communication system?

35. There is one overriding reason why down-link power is limited on a satellite system. What is that reason? What is a second important reason?

36. Give at least three reasons why frequencies between 1 and 10 GHz are preferred for satellite communications.

37. Give the two popular methods of satellite multiple access. Define each in no more than two sentences.

38. What are two of the major advantages of TDMA?

39. Define G/T. What are the two components of receiving system noise temperature for a satellite receiving system?

40. Excess attenuation due to rainfall for a satellite down-link varies with at least four major factors. Give three of them.

41. Give two major reasons why a corporation might opt for a VSAT system.

42. Discuss three generic protocols that may be used for VSAT access.

43. What is the principal overriding advantage of fiber-optic cable?

44. There are three wavelength bands currently used on fiber-optic systems. Identify these bands by wavelength and compare each with loss per unit length of fiber.

45. A glass fiber consists of a _____ and a covering called a _____. Relate these two with indices of refraction.

46. Name the three basic components of a fiber-optics link.

47. How many modes propagate in monomode fiber?

48. There are two basic types of light sources and two basic types of light detectors. Name each and compare.

49. What is the standard BER for a fiber-optic link?

50. What is a pigtail?

51. Give the two common locations where we would install fiber-optic amplifiers.

52. Give pros and cons for having a metallic pair in a fiber-optic cable.

53. A fiber-optic link can be either _____ limited or _____ limited.

54. In the link budget for a fiber-optic link, we have a loss item called *link margin*. What are the three factors we should consider when assigning a dB value to link margin?

55. A fiber-optic link with a bit rate of 622 Mbps has a laser diode with an output of 0 dBm and an APD-based receiver with a threshold of −40 dBm. Design a link around these parameters. What is the maximum link length achievable without repeaters? Show rationale.

REFERENCES

1. *BOC Notes on the LEC Networks—1994*, SR-TSV-002275, Bellcore, Piscataway, NJ, 1994.

2. *Error Performance of an International Digital Connection Forming Part of an Integrated Services Digital Network*, CCITT Rec. G.821, Fascicle III.5, IXth Plenary Assembly, Melbourne, 1988.

3. *Error Performance Parameters and Objectives for International Constant Bit Rate Digital Paths At or Above the Primary Rate*, ITU-T Rec. G.826, Geneva, 1993 (Draft).

4. *Allowable Bit Error Ratios at the Output of the Hypothetical Reference Digital Path for Radio-Relay Systems Which May Form Part of an Integrated Services Digital Network*, CCIR Rec. 594-3, CCIR RF Series, Fixed Service, ITU Geneva, March 1992.

5. *Principles of Digital Transmission*, Raytheon Co., ER79-4307, under U.S. Government contract MDA-904-79-C-0470, Sudbury, MA, 1979 (limited circulation).

6. *Engineering Considerations for Microwave Communications Systems*, GTE-Lenkurt, San Carlos, CA, 1975.

7. INTELSAT Earth Station Standards, *INTELSAT Space Segment Leased Transponder Definition and Associated Operating Conditions* [INTELSAT V, VA, VA(IBS) and VI Satellites], INTELSAT, Washington, DC, March 15, 1988.

8. *Radio Emission from Natural Sources in the Frequency Range Above About 50 MHz*, CCIR Rep. 720, Annex to Vol. V, XVIIth Plenary Assembly, Dusseldorf, 1990.

9. *Satellite Communication Reference Data Handbook*, Defense Communications Agency (NTIS), Washington, DC, July 1972.

10. INTELSAT Earth Station Standards (IESS), IESS 300 series, IESS-308, INTELSAT, Washington, DC, 1994.

11. J. Everett, *VSATS Very Small Aperture Terminals*, IEE/Peter Peregrinus, Stevenage, Herts, UK, 1992.

12. R. L. Freeman, *Telecommunication Transmission Handbook*, 3rd ed., John Wiley & Sons, New York, 1991.

13. *Optical Fibres System Planning Guide*, CCITT/ITU, Geneva, 1989.

14. *1993 Lightwave Symposium*, Hewlett-Packard, Burlington, MA, March 23, 1993.

15. G. P. Agrawal, *Fiber-Optic Communication Systems*, John Wiley & Sons, New York, 1992.

16. *Telecommunication Transmission Engineering*, 3rd ed., Vol. 2, Bellcore, Piscataway, NJ, 1991.

17. R. L. Freeman, *Radio System Design for Telecommunications*, John Wiley & Sons, New York, 1987.

18. *Digital Radio Theory and Measurements*, H-P Application Note 355A, Hewlett-Packard, San Carlos, CA, 1992.

8

DIGITAL TRANSMISSION SYSTEMS

1 DIGITAL VERSUS ANALOG TRANSMISSION

The IEEE dictionary [24] contrasts analog and digital transmission as follows:

> An analog signal implies *continuity*, as contrasted to a digital signal that is concerned with *discrete* states. Often the means of carrying information is the distinguishing feature between analog and digital. The information content of an analog signal is conveyed by the value or magnitude of some characteristics of the signal such as phase, amplitude, frequency of a voltage, the amplitude or duration of a pulse, and so on. To extract the information, it is necessary to compare the value or magnitude of the signal to a standard. The information content of a digital signal is concerned with discrete states of the signal, such as the presence or absence of a voltage, a contact in the open or closed position, or a hole or no hole in certain positions on a card. The digital signal is given meaning by assigning numerical values or other information to the various possible combinations of the discrete states of the signal.

There are three notable advantages to digital transmission that make it extremely attractive to the telecommunication system engineer when compared to its analog counterpart. Dealing in generalities, we can say:

1. Noise does not accumulate as it does in an analog system. Noise accumulation stops at each regenerative repeater where the digital signal is fully regenerated. Noise accumulation was the primary consideration in analog network design.

2. The digital format lends itself ideally to solid-state technology and, in particular, to integrated circuits.

339

3. It is theoretically compatible with digital data, telephone signaling, and computers.

At this juncture, about 85% of the traffic carried on the North American PSTN is voice (analog) traffic. These analog signals must be converted to a digital format compatible with the digital network.

Our first concern in this chapter is the conversion of the nominal 4-kHz analog voice channel to a PCM format. There are two different conversion methods in common use. These are pulse-code modulation (PCM), which is used almost exclusively on the digital PSTN, and delta modulation, which is widely employed by many of the armed forces of the world.

The North American long-distance network is 100% digital, and the local exchange carriers in North America expect to be nearly 100% digital by 1997. This means that not only the transmission facilities are digital, but also the switches are digital. Nearly all major digital switch vendors are marketing third-generation digital switches. These switches are discussed in Chapter 9.

2   BASIS OF PULSE-CODE MODULATION

Pulse-code modulation is a method of modulation in which a continuous analog wave is transmitted in an equivalent digital mode. The cornerstone of an explanation of the functioning of PCM is the Nyquist sampling theorem, which states [Ref. 1, Section 21]:

> If a band-limited signal is sampled at regular intervals of time and at a rate equal to or higher than twice the highest significant signal frequency, then the sample contains all the information of the original signal. The original signal may then be reconstructed by use of a low-pass filter.

As an example of the sampling theorem, the nominal 4-kHz voice channel would be sampled at a rate of 8000 samples per second (i.e., 4000×2). A high-fidelity 15-kHz program channel* would be sampled at 30,000 times per second (i.e., $15,000 \times 2$).

To develop a PCM signal from an analog signal, three processing steps are required: *sampling, quantization*, and *coding*. The result is a serial binary signal or bit stream,† which may or may not be applied to the line without additional modulation or conditioning steps. One major advantage of digital transmission is that signals may be regenerated at intermediate points along a transmission path as well as at the end points. One price for this advantage is the increased bandwidth required for PCM. Conventional PCM systems as

---

*A program channel is a communications channel that carries broadcast material such as music and commentary. This is a facility provided to broadcasters and cable TV operators.
†A bit stream is a continuous series of 1s and 0s.

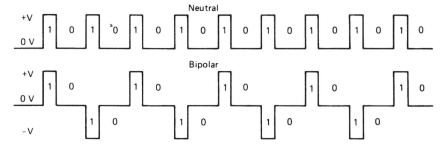

Figure 8.1. Neutral versus bipolar bit streams. The upper diagram illustrates alternating 1s and 0s transmitted in the neutral mode; the lower diagram illustrates the equivalent in a bipolar mode, which is also called AMI or alternate mark inversion.

found in our digital network require 16 times the bandwidth of their analog counterpart (i.e., a 4-kHz analog voice channel requires 16×4 or 64 kHz when transmitted by PCM), assuming 1 bit per hertz of bandwidth. Regeneration of a digital signal is simplified and particularly effective when the transmitted line signal is binary, whether neutral, polar, or bipolar. An example of a bipolar bit stream is shown in Figure 8.1.

Binary transmission tolerates considerably higher noise levels (i.e., degraded signal-to-noise ratio), when compared to its analog counterpart (i.e., FDM in Chapter 5). This fact, in addition to the regeneration capability, is a great step forward in transmission engineering. The regeneration that takes place at each repeater and switch by definition re-creates a new signal; therefore noise, as we know it, does not accumulate.

Error rate is another important factor in the design of PCM systems. Intelligibility is maintained even with an error rate as poor as 1 bit in error in 100 bits (BER = 1×10^{-2}). However, because of line or supervisory signaling concerns, digital networks maintain error rates no worse than 1×10^{-3}. Because present digital networks also carry data, far better error performance is specified for at least 80% of the time.

Another important factor in the design of PCM cable installations is crosstalk, which can degrade error performance. This is crosstalk spilling from one PCM system into another or into the same system from the send path to the receive path inside the same cable sheath.

3 DEVELOPMENT OF A PULSE-CODE MODULATION SIGNAL

3.1 Sampling

Consider the sampling theorem given previously. If we now sample the standard CCITT voice channel, 300–3400 Hz (a bandwidth of 3100 Hz), at a rate of 8000 samples per second, we will have complied with the Nyquist sampling theorem and can expect to recover all the information in the

Figure 8.2. A PAM wave as a result of sampling a single sinusoid.

original analog signal. Therefore a sample is taken every 1/8000 s, or every 125 µs. These are key parameters for our future argument.

Another example may be a 15-kHz program channel. Here the sampling rate would be 30,000 times per second. Samples would be taken at 1/30,000-s intervals, or at 33.3 µs.

3.1.1 The Pulse Amplitude Modulation Wave.
With several exceptions (e.g., SPADE, Chapter 7, Section 5.7.3), practical PCM systems involve time-division multiplexing. Sampling in these cases does not involve just one voice channel but several. In practice, one system (see following paragraph) samples 24 voice channels in sequence, and another samples 30 channels. The result of the multiple sampling is a pulse amplitude modulation (PAM) wave. A simplified PAM wave is shown in Figure 8.2, in this case a single sinusoid. A simplified diagram of the processing involved to derive a multiplexed PAM wave is shown in Figure 8.3.

If the nominal 4-kHz voice channel must be sampled 8000 times per second and a group of 24 such voice channels are to be sampled sequentially to interleave them, forming a PAM multiplexed wave, this could be done by gating. The gate should be open for 5.2 µs (125/24) for each voice channel to be sampled successively from channels 1 through 24. This full sequence must be done in a 125-µs period $\frac{1}{8000}$. We call this 125-µs period a frame, and inside the frame all 24 channels are successively sampled once. This 24-channel system is popularly called T1, but we will call it DS1.

Another system widely used outside of the United States and Canada is E1, which is a 30-voice-channel system plus an additional two service channels for a total of 32 channels. By definition, this system must sample 8000 times per second because it is also optimized for voice operation, and thus its frame period is 125 µs. To accommodate the 32 channels, the gate is open 125/32 or about 3.906 µs.

3.2 Quantization

Our goal is to assign a binary sequence to each voltage sample. For argument's sake, we will contain the maximum excursion of the PAM wave

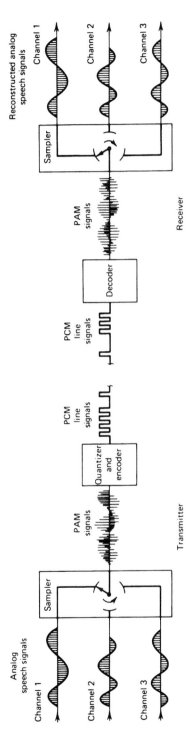

Figure 8.3. A simplified analogy of formation of a PAM wave. Courtesy of GTE Lenkurt Demodulator, San Carlos, California [2].

343

to within $+1$ to -1 volts. In the PAM waveform there could be an infinite number of different values of voltage between $+1$ and -1 volts. For instance, one value could be -0.3875631 volts. To assign a different binary sequence to each voltage value, we would have to construct a code of infinite length. So we must limit the number of voltage values between $+1$ and -1 volts, and the values must be discrete. For example, we could set 20 discrete values between $+1$ and -1 volts, each value at a 0.1-volt increment.

Because we are working in the binary domain, we select the total number of discrete values to be a binary number multiple (i.e., 2, 4, 8, 16, 32, 64, 128, etc.). This facilitates binary coding. For instance, if there were four values, they would be as follows: 00, 01, 10, and 11. This is a 2-bit code. A 3-bit code would yield eight different binary numbers. We find, then, that the number of total possible different binary combinations given a code of n binary symbols (bits) is 2^n. A 7-bit code has 128 different binary combinations (i.e., $2^7 = 128$).

For the quantization process, we want to present to the coder a discrete voltage value. Suppose our quantization steps were on 0.1-volt increments and our voltage measure for one sample was 0.37 volts. That would have to be rounded off to 0.4 volts, the nearest discrete value. Note here that there is a 0.03-volt error, the difference between 0.37 and 0.40 volts.

Figure 8.4 shows one cycle of the PAM wave of Figure 8.2 where we use a 4-bit code. In the figure a 4-bit code is used which allows 16 different binary-coded possibilities or levels between $+1$ and -1 volts. Thus we can assign eight possibilities above the origin and eight possibilities below the origin. These 16 quantum steps are coded as follows:

Step Number	Code	Step Number	Code
0	0000	8	1000
1	0001	9	1001
2	0010	10	1010
3	0011	11	1011
4	0100	12	1100
5	0101	13	1101
6	0110	14	1110
7	0111	15	1111

Examination of Figure 8.4 shows that step 12 is used twice. Neither time it is used is it the true value of the impinging sinusoid. It is a rounded-off value. These rounded-off values are shown with the dashed line in Figure 8.4, which follows the general outline of the sinusoid. The horizontal dashed lines show the point where the quantum changes to the next higher or next lower level if

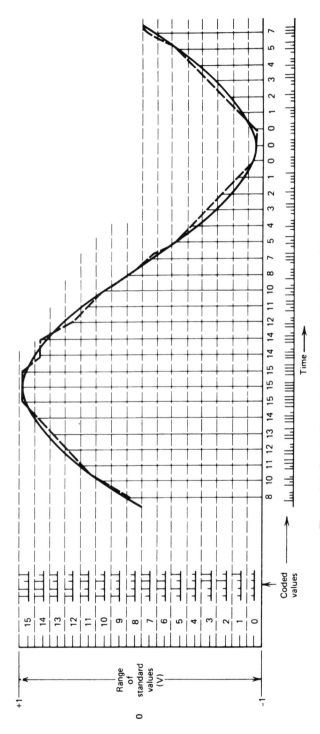

Figure 8.4. Quantization and resulting coding using 16 quantizing steps.

the sinusoid curve is above or below that value. Take step 14 in the curve, for example. The curve, dropping from its maximum, is given two values of 14 consecutively. For the first, the curve is above 14, and for the second, below. That error, in the case of 14, from the quantum value to the true value, is called *quantizing distortion*. This distortion is the major source of imperfection in PCM systems.

In Figure 8.4, maintaining the $-1, 0, +1$ V relationship, let us double the number of quantum steps from 16 to 32. What improvement would we achieve in quantization distortion? First determine the step increment in millivolts in each case. In the first case the total range of 2000 mV would be divided into 16 steps, or 125 mV/step. The second case would have 2000/32 or 62.5 mV/step. For the 16-step case, the worst quantizing error (distortion) would occur when an input to be quantized was at the half-step level or, in this case, 125/2 or 62.5 mV above or below the nearest quantizing step. For the 32-step case, the worst quantizing error (distortion) would again be at the half-step level, or 62.5/2 or 31.25 mV. Thus the improvement in decibels for doubling the number of quantizing steps is

$$20 \log \frac{62.5}{31.25} = 20 \log 2 \text{ or 6 dB (approximately)}$$

This is valid for linear quantization only (see Section 3.2 of this chapter). Thus increasing the number of quantizing steps for a fixed range of input values reduces quantizing distortion accordingly.

Voice transmission presents a problem. It has a wide dynamic range, on the order of 50 dB. That is the level range from the loudest syllable of the loudest talker to lowest-level syllable of the quietest talker. Using linear quantization, we find it would require 2048 discrete steps to provide any fidelity at all. 2048 is 2^{11}. This means we would need an 11-bit code. Such a code sampled 8000 times per second leads to 88,000-bps equivalent voice channel and an 88-kHz bandwidth, assuming 1 bit per Hz. Designers felt this was too great a bit rate/bandwidth.

They turned to an old analog technique of companding. *Companding* stands for two words: compression–expansion. Compression takes place on the transmit side of the circuit; expansion on the receive side. Compression reduces the dynamic range with little loss of fidelity, and expansion returns the signal to its normal condition.

This is done by favoring low-level speech over higher-level speech. In other words, more code segments are assigned to speech bursts at low level than at the higher levels, progressively more as level goes down. This is shown graphically in Figure 8.5, where eight coded sequences are assigned to each level grouping. The smallest range rises only 0.0666 volts from the origin (0 volts). The largest range extends over 0.5 volts, and it is assigned only eight coded sequences [3].

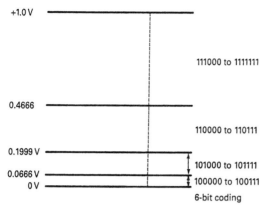

+1.0 V

111000 to 1111111

0.4666

110000 to 110111

0.1999 V

101000 to 101111

0.0666 V

100000 to 100111

0 V

6-bit coding

Figure 8.5. A simple graphic representation of compression. Six-bit coding, eight six-bit sequences per segment.

3.3   Coding

Older PCM systems used a 7-bit code, and modern systems use an 8-bit code with its improved quantizing distortion performance. The companding and coding are carried out together, simultaneously. The compression and later expansion functions are logarithmic. A pseudologarithmic curve made up of linear segments imparts finer granularity to low-level signals and less granularity to the higher-level signals. The logarithmic curve follows one of two laws, the A-law and the μ-law. The curve for the A-law may be plotted from the formula

$$F_A(x) = \left(\frac{A|x|}{1 + \ln(A)}\right) \qquad 0 \leqslant |x| \leqslant \frac{1}{A}$$
$$F_A(x) = \left(\frac{1 + \ln |Ax|}{1 + \ln(A)}\right) \qquad \frac{1}{A} \leqslant |x| \leqslant 1$$

where $A = 87.6$. The curve for the μ law may be plotted from the formula

$$F_\mu(x) = \frac{\ln(1 + \mu|x|)}{\ln(1 + \mu)}$$

where x is the signal input amplitude and $\mu = 100$ for the original North American T1 system (now outdated) and 255 for later North American (DS1) systems and the CCITT 24-channel system (CCITT Rec. G.733). Note the use of the natural logarithms (ln) in these formulas [4].

A common expression used in dealing with the "quality" of a PCM signal is *signal-to-distortion* ratio (expressed in dB). Parameters A and μ, for the respective companding laws, determine the range over which the signal-to-distortion ratio is comparatively constant, about 26 dB. For A-law

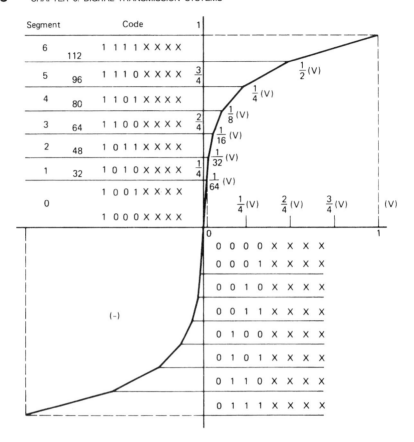

Figure 8.6. The 13-segment approximation of the A-law curve used with E1 PCM equipment.

companding, an $S/D = 37.5$ dB can be expected ($A = 87.6$) and for μ-law companding, an $S/D = 37$ dB ($\mu = 225$) [5].

Turn now to Figure 8.6, which shows the companding curve and resulting coding for the European E1 system. Note that the curve consists of linear piecewise segments, seven above and seven below the origin. The segment just above and the segment just below the origin each consists of two linear elements. Counting the collinear elements by the origin, there are 16 segments. Each segment has 16 8-bit PCM codewords assigned. These are the codewords that identify the voltage level of a sample at some moment in time. Each codeword, often called a PCM "word," consists of 8 bits. The first bit (most significant bit) tells the distant-end receiver if that sample is a positive or a negative voltage. Note that all the PCM words above the origin start with a binary 1, and those below the origin start with a binary 0. The next three bits identify the segment. There are 8 segments (or collinear equivalents) above the origin and 8 below ($2^3 = 8$). The last 4 bits, shown in

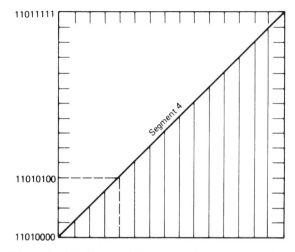

Figure 8.7. The European E1 system, coding of segment 4 (positive).

the figure as XXXX, identify where in the segment that voltage line is located.

Suppose the distant end received the binary sequence 11010100 in an E1 system. The first bit indicates that the voltage is positive (i.e., above the origin in Figure 8.6). The next three bits, 101, indicate that the sample is in segment 4. The last 4 bits, 0100, tell us where it is in that segment as illustrated in Figure 8.7. Note that the 16 steps inside the segment are linear.

Figure 8.8 shows an equivalent logarithmic curve for the North American DS1 system. It uses a 15-segment approximation of the logarithmic μ-law curve ($\mu = 255$). The segments cutting the origin are collinear and are counted as one. So, again, we have a total of 16 segments.

The coding process in PCM systems utilizes straightforward binary codes. Examples of such codes are shown in Figure 8.6, which is expanded in Figure 8.7 and in Figure 8.8.

The North American DS1 (T1) PCM system uses a 15-segment approximation of the logarithmic μ-law ($\mu = 255$), shown in Figure 8.8. The segments cutting the origin are collinear and are counted as one. Figure 8.9 shows some sample DS1 code sequences for particular code levels, for both positive and negative voltages. As can be seen in Figure 8.8, similar to Figure 8.6, the first code element (bit), whether a 1 or a 0, indicates to the distant end whether the sample voltage is positive or negative, above or below the horizontal axis. The next three elements (bits) identify the segment, and the last four elements (bits) identify the actual quantum level inside the segment.

3.3.1 The Concept of Frame. As is shown in Figure 8.3, PCM multiplexing is carried out in the sampling process, sampling the analog sources sequentially. These sources may be the nominal 4-kHz voice channels or other information sources that have a 4-kHz bandwidth, such as data or

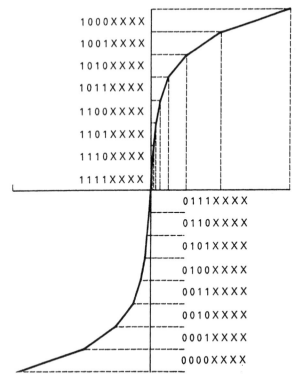

Figure 8.8. Piecewise linear approximation of the μ-law logarithmic curve. Coding based on CCITT Rec. G.711 [6].

freeze-frame video. The final result of the sampling and subsequent quantization and coding is a series of electrical pulses, a serial bit stream of 1s and 0s that requires some identification or indication of the beginning of a scanning sequence. This identification is necessary so that the far-end receiver knows exactly when the sampling sequence starts. The receiver knows a priori (in the case of DS1) that 24 eight-bit slots follow. It synchronizes the receiver. Such identification is carried out by a *framing bit*, and one full sequence or cycle of samples is called a *frame* in PCM terminology.

Consider the framing structure of the two widely implemented PCM systems: the North American DS1 and the European E1.* The North American DS1 system is a 24-channel PCM system using 8-level coding (e.g., $2^8 = 256$ quantizing steps or distinct PCM code words). Supervisory signaling is "in-band" where bit 8 of every sixth frame† is "robbed" for supervisory

---

*Previously, the European system was called CEPT30 + 2, where CEPT stands for Conference European Post & Telegraph. The 30 + 2 means that it has 30 traffic channels and two service channels.

†Note that on each frame that has bit 8 "robbed," 7-bit coding is used versus 8-bit coding used on the other five frames.

Code Level		Digit Number							
		1	2	3	4	5	6	7	8
255	(Peak positive level)	1	0	0	0	0	0	0	0
239		1	0	0	1	0	0	0	0
223		1	0	1	0	0	0	0	0
207		1	0	1	1	0	0	0	0
191		1	1	0	0	0	0	0	0
175		1	1	0	1	0	0	0	0
159		1	1	1	0	0	0	0	0
143		1	1	1	1	0	0	0	0
127	(Center levels)	1	1	1	1	1	1	1	1
126	(Nominal zero)	0	1	1	1	1	1	1	1
111		0	1	1	1	0	0	0	0
95		0	1	1	0	0	0	0	0
79		0	1	0	1	0	0	0	0
63		0	1	0	0	0	0	0	0
47		0	0	1	1	0	0	0	0
31		0	0	1	0	0	0	0	0
15		0	0	0	1	0	0	0	0
2		0	0	0	0	0	0	1	1
1		0	0	0	0	0	0	1	0
0	(Peak negative level)	0	0	0	0	0	0	1[a]	0

[a]One digit is added to ensure that the timing content of the transmitted pattern is maintained.

Figure 8.9. Eight-level coding of the North American DS1 PCM system. Note that there are actually only 255 quantizing steps because steps 0 and 1 use the same bit sequence, thus avoiding a code sequence with no transitions (i.e., all 0s).

signaling. The DS1 signal format, shown in Figure 8.10, has one bit added as a framing bit called an "S" bit. The DS1 frame thus consists of

$$(8 \times 24) + 1 = 193 \text{ bits}$$

making up a full sequence or frame. By definition, 8000 frames are transmitted per second (i.e., 4000×2, the Nyquist sampling rate), so the bit rate for DS1 (T1) is

$$193 \times 8000 = 1,544,000 \text{ bps or } 1.544 \text{ Mbps}$$

The DS1 frame structure is further clarified in Figure 8.11.

The E1 European PCM system is a 32-channel system. Of the 32 channels, 30 transmit speech (or data) derived from incoming telephone trunks and the remaining 2 channels transmit synchronization-alignment and signaling information. Each channel is allotted an 8-bit time slot (TS), and we tabulate TS 0 through 31 as follows:

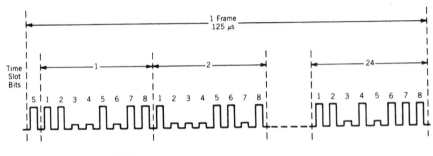

Sampling frequency	8000 Hz
Output bit rate	1.544 Mbps ± 50 bps
Bits/Frame	193
Time slots/Frame	24 (Sequential assignment)
Signaling	Eight bit of every sixth frame

The S-bit is time-shared between terminal framing (F_t) and signal framing (F_S).·

Figure 8.10. DS1 signal format.

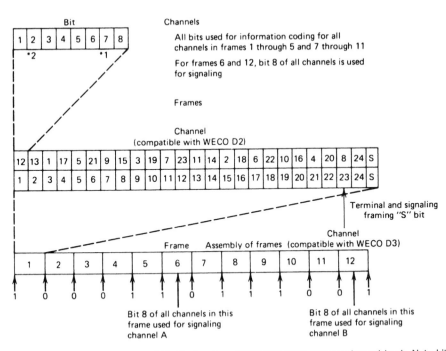

Figure 8.11. Frame structure of North American (ATT) DS1 PCM system channel bank. Note bit "robbing" technique used on each sixth frame to provide signaling information. Courtesy of ITT Telecommunications, Raleigh, N.C. [*Notes*: (1) If bits 1 to 6 and 8 are 0, then bit 7 is transmitted as 1; (2) bit 2 is transmitted as 0 on all channels for transmission of end-to-end alarm; (3) composite pattern 000110111001, etc.].

TS	Type of Information
0	Synchronizing (framing)
1–15	Speech
16	Signaling
17–31	Speech

In TS 0 a synchronizing code or word is transmitted every second frame, occupying digits 2 through 8 as follows:

$$0011011$$

In those frames without the synchronizing word, the second bit of TS 0 is frozen at a 1 so that in these frames the synchronizing word cannot be imitated. The remaining bits of time slot 0 can be used for the transmission of supervisory information signals [8].

As we said, E1 in its primary rate format transmits 32 channels of 8-bit time slots. An E1 frame therefore has $8 \times 32 = 256$ bits. There is no framing bit. Framing alignment is carried out in TS 0. The E1 bit rate to the line is

$$256 \times 8000 = 2,048,000 \text{ bps or } 2.048 \text{ Mbps}$$

Framing and basic timing should be distinguished. "Framing" ensures that the PCM receiver is aligned regarding the beginning (and end) of a bit sequence or frame; "timing" refers to the synchronization of the receiver clock, specifically, that it is in step with its companion far-end transmit clock. Timing at the receiver is corrected via the incoming "1"-to-"0" and "0"-to-"1" transitions.* It is important that periods of no transitions do not occur. This point is discussed later in reference to line codes and digit inversion.

3.3.2 *Quantization Distortion.*

Quantizing distortion has been defined as the difference between the signal waveform as presented to the PCM multiplex (codec) and its equivalent quantized value. For a linear codec with n binary digits per sample, the ratio of the full-load sine wave power to quantizing distortion power (S/D) is [25]

$$\frac{S}{D} = 6n + 1.8 \text{ dB}$$

*A transition in this context is a change of electrical state. We often use the term "mark" for a binary 1 and "space" for a binary 0. The terms mark and space come from old-time telegraphy and have been passed on through the data world to the parlance of digital communications technology.

where n is the number of bits per PCM word, the word expressing the sample. For instance, the older ATT D1 system uses a 7-bit word to express a sample (level), and the 30 + 2 and DS1 systems use essentially 8 bits. If we had a 7-bit word and uniform quantizing, S/D would be 43.8 dB. Each binary digit added to the PCM code word increases the S/D ratio 6 dB for linear quantization. Practical S/D values range in the order of 33–38 dB, depending largely on the talker levels (using 8-bit words).

4 PULSE-CODE MODULATION SYSTEM OPERATION

Pulse-code modulation (PCM) equipment operates on a four-wire basis. Voice-channel inputs and outputs to and from a PCM multiplex channel bank are four-wire, or must be converted to four-wire in the channel bank. The term "codec" is a contraction of the word group *coder–decoder* even though the equipment carries out more functions than just coding and decoding. A block diagram of a typical codec (PCM channel bank) is shown in Figure 8.12.

A codec accepts 24 or 30 voice channels, depending on the system used; digitizes and multiplexes the information; and delivers a serial bit stream to the line of 1.544 Mbps or 2.048 Mbps. It accepts a serial bit stream at one or the other modulation rate, demultiplexes the digital information, and performs digital-to-analog conversion. Output to the analog telephone network

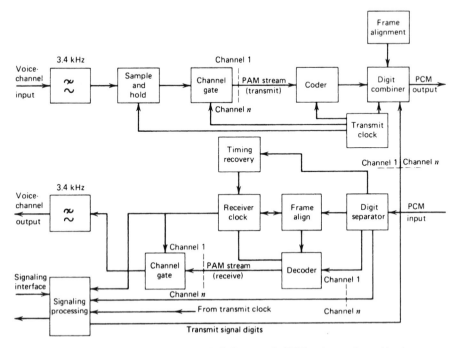

Figure 8.12. Simplified functional block diagram of a PCM codec or channel bank.

is the 24 or 30 nominal 4-kHz voice channels. Figure 8.12 illustrates the processing of a single analog voice channel through a codec. The voice channel to be transmitted is passed through a 3.4-kHz low-pass filter. The output of the filter is fed to a sampling circuit. The sample of each channel of a set of n channels (n usually equals 24 or 30) is released in turn to the pulse amplitude modulation (PAM) highway. The release of samples is under control of a channel gating pulse derived from the transmit clock. The input to the coder is the PAM highway. The coder accepts a sample of each channel in sequence and then generates the appropriate 8-bit signal character corresponding to each sample presented. The coder output is the basic PCM signal that is fed to the digit combiner where framing-alignment signals are inserted in the appropriate time slots, as well as the necessary supervisory signaling digits corresponding to each channel (European approach), and are placed on a common signaling highway that makes up one equivalent channel of the multiplex serial bit stream transmitted to the line. In North American practice supervisory signaling is carried out somewhat differently by "bit robbing," such as bit 8 in frame 6 and bit 8 in frame 12. Thus each equivalent voice channel carries its own signaling (see Figure 8.11).

On the receive side the codec accepts the serial PCM bit stream, inputting the digit separator where the signal is regenerated and split, delivering the PCM signal to four locations to carry out the following processing functions: (1) timing recovery, (2) decoding, (3) frame alignment, and (4) signaling (supervisory). Timing recovery keeps the receive clock in synchronism with the far-end transmit clock. The receive clock provides the necessary gating pulses for the receive side of the PCM codec. The frame-alignment circuit senses the presence of the frame-alignment signal at the correct time interval, thus providing the receive terminal with frame alignment. The decoder, under control of the receive clock, decodes the code character signals corresponding to each channel. The output of the decoder is the reconstituted pulses making up a PAM highway. The channel gate accepts the PAM highway, gating the n-channel PAM highway in sequence under control of the receive clock. The output of the channel gate is fed in turn to each channel filter, thus enabling the reconstituted analog voice signal to reach the appropriate voice path. Gating pulses extract signaling information in the signaling processor and apply this information to each of the reconstituted voice channels with the supervisory signaling interface as required by the analog telephone system in question.

5 PRACTICAL APPLICATIONS

5.1 General

In an early application, pulse-code modulation found wide application in expanding interoffice trunks (junctions) that have reached or will reach

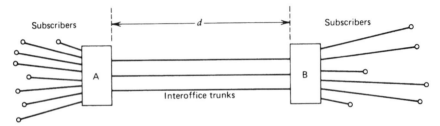

Figure 8.13. Simplified application drawing of PCM as applied to interexchange plant in the local area. A and B are two local serving switches separated by some distance *d*.

exhaust* in the near future. An interoffice trunk is one pair of a circuit group that connects two switching points (exchanges). Figure 8.13 sketches the interoffice trunk concept. Depending on the particular application, at some point where distance *d* is exceeded it will be more economical to install PCM on existing VF cable plant than to rip up streets and add more VF cable pairs. For the planning engineer the distance *d* where PCM becomes an economic alternative is called the "prove-in" distance. The distance *d* may vary from 8 km to 16 km (5 mi to 10 mi), depending on the location and other circumstances. For distances less than *d*, additional VF cable pairs should be used for expanding plant.

The general rule for measuring expansion capacity of a given VF cable is as follows:

- For ATT DS1/DS2 channelizing equipment, two VF pairs will carry 24 PCM channels.
- For the CEPT 30 + 2 system as configured by ITT, two VF pairs plus a phantom pair will carry 30 PCM speech channels.

All pairs in a VF cable may not necessarily be usable for PCM transmission, partly because there is a possibility of excessive crosstalk between PCM carrying pairs. The effect of high crosstalk levels is to introduce digital errors in the PCM bit stream. Error rate may be related on a statistical basis to crosstalk, which, in turn, is dependent on the characteristics of the cable and the number of PCM carrying pairs.

One method for reducing crosstalk and thereby increasing VF pair usage is to turn to two-cable working, rather than have the "go" and "return" PCM cable pairs in the same cable. Another factor that can limit cable pair usage is the incompatibility of FDM and PCM carrier systems in the same cable. On the cable pairs that will be used for PCM, the following should be taken into consideration:

- All load coils must be removed.

---

*"Exhaust" is an outside-plant term meaning that the useful pairs of a cable have been used up (assigned) from a planning point of view.

• Build-out networks and bridged taps must also be removed.

• No crosses, grounds, splits, high-resistance splices, or moisture are permitted.

5.2 Practical System Block Diagram

A block diagram showing the elemental blocks of a PCM transmission link used to expand installed VF cable capacity is shown in Figure 8.14. Most telephone administrations (companies) distinguish between the terminal area of a PCM system and the repeatered line. The term "span" comes into play here. A span line is composed of a number of repeater sections permanently connected in tandem at repeater apparatus cases mounted in manholes or on pole lines along the span. A "span" is defined as the group of span lines that extend between two exchange (switching center) repeater points.

A typical span is shown in Figure 8.14. The spacing between regenerative repeaters is important. Section 5.1 mentioned the necessity of removing load coils from those trunk (junction) cable pairs that were to be used for PCM transmission. It is at these load points that the PCM regenerative repeater should be installed. On a VF line with H-type loading (see Chapter 2, Section 2.4.4), spacing between load points is normally 6000 ft (1830 m). It will be remembered from Chapter 2 that the first load coil out from the exchange on a trunk pair is at half-distance or 3000 ft (915 m). This is provident, because a regenerative repeater also must be installed at this point. Such spacing is shown in Figure 8.14 (1 space = 1000 ft). The purpose of installing a repeater at this location is to increase the pulse level before entering the environment of an exchange area where the levels of impulse noise may be quite high. High levels of impulse noise introduced into the system may cause significant increases in digital error rate of the incoming PCM bit streams, particularly when the bit stream is of a comparatively low level. Generally, the amplitude of a PCM pulse output of a regenerative repeater is of the order of 3 V. Likewise, 3 V is the voltage on the PCM line cross connect field at the exchange (terminal area).

A guideline used by Bell Telephone Manufacturing Company (BTM) (Belgium) is that the maximum distance separating regenerative repeaters is that corresponding to a cable-pair attenuation of 36 dB at 1024 kHz at the maximum expected temperature. This frequency is equivalent to the half-bit rate for the E1 systems (e.g., 2048 kbps). Actually, repeater design permits operation on lines with attenuations anywhere from 4 dB to 36 dB, allowing considerable leeway in placing repeater points. Table 8.1 gives some other practical repeater-spacing parameters for the CEPT-ITT-BTM E1 system. The maximum distance is limited by the maximum number of repeaters, which in this case is a function of power feeding and supervisory considerations. For instance, the fault-location (i.e., troubleshooting) system can handle up to a maximum of 18 tandem repeaters for the BTM (ITT) configuration.

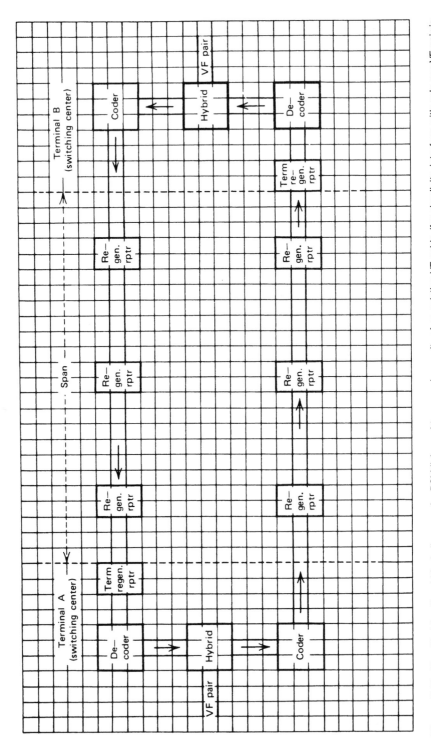

Figure 8.14. Simplified functional block diagram of a PCM link used to expand capacity of an existing VF cable (for simplicity, interface with only one VF pair is shown). Note the spacing between repeaters in the span line.

TABLE 8.1 Line Parameters for ITT/BTM PCM Configuration

Pair Diameter (mm)	Loop Attenuation at 1 MHz (dB/km)	Loop Resistance (Ω/km)	Voltage Drop (V/km)	Maximum Distance^a (km)	Total Repeaters	Maximum Distance System (km)
0.9	12	60	1.5	3	18	54
0.6	16	100	2.6	2.25	16	36

^{a}Between adjacent repeaters.

Power for the BTM system is fed through a constant-current feeding arrangement over a phantom pair serving both the "go" and related "return" repeaters, providing up to 150 V dc at the power feed point. The voltage drop per regenerative repeater is 5.1 V; thus for a "go" and "return" repeater configuration the drop is 10.2 V. For example, let us determine the maximum number of regenerative repeaters in tandem that may be fed from one power feed point by this system, using 0.8-mm-diameter pairs with a 3-V IR drop in an 1830-m spacing between adjacent repeaters:

$$\frac{150}{(10.2 + 3)} = 11$$

Assuming power fed from both ends and an 1800-m "dead" section in the middle, the maximum distance between power feed points is approximately

$$(2 \times 11 + 1)1.8 \text{ km} = 41.4 \text{ km}$$

Fault tracing for the North American (ATT) T1 system is carried out by means of monitoring the framing signal, the 193d bit (Section 3.3.1). The framing signal (amplified) normally holds a relay closed when the system is operative. With loss of framing signal, the relay opens actuating alarms, and thus a faulty system is identified, isolated, and dropped from "traffic."

To locate a defective regenerator on the BTM (Belgium) E1 system, traffic is removed from the system and a special pattern generator is connected to the line. The pattern generator transmits a digital pattern with the same bit rate as does the 30 + 2 PCM signal, but the test pattern can be varied to contain selected low-frequency spectral elements. Each regenerator on the repeatered line is equipped with a special audio filter, each with a distinctive passband. Up to 18 different filters may be provided in a system. The filter is bridged across the output of the regenerator, sampling the output pattern. The output of the filter is amplified and transformer-coupled to a fault-trans-mission pair, which is normally common to all PCM systems on the route, span, or section. To determine which regenerator is faulty, the special test pattern is tuned over the spectrum of interest. As the pattern is tuned through the frequency of the distinct filter of each operative repeater, a return signal will derive from the fault-transmission pair at a minimum specified level. Defective repeaters will be identified by absence of return signal or a return level under specification. The distinctive spectral content of the return signal is indicative of the regenerator undergoing test.

5.3   The Line Code

Pulse-code modulation signals are transmitted to the cable and are in the bipolar mode, as shown in Figure 8.1. The marks, or 1s, have only a 50% duty cycle. There are several advantages to this mode of transmission:

- No dc return is required; thus transformer coupling can be used on the line.
- The power spectrum of the transmitted signal is centered at a frequency equivalent to half the bit rate.

It will be noted in bipolar transmission that the 0s are coded as absence of pulses and 1s are alternately coded as positive and negative pulses, with the alternation taking place at every occurrence of a 1. This mode of transmission is also called *alternate mark inversion* (AMI).

One drawback to straightforward AMI transmission is that when a long string of 0s is transmitted (e.g., no transitions), a timing problem may arise because repeaters and decoders have no way of extracting timing without transitions. The problem can be alleviated by forbidding long strings of 0s. Codes have been developed that are bipolar but with N 0s substitution; they are called "BNZS" codes. For instance, a B6ZS code substitutes a particular signal for a string of six 0s. B8ZS is used on subscriber loop carrier.

Another such code is the HDB3 code (high-density binary 3), where the 3 indicates substitution for binary sequences with more than three consecutive 0s. With HDB3, the second and third 0s of the string are transmitted unchanged. The fourth 0 is transmitted to the line with the same polarity as the previous mark sent, which is a "violation" of the AMI concept. The first 0 may or may not be modified to a 1 to ensure that the successive violations are of opposite polarity. HDB3 is used with European E series PCM systems and is similar to B3ZS.

5.4   Signal-to-Gaussian-Noise Ratio on Pulse-Code Modulation Repeater Lines

As we mentioned earlier, noise accumulation on PCM systems is not an important consideration. However, this does not mean that Gaussian noise*

---

*Same as thermal noise.

**TABLE 8.2   Error Rate of a Binary Transmission System
Versus Signal-to-rms-Noise Ratio**

Error Rate	S/N (dB)	Error Rate	S/N (dB)
10^{-2}	13.5	10^{-7}	20.3
10^{-3}	16.0	10^{-8}	21.0
10^{-4}	17.5	10^{-9}	21.6
10^{-5}	18.7	10^{-10}	22.0
10^{-6}	19.6	10^{-11}	22.2

(or crosstalk or impulse noise) is unimportant. Indeed, it may affect error performance expressed as error rate. Errors are cumulative, as is the error rate. A decision in error, whether 1 or 0, made anywhere in the digital system is not recoverable. Thus such an incorrect decision made by one regenerative repeater adds to the existing error rate on the line, and errors taking place in subsequent repeaters further down the line add in a cumulative manner, thus tending to deteriorate the received signal.

In a purely binary transmission system, if a 20-dB signal-to-noise ratio is maintained, the system operates nearly error free. In this respect, consider Table 8.2.

As discussed in Section 5.3, PCM, in practice, is transmitted on-line with alternate mark inversion. The marks have a 50% duty cycle, permitting energy concentration at a frequency of half the transmitted bit rate. Thus it is advisable to add 1 or 2 dB to the values shown in Table 8.2 to achieve a desired error rate in a practical system.

5.5   Regenerative Repeaters

As we are probably aware, pulses passing down a digital transmission line suffer attenuation and are badly distorted by the frequency characteristic of the line. A regenerative repeater amplifies and reconstructs such a badly distorted digital signal and develops a nearly perfect replica of the original at its output. Regenerative repeaters are an essential key to digital transmission in that we could say that the "noise stops at the repeater."

Figure 8.15 is a simplified block diagram of a regenerative repeater and shows typical waveforms corresponding to each functional stage of signal processing. As shown in the figure, the first stage of signal processing is amplification and equalization. Equalization is often a two-step process. The first is a fixed equalizer that compensates for the attenuation–frequency characteristic of the nominal section, which is the standard length of transmission line between repeaters (often 6000 ft). The second equalizer is variable and compensates for departures between nominal repeater section length and the actual length and loss variations due to temperature. The

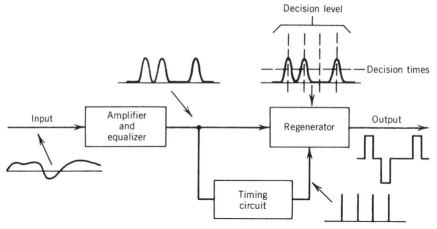

Figure 8.15. Simplified functional block diagram of a regenerative repeater for use on PCM cable systems [3].

adjustable equalizer uses automatic line build-out (ALBO) networks that are automatically adjusted according to characteristics of the received signal.

The signal output of the repeater must be accurately timed to maintain accurate pulse width and space between the pulses. The timing is derived from the incoming bit stream. The incoming signal is rectified and clipped, producing square waves that are applied to the timing extractor, which is a circuit tuned to the timing frequency. The output of the circuit controls a clock-pulse generator that produces an output of narrow pulses that are alternately positive and negative at the zero crossings of the square wave input.

The narrow positive clock pulses gate the incoming pulses of the regenerator, and the negative pulses are used to run off the regenerator. Thus the combination is used to control the width of the regenerated pulses.

Regenerative repeaters are the major source of timing jitter in a digital transmission system. Jitter is one of the principal impairments in a digital network, giving rise to pulse distortion and intersymbol interference. Jitter is discussed in more detail in Section 8.2.

Most regenerative repeaters transmit a bipolar (AMI) waveform (see Figure 8.1). Such signals can have one of three possible states in any instant in time, positive, zero, or negative, and are often designated +, 0, −. The threshold circuits are gated to admit the signal at the middle of the pulse interval. For example, if the signal is positive and exceeds a positive threshold, it is recognized as a positive pulse. If it is negative and exceeds a negative threshold, it is recognized as a negative pulse. If it has a value between the positive and negative thresholds, it is recognized as a 0 (no pulse).

When either threshold is exceeded, the regenerator is triggered to generate a pulse of the appropriate duration, polarity, and amplitude. In this manner the distorted input signal is reconstructed as a new output signal for transmission to the next repeater [3].

5.6 PCM System Enhancements

5.6.1 North American DS1

5.6.1.1 Frame and Superframe. In Section 3.3.1 we defined a frame. It was pointed out in that section the major difference between the North American DS1 framing strategy and European E1 framing. The North American system inserted one S-bit in each frame.

A superframe consists of 12 consecutive frames. Thus we have developed 12 S-bits and they are all used for frame alignment/synchronization. Thus there is a 12-bit sequence, one S-bit from each frame. This 12-bit sequence is subdivided into two sequences. The frame-alignment pattern is 101010 and is located in the odd-numbered frames. The superframe-alignment pattern is 001110 and is located in the even-numbered frames. The superframe pattern is shown in Table 8.3.

5.6.1.2 Extended Superframe. Table 8.4 shows the extended superframe structure. With modern processing technology, it was not necessary to tell the distant-end PCM receiver 8000 times per second where a frame started. Thus

TABLE 8.3 Multiframe Structure

Frame Number	Frame-Alignment Signal[a]	Multiframe-Alignment Signal (S-bit)	Bit Number(s) in Each Channel Time Slot		Signaling Channel Designation[b]
			For Character Signal	For Signaling	
1	1	—	1–8	—	
2	—	0	1–8	—	
3	0	—	1–8	—	
4	—	0	1–8	—	
5	1	—	1–8	—	
6	—	1	1–7	8	A
7	0	—	1–8	—	
8	—	1	1–8	—	
9	1	—	1–8	—	
10	—	1	1–8	—	
11	0	—	1–8	—	
12	—	0	1–7	8	B

[a]When the S-bit is modified to signal the alarm indications to the remote end, the S-bit in frame 12 is changed from state 0 to 1.

[b]Channel-associated signaling provides two independent 667-bit/s signaling channels designated A and B or one 1333-bit/s signaling channel.

Source: Table 5/G.704, page 13, ITU Geneva, 1991 (Ref. 9).

TABLE 8.4 Extended Superframe Structure

Frame Number Within Multiframe	Bit Number Within Multiframe	FAS[a]	DL[b]	CRC[c]	For Character Signal[d]	For signaling[d]	Signaling Channel Designation[d]
		F-bit Assignments			Bit Number(s) in Each Channel Time Slot		
1	1	—	m	—	1–8	—	
2	194	—	—	e_1	1–8	—	
3	387	—	m	—	1–8	—	
4	580	0	—	—	1–8	—	
5	773	—	m	—	1–8	—	
6	966	—	—	e_2	1–7	8	A
7	1159	—	m	—	1–8	—	
8	1352	0	—	—	1–8	—	
9	1545	—	m	—	1–8	—	
10	1738	—	—	e_3	1–8	—	
11	1931	—	m	—	1–8	—	
12	2124	1	—	—	1–7	8	B
13	2317	—	m	—	1–8	—	
14	2510	—	—	e_4	1–8	—	
15	2703	—	m	—	1–8	—	
16	2896	0	—	—	1–8	—	
17	3089	—	m	—	1–8	—	
18	3282	—	—	e_5	1–7	8	C
19	3475	—	m	—	1–8	—	
20	3668	1	—	—	1–8	—	
21	3861	—	m	—	1–8	—	
22	4054	—	—	e_6	1–8	—	
23	4247	—	m	—	1–8	—	
24	4440	1	—	—	1–7	8	D

[a]FAS   Frame-alignment signal (...001011...).
[b]DL   4-kbit/s data link (message bits m).
[c]CRC CRC-6 block check field (check bits e_1...e_6).
[d]Only applicable in the case of channel-associated signaling.
Source: Table 1/G.704, page 2, CCITT Rec. G.704, ITU Geneva, 1991 (Ref. 9).

the extended superframe (ESF) was developed. It consists of 24 sequential frames with 24 available S-bit positions (see Figure 8.10). Every fourth position in the ESF was dedicated to the frame-alignment signal (FAS). The remaining 18-bit positions in the 24-frame ESF were put to good use.

Six of the 18 bits developed a cyclic redundancy check (CRC) pattern, permitting system operators to monitor link error performance in quasi-real time. The generating polynomial for the pattern is $X^6 + X + 1$ based on CRC-6. A complete description of how CRC works is given in Chapter 10.

The remaining 12 bits are used to form a 4-kbps data link. This data link provides a communication path between primary hierarchical level terminals and contains data, an idle data link sequence or a loss of frame alignment alarm sequence. A loss of frame alignment alarm sequence is used when a loss of frame alignment (LFA) condition has been detected. After a loss of frame alignment condition is detected at local end A, one 16-bit LFA sequence of eight 1s and eight 0s (1111111100000000) will be transmitted in the *m*-bits of the 4-kbps data link continuously to remote end B [7, 9].

5.6.2 Enhancements to E1. Allocation of bits 1 to 8 of the E1 frame, time slot (channel) 0, is shown in Table 8.5. Table 8.6 shows the complete CRC-4 multiframe structure of E1, also TS 0.

Each CRC-4 multiframe, which is composed of 16 frames numbered 0 through 15, is divided into two 8-frame sub-multiframes (SMF), designated SMF I and SMF II, which signifies their respective order of occurrence

TABLE 8.5   Allocation of Bits 1 to 8 in TS 0, E1 Frame

Bit Number Alternate Frames	1	2	3	4	5	6	7	8
Frame containing the frame-alignment signal	S_i	0	0	1	1	0	1	1
	Note 1	Frame-alignment signal						
Frame not containing the frame-alignment signal	S_i	1	A	S_{a4}	S_{a5}	S_{a6}	S_{a7}	S_{a8}
	Note 1	Note 2	Note 3	Note 4				

Note 1: S_i = bits reserved for international use. One specific use is described in the text above. Other possible uses may be defined at a later stage. If no use is realized, these bits should be fixed at 1 on digital paths crossing an international border. However, they may be used nationally if the digital path does not cross a border.

Note 2: This bit is fixed at 1 to assist in avoiding simulations of the frame-alignment signal.

Note 3: A = remote alarm indication. In undisturbed operation, set to 0; in alarm condition, set to 1.

Note 4: S_{a4} to S_{a8} = additional spare bits whose use may be as follows:
  (i) Bits S_{a4} to S_{a8} may be recommended by CCITT for use in specific point-to-point applications (e.g., transcoder equipments conforming to Rec. G.761;
  (ii) Bit S_{a4} may be recommended by CCITT as a message-based data link for operations, maintenance, and performance monitoring. This channel originates at the point where the frame is generated and terminates where the frame is split up. This requires further study;
  (iii) Bits S_{a5} to S_{a7} are for national usage where there is no demand on them for specific point-to-point applications (see i) above. Bits S_{a4} to S_{a8} (where these are not used) should be set to 1 on links crossing an international border.

Source: Table 4a/G.704, page 7, CCITT Rec. G.704, ITU Geneva, 1991 (Ref. 9).

TABLE 8.6 CRC-4 Multiframe Structure

Sub-multiframe (SMF)	Frame Number	Bits 1 to 8 of the Frame							
		1	2	3	4	5	6	7	8
Multiframe I	0	C_1	0	0	1	1	0	1	1
	1	0	1	A	S_{a4}	S_{a5}	S_{a6}	S_{a7}	S_{a8}
	2	C_2	0	0	1	1	0	1	1
	3	0	1	A	S_{a4}	S_{a5}	S_{a6}	S_{a7}	S_{a8}
	4	C_3	0	0	1	1	0	1	1
	5	1	1	A	S_{a4}	S_{a5}	S_{a6}	S_{a7}	S_{a8}
	6	C_4	0	0	1	1	0	1	1
	7	0	1	A	S_{a4}	S_{a5}	S_{a6}	S_{a7}	S_{a8}
II	8	C_1	0	0	1	1	0	1	1
	9	1	1	A	S_{a4}	S_{a5}	S_{a6}	S_{a7}	S_{a8}
	10	C_2	0	0	1	1	0	1	1
	11	1	1	A	S_{a4}	S_{a5}	S_{a6}	S_{a7}	S_{a8}
	12	C_3	0	0	1	1	0	1	1
	13	E	1	A	S_{a4}	S_{a5}	S_{a6}	S_{a7}	S_{a8}
	14	C_4	0	0	1	1	0	1	1
	15	E	1	A	S_{a4}	S_{a5}	S_{a6}	S_{a7}	S_{a8}

Note 1: E = CRC-4 error indication bits.
Note 2: S_{a4} to S_{a8} = spare bits (see Note 4 to Table 8.5).
Note 3: C_1 to C_4 = cyclic redundancy check-4 (CRC-4) bits.
Note 4: A = remote alarm indication (see Table 8.5).
Source: Table 4b/G.704, page 8, CCITT Rec. G.704, ITU Geneva, 1991. (Ref. 9).

within the CRC-4 multiframe structure. The SMF is the cyclic redundancy check-4 (CRC-4) block size (i.e., 2048 bits). In those frames containing the frame-alignment signal, bit 1 is used to transmit the CRC-4 bits. There are four CRC-4 bits designated C_1, C_2, C_3, and C_4 in each SMF [9].

In those frames not containing the frame-alignment signal as defined above, bit 1 is used to transmit the 6-bit CRC-4 multiframe-alignment signal and two CRC-4 error indication bits (E). The CRC-4 multiframe-alignment signal has the form 001011.

The E-bits are used to indicate received errored sub-multiframes by setting the binary state of one E-bit from 1 to 0 for each errored sub-multiframe. Any delay between detection of an errored sub-multiframe and the setting of the E-bit that indicates the error state must be less than 1 s. The E-bits are always taken into account even if the SMF which contains them is found to be errored, since there is little likelihood that the E-bits themselves will be errored. The generating polynomial for CRC-4 is $X^4 + X + 1$ [9].

6 HIGHER-ORDER PCM MULTIPLEX SYSTEMS

6.1 Introduction

Higher-order PCM multiplex is developed out of several primary multiplex sources. Primary multiplex is typically DS1 in North America and E1 in Europe; some countries have standardized on E1, such as most of Hispanic America. Not only are E1 and DS1 incompatible, the higher-order multiplexes, as one might imagine, are also incompatible. First we introduce *stuffing*, describe some North American higher-level multiplex, and then discuss European multiplexes based on the E1 system.

6.2 Stuffing and Justification

Stuffing (justification) is common to all higher-level multiplexers that we describe below. Consider the DS2 higher-level multiplex. It derives from an M12 multiplexer, taking inputs from four 24-channel channel banks. The clocks in these channel banks are free running. The transmission rate output of each channel bank is *nominally* 1,544,000 bps. However, there is a tolerance of ±50 ppm (±77 bps). Suppose all four DS1 inputs were operating on the high side of the tolerance or at 1,544,077 bps. The input to the M12 multiplexer is a buffer. It has a finite capacity. Unless bits are read out of the buffer faster than they are coming in, at some time the buffer will overflow. This is highly undesirable. Thus we have bit stuffing.

Stuffing in the output aggregate bit stream means adding extra bits. It allows us to read out of a buffer faster than we write into it.

In Ref. 24 the IEEE defines *stuffing bits* as "bits inserted into a frame to compensate for timing differences in constituent lower rate signals." CCITT uses the term *justification*.

Figure 8.16 illustrates the stuffing concept.

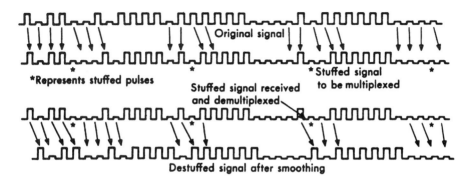

Figure 8.16. Pulse-stuffing synchronization. From Figure 29-2, Ref. 3.

6.3 North American Higher-Level Multiplex

The North American PCM digital hierarchy is shown in Figure 8.17. The higher-level multiplexers are type-coded in such a way that we know the DS levels which are being combined. For example, an M34 has inputs from level 3 (DS3) and the output is at level 4 (DS4). We describe the operation of the M12 multiplexer because it is typical of this series.

The formation of the second-level North American multiplex, DS2, operating from four DS1 inputs is shown in Figure 8.18. There are four inputs, each operating at a nominal 1.544 Mbps. The output bit rate is 6.312 Mbps. Multiply 1.544 by 4 and we get 6.176 Mbps. In other words, the output of the M12 multiplexer is operating 136 kbps faster than the aggregate of the four inputs. Some of these extra bits are overhead bits and the remainder are stuff bits. Figure 8.19 shows the makeup of a DS2 frame.

The M12 multiplex frame consists of 1176 bits. The frame is divided into four 294-bit subframes as illustrated in Figure 8.18. There is a control bit word that is distributed throughout the frame and that begins with an M bit. Thus each subframe begins with an M bit. There are four M bits forming the series 011X, where the fourth bit (X), which may be a 1 or a 0, may be used as an alarm indicator bit. When transmitted as a 1, no alarm condition exists. When it is transmitted as a 0, an alarm is present. The 011 sequence for the first three M bits is used in the receiving circuits to identify the frame.

It is noted in Figure 8.19 that each subframe is made up of six 49-bit blocks. Each block starts with a control bit which is followed by a 48-bit

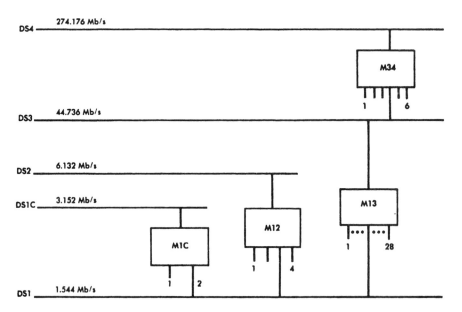

Figure 8.17. The North American digital hierarchy [3].

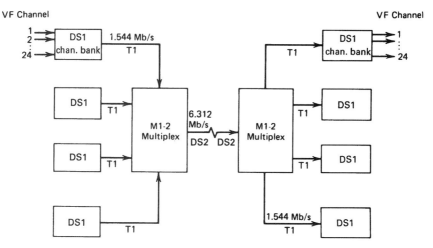

Figure 8.18. The formation of the DS2 signal from four DS1 signals in the M12 multiplexer [3].

block of information. Of these 48 information bits, 12 bits are taken from each of the four DS1 signals. These are interleaved sequentially in the 48-bit block. The first bit in the third and sixth block is designated an F bit. The F bits are a 0101 ... sequence used to identify the location of the control bit sequence and the start of each block of information bits.

The stuff-control bits are transmitted at the beginning of each of the 48-bit blocks numbered 2, 4, and 5 within each subframe. When these control bits, designated C, are 000, no stuff pulse is present; when the C bits are 111, a stuff pulse is added in the stuff position.

The stuff bit positions are all assigned to the sixth 48-bit block in each subframe. In subframe No. 1, the stuff bit is the first bit after the F1 bit. In subframe No. 2, the stuff bit is the second bit after the F1 bit, and so on through the fourth subframe. The nominal stuffing rate is 1796 bps for each DS1 input signal. The maximum is 5367 bps.

Prior to multiplexing at the M12 multiplex unit, input signals 2 and 4 are logically inverted. This is done to improve the statistical properties of the output DS2 signal.

6.4 The European E1 Digital Hierarchy

The E1 hierarchy is identified in a similar manner as DS1. E1 (30 voice channels) is the primary multiplex; E2 is the second level and is derived from four E1s. Thus E2 contains 120 equivalent voice channels. E3 is the third level and is derived from four E2 inputs and contains 480 equivalent voice channels. E4 derives from four E3 formations and contains the equivalent of 1920 voice channels. International digital hierarchies are compared in Table 8.7.

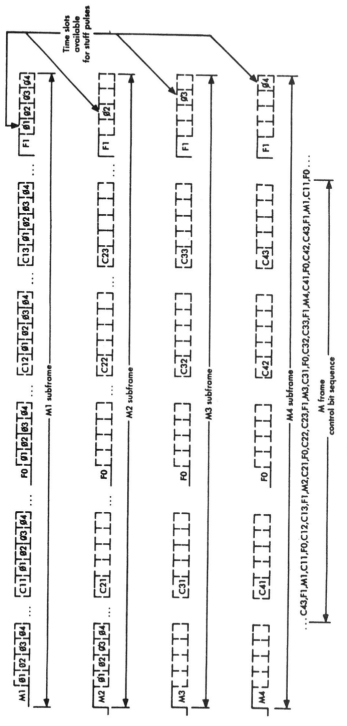

Figure 8.19. Makeup of a DS2 frame [10].

370

TABLE 8.7 Higher-Level PCM Multiplex Comparison

System Type	Level				
	1	2	3	4	5
North American T/D type	1	2	3	4	
Number of voice channels	24	96	672	4032	
Line bit rate (Mbps)	1.544	6.312	44.736	274.176	
Japan					
Number of voice channels	24	96	480	1440	5760
Line bit rate (Mbps)	1.544	6.312	32.064	97.728	400.352
Europe					
Number of voice channels	30	120	480	1920	
Line bit rate (Mbps)	2.048	8.448	34.368	139.264	

Source: Ref. 11.

Table 8.8 gives the basic parameters of the formation of the E2 level in the European digital hierarchy.

CCITT Rec. G.745 recommends cyclic bit interleaving in the tributary (i.e., E1 inputs) numbering order and positive/zero/negative justification with two-command control. The justification control signal is distributed and the C_{jn} bits ($n = 1, 2, 3$; see Table 8.8) are used for justification control bits.

Positive justification is indicated by the signal 111, transmitted in each of two consecutive frames. Negative justification is indicated by the signal 000, also transmitted in each of two consecutive frames. No-justification is indicated by the signal 111 in one frame and 000 in the next frame. Bits 5, 6, 7, and 8 in Set IV (Table 8.8) are used for negative justification of tributaries 1, 2, 3, and 4, respectively, and bits 9 to 12 for positive justification of the same tributaries.

Besides, when information from tributaries 1, 2, 3, and 4 is not transmitted, bits 5, 6, 7, and 8 in Set IV are available for transmitting information concerning the type of justification (positive or negative) in frames containing commands of positive justification and intermediate amount of jitter in frames containing commands of negative justification. The maximum amount of justification rate per tributary is shown in Table 8.8 [12].

7 LONG-DISTANCE PCM TRANSMISSION

7.1 Transmission Limitations

Digital waveforms lend themselves to transmission by wire pair, coaxial cable, fiber-optic cable, and wideband radio media. The PCM multiplex format was first applied to wire-pair cable as described in Section 5.3. Its use

TABLE 8.8 8448-kbps Digital Multiplexing Frame Structure Using Positive/Zero/ Negative Justification

Tributary bit rate (kbit/s)	2048
Number of tributaries	4
Frame Structure	**Bit Number**
	Set I
Frame-alignment signal (11100110)	1 to 8
Bits from tributaries	9 to 264
	Set II
Justification control bits C_{j1} (see Note)	1 to 4
Bits for service functions	5 to 8
Bits from tributaries	9 to 264
	Set III
Justification control bits C_{j2} (see Note)	1 to 4
Spare bits	5 to 8
Bits from tributaries	9 to 264
	Set IV
Justification control bits C_{j3} (see Note)	1 to 4
Bits from tributaries available for negative justification	5 to 8
Bits from tributaries available for positive justification	9 to 12
Bits from tributaries	12 to 264
Frame length	1056 bits
Frame direction	125 µs
Bits per tributary	256 bits
Maximum justification rate per tributary	8 kbps

Note: C_{jn} indicates nth justification control bit of the jth tributary.

Source: Table 1/G.745, CCITT Rec. G.745, page 437, Fascicle III.4, IXth Plenary Assembly, Melbourne, 1988 (Ref. 12).

on coaxial cable will be briefly described with one application. Such use is deprecated because of such excellent advantages of the competing fiber-optic medium.

Each medium has transmission limitations brought about by impairments. In one way or another each limitation is a function of length and transmission rate (bit rate). We have discussed loss, for example. As loss increases (i.e., between regenerative repeaters), signal-to-noise ratio suffers, directly impacting bit error performance. Dispersion is another impairment that limits circuit length over a particular medium, especially as transmission rate increases. The following transmission impairments to PCM transmission are covered: jitter, distortion, noise, crosstalk, and echo.

7.2 Jitter

In the context of digital transmission, *jitter* is defined as short-term variation of the sampling instant from its intended position in time or phase. Longer-term variation of the sampling instant is called *wander*. Jitter can cause transmission impairments such as:

- Displacement of the ideal sampling instant. This leads to a degradation in system error performance.
- Slips in timing recovery circuits manifesting itself in degraded error performance.
- Distortion of the resulting analog signal after decoding at the receive end of the circuit.

The random phase modulation, or *phase jitter*, introduced at each repeater accumulates in a repeater chain and may lead to crosstalk and distortion in the reconstructed analog signal. In digital switching systems, jitter on the incoming lines is a potential source of slips.* The sources of timing jitter may be classified as systematic or nonsystematic according to whether or not they are related to the pulse pattern. Systematic jitter sources lead to jitter which degrades the bit stream in the same way at each repeater in the chain. Systematic sources include intersymbol interference, finite pulse width, and clock threshold effects. Nonsystematic jitter sources such as mistuning and crosstalk result in timing degradations which are random from repeater to repeater. Thermal and impulse noise are not serious contributors to timing jitter. That is, if the total noise at the regenerator input is low enough to permit the regenerator to operate with an acceptably low error rate, the noise passed by the narrow-band timing extractor is orders of magnitude less and therefore negligible. In a long repeater chain, the total accumulated jitter is dominated by components produced by systematic sources [3].

Jitter accumulation is a function of the number of regenerative repeaters in tandem. Keep in mind that switches, fiber-optic receivers, and radios are also regenerative repeaters. The mean square value of jitter in a long chain of repeaters increases with N (the number of repeaters), and the rms value of jitter increases with $(N)^{1/2}$. Jitter is also proportional to the timing filter bandwidth, which leads to the conclusion that higher-Q tuned circuits in the repeater reduce jitter.

Certainly by reducing the number of regenerative repeaters in tandem, we reduce jitter accordingly. Wire-pair systems have repeaters every 6000 ft, and coaxial cable has repeaters approximately every mile. If we are to reduce jitter, these transmission media are not good candidates on long circuits. Fiber-optic systems, depending on design and bit rate, have repeaters every 40–200 mi. This is another reason why fiber-optic systems are favored for

---

*Slips will be discussed in Chapter 9 as a principal impairment in digital networks.

digital transmission. Microwave radio, strictly for budgeting purposes, may have repeaters every 30 mi, so it too is a candidate for long systems. Satellite links have the least repeaters, at least one in a long circuit (say 4000 mi).

The principal effect of jitter on the resulting analog signal after decoding is to distort the signal. The analog signal derives from a PAM pulse train, which is then passed through a low-pass filter. Jitter displaces the PAM pulses from their proper location, showing up as undesired pulse-position modulation (PPM) [3].

7.3   Distortion

On metallic transmission links, such as coaxial cable and wire-pair cable, line characteristics distort and attenuate the digital signal as it traverses the medium. There are three cable characteristics that create this distortion: loss, amplitude distortion (amplitude–frequency response), and delay distortion. Thus the regenerative repeater must provide amplification and equalization of the incoming digital signal before regeneration. There are also trade-offs between loss and distortion on the one hand and repeater characteristics and repeater section length on the other.

7.4   Thermal Noise

As in any electrical communication system, thermal noise, impulse noise, and crosstalk affect system design. Because of the nature of a digital system, these impairments need only be considered on a per-repeater-section basis because noise does not accumulate due to the regenerative process carried out at repeaters and nodes. Bit errors do accumulate, and this impairment family is one of several that create these errors. One way to limit error accumulation is to specify a stringent BER for each repeater section. Repeater sections are often specified with a median BER of 1 in 10^{-9}.

It is interesting to note that PCM provides reasonable voice performance for a BER as poor as 1 in 10^2. However, the worst tolerable BER is 1 in 10^3 at system end points. This value is required to ensure the correct operation of supervisory signaling. The reader should appreciate that such degraded BER values are completely unsuitable for data transmission.

7.5   Crosstalk

Crosstalk is a major impairment in PCM wire-pair systems, particularly when "go" and "return" channels are carried in the same cable sheath. The major offender of single-cable operation is near-end crosstalk (NEXT). When the two directions of transmission are carried in separate cables or use shielded pairs in a common cable, far-end crosstalk (FEXT) becomes dominant.

One characteristic has been found to be a major contributor to poor crosstalk coupling loss. This is the capacitance imbalance between wire pairs.

Stringent quality control during cable manufacture is one measure taken to ensure that minimum balance values are met.

7.6 Echo

Echo is caused by impedance discontinuities in the transmission line, including repeaters and terminations (MDFS, codecs, switch ports). Good impedance match across the entire system eliminates the cause of echo or reduces its level. On a PCM transmission system there are many causes of echo, such as gas plugs and splices.

Gas plugs are used on cable systems to allow gas to be applied under pressure to the cable to prevent moisture buildup. The plug tends to add capacitance to the line. To compensate for this, repeater sections that incorporate plugs are made short to accommodate the added capacitance.

Other sources of echo are where gauge and insulation changes take place along the cable run. Bridged taps are still another potential source of mismatch.

8 DIGITAL LOOP CARRIER (DLC)

Digital subscriber loop carrier is a method of extending the metallic subscriber plant by using one or more DS1 configurations. As an example, the SLC-96 uses four DS1 configurations to derive an equivalent of 96 voice channels.

The digital transmission facility used by a DLC system may be repeatered wire-pair cable, optical fibers, either or both combined with digital multiplexers, or other appropriate media. In Bellcore terminology, the central office termination (COT) is the digital terminal colocated with the local serving switch. The RT is the remote terminal. The RT must provide all of the features to a subscriber loop that the local serving switch normally does, such as supervision, ringing, address signaling, both dial pulse and touch tone, and so on [13].

9 SONET AND SDH

9.1 Introduction

SONET is an acronym for *Synchronous Optical Network*. SONET provides digital formats extending up to 9 Gbps, although our references cover only up to 2.4 Gbps. SONET is a North American development. The equivalent European format is called SDH or *synchronous digital hierarchy*. The two are very similar. Either one can accommodate the standard DS (i.e., 1.544 Mbps, etc.) and E (i.e., 2.048 Mbps, etc.) line rates.

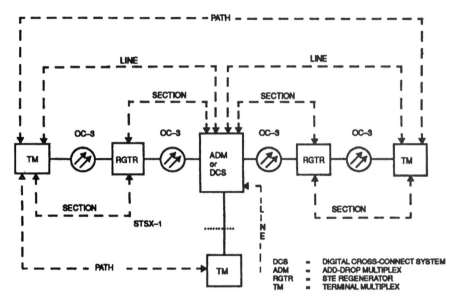

Figure 8.20. SONET: section, line, and path definitions [14].

9.2   SONET

SONET's higher-level digital format was originally intended for transmission over optical fiber facilities. It can, however, be accommodated on any transmission medium that meets the bandwidth requirements. Figure 8.20 is a functional diagram depicting SONET section, line, and path for the purpose of definition.

9.2.1   SONET Rates and Formats. SONET was designed to have a synchronous hierarchy that has sufficient flexibility to carry many different capacity signals such as DS1, DS3, E1, and ATM. This is realized by defining a basic module with a bit rate of 51.840 Mbps, as well as defining a byte-interleaved multiplex scheme that results in a family of signals with N times 51.840 Mbps, where N is an integer [15].*

The basic SONET module is divided into a portion assigned to overhead and a portion that carries the payload. The payload can be used to transport DS3 signals or a variety of sub-DS3 signals. Because some signals requiring transport have rates greater than the basic rate (e.g., broadband-ISDN), a technique of linking several basic modules together to build a transport signal of increased capacity is provided. To maintain a consistent payload structure while providing for the transport of a variety of lower-rate payloads

---

*Note the difference with plesiochronous digital hierarchy (PDH), what the industry calls the DS1 multiplex hierarchy (DS1C, DS2, etc.). These are *not* integer multiples of DS1.

(e.g., DS1, DS1C, and DS2 signals), a structure called the *virtual tributary* (VT) is defined. Payloads below the DS3 rate are transported with a VT structure.

9.2.1.1 Synchronous Hierarchical Rates. The synchronous transport signal—level 1 (STS-1) is the basic SONET module. It has a bit rate of 51.840 Mbps. The optical counterpart of STS-1 is optical carrier—level 1 signal (OC-1).

Higher-level SONET signals are obtained by synchronous multiplexing lower-level modules. When these lower-level modules are multiplexed, the result is denoted STS-N, where N is an integer. STS-N can then be converted to an OC-N or STS-N electrical signal. The rate of OC-N is N times the rate of STS-1. The popular SONET line rates are shown in Table 8.9.

An STS-1 is a specific sequence of 810 bytes (6480 bits), which includes various overhead bytes and an envelope capacity for transporting payloads. STS-1 is shown in Figure 8.21. It has a 90-column by 9-row structure. The frame duration is 125 μs (i.e., 8000 frames per second), deriving a bit rate of 51.840 Mbps. The order of transmission of bytes in Figure 8.21 is row by row, from left to right [14].

TABLE 8.9   Line Rates for Standard SONET Interface Signals

OC-N Level	STS-N Electrical Level	Line Rate (Mbps)
OC-1	STS-1 electrical	51.84
OC-3	STS-3 electrical	155.52
OC-12	STS-12 electrical	622.08
OC-24	STS-24 electrical	1244.16
OC-48	STS-48 electrical	2488.32
OC-192	STS-192 electrical	9953.28

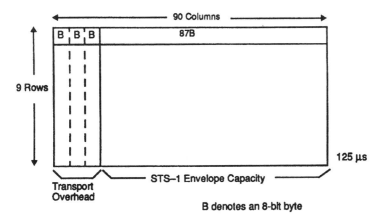

Figure 8.21. SONET STS-1 frame structure.

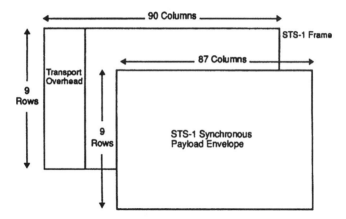

Figure 8.22. STS-1 synchronous payload envelope (SPE) [14].

Transport overhead occupies the first three columns of the STS-1 frame for a total of 27 bytes. The remaining 87 columns of the STS-1 frame, a total of 783 bytes, are allocated to the synchronous payload envelope (SPE) signal. This provides a channel capacity of 50.11 Mbps in the STS-1 signal structure for carrying tributary payloads intact across the synchronous network.

It should be noted that at 8000 frames per second, each byte within the SONET signal structure represents a channel bandwidth of 64 kbps (i.e., 8 bits/byte × 8000 bytes/second = 64 kbps). This is the same bit rate of a PCM voice channel or a DS0/E0 time slot.

Figure 8.22 shows the SPE providing 87 columns and 9 rows of payload. The figure gives the appearance that an SPE is wholly contained in one STS-1 frame. This is not necessarily true. In fact the STS-1 SPE can begin anywhere in the STS-1 envelope capacity, as shown in Figure 8.23. Of course, if it

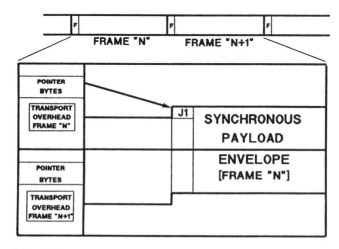

Figure 8.23. Link between transport overhead and SPE. Courtesy of Hewlett-Packard [16].

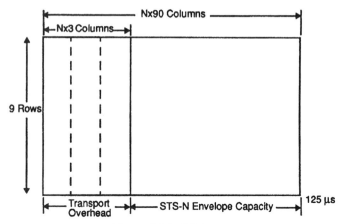

Figure 8.24. STS-N frame.

begins in one STS-1 frame and is not wholly contained in that frame, the remainder of the SPE appears in the following contiguous frame. Allowing the SPE to begin anywhere in the STS-1 frame facilitates efficient multiplexing (especially add–drop multiplexing) and cross-connection of signals in the synchronous network.

When an SPE is assembled into the transport frame, additional bytes, referred to as the *payload pointer*, are made available in the transport overhead. These bytes contain a pointer value which indicates the location of the first byte (J1) of the STS-1 SPE. The SPE is allowed to float freely within the space made available for it in the transport frame so that timing phase adjustments can be made as required between the SPE and the transport frame. The payload pointer identifies the first byte location of the SPE [14].

An STS path overhead (POH) associated with each payload is used to communicate various information from the point where a payload is mapped into the STS-1 SPE to where it is delivered. The POH is contained in the first column of the SPE and thus consists of 9 bytes. This signal capacity provides such facilities as alarm and performance monitoring required to support and maintain the transport of the SPE between *path terminations*. A path termination is where the SPE is either assembled or disassembled.

The frame structure of the STS-N is shown in Figure 8.24. It consists of N × 810 bytes. The STS-N is formed by byte-interleaving STS-1 and STS-M (M < N) modules. The transport overhead of the individual STS-1 and STS-M modules are frame-aligned before interleaving, but the associated STS SPEs are not required to be aligned because each STS-1 has a payload pointer to indicate the location of the SPE. The concept of byte interleaving is shown in Figure 8.25.

9.2.1.2 Virtual Tributaries. The virtual tributary (VT) structure is designed for transport and switching of sub-STS-1 payloads. There are four sizes of VTs:

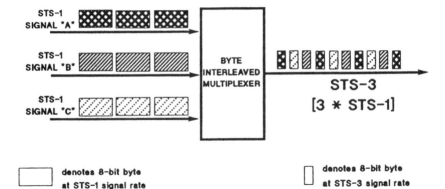

Figure 8.25. The concept of byte-interleaved multiplexing. Courtesy of Hewlett-Packard [16].

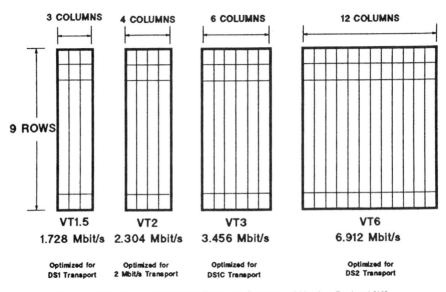

Figure 8.26. The four different VT frames. Courtesy of Hewlett-Packard [16].

VT1.5 (1.728 Mbps), VT2 (2.304 Mbps), VT3 (3.456 Mbps), and VT6 (6.912 Mbps). These are illustrated in Figure 8.26.

A VT1.5 packed in an STS-1 SPE is shown in Figure 8.27.

9.2.1.3 SPE Assembly/Disassembly Process.   Fundamental to the SONET format of transmission is the concept of a tributary signal (such as a DS3) being assembled into an SPE to be transported end-to-end across the synchronous network. The assembly process is called *payload mapping*. The payload capacity provided for each tributary signal is always slightly greater than that required by the tributary signal. This provides uniformity across all SONET transport facilities. The mapping process synchronizes the tributary

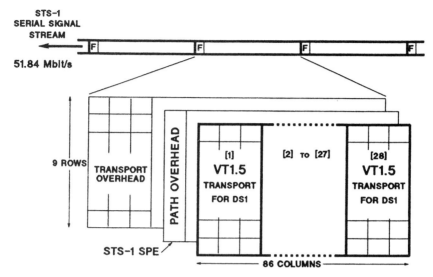

Figure 8.27. VT1.5 packaged in an STS-1 SPE. Courtesy of Hewlett-Packard [16].

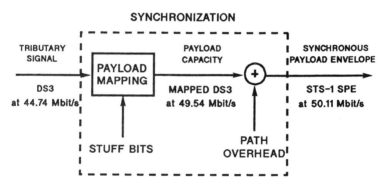

Figure 8.28. The SPE assembly process. Courtesy of Hewlett-Packard [16].

signal with the payload capacity. This is achieved by stuffing bits to the signal stream as part of the mapping process. The assembly process is illustrated in Figure 8.28. In this example, a DS3 tributary signal enters at its nominal 44.736 Mbps and needs to be synchronized with the payload capacity of 49.54 Mbps provided by the STS-1 SPE. This bit rate is increased to 50.11 Mbps by addition of the path overhead (POH).

At the point of exit from the synchronous network, the payload tributary signal that has been transported over the network needs to be recovered from the SPE. The process of disassembling the tributary signal from the SPE is referred to as *payload demapping*. This disassembly process is shown in Figure 8.29. [16].

DESYNCHRONIZATION

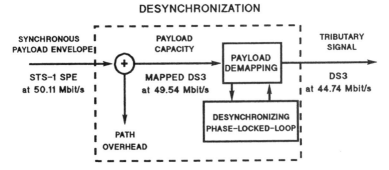

Figure 8.29. The SPE disassembly process. Courtesy of Hewlett-Packard [16].

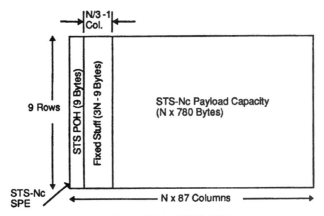

Figure 8.30. STS-Nc SPE.

9.2.1.4 Super Rate Payloads.   Multiple STS-1 SPEs are required to transport super rate payloads as might be encountered with broadband-ISDN ATM. To accommodate such a payload, an STS-Nc module is formed by linking N constituent STS-1s together in fixed phase alignment. The super rate payload is then mapped into the resulting STS-Nc SPE for transport. The STS-Nc can be carried by an OC-N, STS-N electrical, or higher. Concatenation indicators contained in the second through Nth STS payload pointers are used to show that the STS-1s of an STS-Nc are linked together. The STS-Nc SPE is shown in Figure 8.30.

The STS-Nc consists of $N \times 783$ bytes and can be depicted as an $N \times 97$ column by 9-row structure. Only one set of STS POH is required in the STS-Nc SPE. The STS-Nc SPE is carried within the STS-Nc so that the STC POH will always appear in the first of the N STS-1s that make up the STS-Nc.

In all the super rate payload mappings, the first $(N/3) - 1$ columns of the STS-Nc SPE following the STS POH are not used for payload, but are designated fixed stuff columns (i.e., columns of undefined bytes). Only mappings into STS-3c and STS-12c have been defined by Bellcore [14]. Other mapping may be defined in the future.

9.3   Synchronous Digital Hierarchy (SDH)

9.3.1   Introduction. Synchronous digital hierarchy (SDH) was a European development, whereas SONET was a North American development. They are very similar. The principal difference is that their level 1 rates are dissimilar. For SONET it is STS-1 at 51.84 Mbps, and for SDH it is STM-1 at 155.52 Mbps. The STM-1 rate is the same as the SONET STS-3 rate. SDH is generally understood to be more flexible than SONET and was designed more with the PTT (government-owned telecommunication administrations) in mind. The SONET design, on the other hand, has more flavor of the private network. SDH is also several years behind SONET in implementation.

9.3.2   SDH Standard Bit Rates. The standard SDH bit rates are shown in Table 8.10. ITU-T Rec. G.707 [17] states "that the first level of the digital hierarchy shall be 155,520 kbps ... and ... that high synchronous digital hierarchy bit rates shall be obtained as integer multiples of the first level bit rate."

TABLE 8.10   SDH Bit Rates with SONET Equivalents

SDH Level	SDH Bit Rate (kbps)	SONET Equivalent Line Rate
1	155,520	STS-3/OC-3
4	622,080	STS-12/OC-12
16	2,488,320	STS-48/OC-48

Two other SDH hierarchical levels are under consideration by CCITT [17]:

Level 8:   1,244,160 kbps
Level 12:   1,866,240 kbps

9.3.3   Interface and Frame Structure of SDH. Figure 8.31 illustrates the relationship between various multiplexing elements that are given below and shows generic multiplexing structures. Figures 8.32, 8.33, and 8.34 show specific derived multiplexing methods.

Definitions

Synchronous Transport Module (STM): An STM is the information structure used to support section layer connections in the SDH. It is analogous to STS in the SONET regime. STM consists of information payload and

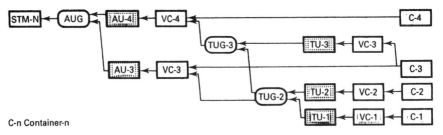

C-n Container-n

Figure 8.31. Generalized SDH multiplexing structure. From ITU-T Rec. G.708, Figure 2-1/G.708, page 3, 3/93 (Ref. 18).

section overhead (SOH) information fields organized in a block frame structure which repeats every 125 µs. The information is suitably conditioned for serial transmission on selected media at a rate which is synchronized to the network. A basic STM (STM-1) is defined at 155,520 kbps. Higher-capacity STMs are formed at rates equivalent to N times multiples of this basic rate. STM capacities for $N = 4$ and $N = 16$ are defined, and higher values are under consideration by ITU-T. An STM comprises a single administrative unit group (AUG) together with the SOH. STM-N contains N AUGs together with SOH.

Container, C-n ($n = 1$ to $n = 4$): This element is a defined unit of payload capacity which is dimensioned to carry any of the levels currently defined in Section 9.3.2 and may also provide capacity for transport of broadband signals which are not yet defined by CCITT (ITU-T organization) [18].

Virtual Container-n (VC-n): A virtual container is the information structure used to support path layer connection in the SDH. It consists of information payload and path overhead (POH) information fields organized in a block frame which repeats every 125 or 500 µs. Alignment information to identify VC-n frame start is provided by the server network layer. Two types of virtual container have been identified:

1. *Lower-Order Virtual Container-n, VC-n* ($n = 1, 2$). This element comprises a single C-n ($n = 1, 2$), plus the basic virtual container path overhead (POH) appropriate to that level.
2. *Higher-Order Virtual Container-n, to VC-n* ($n = 3, 4$). This element comprises a single C-n ($n = 3, 4$), an assembly of tributary unit groups (TUG-2s), or an assembly of TU-3s, together with virtual container POH appropriate to that level.

Administrative Unit-n, AU-n: An administrative unit is the information structure which provides adaptation between the higher-order path layer and the multiplex section. It consists of an information payload (the higher-

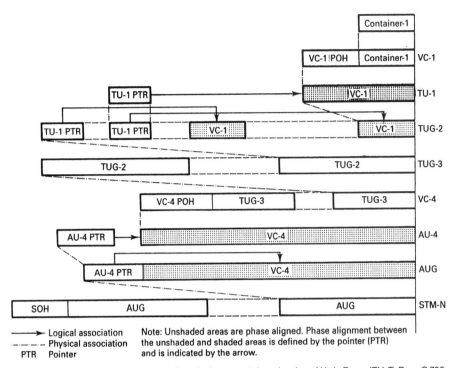

Figure 8.32. Multiplexing method directly from container-1 using AU-4. From ITU-T Rec. G.708, Figure 2-2/G.708, page 3, 3/93 (Ref. 18).

order virtual container) and an administrative unit pointer which indicates the offset of the payload frame start relative to the multiplex section frame start. Two administrative units are defined. The AU-4 consists of a VC-4 plus an administrative unit pointer which indicates the phase alignment of the VC-4 with respect to the STM-N frame. The AU-3 consists of a VC-3 plus an administrative unit pointer which indicates the phase alignment of the VC-3 with respect to the STM-N frame. In each case the administrative unit pointer location is fixed with respect to the STM-N frame [18].

One or more administrative units occupying fixed, defined positions in an STM payload is termed an *administrative unit group* (AUG). An AUG consists of a homogeneous assembly of AU-3s or an AU-4.

Tributary Unit-*n*, TU-*n*: A tributary unit is an information structure which provides adaptation between the lower-order path layer and the higher-order path layer. It consists of an information payload (the lower-order virtual container) and a tributary unit pointer which indicates the offset of the payload frame start relative to the higher-order virtual container frame start.

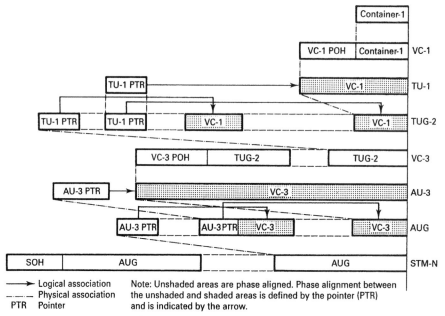

Figure 8.33. SDH multiplexing method directly from container-1 using AU-3. From ITU-T Rec. G.708, Figure 2-3/G.708, page 4, 3/93 (Ref. 18).

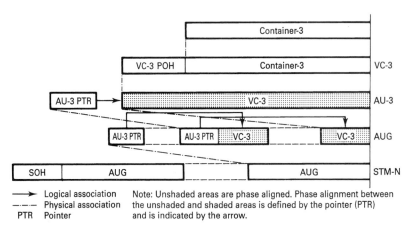

Figure 8.34. Multiplexing method directly from container-3 usaing AU-3. From ITU-T Rec. G.708, Figure 2-4/G.708, page 5, 3/93 (Ref. 18).

The TU-n ($n = 1, 2, 3$) consists of a VC-n together with a tributary unit pointer.

One or more tributary units occupying fixed, defined positions in a higher-order VC-n payload is termed a *tributary unit group* (TUG). TUGs are defined in such a way that mixed-capacity payloads made up of different-size tributary units can be constructed to increase flexibility of the transport network.

A TUG-2 consists of a homogeneous assembly of identical TU-1s or a TU-2. A TUG-3 consists of a homogeneous assembly of TUG-2s or a TU-3 [18].

Container-*n* (*n* = 1–4): A container is the information structure which forms the network synchronous information payload for a virtual container. For each of the defined virtual containers there is a corresponding container. Adaptation functions have been defined for many common network rates into a limited number of standard containers [18]. These include standard E-1/DS-1 rates defined in ITU-T Rec. G.702 [19].

9.3.3.1 Frame Structure. The basic frame structure, STM-N, is shown in Figure 8.35. The three main areas of the STM-1 frame are section overhead, AU pointers, and STM-1 payload.

Section Overhead. Section overhead is shown in rows 1–3 and 5–9 of columns 1–9 × *N* of the STM-N in Figure 8.35.

Administrative unit (AU) pointers. Row 4 of columns 1–9 × *N* in Figure 8.35 is available for AU pointers. The positions of the pointers of the AUs for different organizations of the STM-1 payload are shown in Table 8.11. See ITU-T Rec. G.709 for application of pointers and their detailed specifications. The rules for interpreting the AU-*n* pointers are summarized below [20]:

1. During normal operation, the pointer locates the start of the VC-*n* within the AU-*n* frame.

2. Any variation from the current pointer value is ignored unless a consistent new value is received three times consecutively or it is

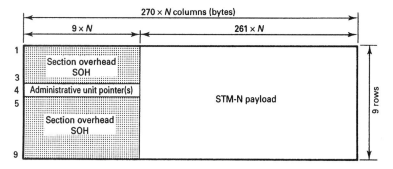

Figure 8.35. STM-N frame structure. From ITU-T Rec. G.708, Figure 3-1/G.708, page 7, 3/93 (Ref. 18).

TABLE 8.11   Position of AU Pointers

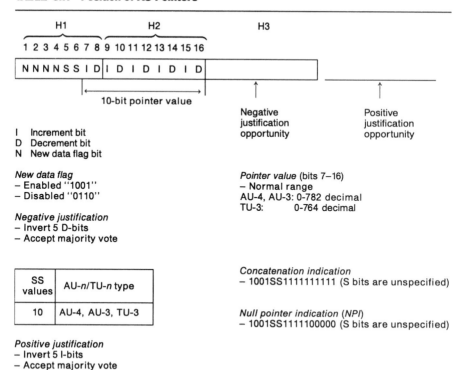

I   Increment bit
D   Decrement bit
N   New data flag bit

New data flag
– Enabled "1001"
– Disabled "0110"

Negative justification
– Invert 5 D-bits
– Accept majority vote

SS values	AU-*n*/TU-*n* type
10	AU-4, AU-3, TU-3

Positive justification
– Invert 5 I-bits
– Accept majority vote

Pointer value (bits 7–16)
– Normal range
AU-4, AU-3: 0-782 decimal
TU-3:   0-764 decimal

Concatenation indication
– 1001SS1111111111 (S bits are unspecified)

Null pointer indication (NPI)
– 1001SS1111100000 (S bits are unspecified)

Notes:
1. NPI value applies only to TU-3 pointers.
2. The pointer is set to all "1"'s when an AIS occurs.

Source: Ref. 20.

preceded by one of the rules 3, 4, or 5. Any consistent new value received three times consecutively overrides (i.e., takes priority over) rules 3 or 4.

3. If the majority of the I-bits of the pointer word are inverted, a positive justification operation is indicated. Subsequent pointer values shall be incremented by one.

4. If the majority of the D-bits of the pointer word are inverted, a negative justification operation is indicated. Subsequent pointer values shall be decremented by one.

5. If the NDF is set to "1001", then the coincident pointer value shall replace the current one at the offset indicated by the new pointer value unless the receiver is in a state that corresponds to a loss of pointer.

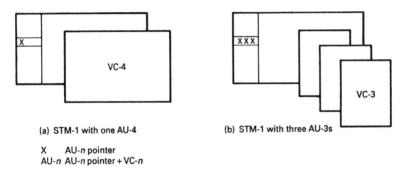

(a) STM-1 with one AU-4   (b) STM-1 with three AU-3s

X   AU-*n* pointer
AU-*n* AU-*n* pointer + VC-*n*

Figure 8.36. Administrative units in the STM-1 frame. From ITU-T Rec. G.708, Figure 3-2/G.708, page 8, 3/93 (Ref. 18).

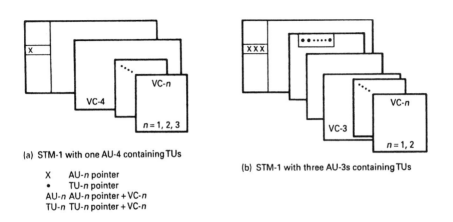

(a) STM-1 with one AU-4 containing TUs

X   AU-*n* pointer
•   TU-*n* pointer
AU-*n* AU-*n* pointer + VC-*n*
TU-*n* TU-*n* pointer + VC-*n*

(b) STM-1 with three AU-3s containing TUs

Figure 8.37. Two-stage multiplex. From ITU-T Rec. G.708, Figure 3-3/G.708, page 9, 3/93 (Ref. 18).

Administrative units in the STM-N. The STM-N payload can support N AUGs where each AUG may consist of one AU-4 or three AU-3s. The VC-*n* associated with each AU-*n* does not have a fixed phase with respect to the STM-N frame. The location of the first byte of the VC-*n* is indicated by the AU-*n* pointer. The AU-*n* pointer is in a fixed location in the STM-*N* frame. This is shown in Figures 8.32, 8.33, 8.34, 8.35, 8.36, and 8.37. The AU pointer bytes H1 and H2 with bit meaning are shown in Table 8.11.

The AU-4 may be used to carry, via the VC-4, a number of TU-*n*s ($n = 1, 2, 3$) forming a two-stage multiplex. An example of this arrangement is illustrated in Figures 8.32 and 8.36a. The VC-*n* associated with each TU does not have a fixed-phase relationship with respect to the start of the VC-4. The TU-*n* pointer is in a fixed location in the VC-4, and the location of the first byte of the VC-*n* is indicated by the TU-*n* pointer.

The AU-3 may be used to carry, via the VC-3, a number of TU-*n*s ($n = 1, 2$) forming a two-stage multiplex. An example of this arrangement is illustrated in Figures 8.33 and 8.37b. The VC-*n* associated with each TU-*n* does not have a fixed-phase relationship with respect to the start of the VC-3. The TU-*n* pointer is in a fixed location in the VC-3, and the location of the first byte of the VC-*n* is indicated by the TU-*n* pointer [18].

9.3.4 Interconnection of STM-1s. SDH has been designed to be universal, allowing transport of a large variety of signals including those specified in ITU-T Rec. G.702, such as North American 1.544-Mbps and European 2.048-Mbps regimes. However, different structures can be used for the transport of virtual containers. The following interconnection rules are used:

1. The rule for interconnecting two AUGs based upon two different types of administrative unit, namely AU-4 and AU-3, is to use the AU-4 structure. Therefore, the AUG based upon AU-3 is demultiplexed to the TUG-2 or VC-3 level according to the type of the payload and is remultiplexed within an AUG via the TUG-3/VC-4/AU-4 route.

2. The rule for interconnecting VC-11s transported via different types of tributary unit, namely TU-11 and TU-12, is to use the TU-11 structure. VC-11, TU-11, and TU-12 are described in ITU-T Rec. G.709 [20].

9.3.5 Basic SDH Multiplexing Structure. The SDH multiplexing structure is shown in Figure 8.38.

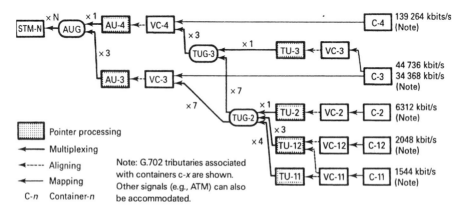

Figure 8.38. SDH multiplexing structure. From ITU-T Rec. G.709, Figure 1/G.709, page 3, 3/93 (Ref. 20).

10 DELTA MODULATION

10.1 Introduction

Delta modulation is another method of transmitting an analog signal such as voice in a digital format. It is quite different from PCM in that coding is carried out before multiplexing and the code is far more elemental, actually coding at only 1 bit at a time. Delta modulation exploits the sample-to-sample redundancy typical in a speech or video waveform.

The delta modulation code is a one-element code and is differential in nature, providing 1 bit per sample of the difference signal. That single bit specifies the polarity of the difference sample. It thereby indicates whether the signal has increased in amplitude or decreased since the last sample.

An approximation of the input waveform is constructed in a feedback path by stepping up one quantization level when the difference is positive (1) and stepping down when the difference is negative (0). Here we mean that the derivative of the analog input is transmitted rather than the instantaneous amplitude as in PCM. This is done by integrating the analog input to decide which of the two has the larger amplitude. The polarity of the next digit placed on the line is either plus or minus to reduce the amplitude of the two waveforms [i.e., analog input and integrated digital output (previous digit)]. We thus see the delta encoder basically as a feedback circuit as shown in Figure 8.39.

Most delta encoders sample at a rate greater than the Nyquist rate, which, for a 4-kHz channel, is 8000 samples per second. However, since each encoded sample contains relatively little information (i.e., 1 bit), delta modulation systems require a higher sampling rate than conventional PCM. These rates are typically 16 kbps and 32 kbps. Up to a certain point, the higher the sampling rate in this case, the better the signal-to-quantizing-noise ratio.

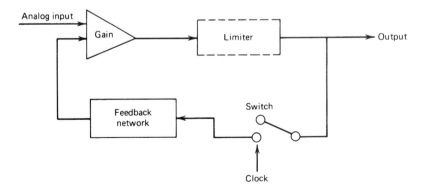

Figure 8.39. Basic electronic feedback circuit used in delta modulation.

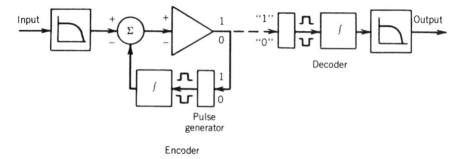

Figure 8.40. Simplified functional block diagram of a delta encoder/decoder.

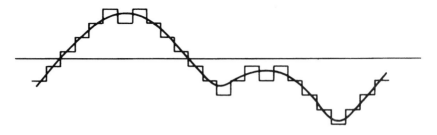

Figure 8.41. A delta encoding waveform.

A simplified functional block diagram of a delta coder/decoder is shown in Figure 8.40, and Figure 8.41 shows a typical delta waveform superimposed in a simple sinusoid audio input signal.

From the Figure 8.40 block diagram we can see that the digital output signal of the delta coder is indicative of the slope of the analog input signal (its derivative of the slope)—1 for a positive slope and 0 for a negative slope. But the 1 and 0 give no idea of the instantaneous or even semi-instantaneous steepness of the slope. This leads to the basic weakness found in delta modulation systems, namely, poor dynamic range or poor dynamic response given a satisfactory signal-to-quantizing-noise ratio.

With rapid changes of input level to the delta encoder, the output digital signal tends to lag behind these changes. This is called *slope overload*. Slope overload, shown diagrammatically in Figure 8.42, occurs when the rate of change of the input signal exceeds the maximum rate of change of the feedback loop. Thus a linear delta modulator (encoder) has severe dynamic range limitations.

10.2 Continuous Variable-Slope Delta Modulation

Continuous variable-slope delta modulation (CVSD) is a method of adaptive delta modulation using a form of digitally controlled companding, (com-

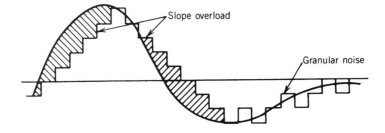

Figure 8.42. Typical slope overload of a delta modulator.

pression and expansion). It derives its step size from the transmit bit stream. As shown in Figure 8.43, adaptive logic monitors occurrence of three or four successive 1s or three or four successive 0s. A string of 1s indicates that the feedback is probably not rising as fast as the input, and conversely a string of 0s indicates that the feedback is not falling as fast as the input. This all 1s or all 0s condition enables control of a pulse generator, increasing step size voltage. Through resistor leak-off, normal operation returns unless reenabled. Figure 8.43 shows a CVSD coder/decoder, and we see that the circuitry in the decoder is similar to that in the coder.

Turning to Figure 8.43, the analog input signal to the CVSD encoder is band-limited by the input band-pass filter. The encoder compares the band-limited input signal S_i with the analog feedback approximation signal S_f, which is generated at the reconstruction integrator output. The digital output signal X_o of the encoder is the output of the first register in the "run-of-three" counter. The digital output signal is transmitted at the clock (sample) rate and will equal 1 if $S_i \geq S_f$ at the instant of sampling. For this value of X_o the pulse amplitude modulator (PAM) applies a positive feedback pulse $(+H)$ to the reconstruction integrator; otherwise, a negative pulse $(-H)$ is applied. This function is accomplished by the polarity control signal, which is equal to the digital encoder output X_o. The amplitude of the feedback pulse is derived by means of a 3-bit shift register, logic sensing for overload, and a syllabic low-pass filter. When a string of three consecutive 1s or 0s appears at the output of X_o, a discrete voltage level F is applied to the syllabic filter and the feedback pulse amplitude $(+H)$ increases until the overload string is broken. In such an event ground potential is fed to the filter by the overload logic, forcing a decrease in the amplitude of the slope voltage V_s.

The encoder and decoder, as mentioned, have identical characteristics, except for the comparator and filter functions. The CVSD decoder consists of the shift register, overload logic, syllabic filter, PAM, and reconstructed integrator used in the encoder, followed by a 4-kHz low-pass filter. The decoder performs the inverse function of the encoder and regenerates speech by passing the output signal S_f of the reconstruction integrator through the low-pass filter. Other characteristics optimize the CVSD technique for voice signals. These characteristics include:

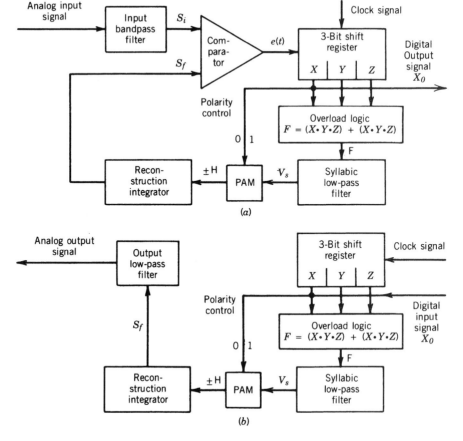

Figure 8.43. Block diagram of CVSD encoder (a) and decoder (b) [21].

- Changes in the slope of the analog input signal determine the step size changes of the digital output signal.
- The feedback loop is adaptive to the extent that the loop provides continuous or smoothly incremental changes in step size.
- Companding is performed at a syllabic rate to extend the dynamic range of the analog input signal.
- The reconstruction integrator of the CVSD device is of the exponential (leaky) type to reduce the effects of digital errors.

Table 8.12 shows typical signal-to-quantizing-noise ratio values. Figure 8.44 shows the U.S. Department of Defense TRI-TAC digital multiplexer hierarchy where the user (subscriber) input signals are 16 kbps or 32 kbps CVSD waveforms [21, 22].

TABLE 8.12 Signal-to-Quantizing-Noise Ratios for Typical Military CVSD Operation

| | Minimum Test Signal-to-Quantizing-Noise Plus Idle Channel Noise Ratio (dB) | |
Input Signal (dBm)	32 kbps	16 kbps
+2 to −3	22	15
−4 to −8	25	14
−9 to −13	26	14
−14 to −18	24	14
−19 to −23	22	14
−24 to −28	21	14
−29 to −33	19	13
−34 to −39	16	13

Source: MIL-STD-188-200 (Ref. 22).

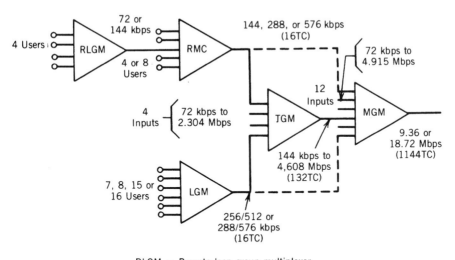

RLGM = Remote loop group multiplexer
RMC = remote multiplexer combiner
LGM = Loop group multiplexer
TGM = Trunk group multiplexer
MGM = Master group multiplexer
TC   = Traffic channels (maximum capability shown)
User = 16 or 32 kbps digital subscriber

Figure 8.44. U.S. Department of Defense TRI-TAC digital hierarchy. Courtesy of Raytheon Company (DGM data) [21].

11   SUMMARY OF ADVANTAGES AND DISADVANTAGES OF DIGITAL TRANSMISSION

The advantages of digital transmission tend to far outweigh the dis-advantages. This is being borne out with the rapid conversion of the trunk plant to all digital. Because of the large investment in the subscriber loop plant, conversion to all digital may take more time. However, this time may be shortened if ISDN (Chapter 13) implementation is accelerated.

Some advantages and disadvantages of digital transmission are listed below.

Advantages

1. System noise is controlled by the design of the terminal (quantization noise) and is essentially independent of the length of the system or of line noise and distortion. This assumes that the full length of the system is digital.

2. The signal-to-distortion performance of the system increases linearly with the number of bits per sample, giving a more efficient noise/band-width trade-off than other *bandwidth expansion* techniques such as FM. This efficiency, combined with the ruggedness of digital transmission, gives better utilization of noisy media such as wire-pair cable.

3. Increases in device speed (i.e., ICs, VLSI) allow common circuit com-ponents to be shared by many channels, thus lowering the per-channel cost of a PCM terminal. This is one major factor that makes the per-channel cost of PCM more cost-effective than FDM.

4. Digital systems are insensitive to traffic loading up to their full capacity. FDM is highly sensitive to traffic loading.

5. Likewise, digital time-division multiplex treats all channels alike in con-trast to FDM regarding phase and amplitude distortion (and resulting noise degradation) suffered by channels at band edge.

6. There is no appreciable degradation incurred in multiplexing/demultiplexing so that facility arrangements do not need to take into account the number of previous multiplex/demultiplex operations.

7. Digital transmission gives complete freedom to multiplex digital data, voice, video, facsimile, etc., on the same facility, whereas analog trans-mission does not.

8. Digital systems are more efficient than analog in the transmission of digital data in that fewer voice channels must be displaced to obtain a given digital capacity.

9. Digital transmission provides the most economically possible interface to digital switching systems. Analog systems, on the other hand, require full demultiplexing/remultiplexing at switching nodes. Digital switches have inlets/outlets at the digital multiplex rates.

10. Fiber-optics transmission systems tend to favor digital transmission in that its attenuation is relatively independent of frequency. This makes bandwidth expansion transmission techniques to reduce baseband noise particularly attractive since the added bandwidth on fiber-optic systems is almost "free," and PCM is one of the most efficient of such schemes. In addition, light sources used in optical transmission exhibit nonlinearities that make them better suited to nonlinear modulation techniques such as PCM than to linear modulation methods such as AM.

11. Signaling is digital. Signaling on analog systems has to be converted to something compatible, such as a tone or multitone format. On digital systems, only a bit has to be changed in state for supervisory signaling and a bit sequence for address signaling (SSN No. 7). A digital system is also compatible with digital processors used in SPC switches.

Disadvantages

1. Bit errors accumulate across a digital system. These are not recoverable unless we resort to an error-correction system such as ARQ or FEC, both of which require still additional bandwidth.

2. System timing is a major issue and is discussed in the following chapter.

3. Although digital terminals tend to be less expensive than their analog counterparts on a per-channel basis, digital transmission lines (metallic media) tend to be more expensive than their analog counterparts. With the cost of fiber-optic systems dropping, digital transmission on fiber is less expensive than on metallic media [26].

REVIEW QUESTIONS

1. Give the three principal advantages of digital transmission.

2. What are the three basic steps in the development of a PCM signal from an analog source, typically voice?

3. Following the Nyquist sampling theorem, what is the sampling rate of a 4-kHz voice channel? Of a 7.5-kHz program channel? Of a 4.2-MHz video channel?

4. What is the polarity of a mark (1) and of a space (0) in AMI (bipolar) transmission?

5. For a 24-channel PCM system, calculate the period of one frame.

6. Define *quantization distortion.*

7. If the number of quantization steps is doubled in a particular PCM design using linear quantization, what improvement in quantization distortion (noise) is achieved? Express the answer in decibels.

8. Why is it desirable to reduce the size (number of bits or elements) of a PCM code word as much as possible and yet maintain reasonable voice quality?

9. There are only 16 quantization steps in a particular PCM system. What minimum length (bits) code word is required to accommodate these 16 discrete levels?

10. Identify the two distinct logarithmic companding laws used in modern PCM systems.

11. What key piece of information does the first significant bit in a PCM code word tell us?

12. Derive the value 1.544 Mbps in accordance with DS1 format.

13. Name at least five differences between European and North American PCM systems.

14. How is supervisory signaling carried out in the North American PCM system? In the European system? Argue pros and cons of each approach.

15. In the earliest implementations of PCM, where was it applied and why? (This does not imply that today it is not applied for the same reason in various situations.)

16. In the North American T1/DS1 system, what is the repeater spacing in feet? Going out from a switching node, why is there always a half-section rather than a full repeater section? For extra credit, what additional benefit can we get from that particular repeater spacing distance? (*Hint*: Turn back to subscriber/trunk loop design.)

17. Give at least two reasons why AMI (bipolar) waveform is used on T1/ T1C (DS1/DS1C) cable transmission systems.

18. In a simplified functional block diagram of a regenerative repeater there are three basic functional blocks. Name and describe the function of each.

19. Describe BNZS. Why do we use it? What would B3ZS be?

20. How do we derive "enhancements" to a DS1 PCM system? Talk about framing bits. (I tell my classes: "Do you mean that I have to tell you 8000 times a second where a frame begins?")

21. In the DS1 extended superframe, what are the two basic enhancements (besides frame alignment, which we have had right along)?

22. Of what use is stuffing (justification) in higher-order PCM multiplex systems?

23. Jitter accumulation, a major impairment on long-distance digital transmission, is a function of what?

24. Besides jitter, give at least two other transmission impairments for digital transmission.

25. SONET, synchronous optical network, implies transmission on optical fiber. Can the SONET format be transmitted on other media? What would be the principal constraint, say, for transmitting SONET format on digital LOS radio?

26. Compare the first level digital line rate of SONET and SDH. The only concern in this question is the transmission rate in bps.

27. What is the payload capacity of STS-1?

28. What is the primary purpose of the payload pointer? Describe another benefit we get from the payload pointer.

29. How do we derive other line rates from STM-1?

30. Comment on the superiority of DS1 over E1, or the superiority of E1 over DS1.

31. If PCM is performance limited by the number of regenerative repeaters in tandem, then how can we get around this problem?

32. If the fact that noise does not accumulate on a digital network, and we see that as the principal advantage, then what is the principal disadvantage of, say, PCM transmission? (*Hint*: It deals with accumulation.)

33. Give two primary advantages of delta modulation and one principal disadvantage.

34. What causes slope overload in a delta modulator?

35. What is the advantage of CVSD compared to conventional delta modulation?

36. Why is the Nyquist sampling rate exceeded in delta/CVSD systems?

REFERENCES

1. *Reference Data for Radio Engineers*, 6th ed., ITT/Howard W. Sams, Indianapolis, 1976.

2. GTE Lenkurt Demodulator, *PCM Update, Parts 1 and 2*, GTE-Lenkurt Electric Company, San Carlos, CA, February 1975.

3. *Transmission Systems for Communications*, 5th ed., Bell Telephone Laboratories, Holmdel, NJ, 1982.

4. J. Bellamy, *Digital Telephony*, 2nd ed., John Wiley & Sons, New York, 1991.

5. D. R. Smith, *Digital Transmission Systems*, 2nd ed., Van Nostrand Reinhold, New York, 1993.

6. *Pulse Code Modulation (PCM) of Voice Frequencies*, CCITT Rec. G.711, Fascicle III.4, IXth Plenary Assembly, Melbourne, 1988.

7. *Digital Channel Bank—Requirements and Objectives*, Bell System Technical Reference, Publication 43801, American Telephone & Telegraph Co., Basking Ridge, NJ, 1982.

8. *Physical/Electrical Characteristics of Hierarchical Digital Interfaces*, CCITT Rec. G.703, ITU, Geneva, 1991.

9. *Synchronous Frame Structures Used at Primary and Secondary Hierarchical Levels*, CCITT Rec. G.704, ITU Geneva, 1991.

10. R. L. Freeman, *Reference Manual for Telecommunications Engineering*, 2nd ed., John Wiley & Sons, New York, 1994.

11. R. L. Freeman, *Telecommunication Transmission Handbook*, 3rd ed., John Wiley & Sons, New York, 1991.

12. *Second Order Digital Multiplex Equipment Operating at 8448 kbps Using Positive/Zero/Negative Justification*, CCITT Rec. G.745, Fascicle III.4, IXth Plenary Assembly, Melbourne, 1988.

13. *Functional Criteria for Digital Loop Carrier Systems*, Bellcore Technical Reference TR-NWT-000057, Issue 2, Bellcore, Piscataway, NJ, 1993.

14. *Synchronous Optical Network (SONET) Transport Systems: Common Generic Criteria*, Bellcore GR-253-CORE, Issue 1, Bellcore, Piscataway, NJ, December 1994.

15. *BOC Notes on the LEC Networks—1994*, Bellcore Special Report SR-TSV-002275, Issue 2, Piscataway, NJ, April 1994.

16. "Introduction to SONET," a Hewlett-Packard Seminar, Burlington, MA, 1993.

17. *Synchronous Digital Hierarchy Bit Rates*, ITU-T Rec. G.707, ITU Helsinki, March 1993.

18. *Network Node Interface for Synchronous Digital Hierarchy*, ITU-T Rec. G.708, ITU-T Helsinki, March 1993.

19. *Digital Hierarchy Bit Rates*, CCITT Rec. G.702, Fascicle III.4, IXth Plenary Assembly, Melbourne, 1988.

20. *Synchronous Multiplexing Structure*, ITU-T Rec. G.709, ITU-T Helsinki, March 1993.

21. *Tri-Tac Digital Group Multiplex (DGM) Equipments: Characteristics, Specifications and System Interfaces*, Raytheon Company brochure, Raytheon Co., Marlborough, MA, 1987.

22. MIL-STD-188-200, Tactical Communication Systems, U.S. Department of Defense, Washington, DC, June 1983.

23. R. L. Freeman, "An Overview of Digital Transmission and Multiplexing," a tutorial presentation to the MITRE Institute, Bedford, MA, October 1987.

24. *The New IEEE Standard Dictionary of Electrical and Electronic Terms*, 5th ed., IEEE Std. 100–1992, IEEE Press, New York, 1992.

25. K. W. Catermole, *Principles of Pulse Code Modulation*, Illiffe, London, UK, 1969.

26. *Telecommunications Transmission Engineering*, ATT–Western Electric Co., Winston-Salem, NC, 1977.

9

DIGITAL SWITCHING AND NETWORKS

1 INTRODUCTION

In Chapter 3 we dealt with analog space-division switching in which a metallic path is set up between calling and called subscriber. "Space division" in this context refers to the fact that speech paths are physically separated (in space). Figure 9.1A illustrates this concept by showing a representative cross-point matrix. Time-division switching (Figure 9.1B) permits a single common metallic path to be used by many calls separated one from the other in the time domain. In this context, with time-division switching the speech or other information to be switched is digital in nature, either PCM or delta modulation (DM). Samples of each telephone call are assigned time slots, as described in Chapter 8. PCM or DM switching involves the distribution of these slots in sequence to the desired destination port(s) of the switch. Internal functional connectivities in the switch are carried out by digital "highways." A highway consists of sequential speech path time slots.

This chapter describes PCM switching in a simplified step-by-step fashion. It covers generic switch architectures and explains basic functional operations. This is followed by a walk-through of two of the most popular telephone switches: the Northern Telecom DMS-100 family of switches and the AT&T 5ESS.* We will discuss the trend toward higher internal bit rates in the advanced versions of these switches. This is followed by an overview of digital networks and their topologies, where the switch is probably the most important element. The discussion of digital networks includes network synchronization and timing, network performance, and digital network impairments such as jitter, wander, and slips.‡

---

*5ESS is an AT&T trademark.
‡Ref. 13 is useful for definitions of some of the terms given in this chapter.

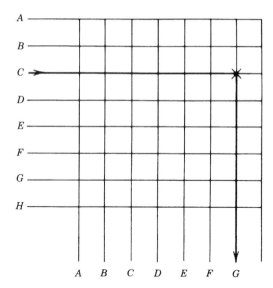

Figure 9.1A. A space-division switch showing connectivity from user C to user G.

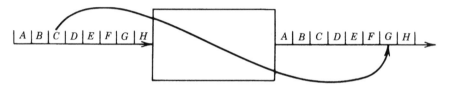

Figure 9.1B. A time-division switch which is a time-slot interchanger (TSI). Connectivity is from user C (in incoming time slot C) to user G (in outgoing time slot G).

2 ADVANTAGES AND ISSUES OF PCM SWITCHING

There are both economic and technical advantages to digital switching; in this context we refer to PCM switching (or course, most of these same arguments hold for delta/CVSD switching as well). The economic advantages of time-division PCM switching include the following:

- There are notably fewer equivalent cross-points for a given number of lines and trunks than in a space-division switch.
- A PCM switch is of considerably smaller size.
- It has more common circuitry (i.e., common modules).
- It is easier to achieve full availability within economic constraints.

The technical advantages include the following:

- It is regenerative (i.e., the switch does not distort the signal; in fact, the output signal is "cleaner" than the input).
- It is noise-resistant.
- It is computer-based and thus incorporates all the advantages of SPC.
- The binary message format is compatible with digital computers. It is also compatible with signaling.
- A digital exchange is lossless. There is no insertion loss as a result of a switch inserted in the network.
- It exploits the continuing cost erosion of digital logic and memory; LSI, VLSI, and VHSIC* insertion.

Two technical issues may be listed as disadvantages:

- A digital switch deteriorates error performance of the system. A well-designed switch may only impact network error performance minimally, but it still does it.
- Switch and network synchronization and the reduction of wander and jitter can be gating issues in system design.

3 APPROACHES TO PCM SWITCHING

3.1 General

A digital switch's architecture is made up of two elements called T and S for time-division switching (T) and space-division switching (S) and can be made up of sequences of T and S. For example, the AT&T No. 4 ESS is a TSSSST switch, No. 3 EAX is an SSTSS, and the classical Northern Telecom DMS-100 is TSTS-folded. Many of these switches (e.g., DMS-100) are still available.

One thing in common with these switches is that they had multiple space (S) stages. This has now changed. Many of the new switches or enhanced versions of the switches just mentioned have very large capacities (e.g., 100,000 lines) and are simply TST or STS switches.

We will describe a simple time switch, a space switch, and methods of making up an architecture combining T and S stages. We will show that designing a switch with fairly high line and trunk capacity requires multiple stages. Then we will discuss the "new look" at the time stage.

3.2 Time Switch

In a most simplified way, Figure 9.1B is a time-switch or time-slot interchange (TSI). From Chapter 8 we know that a time slot in conventional

*VHSIC = Very high speed integrated circuit.

PCM contains 8 bits and that a basic frame is 125 μs in duration. For the North American DS1 format the basic frame contains 24 time slots, and for the European E1 it has 32 time slots. The time duration of an 8-bit time slot in each case is $125/24 = 5.2083$ μs for the DSI case and $125/32 = 3.906$ μs for the E1 case. Time-slot interchanging involves moving the data contained in each time slot from the incoming bit stream to an outgoing bit stream but with a different time-slot arrangement in accordance with the destination of each time slot. What is done, of course, is to generate a new frame for transmission at the appropriate switch outlet.

Obviously, to accomplish this, at least one time slot must be stored in memory (write) and then called out of memory in a changed position (read). The operations must be controlled in some manner, and some of these control actions must be kept in memory together with the software managing such actions. Typical control functions are time-slot "idle" or "busy." Now we can identify three of the basic functional blocks of a time switch:

1. Memory for speech
2. Memory for control
3. Time-slot counter or processor

These three blocks are shown in Figure 9.2. There are two choices in handling the time switch: (1) sequential write, random read as shown in

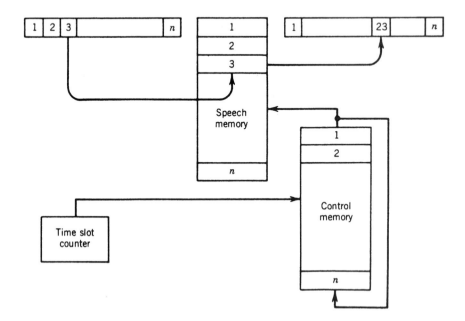

Figure 9.2A. Time-slot interchange: time switch (T). Sequential write, random read.

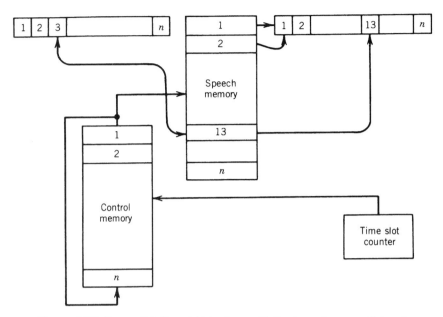

Figure 9.2B. Time-switch, time-slot interchange (T). Random write, sequential read.

Figure 9.2A and (2) the reverse, namely, random write, sequential read. In the first case, sequential write, the time slots are written into the speech memory as they appear in the *incoming* bit stream. For the second case, random write, the incoming time slots are written into memory in the order of appearance in the *outgoing* bit stream. This means that the incoming time slots are written into memory in the desired *output* order. The writing of incoming time slots into the speech memory can be controlled by a simple time-slot counter and can be sequential (e.g., in the order in which they appear in the incoming bit stream, Figure 9.2A). The readout of the speech memory is controlled by the control memory. In this case the readout is random where the time slots are read out in the desired output order. The memory has as many cells as there are time slots. For the DS1 example, there would be 24 cells. This time switch, as shown, works well for a single inlet–outlet switch. With just 24 cells it could handle 23 stations besides the calling subscriber, not an auspicious number.

How can we increase a switch's capacity? Enter the space switch (S). Figure 9.3 affords a simple illustration of this concept. For example, time slot B_1 on the B trunk is moved to the Z trunk into time slot Z_1, and time slot C_n is moved to trunk W into time slot W_n. However, we see that there is no change in time-slot position.

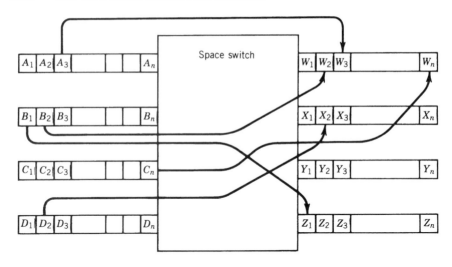

Figure 9.3. Space switch connects time slots in a spatial configuration.

3.3 Space Switch

A typical time-division space switch is shown in Figure 9.4. It consists of a cross-point matrix made up of logic gates that allow the switching of time

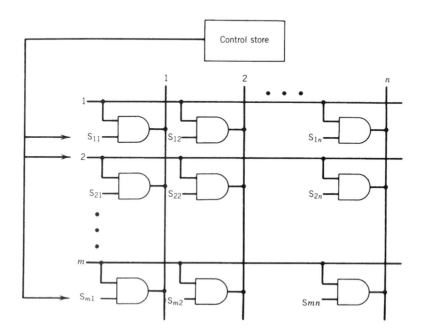

Figure 9.4. Time-division space switch cross-point array showing enabling gates.

slots in the spatial domain. These PCM time-slot bit streams are organized by the switch into a pattern determined by the required network connectivity. The matrix consists of a number of input horizontals and a number of output verticals with a logic gate at each cross-point. The array, as shown in the figure, has M horizontals and N verticals, and we call it an $M \times N$ array. If $M = N$, the switch is nonblocking. If $M > N$, the switch concentrates; and if $N > M$, the switch expands.

Return to Figure 9.4. The array consists of a number of (M) input horizontals and (N) output verticals. For a given time slot, the appropriate logic gate is enabled and the time slot passes from the input horizontal to the desired output vertical. The other horizontals, each serving a different serial stream of time slots, can have the same time slot (e.g., a time slot from time slots number 1–24, 1–30, or 1–n; for instance, time slot 7 on each stream) switched into other verticals enabling their gates. In the next time-slot position (e.g., time slot 8), a completely different path configuration could occur, again allowing time slots from horizontals to be switched to selected verticals. The selection, of course, is a function of how the traffic is to be routed at that moment for calls in progress or being set up.

The space array (cross-point matrix) does not switch time slots as does a time switch (time-slot interchanger). This is because the occurrences of time slots are identical on the horizontal and on the vertical. It switches in the space domain, not in the time domain. The control memory in Figure 9.4 enables gates in accordance with its stored information.

If an array has M inputs and N outputs, M and N may be equal or unequal depending on the function of the switch on that portion of the switch. For a tandem or transit switch we would expect $M = N$. For a local switch requiring concentration and expansion, M and N would be unequal.

If, in Figure 9.4, it is desired to transmit a signal from input 1 (horizontal) to output 2 (vertical), the gate at the intersection would be activated by placing an enable signal on S_{12} during the desired time-slot period. Then the eight bits of that time slot would pass through the logic gate onto the vertical. In the same time slot, an enable signal on S_{M1} on the Mth horizontal would permit that particular time slot to pass to vertical 1. From this we can see that the maximum capacity of the array during any one time-slot interval measured in simultaneous call connections is the smaller value of M or N. For example, if the array is 20×20 and a time-slot interchanger is placed on each input (horizontal) line and the interchanger handles 30 time slots, the array then can serve $20 \times 30 = 600$ different time slots. The reader should note how the TSI (time-slot interchanger) multiplies the call-handling capability of the array when compared to its analog counterpart.

3.4 Time–Space–Time Switch

Digital switches are composed of time and space switches in any order or in time switches only. We use the letter T to designate a time-switching stage

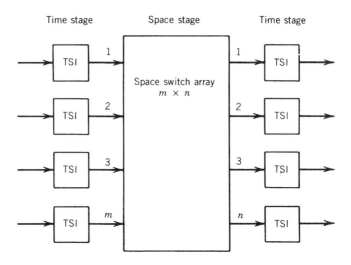

Figure 9.5. A time–space–time (TST) switch (TSI = time-slot interchanger).

and use S to designate a space-switching stage. For instance, a switch that consists of a sequence of a time-switching stage, a space-switching stage, and a time-switching stage is called a TST switch. A switch consisting of a space-switching stage, a time-switching stage, and a space-switching stage is designated an STS switch. There are other combinations of T and S. The ATT No. 4 ESS switch is an example. It is a TSSSST switch.

Figure 9.5 illustrates the time–space–time (TST) concept. The first stage of the switch is the time-slot interchanger (TSI) or time stages that interchange time slots (in the time domain) between external incoming digital channels and the subsequent space stage. The space stage provides connectivity between time stages at the input and output. It is a multiplier of call-handling capacity. The multiplier is either the value for M or value for N, whichever is smaller. We also saw earlier that space-stage time slots need not have any relation to either external incoming or outgoing time slots regarding number, numbering, or position. For instance, incoming time slot 4 can be connected to outgoing time slot 19 via space network time slot 8.

If the space stage of a TST switch is nonblocking, blocking in the overall switch occurs if there is no internal space-stage time slot during which the link from the inlet time stage and the link to the outlet time stage are both idle. The blocking probability can be minimized if the number of space-stage time slots is large. A TST switch is strictly nonblocking if

$$l = 2c - 1 \tag{9.1}$$

where l is the number of space-stage time slots and c is the number of external TDM time slots [3].

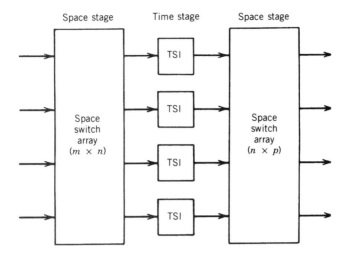

Figure 9.6. A space–time–space (STS) switch.

3.5 Space–Time–Space Switch

A space–time–space (STS) switch reverses the architecture of a TST switch. The STS switch consists of a space cross-point matrix at the input followed by an array of time-slot interchangers whose ports feed another cross-point matrix at the output. Such a switch is shown in Figure 9.6. Consider this operational example with an STS. Suppose that an incoming time slot 5 on port No. 1 must be connected to an output slot 12 at outgoing port 4. This can be accomplished by time-slot interchanger No. 1 which would switch it to time slot 12; then the outgoing space stage would place that on outgoing trunk No. 4. Alternatively, time slot 5 could be placed at the input of TSI No. 4 by the incoming space switch where it would be switched to time slot 12, thence out port No. 4.

3.6 TST Compared to STS

Both TST and STS switches can be designed with identical call-carrying capacities and blocking probabilities. It can be shown that a direct one-to-one mapping exists between time-division and space-division networks [2].

The architecture of TST switching is more complex than STS switching with space concentration. The TST switch becomes more cost-effective because time expansion can be achieved at less cost than space expansion. Such expansion is required as link utilization increases because less concentration is acceptable as utilization increases.

It would follow, then, that TST switches have a distinct implementation advantage over STS switches when a large amount of traffic must be handled. Bellamy [3] states that for small switches STS is favored due to reduced implementation complexities. The choice of a particular switch architecture may be more dependent on such factors as modularity, testability, and expandability.

One consideration that generally favors an STS implementation is the relatively simpler control requirements. However, for large switches with heavy traffic loads, the implementation advantage of the TST switch and its derivatives is dominant. A typical large switch is the ATT No. 4 ESS, which has a TSSSST architecture and has the capability of terminating 107,520 trunks with a blocking probability of 0.5% and channel occupancy of 0.7.

4   REVIEW OF SOME DIGITAL SWITCHING CONCEPTS

4.1   Early Ideas

In Section 3, the reader was probably led to believe that the elemental time-switching stage, the time-slot interchanger, would have 24 or 30 time-slot capacity to match the North American DS1 rate or the European E1 rate, respectively. That means a manufacturer would have to make two distinct switches, one to satisfy the North American market and one for the European market. Most switch manufacturers made just one switch with a common internal switching network, the time and space arrays we just discussed. For one thing, they could map 5 DS1 groups into 4 E1 groups, the common denominator being 120 DS0/E0 (64-kbps channels). Peripheral modules cleaned up any differences remaining, such as signaling. The "120" is a number used in ATT's 4ESS. It maps 120 8-bit time slots into 128 time slots. The 8 time slots of the remainder are used for diagnostic and maintenance purposes [2].

Another early concept was a common internal bit rate, to be carried on those "highways" we spoke about, or on junctors.* At the points of interface that a switch has with the outside world, it must have 8-bit time slots in DS1 (or higher) or E1 (or higher, E2, E3) frames each 125 µs in duration. Inside the switch was another matter. For instance, with the Northern Telecom's DMS-100 the incoming 8-bit time slot was mapped into a 10-bit time slot as shown in Figure 9.7. The example used in the figure is DS1.

Note in Figure 9.7 that one bit is a parity bit (bit 0) and the other appended bit (bit 1) carries the supervisory signaling information: The line is idle or busy. Bits 2 through 9 are the bits of the original 8-bit time slot. Because Northern Telecom in their DMS-100 wanted a switch that was simple to convert from E1 to DS1, they built up their internal bit rate to 2.560 Mbps as follows: 10 bits per time slot, 32 time slots × 8000 (the frame

---

*A junctor is a path connecting switching networks internal to a switch.

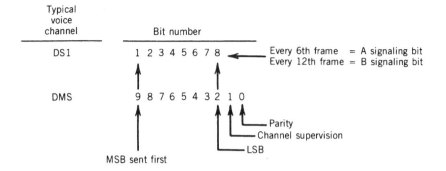

Figure 9.7. Bit mapping in the DMS-100, DS1 to DMS.

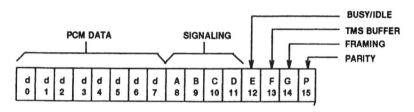

Figure 9.8. The composition of the AT&T 5ESS internal 16-bit time slot.

rate)* or 2.560 Mbps. This now accommodates E1, all 32 channels. As we mentioned, 5 DS1 are now easily mapped into 4 E1s and vice versa.

AT&T's 5ESS maps each 8-bit time slot into a 16-bit internal PCM word. It actually appends 8 additional bits onto the 8-bit PCM word as shown in Figure 9.8.

4.2   Higher-Level Multiplex Structures Internal to a Digital Switch

We pictured a simple time-slot interchanger switch with 24 8-bit time slots to satisfy DS1 requirements. It would meet the needs of a population of 24 subscribers without blocking. There is no reason why we couldn't build a TSI with greater capacity. Suppose that instead of the DS1 rate, we built the TSI with a DS3 rate. The basic TSI then could handle 672 subscribers (i.e., 672 time slots). If we built a concentrator in front of it for 4:1 concentration, then it could handle 4×672 or 2688 subscribers. But note an important aspect, the gating time per channel. In a DS3 frame there are 672 time slots available for traffic. Each time slot is visited once for each frame, and thus the time of dwell is $125 \ \mu s/672 = 186$ ns. This is still inside the state of the art. But as we

---

*The 8000 frames per second or frame rate is common on all conventional PCM systems. As the reader will recall from Chapter 8, this is the Nyquist sampling rate for the 4-kHz analog voice channel on converting it to a PCM equivalent.

increase the bit rate, that gating time per time slot can get very small indeed. A 50,000 time-slot switch would have only 2.5-ns gating time.

The AT&T 5ESS is a TST switch. It has the capacity for 100,000 or more lines. They are able to accomplish this simpler architecture by using larger-capacity time-slot interchangers (TSIs) and higher bit rates on the space stage. A 5ESS TSI handles 512 time slots.* However, each TSI port has an incoming/outgoing time-slot rate of 256 time slots. Two ports are required (in one direction) to handle the 512 time slots: one for odd-numbered channels and one for even-numbered channels. Thus the bit rate at a TSI port is $256 \times 16 \times 8000 = 32.768$ Mbps. This odd-channel, even-channel arrangement carries through the entire switching fabric with each port handling 256 time slots or 32.768 Mbps.

Another example of a widely implemented modern digital switch is the Northern Telecom DMS-100 with supernode/ENET. They modified the older DMS100 conventional switch, which had a TSTS-folded architecture. Like the 5ESS, they also moved into the 2048-time-slot domain in the ENET (extended network). But their time slot is 10 bits, and the ENET uses a 10-bit parallel format, so each line (i.e., there are 10 lines) has 2048×8000 or 16.384 Mbps.

4.3   An Overview of the AT&T 5ESS

The AT&T 5ESS latest version is the 5ESS-2000, which can serve as a local switch or a combined local/tandem switch. It can also serve as an equal-access tandem switch. Depending on the software release, the 5ESS-2000 maximum capacity is from 300,000 to 600,000 call completions per hour. If we incorporate 10:1 concentration, the 5ESS line unit (LU) can handle 1700 ccs for an ABS (average busy season) blocking probability of 1.5%. A line unit handles 64 channels. Thus, with 10:1 concentration an LU can handle 640 lines and a 5ESS can have a large number of line units. The final maximum capacity of the 5ESS switch depends on how it is configured. There are variables as to whether some of the switch is devoted to tandem operation, the blocking probability specified and the concentration ratio. The user can select concentration ratios of 10, 8, 6, and 4:1. The architecture of the 5ESS is TST (time–space–time) as illustrated in Figure 9.9.

Figure 9.10 illustrates standard switching operations in the 5ESS switch. The 5ESS consists of three major modules: administrative module (not shown), communication module (CM) which contains the space-switching stage (TMS), and the switching module (SM) which contains the time-switching stages (TSI or time-slot interchange). The architecture is TST because time slots pass through the TSI twice.

---

*Remember that a time slot here has 16 bits. See Figure 9.8 for a time-slot layout.

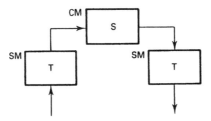

Figure 9.9. Simplified functional block diagram showing the TST architecture of a 5ESS switch. CM, communication module; SM, switching module.

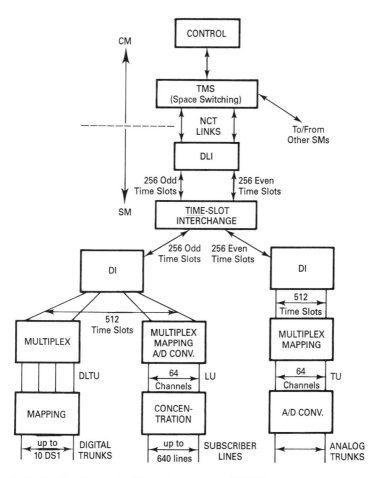

Figure 9.10. Functional/conceptual block diagram of the AT&T 5ESS switch. Only one SM (switching module) is shown. CM, communications module; TU, trunk unit; LU, line unit; DLTU, digital line and trunk unit; DI, data interface; DLI, dual link interface; NCT, network control and timing; TMS, time-multiplexed switch.

415

4.3.1 Communications Path. Each SM of the 5ESS contains a number of interface units shown at the bottom of Figure 9.10. These interface units (IU) provide the actual physical terminations of lines and trunks. For the case of a typical line-to-line POTS* call, each interface unit contains a space-division concentrator which provides up to 640 analog lines with access to 64 digital, time-multiplexed time slots. Each time slot contains 16 data bits, 8 of which are the PCM code for the analog signal which is sampled 8000 times per second. These 64 time slots are organized into groups of 32 on two PIDBs (peripheral interface data buses). Each PIDB carries a serial data stream made up of frames of 32 data words repeated 1/8000 s (125 μs). The PIDBs from all interface units in an SM converge on the peripheral side of the TSI.

The TSI has the ability to reorder any of the 512 time slots arriving at its peripheral side to any of the 512 time slots leaving its network side (out of the SM) for transmission to another SM. This time-switching property allows the calling and called parties to occupy different time slots on the peripheral side of their respective SMs (for intra-SM calls, the time slots are simply transferred from one PIDB to another through the TSI via a similar process).

The network side of the TSI is connected to the NCT† fiber-optic links which carry the time slots to another SM via the CM (actually the TMS in the CM). One NCT link carries the 256 odd time slots, and the other carries the 256 even time slots. The CM also provides the timing interface for switching center synchronization.

The pairs of NCT links from each SM terminate on the TMS. The TMS, the space switch, has the ability to connect time slots (odd-to-odd, even-to-even) from one SM to another via electronic cross-points. The space switch can be configured differently for each of the 512 time slots. This is the way full interconnection can be accomplished between SMs.

At the terminating SM, the TSI reorders the arriving NCT time slots for proper sequence as they are distributed to the PIDBs. The remote interface unit then demultiplexes the PIDB time slots, performs digital-to-analog conversion, and completes the path to the remote termination.

At the bottom of Figure 9.10 the digital lines and trunks enter (and leave) the switch. The DLTU2 provides inlet/outlet ports for up to 20 DS1 trunks. It can also interface with E1 trunks. It accepts the standard 8-bit time slot and maps that into a 16-bit time slot as shown in Figure 9.8. It then multiplexes the result into 32 16-bit time slots. The aggregate 512 time slots is fed to the DI (digital interface), where it is reduced to 256 time slots, using a 2:1 concentrator, for delivery to the TSI.

The line unit (LU), depending on the concentration ratio, can terminate up to 640 subscriber loops as mentioned above. It concentrates with concentration ratios selectable up to 10:1. It then performs A/D (analog-to-

---

*POTS, meaning "plain old telephone service," is a term used commonly in the North American telephone world.
†NCT stands for network control and timing.

digital) conversion as any PCM channel bank (codec) would. The remainder of the processing, switching, and transmission is similar to the DLTU.

The NCT links connecting the SM to the CM use fiber-optic cable as the transmission medium. Each link uses an LED transmitter and a PIN receiver. The bit rate on these NCT links carrying 256 time slots is (16 bits per slot) $\times$ (256 slots) $\times$ 8000 = 32.768 Mbps.

The interface units and TSI are controlled by the SMPU (switching module processing unit), and the TMS is controlled via messages from the AM/CMP processor. All switching elements and links have a receive half and a transmit half that operate simultaneously.

4.3.2 Control Path. Call processing and many other 5ESS switch/5ESS-2000 switch functions require coordinated activities between SM processors and between the AM/CMP (administrative module/communications module processor) and SM processors. To facilitate this coordination, special message channels are provided for interprocessor communication. These channels, which share the call-carrying inter-SM links, are designed to transmit packets of information within X.25 level 2 frames. This technique is consistent with the advanced, process-oriented nature of the 5ESS switch/5ESS-2000 switch operational software and supports the loosely coupled nature of the distributed processing complex.

To avoid using a dedicated control bus for SMs, the 5ESS switch/5ESS-2000 switch uses one time slot of each NCT link to carry control information. The control time slot is fixed by hardware strapping for each SM and cannot be used for call data. The control time slots transport bi-level messages between SMs and between SMs and the AM/CMP.

Section 4.3 is based on AT&T *5ESS Switch and 5ESS = 2000 Switch System Description*, AT&T 235-100-125, Issue 7, November 1994 (Ref. 7).

4.4   An Overview of the NTI DMS-100 with Supernode/ENET

Northern Telecom uses the same sort of thinking on the DMS-100 supernode/ENET* switch family that AT&T employed on their 5ESS-2000 switch family. Why not use higher-capacity time switches? NTI migrated from DS-30 links to DS-512 links, and from a 512-time-slot bus to a 2048-time-slot bus. The final product, the ENET, is a single-stage time/space switch. This is a single-stage switch that accomplishes both time-slot interchanging and space switching in one stage. Its practical maximum capacity is 128,000 lines in a two-cabinet configuration with full redundancy (to improve availability). The switch is nonblocking.

*ENET stands for enhanced network.

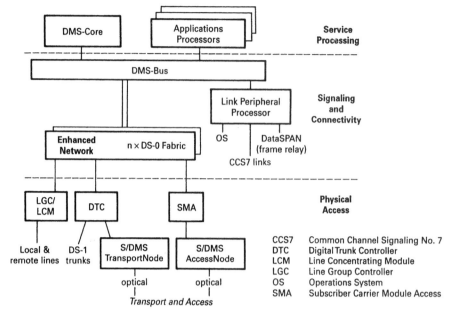

Figure 9.11. ENET's place in the DMS-100 superNode signaling and connectivity layer. Courtesy of Northern Telecom.

The new DMS-100 with ENET uses a standard internal link bit rate, time slot and frame structure. It is based on 512 time slots where each time slot has 10 bits; by applying the Nyquist sampling rate of 8000/s, the bit rate is $512 \times 10 \times 8000 = 40.960$ Mbps.

Internal to the ENET, each group of four 512-channel serial links is multiplexed to a single 2048-channel parallel bus 10 bits wide running at a bus rate of 16,384 MHz (i.e., 2048×8000).

Figure 9.11 is a functional block diagram of a DMS-100 with supernode, equipped with ENET. Figure 9.12 shows the ENET architecture. Figure 9.13 is a mechanical overview of the crosspoint matrix, H for horizontal and V for vertical. Figure 9.14 is a functional block diagram of a two-bus by two-bus matrix where each horizontal bus is equipped with two time switches.

The ENET, which is the revolutionary key to the DMS-100, consists of a 64×64 matrix. There are 64 rows, each with 64 time switches. These have the ability to connect, time slot by time slot, to 64 columns. The 64 horizontal buses have access to every time switch on that row, but one at a time. This selection is determined by a control processor. Whereas with conventional digital switches, incoming time slots are stored in N memories, with the DMS-100 ENET, this information is stored in N^2 memories. In this case, there are 64×64 or 4096 memories. For a specific connectivity, time slots related to a particular call are available at the time switches on each row.

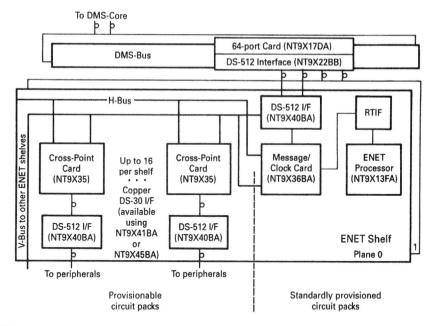

Figure 9.12. DMS-100 ENET architecture. Courtesy of Northern Telecom, "Enhanced Network," Figure 4, page 10 (Ref. 8).

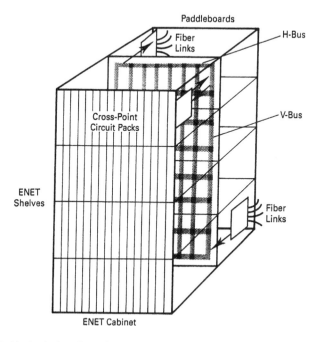

Figure 9.13. Mechanical outline of key elements of ENET single-stage switching. Courtesy of Northern Telecom.

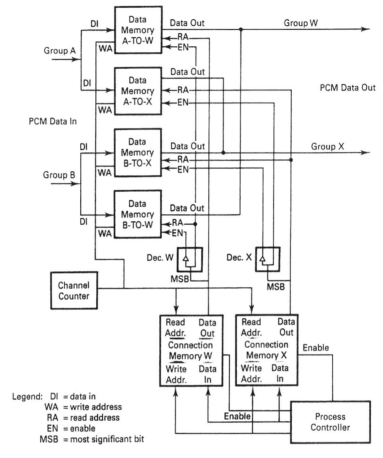

Figure 9.14. Block diagram of a sample ENET 2 × 2 switch matrix and time switch. Courtesy of Northern Telecom. From NTI patent 4,470,139, Fig. 3 (Ref. 10).

Each row carries 2048 channels deriving a theoretical maximum capability for ENET of 64 × 2048 or 131,072 channels (time slots*).

Turn now to Figure 9.14, which shows a very simplified block diagram of an ENET 2 × 2 matrix where serial 10-bit time slots enter in two streams, which we call Group A and Group B. These time slots appear in two outgoing streams, which we call Group W and Group X, rearranged in time and in space depending on their destinations. Note the duplication in data memories. In this example there are only two incoming bit streams and two outgoing bit streams and four data memories are required. If there were four incoming/outgoing bit streams, 16 data memories would be required. For N streams in/out, N^2 memories would be required.

---

*A time slot represents a virtual channel, which we have shortened to just "channel." In the context of these paragraphs, the meaning of "channel" and "time slot" can be interchanged.

The incoming bit streams Group A and Group B contain PCM frames of 2048 time slots. The outgoing bit streams also contain PCM frames of 2048 time slots. The incoming Group A bus applies its 2048 time slots, sequentially and in parallel to both memories A-to-W and A-to-X. Likewise, incoming Group B bus carries its 2048 time slots and applies them to both memories B-to-X and B-to-W. Each data memory has a 4096-time-slot memory, where each time slot has 10 bits.

Note the similarity with the time switching in Figures 9.2A and 9.2B; there is a time-slot counter and control memory, and in this case we have a channel counter and process controller. In other words, time switching or time-slot interchanging is carried out with the ENET. In this simplified ENET switch, the Group W or Group X output path can be selected for a particular time slot.

Consider the following example. Assume that a time slot on the Group A input bus is to be connected to a time slot on the Group W output bus. As data from all time slots on the Group A bus are stored in the A-to-W data memory (at least temporarily), the outgoing Group W bus can receive the data from the required time slot, as stored in data memory A-to-W, under the control of connection memory W. In a similar manner, if a time slot on the Group A incoming bus is to be connected to a time slot on the Group X outgoing bus, data memory A-to-X is accessed, under control of connection memory X. Since data memory A-to-X stores (at least temporarily) data from all the time slots of the Group A incoming bus, all the data from those time slots are available to be applied to the Group X outgoing bus as required.

In a similar manner, any time slot on incoming Group B bus can be applied to outgoing Group X bus (via data memory B-to-X), and any time slot on the Group B incoming bus can be applied to outgoing Group W bus (via data memory B-to-W). As a result, the Group W outgoing bus has access to any one of the 2048 time slots received on the Group A incoming bus and to any one of the 2048 time slots received on the Group B incoming bus. Thus this same time-switching stage can also switch in the space domain.

The switching control in Figure 9.14 resides in the lower four blocks in the drawing: channel counter, connection memory W, connection memory X, and the process controller. The control is supported by blocks Dec. W (decoder W) and Dec. X (decoder X) which actuate enablers abbreviated EN.

The channel counter (i.e., time-slot counter) provides an 11-bit address at its output which feeds the four data memories and the two connection memories, as shown. It counts in increasing order from 0 to 2047 in step with the incoming time slots, and then repeating its count over and over again. Each one of the four data memories is written sequentially as the data appear on the respective data input (DI). The address (memory location) at which the data are stored is, of course, defined by the binary address from the channel counter and is applied, via the address bus (i.e., the output of the channel counter) to the respective WA (write address) inputs. A specific PCM time slot, from a given channel, is stored in two data memories. For example, a PCM time slot from a channel on bus Group A is stored in both data memories A-to-W and A-to-X at an address corresponding to the incoming channel number.

There are four data out ports, one from each data memory. Each port provides an output in response to both an enable signal on the respective EN and to a read address applied to the respective read address input RA of each of the four data memories. The read address derives from the two connection memories W and X. In response to the address provided on the address bus (i.e. , the output of the channel counter), connection memory W produces on its "data out" terminal a 12-bit address indicative of which channel on either Group A or Group B (as stored in data memories A-to-W and B-to-W) is to be connected to the current channel on the output bus Group W. It should be noted that the MSB (most significant bit) on the "data out" terminal on the connection memory W is used to determine which one of the data memories, A-to-W or B-to-W, will be enabled (via the enable input EN of each data memory), and the remaining 11 bits on the "data out" bus of connection memory W address a storage location in the enabled memory (i.e., either data memory A-to-W or B-to-W) containing data from a specific channel (time slot) from either Group A or Group B input buses; these data are then read and applied to data out(put) DO of the enabled data memory (i.e., either data memory A-to-W or B-to-W). We can follow through with the same reasoning for connection memory X which deals with data memories A-to-X and B-to-X.

The process controller in the lower right of Figure 9.14 functions in a similar manner to that of the control memory in Figures 9.2A and 9.2B. Write address bus output entering from the control processor defines a location in connection memory W (or connection memory X) corresponding to a time slot (i.e., a channel) on bus Group W (or bus Group X). The process controller output to "data in" in the connection memories W and X, is stored in these memories as data which consists of the addresses of the time slots (channels deriving from input buses Group A and Group B to be connected to the output bus Group W or Group X). These addresses are, of course, the addresses of locations in the four data memories (A-to-W, A-to-X, etc.) into which data are carried by time slots on bus Group A (input) and bus Group B (input).

As a consequence, outgoing bus Group A can convey data from up to 2048 channels (time slots) originating from the 4096 time slots (channels) on both buses Group A and Group B. In a similar fashion, outgoing bus Group X can convey data from up to 2048 time slots (channels) originating from the 4096 time slots (channels) on both input bus Group A and input bus Group B.

This in a nutshell provides a description of how just one switching stage can switch in both time and space without blocking.

Return to Figure 9.11. The lower portion of this figure shows the "physical access." It is this portion of the DMS-100 that prepares the various lines and trunks for processing in the "signaling and connectivity layer." The physical access consists of peripheral modules (PMs). Each PM performs two basic functions:

1. It converts voice and data signals into the digital format used by the DMS-100 supernode/ENET system.
2. It performs low-level processing tasks to relieve the DMS-Core of routine processing chores such as loop supervision, loop loss control, and transmission of ringing and tones to telephone equipment.

One such PM is the line concentrating module (LCM) which can support up to 640 line cards (i.e., 640 subscriber loops) and provide various preset line concentrating ratios. The LCM converts subscriber line analog signals to the standard PCM format (either DS1 or E1). Before entering the supernode/ENET, the output of an LCM is fed to a line group controller (LGC). This unit converts the 24-channel PCM format to a 32-channel format.

There is a trunk module (TM) peripheral and a digital trunk controller (DTC). The TM handles analog trunks, converting them to the standard digital format; the DTC handles digital trunks. An important function of all the PMs is to convert from an 8-bit format to the standard 10-bit DMS-100 format. Of course, for the ENET operation, digital channels must be multiplexed to the 2048-time-slot PCM frame.

Section 4.4 is based on Refs. 8, 9, and 10 and private communication with NTI design engineers [11].

4.5 Remote Switching Capabilities

Both the AT&T 5ESS and the NTI DMS-100 offer remote switching capabilities in several configurations. Remote switching provides a number of advantages:

• It can serve as a CDO (community dial office) where a full-blown switch would not be justified.
• It can dramatically extend the operational area of a switch.
• Variants of remote switching can bring conventional telephone service to rural areas that might not otherwise have this service.
• It can bring digital service that much closer to the subscriber, discussed in Section 5.2 below.

4.5.1 *Remote Switching Capabilities with AT&T 5ESS.* Three of the 5ESS remote operation options are described below. These are:

- RSM (remote switching module)
- ORM (optical remote module)
- Extended switch module-2000

The RSM provides the switching capabilities of the 5ESS switch/5ESS-2000 switch to areas that would not economically support a 5ESS switch. The RSM offers custom calling features to areas normally served by community dial offices (CDOs) or small wire centers. The RSM may also be used to provide toll and operator assistance services in these areas. It can have an independent or shared exchange code (numbering plan office code designation) with its host switch.

An RSM offers capabilities to terminate analog lines and pair gain systems such as the SLC carrier system and provides all the features of the host SM. It is a zero-resistance point for the loop plant and can switch intra-RSM local calls. The host-to-RSM umbilical usually consists of DS1 or E1 configurations on wire-pair, digital radio, or fiber-optic cable. The RSM can be located up to 104 miles from its host switch. The primary distance-limiting factor is echo delay. When an SLC carrier system is hosted by an RSM, the total distance may be increased by 82 miles (i.e., SLC distance plus RSM distance to host). When fiber-optic cable is used by the umbilical, a DS3 configuration is preferred.

An ORM is an SM that is remoted from the 5ESS switch/5ESS-2000 switch via DS3 transmission facilities. The ORM can be located up to 100 miles from the host 5ESS switch over standard transmission facilities. Similar extended distance rules hold as with the RSM described above.

The ORM (optical remote module) provides the same services provided by an SM at the host 5ESS switch at a geographically different location. It is similar in function to the RSM, but it requires no host SM or any of the RSM-associated interface hardware. The ORM is connected to the CM via a DS3 interface.

Each DS3 channel carries 256 time slots (as configured for this application). This is one type of NCT link equivalent information. Therefore, each ORM requires four DS3 channels, two active and two standby. These links carry control information between the ORM and the host 5ESS switch as well as the call traffic on inter-SM calls.

The extended switch module-2000 (EXM-2000) is an optically remote SM-2000 somewhat analogous to the ORM. It has an optional stand-alone capability and an optional stratum 3 clock capability (stratum clocks are discussed below in Section 5.5) and is remoted from a CM2 (an improved communication module) using QLIs (quad link interfaces) and NCT link-formatted optical fiber. The EXM-2000 accesses the functions provided by the local classic SMs during normal operations. It supports analog lines, digital trunks for low-speed interfaces, and a SONET high-speed interface [7].

4.5.2 *Remote Switching Capabilities with NTI DMS-100/ENET.* The
DMS-100 has essentially four types of remote switching/concentration units:

1. Remote line concentrating module (RLCM) supporting up to 640 lines.
2. Outside plant module (OPM) for outside plant deployment, supporting
   up to 640 lines.
3. Remote switching center (RSC), supporting up to 7600 lines and
   distances greater than 100 mi (160 km).
4. Remote switching center-SONET (RSC-S), supporting up to 12,000
   lines.

The RLCM is similar to the DMS-100 internal LCM (line concentrating
module). It provides full-feature transparency with the host DMS-100 switch,
including digital multiplexing, forming groups of DS1 (T1) in North America
and E1 in countries under European hegemony. Calls between subscribers
connected to an RLCM have to be returned to the host DMS-100 for
switching.

The OPM is an outside plant version of the RLCM and is used in situations that require outdoor placement. An OPM can support up to 640
subscriber lines (loops).

An RSC is used for medium-size remote switching applications. It delivers
the full complement of DMS-100 features such as Meridian Centrex, data,
and ACD (automatic call distribution). Based on 3 ccs (0.083 erlangs) per
line, an RSC can support 3800 subscriber lines when using a single remote
cluster controller (RCC). With a dual RCC, an RSC can support 7600
subscriber lines. The RSC has full satellite switch capabilities. It can switch
local calls. In other words, local subscribers desiring connectivity one with
another do not require the assistance of the host serving switch.

A remote switching center-SONET (RSC-S) uses a SONET on fiber
configuration for the umbilical with its host switch. Fully configured, it can
serve up to 12,000 lines and can be located up to 100 miles (160 km) from its
host switch [9].

5   THE DIGITAL NETWORK

5.1   Introduction

The North American public switched telecommunication network should be
nearly 100% digital by 1996, with some possible holdouts in the local area
with small, independent telephone companies. The interexchange (IXC)
carrier portion is nearly 100% digital today. We imagine that the world
network will be fully digital in the first decade of the 21st century. It is still
basically hierarchical, and the structure changes slowly. There are possibly
only two factors that change network structure:

1. Political
2. Technological

In the United States, certainly divestiture of the Bell System affected network structure with the formation of LECs (local exchange carriers) and IXCs. Outside of North America, the movement toward privatization of government telecommunication monopolies in one way or another will affect structure. As we mentioned in Chapter 6, there is a trend away from strict hierarchical structures, particularly in routing schemes, less so in topology.

Technology and its advances certainly may be equally or even more important than political causes. Satellite communications, we believe, brought about the move by CCITT away from any sort of international network hierarchy. International high-usage and direct routes became practical. We should not lose sight of the fact that every digital exchange has powerful computer power, permitting millisecond routing decisions for each call. This was greatly aided by the implementation of CCITT Signaling System No. 7 (Chapter 16). Another evident factor certainly is fiber-optic cable for a majority of trunk routes. It also has forced the use of geographic route diversity to improve survivability and availability. What will be the impact of ATM (Chapter 15) on the evolving changes in network structure (albeit slowly)?

We must also take into account the influence of cellular radio as an adjunct and subscriber traffic feed source to the network. PCN/PCS will be another factor. Motorola's IRIDIUM is one example. This is a worldwide coverage PCN/cellular system of 66 low earth orbit (LEO) satellites operating at L-band (1–2 GHz).

We believe that "alternative access," an expression unheard of 5 years ago, may be another important element that will bring about change in structure. The group that is best positioned to bring competitive local telephone service to the home and office is the cable TV industry. Two other groups, less well-positioned except in certain metropolitan areas, are the teleport companies and metropolitan fiber companies. Even some of the power companies are getting into the act. This causal influence factor may be classified both as political and technical.

Fiber-optic cable will more and more pervade the trunk plant. Its extension directly to the user remains a moot point. When copper pairs can no longer provide the necessary bit rate capacity (ergo bandwidth), fiber to the home or office desk may be the only alternative. Internet services, digital TV, ISDN, and new services brought about by ATM may bring the conversion of the "last mile" of local plant sooner than we ever imagined several years ago.

All of the above will affect network structure including topology. However, idealism is tempered by the reality of current investment in existing plant yet to be amortized. This forces the macroplanner to employ gradual conversion rather than to employ rapid, radical change.

Our first consideration in this section is digital extension to the subscriber or user. We then cover some of the design issues of an all-digital PSTN. Among these issues are:

- Change in profile of services that digital brings about
- New technology, specifically SONET and SDH
- Digital network performance and performance requirements
- International interface (European versus North American PCM standards)
- Signaling: CCITT Signaling System No. 7 (covered in Chapter 16)

5.2 Digital Extension to the Subscriber

The driving factors of digital extension directly to the subscriber are cost and demand. The demand is growing and the cost is dropping. We still must consider, however, the existing outside plant from the main distribution frame (MDF) at the serving switch to literally thousands of subscribers. This outside plant can represent 30–50% of the total telecommunication plant investment. The greater portion of this investment is to provide one wire pair to each telephone subscriber.

If we are to assume that we will remain with the present basic telephone subset, several other problems then arise, deriving from the fundamental features that must be provided the subscriber. These are:

- Transmitter DC feed voltage (talk battery); supervision (i.e., on hook, off hook)
- Telephone alerting: bell, ringing voltage
- Subset sidetone
- Overvoltage protection

We can assume either two-wire* or four-wire operation. With four-wire operation, the hybrid in the subset is removed and the subset will operate without sidetone. With no sidetone, the user will think that the set is not operational. Therefore some method must be devised to reinsert sidetone.

The carbon microphone requires, as a minimum, 3–5 VDC for its operation at some 25–80 mA. This emf† is derived from the central battery (−48 VDC) at the serving switch. The *IR* drop of the loop and other resistive elements bring the voltage down to a considerably lower range, but must

---

*This would be similar to the ISDN BRI with its "U" interface, which allows simultaneous two-way digital operation on a two-wire subscriber line. The topic is discussed in Chapter 13.
†emf stands for electromotive force. In this context it has the same meaning as voltage.

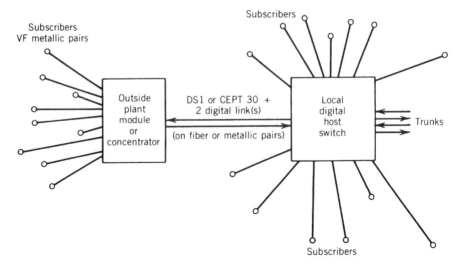

Figure 9.15. Remote concentrator, outside plant module, or switch. Remote unit performs the SLIC functions on each subscriber line.

always be > 3 V across the telephone transmitter terminals for the transmitter to operate properly.

To ring the subset bell, which we call alerting, a 105- to 140-V ringing voltage source is used at 16–66 Hz, requiring about 1 W of power. These characteristics are alien to solid-state devices such as a codec on an IC chip.

Rather than extend directly to the subscriber, an alternative approach is more feasible with today's technology. This is the use of a concentrator, satellite switch, outside plant module (OPM), remote switch, and so on, bringing digital service to the vicinity of the subscriber. Depending on subscriber density, the distance may be 100 ft (30 m) up to a mile (1.6 km). We describe some kind of device which can concentrate greater than 24 up to 240 subscribers to derive a T1 or E1 configuration which extends to the local serving switch. The concept is shown in Figure 9.15. The SLC-96 carries out a similar function but can be considered merely an appendage to a digital local serving switch. The SLC-96,* as the name infers, is a 96-channel subscriber carrier system which can provide the necessary features for an operational system.

Turning to Figure 9.15, the OPM or concentrator accepts analog VF inputs from two-wire analog pairs connected through a cable route to subscriber subsets. The OPM/concentrator carries out the SLIC (subscriber line interface card) functions on each active analog subscriber line. A typical SLIC functional block diagram is shown in Figure 9.16.

*The SLC-96 is an AT&T trademark.

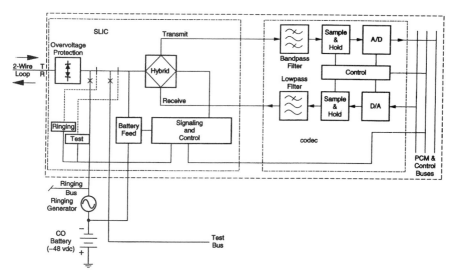

Figure 9.16. Functional block diagram of an SLIC. Reprinted with permission, courtesy of the IEEE [12].

The important concept here is that an SLIC must carry out certain functions. The term BORSCHT describes the seven functions as follows:

B battery feed
O overvoltage protection
R ringing (ringing voltage source)
S signaling (on-hook, off-hook supervision)
C coding (A/D and D/A conversion to/from PCM format)
H hybrid (two-wire to four-wire conversion)
T test

All of these functions can be identified in Figure 9.16.

There are two comparatively new digital subscriber carrier systems that should be mentioned. These are the integrated digital loop carrier (IDLC) and the universal digital loop carrier (UDLC). Neither carry digital service directly to the subscriber. The UDLC can operate with either an analog switch or a digital switch. The IDLC is designed to operate exclusively with a digital serving switch. They can be used with or without concentration at the remote terminal. Both use a multiplexed digital transport such as repeatered T1. Another digital multiplexed transport system is high-bit-rate digital subscriber line (HDSL). HDSL provides equivalent T1 service with a payload bit rate of 1.536 Mbps (24 DSO channels). It is repeaterless with a maximum extension of 12,000 ft (3692 m). HDSL is designed for two-wire duplex operation with echo suppressors (much like the U-interface of ISDN described in Chapter 13). Often the payload is split in half with 784 kbps on each wire pair. The 784 kbps includes 768-kbps payload plus overhead bits. Its remote terminal provides a DS1 (T1) output to the user.

Several advantages accrue from the use of these remote devices. Their use provides an interim solution of extending the digital network toward the subscriber. It is also a convenient means of loop extension, in some applications up to 100 mi (160 km) or more, if necessary. Such techniques also reduce the size of the intervening plant. When using the traditional analog wire-pair techniques, each subscriber pair required individual conditioning for long loops (e.g., loading), possibly with range extension. With the use of such digital plant extension techniques, the total pairs back to the host serving switch are reduced: two pairs for each 24 active subscribers or 30 active subscribers, depending on the system being employed. As an example, suppose there was a community in Spain with 60 telephones and the community was 48 km (30 mi) from the local serving switch. A remote concentrator with codecs could be installed where the concentration was 2:1. Thus a single E1 circuit could serve that entire community on two-wire pair (four-wire) with regenerators every 1830 m (6000 ft). Ordinarily this would require a 30-pair cable with incumbent VF repeaters or some sort of analog subscriber carrier system.

If the 60 telephones were at a business facility, a modern PABX could essentially carry out the same function.

5.3 Change of Profile of Services

Customers demand a greater variety of services than ever before, and that demand is growing. We are shortly reaching a point where the entire network is digital. By definition, digital switches are SPC switches (i.e., they are computer-controlled). Such switches offer an extremely long list of available customer services such as call waiting, call forwarding, voice mail, and many more.

Data communications is increasingly in demand with ever-escalating requirements for higher bit rates and special service offerings. One typical offering is ISDN (Chapter 13), which in many locations is available now. ATM (Chapter 15) is another, which is also available although many of its specifications are still in the formative stages. In Chapter 14 we discuss frame relay and SMDS, which are available now in most locations in North America. In many locations, perhaps most locations, the "analog constraint" remains. The constraint, we believe, is overstated. Group 3 facsimile works well at 9600 bps. Service is a subminute page. However, when the digital 64-kbps channel is available, Group 4 facsimile becomes an option. In this case we can talk about a 1-s page. Another service made available by digital operation is video conferencing.* Many organizations use video conferencing in lieu of traveling to a meeting.

---

*Video conferencing can also be implemented with analog techniques. Video, by its very nature, requires wide bandwidth, 1–4 MHz or more. With digital techniques, video compression is much more feasible. A good picture requires only 384 kbps.

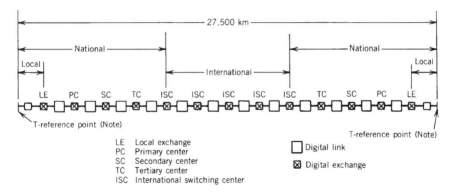

Figure 9.17. The standard CCITT (ITU-T) hypothetical reference connection (HRX), longest length. From CCITT Rec. G.801, Figure 1/G.801, page 5, Fascicle III.5, IXth Plenary Assembly, Melbourne, 1988 (Ref. 14).

5.4 Digital Transmission Network Models: CCITT

Digital transmission network models are hypothetical entities of defined length and composition for use in the study of digital transmission impairments, such as bit and block errors, jitter, wander, transmission delay, and slips. CCITT (ITU-T) provides a *standard hypothetical reference connection* (HRX) based on an all-digital 64-kbps circuit. The standard HRX is shown in Figure 9.17.

The implementation of the standard HRX for a particular application must be tailored for that application. CCITT has two additional models for shorter connections and where a local exchange connects directly to an international switching center (ISC).

The HRX is further broken down into hypothetical reference digital links (HRDLs) 2500 km in length. A digital link is defined in CCITT Rec. G.701 [13] as "the whole of the means of digital transmission of a digital signal of a specified rate between two digital distribution frames (or equivalent)." The HRDL is made up of hypothetical reference digital sections with lengths of 50 and 280 km.

5.5 Digital Network Synchronization

5.5.1 Need for Synchronization. When a PCM bit stream is transmitted over a telecommunication link, there must be synchronization at three different levels: bit, time slot, and frame. Bit synchronization refers to the need for the transmitter (coder) and receiver (decoder) to operate at the same bit rate. It also refers to the requirement that the receiver decision point be exactly at the mid-position of the incoming bit. Bit synchronization assures that the bits will not be misread by the receiver.

Obviously a digital receiver must also know where a time slot begins and ends. If we can synchronize a frame, time-slot synchronization can be assured. Frame synchronization assumes that bit synchronization has been achieved. We know where a frame begins (and ends) by some kind of marking device. With DS1 it is the framing bit. In some frames it appears as a 1 and in others it appears as a 0. If the 12-frame superframe is adopted, it has 12 framing bits, one in each of the 12 frames. This provides the 000111 framing pattern [3]. In the case of the 24-frame extended superframe, the repeating pattern is 001011, and the framing bit occurs only once in four frames.

E1, as we remember in Chapter 8, has a separate framing and synchronization channel, namely channel 0. In this case the receiver looks in channel 0 for the framing sequence in bits 2 through 8 (bit 1 is reserved) of every other frame. The framing sequence is 0011011. Once the framing sequence is acquired, the receiver knows exactly where frame boundaries are. It is also time-slot-aligned.

All digital switches have a master clock. Outgoing bit streams from a switch are slaved to the switch's master clock. Incoming bit streams to a switch derive timing from bit transitions of that incoming bit stream. It is mandatory that each and every switch in a digital network generate outgoing bit streams whose bit rate is extremely close to the nominal bit rate. To achieve this, network synchronization is necessary. Network synchronization can be accomplished by synchronizing all switch (node) master clocks so that transmissions from these nodes have the same average line bit rate. Buffer storage devices are judiciously placed at various transmission interfaces to absorb differences between the actual line bit rate and the average rate. Without this network-wide synchronization, *slips* will occur. Slips are a major impairment in digital networks. Slip performance requirements are discussed in Section 5.6.5. A properly synchronized network will not have slips (assuming negligible phase wander and jitter). In the next paragraph we explain the fundamental cause of slips.

As we mentioned above, timing of an outgoing bit stream is governed by the switch clock. Suppose a switch is receiving a bit stream from a distant source and expects this bit stream to have a transmission rate of $F(0)$ in Mbps. Of course this switch has a buffer of finite storage capacity into which it is streaming these incoming bits. Let's further suppose that this incoming bit stream is arriving at a rate slightly greater than $F(0)$, yet the switch is draining the buffer at exactly $F(0)$. Obviously, at some time, sooner or later, that buffer must overflow. That overflow is a *slip*. Now consider the contrary condition: The incoming bit stream has a bit rate slightly less than $F(0)$. Now we will have an underflow condition. The buffer has been emptied and for a moment in time there are no further bits to be streamed out. This must be compensated for by the insertion of idle bits, false bits, or frame. However, it is more common just to repeat the previous frame. This is also a slip. We may remember the discussion of stuffing in Chapter 8 in the description of higher-order multiplexers. Stuffing allowed some variance of incoming bit rates without causing slips.

When a slip occurs at a switch port buffer, it can be controlled to occur at frame boundaries. This is much more desirable than to have an uncontrolled slip that can occur anywhere. Slips occur for two basic reasons:

1. Lack of frequency synchronization among clocks at various network nodes

2. Phase wander and jitter on the digital bit streams

Thus, even if all the network nodes are operating in the synchronous mode and synchronized to the network master clock, slips can still occur due to transmission impairments. An example of environmental effects that can produce phase wander of bit streams is the daily ambient temperature variation affecting the electrical length of a digital transmission line.

Consider this example. A 1000-km coaxial cable carrying 300 Mbps (3×10^8 bps) will have about 1 million bits in transit at any given time, each bit occupying about 1 meter of the cable. A 0.01% increase in propagation velocity, as would be produced by a 1°F decrease in temperature, will result in 100 fewer bits in the cable; these bits must be absorbed to the switch's incoming elastic store buffer. This may end up causing an underflow problem forcing a controlled slip. Because it is underflow, the slip will be manifested by a frame repeat; usually the last frame just before the slip occurs; is repeated.

In speech telephony, a slip only causes a click in the received speech. For the data user, the problem is far more serious. At least one data frame or packet will be corrupted.

Slips due to wander and jitter can be prevented by adequate buffering. Therefore adequate buffer size at the digital line interfaces and synchronization of the network node clocks are the basic means by which to achieve the network slip rate objective [15].

5.5.2 Methods of Network Synchronization. There are a number of methods that can be employed to synchronize a digital network. Six such methods are shown graphically in Figure 9.18.

Figure 9.18a illustrates plesiochronous operation. In this case each switch clock is free running (i.e., it is *not* synchronized to a network master clock). Each network nodal switch has identical high-stability clocks operating at the same nominal rate. When we say high-stability, we mean a stability range from 1×10^{-11} to about 5×10^{-13} per month. This implies an atomic clock, rubidium or cesium. The accuracy and stability of each clock are such that there is almost complete coincidence in time-keeping, and the phase drift among many clocks is, in theory, avoided or the slip rate between network nodes is acceptably low. This requires that all switching nodes, no matter

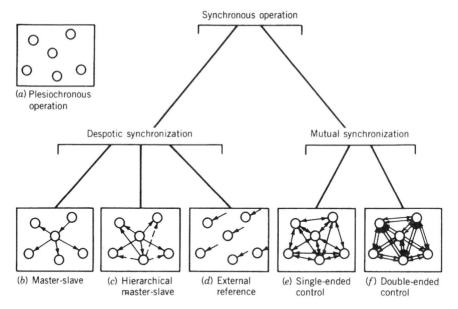

Figure 9.18. Digital network synchronization methods.

how small, have such high-precision clocks. For commercial networks this causes a high cost burden. However, for military networks it is very attractive for survivability because there is no mutual network synchronization. Thus the loss of a node and its clock does not affect the rest of the network timing. CCITT (ITU-T) recommends plesiochronous operation on transnational connectivities (i.e., at international switching centers). (Refer to ITU-T Rec. G.811, Ref. 16.)

Another general synchronization scheme is mutual synchronization, which is shown in Figures 9.18e and 9.18f. Here all nodes in the network exchange frequency references, thereby establishing a common network clock frequency. Each node averages the incoming references and uses the result to correct its local transmitted clock. After an initialization period, the network aggregate clock normally converges to a single stable frequency.

A number of military systems as well as a growing number of private entities use external synchronization.* This is illustrated in Figure 9.18d. Switch clocks use "disciplined" oscillators slaved to an external radio source. One of the most popular today is GPS (Geographical Positioning System) which disseminates coordinated universal time called UTC, an acronym deriving from the French. It is a multiple satellite system such that there are always three or four satellites in view at once anywhere on the earth's surface. Its time transfer capability is in the 10- to 100-ns range from UTC.

---

*We believe that by 1998 all U.S. LEC switches will change over to this type of synchronization using GPS. Each switch then will have a stratum 2 clock. See Table 9.1.

TABLE 9.1 Stratum-Level Specifications

Stratum Level	Free-Run Accuracy	Holdover Stability	Pull-in/Hold-in
1	$\pm 10^{-11}$	N/A	N/A
2	$\pm 1.6 \times 10^{-8}$	$\pm 1 \times 10^{-10}$ per day	$\pm 1.6 \times 10^{-8}$
3E	$\pm 4.6 \times 10^{-6}$	$\pm 1 \times 10^{-8}$ day 1	4.6×10^{-6}
3	$\pm 4.6 \times 10^{-6}$	<255 slips during first day of holdover	4.6×10^{-6}
4	$\pm 32 \times 10^{-6}$	No holdover	32×10^{-6}

Source: Ref. 15, Table 3-1, page 3-3.

Other time dissemination systems by radio are also available, such as satellite-based Transit and GOES, terrestrially based Omega (worldwide), and Loran-C with spotty worldwide coverage. HF radio time transfer techniques are not recommended.

5.5.3 *North American Synchronization Plan as Specified by ANSI/ Bellcore.* The North American network uses a hierarchical timing distribution system as shown in Figure 9.18c. It is based on a four-level hierarchy and these levels are called strata (stratum in the singular). The North American synchronization (Bellcore) arrangement is shown in Figure 9.19.

Timing requirements for each stratum level are shown in Table 9.1. The parameters given in the table are described below.

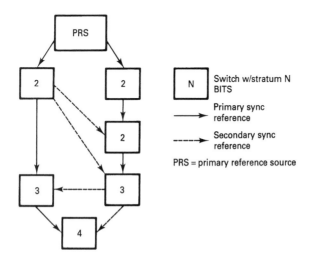

Figure 9.19. North American hierarchical network synchronization. From Ref. 15, Figure 11-2.

The stratum levels for synchronized clocks are based on three parameters:

1. *Free-Run Accuracy.* This is the maximum fractional frequency offset that a clock may have when it has never had a reference or has been in holdover for an extended period, greater than several days or weeks.

2. *Holdover Stability.* This is the amount of frequency offset that a clock experiences after it has lost its synchronization reference. Holdover is specified for stratum 2. The stratum 3 holdover extends beyond one day and it breaks the requirement up into components for initial offset, drift, and temperature. See Bellcore document TR-1244 [17].

3. *Pull-in/Hold-in.* This is a clock's ability to achieve or maintain synchronization with a reference that may be off-frequency. A clock is required to have a pull-in/hold-in range at least as wide as its free-run accuracy. This ensures that a clock of a given stratum level can achieve and maintain synchronization with the clock of the same or higher stratum level.

A stratum 1 clock, which is the PRS (primary reference source), is required to have a timing signal whose long-term accuracy is maintained at 1×10^{-11} or better and completely autonomous of other references. Currently, cesium beam atomic references are the only type clocks that are true stratum 1 references in that they are autonomous and achieve the desired stability. Stratum 1 clocks must be capable of being verified with Universal Coordinated Time (UTC). Alternatively, the PRS source may not be a completely autonomous implementation, in which case it may employ direct control from UTC-derived frequency and time dissemination services such as Loran C or GPS (mentioned above).

Stratum 2 clocks are typically based on either double-oven crystal oscillators or rubidium oscillators. To take advantage of their stable oscillators and provide the best holdover estimate possible, stratum 2 clocks usually have long time constants for averaging their input frequency reference. Stratum 2 clocks are historically deployed as part of tandem and transit switches. To improve reliability, stratum 2 clocks often use two input references with automatic protection switching.

Stratum 3 is a clock with reduced capability in terms of holdover performance. These clocks are typically based on temperature-compensated crystal oscillators (TCXO). Stratum 3 clocks are commonly used by local serving switches. A stratum 3E clock filters its reference timing input to clean up the large amounts of wander to create a timing signal output with low levels of wander. 3E clocks provide significantly better holdover performance.

Stratum 4 clocks are commonly used for digital channel banks. Digital PABXs use a special stratum 4E clock which has two input references [15].

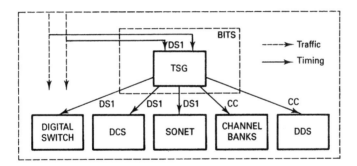

Figure 9.20. Recommended BITS implementation (simplified). From Ref. 15, reprinted with permission.

5.5.3.1 Building Integrated Timing Supply (BITS) Plan. In this plan each switching center has one master clock called the BITS. All other synchronized clocks in the switching center are timed from the BITS clock. The BITS clock is the same or higher stratum level than all other clocks in the switching center and is the only clock that has an external reference.

With the introduction of DS0 dataports, all channel banks must derive timing from the BITS clock.

The recommended BITS implementation is to have one network element (NE) which is a TSG (timing signal generator). It provides DSI and CC (composite clock) timing to all synchronized clocks in the switching center. The TSG is timed by two DS1 signals using any of the following methods:

- Bridging off traffic DS1s coming into the switching center
- Terminating a DS1 dedicated to synchronization distribution
- Having SONET network elements derive DS1s as described in Bellcore specification GR-253 [18]
- Receiving a DS1 from co-located PRS

All equipment in the switching center takes its timing (DS1 or CC) from the TSG. This is shown conceptually in Figure 9.20.

5.5.3.2 Holdover and Slip Performance. When a BITS clock loses its references, it enters holdover and drifts off frequency. The magnitude of this frequency drift determines the average slip rate experienced by equipment that depends on that clock timing source. Table 9.2 shows the number of slips expected after 1 day and 1 week of holdover given limited ambient temperature variations of $\pm 1°F$ in the switching center. The table shows the difference between stratum levels for performance during holdover. If maintenance actions are prompt when that unusual holdover occurs and we

TABLE 9.2 Expected Slip Performance in Holdover

Stratum Level	Slips in Day 1	Slips in Week 1
2	1 or less	2
3E	1 or less	13
3	17	266

Source: Ref. 15, Table 5-1, page 5-2.

base a network on stratum 2 or 3E clocks, a virtually slip-free network can be expected [15].

5.5.3.3 Timing Distribution. The synchronization network has historically been tied to the switching network, using traffic-carrying DS1s between digital switches for synchronization distribution. The switching network includes local serving exchanges and tandem switches (and transit switches in the long-distance network). The tandem switches (and transit switches), which are connected to many downstream switches, provide a convenient point to distribute synchronization. In the future, as synchronization distribution evolves to become SONET-based, SONET facility hubs may become synchronization distribution hubs. Thus Bellcore recommends that synchronization hubs that distribute timing to several downstream switches have stratum 2 BITS clocks, and consideration should be given to making them PRS sites.

5.5.4 Synchronization between Autonomous Networks. When two disparate networks must interface, we are up against what some call the *two-clock problem*. In the United States, typically this is the interface between a local exchange carrier (LEC) and one or several interexchange carriers (IECS). Each network is autonomous for timing synchronization, and each has a clock traceable to some sort of PRS. This also can be an interface between a private digital network and a public network. When two disparate networks have traceability each to their own PRS, it is called *plesiochronous operation*. This concept is shown in Figure 9.21.

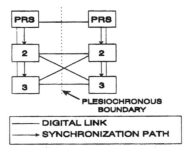

Figure 9.21. Plesiochronous operation. From Ref. 15, Figure 8-1, page 8-1.

The effect of the frequency offset due to the separate stratum 1 traceability is a low slip rate. In the worst case when each PRS is operating at stratum 1 requirement extremes of 1×10^{-11}, the expected slip rate between the networks is less than 1 slip in 72 days.

5.5.5 CCITT Synchronization Plans.

CCITT Rec. G.811 [16] deals with synchronization of international links. Plesiochronous operation is preferred (see Section 5.5.2). The recommendation states the problem at the outset:

International digital links will be required to interconnect a variety of national and international networks. These networks may be of the following form:

(a) a wholly synchronized network in which the timing is controlled by a single reference clock.

(b) a set of synchronized subnetworks in which the timing of each is controlled by a reference clock but with plesiochronous operation between the subnetworks.

(c) a wholly plesiochronous network (i.e., a network where the timing of each node is controlled by a separate reference clock).

Plesiochronous operation is the only type of synchronization that can be compatible with all three types listed. Such operation requires high-stability clocks. Thus Rec. G.811 states that all clocks at network nodes that terminate international links will have a long-term frequency departure of not greater than 1×10^{-11}. This is further described in what follows.

The theoretical long-term mean rate of occurrence of controlled frame or octet (time slot) slips under ideal conditions in any 64-kbps channel is consequently not greater than *1 in 70 days* per international digital link.

Any phase discontinuity due to the network clock or within the network node should result only in the lengthening or shortening of a time signal interval and should not cause a phase discontinuity in excess of one-eighth of a unit interval on the outgoing digital signal from the network node.

Rec. G.811 states that when plesiochronous and synchronous operation coexist within the international network, the nodes will be required to provide both types of operation. It is therefore important that the synchronization controls do not cause short-term frequency departure of clocks, which is unacceptable for plesiochronous operation. The magnitude of the short-term frequency departure should meet the requirements specified in Section 5.5.5.1.

5.5.5.1 Time Interval Error and Frequency Departure. Time interval error (TIE) is based on the variation of ΔT, which is the time delay of a given timing signal with respect to an ideal timing signal, such as UTC (universal coordinated time). The TIE over a period of S seconds is defined to be the

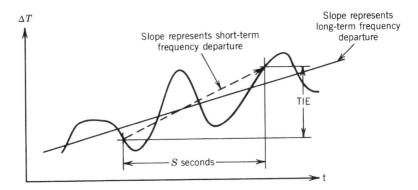

Figure 9.22. Definition of time interval error (TIE). From CCITT Rec. G.811, page 298, Fascicle III.5, Red Books (Ref. 16).

magnitude of difference between time delay values measured at the end and at the beginning of the period:

$$\text{TIE}(S) = |\Delta T(t + S) - \Delta T(t)| \tag{9.2}$$

This is shown diagrammatically in Figure 9.22. The corresponding normalized frequency departure $\Delta f/f$ is the TIE divided by the duration of the period (i.e., S seconds).

The TIE at the output of a reference clock is specified in CCITT Rec. G.811 for three values of frequency as follows:

The TIE over a period of S seconds shall not exceed the following limits:

(a) $(100S)$ ns $+ 1/8$ unit interval. Applicable to S less than 5. These limits may be exceeded during periods of internal clock testing and re-arrangements. In such cases the following conditions should be met: TIE over any period up to 2 UI (unit intervals) should not exceed 1/8 of a UI. For periods greater than 2 UI, the phase variation for each interval of 2 UI should not exceed 1/8 UI up to a total maximum TIE of 500 ns.

(b) $(5S + 500)$ ns for values of S between 5 and 500.

(c) $(10^{-2S} + 3000)$ ns for values of S greater than 500.

The allowance in (c) of 3000 ns is for component aging and environmental effects.

For clarification, CCITT defines UI in Rec. G.701 [13] as the nominal difference in time between consecutive significant instants of an isochronous signal. With NRZ coding we can think of a unit interval (UI) as the duration of 1 bit or the bit period. The bit period in NRZ is the inverse of the bit rate.

TABLE 9.3   Maximum Permissible Degradation of Timing at a Network Node

Performance Category[a]	Frequency Departure $\left\|\dfrac{\Delta f}{f}\right\|$ of Node Timing[b]		Proportion of Time During Which Degradation May Occur, Referred to Total Time[d]	
	Local[c]	Transit	Local[c]	Transit
Nominal	See Section 5.5.5.1	See Section 5.5.5.1	≥98.89%	≥99.945%
(a)	$10^{-11} < \left\|\dfrac{\Delta f}{f}\right\| \le 10^{-8}$	$10^{-11} < \left\|\dfrac{\Delta f}{f}\right\| \le 2.0 \times 10^{-9}$	≤1%	≤0.05%
(b)	$10^{-8} < \left\|\dfrac{\Delta f}{f}\right\| \le 10^{-6}$	$2.0 \times 10^{-9} < \left\|\dfrac{\Delta f}{f}\right\| \le 5.0 \times 10^{-7}$	≤0.1%	≤0.005%
(c)	$\left\|\dfrac{\Delta f}{f}\right\| > 10^{-6}$	$\left\|\dfrac{\Delta f}{f}\right\| > 5.0 \times 10^{-7}$	≤0.01%	≤0.0005%

[a]The performance categories (b) and (c) correspond to (b) and (c) in Rec. G.822 while category (a) in Rec. G.822 corresponds to ''Nominal'' and (a) in Rec. G.811, combined.
[b]All values are provisional.
[c]The values for local nodes are given for guidance only, and administrations are free to adopt other performance levels provided the overall controlled slip performance objective of Rec. G.822 are met.
[d]These values are more stringent than would be strictly required by Rec. G.822 for a 64-kbps connection, to allow for the future introduction of services at higher bit rates that may require a better slip performance. They also allow a margin for possible network effects.
Source: From CCITT Rec. G.811, page 298, *Red Book*, vol. III, Fascicle III.5 (Ref. 16).

In the case of AMI coding, the mark and space are usually not of equal duration.

Table 9.3 shows the permissible degradation of the timing of a network node. The performance category refers to slip rate performance objectives given in Section 5.6.3.

5.6   Digital Network Performance Requirements

5.6.1   Blocking Probability. Based on Bellcore Ref. 19, a block probability of $B = 0.01$ is the quality-of-service objective. With judicious use of alternative routing, a blocking probability of $B = 0.005$ might be expected.

5.6.2   Error Performance—Bellcore Perspective
Definitions

BER: The BER is the ratio of the number of bits in error to the total number of bits transmitted during a measurement period.

Errored Seconds (ES): An errored second is any 1-s interval containing at least one error.

Burst Errored Seconds: A burst errored second is any errored second containing at least 100 errors.

1. The BER at the interface levels DSX-1, DSX-1C, DSX-2, and DSX-3* shall be less than 2×10^{-10}, excluding all burst errored seconds in the measurement period. During a burst errored second, neither the number of bit errors nor number of bits is counted. This requirement applies in a normal operating environment, and it shall be met by every channel in each protection switching section.

2. The frequency of burst errored seconds, other than those caused by protection switching induced by hard equipment failures, shall average no more than four per day at each of the interface levels DSX-1, DSX-1C, DSX-2, and DSX-3.† This requirement applies in a normal operating environment and must be met by every channel in each protection switching system.

3. For systems interfacing at the DS1 level, the long-term percentage of errored seconds (measured at the DS1 rate) shall not exceed 0.04%. This is equivalent to 99.96% error-free seconds (EFS). This requirement applies in a normal operating environment and is also an acceptance criterion. It is equivalent to no more than 10 errored seconds during a 7-h, one-way (loopback) test.

4. For systems interfacing at the DS3 level, the long-term percentage of errored seconds (measured at the DS3 rate) shall not exceed 0.4%. This is equivalent to 99.6% error-free seconds. This requirement applies in a normal operating environment and is also an acceptance criterion. It is equivalent to no more than 29 errored seconds during a 2-h, one-way (loopback) test.

Based on "Transport Systems Generic Requirements (TS6R): Common Requirements," TR-NWT-000499, Issue 5, Bellcore, Piscataway, NJ, Dec. 1993, Ref. 20.

5.6.3 Error Performance—CCITT Perspective. CCITT Rec. G.821 [21] error performance objectives are based on a 64-kbps circuit-switched connection used for voice traffic or as a "bearer channel" for data traffic. The error performance parameters given refer to the HRX discussed above in Section 5.4.

The CCITT error performance parameters are defined as follows (CCITT Rec. G.821): "The percentage of averaging periods each of time interval $T(0)$ during which the bit error rate (BER) exceeds a threshold value. The percentage is assessed over a much longer time interval $T(L)$." A suggested interval for $T(L)$ is 1 month.

*DSX means digital system cross-connect.
†This is a long-term average over many days. Due to day-to-day variation, the number of burst errored seconds occurring on a particular day may be greater than the average.

TABLE 9.4 CCITT Error Performance Objectives for International ISDN Connections

Performance Classification	Objective[c]
(a) (Degraded minutes)[a,b]	Fewer than 10% of 1-min intervals to have a bit error ratio worse than 1×10^{-6} [d]
(b) (Severely errored seconds)[a]	Fewer than 0.2% of 1-s intervals to have a bit error ratio worse than 1×10^{-3}
(c) (Errored seconds)[a]	Fewer than 8% of 1-s intervals to have any errors (equivalent to 92% error-free seconds)

[a]The terms "degraded minutes," "severely errored seconds," and "errored seconds" are used as a convenient and concise performance objective "identifier." Their usage is not intended to imply the acceptability, or otherwise, of this level of performance.
[b]The 1-min intervals mentioned in the table and in the notes are derived by removing unavailable time and severely errored seconds from the total time and then consecutively grouping the remaining seconds into blocks of 60. The basic 1-s intervals are derived from a fixed time pattern.
[c]The time interval $T(L)$, over which the percentages are to be assessed, has not been specified since the period may depend on the application. A period of the order of any one month is suggested as a reference.
[d]For practical reasons, at 64 kbps, a minute containing four errors (equivalent to an error ratio of 1.04×10^{-6}) is not considered degraded. However, this does not imply relaxation of the error ratio objective of 1×10^{-6}.
Source: CCITT Rec. G.821, *Blue Book*, Fascicle III. (Ref. 21).

It should be noted that total time $T(L)$ is broken down into two parts:

- Time that the connection is available.
- Time that the connection is unavailable.

The following BERs and intervals are used in CCITT Rec. G.821 in the statement of objectives [21]:

- A BER of less than 1×10^{-6} for $T(0) = 1$ min.
- A BER of less than 1×10^{-3} for $T(0) = 1$ s.
- Zero errors for $T(0) = 1$ s.

Table 9.4 gives CCITT error performance objectives. Table 9.5 gives some guidelines for interpreting Table 9.4.

5.6.3.1 ITU-T* Rec. G.826. ITU-T Rec. G.826 makes use of block-based error measurements so that in-service (error) measurements (ISMs) are easier to carry out. First, we review some terminology of ITU-T Rec. G.826 [22]. Consider Figure 9.23, which shows some of the relationships to be discussed. The following notation is defined:

---

*Recall that on January 1, 1993, CCITT became the ITU-T (organization).

TABLE 9.5 Guidelines for the Interpretation of Table 9.4

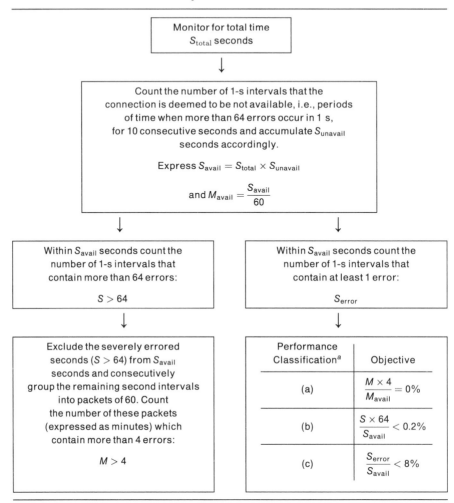

[a]See Table 10.3. From CCITT Rec. G.821, Annex B, page 316, *Red Book*, Ref. 26.

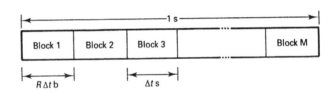

Figure 9.23. Block representation of signals. From Ref. 22, Figure 1, page 57. Reprinted with permission of the IEEE.

$$M = \text{number of blocks per second}$$
$$R = \text{number of bits per second}$$
$$\Delta t = 1/M$$

We introduce the following definitions:

Errored Block (EB): A block in which one or more bits are in error.

Errored Second (ES): A 1-s period with one or more errored blocks.

Severely Errored Second (SES): A 1-s period that contains $\geq 30\%$ EBs or at least one severely disturbed period (SDP). For out-of-service measurements, an SDP occurs when, over a minimum period of time equivalent to four contiguous blocks, either (a) all the contiguous blocks are affected by a high binary error density of 10^{-2} or (b) a loss of signal information is observed. For in-service monitoring purposes, an SDP is estimated by the occurrence of a network defect. CCITT defines *network defect* in the annexes to Rec. G.826 for different network formats such as SDH, PDH, or cell-based (see Chapter 15, ATM).

Background Block Error (BBE): An EB not occurring as part of an SES.

ES Ratio (ESR): The ratio of ESs to total seconds available time during a fixed measurement interval.

SES Ratio (SESR): The ratio of SESs to total seconds in available time during a fixed measurement interval.

BBE Ratio (BBER): The ratio of EBs to total blocks during a fixed measurement interval, excluding all blocks during SESs and unavailable time.

ERROR PERFORMANCE OBJECTIVES (EPOS). The EPOs are summarized in Table 9.6. The objectives are measured over *available* time in a fixed measurement interval (e.g., one month recommended). All three objectives (i.e., ESR,

TABLE 9.6   Error Performance Objectives (EPOs) for G.826

Rate (Mb/s)	Bits/block	ESR	SESR	BBER
1.5–5	2000–8000	0.04	0.002	3×10^{-4}
>5–15	2000–8000	0.05	0.002	2×10^{-4}
>15–55	4000–20,000	0.075	0.002	2×10^{-4}
>55–160	6000–20,000	0.16	0.002	2×10^{-4}
>160–3500	15,000–30,000	—[a]	0.002	10^{-4}
>3500	FFS[b]	FFS	FFS	FFS

[a]No objective given due to the lack of available information.
[b]FFS, for further study.

Source: Ref. 22, Table 2, page 58. Reprinted with permission, courtesy of *IEEE Communications Magazine*.

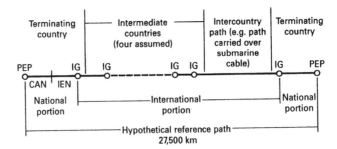

PEP: Path end point
IG: International gateway

Figure 9.24. Hypothetical reference path (HRP). From Ref. 22, Figure 2, page 57. Reprinted with permission of the IEEE.

SESR, and BBER) must hold concurrently to satisfy G.826, and they apply end-to-end for a 27,500-km hypothetical reference path (HRP) (see Section 5.4, Figure 9.17 for a hypothetical reference connection), which is shown in Figure 9.24.

Availability. The concept of available time is the same as was defined in ITU-T Rec. G.821, except that the BER $> 1 \times 10^{-3}$ criterion is replaced by an SES criterion. Thus, unavailable time begins at the start of a block of 10 consecutive SESs. The unavailable time stops (and, of course, available time begins) at the start of 10 consecutive Non-SES events. These 10 seconds are considered part of available time.

Apportionment. The apportionment rules contained in G.826 are different from those used in G.821. These rules are given in Table 9.7, and one should refer to Figure 9.24 for the HRP.

TABLE 9.7 Apportionment Rules

Percent of EPO			
National Portion[a]		International Portion[a]	
Block Allowance	Distance Allowance	Transit Allowance[b]	Distance Allowance
17.5% to both terminating countries	1% per 50 km	2% per intermediate country 1% per terminating country	1% per 500 km

[a]Satellite hops each receive 35%, but the distance of the hop is removed from the distance allowance.
[b]Four intermediate countries are assumed.

Source: Ref. 22, Table 3, page 59. Reprinted with permission, courtesy of *IEEE Communications Magazine*.

The IEEE reference article (23) notes that it is not immediately obvious that the EPO percentage figures of Table 9.7 yield 100%. However, under the assumption of four intermediate countries and no satellite hops, the following breakdown can be obtained:

Terminating countries:

$$2 \times 17.5\% + 2 \times 1\% = 37\%$$

Intermediate countries:

$$4 \times 2\% = 8\%$$

Distance allowance:

$$27,500 \text{ km} = 55 \times 500 \text{ km} (55 \times 1\%) = \underline{55\%}$$
$$\text{Total} \qquad \overline{100\%}$$

If one satellite hop is used, it uses 35%, corresponding to a nominal hop distance of 17,500 km.

The identification of an SES event during ISM is not so straightforward. This is because the definition of an SES involves SDPs. SDP events can only be measured during an out-of-service condition. Thus, "equivalent" ISM events need to be defined if SDPs are to be detected while in-service. It is recognized that there is not an exact 1:1 correspondence between SDP events measured in-service and measured out-of-service. The objective of annexes 2, 3, and 4 to G.826 is to provide ISM events that are reasonably close to the G.826 error events.

It should be noted that G.826 generally applies to underlying transport systems such as E1/T1 (PDH), SDH/SONET, and cell-based transport (see Chapter 15).

The material presented in Section 5.6.3.1 was based on portions of Refs. 22 and 23.

5.6.4 Jitter

5.6.4.1 From a Bellcore Perspective. Jitter was discussed in Chapter 8, where we stated it to be a major digital transmission system impairment. We also said that jitter magnitude was a function of the number of regenerative repeaters there were in tandem. In this section we look at jitter as an overall digital network impairment. The following presents some Bellcore thinking on the matter.

Definition of Timing Jitter. Timing jitter is the short-term variations of a digital signal's significant instants (e.g., optimum sampling instants) from their ideal positions in time. Short-term variations are phase oscillations of

TABLE 9.8 Maximum Permissible Timing Jitter at Non-SONET Network Interfaces

Bit Rate Mb/s	Network Limit UI Peak-to-Peak[a]		Measurement Filter: Bandpass Filter Having a Lower Cutoff Frequency F_1 or F_3 and Minimum Upper Cutoff Frequency F_4		
	B_1 $(F_1 - F_4)$	B_2 $(F_3 - F_4)$	F_1	F_3	F_4
1.544	5	0.1	10 Hz	8 kHz	40 kHz
3.152	5	0.1	10 Hz	1.5 kHz	40 kHz
6.312	3	0.1	10 Hz	3 kHz	60 kHz
44.736	5	0.1	10 Hz	30 kHz	400 kHz

[a]UI = unit interval
For 1.544 Mb/s: 1 UI = 648.00 ns
For 3.152 Mb/s: 1 UI = 317.00 ns
For 6.312 Mb/s: 1 UI = 158.00 ns
For 44.736 Mb/s: 1 UI = 22.35 ns
Source: Ref. 20, Table 7-1, page 7-2. Reprinted with permission.

frequency greater than a demarcation point that is specified for each interface rate (e.g., DS1, phase modulation that, after demodulation, passes through a high-pass filter with a cutoff frequency of 10 Hz, and a 20-dB/decade roll-off). Table 9.8 shows demarcation frequencies for DS1 through DS3 as frequency F.

5.6.4.2 Jitter and Wander from a CCITT Perspective. Jitter is defined as the short-term variations of the significant instants of a digital signal from their ideal positions in time. Excessive accumulating jitter can cause the following impairments:

(a) An increase in error rate at points of signal regeneration as a result of timing signals being displaced from their optimum position in time.

(b) The introduction of uncontrolled slips into digital signals through store spillage and depletion in certain types of terminal equipment incorporating buffer stores and phase comparators, typically in jitter reducers and certain digital multiplex equipment.

(c) A degradation of digitally encoded analog information as a result of phase modulation of the reconstructed samples in the D/A conversion device at the end of a connection.

Jitter can be reduced in magnitude by the use of jitter reducers.

Wander is defined as long-term variations of the significant instants of a digital signal from their ideal positions in time. It can arise as a result of changes in propagation delay of transmission media and equipments. It is

TABLE 9.9A   Maximum Permissible Jitter at Hierarchical Interfaces—E1 Hierarchy

Parameter Value	Network Limit		Measurement Filter Bandwidth Bandpass Filter Having a Lower Cutoff Frequency F_1 or F_3 and an Upper Cutoff Frequency F_4		
Digit Rate (kbit/s)	B_1 Unit Interval[a] Peak-to-Peak	B_2 Unit Interval Peak-to-Peak	F_1	F_3	F_4
64[b]	0.25	0.05	20 Hz	3 kHz	20 kHz
2,048	1.5	0.2	20 Hz	18 kHz (700 Hz)[c]	100 kHz
8,448	1.5	0.2	20 Hz	3 kHz (80 kHz)	400 kHz
34,368	1.5	0.15	100 Hz	10 kHz	800 kHz
139,264	1.5	0.075	200 Hz	10 kHz	3500 kHz

[a]Unit interval (UI):

For 64 kbit/s	1 UI = 15.6 μs
For 2,048 kbit/s	1 UI = 488 ns
For 8,448 kbit/s	1 UI = 118 ns
For 34,368 kbit/s	1 UI = 29.1 ns
For 139,264 kbit/s	1 UI = 7.18 ns

[b]For the codirectional interface only.

[c]The frequency values shown in parentheses only apply to certain national interfaces.

Source: CCITT Rec. G.823, Table 1/G.823 , ITU Geneva, 1993 (Ref. 25).

necessary to accommodate wander at the input ports of digital equipments if uncontrolled slips are to be minimized.

Network Limits of Jitter. Tables 9.9A and 9.9B give limits of maximum permissible levels of jitter as hierarchical interfaces within a digital network. A test setup for measuring output jitter at a digital interface is shown in Figure 9.25. The frequency response of the filters associated with the measuring equipment should have a roll-off of 20 dB per decade.

Network Limits of Wander. A maximum network limit for wander at all hierarchical interfaces has not been defined. Actual magnitudes of wander, which are largely dependent on the propagation characteristics of the intervening transmission media and the aging of the clock circuitry, can be predicted.

5.6.5   Slips

5.6.5.1   From a Bellcore Perspective. Slips are a major impairment in digital networks. The basic cause of slips was explained in Section 5.5.1.

When stratum-3 slip conditions are trouble-free, the nominal clock slip rate is 0. If there is trouble with the primary reference, a maximum of one slip on any trunk will result from a switched reference or any other

TABLE 9.9B   Maximum Permissible Output Jitter at Hierarchical Interfaces—DS1 Hierarchy

	Network Limit (UI Peak-to-Peak)[a]		Bandpass Filter Having a Lower Cutoff Frequency F_1 or F_3 and a Minimum Upper Cutoff Frequency F_4		
Digital Rate (kbit/s)	B_1	B_2	F_1 (Hz)	F_3 (kHz)	F_4 (kHz)
1,544	5.0	0.1[b]	10	8	40
6,312	3.0	0.1[b]	10	3	60
32,064	2.0	0.1[b]	10	8	400
44,736	5.0	0.1	10	30	400
97,728	1.0	0.05	10	240	1000

[a]UI = unit interval.
[b]This value requires further study.

Source: CCITT Rec. G.824, Table 1/G.824, page 51, Fascicle III.5, IXth Plenary Assembly, Melbourne, 1988 (Ref. 26).

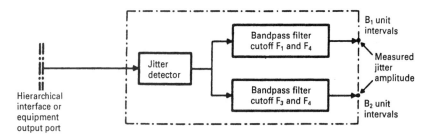

Figure 9.25. Measurement arrangements for output jitter from a hierarchical interface or an equipment output port. From CCITT Rec. G.823, Figure 1/G.823, page 40, Fascicle III.5, IXth Plenary Assembly, Melbourne, 1988 (Ref. 25).

rearrangement. If there is a loss of all references, the maximum slip rate is 255 slips the first day for any trunk. This occurs when the stratum-3 clocks drift a maximum of 0.37 parts per million from their referenced frequency.

From *BOC Notes on the LEC Networks—1990*, Issue 1, pages 11–14, Table 11-2 (Ref. 27).

5.6.5.2   From a CCITT Perspective.   With plesiochronous operation, the number of slips on the international links will be governed by the sizes of buffer stores and the accuracies and stabilities of the interconnecting national clocks. The end-to-end slip performance should satisfy the service requirements for telephone and nontelephone services on a 64-kbit/s digital connection in an ISDN.

TABLE 9.10   Controlled Slip Performance on a 64-kbps International Connection Bearer Channel

Performance Category	Mean Slip Rate	Proportion of Time[a]
(a)[b]	≤5 slips in 24 h	>98.9%
(b)	>5 slips in 24 h and ≤30 slips in 1 h	< 1.0%
(c)	>30 slips in 1 h	< 0.1%

[a]Total time ≥1 year.
[b]The nominal slip performance due to plesiochronous operation alone is not expected to exceed 1 slip in 5.8 days.
Source: CCITT Rec. G.822, Table 1/G.822, page 37, Fascicle III.5, IXth Plenary Assembly, Melbourne, 1988 (Ref. 28).

The slip rate objectives for an international end-to-end connection are stated with reference to the standard hypothetical reference connection (HRX) (see Section 5.4) of 27,500 km in length.

The theoretical slip rate is one slip in 70 days per plesiochronous interexchange link assuming clocks with specified accuracies (see Section 5.5.1) and provided that the performance of the transmission and switching requirements remain within their design limits.

In the case where the international connection includes all of the 13 nodes identified in the HRX and those nodes are all operating together in a plesiochronous mode, the nominal slip performance of a connection could be 1 in 70/12 days (12 links in tandem) or 1 in 5.8 days. In practice, however, some nodes in such a connection would be part of the same synchronized network. Therefore, a better nominal slip performance can be expected (e.g., where the national networks at each end are synchronized). The nominal slip performance of the connection would be 1 in 70/4 or 1 in 17.5 days. Note that these calculations assume a maximum of four international links.

The performance objectives for the rate of *octet* slips on an international connection of 27,500 km in length of a corresponding bearer channel are given in Table 9.10. CCITT [28] adds that further study is required to confirm that these values are compatible with other objectives such as error performance given in Section 5.6.3.

ALLOCATION OF SLIPS. Because the impact of slips occurring in different parts of a connection varies in importance depending on the type of service and level of traffic affected, the allocation process includes placing higher limits on slips detected at international and national transit exchanges and less stringent limits on small local exchanges. The allocation process is based on subdividing the percentage of time objective for performance categories (b) and (c) in Table 9.10. Table 9.11 shows these allocations assigned to the

TABLE 9.11   Allocation of Controlled Slip Performance

Portion of HRX Derived from Figure 9.26[a]	Allocated Proportion of Each Objective in Table 9.10[b]	Objectives as Proportion of Total Time[c]	
		(b)	(c)
International transit portion	8.0%	0.08%	0.008%
Each national transit portion[d]	6.0%	0.06%	0.006%
Each local portion[d]	40.0%	0.4%	0.04%

[a]The portions of the HRX are defined in Figure 9.26. They are derived from, but not identical to, Rec. G.801.
[b]Performance levels are defined in Table 9.10.
[c]Total time ≥1 year.
[d]The allocation between national transit portion and local portion is given for guidance only. Administrations are free to adopt a different apportionment provided the total for each national portion (local plus transit) does not exceed 46%.

Source: CCITT Rec. G.822, Table 2/G.822, page 38, Fascicle III.5, IXth Plenary Assembly, Melbourne, 1988 (Ref. 28).

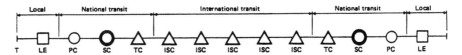

Figure 9.26. Subdivision of the HRX for the purpose of allocation of slip performance objectives. From CCITT Rec. G.822, Figure 1/G.822, page 38, Fascicle III.5, IXth Plenary Assembly, Melbourne, 1988 (Ref. 28).

various portions of the HRX (see Section 5.4). CCITT [28] states that these allocations are provisional. Figure 9.26 shows the subdivision of the HRX for slip allocation.

5.7   A-Law Conversion to μ-Law; Digital Loss

Chapter 8 discussed two quite different PCM systems. In North America, there is the DS1 or T1* system; in Europe and in those countries under European hegemony there is the E1 system, previously called CEPT30 + 2. The logarithmic companding curve used for the development of the DS1 signal is the μ-law, where $\mu = 255$. E1 uses the A-law, where $A = 87.6$. An international switching center is probably the most convenient location to convert from one to the other. To carry out such conversion, a processor is required with a look-up table. The processor forces some small changes in the binary value of time-slot samples based on look-up table values. As we

*T1 and DS1 are synonymous.

mentioned earlier, to match E1 to DS1 channel-by-channel, five DS1s are mapped into four E1s, and during this mapping the companding conversion can also take place. The conversion equipment must also handle the different signaling and framing strategies used by each system. This can be carried out in fairly straightforward software.

Like its analog counterpart, the digital network must be lossy to control echo and singing. Specific loss plans were described in Chapter 6. In North America, 3 dB is a common value for loss on one end of a long-distance connection. It was a fairly simple matter to switch in a 3-dB pad (attenuator) in an analog switch. It is not so straightforward if we wish to insert such a loss on the digital side. Remember that loss equates to reducing level. To do this on the digital side, we again can resort to the use of a processor and a look-up table.

REVIEW QUESTIONS

1. Give a simple definition of time-division switching. Compare space-division switching with time-division switching.

2. Give at least five advantages of time-division switching compared to space-division switching.

3. Give two technical issues relating to time-division switching and digital networks. (*Hint*: These may be listed as disadvantages in relation to question 2.)

4. What are the two principal functional elements of digital switching? These elements carry out the actual switching of digital connections (e.g., not control).

5. Define the terms *read* and *write* as used in this chapter. Differentiate between sequential read–random write and random read–sequential write.

6. What are the three basic building blocks of a time switch?

7. What are the limitations of a time (T) switch? How are these solved by the addition of space (S) switching capability?

8. A space array has M horizontals and N verticals in its matrix. Relate M to N for nonblocking tandem switching, for local switch expansion, and for local switch concentration.

9. We have a 30×30 space array with a time-slot interchange (TSI) at each input. The TSI is designed for DS1 operation. How many total time slots can the array handle?

10. Why is a TST switch more desirable than an STS switch? Where would an STS switch have application?

11. How can blocking probability be reduced in a TST switch?

12. What is the function of "junctors" in a switch (e.g., a DMS-100)?

13. Internal bit rates of switches differ from the external interface of 8-bit time slots. Northern Telecom DMS-100 maps the 8-bit PCM "word" into a _____-bit time slot. What is/are the function(s) of the extra two bits?

14. To make a switch "universal"—that is, easy to change over from DS1 format to E1 format and vice versa—often internal bit rates were also based on an X/Y ratio between DS1 and E1. What are the values of X and Y?

15. The present AT&T 5ESS has 100,000-line capacity. How does it accomplish this capacity with only a TST architecture (i.e., only one space stage)?

16. Suppose a digital switch were based solely on a time stage consisting of an E3 format. How many lines could it handle?

17. Why would fiber-optic transmission links find application internally to these third-generation PCM switches?

18. Give at least two applications of remote switching, particularly in light of Chapter 2.

19. What are the two principal factors that change network structure?

20. In your opinion, what is delaying "fiber to the home"?

21. What is the effect on the subscriber when his/her telephone subset operates without sidetone?

22. What is the meaning of each letter in BORSCHT?

23. Of what use to the telecommunication system engineer is a hypothetical reference circuit or hypothetical reference connection?

24. In the digital network, digital bit streams require synchronization at three levels. What are they?

25. Why is it imperative that every outgoing bit stream from a digital switch have its bit rate *very* close to the nominal? If not, what impairment results?

26. What is the cause of slips?

27. Which type of user of the digital network suffers the most when a slip occurs?

28. What is the most common boundary for a controlled slip? Why there?

29. The velocity of propagation for optical fiber is 2×10^8 m/s. A data source is transmitting at 1000 Mbps. How many bits will there be on 1000 km of fiber-optic cable?

30. Six methods of synchronizing a digital network were given in the text. Name four of them.

31. Give a value of frequency stability when we mean *high* stability regarding synchronization of the digital network.

32. What is the most common way for a digital network to disseminate timing?

33. We commonly meet what is termed the "two-clock problem." An international switch is an example, where it is more of a multiclock problem. How can a switch handle such a problem?

34. Define plesiochronous operation.

35. What is a unit interval (UI)? Define it.

36. What is the end-to-end BER specified by CCITT for an ISDN?

37. Why did CCITT come up with a new way of specifying error performance on the digital network?

38. What is the basic performance measure of a digital network?

39. What primary performance parameter does timing jitter affect?

40. If network synchronization is working properly, what slip rate can be expected?

41. What is the most common method of converting A-law to μ-law and vice versa? (We face this when E1 interfaces with DS1 and vice versa.)

42. To control echo, we can insert loss in a voice channel. How can loss be inserted on the digital side?

REFERENCES

1. J. P. Ronadyne, *Introduction to Digital Communications Switching*, Howard W. Sams & Co., Indianapolis, Indiana, 1986.

2. J. C. McDonald, ed., *Fundamentals of Digital Switching*, 2nd ed., Plenum Press, New York, 1990.

3. J. C. Bellamy, *Digital Telephony*, 2nd ed., John Wiley & Sons, New York, 1991.

4. *Planning Guide: DMS-100/200 Family*, Northern Telecom, Research Triangle Park, NC, 1985.

5. *Network Frame*, General Specification, Northern Telecom, Research Triangle Park, NC, 1983/1988.

6. R. L. Freeman, *Reference Manual for Telecommunication Engineering*, 2nd ed., John Wiley & Sons, New York, 1994.

7. *5ESS Switch and 5ESS-2000 Switch System Description*, AT&T 235-100-125, Issue 7.00, Winston-Salem, NC, November 1994.

8. *Enhanced Network*, a brief description of ENET, Northern Telecom, Research Triangle Park, NC, Issue 2, August 1992.

9. *DMS-100 Advantage*, Issue 2, Northern Telecom, Research Triangle Park, NC, November 1993.

10. Northern Telecom US Patent 4,470,139, issued September 4, 1984.

11. Private communication with Ernst Munter, Northern Telecom, Ottawa, March 25, 1995.

12. W. D. Reeve, *Subscriber Loop Signaling and Transmission Handbook, Digital*, IEEE Press, New York, 1995.

13. *Vocabulary of Digital Transmission and Muliplexing and Pulse Code Modulation (PCM) Terms*, CCITT Rec. G.701, Fascicle III.4, IXth Plenary Assembly, Melbourne, 1988.

14. *Digital Transmission Models*, CCITT Rec. G.801, Fascicle III.5, IXth Plenary Assembly, Melbourne, 1988.

15. *Digital Network Synchronization Plan*, Bellcore Generic Requirements, GR-436-CORE, Bellcore, Piscataway, NJ, June 1994.

16. *Timing Requirements at the Outputs of Reference Clocks and Network Nodes Suitable for Plesiochronous Operation of International Digital Links*, CCITT Rec. G.811, Fascicle III.5, IXth Plenary Assembly, Melbourne, 1988.

17. *Clocks for the Synchronized Network: Common Generic Criteria*, TR-NWT-001244, Issue 1, Bellcore, Piscataway, NJ, June 1993.

18. *Synchronous Optical Network (SONET) Transport Systems: Common Generic Criteria*, GR-253-CORE, Bellcore, Piscataway, NJ, December 1994.

19. *BOC Notes on the LEC Networks—1994*, TR-TSV-002275, Issue 2, Bellcore, Piscataway, NJ, April 1994.

20. *Transport Systems Generic Requirements (TSGR): Common Requirements*, TR-NWT-000499, Issue 5, Bellcore, Piscataway, NJ, December 1993.

21. *Error Performance of an International Digital Connection Forming Part of an ISDN*, CCITT Rec. G.821, Fascicle III.5, IXth Plenary Assembly, Melbourne, 1988.

22. *Error Performance Parameters and Objectives for International Constant Bit Rate Digital Paths at or above the Primary Rate*, ITU-T Rec. G.826, Geneva, 1994.

23. M. Shafi and P. J. Smith, "The Impact of G.826," *IEEE Communications Magazine*, September 1993, IEEE, New York, 1993.

24. *The New IEEE Standard Dictionary of Electrical and Electronics Terms*, 5th ed., IEEE Std 100-1992, IEEE, New York, 1992.

25. *The Control of Jitter and Wander within Digital Networks Which Are Based on the 2048 kbps Hierarchy*, CCITT Rec. G.823, ITU Geneva, 1993.

26. *The Control of Jitter and Wander within Digital Networks Which Are Based on the 1544 kbps Hierarchy*, CCITT Rec. G.824, ITU Geneva, 1993.

27. *BOC Notes on the LEC Networks—1990*, Issue 1, Bellcore, Piscataway, NJ, 1990.

28. *Controlled Slip Rate Objectives on an International Digital Connection*, CCITT Rec. G.822, Fascicle III.5, IXth Plenary Assembly, Melbourne, 1988.

10

INTRODUCTION TO DATA COMMUNICATIONS

1 OVERVIEW

Data communications emerged as a technology slightly delayed from the development of the computer. Today it is probably the most rapidly growing branch of telecommunications. My initial experience with data communications was in 1961 with DoD's LogComNet, which became AutoDiN. Many of its concepts, technology, and certainly language derived from the much older telegraph transmission.

The IEEE [1] defines data communications (data transmission) as "The movement of encoded information by means of communications techniques." In this chapter we will discuss the *encoding of data* and then the *techniques of its communication*. In Chapter 11, we will address data networks and their operation. This discussion will be a general treatment of wide area (data) networks or WANS. Local area networks (LANS) are covered in Chapter 12. Chapter 13 deals with integrated services digital networks (ISDN), and Chapter 14 covers emerging broadband data communication technologies such as frame relay. Chapter 15, dealing with the asynchronous transfer mode, is the final chapter on the subject.

The fact that this book covers data in 5 chapters out of 18 shows its significance in the world of telecommunication systems. The present North American public switched telecommunication network (PSTN) is optimized for voice communication. It is estimated that 80% of the traffic on that network is voice traffic. On other networks in the world, the percentage is considerably higher.

This begs the question: What is the other 20%? It is a mix of data, facsimile, broadcast, transport, and conference television. However, facsimile as transmitted on the network is data, and conference television is transmitted

459

in a digital format. If we said that the remaining 20% was data, we would be pretty much correct.

In this chapter we address data transmission on wide area networks. Facsimile will be considered as just another data transmission circuit. Wide area networks often use the underlying PSTN for transport. But not always. Private networks that have nothing to do with the PSTN* are becoming ever more prevalent. VSAT networks (Chapter 7) are just one example. Often such networks are 100% dedicated to data communication. Other private networks carry a mix of voice, data, and possibly conference television.

2 THE BIT

The *bit* is often called the most elemental unit of information. The IEEE [1] calls it a contraction of *binary digit*, a unit of information represented by either a zero or a one. These are the same bits we were dealing with in Chapters 8 and 9. In those chapters the primary purpose of those bits was to signal the distant end the voltage level of an analog voice channel at some moment in time. Here we will be assembling bit groupings which will represent the letters of the alphabet, numerical digits 0 through 9, punctuation, graphic symbols, or just operational sequences that have meaning to a machine but have no ostensible outward meaning to us.

From old-time telegraphy the terminology has migrated to data communications. A *mark* is a binary 1 and a *space* is a binary 0. A space or 0 is represented by a positive-going voltage, and a mark or 1 is represented by a negative-going voltage. Now I am getting confused. When I was growing up in the industry, a 1 or mark was a positive-going voltage (and so forth).

3 REMOVING AMBIGUITY—BINARY CONVENTION

To remove ambiguity of the various ways we can express a 1 and a 0, CCITT in Rec. V.1 [2] states clearly how to represent a 1 and a 0. This is summarized in Table 10.1 with several additions from other sources. Table 10.1 defines the *sense* of transmission so that the mark and space, the 1 and 0, respectively, will not be inverted. Inversion can take place by just changing the voltage polarity. We call it reversing the sense. Some data engineers often refer to such a table as a "table of mark-space convention."

---

*Some private networks lease circuits from, or overlay on, the PSTN. That is why the statement is made as it is.

TABLE 10.1   Equivalent Binary Designations: Summary of Equivalence

Symbol 1	Symbol 0
Mark or marking	Space or spacing
Current on	Current off
Negative voltage	Positive voltage
Hole (in paper tape)	No hole (in paper tape)
Condition Z	Condition A
Tone on (amplitude modulation)	Tone off
Low frequency (frequency shift keying)	High frequency
Inversion of phase	No phase inversion (differential phase shift keying)
Reference phase	Opposite to reference phase

Source: CCITT Recs. V.1 [2], V.10, and V.11.

4 CODING

4.1 Introduction to Binary Coding Techniques

Written information must be coded before it can be transmitted over a data network. One bit carries very little information. There are only those two possibilities: the 1 and the 0. It serves good use for supervisory signaling where a telephone line could only be in one of two states. It is either idle or busy. As a minimum we would like to transmit every letter of the alphabet and the 10 basic decimal digits plus some control characters, such as a space and carriage return, and some punctuation.

Suppose we stick two bits together for transmission. There are four possibilities:

$$00 \quad 10$$
$$01 \quad 11 *$$

or four pieces of information. Suppose 3 bits are transmitted in sequence. Now there are eight possibilities:

$$000 \quad 100$$
$$001 \quad 101$$
$$010 \quad 110$$
$$011 \quad 111$$

We can now see that for a binary code, the number of distinct information characters available is equal to two raised to a power equal to the number of

---

*To a degree this is a review of an argument in Chapter 8.

Characters				Code Elements [a]							
Letters Case	Communications	Weather	CCITT #2[b]	START	1	2	3	4	5	STOP
A	–	↑		X	X	X				X
B	?	⊕		X	X			X	X	X
C	:	○		X		X	X	X		X
D	$	↗	WRU	X	X			X		X
E	3	3		X	X					X
F	!	→	Unassigned	X	X		X	X		X
G	&	↘	Unassigned	X		X		X	X	X
H	STOP[c]	↓	Unassigned	X			X		X	X
I	8	8		X		X	X			X
J	'	↗	Audible signal	X	X	X		X		X
K	(	←		X	X	X	X	X		X
L	)	↖		X		X			X	X
M	.	.		X			X	X	X	X
N	,	⊕		X			X	X		X
O	9	9		X				X	X	X
P	0	0		X		X	X		X	X
Q	1	1		X	X	X	X		X	X
R	4	4		X		X		X		X
S	BELL	BELL	,	X	X		X			X
T	5	5		X					X	X
U	7	7		X	X	X	X			X
V	;	⊙	=	X		X	X	X	X	X
W	2	2		X	X	X			X	X
X	/	/		X	X		X	X	X	X
Y	6	6		X	X		X		X	X
Z	"	+	+	X	X				X	X
BLANK		–		X						X
SPACE				X			X			X
CAR. RET.				X				X		X
LINE FEED				X		X				X
FIGURE				X	X	X		X	X	X
LETTERS				X	X	X	X	X	X	X

[a] Blank, spacing element; crosshatched, marking element.
[b] This column shows only those characters that differ from the American "communications" version
[c] Figures case H(COMM) may be STOP or +.

Figure 10.1. Communication and weather codes. CCITT Alphabet No. 2 (ITA No. 2 code).

elements or bits per character. For instance, the last example was based on a three-element code giving eight possibilities or information characters, or 2^3.

Another, more practical example is the CCITT ITA No. 2 teleprinter code (Figure 10.1), which has 5 bits or information elements per character. Therefore the number of different graphics* and characters available is $2^5 = 32$. The American Standard Code for Information Interchange (ASCII)

*In this context a graphic is a printing character other than a letter or number. Typical graphics are asterisks, punctuation, parentheses, dollar signs, and so forth.

has seven information elements per character, or $2^7 = 128$; thus it has 128 distinct combinations of marks and spaces that are available for assignment as characters or graphics.

The number of distinct characters for a specific code may be extended by establishing a bit sequence (a special character assignment) to shift the system or machine to uppercase (as is done with a conventional typewriter). Uppercase is a new character grouping. A second distinct bit sequence is then assigned to revert to lowercase. For example, the CCITT ITA No. 2 code (Figure 10.1) is a five-unit code with 58 letters, numbers, graphics, and operator sequences. The additional characters and graphics (additional above $2^5 = 32$) originate from the use of uppercase. Operator sequences appear on a keyboard as "space" (spacing bar), "figures" (uppercase), "letters" (lowercase), "carriage return," "line feed" (spacing vertically), and so on. When we refer to a 5-unit, 6-unit, or 12-unit code, we refer to the number of information units or elements that make up a single character or symbol. That is, we refer to those elements assigned to each character that carry information and that distinguish it from all other characters or symbols of the code.

4.2 Hexadecimal Representation and BCD Code

The hexadecimal system is a numeric representation in the number base 16. This number base uses 0 through 9 as in the decimal number base and uses the letters A through F to represent the decimal numbers 10 through 15. The "hex" numbers can be translated to the binary base as follows:

Hex	Binary	Hex	Binary
0	0000	8	1000
1	0001	9	1001
2	0010	A	1010
3	0011	B	1011
4	0100	C	1100
5	0101	D	1101
6	0110	F	1110
7	0111	F	1111

Two examples of the hexadecimal notation are as follows:

Number Base 10	Number Base 16
21	15
64	40

The BCD (binary-coded decimal) is a compromise code assigning 4-bit binary numbers to the digits between 0 and 9. The BCD equivalents to decimal digits appear as follows:

Decimal Digit	BCD Digit	Decimal Digit	BCD Digit
0	1010	5	0101
1	0001	6	0110
2	0010	7	0111
3	0011	8	1000
4	0100	9	1001

To cite some examples, consider the number 16; it is broken down into 1 and 6. Thus its BCD equivalent is 0001 0110. If it were written in straight binary notation, it would appear as 10000. The number 25 in BCD combines the digits 2 and 5 above as 0010 0101.

4.3 Some Specific Binary Codes for Information Interchange

The most commonly used binary source code for data communications is the ASCII code. ASCII stands for American Standard Code for Information Interchange. It is used worldwide. ASCII is a 7-unit or 7-level or 7-bit code and is illustrated in Figure 10.2. Almost universally ASCII has a parity bit appended, making it an 8-level code. Figure 10.3 shows EBCDIC (extended binary-coded decimal interchange code) developed by IBM. It is a true 8-level code with 2^8 or 256 coded character possibilities. As shown in Figure 10.3, some of the bit sequences have their meaning left in blank (i.e., no meaning has been assigned). Table 10.2 gives a description of ASCII control characters.

5 ERRORS IN DATA TRANSMISSION

5.1 Introduction

In data transmission one of the most important design goals is to minimize the error rate. Error rate may be defined as the ratio of the number of bits incorrectly received to the total number of bits transmitted. On many data circuits the design objective is an error rate better than one error in 1×10^6 bits (often expressed 1×10^{-6}), and for telegraph circuits, one error in 1×10^4.

One method for minimizing the error rate would be to provide a "perfect" transmission channel, one that will introduce no errors in the transmitted information at the output of the receiver. However, that perfect channel can

b7				0	0	0	0	1	1	1	1
b6				0	0	1	1	0	0	1	1
b5				0	1	0	1	0	1	0	1
b4	b3	b2	b1		0	1	2	3	4	5	6	7
0	0	0	0	0	NUL	DLE	SP	0	@	P	`	p
0	0	0	1	1	SOH	DC1	!	1	A	Q	a	q
0	0	1	0	2	STX	DC2	"	2	B	R	b	r
0	0	1	1	3	ETX	DC3	#	3	C	S	c	s
0	1	0	0	4	EOT	DC4	$	4	D	T	d	t
0	1	0	1	5	ENQ	NAK	%	5	E	U	e	u
0	1	1	0	6	ACK	SYN	&	6	F	V	f	v
0	1	1	1	7	BEL	ETB	'	7	G	W	g	w
1	0	0	0	8	BS	CAN	(	8	H	X	h	x
1	0	0	1	9	HT	EM	)	9	I	Y	i	y
1	0	1	0	10	LF	SUB	*	:	J	Z	j	z
1	0	1	1	11	VT	ESC	+	;	K	[	k	{
1	1	0	0	12	FF	FS	,	<	L	\	l	|
1	1	0	1	13	CR	GS	-	=	M	]	m	}
1	1	1	0	14	SO	RS	.	>	N	^	n	~
1	1	1	1	15	SI	US	/	?	O	_	o	DEL

NOTE: The font used in this code table is OCR-B. It is intended only as an example of a conforming font and is not intended to indicate preference for OCR-B.

Figure 10.2. American Standard Code for Information Interchange (ASCII). From ANSI X3.4-1986 [3].

B I T S		4	0	0	0	0	0	0	0	0	1	1	1	1	1	1	1	1
		3	0	0	0	0	1	1	1	1	0	0	0	0	1	1	1	1
		2	0	0	1	1	0	0	1	1	0	0	1	1	0	0	1	1
		1	0	1	0	1	0	1	0	1	0	1	0	1	0	1	0	1
8	7	6	5																
0	0	0	0	NUL				PF	HT	LC	DEL								
0	0	0	1					RES	NL	BS	IL								
0	0	1	0					BYP	LF	EOB	PRE			SM					
0	0	1	1					PN	RS	UC	EOT								
0	1	0	0	SP										¢	.	<	(	+	|
0	1	0	1	&										!	$	*	)	;	¬
0	1	1	0	-	/									^	,	%	_	>	?
0	1	1	1											⌿:	#	@	'	=	"
1	0	0	0		a	b	c	d	e	f	g	h	i						
1	0	0	1		j	k	l	m	n	o	p	q	r						
1	0	1	0			s	t	u	v	w	x	y	z						
1	0	1	1																
1	1	0	0		A	B	C	D	E	F	G	H	I						
1	1	0	1		J	K	L	M	N	O	P	Q	R						
1	1	1	0			S	T	U	V	W	X	Y	Z						
1	1	1	1	0	1	2	3	4	5	6	7	8	9						⌷

PF – Punch Off	RES – Restore	BYP – Bypass
HT – Horiz. Tab	NL – New Line	LF – Line Feed
LC – Lower Case	BS – Backspace	EOB – End of Block
DEL – Delete	IL – Idle	PRE – Prefix
SP – Space	PN – Punch On	RS – Reader Stop
UC – Upper Case	EOT – End of Transmission	SM – Start Message

Figure 10.3. Extended binary-coded decimal interchange code (EBCDIC).

never be achieved. Besides improvement of the channel transmission parameters themselves, error rate can be reduced by forms of a systematic redundancy. In old-time Morse code, words on a bad circuit were often sent twice; this is redundancy in its simplest form. Of course, it took twice as long to send a message; this is not very economical if the number of useful words per minute received is compared to channel occupancy.

This illustrates the trade-off between redundancy and channel efficiency. Redundancy can be increased such that the error rate could approach zero. Meanwhile, the information transfer across the channel would also approach zero. Thus unsystematic redundancy is wasteful and merely lowers the rate of useful communication. On the other hand, maximum efficiency could be obtained in a digital transmission system if all redundancy and other code elements, such as "start" and "stop" elements, parity bits, and other "overhead" bits, were removed from the transmitted bit stream. In other words, the channel would be 100% efficient if all bits transmitted were information bits. Obviously, there is a trade-off of cost and benefits somewhere between maximum efficiency on a data circuit and systematically added redundancy (see Chapter 11, Section 5).

TABLE 10.2 ASCII Control Characters

ACK (Acknowledge). A transmission control character transmitted by a receiver as an affirmative response to the sender.

BEL (Bell). A control character that is used when there is a need to call for attention; it may control alarm or attention devices.

BS (Backspace). A format effector that causes the active position to move one character position backwards.

CAN (Cancel). A character, or the first character of a sequence, indicating that the data preceding it is in error. As a result, this data is to be ignored. The specific meaning of this character shall be defined for each application and/or defined between sender and recipient.

CR (Carriage Return). A format effector that causes the active position to move to the first character position on the same line.

DC1 (Device Control One). A device control character that is primarily intended for turning on or starting an ancillary device. If it is not required for this purpose, it may be used to restore a device to the basic mode of operation (see also DC2 and DC3), or for any other device control function not provided by other DCs.

DC2 (Device Control Two). A device control character that is primarily intended for turning on or starting an ancillary device. If it is not required for this purpose, it may be used to set a device to a special mode of operation (in which case DC1 is used to restore the device to the basic mode), or for any other device control function not provided by other DCs.

DC3 (Device Control Three). A device control character that is primarily intended for turning off or stopping an ancillary device. This function may be a secondary level stop—for example, wait, pause, standby, or halt (in which case DC1 is used to restore normal operation). If it is not required for this purpose, it may be used for any other ancillary device control function not provided by other DCs.

DC4 (Device Control Four). A device control character that is primarily intended for turning off, stopping, or interrupting an ancillary device. If it is not required for this purpose, it may be used for any other device control function not provided by other DCs.

DEL (Delete). A character used primarily to erase or obliterate an erroneous or unwanted character in punched tape. DEL characters may also serve to accomplish media-fill or time-fill. They may be inserted into or removed from a stream of data without affecting the information content of that stream, but such action may affect the information layout and/or the control of equipment. If media-fill or time-fill is required, it is preferred that the NUL character be used.

DLE (Data Link Escape). A transmission control character that changes the meaning of a limited number of contiguously following bit combinations. It is used exclusively to provide supplementary transmission control functions. Only graphic characters and transmission control characters may be used in DLE sequences. Appropriate sequences are defined in ANSI X3.28-1976.

EM (End of Medium). A control character that may be used to identify the physical end of a medium, the end of the used portion of a medium, or the end of the wanted portion of data recorded on a medium. The position of this character does not necessarily correspond to the physical end of the medium.

ENQ (Enquiry). A transmission control character used as a request for a response from a remote station—the response may include station identification and/or station status. When a "Who are you" function is required on a switched transmission network, the first use of ENQ after the connection is established shall have the meaning "Who are you" (station identification). Subsequent use of ENQ may or may not include the function "Who are you," as determined by agreement.

EOT (End of Transmission). A transmission control character used to indicate the conclusion of the transmission of one or more texts.

ESC (Escape). A control character that is used to provide additional characters (code extension). It alters the meaning of a limited number of contiguously following bit combinations. The use of this character is specified in ANSI X3.41-1974.

ETB (End of Transmission Block). A transmission control character used to indicate the end of a transmission block of data where data

are divided into such blocks for transmission purposes.

ETX (End of Text). A transmission control character that terminates a text.

FF (Form Feed). A format effector that causes the active position to advance to the corresponding character position on a predetermined line of the next form or page.

FS (File Separator) (Information Separator Four). A control character used to separate and qualify data logically; its specific meaning has to be defined for each application. If this character is used in hierarchical order, as specified in the general definition of the information separators, it delimits a data item called a "file."

GS (Group Separator) (Information Separator Three). A control character used to separate and qualify data logically; its specific meaning has to be defined for each application. If this character is used in hierarchical order, as specified in the general definition of the information separators, it delimits a data item called a "group."

HT (Horizontal Tabulation). A format effector that causes the active position to advance to the next predetermined character position.

LF (Line Feed). A format effector that causes the active position to advance to the corresponding character position of the next line.

NAK (Negative Acknowledge). A transmission control character transmitted by a receiver as a negative response to the sender.

NUL (Null). A control character used to accomplish media-fill or time-fill. NUL characters may be inserted into or removed from a stream of data without affecting the information content of that stream, but such action may affect the information layout and/or the control of equipment.

RS (Record Separator) (Information Separator Two). A control character used to separate and qualify data logically; its specific meaning has to be defined for each application. If this character is used in hierarchical order, as specified in the general definition of the information separators, it delimits a data item called a "record."

SI (Shift-In). A control character that is used in conjunction with SO and ESC
(continued)

TABLE 10.2 *(continued)*

to extend the graphic character set of the code. It may reinstate the standard meanings of the bit combinations that follow it. The effect of this character is described in ANSI X3.41-1974.

SO (Shift-Out). A control character that is used in conjunction with SI and ESC to extend the graphic character set of the code. It may alter the meaning of the bit combinations that follow it until an SI character is reached. The effect of this character is described in ANSI X3.41-1974.

SOH (Start of Heading). A transmission control character used as the first character of a heading of an information message.

STX (Start of Text). A transmission control character that precedes a text and that is used to terminate a heading.

SUB (Substitute Character). A control character used in the place of a character that has been found to be invalid or in error. SUB is intended to be introduced by automatic means, as, for example, when a transmission error is detected.

SYN (Synchronous Idle). A transmission control character used by a synchronous transmission system in the absence of any other character (idle condition) to provide a signal from which synchronism may be achieved or retained between data terminal equipment.

US (Unit Separator) (Information Separator One). A control character used to separate and qualify data logically; its specific meaning has to be defined for each application. If this character is used in hierarchical order, as specified in the general definition of the information separators, it delimits a data item called a "unit."

VT (Vertical Tabulation). A format effector that causes the active position to advance to the corresponding character position on the next predetermined line.

Source: Ref. 3.

5.2   Throughput

Throughput of a data channel is the expression of how much data are put through. In other words, throughput is an expression of channel efficiency. The term gives a measure of *useful* data put through the communication link. These data are directly useful to the computer or DTE (data-terminal equipment).

Therefore, on a specific circuit, throughput varies with the raw-data rate, is related to the error rate and the type of error encountered (whether burst or random), and varies according to the type of error detection and correction system used, the message-handling time, and the block length from which we must subtract overhead bits such as parity, flags, and cyclic redundancy checks. Throughput and the operational features of data circuits are described in detail in Chapter 11.

5.3   The Nature of Errors

In binary transmission an error is a bit that is incorrectly received. For instance, suppose a 1 is transmitted in a particular bit location and at the receiver the bit in that same location is interpreted as a 0. Bit errors occur either as single random errors or as bursts of errors.

Random errors occur when the signal-to-noise ratio deteriorates. This assumes, of course, that the noise is thermal noise. In this case noise peaks, at certain moments of time, are of sufficient level as to confuse the receiver's decision, whether a 1 or a 0.

Burst errors are commonly caused by fading on radio circuits. Impulse noise can also cause error bursts. Impulse noise can derive from lightning, car ignitions, electrical machinery, and certain electronic power supplies, to name a few sources.

5.3.1 *Error Performance.* The principal measure of quality of service (QoS) of a data circuit is its error performance. The most common method to express error performance is *bit error rate* (BER). It is also called *bit error ratio*. BER is almost universally expressed in the powers of 10. If the bit error rate of a circuit was 1 error in 1000 bits, we would express it as 1×10^{-3}. This means that if I receive 1000 bits, I should expect 1 bit to be in error of the 1000.

CCITT Rec. G.821 [4] recommends a BER of 1×10^{-6}; certain service providers in the United States and Canada offer a BER of 1×10^{-7} to the end user. In either case, these performance values are given in some sort of statistical distribution, for example, a BER of 1×10^{-7} for 97% of connections. If data bits are being mapped into a digital circuit such as a 64-kbps channel, then the underlying digital circuit should have the same or better error performance.

5.4 Error Detection and Error Correction

Error detection just identifies that a bit (or bits) has been received in error. Error correction corrects errors at a far-end receiver. Both require a certain amount of redundancy to carry out the respective function. Redundancy, in this context, means those added bits or symbols that carry out no other function than as an aid in the error detection or error correction process.

One of the earliest methods of error detection was the *parity check*. With the 7-bit ASCII code, a bit was added for parity, making it an 8-bit code. This is character parity. It is also referred to as *vertical redundancy checking* (VRC).

We speak of *even parity* and *odd parity*. One system or the other may be used. Either system is based on the number of marks or 1s in a 7-bit character, and the eighth bit is appended accordingly, either a 0 or a 1. Let's assume even parity and we transmit the ASCII bit sequence 1010010. There are three 1s, an odd number. Thus a 1 is appended as the eighth bit to make it an even number.

Suppose we use odd parity and transmit the same character. There is an odd number of 1s (marks) so we append a 0 to leave the total number of 1s an odd number. With odd parity, try 1000111. If you added a 1 as the eighth bit, you'd be correct.

Character parity has the weakness that a lot of errors can go undetected. Suppose two bits are changed in various combinations and locations. Suppose a 10 became a 01; a 0 became a 1 and a 1 became a 0; and two 1s became two 0s. All would get by the system undetected.

To strengthen this type of parity checking, the *longitudinal redundancy check* (LRC) was included as well as the VRC. This is a summing of the 1s in a vertical column of all characters, including the 1s in the 8th bit location. The sum is now appended at the end of a message block or frame which is a special field for error detection, which we can call the block check count or BCC. Later, we will call that field the *frame check sequence* (FCS). At the distant-end receiver, the same addition is carried out and if the sum agrees with the BCC value received, the block is accepted as error-free. If not, it contains at least one error, and a request is sent to the transmit end to re-transmit the block.*

Even with the addition of LRC, errors can get through. In fact no error detection system is completely foolproof. There is another method, though, that has excellent error detection properties. This is the *cyclic redundancy check* (CRC). It comes in a number of varieties.

5.4.1 Cyclic Redundancy Check (CRC). In very simple terms the CRC error detection technique works as follows. A data block or frame is placed in storage. We can call it a k-bit sequence and it can be represented by a polynomial which is called $G(x)$. Various modulo-2 arithmetic† operations are carried out on $G(x)$ and the result is divided by a known generator polynomial called $P(x)$. This results in a quotient $Q(x)$ and a remainder $R(x)$. The remainder is appended to the frame as an FCS (frame check sequence), and the total frame with FCS is transmitted to the distant-end receiver where the frame is stored, then divided by the same generating polynomial $P(x)$. The calculated remainder is compared to the received remainder (i.e., the FCS). If the values are the same, the frame is error free. If they are not, there is at least one bit in error in the frame.

For many wide area network applications the FCS is 16 bits long; on LANs it is often 32 bits long. In Chapter 8 we dealt with FCS of length 4 and 6 bits. Generally speaking the greater the number of bits, the more powerful the CRC is for catching errors.

The following are two common generating polynomials:

- ANSI CRC-16: $X^{16} + X^{15} + X^2 + 1$
- CRC-CCITT: $X^{16} + X^{12} + X^5 + 1$

producing a 16-bit FCS.

CRC-16 provides error detection of error bursts up to 16 bits in length. Additionally, 99.955% of error bursts greater than 16 bits can be detected [43].

---

*A *block* is a group of bits, bytes, or octets transmitted as a unit over which an error control procedure is applied. We may also use the terms packet or frame synonymously for block. The more popular term today is frame.

†Modulo-2 arithmetic is the same as binary arithmetic but without carries or borrows.

5.4.2 Forward-Acting Error Correction (FEC)

Forward-acting error correction (FEC) uses certain binary codes that are designed to be self-correcting for errors introduced by the intervening transmission media. In this form of error correction the receiving station has the ability to reconstitute messages containing errors.

The codes used in FEC can be divided into two broad classes: block codes and convolutional codes. In block codes information bits are taken k at a time, and c parity bits are added, checking combinations of the k information bits. A block consists of $n = k + c$ digits. When used for the transmission of data, block codes may be systematic. A systematic code is one in which the information bits occupy the first k positions in a block and are followed by the $(n - k)$ check digits.

Still another block code is the group code, where the modulo-2 sum of any two n-bit code words is another code word. Modulo-2 addition is denoted by the symbol $\oplus$. It is a binary addition without the "carry" or $1 + 1 = 0$, and we do not carry the 1. Summing 10011 and 11001 in modulo-2, we get 01010.

The minimum Hamming distance is a measure of the error detection and correction capability of a code. This "distance" is the minimum number of digits in which two encoded words differ. For example, to detect E digits in error, a code of a minimum Hamming distance of $(E + 1)$ is required. To correct E errors, a code must display a minimum Hamming distance of $(2E + 1)$. A code with a minimum Hamming distance of 4 can correct a single error *and* detect two digits in error.

A convolution(al) code is another form of coding used for error correction. As the word "convolution" implies, this is one code wrapped around or convoluted on another. It is the convolution of an input-data stream and the response function of an encoder. The encoder is usually made up of shift registers. Modulo-2 adders are used to form check digits, each of which is a binary function of a particular subset of the informations digits in the shift register.

Error performance can also be improved by the assistance of a microprocessor in the decoder. Mark or space decisions made by a demodulator are not hard or irrevocable decisions; rather, these are called "soft" decisions. In this case a tag of 3 bits is attached to each received digit to indicate the confidence level of the decision in the demodulator before processing. After processing, when errors are indicated, the bits with the lowest confidence level are changed from 0 to 1 or 1 to 0, as the case may be.

FEC as is works on random errors only. It does not help with burst errors. But we can fool the system by using an interleaver–deinterleaver. This concept is shown in Figure 10.4. An interleaver first stores a stream of x number of serial bits. It then shuffles the bits into a pseudorandom sequence and releases those x number of bits for transmission. The deinterleaver operates at the receive end using the same randomizing polynomial, but de-randomizes, placing the bits back in their original order before they entered the interleaver. Of course, the deinterleaver has to be time-synchronized with the

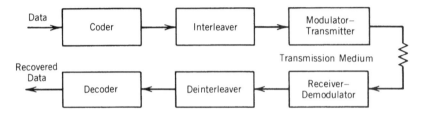

Figure 10.4. FEC scheme for a channel with burst errors.

interleaver. The value of x is important. x can be related to time. If we transmit 1200 bps and we wish time to be 1 s, then x must be 1200 bits. x must be greater than the expected duration of a burst. If we can meet these conditions, burst errors will be treated as though they are random errors.

5.5 Error Correction with Feedback Channel

Two-way or feedback error correction is used widely today on data and some telegraph circuits. Such a form of error correction is called ARQ. The letter sequence ARQ derives from the old Morse and telegraph signal, "automatic repeat request."

There are three varieties of ARQ:

- Stop-and-wait ARQ
- Selective or continuous ARQ
- Go-back-n ARQ

Stop-and-wait ARQ is simple to implement and may be the most economic in the short run. It works on a frame-by-frame basis. A frame is generated; it goes through CRC processing and an FCS is appended. It is transmitted to the distant end where the frame runs through CRC processing. If no errors are found, an acknowledgment signal (ACK) is sent to the transmitter, which now proceeds to send the next frame—and so forth. If a bit error is found, a negative acknowledgment (NACK) is sent to the transmitter, which then proceeds to repeat that frame. It is the waiting time of the transmitter as it waits for either acknowledgment or negative acknowledgment signals. Many point to this wait time as wasted time. It could be costly on high-speed circuits. However, the control software is simple and the storage requirements are minimal (i.e., only one frame).

Selective ARQ, sometimes called *continuous ARQ*, eliminates the waiting. The transmit side pours out a continuous stream of contiguous frames. The receive side stores and CRC processes as before, but it is processing a continuous stream of frames. When a frame is found in error, it informs the transmit side on the return channel. The transmit side then picks that frame out of storage and places it in the transmission queue. Several points become obvious to the reader. First, there must be some way to identify frames. Second, there must be a better way to acknowledge or "negative-acknowledge." The two problems are combined and solved by the use of send sequence numbers and receive sequence numbers. The header of a frame has bit positions for a send sequence number and a receive sequence number. The send sequence number is inserted by the transmit side, whereas the receive sequence number is inserted by the receive side. The receive sequence numbers forwarded back to the transmit side are the send sequence of frame numbers acknowledged by the receive side. Of course, the receive side has to insert the corrected frame in its proper sequence before passing the data message to the end user.

Continuous or *selective ARQ* is more costly in the short run compared to stop-and-wait ARQ. It requires more complex software and notably more storage on both sides of the link. However, there are no gaps in transmission and no time is wasted waiting for the ACK or NACK.

Go-back-n ARQ is a compromise. In this case the receiver does not have to insert the corrected frame in its proper sequence, thus less storage is required. It works this way. When a frame is received in error, the receiver informs the transmitter to "go-back-*n*," *n* being the number of frames back to where the errored frame was. The transmitter then repeats all *n* frames, from the errored frame forward. Meanwhile, the receiver has thrown out all frames from the errored frame forward. It replaces this group with the new set of *n* frames it received, all in proper order.

6 THE dc NATURE OF DATA TRANSMISSION

6.1 Loops

Binary data are transmitted on a dc loop. More correctly, the binary data end instrument delivers to the line and receives from the line one or several dc loops. In its most basic form a dc loop consists of a switch, a dc voltage, and a termination. A pair of wires interconnects the switch and termination. The voltage source in data and telegraph work is called the *battery*, although the device is usually electronic, deriving the dc voltage from an ac power line source. The battery is placed in the line to provide voltage(s) consistent with the type of transmission desired. A simplified dc loop is shown in Figure 10.5.

6.2 Neutral and Polar dc Transmission Systems

Older telegraph and data systems operated in the *neutral* mode. Nearly all present data transmission systems operate in some form of *polar* mode. The words "neutral" and "polar" describe the manner in which battery is applied to the dc loop. On a "neutral" loop, following the convention of Table 10.1, battery is applied during spacing (0) conditions and is switched off during

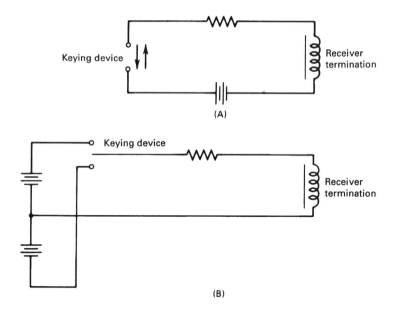

Figure 10.5. Simplified diagram illustrating a dc loop with (A) neutral and (B) polar keying.

marking (1). Current therefore flows in the loop when a space is sent and the loop is closed. Marking is indicated on the loop by a condition of no current. Thus we have two conditions for binary transmission, an open loop (no current flowing) and a closed loop (current flowing). Keep in mind that we could reverse this, namely, change the convention and assign marking to a condition of current flowing or closed loop and spacing to a condition of no current or an open loop.* As we mentioned, this is called "changing the sense." Either way, a neutral loop is a dc loop circuit where one binary condition is represented by the presence of voltage and the flow of current, and the other condition is represented by the absence of voltage and current. Figure 10.5A illustrates a neutral loop.

Polar transmission approaches the problem differently. Two battery sources are provided, one "negative" and the other "positive." Following the convention in Table 10.1, during a condition of spacing (binary 0), a positive battery (i.e., positive voltage) is applied to the loop, and a negative battery is applied during marking (binary 1). In a polar loop, current is always flowing. For a mark or binary "1" it flows in one direction and for a space or binary 0 it flows in the opposite direction. Figure 10.5B shows a simplified polar loop. Notice that the switch used to select the voltage we called a *keying device*. Figure 10.6 shows the two electrical waveforms.

*In fact this was the older convention, about prior to 1960.

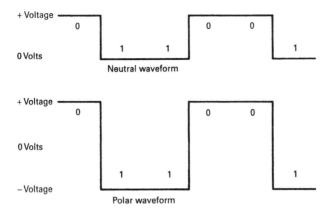

Figure 10.6. Neutral and polar electrical waveforms.

7 BINARY TRANSMISSION AND THE CONCEPT OF TIME

7.1 Introduction

Time and timing are most important factors in digital transmission. For this discussion, consider a binary end instrument sending out in series a continuous run of marks and spaces. Those readers who have some familiarity with the Morse code will recall that the spaces between dots and dashes told the operator where letters ended and where words ended. With the sending device or transmitter delivering a continuous series of characters to the line, each consisting of five, six, seven, eight, or nine elements (bits) per character, a receiving device that starts its print cycle when the transmitter starts sending and subsequently is perfectly in step with the transmitter can be expected to provide good printed copy and few, if any, errors at the receiving end.

It is obvious that when signals are generated by one machine and received by another, the speed of the receiving machine must be the same or very close to that of the transmitting machine. When the receiver is a motor-driven device, timing stability and accuracy are dependent on the accuracy and stability of the speed of rotation of the motors used. Most simple data–telegraph receivers sample at the presumed center of the signal element. It follows, therefore, that whenever a receiving device accumulates timing error of more than 50% of the period of one bit, it will print in error.

The need for some sort of synchronization is illustrated in Figure 10.7. A five-unit code is employed, and three characters transmitted sequentially are shown. Sampling points are shown in Figure 10.7 as vertical arrows. Receiving timing begins when the first pulse is received. If there is a 5% timing difference between the transmitter and receiver, the first sampling at the receiver will be 5% away from the center of the transmitted pulse. At the end of the tenth pulse or signal element the receiver may sample in error. The

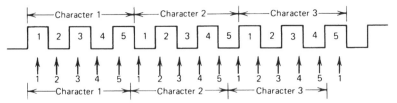

Figure 10.7. Five-unit synchronous bit stream with timing error.

eleventh signal element will, indeed, be sampled in error, and all subsequent elements will be errors. If the timing error between transmitting machine and receiving machine is 2%, the cumulative error in timing would cause the receiving device to print all characters in error after the 25th bit.

7.2 Asynchronous and Synchronous Transmission

In the earlier days of printing telegraphy, "start–stop" transmission, or asynchronous operation, was developed to overcome the problem of synchronism. Here timing starts at the beginning of a character and stops at the end. Two signal elements are added to each character to signal the receiving device that a character has begun and ended.

For example, consider a five-element code such as CCITT No. 2 (see Figure 10.1). In the front of a character an element called a "start space" is added, and a stop mark is inserted at the end of each character. To send the letter Y in Figure 10.1, the receiving device starts its timing sequence on the first signal element, which is a space or 0, followed by 10101, which is the code sequence for the character Y, followed by a stop mark, which terminates the timing sequence, as shown in Figure 10.8. In such an operation, timing errors can accumulate only inside each character. Suppose the receiving device is again 5% slower or faster than its transmitting counterpart; now the fifth information element will be no more than 30% displaced in time from the transmitted pulse and well inside the 50% or halfway point for correct sampling to take place.

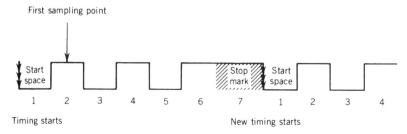

Figure 10.8. Five-unit start–stop stream of bits with a 1.5-unit stop element.

In start–stop transmission, information signal elements are each of the same duration, which is the duration or pulse width of the start element. The stop element has an indefinite length or pulse width beyond a certain minimum. If a steady series of characters is sent, the stop element is always of the same width or has the same number of unit intervals. Consider the transmission of two Y's, $0101011010101111 \rightarrow 11111$. The start space (0) starts the timing sequence for six additional elements, which are the five code elements in the letter Y and the stop mark. Timing starts again on the mark-to-space transition between the stop mark of the first Y and the start of the second. Sampling is carried out at pulse center for most asynchronous systems. Note that a continuous series of marks is sent at the end of the second Y; thus the signal is a continuation of the stop element or just a continuous mark. It is the mark-to-space transition of the start element that tells the receiving device to start timing a character.

Minimum lengths of stop elements vary. The preceding example above shows a stop element of one-unit interval duration (1 bit). Some are 1.42-unit intervals, others are of 1.5- and 2-unit interval duration. The proper semantics of data–telegraph transmission would describe the code of the previous paragraph as a five-unit start–stop code with a one-unit stop element.

A primary objective in the design of data systems is to minimize errors received or to minimize the error rate. Two of the prime causes of errors are noise and improper timing relationships. With start–stop systems a character begins with a mark-to-space transition at the beginning of the start space. Then 1.5-unit intervals later, the timing causes the receiving device to sample the first information element, which simply is a mark or space decision. The receiver continues to sample at one-bit intervals until the stop mark is received. In start–stop systems the last information bit is most susceptible to cumulative timing errors. Figure 10.8 is an example of a five-unit start–stop bit stream with a 1.5-unit stop element.

Another problem in start–stop systems is the mutilation of the start element. Once this happens, the receiver starts a timing sequence on the next mark-to-space transition it sees and then continues to print in error until, by chance, it cycles back properly on a proper start element.

Synchronous data systems do not have start and stop elements, but consist of a continuous serial stream of information elements or bits such as shown in Figure 10.7. With start–stop systems, timing error could only accumulate inside a character, those five or eight bits of character length. This is not so for synchronous systems. Timing error can accumulate for the entire length of a frame.

With start–stop systems, the receiving device knows when a character starts by the mark-to-space transition at the start space. In a synchronous transmission system, some marker must be provided to tell the receiver when a frame starts. This "marker" is the *unique field*. Every data frame starts with a unique field. A generic data frame is shown in Figure 10.9. View the frame from left to right. The first field is the unique field or flag, and it generally

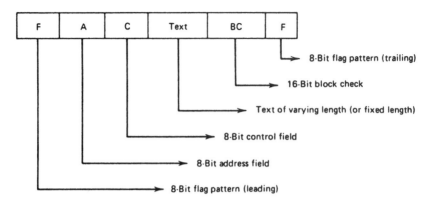

Figure 10.9. A generic data frame. Flag pattern = unique field. Text often called "info" field or information field.

consists of the binary sequence 01111110. A frame always starts with this field and ends with the same field. If one frame follows another contiguously, the unique field ending frame No. 1 is the unique field starting frame No. 2, and so forth. Once a frame knows where it starts, it will know a priori where the following fields begin and end by simple bit/octet counting. However, with some data link protocols, the information field is of variable length. In this case, the information field length will appear as a subfield in the control field.

The unique field *aligns* the frame. Suppose the unique field binary sequence (01111110) occurred inside the frame. The receiver, of course, would misalign, and the entire frame would be in error. We don't want our hands tied in the use of a particular binary sequence. A frame should be *transparent*, meaning that we can use any 8-bit sequence we desire, even 01111110. Thus care must be taken with sequences of contiguous 1s. To avoid interpreting that unique field (or flag), *bit stuffing*, sometimes called *zero insertion*, is used.

The rule for bit stuffing is to insert (stuff) a 0 into the data stream of the frame proper after each successive appearance of five 1s. This concept is shown in Figure 10.10. Thus, the frame, after stuffing, never contains more

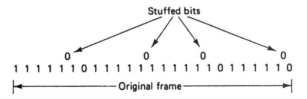

Figure 10.10. The concept of bit stuffing (zero insertion). A 0 bit is stuffed after each consecutive five 1s inside the original frame. The unique field (flag) has no stuffing.

than five consecutive 1s, and the unique field (flag) at the end of the frame is singularly recognizable. At the receiving end of the link, the first 0 after each string of five consecutive 1s is deleted. If, however, a string of five 1s is followed by a 1, the frame is declared to be finished.

Bit stuffing is used for other purposes beyond eliminating flags within the frame. Most data link protocols have an abort capability in which a frame can be aborted by transmitting seven or more 1s in a row. In addition, a link is regarded as idle if 15 or more 1s in a row are received.

Synchronous data transmission systems not only require frame alignment, but must be bit-aligned as well. This was an easy matter on start–stop systems because we could let the receive clock run freely inside the 5 or 8 bits of a character. There is no such freedom with synchronous systems. Suppose we assume a free-running receive clock. However, if there was a timing error of 1% between the transmit and receive clocks, not more than 100 bits could be transmitted until the synchronous receiving device would be off in timing by the duration of 1 bit from the transmitter, and all bits received thereafter would be in error. Even if the timing accuracy of one relative to the other was improved to 0.05%, the correct timing relationship between transmitter and receiver would exist for only the first 2000 bits transmitted. It follows, therefore, that no timing error whatsoever can be permitted to accumulate since anything but absolute accuracy in timing would cause eventual malfunctioning. In practice, the receiver is provided with an accurate clock that is corrected by small adjustments based on the transitions of the received bit stream, as explained in Section 7.3.

7.3 Timing

All currently used data transmission systems are synchronized in phase and symbol rate in some manner. Start–stop synchronization has already been discussed. All fully synchronous transmission systems have timing generators or clocks to maintain stability. The transmitting device and its companion receiver at the far end of the circuit must maintain a timing system. In normal practice, the transmitter is the master clock of the system. The receiver also has a clock that in every case is corrected by some means to its transmitter's master clock equivalent at the far end.

Another important timing factor is the time it takes a signal to travel from the transmitter to the receiver. This is called *propagation time*. With velocities of propagation as low as 20,000 mi/s, consider a circuit 200 mi in length. The propagation time would then be 200/20,000 s or 10 ms. Ten milliseconds is the time duration of 1 bit at a data rate of 100 bps; thus the receiver in this case must delay its clock by 10 ms to be in step with its incoming signal. Temperature and other variations in the medium may also affect this delay, as well as variations in the transmitter master clock.

There are basically three methods of overcoming these problems. One is to provide a separate synchronizing circuit to slave the receiver to the transmitter's master clock. However, this wastes bandwidth by expending a voice channel or subcarrier just for timing. A second method, which was quite widely used until several years ago, was to add a special synchronizing pulse for groupings of information pulses, usually for each character. This method was similar to start–stop synchronization and lost its appeal largely because of the wasted information capacity for synchronizing. The most prevalent system in use today is one that uses transition timing, where the receiving device is automatically adjusted to the signaling rate of the transmitter by sampling the transitions of the incoming pulses. This type of timing offers many advantages, particularly automatic compensation for variations in propagation time. With this type of synchronization the receiver determines the average repetition rate and phase of the incoming signal transition and adjusts its own clock accordingly by means of a phase-locked loop.

In digital transmission the concept of a transition is very important. The transition is what really carries the information. In binary systems the space-to-mark and mark-to-space transitions (or lack of transitions) placed in a time reference contain the information. In sophisticated systems, decision circuits regenerate and retime the pulses on the occurrence of a transition. Unlike decision circuits, timing circuits that reshape a pulse when a transition takes place must have a memory in case a long series of marks or spaces is received. Although such periods have no transitions, they carry meaningful information. Likewise, the memory must maintain timing for reasonable periods in case of circuit outage. Note that synchronism pertains to both frequency and phase and that the usual error in high-stability systems is a phase error (i.e., the leading edges of the received pulses are slightly advanced or retarded from the equivalent clock pulses of the receiving device). Once synchronized, high-stability systems need only a small amount of correction in timing (phase). Modem internal timing systems may have a long-term stability of 1×10^{-8} or better at both the transmitter and receiver. At 2400 bps, before a significant timing error can build up, the accumulated time difference between transmitter and receiver must exceed approximately 2×10^{-4} s. Whenever the circuit of a synchronized transmitter and receiver is shut down, their clocks must differ by at least 2×10^{-4} s before significant errors take place once the clocks start back up again. This means that the leading edge of the receiver-clock equivalent timing pulse is 2×10^{-4} in advance or retarded from the leading edge of the pulse received from the distant end. Often an idling signal is sent on synchronous data circuits during periods of no traffic to maintain the timing. Some high-stability systems need resynchronization only once a day.

Note that thus far in our discussion we have considered dedicated data circuits only. With switched (dial-up) synchronous circuits, the following problems exist:

- No two master clocks are in perfect phase synchronization.

- The propagation time on any two paths may not be the same.

Thus such circuits will need a time interval for synchronization for each call setup before traffic can be passed.

To summarize, synchronous data systems use high-stability clocks, and the clock at the receiving device is undergoing constant but minuscule corrections to maintain an in-step condition with the received pulse train from the distant transmitter, which is accomplished by responding to mark-to-space and space-to-mark transitions. The important considerations of digital network timing were also discussed in Chapter 9.

7.4 Distortion

It has been shown that the key factor in data transmission is timing. Although the signal must be either a mark or space, this alone is not sufficient. The marks and spaces (or 1s and 0s) must be in a meaningful sequence based on a time reference.

In the broadest sense, distortion may be defined as any deviation of a signal in any parameter, such as time, amplitude, or wave shape, from that of the ideal signal. For binary data transmission, distortion is defined as a displacement in time of a signal transition from the time that the receiver expects to be correct. In other words, the receiving device must make a decision as to whether a received signal element is a mark or a space. It makes the decision during the sampling interval, which is usually at the center of where the received pulse or bit should be; thus it is necessary for the transitions to occur between sampling times and preferably halfway between them. Any displacement of the transition instants is called "distortion." The degree of distortion suffered by a data signal as it traverses the transmission medium is a major contributor in determining the error rate that can be realized.

Telegraph and data distortion is broken down into two basic types, systematic and fortuitous. Systematic distortion is repetitious and is broken down into bias distortion, cyclic distortion, and end distortion, which is more common in start–stop systems. Fortuitous distortion is random and characterized by a displacement of a transition from the time interval in which it should have occurred. Distortion caused by noise spikes or other transients in the transmission medium may be included in this category. Characteristic distortion is still another type and is caused by transients in the modulation process that then appear in the demodulated signal.

Figure 10.11 shows some examples of distortion. Figure 10.11A is a binary signal without distortion, and Figure 10.11B shows the sampling instants, which should ideally occur in the center of the pulse to be sampled. From this we can see that the displacement tolerance is nearly 50%; that is, the point of sample could be displaced by up to 50% of a pulse width and still record the mark or space condition present without error. However, the sampling interval does require a finite amount of time; thus in actual practice the permissible displacement is somewhat less than 50%. Figures 10.11C and

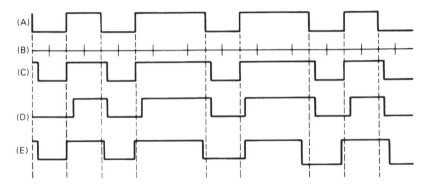

Figure 10.11. Three typical distorted data signals.

10.11D show the two typical types of bias distortion. Spacing bias is shown in Figure 10.11C, where all the spacing impulses are lengthened at the expense of the marking impulses. Figure 10.11D shows marking bias, which is the reverse; the marking impulses are lengthened at the expense of the spaces. Figure 10.11E shows fortuitous distortion, which is random.

Figure 10.12 shows distortion that is more typical of start–stop transmission. Figure 10.12A is an undistorted start–stop signal. Figure 10.12B shows cyclic or repetitive distortion typical of mechanical transmitters. In this type of distortion the marking elements may increase in length for a period of time and then the spacing elements will increase in length. Figure 10.12C shows peak distortion. Identifying the type of distortion present on a signal often gives a clue to the source or cause of distortion. Distortion-measurement equipment measures the displacement of the mark-to-space transition from the ideal of the digital signal. If a transition occurs too near to the sampling point, the signal element is liable to be in error.

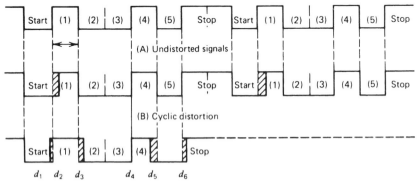

Figure 10.12. Distorted telegraph signals illustrating cyclic and peak distortion. Peak distortion appears at transition d_5.

7.5 Bits, Bauds, and Symbols

There is much confusion among professionals in the telecommunication industry over terminology, especially in differentiating bits, bauds, and symbols. The bit, a binary digit, has been defined previously.

The baud is a unit of transmission rate or modulation rate. It is a measure of transitions per second. A transition is a change of state. In *binary* systems, bauds and bits per second (bps) are synonymous. In higher-level systems, typically *m*-ary systems, bits and bauds have different meanings. For example, we will be talking about a type of modulation called QPSK. In this case, every transition carries two bits. Thus the modulation rate in bauds is half the bit rate.

The industry often uses symbols per second and bauds interchangeably. It would be preferable, in our opinion, to use "symbols" for the output of a coder or other conditioning device. For the case of a channel coder (or encoder), bits go in and symbols come out. There are more symbols per second in the output than bits per second in the input. They differ by the coding rate. For example, a 1/2 rate coder (used in FEC) may have 4800 bps at the input and then would have 9600 symbols per second at the output.

7.5.1 *The Period of a Bit, Symbol, or Baud.* The period of a bit is the time duration of a bit pulse. When we use NRZ coding (discussed in Section 7.6), the period of a bit, baud, or symbol is simply 1/(bit rate) or 1/(symbol rate). For example, if we are transmitting 2400 bps, what is the period of a bit? It is 1/2400 or 416 μs; for 19.2 kbps it is $1/19,200 = 52.08$ μs; and for 1.544 Mbps it is 647 ns.

7.6 Digital Data Waveforms

Digital symbols may be represented in many different ways by electrical signals to facilitate data transmission. All these methods for representing (or coding) digital symbols assign electrical parameter values to the digital symbols. In binary coding, of course, these digital symbols are restricted to two states, space (0) and mark (1). The electrical parameters used to code digital signals are levels (or amplitudes), transitions between different levels, phases (normally 0° and 180° for binary coding), pulse duration, and frequencies or a combination of these parameters. There is a variety of coding techniques for different areas of application, and no particular technique has been found to be optimum for all applications, considering such factors as implementing the coding technique in hardware, type of transmission technique employed, decoding methods at the data sink or receiver, and timing and synchronization requirements.

In this section we discuss several basic concepts of *electrical* coding of binary signals. In this discussion reference is made to Figure 10.13, which graphically illustrates several line coding techniques.

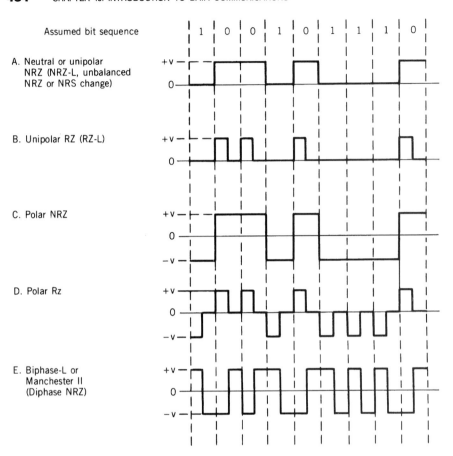

Figure 10.13. Digital data transmission waveforms.

Figure 10.13A shows what is still called by many today "neutral transmission." This was the principal method of transmitting telegraph signals until about 1960. In many parts of the world, neutral transmission is still widely used. First, this waveform is a non-return-to-zero (NRZ) format in its simplest form. "Non-return-to-zero" simply means that if a string of 1s (marks) are transmitted, the signal remains in the mark state with no transitions. Likewise, if a string of 0s is transmitted, there is no transition and the signal remains in the 0 state until a 1 is transmitted. As we can now see, with NRZ transmission, we can transmit information without transitions.

Figures 10.13B and 10.13D show the typical "return-to-zero" (RZ) waveform, where, when a continuous string of marks (or spaces) is transmitted, the signal level (amplitude) returns to the zero voltage condition at each element or bit. Obviously RZ transmission is much richer in transitions than NRZ.

In Section 6.2 we discussed neutral and polar dc transmission systems. Figure 10.13A shows a typical neutral waveform where the two state conditions are 0 V for the mark or 1 condition and some positive voltage for the space or 0 condition. On the other hand, in polar transmission, as shown in Figures 10.12C and 10.12D, a positive voltage represents a space and a negative voltage, a mark. With NRZ transmission, the pulse width is the same as the duration of a unit interval or bit. Not so with RZ transmission, where the pulse width is less than the duration of a unit interval. This is because we have to allow time for the pulse to return to the zero condition.

Bi-phase-L or Manchester coding (Figure 10.13E) is a code format that is being used ever more widely on digital systems such as wire pair, coaxial cable, and fiber optics. Here the binary information is carried in the transition. By convention a logic 0 is defined as a positive-going transition and a logic 1 as a negative-going transition. Note that Manchester coding has a signal transition in the middle of each unit interval (or bit). Manchester coding is a form of phase coding.

The reader should be cognizant of and be able to differentiate between two sets or ways of classifying binary digital waveforms. The first set is *neutral* and *polar*. The second set is *NRZ* and *RZ*. Manchester coding is still another way to represent binary digital data where the transition takes place in the middle of the unit interval. In Chapter 8 one more class of waveform was introduced: alternate mark inversion (AMI).

8 DATA INTERFACE—THE PHYSICAL LAYER

When we wish to transmit data over a conventional analog network, the electrical representation of the data signal is essentially direct current. As such it is incompatible with that network that accepts information channels in the band 300–3400 Hz.* A data modem is a device that brings about this compatibility. It translates the electrical data signal into a modulated frequency tone in the range of 300–3400 Hz, often 1800 Hz. Also, more often than not, the digital network (described in Chapters 8 and 9) extensions are analog and require the same type(s) of modem. If the digital network extends to the user's premise, a digital conditioning device (CSU/DSU) is required for bit rate and waveform compatibility.

For this discussion of data interface, we will call the modem or digital conditioning device *data communication equipment (DCE)*. This equipment has two interfaces, one on each side as shown in Figure 10.14. The first, which is discussed in this section, is on the user side, which is called *data terminal equipment (DTE)*, and the applicable interface is the DTE–DCE interface. The second interface is on the line side, which is covered in Section 9. It should be noted that the DTE–DCE interface is well defined.

---

*This is the traditional analog voice channel.

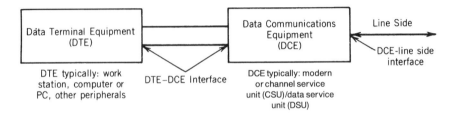

Figure 10.14. Data circuit interfaces, physical layer.

The most well-known DTE–DCE standard was developed by the (US) Electronics Industries Association (EIA) and is called EIA-232E [5]. It is essentially equivalent to international standards covered by CCITT Recs. V.24, V.28, and ISO IS2110.

EIA-232E and most of the other standards discussed are applicable to the DTE–DCE interface employing serial binary data interchange. It defines signal characteristics, mechanical interface characteristics, and functional descriptions of the interchange circuits. EIA-232E is applicable for data transmission rates up to 20,000 bps and for synchronous/asynchronous serial binary data systems.

Section 2.1.3 is quoted from EIA-232E. It is crucial to the understanding of signal state conventional and level:

> For data interchange circuits, the signal shall be considered in the marking condition when the voltage (V_1) on the interchange circuit, measured at the interface point [Figure 10.15], is more negative than minus three volts with respect to circuit AB (signal ground), The signal shall be considered in the spacing condition when the voltage V_1 is more positive than plus three volts with respect to circuit AB. . . . The region between plus three volts and minus three volts is defined as the transition region. The signal state is not uniquely defined when the voltage (V_1) is in this transition region.

During the transmission of data, the marking condition is used to denote the binary state *ONE* and the spacing condition is used to denote the binary state *ZERO*.

Figure 10.15 shows the interchange equivalent circuit.

Besides EIA-232 there are many other interface standards issued by EIA, CCITT, U.S. federal standards, U.S. military standards, and ISO. Each defines the DTE–DCE interface. Several of the more current standards are briefly described below.

EIA-530* [6] is a comparatively recent standard developed by the EIA. It provides for all data rates below 2.1 Mbps and it is intended for all applications requiring a balanced electrical interface. It can also be used for unbalanced operation.

*More properly called ANSI/EIA/TIA-530-A.

Interchange Circuit

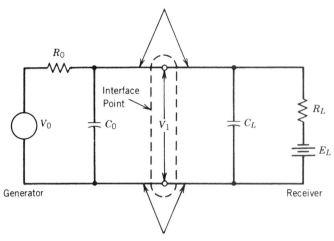

Circuit AB, Signal Ground

Figure 10.15. Interchange equivalent circuit. Courtesy of Electronics Industries Association, from EIA RS-232E (Ref. 5).

V_0 is the open-circuit generator voltage.

R_0 is the generator internal dc resistance.

C_0 is the total effective capacitance associated with the generator, measured at the interface point and including any cable to the interface point.

V_1 is the voltage at the interface point.

C_L is the total effective capacitance associated with the receiver, measured at the interface point and including any cable to the interface point.

R_L is the receiver load dc resistance.

E_L is the open-circuit receiver voltage (bias).

Let us digress for a moment. An unbalanced electrical interface is where one of the signal leads is grounded; for a balanced electrical interface, no ground is used.

EIA-530 applies for both synchronous and nonsynchronous (i.e., start–stop) operation. It uses a standard 25-pin connector; alternatively it can use a 26-pin connector. A list of interchange circuits showing circuit mnemonic, circuit name, circuit direction, and circuit type is presented in Table 10.3.

The electrical characteristics of EIA-530 are described in EIA/TIA-422 and 423 [7, 8].

EIA-422 deals with a balanced electrical interface, and EIA-423 deals with an unbalanced electrical interface. The state transition region for these two standards is between $+2$ V and -2 V at the generator side. The sense is the same as EIA-232.

The applicable U.S. military standard is MIL-STD-188-114. This provides a potpourri of EIA-232 and EIA-422/423 parameters. One significant difference reflected in MIL-STD-188-114 [9] is that the receiver has a balanced

TABLE 10.3   EIA-530 Interchange Circuits

Circuit Mnemonic	CCITT Number	Circuit Name	Circuit Direction	Circuit Type
AB	102	Signal Common		Common
AC	102B	Signal Common		
BA	103	Transmitted Data	To DCE	Data
BB	104	Received Data	From DCE	
CA	105	Request to Send	To DCE	
CB	106	Clear to Send	From DCE	
CF	109	Received Line Signal Detector	From DCE	
CJ	133	Ready for Receiving	To DCE	
CE	125	Ring Indicator	From DCE	Control
CC	107	DCE Ready	From DCE	
CD	108/1, /2	DTE Ready	To DCE	
DA	113	Transmit Signal Element Timing (DTE Source)	To DCE	
DB	114	Transmit Signal Element Timing (DCE Source)	From DCE	Timing
DD	115	Receiver Signal Element Timing (DCE Source)	From DCE	
LL	141	Local Loopback	To DCE	
RL	140	Remote Loopback	To DCE	
TM	142	Test Mode	From DCE	

Source: ANSI/EIA/TIA-530-A, EIA, June 1992 (Ref. 6).

input even though it is used with an unbalanced generator. The choice of a balanced receiver was done deliberately for the following reasons:

1. Noise immunity and reducing problems of ground potential differences between generator and receivers.

2. Convenience of inverting mark and space signaling sense.

3. Uniformity of receiver design for economic advantages in mass production.

CCITT has issued a number of recommendations for the DTE–DCE interface. The equivalent of EIA-232 is CCITT Rec. V.24 [12]. The following are other pertinent CCITT recommendations:

Rec. V.10, *Electrical Characteristics for Unbalanced Double-Current Interchange Circuits for General Use with Integrated Circuit Equipment in the Field of Data Communications.* The term *double-current* is synonymous with polar transmission discussed in Sections 6.2 and 7.6 [10].

Rec. V.11, *Electrical Characteristics for Balanced Double-Current Interchange Circuits for General Use with Integrated Circuit Equipment in the Field of Data Communications* [11].

Rec. V.28, *Electrical Characteristics for Unbalanced Double-Current Interchange Circuits* [13].

Rec. V.31, V.31 bis, *Electrical Characteristics for Single-Current Interchange Circuits Controlled by Contact Closure.* Single-current refers to *neutral transmission* [14].

9 DIGITAL TRANSMISSION ON AN ANALOG CHANNEL

9.1 Introduction

Two fundamental approaches to the practical problem of data transmission are (1) to design and construct a complete, new network expressly for the purpose of data transmission and (2) to adapt the many existing telephone facilities for data transmission. The following paragraphs deal with the latter approach.

Analog transmission facilities designed to handle voice traffic have characteristics that hinder the transmission of dc binary digits or bit streams. To permit the transmission of data over voice facilities (i.e., the telephone network), it is necessary to convert the dc data into a signal within the voice-frequency range. The equipment that performs the necessary conversion to the signal is generally called a *modem*, an acronym for *mo*dulator–*dem*odulator.

9.2 Modulation–Demodulation Schemes

A modem modulates and demodulates a carrier signal with digital data signals. The types of modulation used by present-day modems may be one or a combination of the following:

Amplitude modulation, double sideband (DSB).

Amplitude modulation, vestigial sideband (VSB).

Frequency shift modulation, commonly called frequency shift keying (FSK).

Phase shift modulation, commonly called phase shift keying (PSK).

9.2.1 Amplitude Modulation: Double Sideband. With the double-sideband (DSB) modulation technique, binary states are represented by the presence or absence of an audio tone or carrier. More often it is referred to as "on–off telegraphy." For data rates up to 1200 bps, one such system uses a carrier frequency centered at 1600 Hz. For binary transmission, amplitude

modulation has significant disadvantages, which include (1) susceptibility to sudden gain change and (2) inefficiency in modulation and spectrum utilization, particularly at higher modulation rates (see CCITT Rec. R.70).

9.2.2 Amplitude Modulation: Vestigial Sideband.

An improvement in the amplitude modulation double-sideband (DSB) technique results from the removal of one of the information-carrying sidebands. Since the essential information is present in each of the sidebands, there is no loss of content in the process. The carrier frequency must be preserved to recover the dc component of the information envelope. Therefore digital systems of this type use VSB modulation in which one sideband, a portion of the carrier, and a "vestige" of the other sideband are retained. This is accomplished by producing a DSB signal and filtering out the unwanted sideband components. As a result, the signal takes only about 75% of the bandwidth required for a DSB system. Typical VSB data modems are operable up to 2400 bps in a telephone channel. Data rates up to 4800 bps are achieved using multilevel (*M*-ary) techniques. The carrier frequency is usually located between 2200 Hz and 2700 Hz.

9.2.3 Frequency Shift Modulation.

Many data transmission systems utilize frequency shift modulation (FSK*). The two binary states are represented by two different frequencies and are detected by using two frequency-tuned sections, one tuned to each of the 2 bit frequencies. The demodulated signal is then integrated over the duration of 1 bit, and a binary decision is based on the result.

Digital transmission using FSK modulation has the following advantages: (1) The implementation is not much more complex than an AM system; and (2) since the received signals can be amplified and limited at the receiver, a simple limiting amplifier can be used, whereas the AM system requires sophisticated automatic gain control for operation over a wide level range. Another advantage is that FSK can show a 3 to 4-dB improvement over AM in most types of noise environment, particularly at distortion threshold (i.e., at the point where the distortion is such that good printing is about to cease). As the frequency shift becomes greater (i.e., a greater frequency separation between the mark and space frequencies), the advantage over AM improves in a noisy environment.

Another advantage of FSK is its immunity from the effects of nonselective level variations, even when they occur extremely rapidly. Thus a major application is on worldwide high-frequency radio transmission where rapid fades are a common occurrence. In the United States, FSK has nearly universal application for the transmission of data at the lower data rates (i.e., $\leq$1200 bps).

*FSK meaning frequency shift keying. "Keying" derives from old-time telegraph transmission.

9.2.4 *Phase Shift Modulation.* For systems using higher data rates, phase modulation becomes more attractive. Various forms are used, such as two-phase, relative phase, and quadrature phase. A two-phase system uses one phase of the carrier frequency for one binary state and the other phase for the other binary state. The two phases are ideally 180° apart and are detected by a synchronous detector using a reference signal at the receiver that is of known phase with respect to the incoming signal. This known signal operates at the same frequency as the incoming signal carrier and is arranged to be in phase with one of the binary signals. In the relative-phase system a binary 1 is represented by sending a signal burst of the same phase as that of the previous signal burst sent. A binary 0 is represented by a signal burst of a phase opposite to that of the previous signal transmitted. The signals are demodulated at the receiver by integrating and storing each signal burst of 1-bit period for comparison in phase with the next signal burst. In the quadrature-phase system (QPSK), two binary channels (2 bits) are phase multiplexed onto one tone by placing them in phase quadrature, as shown in the following sketch. An extension of this technique places two binary channels on each of several tones spaced across the voice channel of a typical telephone circuit. Almost all modems operating at 2400 bps use QPSK. The baud rate in the case is one-half of the bit rate.

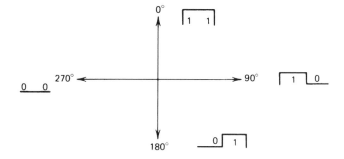

Some of the advantages of phase modulation are as follows:

1. All available power is utilized for intelligence conveyance.
2. The demodulation scheme has good noise-rejection capability.
3. The system yields a smaller noise bandwidth.

A disadvantage of such a system is the complexity of equipment required, compared to FSK systems.

9.3 Critical Parameters

The effect of the various telephone-circuit parameters on the capability of a circuit to transmit data is a most important consideration. The following discussion is intended to familiarize the reader with the problems most likely to be encountered in the transmission of data over analog circuits (e.g., the analog telephone network) and to make certain generalizations in some cases, which can be used to facilitate planning the implementation of data systems.

9.3.1 Phase Distortion. Phase distortion "constitutes the most limiting impairment to data transmission, particularly over telephone voice channels" [17]. When specifying phase distortion, the terms "envelope delay distortion" (EDD) and "group delay" are often used. The IEEE Standard Dictionary states that "envelope delay is often defined the same as group delay, that is the rate of change, with angular frequency, of the phase shift between two points in a network." (See Chapter 5, Section 2.2.)

The problem is that in a band-limited analog system, such as the typical telephone voice channel, not all frequency components of the input signal will propagate to the receiving end in exactly the same elapsed time, particularly on loaded cable circuits and FDM carrier systems. In carrier systems it is the cumulative effect of the many filters used in the FDM equipment. On long-haul circuits the magnitude of delay distortion is generally dependent on the number of carrier modulation stages that the circuit must traverse rather than the length of the circuit. Figure 10.16 shows a typical frequency–delay response curve in milliseconds of a voice channel due to FDM equipment only. For the voice channel (or any symmetrical passband, for that matter), delay increases toward band edge and is minimum around the center portion (around 1800 Hz).

Phase or delay distortion is the major limitation to modulation rate. The shorter the pulse width (the width of 1 bit in binary systems), the more critical will be the EDD parameters. As we discuss in Section 9.5, it is desirable to keep the delay distortion in the band of interest below the period of 1 bit.

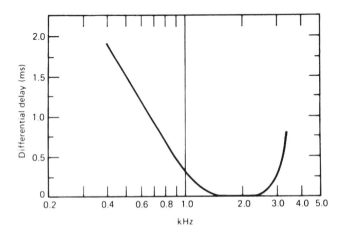

Figure 10.16. Typical differential delay across a voice channel, FDM equipment back to back.

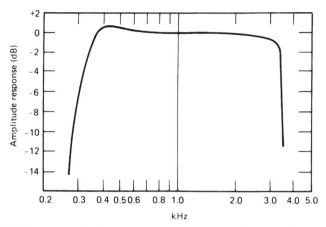

Figure 10.17. Typical amplitude–frequency response across a voice channel; channel modulator, demodulator back to back, FDM equipment.

9.3.2 ***Attenuation Distortion (Amplitude Response).*** Another parameter that seriously affects the transmission of data and can place very definite limits on the modulation rate is amplitude response, also called "attenuation distortion." Ideally, all frequencies across the passband of a channel of interest should undergo the same attenuation. For example, let a −10-dBm signal enter a channel at any frequency between 300 Hz and 3400 Hz. If the channel has 13 dB of flat attenuation, we would expect an output of −23 dBm at any and all frequencies in the band. This type of channel is ideal but unrealistic in a real working system.

In Rec. G.132 [15], the CCITT recommends no more than 9 dB of attenuation distortion relative to 800 Hz between 400 Hz and 3000 Hz. This figure, 9 dB, describes the maximum variation that may be expected from the reference level at 800 Hz. This variation of amplitude response is often called attenuation distortion. A conditioned channel, such as a Bell System C-4 channel, will maintain a response of −2 dB to +3 dB from 500 Hz to 3000 Hz and −2 dB to +6 dB from 300 Hz to 3200 Hz. Channel conditioning is discussed in Section 9.6.

Considering tandem operation, the deterioration of amplitude response is arithmetically cumulative when sections are added. This is particularly true at band edge in view of channel unit transformers and filters that account for the upper and lower cutoff characteristics. Figure 10.17 illustrates a typical example of amplitude response across FDM carrier equipment (see Chapter 5) connected back to back at the voice channel input–output. For additional discussion of attenuation distortion, see Chapter 5, Section 2.1.

9.3.3 ***Noise.*** Another important consideration in the transmission of data is noise. All extraneous elements appearing at the voice channel output that were not due to the input signal are considered to be noise. For convenience, noise is broken down into four categories: (1) thermal, (2) crosstalk, (3) inter-modulation, and (4) impulse [16]. Thermal noise, often called "resistance noise," "white noise," or "Johnson noise," is of a Gaussian nature or completely random. Any system or circuit operating at a temperature above absolute zero inherently will display thermal noise. The noise is caused by the random motions of discrete electrons in the conduction path. Crosstalk is a form of noise caused by unwanted coupling from one signal path into another. It may be caused by direct inductive or capacitive coupling between conductors or between radio antennas. Intermodulation noise is another form of unwanted coupling, usually caused by signals mixing in nonlinear elements of a system. Carrier and radio systems are highly susceptible to inter-modulation noise, particularly when overloaded. Impulse noise can be a primary source of errors in the transmission of data over telephone networks. It is sporadic and may occur in bursts or discrete impulses called "hits." Some types of impulse noise are natural, such as that from lightning. However, man-made impulse noise is ever increasing, such as that from automobile ignition systems and power lines. Impulse noise may be of a high level in conventional analog telephone switching centers as a result of dialing, supervision, and switching impulses that may be induced or otherwise coupled into the data transmission channel. The worst offender in the switching area is the step-by-step exchange, some of which are still around. Impulse noise in digital exchanges is almost nonexistent.

For our discussion of data transmission, two types of noise are considered, random (or Gaussian) noise and impulse noise. Random noise measured with a typical transmission measuring set appears to have a relatively constant value. However, the instantaneous value of the noise fluctuates over a wide range of amplitude levels. If the instantaneous noise voltage is of the same magnitude as the received signal, the receiving detection equipment may yield an improper interpretation of the received signal and an error or errors will occur. Thus we need some way of predicting the behavior of data trans-mission in the presence of noise. Random noise or white noise has a Gaussian distribution and is considered representative of the noise en-countered on the analog telephone channel (i.e., the voice channel). From the probability distribution curve for Gaussian noise shown in Figure 10.18, we can make some statistical predictions. It may be noted from this curve that the probability of occurrence of noise peaks that have amplitudes 12.5 dB above the rms level* is 1 in 10^5. Hence, if we wish to ensure an error rate of 10^{-5} in a particular system using binary polar modulation, the rms noise should be at least 12.5 dB below the signal level [Ref. 17, p. 114]. This simple analysis is only valid for the type of modulation used (i.e., binary polar baseband modulation), assuming that no other factors are degrading the operation of the system and that a cosine-shaped receiving filter is used. If we were to interject distortion such as EDD into the system, we could translate

---

*In a Nyquist bandwidth or 9.5 dB in a bit-rate bandwidth.

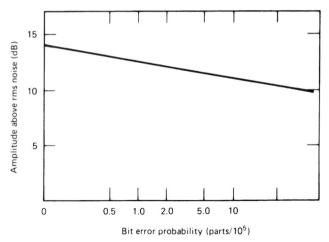

Figure 10.18. Probability of bit error in Gaussian noise; binary polar transmission with a Nyquist bandwidth.

the degradation into an equivalent signal-to-noise ratio improve-ment necessary to restore the desired error rate. For example, if the delay distortion were the equivalent of one pulse width, the signal-to-noise ratio improvement required for the same error rate would be about 5 dB, or the required signal-to-noise ratio would now be 17.5 dB.

Let us assume a telephone system where the signal level is −10 dBm at the zero transmission-level point of the system; the rms noise measured at the same point would be −27.5 dBm to retain the error rate of 1 in 10^5. This figure can be significant only if it is related to the actual noise found in a channel. The CCITT recommends no more than 50,000 pW of noise pso-phometrically weighted (−43 dBmp) on an international connection made up of six circuits in a chain. However, the CCITT states (Rec. G.143 [18]) that for data transmission at as high a modulation rate as possible without significant error rate, a reasonable circuit objective for maximum random noise would be −40 dBm0p for leased circuits (impulse noise not included) and −36 dBm0p for switched circuits without compandors. This figure obviously appears quite favorable when compared to the −27 dBm0 (−29.5 dBm0p) required in the preceding example. However, other factors developed later will consume much of the noise margin that appears to be available.

Unlike random noise, which is measured by its rms value when we measure level, impulse noise is measured by the number of "hits" or "spikes" per interval of time above a certain threshold. In other words, it is a measure-ment of the recurrence rate of noise peaks over a specified level. The word "rate" should not mislead the reader. The recurrence is not uniform per unit time, as the word "rate" may indicate, but we can consider a sampling and convert it to an average.

Bellcore [19] is quoted:

Impulse noise is any burst of noise, usually less than 5 but up to 10 ms in duration, that produces a voltage exceeding the rms (root mean square) noise voltage (i.e., the mean noise as measured with an AT&T Technologies 3A-type noise measuring set using C-message weighting or its equivalent) by a given magnitude. The impulse noise threshold is nominally 12 dB above the rms value for 3-kHz bandwidth systems, but in some systems, particularly microwave systems, it may increase to 16 dB. Impulse noise objectives and requirements are normally stated in terms of threshold noise setting and the number, or "count," of times impulse noise exceeds the threshold during a given time interval.

The performance objective is that there should be no more than five counts in 5 minutes on at least 50 percent of the switching system connections of each connection type. The objectives apply to a working (installed EO*) environment and over the life of the office.† These objectives are associated with the following noise thresholds:

- 47 dBrnC0 for analog and digital offices on connections between voice frequency interfaces including remote LI‡-to-remote LI (intra-RSU§), but excluding remote LI to host.
- 60 dBrnC0 with a −23 dBm0 holding tone of 1004 Hz and 66 dBrnC0 with a −13 dBm0 holding tone at 1004 Hz for remote LI-to-host voice frequency connections. The requirement is that 99 percent or more of the connections of each connection type between voice frequency interfaces (excluding the carrier facility used in RSU-host connections) should have zero counts in 5 minutes above 54 dBrnC0 for any possible combination of laboratory-simulated switching system operations. It is desirable that there be no counts in 5 minutes above 47 dBrnC0 under these same conditions.

CCITT Rec. Q.45 [20] states that "in any four-wire international exchange the busy hour impulsive noise counts should not exceed 5 counts in 5 minutes at a threshold level of −35 dBm0."

Remember that random noise has a Gaussian distribution and will produce peaks at 12.5 dB over the rms value (unweighted) 0.001% of the time on a data-bit stream for an equivalent error rate of 1×10^{-5}. It should be noted that some references use 12 dB, some use 12.5 dB, and others use 13 dB. The 12.5 dB above the rms random noise floor should establish the impulse noise threshold for measurement purposes. We should assume that in a well-designated data transmission system traversing the telephone network, the signal-to-noise ratio of the data signal will be well in excess of 12.5 dB. Thus impulse noise may well be the major contributor to degradation of the error rate.

*EO stands for end office, meaning a local switch.
†"Office" in North America means a switching center.
‡LI stands for line interface.
§RSU stands for remote switching unit.

When an unduly high error rate has been traced to impulse noise, there are some methods for improving conditions. Noisy areas may be bypassed, repeaters may be added near the noise source to improve signal-to-impulse-noise ratio, or in special cases pulse smearing techniques may be used. This latter approach uses two delay-distortion networks that complement each other such that the net delay distortion is zero. By installing the networks at opposite ends of the circuit, impulse noise passes through only one network* and hence is smeared because of the delay distortion. The signal is unaffected because it passes through both networks.

Signal-to-noise ratio may be traded for the implementation of forward-acting error correction (FEC). This trade-off may be economically feasible or even mandatory, such as in certain digital satellite circuits, to reduce power output of a transmitter or reduce level out of a modem. Reducing power by 3 dB (Figure 10.18) on a circuit displaying a bit error rate of 1×10^{-5} would deteriorate error rate to some 5×10^{-2}. The error rate could be recovered by selecting the proper method of FEC, decoding algorithm and possibly implementing soft mark–space decisions. Soft decision decoding can improve error performance by an equivalent 2 dB. For additional discussion of FEC, see Section 5 of this chapter.

9.3.4 *Levels and Level Variations.* The design signal levels of telephone networks traversing FDM carrier systems are determined by average talker levels, average channel occupancy, permissible overloads during busy hours, and so on. Applying constant amplitude digital data tone(s) over such an equipment of 0 dBm0 on each channel would result in severe overload and intermodulation within the system.

Loading does not affect (metallic) wire systems except by increasing crosstalk. However, once the data signal enters FDM carrier multiplex (voice) equipment, levels must be carefully considered, and the resulting levels most probably have more impact on the final signal-to-noise ratio at the far end than anything else. The CCITT (Rec. V.2 [21]) recommends that signal power shall not exceed −10 dBm0 in 10 Hz of bandwidth and −13 dBm0 when the portion of nonspeech circuits on an international carrier circuit exceeds 10% or 20%. For multichannel telegraphy −8.7 dBm0 for the composite level, or for 24 channels, each individual telegraph channel would be adjusted for −22.5 dBm0. Even this loading may be too heavy if a large portion of the voice channels is loaded with data. Depending on the design of the carrier equipment, cutbacks to −13 dBm0 or less may be advisable.

In a properly designed transmission system the standard deviation of the variation in level should not exceed 1.0 dB/circuit. However, data communication equipment should be able to withstand level variations in excess of 4 dB.

Levels are a secondary issue on digital (TDM) networks.

---

*This assumes that impulse noise enters the circuit at some point beyond the first network.

9.3.5 Frequency-Translation Errors.

Total end-to-end frequency-translation error on a voice channel being used for data or telegraph transmission must be limited to 2 Hz (CCITT Rec. G.135 [22]); this is an end-to-end requirement. Frequency translations occur largely because of FDM carrier equipment modulation and demodulation steps. Frequency-division multiplex carrier equipment widely uses single-sideband suppressed-carrier techniques. Nearly every case of error can be traced to errors in frequency translation (we refer here to deriving the group, supergroup, mastergroup, and its reverse process; see also Chapter 5) and carrier reinsertion frequency offset, where the frequency error is exactly equal to the error in translation and offset or the sum of several such errors. Frequency-locked (e.g., synchronized) or high-stability master carrier generators (1×10^{-8} or 1×10^{-9}, depending on the system), with all derived frequency sources slaved to the master source, are usually employed to maintain the required stability.

On digital networks, frequency-translation errors can be disregarded.

Although 2 Hz seems to be a very rigid specification, when added to the possible back-to-back error of the modems themselves, the error becomes more appreciable. Much of the trouble arises with modems that employ sharply tuned filters. This is particularly true of telegraph equipment. But for the more general case, some high-speed data modems occupying the whole voice channel can be designed to withstand greater carrier shifts than can their slower-speed counterparts used for asynchronous data or printing telegraphy. On the other hand, V.22-, V.27-, and V.29-type modems standardized by CCITT require ±1 Hz stability.

9.3.6 Phase Jitter.

The unwanted change in phase or frequency of a transmitted signal caused by modulation by another signal during transmission is defined as "phase jitter." If a simple sinusoid is frequency or phase modulated during transmission, the received signal will have sidebands. The amplitude of these sidebands compared to the received signal is a measure of the phase jitter imparted to it during transmission.

Phase jitter is measured in degrees of variation peak to peak for each hertz of transmitted signal. Phase jitter is manifested as unwanted variations in zero crossings of a received signal. It is the zero crossings that data modems use to distinguish marks from spaces. Thus the higher the data rate, the more jitter can affect error rate on the receive bit stream.

The greatest cause of phase jitter in the telephone network is FDM carrier equipment, where it is manifested as undesired incidental phase modulation. Modern FDM equipment derives all translation frequencies from one master frequency source by multiplying and dividing its output. To maintain stability, phase-lock techniques are used; thus the low-jitter content of the master oscillator may be multiplied many times. It follows, then, that there will be more phase jitter in the voice channels occupying the higher baseband frequencies.

Jitter most commonly appears on long-haul systems at rates related to the power line frequency (e.g., 60 Hz and its harmonics and submultiples) or is derived from 20-Hz ringing frequency. Modulation components that we define as "jitter" usually occur close to the carrier ±300 Hz maximum.

9.4 Channel Capacity

A leased or switched voice channel represents a financial investment. Therefore one goal of the system engineer is to derive as much benefit as possible from the money invested. For the case of digital transmission, this is done by maximizing the information transfer across the system. This section discusses how much information in bits can be transmitted, relating information to bandwidth, signal-to-noise ratio, and error rate. These matters are discussed empirically in Section 9.5.

First, looking at very basic information theory, Shannon stated in his classic paper [23] that if input information rate to a band-limited channel is less than C (bps), a code exists for which the error rate approaches zero as the message length becomes infinite. Conversely, if the input rate exceeds C, the error rate cannot be reduced below some finite positive number.

The usual voice channel is approximated by a Gaussian band-limited channel (GBLC) with additive Gaussian noise. For such a channel, consider a signal wave of mean power of S watts applied at the input of an ideal low-pass filter that has a bandwidth of W (Hz) and contains an internal source of mean Gaussian noise with a mean power of N watts uniformly distributed over the passband. The capacity in bits per second is given by

$$C = W \log_2 \left(1 + \frac{S}{N} \right)$$

Applying Shannon's "capacity" formula to an ordinary voice channel (GBLC) of bandwidth (W) 3000 Hz and a signal-to-noise (S/N) ratio of 1023, the capacity of the channel is 30,000 bps. (Remember that bits per second and bauds are interchangeable in binary systems.) Neither S/N nor W is an unreasonable value. Seldom, however, can we achieve a modulation rate greater than 3000 bauds. The big question in advanced design is how to increase the data rate and keep the error rate reasonable.

One important item not accounted for in Shannon's formula is intersymbol interference. A major problem of a pulse in a band-limited channel is that the pulse tends not to die out immediately, and a subsequent pulse is interfered with by "tails" from the preceding pulse (see Figure 10.19).

Nyquist [24] provided another approach to the data-rate problem, this time using intersymbol interference (the tails in Figure 10.19) as a limit. This resulted in the definition of the so-called Nyquist rate = $2W$ symbols/s, where W is the bandwidth (Hz) of a band-limited channel. In binary transmission we are limited to $2W$ bps, where a symbol is 1 bit. If we let $W = 3000$ Hz, the

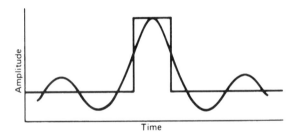

Figure 10.19. Pulse response through a Gaussian band-limited channel (GBLC).

TABLE 10.4 Medium- and High-Rate Modems for a Telephone Voice Channel

Data Rate (bps)	Modulation Rate (Bauds)	Modulation	Bits per Hertz[a]	Bandwidth Required (Hz)[a]
1. 2400 Synchronous (e.g., Rec. V.26)	1200	Differential four-phase	2	1200
2. 4800 Synchronous (e.g., Rec. V.27)	1600	Differential eight-phase	3	1600
3. 3600 Synchronous	1200	Differential four-phase, two-level (combined PSK-AM)	3	1200
4. 2400 Synchronous	800	Differential eight-phase	3	800
5. 9600 Synchronous (e.g., Rec. V.29)	2400	Differential four-phase, two-level	4	2400[b]
6. 14,400 Synchronous (e.g., Rec. V.33)	2400	Differential four-phase, six-level, trellis coding	6	2400

[a]Theoretical values.
[b]Uses automatic equalizer.

maximum data rate attainable is 6000 bps. Some refer to this as "the Nyquist 2-bit rule."

The key here is that we have restricted ourselves to binary transmission and are limited to $2W$ bps no matter how much we increase the signal-to-noise ratio. The Shannon GBLC equation indicates that we should be able to increase the information rate indefinitely by increasing the signal-to-noise ratio. The way to attain a higher C value is to replace the binary transmission system with a multilevel system, often termed an M-ary transmission system, with $M > 2$. An M-ary channel can pass $2W \log_2 M$ bps with an acceptable error rate. This is done at the expense of signal-to-noise ratio. As M increases (as the number of levels increases), so must S/N increase to maintain a fixed error rate [17].

Table 10.4 shows some typical modem bit rates, corresponding baud rates, and bandwidths.

9.5 Some Modem Selection Considerations

The critical parameters that affect data transmission have been discussed; these are amplitude–frequency response (sometimes called "amplitude distortion"), envelope delay distortion, and noise. Now we relate these parameters to the design of data modems to establish some general limits or "boundaries" for equipment of this type. The discussion that follows purposely avoids HF radio considerations.

As stated earlier in the discussion of envelope delay distortion, it is desirable to keep the transmitted pulse (bit) length equal to or greater than the residual differential EDD. Since about 1.0 ms is assumed to be reasonable residual delay after equalization (conditioning), the pulse length should be no less than approximately 1 ms. This corresponds to a modulation rate of 1000 pulses per second (binary). In the interest of standardization (CCITT Rec. V.22 [25]), this figure is modified to 1200 bps.

The next consideration is the usable bandwidth required for the transmission of 1200 bps. This figure is approximately 1800 Hz, using modulation methods such as phase shift (PSK), frequency shift (FSK) or double-sideband AM (DSB–AM), and somewhat less for vestigial-sideband AM (VSB–AM). Since delay distortion of a typical voice channel is at its minimum between 1700 Hz and 1900 Hz, the required band, when centered about these points, extends from 800 Hz to 2600 Hz or 1000 Hz to 2800 Hz. From the previous discussion and with reference to Figure 10.16, we can see that the EDD requirement is met easily over the range of 800–2800 Hz.

Bandwidth limits modulation rate. However, the modulation rate in bauds and the data rate in bits per second may not necessarily be the same. This is a very important concept. Suppose a modulator looked at the incoming serial bits stream 2 bits at a time rather than the conventional 1 bit at a time. Now let four discrete signal amplitudes be used to define each of four possible combinations of two consecutive bits such that

$$A_1 = 00$$
$$A_2 = 01$$
$$A_3 = 11$$
$$A_4 = 10$$

where A_1, A_2, A_3, and A_4 represent the four pulse amplitudes. This form of treating 2 bits at a time is called "di-bit coding" (see Section 9.2).

Similarly, we could let eight amplitude levels cover all the possible combinations of three consecutive bits so that with a modulation rate of 1200 bauds it is possible to transmit information at a rate of 3600 bps. Rather than vary amplitude to four or eight levels, phase can be varied. A four-phase system (PSK) could be coded as follows.

$$F_1 = \quad 0° = 00$$
$$F_2 = \quad 90° = 01$$
$$F_3 = 180° = 11$$
$$F_4 = 270° = 10$$

Again, with a four-phase system using di-bit coding, a tone with a modulation rate of 1200 bauds PSK can be transmitting 2400 bps. An eight-phase PSK system at 1200 bauds could produce 3600 bps of information transfer. Obviously, this process cannot be extended indefinitely. The limitation comes from channel noise. Each time the number of levels or phases is increased, it is necessary to increase the signal-to-noise ratio to maintain a given error rate.

Sufficient background has been developed to appraise the data modem for the voice channel. Now consider a data modem for a data rate of 2400 bps. By using quaternary phase shift keying (QPSK or 4-ary PSK) as described earlier, 2400 bps is transmitted with a modulation rate of 1200 bauds. Assume that the modem uses differential phase detection wherein the detector decisions are based on the change in phase between the last transition and the preceding one. Assume the bandwidth of the data modem under consideration to be 1800 Hz. It is now possible to determine whether the noise requirements can be satisfied. Figure 10.17 shows that a 12.5-dB signal-to-noise ratio (Gaussian noise*) is required to maintain an error rate 1×10^{-5} for a binary polar (AM) system. It is well established that PSK systems have about 3-dB improvement. In this case only a 9.5-dB† signal-to-noise ratio would be needed, if all other factors were held constant (no other contributing factors). Assume the input from the line to be −10 dBm0 to satisfy loading conditions. To maintain the proper signal-to-noise ratio, the channel noise must be down to −19.5 dBm0. To improve the modulation rate without expense of increased bandwidth, quaternary phase shift keying (four-phase) is used. This introduces a 3-dB noise degradation factor, bringing the required noise level down to −22.5 dBm0.

Now consider the effects of EDD. It has been found that for a four-phase differential system, this degradation will amount to 6 dB if the permissible delay distortion is one pulse width. This impairment brings the noise requirement down to −28.5 dBm0 of average noise power in the voice channel. Allow 1 dB for frequency-translation error or other factors, and the noise requirement is now down to −29 dBm0. If the transmission level were reduced by 3 dB to −13 dBm0 (see Section 9.3.4), the noise level must be reduced downward another 3 dB to −32.5 dBm0 (or 19.5 dB signal-to-noise ratio). Thus it can be seen that to achieve a certain error rate for a given modulation rate, several modulation schemes should be considered. It is safe to say that the noise requirement will fall somewhere between −25 dBm0 and

---

*Thermal noise.
†Actually 9.6-dB theoretical ratio (E_b/N_0).

−40 dBm0. This is well inside the CCITT figure of −43 dBmp referenced in Section 9.3.

9.6 Equalization

Of the critical circuit parameters mentioned in Section 9.3, two that have severely deleterious effects on data transmission can be reduced to tolerable limits by *equalization*. These two are amplitude–frequency response (amplitude distortion) and EDD (delay distortion).

The most common method of performing equalization is the use of several networks in tandem. Such networks tend to flatten response and, in the case of amplitude response, add attenuation increasingly toward channel center and less toward its edges. The overall effect is one of making the amplitude response flatter. The delay equalizer operates in a similar manner. Delay increases toward channel edges parabolically from the center. To compensate, delay is added in the center much like an inverted parabola, with less and less delay added as the band edge is approached. Thus the delay response is flattened at some small cost to absolute delay, which has no effect in most data systems. However, care must be taken with the effect of a delay equalizer on an amplitude equalizer and, conversely, of an amplitude equalizer on the delay equalizer. Their design and adjustment must be such that the flattening of the channel for one parameter does not entirely distort the channel for the other.

Another type of equalizer is the transversal type of filter, which is useful where it is necessary to select among, or to adjust, several attenuation (amplitude) and phase characteristics. The basis of the filter is a tapped delay line to which the input is presented. The output is taken from a summing network that adds or sums the outputs of the taps. Such a filter is adjusted to the desired response (equalization of both phase and amplitude) by adjusting the contributions for each tap.

If the characteristics of a line are known, a common method of equalization is predistortion of the output signal of the data set. Some devices use a shift register and a summing network. If the equalization needs to be varied, a feedback circuit from the receiver to the transmitter would be required to control the shift register. This type of dynamic predistortion is practical for binary transmission only.

A major drawback of all the equalizers discussed (with the exception of the latter one with the feedback circuit) is that they are useful only on dedicated or leased circuits where the circuit characteristics are known and remain fixed. Obviously, a switched circuit would require a variable automatic equalizer, or conditioning would be required on every circuit in the switched system that would be transmitting data.

Circuits are usually equalized on the receiving end. This is called *postequalization*. An equalizer must be balanced and must present the proper impedance to the line. Administrations may choose to condition trunks and attempt to eliminate the need to equalize station lines; the economy of considerably fewer equalizers is obvious. In addition, each circuit that would possibly carry high-speed data in the system would have to be equalized, and the equalization must be sufficient for any possible combination to meet the overall requirements. If equalization requirements become greater (i.e., parameters more stringent), the maximum number of circuits (trunks) in tandem may have to be restricted still further (i.e., < 12).

Equalization to meet amplitude–frequency response requirements is less exacting in the overall system than is envelope delay.* Equalization for envelope delay and its associated measurements are time consuming and expensive. In general, envelope delay is arithmetically cumulative. If there is a requirement of overall envelope delay distortion of 1 ms for a circuit between 1000 Hz and 2600 Hz, then for 3 links in tandem, each link must be better than 333 μs between the same frequency limits. For 4 links in tandem, each link would have to be at least 250 μs. In practice, accumulation of delay distortion is not entirely arithmetical, because it results in a reduction of requirements by about 10%. Delay distortion tends to be inversely proportional to the velocity of propagation. Loaded cables display greater delay distortion than do nonloaded cables. Likewise, with sharp filters a greater delay is experienced for frequencies approaching band edge than for filters with a more gradual cutoff.

In carrier multiplex systems, channel banks contribute more to the overall EDD than does any part of the system. Because channels 1 and 12 of the standard CCITT modulation plan (those nearest the group band edge) suffer additional delay distortion because of group filter effects and, in some cases, supergroup filters, the system engineer should allocate channels for data transmission near group and supergroup center. On long-haul critical-data systems, the data channels should be allocated to through-groups and through-supergroups, minimizing as much as possible the steps of demodulation back to voice frequencies (channel demodulation).

Automatic equalization for both amplitude and delay are effective, particularly for switched data systems. Such devices are self-adaptive and require a short adaptation period after switching, on the order of < 1 s [42]. This can be carried out during synchronization. Not only is the modem clock being "averaged" for the new circuit on transmission of a synchronous idle signal, but the self-adaptive equalizers adjust for optimum equalization as well. The major drawback of adaptive equalizers is cost.

9.7   Practical Modem Applications

9.7.1   Voice-Frequency Carrier Telegraph.
Narrow-shifted FSK transmission of digital data is commonly referred to as voice-frequency telegraph (VFTG) and voice-frequency carrier telegraph (VFCT).

---

*Remember "envelope delay distortion" (EDD) is a measure of phase distortion.

In practice, VFCT techniques handle data rates of up to 1200 bps by a simple application of FSK modulation. The voice channel is divided into segments or frequency-bounded zones or bands. Each segment represents a data or telegraph channel, each with a frequency-shifted subcarrier.

For proper end-to-end system interface, it is convenient to use standard modulation plans, particularly on international circuits. For the far-end demodulator to operate with the near-end modulator, the former must be tuned to the same center frequency and accept the same shift. Center frequency is that frequency in the center of the passband of the modulator–demodulator. The shift is the number of hertz that the center frequency is shifted up and down in frequency for the mark–space condition. From Table 10.1, by convention, the mark condition is the center frequency shifted downward and the space, upward. For modulation rates below 80 bps, bandpasses have either 170-Hz (CCITT Rec. R.39) or 120-Hz bandwidths, with frequency shifts of ±42.5 Hz or ±30 Hz, respectively. The CCITT recommends (Rec. R.31 [26]) the 120-Hz channels for operating at 50 bps and below; however, some administrations operate these channels at higher modulation rates.

The number of tone telegraph or data channels that can be accommodated on a voice channel depends partly on the usable voice-channel bandwidth. For high-frequency radio with a voice-channel limit on the order of 3 kHz, 16 channels may be accommodated using 170-Hz spacing (170 Hz between center frequencies). On the nominal 4-kHz voice channel, 24 VFCT channels may be accommodated between 390 Hz and 3210 Hz with 120-Hz spacing, or 12 channels with 240-Hz spacing. This can easily meet standard telephone FDM carrier channels of 300–3400 Hz.

Some administrations use a combination of voice and telegraph data simultaneously on a telephone channel. This technique is commonly referred to as "voice plus" or $S + D$ (speech plus derived), There are two approaches to this technique. The first is recommended by CCITT and is used widely by INTELSAT order wires. It places five telegraph channels (channels 20 through 24) above a restricted voice band with a roofing filter* near 2500 Hz. Speech occupies a band between 300 Hz and 2500 Hz. Up to five 50-bps telegraph channels appear above 2500 Hz. The second approach removes a slot from the center of the voice channel into which up to two telegraph channels may be inserted. The slot is a 500-Hz band centered on 1275 Hz.

However, some administrations use a slot for telegraphy of frequencies 1680 Hz and 1860 Hz by either amplitude or frequency modulation (FSK) (see CCITT Rec. R.43 [26]). The use of speech plus should be avoided on trunks in large networks because it causes degradation to speech and also precludes the use of the channel for higher-speed data. In addition, the telegraph channels have to be removed before going into two-wire telephone service (i.e., at the hybrid or term set); otherwise, service drops to half-duplex on telegraph.

*This is a low-pass filter.

9.7.2 CCITT Recommended Medium Data Rate Modems. CCITT has

issued a number of modem recommendations under the V series; in other words the recommendation number begins with the letter V. Some of the lower data rate modems are covered under the R recommendations.

In normal practice binary FSK is used for the transmission of data for data rates up to and including 1200 bps. For rates above 75 bps, one approach is to apply CCITT Rec. R.31. For instance, a 150-bps circuit would use 240-Hz spacing, a 300-bps circuit would use 480-Hz spacing, and so forth up to one 1200-bps channel. Of course, there would be concurrent increases for the frequency shift, such as ±60 Hz for the 150-bps channel and ±120 Hz for the 300-bps channel.

The first of the V recommendations of CCITT is V.21 [27], which provides 300-bps service; two such channels can be accommodated on a standard VF channel. The mean frequency of each channel is 1080 Hz and 1750 Hz, and the frequency shift on each channel is ±100 Hz.

CCITT Rec. V.22 [25] standardizes a 1200-bps modem that can operate either in the start–stop or synchronous mode. Two data rates are available— 600 bps and 1200 bps—and the modulation rate is 600 bauds in either case. This modem is an exception to the rule stated about FSK. Rec. V.22 specifies a PSK waveform: BPSK for 600 bps operation and QPSK for 1200 bps operation. The modulation approach is similar to that described in Section 9.5 for PSK and QPSK operation; however, the phase values are different for the appropriate di-bit values. For instance, bit 00 is 90°. The recommendation also provides for a standardized scrambler/descrambler. Scrambling is not done for security but to randomize bit patterns. This allows for more uniform dispersal of energy and eliminates long strings of 1s or 0s when an NRZ format is used. Long strings of 1s or 0s give no transitions, which is an unfavorable condition for a data receiver that is looking for transitions to keep its timing circuit in sync with the far-end transmitter.

A related CCITT recommendation is Rec. V.22 bis [28], which is for full-duplex 2400-bps operation in a frequency division mode on two-wire circuits. The "go" and "return" channels are separated by frequency division. Modulation is quadrature amplitude modulation utilizing four levels of phase and four levels of amplitude. A scrambler and an adaptive equalizer are also specified. The modulation rate in all cases is 600 bauds. For instance, for 2400-bps operation the data stream to be transmitted is divided into groups of 4 consecutive bits (quadbits). The first two bits of the quadbit are encoded as a phase quadrant change relative to the quadrant occupied by the preceding signal element.

Recommendation V.23 [29] is titled "600/1200-Baud Modem Standardized for Use on the General Switched Telephone Network." It uses FSK and is compatible with synchronous and start–stop bit streams. Center frequencies are 1500 Hz for 600-bps operation and 1700 Hz for 1200-bps operation. Frequency shifts are ±200 Hz and ±400 Hz, respectively.

A 2400-bps operation modem is covered in CCITT Rec. V.26 [30] which uses a conventional QPSK waveform. The center frequency is 1800 Hz; this value is chosen because for most VF channels it sits at about midpoint on the phase delay curve. Often we find envelope delay distortion specified for the band 1000–2600 Hz and, of course, 1800 Hz is just midpoint.

CCITT Rec. V.26 bis is similar in most respects to Rec. V.26. CCITT Rec. V.26 ter is also similar to Rec. V.26, but is specified for half-duplex operation as well as for full duplex. It provides for echo cancellation.

A 4800-bps capability modem is described in CCITT Rec. V-27 [31], employing a manual equalizer for use on leased circuits. It uses 8-ary PSK on an 1800-Hz carrier frequency, and the modulation rate is 1600 bauds. In this case each transition carries 3 bits of information as shown below.

Tribit Value	Phase Change
001	0°
000	45°
010	90°
011	135°
111	180°
110	225°
100	270°
101	315°

As in many similar CCITT recommendations for data modems, there is provision for a 75-bps backward channel for ARQ commands such as ACK (acknowledgment, a message received without error) and NACK (negative acknowledgment, a result of a message received in error requesting repetition). There is provision for a scrambler having a generating polynomial of $1 + X^{-6} + X^{-7}$. Before data traffic exchange, a synchronization signal is specified to sync the demodulator and to establish descrambler synchronization. The synchronization signal consists of a continuous 180° phase reversal on the line for about 9 ms followed by continuous 1s until full synchronization is achieved as indicated on circuit 106 (ready for sending). (Circuit 106, see CCITT Rec. V.24 or Ref. 43).

CCITT Rec. V.27 bis is similar in many respects to Rec. V.27. Some of the differences are that it has provision for 2400-bps operation using quadrature PSK as well as 4800 bps using 8-ary PSK. It is provided with an automatic equalizer, which uses rapid convergence techniques and thus requires a short training period. By a training period we mean the time it takes to adjust its constants in phase and amplitude, providing a best fit amplitude and phase response line equivalent of the automatic equalizer.

The third V.27 recommendation (CCITT Rec. V.27 ter) is a 4800/2400-bps modem standardized for use in the *general switched telephone network.*

This modem is the Rec. V.27 bis equivalent for use on dial-up data connections.

Operation at 9600 bps over a standard VF channel is now covered by two CCITT recommendations: Recs. V.29 [32] and V.32 [33]. Rec. V.29 is titled "9600 BPS Modem Standardized for Use on Point-to-Point 4-Wire Leased Telephone-Type Circuits." Some of its principal characteristics are:

- Fallback rates of 7200 bps and 4800 bps.
- Capable of operating in duplex or half-duplex mode with continuous or controlled carrier.
- Combined amplitude and phase modulation with synchronous mode operation.
- Inclusion of an automatic adaptive equalizer.
- Optional inclusion of a multiplexer for combining data rates of 7200 bps, 4800 bps, and 2400 bps.

The carrier frequency is 1700 Hz, and the modulation is quadrature amplitude modulation (QAM) with 16 distinct phase-amplitude states. Thus the theoretical bit packing is 4 bits per hertz. As in other modems of similar type, a data scrambler/descrambler is provided based on a pseudorandom sequence generator.

CCITT Rec. V.32 is a recommendation for a family of two-wire duplex modems at data signaling rates up to 9600 bps for use on the general switched telephone network and on leased telephone-type circuits. The principal characteristics of these modems are:

- Duplex mode of operation on the GSTN (general switched telephone network) and two-wire point-to-point leased circuits.
- Channel separation by echo cancellation techniques.
- Quadrature amplitude modulation (QAM) for each channel with synchronous line transmission at 2400 bauds.
- Any combination of the following data rates may be implemented: 9600 bps, 4800 bps, and 2400 bps, all synchronous.
- At 9600 bps operation with two alternative modulation schemes, one using 16 carrier states and one using trellis coding with 32 carrier states. It notes that modems providing the 9600-bps data rate shall be capable of interworking using the 16-state alternative.
- Exchange of rate sequences during start-up to establish the data rate, coding, and any other special facilities.

CCITT Rec. V.32 bis [34] is a full-duplex modem that operates at data signaling rates of up to 14,400 bps for use on the general switched telephone network and on leased point-to-point 2-wire telephone-type circuits. It has fallback modes of 12,000, 9600, 7200, and 4800 bps. Its carrier frequency is 1800 Hz ± 1 Hz. The receiver can operate with a maximum frequency offset of up to ± 7 Hz. The modulation rate is 2400 baud $\pm 0.01\%$. In the 14,400-bps data rate mode, there are 128 distinct points in the space diagram (for QPSK there are just 4, and for 8-PSK there are 8). This achieves a bit packing of 6 bits per hertz plus one additional bit called a *redundant bit* used in trellis coding.

CCITT Rec. V.33 [35] is similar in many respects to V.32 bis. it provides for a 14,400 bps. It uses trellis coding with a similar signal space constellation (i.e., 128 points). However, the modem is designed for use with 4-wire leased telephone-type circuits. It has fallback modes all the way to 2400 bps. These fallback modes are useful when circuit conditions deteriorate, such as increased noise and fading. In general one can say that the lower the modulation rate, the more robust the data transmission system is.

9.8 Data Transmission on the Digital Network

9.8.1 The Problem. Chapters 8 and 9 described the digital network, whether based on DS1 or E1; the digital voice channel is 56 and/or 64 kbps. These transmission rates are incompatible with the standard data rates based on CCITT Recs. V.5 [36], and V.6 [37]. The standard data rates are 600, 1200, 2400, 4800, 9600, and 14,400 bps. Other acceptable data rates are 3000, 6000, 7200, and 12,000 bps. EIA-269 [38] reflects the same standard rates.

Two methods are described below to interface standard data rates with the 56/64-kbps digital channel. The first is AT&T's Digital Data System (DDS) and the second is based on CCITT Rec. V.110 [39].

9.8.2 The AT&T Digital Data System (DDS). The AT&T digital data system (DDS) provides duplex point-to-point and multipoint private line digital data transmission at a number of synchronous data rates. This system is based on the standard 1.544-Mbps DS1 PCM line rate, where individual bit streams have data rates that are submultiples of that line rate (i.e., based on 64 kbps). However, pulse slots are reserved for identification in the demultiplexing of individual user bit streams as well as for certain status and control signals and to ensure that sufficient line pulses are transmitted for receive clock recovery and pulse regeneration. The maximum data rate available to a subscriber to the system is 56 kbps, some 87.5% of the 64-kbps theoretical maximum.

The 1.544-Mbps line signal as applied to DDS service consists of 24 sequential 8-bit words (i.e., channel time slots) plus one additional framing bit. This entire sequence is repeated 8000 times per second. Note that again we have $(192 + 1)8000 = 1.544$ Mbps, where the value 192 is 8×24 (see Chapter 8). Thus the line rate of a DDS facility is compatible with the DS1 (T1) PCM line rate and offers the advantage of allowing a mix of voice

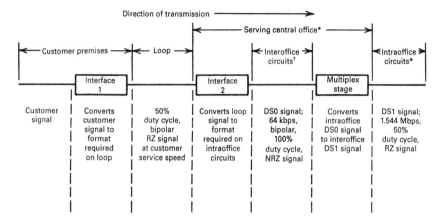

Figure 10.20. Subhierarchy of DDS signals. *Note:* Inverse processing must be provided for the opposite direction of transmission. Four-wire transmission is used throughout. *Exchange; †PCM trunk. Copyright by American Telephone and Telegraph Company.

(PCM) and data where the full dedication of a DS1 facility to data transmission would be inefficient in most cases.

AT&T calls the basic 8-bit word a *byte*. One bit of each 8-bit word is reserved for network control and for stuffing to meet nominal line bit rate requirements. This control bit is called a *C*-bit. With the *C*-bit removed we see where the standard channel bit rate is derived, namely, 56 kbps or 8000×7. Three subrates or submultiple data rates are also available: 2.4, 4.8, and 9.6 kbps. However, when these rates are implemented, an additional bit must be robbed from the basic byte to establish flag patterns to route each subrate channel to its proper demultiplexer port. This allows only 48 kbps out of the original 64 kbps for the transmission of user data. The 48-kbps composite total may be divided down to five 9.6-kbps channels, or ten 4.8-kbps channels, or twenty 2.4-kbps channels [40, 41]. The subhierarchy of DDS signals is shown in Figure 10.20.

Often in practical implementations of DDS, the link connecting the local digital serving switch to the customer premise is via digital loop carrier such as the SLC-96, described in Chapter 8.

9.8.3 Transmitting Data on the Digital Network Based on CCITT Rec.
V.110. This CCITT recommendation covers data rate adaptation for standard rates up through 19.2 kbps through a two-stage process. It also includes adaptation for 48 and 56 kbps to the 64-kbps E0 or clear DS0 channel.

In the two-step case, the first conversion is to take the incoming data rate and convert it to an appropriate intermediate rate expressed by $2^k \times 8$ kbps, where $k = 0$, 1, or 2. The second conversion takes the intermediate rate and converts it to 64 kbps.

Simple division of 64,000 bps by standard data rates shows that as a minimum, a lot of bit stuffing would be required. These would essentially be wasted bits. CCITT makes use of these bits to provide framing overhead, status information, and control information. A frame (step 1 conversion) consists of 10 octets ($10 \times 8 = 80$ bits). Six octets carry user data, the first octet is all 0s for frame alignment, and bit 1 of the remaining 9 octets is set to 1. There are 15 overhead bits. Nine different frames accommodate the various data rates, and each frame has 80 bits (10 octets). The recommendation also covers conversion of start–stop data rates including 50, 75, 110, 150, 300 bps, and the standard rates up through 19.2 kbps [39].

REVIEW QUESTIONS

1. What is the basic element of information in a binary data system? How much information does it contain? (*Hint*: How many distinct states may it have?)

2. How does one extend the information content of that basic information element (question 1), for example, to construct a code that represents our alphabet?

3. There are many ways we can express the binary 1 and the binary 0. Also, we do not want to confuse the 1 and the 0 (i.e., reverse them). Give at least four ways we can express a 1 and a 0, and show that we are in common agreement on how to do this such that all ambiguity is removed.

4. How many distinct characters or symbols can be represented by a 4-unit binary code? a 7-unit binary code? an 8-unit binary code? (*Hint*: For this argument, consider that a unit is a bit.)

5. Name at least three nonprinting characters that might be encountered in a practical binary code.

6. How many information elements (bits) are in an ASCII character?

7. What are some of the more common causes of burst errors?

8. An ASCII character is represented as 1001011 with odd parity assumed. Give the value of the eighth bit. In another situation even parity is assumed and the ASCII character is 0101110. What is the value of the eighth bit?

9. Would you imagine that throughput is at all sensitive to the underlying error rate of a circuit? Explain your answer.

10. Why the insistence on careful definition of throughput?

11. Which is the more powerful error detection scheme: vertical redundancy check, longitudinal redundancy check, or cyclic redundancy check (CRC)?

12. Describe the two common methods of correcting errors that occur on data links.

13. A frame check sequence (FCS) contains either a 16-bit or 32-bit sequence of bits. What is it and how is it derived?

14. When FEC (forward error correction) is implemented, what types of errors can be corrected? How can one use FEC to correct burst errors?

15. Define "Hamming distance."

16. Name the three different types of ARQ. Which is the simplest to implement? Which is the most efficient for maximum "throughput."

17. Differentiate between neutral and polar transmission.

18. A data link transmits 2400 bps synchronously. There is a time base stability difference between transmitter and distant-end receiver of 20 ppm (parts per million). If the link can be considered "error free" except for timing, how many bits will the receiver receive correctly before printing in error? Assume, of course, that the receiver started copying the first bit out of the transmitter and that their two clocks started exactly in synchronism.

19. On a start–stop circuit, where does the receiver start counting information bits? At which signal element?

20. There are three major causes of errors in a data system. Name two of them.

21. The mark-to-space transition for the start space element tells a data receiver when a character is to begin for start–stop systems. How does a synchronous data receiver know when things start so it can begin a bit/byte count?

22. Show and explain the function of each of the 5/6 fields of a generic data frame.

23. How does a synchronous data receiver keep in synchronization with an incoming bit stream?

24. Explain timing distortion and how it can affect error performance.

25. A serial bit stream with an NRZ waveform has a bit rate of 19.2 kbps. What is the period of one bit?

26. Why should we be careful on the use of terms such as baud and bit?

27. The CCITT V.29 modem operates on the standard analog voice channel at 9600 bps. How can it possibly do this if we allow 1 bit per hertz of bandwidth?

28. What is notably richer in transitions per second, RZ or NRZ coding? Why would that be important anyway?

29. Keeping strictly in the electrical domain (i.e., the physical layer), draw a simple functional block diagram of a data circuit identifying the DTE and DCE, and show two major points of interface.

30. On the diagram prepared for question 29, show the EIA-232 interface point.

31. Distinguish between a 1 and a 0 based on EIA-232 using voltage values.

32. Name at least two advantages of using a balanced interface between DTE and DCE.

33. For conventional data transmission for data rates ≤1200 bps, what type of modulation would one expect? This, of course, is in reference to analog voice channel modems.

34. Discuss bit rate, modulation rate, and modulation type of a conventional 2400-bps modem. What is the "baud rate?"

35. What are the three basic impairments for data transmission?

36. Name the four types of noise we must consider for data transmission. One does not normally affect voice communication but can seriously degrade error performance on a data circuit. Which type of noise is that?

37. Phase distortion, in general, has little effect on speech communication. What can we say about it for data transmission?

38. Shannon's formula for capacity (in bps) for a particular bandwidth was based on only one other parameter. What was it?

39. We usually equalize two voice channel impairments. What are they and what does equalization do?

40. Why do higher-speed modems use a center tone frequency around 1700–1800 Hz?

41. Although we said in Chapter 8 that the digital network was compatible with conventional data bit streams, what seems to be the real problem? Discuss three ways we can transmit binary digital data on the digital network.

REFERENCES

1. *The New IEEE Standard Dictionary of Electrical and Electronics Terms,*" 5th ed, IEEE Std 100-1992, IEEE, New York, 1992.

2. *Equivalence Between Binary Notation Symbols and the Significant Conditions of a Two-Condition Code,* CCITT Rec. V.1, Fascicle VIII.1, IXth Plenary Assembly, Melbourne, 1988.

3. *Coded Character Sets—7-Bit American National Standard Code for Information Interchange (7-bit ASCII),* ANSI X.3.4-1986, ANSI, New York, 1986.

4. *Error Performance on an International Digital Connection Forming Part of an Integrated Services Digital Network,* CCITT Rec. G.821, Fascicle III.5, IXth Plenary Assembly, Melbourne, 1988.

5. *Interface Between Data Terminal Equipment and Data Circuit-Terminating Equipment Employing Serial Binary Data Interchange,* EIA/TIA-232E, Electronics Industries Association, Washington, DC, July 1991.

6. *High Speed 25-Position Interface for Data Terminal Equipment and Data Circuit-Terminating Equipment Including Alternative 26-Position Connector,* EIA/TIA-530-A, Electronics Industries Association, Washington, DC, June 1992.

7. *Electrical Characteristics of Balanced Voltage Digital Interface Circuits,* EIA-422A, Electronic Industries Association, Washington, DC, December 1978.

8. *Electrical Characteristics of Unbalanced Voltage Digital Interface Circuits,* EIA-423-A, Electronics Industries Association, Washington, DC, December 1978.

9. *Electrical Characteristics of Digital Interface Circuits,* MIL-STD-188-114A, U.S. Department of Defense, Washington, DC, September 1985.

10. *Electrical Characteristics for Unbalanced Double-Current Interchange Circuits for General Use with Integrated Circuit Equipment in the Field of Data Communications,* ITU-T Rec. V.10, ITU Geneva, 1994.

11. *Electrical Characteristics of Balanced Double-Current Interchange Circuits for General Use with Integrated Circuit Equipment in the Field of Data Communications,* ITU-T Rec. V.11, ITU Geneva, 1994.

12. *List of Definitions of Interchange Circuits Between Data Terminal Equipment (DTE) and Data Circuit-Terminating Equipment (DCE),* ITU-T Rec. V.24, ITU Geneva, 1994.

13. *Electrical Characteristics of Unbalanced Double-Current Interchange Circuits,* ITU-T Rec. V.28, ITU Geneva, 1994.

14. *Electrical Characteristics of Single-Current Interchange Circuits Controlled by Contact Closure,* CCITT Rec. V.31, Fascicle VIII.1, IXth Plenary Assembly, Melbourne, 1988.

15. *Attenuation Distortion,* CCITT Rec. G.132, Fascicle III.1, IXth Plenary Assembly, Melbourne, 1988.

16. *Reference Data for Engineers: Radio, Electronics, Computer and Communications,* 7th ed., Howard W. Sams, Indianapolis, 1985.

17. W. R. Bennett and J. R. Davey, *Data Transmission,* McGraw-Hill, New York, 1965.

18. *Circuit Noise and the Use of Compandors*, CCITT Rec. G.143, Fascicle III.1, IXth Plenary Assembly, Melbourne, 1988.

19. *LATA Switching Systems Generic Requirements, Transmission*, Bellcore TR-TSY-000507, Issue 3, Piscataway, NJ, 1989.

20. *CCITT Recommendations, Blue Books*, Volume VI, Fascicle VI.1, IXth Plenary Assembly, Melbourne, 1988.

21. *Power Levels for Data Transmission over Telephone Lines*, CCITT Rec. V.2, Fascicle VIII.1, IXth Plenary Assembly, Melbourne, 1988.

22. *Error on the Reconstituted Frequency*, CCITT Rec. G.135, Fascicle III.1, IXth Plenary Assembly, Melbourne, 1988.

23. C. E. Shannon, *A Mathematical Theory of Communications*, BSTJ 27, Bell Telephone Laboratories, Holmdel, NJ, 1948.

24. H. Nyquist, *Certain Topics in Telegraph Transmission Theory*, BSTJ, 6-7-644, Bell Telephone Laboratories, Holmdel, NJ, 1928.

25. *1200 BPS Duplex Modem Standardized for Use in the General Switched Telephone Network and on Point-to-Point 2-Wire Leased Telephone-Type Circuits*, CCITT Rec. V.22, Fascicle VIII.1, IXth Plenary Assembly, Melbourne, 1988.

26. *CCITT Recommendations*, Volume VII, *Blue Books*, IXth Plenary Assembly, Melbourne, 1988.

27. *300 BPS Duplex Modem Standardized for Use in the General Switched Telephone Network*, CCITT Rec. V. 21, Fascicle VIII.1, IXth Plenary Assembly, Melbourne, 1988.

28. *2400 BPS Modem Using the Frequency Division Technique Standardized for Use on the General Switched Telephone Network and on Point-to-Point 2-Wire Leased Telephone-Type Circuits*, CCITT Rec. V.22 bis, Fascicle VIII.1, IXth Plenary Assembly, Melbourne, 1988.

29. *600/1200 Baud Modem Standardized for Use on the General Switched Telephone Network*, CCITT Rec. V.23, Fascicle VIII.1, IXth Plenary Assembly, Melbourne, 1988.

30. *2400 BPS Modem Standardized for Use on 4-Wire Leased Telephone-Type Circuits*, CCITT Rec. V.26. Fascicle VIII.1, IXth Plenary Assembly, Melbourne, 1988.

31. *4800 BPS Modem with Manual Equalizer Standardized for Use on Leased Telephone-Type Circuits*, CCITT Rec. V.27, Fascicle VIII.1, IXth Plenary Assembly, Melbourne, 1988.

32. *9600 BPS Modem Standardized for Use on Point-to-Point 4-Wire Leased Telephone-Type Circuits*, CCITT Rec. V.29, Fascicle VIII.1, IXth Plenary Assembly, Melbourne, 1988.

33. *A Family of 2-Wire, Duplex Modems Operating at Data Signaling Rates Up to 9600 BPS for Use on the General Switched Telephone Network and on Leased Telephone-Type Circuits*, ITU-T Rec. V.32, ITU Geneva, 1994.

34. *A Duplex Modem Operating at Data Signaling Rates Up to 14,400 BPS for Use on the General Switched Telephone Network and on Leased Point-to-Point 2-Wire Telephone-Type Circuits*, CCITT Rec. V.32 bis, ITU Geneva, 1991.

35. *14,400 BPS Modem Standardized for Use on Point-to-Point 4-Wire Leased Telephone-Type Circuits*, CCITT Rec. V.33, Fascicle VIII.1, IXth Plenary Assembly, Melbourne, 1988.

36. *Standardization of Data Signaling Rates for Synchronous Data Transmission in the General Switched Telephone Network*, CCITT Rec. V.5, Fascicle VIII.1, IXth Plenary Assembly, Melbourne, 1988.

37. *Standardization of Data Signaling Rates for Synchronous Data Transmission on Leased Telephone-Type Circuits*, CCITT Rec. V.6, Fascicle VIII.1, IXth Plenary Assembly, Melbourne, 1988.

38. *Synchronous Signaling Rates for Data Transmission*, EIA-269A, Electronics Industries Association, Washington, DC, May 1968.

39. *Support of Data Terminal Equipments with V-Series Type Interfaces by an Integrated Services Digital Network*, CCITT Rec. V.110, ITU Geneva, 9/92.

40. *Digital Data System Data Service Unit Interface Specification*, Bell System Technical Reference Pub 41450, AT&T, New York, 1981.

41. *Digital Data System Channel Interface Specification*, Bell System Technical Reference PUB 62310, AT&T, New York, 1983.

42. K. Pahlavan and J. L. Holsinger, "Voice-Band Communication Modems: A Historical Review, 1919–1988," *IEEE Communications Magazine*, **26**(1) (January 1988).

43. R. L. Freeman, *Telecommunication Transmission Handbook*, 3rd ed., John Wiley & Sons, New York, 1991.

DATA NETWORKS AND
THEIR OPERATION

1 INTRODUCTION

Data networking deals with the rapid transporting of information. The transport system should be efficient and cost-effective. There are two generic data network regimes that have developed and evolved over the past 40 years. These are the *wide area network* (WAN) and the *local area network* (LAN). There is also the *metropolitan area network* (MAN) that is written about in the literature but has not been implemented to any great extent.

The WAN appeared on the scene first, providing connectivity among computer and computer-related sources and destinations that are geographically widely dispersed. Data circuits were set up using the ubiquitous telephone network connections either by dial-up or leased lines. This means of establishing data circuits is still very prevalent today. Entire networks are built up from leased telephone network facilities. These could be private networks dedicated completely to a single business enterprise or public sub-scription networks such as Tymnet and Telenet, both of which offer packet data services. An example of an early WAN is ARPANET shown in Figure 11.1, which went on-line in 1969.

Local area networks (LANS) came along later, particularly due to the widespread use of the PC in the enterprise environment. It is interesting to note that the first LAN protocol, Ethernet, was formulated in 1972, some three years before the PC was introduced. Ethernet still remains the most popular LAN access protocol. LANs are discussed in Chapter 12.

In the design of a WAN, what sort of topology will be used? In this context we will be dealing with how to connect up these data assets, be they switches, PCs, printers, servers, mass storage devices, and so on. Topology is defined in Ref. 27 as the physical or logical placement of nodes in a computer

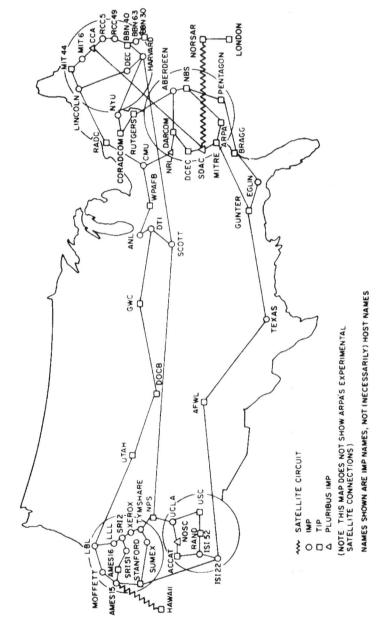

Figure 11.1. The ARPANET geographic map. An example of a very large data network.

∿∿ SATELLITE CIRCUIT
○ IMP
▢ TIP
△ PLURIBUS IMP

(NOTE: THIS MAP DOES NOT SHOW ARPA'S EXPERIMENTAL SATELLITE CONNECTIONS)

NAMES SHOWN ARE IMP NAMES, NOT (NECESSARILY) HOST NAMES

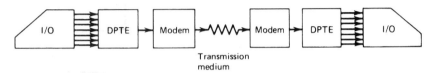

Figure 11.2. Elemental point-to-point data connectivity. The DTE, described in Chapter 10, consists of the DPTE and I/O combined. The DCE is the modem. DPTE, data processing terminal equipment; I/O, input/output device.

network. One very common approach is to use point-to-point connectivity as shown in Figure 11.2.

Cost, quantity of data, urgency, and reliability would drive the designer to other types of topology such as mesh and star, described in Chapters 2 and 6. We will also describe multipoint and ring networks.

There is a disparity between LANs and WANs. LANs operate at megabit data rates and WANs operate at kilobit rates, and they operate with entirely different protocol regimes. Shortly after LANs came into existence, the problem arose interconnecting them over comparatively long distances. Enter frame relay, Switched Multimegabit Data Service (SMDS), and ATM, described in Chapters 14 and 15. VSATs, discussed in Chapter 7, brought a whole new dimension to data networking and bypassing the telephone network.

1.1 Applications

Data and data networks enter every facet of our society, particularly in the business or "enterprise" world. The following is a listing of some applications:

1. Multipoint inventory control.
2. Airline, train, and hotel reservations.
3. Banking transactions, electronic funds transfer (EFT).
4. Truck, train, and ship cargo control.
5. Air traffic control.
6. Weather forecasts, weather warnings.
7. Defense: command, control, and communication networks both strategic and tactical; intelligence, logistics, fire control, remote radar, and other sensors.
8. Resource sharing, particularly in research and development; Internet.
9. Remote text composition (newspapers, magazines, other publishing).
10. Police anticrime, counter-narcotics.
11. Factory automation; CAD/CAM networks, other industrial networks, inventory control.

12. Office automation, payroll, engineering, accounting.

13. Finance (stock market); insurance; taxes.

14. E-mail, teletext.

15. Credit card verification.

The list is not exhaustive. However, each entry on the list, as well as those that are not, requires a distinct approach in data network design. Of course, the distribution of data assets can be an overriding consideration. The material presented in Chapter 10 is the base required for the transmission aspects of data communications.

2 INITIAL DESIGN CONSIDERATIONS

2.1 General

Data network architecture can vary from a complex, widely distributed network shown in Figure 11.1 to the elemental point-to-point network shown in Figure 11.2. Between these extremes there is a large variety of data networks regarding size, configuration, and capability. Some of the more important considerations entering into the network design are:

- Data sources and destinations: locations and geographic dispersal.
- Traffic profile among all connectivities.
- Network organization, topology: existing and planned.
- Private network, use of PSTN, hybrid schemes (i.e., partially private network, partially use of PSTN).
- For PSTN usage, tariffs and tariff structures.
- Type of data communication now in use or planned; compatibility for upgrade.
- Requirements for data perishability (e.g., inquiry–response).

Modern data networks all have distributed processing capability. Certainly the all-pervasive PC was the major instigator for distributed processing. One or more mainframe computers and minicomputers are also connected to most networks. They are just represented as another node.

We have on one hand user needs and on the other hand a "toolbox" of candidate solutions to satisfy the needs. In the sections that follow, we describe what is available in the toolbox and how to put the tools to work. Chapter 10 covered the transmission aspects of data. Here we cover network and operational issues.

2.2 Data Terminals and Workstations

A key element of a data network, whether LAN or WAN, is the data terminal or workstation. It may be no more than a PC or it could be a powerful Sun workstation. Generally, a network consists of workstations and other computer assets, such as CPUs,* high-speed printers, servers, and in some cases mass storage devices. Our interest in this section is how data terminals and workstations can drive network design.

A workstation has a human operator. It is the principal input/output device for the network. As a minimum, it consists of a display, a keyboard or keypad input device, and a processor. To interface a data network, it must also have a communication interface device, which may or may not be physically a part of its processor, such as a plug-in card. It may also have an associated printer. For planning the system, design engineers should determine the following:

- The type of protocol(s) supported.
- Allowance for protocol upgrade.
- Inclusion of automatic polling.
- Standard interface (OSI Layer 1): RS-232E, MIL-STD 188-114B, etc.
- Handling of varying data rates.
- Coprocessing and interrupt capabilities.
- LAN interface and NIU card slot.
- Processor's operating system.
- Buffer memory. Can a portion be used for communications?
- Security device support, if any.
- Asynchronous/synchronous interface.

There are three ways of incorporating terminals into a WAN:

1. Stand-alone terminal with the necessary interface.
2. Cluster of terminals with communications controller.
3. LAN with interconnect through a gateway or router.

Their incorporation into a WAN is illustrated in Figure 11.3.

3 NETWORK TOPOLOGIES AND CONFIGURATIONS

For the wide area network, there are six ways of configuring data assets:

*CPU stands for central processing unit, a computer.

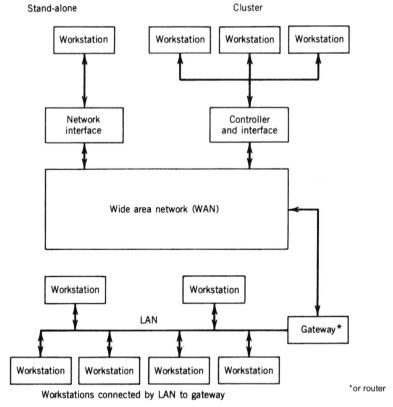

Figure 11.3. Graphic representation of three methods of connecting terminals or workstations to a WAN.

1. Point-to-point (Figure 11.4A)

2. Multipoint or multidrop (Figure 11.4B)

3. Star with host computer at hub (11.4C)

4. Multistar/hierarchical (Figure 11.4D)

5. Ring network (Figure 11.4E)

6. Grid network, packet-switched or circuit-switched (Figure 11.4F)

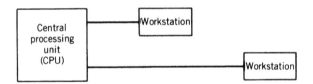

Figure 11.4A. A simplified diagram of a point-to-point network.

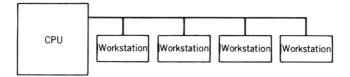

Figure 11.4B. A simplified multipoint or multidrop network.

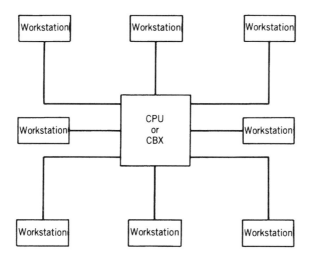

Figure 11.4C. A simplified diagram of a star network with a CPU at the center or hub of the star.

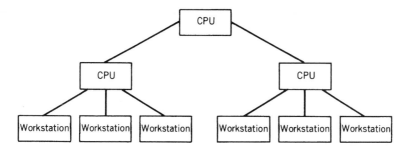

Figure 11.4D. A simplified diagram of a multistar or hierarchical network.

There are advantages and disadvantages of each. An advantage of a particular type of network in one situation may turn out to be a disadvantage in another. For instance, a multidrop (multipoint) network is simple and economical but should be used only when each access has a low traffic intensity. For the high traffic intensity and occupancy case, a point-to-point network is more attractive and its expense is a function of geographic dispersal.

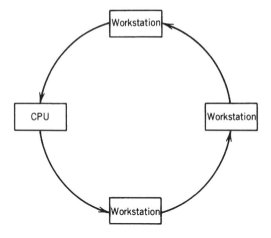

Figure 11.4E. A simplified diagram of a ring network.

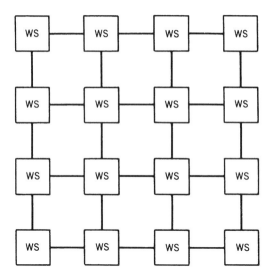

Figure 11.4F. A grid network of workstations (WS). A CPU can replace a workstation at any location, or be colocated with a workstation.

The point-to-point network (Figure 11.4A) has the advantage of simplicity of design. As the number of accesses increases, a port contention device may be required. The disadvantage of point-to-point operation is the cost of the transmission media, particularly as the circuits are extended from feet to miles to hundreds or thousands of miles. For low-usage accesses, dial-up connections may reduce cost; local clustering is another alternative (Figure 11.3).

Multipoint or multidrop operation can solve the transmission media cost problem unless the connectivity route tends to meander a lot. That only one access at a time may use the medium also detracts from its utility. Depending on the operational regime and embedded firmware and software in each terminal, there are two common methods to access the host CPU: contention and polling. The first is simple random access where a user can get access of the medium for an indefinite period. Polling provides an access control mechanism where the host asks each access in turn if it has traffic and takes or passes traffic during the poll interval. These protocol techniques are covered in Section 6.2.

A star network (Figure 11.4C) has a processing and switching capability at its hub. The hub is usually a computer. The star network gives the appearance of an extension of the point-to-point concept. The exception is that not only can workstations access the hub computer, but they can access each other through the hub. VSAT networks are star networks, but only in special VSAT networks can remote terminals access each other. The star network has poor grades for survivability.

Figure 11.4D expands the star network concept shown in Figure 11.4C. Our present telephone network is a modified hierarchical network (Chapters 1 and 6). One application for data communications is to place workstations at the lowest levels in the hierarchy. These workstations access nearby computers, which carry on the brunt of the processing and also provide a concentration capability for data communication to a large mainframe, centrally located computer (at the top of the figure). Some of the same techniques employed in the telephone network can be applied to the hierarchical data network. One is the application of high-usage (HU) and/or direct routes for high traffic intensity data relations.

Figure 11.4E illustrates a ring network. We note in the figure the unidirectionality of data flow. In the LAN arena, this is typical of token ring. If a second ring is added, overlaid on the first, but providing traffic flow in the other direction, survivability and availability are improved. If there is a failure or cut across both rings, traffic can be routed in opposite directions avoiding catastrophic breakdown. If one ring fails, the other still operates. Of course, there is the added cost of the additional ring.

In large networks, survivability and availability* can be enhanced still further with a grid network, shown in Figure 11.4F. This is because of the large number of alternative routing possibilities available with a grid network. Whereas the ring network can withstand one failure, the grid network can withstand multiple failures. A grid network also lends itself well to packet communications, where each node has at least three inlets and outlets. Packets can travel on a variety of routes between any two nodes.

---

*Availability is related to reliability and expresses the percentage of time that a network is up and fully operational.

4 THREE APPROACHES TO DATA SWITCHING

The three approaches to data switching are:

1. Circuit switching
2. Message switching
3. Packet switching

Plain old telephone service, affectionately called POTS, uses circuit switching. A data switch can be set up in a similar manner, providing address information in the message header. The switch responds to the header by setting up a real (physical) or virtual circuit to the desired addressee. AutoDiN, a U.S. Department of Defense digital network, offers circuit switching as one operational mode. The dial-up connection described previously uses the POTS circuit-switching capability of the telephone network to provide the required connectivity for data messages.

Message switching, often called store-and-forward switching, accepts messages from originators, stores the messages, and then forwards each message to the next node or destination when circuits become available. We wish to make two points when comparing circuit and message switching. Circuit switching provides end-to-end connectivity in near real time. Message switching does not. There usually is some delay as the message makes its way through the system to its destination. The second point deals with efficient use of expensive transmission links. A well-designed message-switching system keeps a uniform load throughout the working day and even into the night. Circuit-switching systems, such as the telephone systems, are designed for busy-hour loading, and the system tends to loaf after-hours.

Telex is a good example of message switching. Another example is the older "torn-tape" systems. Such systems were labor-intensive, rather primitive in this day and age. With torn tape, messages arriving at a switching node from an originator or relay node used paper tape to copy the messages in the ITA No. 2 code (Chapter 10). Headers were printed out laterally along the tape holes. Operators at a torn-tape center would read the header, tear off the tape, and insert it in the tape reader accessing the required outgoing circuit. If the reader was busy, the tape was placed in queue in accordance with the message precedence (priority). I've seen clothes-pins and paper clips used to piece the tapes together in such queues.

Store-and-forward systems today use magnetic tape storage or other storage media rather than paper tape. A processor arranges messages in the correct queue. Message processors "read" precedence and header information and automatically pass the data traffic to the necessary outgoing circuits.

Packet switching utilizes some of the advantages of message switching and circuit switching and mitigates some of the disadvantages of both. A data "packet" is a comparatively short block of message data of fixed length.

Complete data messages are broken down into short packages, each with a header. These packets may be sent on diverse routes to their eventual destination, and each packet is governed by an ARQ error-correction protocol. Because packets often travel on diverse routes, they may not arrive at the far-end receiving node in sequential order. Thus the far-end node must have the capability to store incoming packets and rearrange them in sequential order. The destination node then reformats the message as it was sent by the originator and forwards it to the final destination user.

Packet switching can show considerably greater efficiency when compared to circuit and message switching. There is one caveat, however. For efficient packet-switching operation, a multiplicity of paths (more than three) must exist from the originating local node to the destination local switching node. Well-designed packet-switching systems can reduce delivery delays when compared to conventional message switching. Expensive transmission facilities tend to be more uniformly loaded than with other data-switching methods. Adaptive routing becomes possible where a path between nodes has not been selected at the outset. Delivery delay is also reduced when compared to message switching because a packet switch starts forwarding packets before receipt of all the constituent packets making up a message. Path selection is a dynamic function of real-time conditions of the network. Packets advancing through the network can bypass trunks and nodes where congestion or failure exist.

A packet switch is a message processor. Packets are forwarded over optimum routes based on route condition, delay, and congestion. It provides error control and notifies the originator of packet receipt at the destination. Packet switching approaches the intent of real-time switching. Of course, there is no "real" connection from originator to destination. The connection is one form of a "virtual" connection.

Table 11.1 summarizes advantages and disadvantages of the three methods of data switching.

5 CIRCUIT OPTIMIZATION

Data links have sources and destinations with varying requirements regarding *quantity* of data and its urgency. Consider a so-called voice-grade line (analog) that is full-period-leased. It can support 4800 bps, or 9600 bps, or 14,400 bps. It would be uneconomical to underutilize such an expensive facility. Let's consider a number of scenarios. The first is where urgency is not an underlying consideration and the data rate then is selected to be 75 bps. This circuit can be frequency-shared with 20 or more other users. The second is 2400 bps, on a circuit time-shared with three or four other 2400-bps users. The third is a full-period, 14,400-bps circuit and the fourth is a 64-kbps digital DS0/E0 channel. Table 11.2 expresses these results for time periods from 10 s to 24 h. In the table, data flow is treated as continuous with no

TABLE 11.1 Summary of Data-Switching Methods

Switching Method	Advantages	Disadvantages

Circuit switching	Mature technology
	Near real-time connectivity
	Excellent for inquiry and
	  response
	Leased service attractive
		High cost of switch
		Lower system utilization, particularly
		  link utilization
		Privately owned service can only be
		  justified with high traffic volume

Message switching	Efficient trunk utilization Cost-effective for low-volume leased service	Delivery delay may be a problem Not viable for inquiry and response Survivability problematical Requires large storage buffers

Packet switching	Efficiency
	Approaches near real-time
	  connectivity
	Highly reliable, survivable
	Low traffic volume attractive
	  for leased service
		Multiple route and node network
		  expensive
		Processing intensive
		Large traffic volume justifies private
		  ownership

TABLE 11.2 Circuit Capacity in kbps versus Bit Rate versus Time Period

Time Unit	Bits per Second			
	75	2400	14,400	64,000
10 s	0.750	24.000	144.000	640.000
1 min	4.500	144.000	864.000	3,840.000
10 min	45.000	1,440.000	8,640.000	38,400.000
1 h	270.000	8,640.000	51,840.000	230,400.000
8 h	2,160.000	69,120.000	414,720.000	1,843,200.0
24 h	6,480.000	207,360.0	1,244,160.0	5,529,600.0

breaks and no repeats. This would be unlikely in the real world. The table does give a good idea of data quantity, although on the liberal side.

One could say that these are *throughput* values in Table 11.2. Of course, it depends on how we would define throughput. Let's look at a simple case of the ASCII 7-bit code. In this example, it is transmitted in a start–stop format with a 2-bit stop element. Thus there is a start element (bit) and a stop element (2 bits). The sole purpose of these three bits is to facilitate timing and synchronization; they carry no useful information. If Table 11.2 is based on such a start–stop format, only 72% of the bits are useful. There is also a parity bit which is usually appended as the eighth bit. It carries no useful information. It goes along for the ride, making each character 8 bits long, an octet. This is useful for the organization of 8-bit, 16-bit, 32-bit, and so on, processors. It did, at one time, serve an error detection function, but seldom

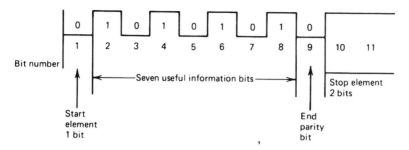

Figure 11.5. An ASCII start–stop bit sequence with a 2-bit stop element and a parity bit.

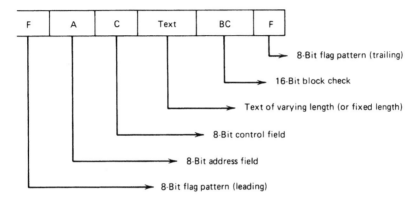

Figure 11.6. A typical data frame, block, or packet.

does now. It is overhead, bearing no useful information. We are now reduced to a 63% efficiency. In other words, based on our start–stop assumption, only 63% of the bits in Table 11.2 are useful to the end-user. This stop–start format is shown in Figure 11.5.

We now turn to synchronous transmission, where the data traffic must be sent in frames, blocks, or packets (and now cells). A typical frame is shown in Figure 11.6. In this case it is the "text" field, often called the "info" field, that contains the "useful" data for the destination user. Based on the figure, we must amortize this "text" over 48 bits of overhead. Obviously, the longer the text field is, the better we amortize the overhead bits, and our apparent efficiency goes up. Suppose the text had 48 bits and there were 48 bits of overhead; we don't count the second flag pattern. There is a total of 96 bits in the frame, 48 of which were overhead. Thus our efficiency is 48/96, or 50%. If the number of bits now are doubled for the text, we have 96/144 or 67% efficiency. One can see that if we had a very long text and maintained the overhead at 48 bits, the efficiency becomes very high.

A high noise level, or even just a few noise spikes, tend to ruin our good record. This is because of repeats required of errored messages. The longer the frame, the more time is spent on a repeat through an ARQ regime. There is a general rule that on noisy channels, short frames lead to greater throughput.

Some links are so overburdened with inefficiencies that, for example, a 4800-bps data link may only afford a net 200 bps of effective, useful bits put through. This poor performance is attributed to large overhead and frame repeats due to errors. As we mentioned, just one noise spike of short duration, a few microseconds, could cause just one bit in error, requiring that the whole frame be repeated.

5.1   Throughput

The preceding discussion leads to the question of throughput. Throughput means different things to different people. We define throughput as the net *useful* bits put through per unit time. "Useful" is the key word. Useful to whom? The start and stop bits are useful to make the transmission system work. They are not useful to the data user who couldn't care less how the system works. The same argument holds for the synchronous overhead bits (not that start and stop systems do not have additional overhead).

Turning back to the frame format in Figure 11.6, the text field contains the "useful" bits. As we will see later, even bits in this field are required for control overhead, which can be argued as *not useful* bits (noninformation characters). We also must try to partition the data communication system in some way to determine where the telecommunication responsibility leaves off and the data user responsibility begins. This point will be discussed in Section 6.

5.2   Some Cost-Effective Options to Meet "Throughput" Requirements

In this section we treat point-to-point service exclusively for off-premises connectivity. In general, cost is a function of the product of bit rate and distance. This assumes a tariff structure that is based on circuit length. Table 11.3 gives a number of options available. In many cases, the smallest segment available is a full-period voice channel. Also, if the data are very perishable, such as on inquiry/response systems, the slow reaction of low bit rate systems may make such systems impractical. Typical inquiry/response systems are credit card verification and hotel and airline reservation systems. In some cases, data compression can speed up transaction time and thus take advantage of the less expensive lower bit rate techniques.

Cost can be shared among users by breaking a voice channel up into segments. One way is to break it up into frequency segments and use voice-frequency carrier techniques. Another is to break it up into short time segments using time-divison multiplexers (TDM) or statistical multiplexers. It

TABLE 11.3 Some Available Data Transmission Options in Order of Increasing Cost

Bit Rate (bps)	Bandwidth (Hz)	Modulation (see Chapter 10)
75	120	VFCT (FSK)
150	240	VFCT (FSK)
300	480	VFCT (FSK)
600	960	VFCT (FSK)
1200	3100	FSK
2400	3100	QPSK
4800	3100	8-ary PSK
9600	3100	16-QAM
14,400	3100	128-QAM/trellis coded
56,000	1/24 T1	Baseband (ATT DDS)
64,000	Approx. 64 kHz	Baseband (ISDN rate)
384,000	6 × 64 kHz approx.	Baseband six 64 kbps ch
1,544,000	Approx. 1.544 MHz	Baseband DS1

is more common today to lease a voice channel and share it among internal users.

Of the two voice-channel sharing techniques, it can be shown that the time-division techniques are more efficient, producing a higher aggregate bit rate. Aggregate bit rates can be 4800, 9600, and 14,400 bps. If we were to break this down into 75-bps users, it would represent 64, 128, and 192 users respectively, assuming each is a full-period user. If they are not, meaning that they have a bursty traffic profile and we use a statistical multiplexer, even more startling numbers of users can be accommodated. In fact, many of these devices accept a mix of input data rates often based on the standard 75×2^n bps. However, overhead is often required to identify the far-end output data ports as the proper destinations.

Commonly, the inputs to time-division multiplexers are start–stop. The multiplexers remove the start and stop bits and transmit the aggregate to the distant end with a synchronous bit stream. The start and stop bits are restored at the receive demultiplexer. This further improves circuit efficiency.

In the United States the Digital Data System (DDS) described in Chapter 10 provides another form of sharing; in this case it may be with "foreign" users.* This is the sharing of a 56-kbps PCM channel which is broken down into 9600-, 4800-, and 2400-bps users.

The 384-kbps option is 1/4 of a DS1 configuration, assuming the availability of clear 64-kbps channels.† In the case of E1, 64-kbps channels are available. Here six contiguous E0 or special DS0 channels make up the 384-kbps configuration. It is a popular bit rate for television conferencing and for frame relay.

*Foreign users meaning data/telecommunication users of a different organization.
†We will recall from Chapter 8 that the North American DS1 commonly provides the user with a 56-kbps channel because of embedded signaling bits. However, clear 64-kbps channels can be provided in many instances at extra cost.

The final entry in Table 11.3 is a full DS1 configuration for North America or E1 for countries using E1. This provides 24 56/64-kbps channels for DS1 and 30 64-kbps channels for E1. In most cases, telephone companies or telecommunication administrations offer the option of "mix and match." In some cases, such as for frame relay, the entire aggregate can be used for data. In other situations, a 384-kbps group can be used for frame relay or other data connectivity, and other channels can be used for voice and/or data. In North America, leasing of a DS3 configuration is available providing 28 DS1s for the user.

In the United States, where private networks are prevalent, organizations, typically power companies, have excess capacity in their network and lease circuits to other users. In many cases such lease revenues not only pay for the private network, but also turn a profit on top of that.

6   DATA NETWORK OPERATION

6.1   Introduction

When data networks involve more than two terminals or nodes, some form of circuit discipline should be invoked. The first case is multidrop or multipoint, illustrated in Figure 11.4B. Here several terminals or nodes share a common transmission medium. They could operate on a "first come, first served" basis. This is known as *contention*, where terminals or nodes compete for access to the medium. There is no discipline. As more terminals are added, such networks become unwieldy.

One simple form of discipline in this situation is *polling*. In this case, one of the terminals is assigned as master station and polls the others periodically, querying them whether they have traffic for transmission. If so, the traffic is transmitted in frames and received by all stations on the network. It is only copied, though, by those stations identified in the frame header.

There are wide area and local area networks with thousands of users. To further complicate the matter, these networks consist of terminals, nodes, servers, routers, and CPUs from different vendors loaded with different software. Thus there is a problem offering equitable access to a network, but there is also an interface problem at many levels. We have already covered the electrical interface, such as EIA-232, which is more formally called the *physical interface* or *physical layer*.

In the late 1960s a term came into popular use, the *protocol*. Here we will slightly amplify the IEEE definition of protocol [1]: "*A formal set of rules and conventions governing the format and relative timing of message exchange among two or more communication terminals.*"

There are literally many dozens of protocols. Some have been developed by standardization bodies such as the ISO, IEEE, the U.S. Department of Defense, ITU-T organization, and to some lesser extent ANSI and EIA/TIA.

Others are proprietary, having been developed by various interests of the large population of data communication and processing equipment manufacturers such as IBM and DEC.

In the next sections we will introduce protocols, discuss the ISO (International Standards Organization) Open Systems Interconnection (OSI), several protocols fitting into the OSI environment, the U.S. DoD TCI/IP, and IBM's SNA.

6.2 Protocols

6.2.1 Basic Protocol Functions. Stallings [2] lists protocol functions in the following categories:

- Segmentation and reassembly (SAR)
- Encapsulation
- Connection control
- Ordered delivery
- Flow control
- Error control
- Multiplexing

A short description of each functional category is provided below.

Segmentation and Reassembly. Segmentation refers to breaking up the data into blocks with some bounded size. Depending on the semantics or system, these blocks may be called frames or packets. Reassembly is the counterpart of segmentation—that is, putting the blocks or packets back into their original order. Another name used for a data block is *protocol data unit* (PDU).

Encapsulation. Encapsulation is the adding of control information on either side of the data *text* of a block. Typical control information is the *header*, which contains address information and sequence numbers. An error control field is appended at the end of a block.

Connection Control. There are three stages of connection control:

1. Connection establishment.
2. Data transfer.
3. Connection termination.

Some of the more sophisticated protocols also provide connection interrupt and recovery capabilities to cope with errors and other sorts of interruptions.

Ordered Delivery. PDUs are assigned sequence numbers to ensure an ordered delivery of the data at the destination. In a large network, especially if it operates in the packet mode, PDUs (packets) can arrive at the destination out of order. With a unique PDU numbering plan using a simple numbering sequence, it is a rather simple task for a long data file to be reassembled at the destination in its original order.

Error Control. Error control is a technique that permits recovery of lost or errored PDUs. There are three possible functions involved in error control:

1. Acknowledgment of each PDU or string of PDUs.

2. Sequence numbering of PDUs (e.g., missing numbers).

3. Error detection (see Chapter 10).

Acknowledgment may be carried out by returning to the source the source sequence number of a PDU. This ensures delivery of all PDUs to the destination. Error detection initiates retransmission of errored PDUs.

Flow Control. Flow control refers to the management of data flow from source to destination such that buffers do not overflow but maintain full capacity of all facility components involved in the data transfer. Flow control must operate at several peer layers of a protocol, as will be discussed later.

Multiplexing. We will be talking about a vertical layered architecture in the next section. With this in mind, multiplexing can be used in one of two directions, upwards and downwards. *Upward multiplexing* can be used when multiple higher-level (layer) connections are multiplexed on a single lower-layer connection. This could be done to make more efficient use of the lower-layer service. This, typically, may be the multiplexing of several transport connections on a single network connection. *Downward multiplexing*, sometimes called *splitting*, is the building up of a single high-layer connection on top of multiple lower-layer connections. Reasons for downward multiplexing are to improve performance, reliability, and/or efficiency.

6.2.2   Open Systems Interconnection

6.2.2.1 Rationale.   Data communication systems can be very diverse and complex. These systems involve very elaborate software which must run on equipment having an ever-increasing processing capacity. Under these conditions, it is desirable to ensure maximum independence between the various software and hardware elements of a system for several reasons:

- To facilitate intercommunication among disparate elements

• To eliminate the "ripple effect" when there is a modification to one software element that may affect all elements

The International Standards Organization set about to make this data inter-communication problem more manageable. It developed its famous Open Systems Interconnection (OSI) reference model [4]. Instead of trying to solve the global dilemma, it decomposed the problem into more manageable parts. This provided standard-setting agencies with an architecture that defines communication tasks. The OSI model provides the basis for connecting open systems for distributed applications processing. The term *open* denotes the ability of any two systems conforming to the reference model and associated standards to interconnect. OSI thus provides a common groundwork for the development of families of standards permitting data assets to communicate.

ISO broke data communications down into seven areas or layers arranged vertically starting at the bottom with layer 1, the input/output ports of a data device. The OSI reference model is shown in Figure 11.7. It takes at least two to communicate. Thus we consider the model in twos, one entity to the left in the figure and one to the right. ISO and the ITU-T organization use the term *peers*. Peers are corresponding entities on either side of Figure 11.7. A peer on one side (system A) communicates with its peer on the other side (system B) by means of a common protocol. For example, the transport layer of system A communicates with its peer transport layer at system B. It is important to note that there is no direct communication between peer layers except at the physical layer (layer 1). That is, above the physical layer, each

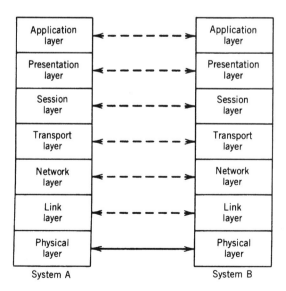

Figure 11.7. The OSI reference model.

protocol entity sends data down to the next lower layer, and so on to the physical layer, then across and up to its peer on the other side. Even the physical layer may not be directly connected to its peer on the other side of the "connection" such as in packet communications. This we call *connectionless service*. However, peer layers must share a common protocol in order to communicate.

There are seven OSI layers, as shown in Figure 11.7. Any layer may be referred to an N layer. Within a particular system there are one or more active entities in each layer. An example of an entity is a process in a multiprocessing system. It could simply be a subroutine. Each entity communicates with entities above and below it across an interface. The interface is at a service access point (SAP). An $(N - 1)$ entity provides services to an N entity by use of primitives. A primitive [1, 2] specifies the function to be performed and is used to pass data and control information.

CCITT Rec. X.200 [3] describes four types of primitive used to define the interaction between adjacent layers of the OSI architecture. A brief description of each of these primitives is given below.

Request. A primitive issued by a service user to invoke some procedure and to pass parameters needed to fully specify the service.

Indication. A primitive issued by a service provider either to invoke some procedure or to indicate that a procedure has been invoked by a service user at the peer service access point.

Response. A primitive issued by a service user to complete at a particular SAP some procedure invoked by an *indication* at that SAP.

Confirm. A primitive issued by a service provider to complete at a particular SAP some procedure previously invoked by a request at that SAP. CCITT Rec. X.200 [3] adds this note: Confirms and responses can be positive or negative depending on the circumstances.

The data that pass between entities are a bit grouping called a *data unit*. We discussed protocol data units (PDUs) earlier. Data units are passed downward from a peer entity to the next OSI layer, called the $(N - 1)$ layer. The lower layer calls the PDU a *service data unit* (SDU). The $(N - 1)$ layer adds control information, transforming the SDU into one or more PDUs. However, the identity of the SDU is preserved to the corresponding layer at the other end of the connection. This concept is shown in Figure 11.8.

When we discussed throughput in Section 5.1, it became apparent that throughput must be viewed from the eyes of the user. With OSI some form of encapsulation takes place at every layer above the physical layer. To a greater or lesser extent OSI is used on every and all data connectivities. The concept of encapsulation, the adding of overhead, from layers 2 through 7 is shown in Figure 11.9.

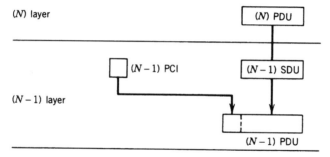

(N) layer

(N) PDU

(N – 1) PCI

(N – 1) SDU

(N – 1) layer

(N – 1) PDU

PCI = protocol control information
PDU = protocol data unit
SDU = service data unit

Figure 11.8. An illustration of mapping between data units in adjacent layers.

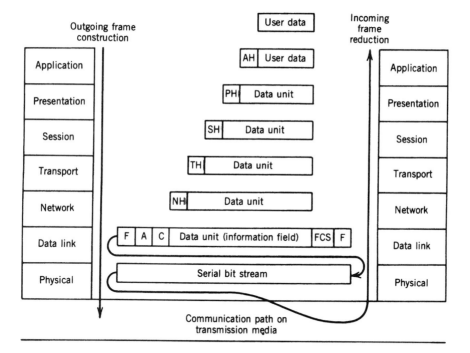

Figure 11.9. Buildup and breakdown of a data message following the OSI model. OSI encapsulates at every layer, except layer 1, adding significant overhead.

6.2.2.2 Functions of OSI Layers

PHYSICAL LAYER. The physical layer is layer 1, the lowest OSI layer. It provides the physical connectivity between two data terminals who wish to communicate. The services it provides to the data-link layer (layer 2) are those required to connect, maintain the connection, and disconnect the physical circuits that form the physical connectivity. The physical layer represents the traditional interface between data terminal equipment (DTE) and data communication equipment (DCE). This was described in Chapter 10, Section 8.

The physical layer has four important characteristics:

1. Mechanical
2. Electrical
3. Functional
4. Procedural

The mechanical aspects include the actual cabling and connectors necessary to connect the communications equipment to the media. Electrical characteristics cover voltage and impedance, balanced and unbalanced. Functional characteristics include connector pin assignments at the interface and the precise meaning and interpretation of the various interface signals and data set controls. Procedures cover sequencing rules that govern the control functions necessary to provide higher-layer services such as establishing a connectivity across a switched network.

Some applicable standards for the physical layer are:

- EIA-232, EIA-422, EIA-423, and EIA-530
- ITU-T Recs. V.10, V.11, V.24, V.28, X.21, and X.21 bis
- ISO 2110, 2593, 4902, and 4903
- U.S. Fed. Stds, 1020A, 1030A, and 1031
- U.S. MIL-STD-188-114

DATA-LINK LAYER. The data-link layer provides services for reliable interchange of data across a data link established by the physical layer. Link-layer protocols manage the establishment, maintenance, and release of data-link connections. These protocols control the flow of data and supervise error recovery. A most important function of this layer is recovery from abnormal conditions. The data-link layer services the network layer or logical link control (LLC; in the case of LANs) and inserts a data unit into the INFO portion of the data frame or block. A generic data frame generated by the link layer is shown in Figure 11.10.

Some of the more common data-link layer protocols are:

- ISO HDLC, ISO 3309, 4375
- CCITT LAP-B and LAP-D
- IBM BSC, SDLC
- DEC DDCMP
- ANSI ADCCP (also a U.S. government standard)

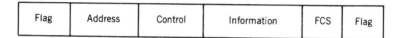

| Flag | Address | Control | Information | FCS | Flag |

Figure 11.10. Generalized data-link layer frame.

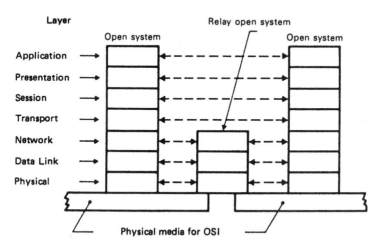

Figure 11.11. Only the first three layers of OSI are required at a relay (switching) point [3].

NETWORK LAYER. The network layer moves data through the network. At relay and switching nodes along the traffic route, layering concatenates. In other words, the higher layers (above layer 3) are not required and are utilized only at user end points.

The concept of relay open system is shown in Figure 11.11. At the relay or switching point, only the first three layers of OSI are required.

The network layer carries out the functions of switching and routing, sequencing, logical channel control, flow control, and error recovery functions. We note the duplication of error recovery in the data-link layer. However, in the network layer error recovery is network-wide, whereas on the data-link layer error recovery is concerned only with the data link involved.

The network layer also provides and manages logical channel connections between points in a network such as virtual circuits across the public switched network (PSN). It will be appreciated that the network layer concerns itself with the network switching and routing function. On simpler data connectivities, where a large network is not involved, the network layer is not required and can be eliminated. Typical of such connectivities are point-to-point circuits, multipoint circuits, and LANS. A packet-switched network is a typical example where the network layer is required.

The best-known layer 3 standard is CCITT Rec. X.25.

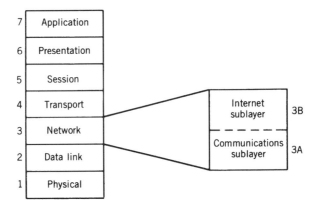

Figure 11.12. Sublayering of OSI layer 3 to achieve internetworking.

LAYER 3.5: INTERNETWORK PROTOCOLS. Layer 3.5 is a suggested sublayer of the OSI network layer.* It carries out the functions of internetworking—that is, the interconnection of two disparate networks. This sublayer is shown in Figure 11.12. The internet function is carried out by routers or gateways.

There are two applicable protocols for internetworking: CCITT Rec. X.75, which may be considered a subset of X.25, and IP (internet protocol), which is often accompanied by TCP, the transmission control protocol. IP and TCP are members of a fairly large family of protocols developed by U.S. Department of Defense ARPA (Advanced Research Projects Agency). There is an equivalent standard developed by ISO (ISO 8473).

The CCITT X.75 protocol is a companion of X.25 and assumes that all networks involved are based on the latter protocol. We will discuss X.25 in Section 8 of this chapter. A brief description of IP and its companion TCP is provided in Section 9.

TRANSPORT LAYER. The transport layer (layer 4) is the highest layer of the services associated with the provider of communication services. One can say that layers 1–4 are the responsibility of the communication system engineer. Layers 5, 6, and 7 are the responsibility of the data end-user. However, we believe that the telecommunication system engineer should have a working knowledge of all seven layers.

The transport layer has the ultimate responsibility for providing a reliable end-to-end data delivery service for higher-layer users. It is defined as an end system function, located in the equipment using network service or services. In this way its operations are independent of the characteristics of all the networks that are involved. Services that a transport layer provides are as follows:

---

*Layer 3.5 is not included in CCITT Rec. X.200. It has been suggested by Ref. 2 and bears a lot of merit.

- *Connection Management.* This includes establishing and terminating connections between transport users. It identifies each connection and negotiates values of all needed parameters.
- *Data Transfer.* This involves the reliable delivery of transparent data between the users. All data are delivered in sequence with no duplication or missing parts.
- *Flow Control.* This is provided on a connection basis to ensure that data are not delivered at a rate faster than the user's resources can accommodate.

The TCP (transport control protocol) was the first working version of a transport protocol and was created by DARPA for DARPANET. All the features in TCP have been adopted in the ISO version. TCP is often lumped with the internet protocol and referred to as TCP/IP.

The ISO transport protocol messages are called *transport protocol data units* (TPDUs). There are connection management TPDUs and data transfer TPDUs. The applicable ISO references are ISO 8073 OSI (*Transport Protocol Specification*) and ISO 8072 OSI (*Transport Service Definition*).

SESSION LAYER. The purpose of the session layer is to provide the means for cooperating presentation entities to organize and synchronize their dialogue and to manage the data exchange. The session protocol implements the services that are required for users of the session layer. It provides the following services for users:

1. The establishment of session connection with negotiation of connection parameters between users.
2. The orderly release of connection when traffic exchanges are completed.
3. Dialogue control to manage the exchange of session user data.
4. A means to define activities between users in a way that is transparent to the session layer.
5. Mechanisms to establish synchronization points in the dialogue and, in case of error, resume from a specified point.
6. Interrupt a dialogue and resume it later at a specified point, possibly on a different session connection.

Session protocol messages are called session protocol data units (SPDUs). The session protocol uses the transport layer services to carry out its function. A session connection is assigned to a transport connection. A transport connection can be reused for another session connection if desired. Transport connections have a maximum TPDU size. The SPDU cannot exceed this size. More than one SPDU can be placed on a TPDU for transmission to the remote session layer.

Reference standards for the session layer are ISO 8327 [*Session Protocol Definition* (CCITT Rec. X.225)] and ISO 8326 [*Session Services Definition* (CCITT Rec. X.215)].

PRESENTATION LAYER. The presentation layer services are concerned with data transformation, data formatting, and data syntax. These functions are required to adapt the information handling characteristics of one application process to those of another application process.

The presentation layer services allow an application to interpret properly the data being transferred. For example, there are often three syntactic versions of the information to be exchanged between end-users A and B as follows:

- Syntax used by the originating application entity A.
- Syntax used by the receiving application entity B.
- Syntax used between presentation entities. This is called the *transfer syntax* [5].

Of course, it is possible that all three or any two of these may be identical. The presentation layer is responsible for translating the representation of information between the transfer syntax and each of the other two syntaxes as required.

The following standards apply to the presentation layer:

- ISO 8822 Connection-Oriented Presentation Service Definition.
- ISO 8823 Connection-Oriented Presentation Service Specification.
- ISO 8824 Specification of Abstract Syntax Notation One.
- ISO 8824 Specification of Basic Encoding Rules for Abstract Syntax Notation One.
- CCITT Rec. X.409 Message Handling Systems: Presentation Transfer Syntax and Notation.

APPLICATION LAYER. The application layer is the highest layer of the OSI architecture. It provides services to the application processes. It is important to note that the applications do not reside in the application layer. Rather, the layer serves as a window through which the application gains access to the communication services provided by the model.

This highest OSI layer provides to a particular application all services related to communication in such a format that easily interfaces with the user application and is expressed in concrete quantitative terms. These include identifying cooperating peer partners, determining the availability of resources, establishing the authority to communicate, and authenticating the communication. The application layer also establishes requirements for data syntax and is responsible for overall management of the transaction.

Of course, the application itself may be executed by a machine, such as a CPU in the form of a program, or by a human operator at a workstation. The following standards apply to the application layer:

- ISO 8449/3 Definition of Common Application Service Elements
- ISO 8650 Specification of Protocols for Common Application Service Elements [2–6].

6.2.3 High-Level Data-Link Control—A Typical Link-Layer Protocol. High-Level Data-Link Control (HDLC) was developed by the International Standards Organization (ISO). It has spawned many related or nearly identical protocols. Among these are ANSI ADCCP, CCITT LAPB and LAPD, and IBM SDLC [7–9].

HDLC DEFINITIONS. Stations, configurations, and three modes of operation.

Primary Station. A logical primary station is an entity that has primary link control responsibility. It assumes responsibility for organization of data flow and for link level error recovery. Frames issued by the primary station are called *commands*.

Secondary Station. A logical secondary station operates under control of a primary station. It has no direct responsibility for control of the link but instead responds to primary station control. Frames issued by a secondary station are called *responses*.

Combined Station. A combined station combines the features of primary and secondary stations. It may issue both commands and responses.

Unbalanced Configuration. An unbalanced configuration consists of a primary station and one or more secondary stations. It supports full-duplex and half-duplex operation, point-to-point and multipoint circuits. An unbalanced configuration is shown in Figure 11.13a.

Balanced Configuration. A balanced configuration consists of two combined stations in which each station has equal and complementary responsibility of the data link. A balanced configuration operates only in the point-to-point mode and supports full-duplex and half-duplex operation. Figure 11.13b shows a balanced configuration.

Modes of Operation. With *normal response mode* (*NRM*) a primary station initiates data transfer to a secondary station. A secondary station transmits data only in response to a poll from the primary station. This mode of operation applies to an unbalanced configuration. With *asynchronous response mode (ARM)* a secondary station may initiate transmission without receiving a poll from a primary station. It is useful on a circuit where there is only one active secondary station. The overhead of continuous polling is thus eliminated. *Asynchronous balanced mode (ABM)* is

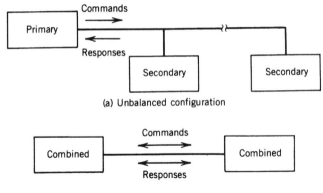

(a) Unbalanced configuration

(b) Balanced configuration

Figure 11.13. HDLC link configurations.

a balanced mode that provides symmetric data transfer capability between combined stations. Each station operates as if it were a primary station, can initiate data transfer, and is responsible for error recovery. One application of this mode is hub polling, where a secondary station needs to initiate transmission.

6.2.3.1 The HDLC Frame. Figure 11.14 shows the HDLC frame format. It has a similar format of generic data frames shown previously in this chapter and in Chapter 10. Moving from left to right in the figure, we have the flag field (F) which delimits the frame at both ends with the unique bit pattern 01111110. If frames are sent sequentially, the closing flag of the first frame is the opening flag of the next frame.

We called the flag sequence unique. Receiving stations constantly search for the flag bit sequence to mark the beginning and end of a frame. A data

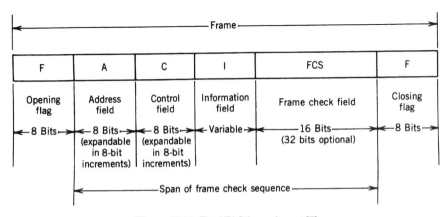

Figure 11.14. The HDLC frame format [7].

frame should be transparent to any 8-bit sequence of bits. Suppose, then, that the sequence 01111110 appeared in the middle of a frame. This would incorrectly tell a receiving station that the frame has ended and a new frame begun. Of course, this corrupts the whole frame and probably some subsequent frames. To avoid the problem, no sequence of six consecutive 1s bracketed by 0s is permitted. This is accomplished by what some call *bit stuffing* and others call *zero insertion*. With the exception of the flags, a transmitting station inserts an extra 0 bit after each occurrence of five 1s in the frame.

After detecting the opening flag, a receiver monitors the bit stream for a pattern of five contiguous 1s. If such a pattern occurs, the sixth bit is examined. If that bit is a 0, it is deleted. If the sixth bit is a 1 and the seventh bit is a 0, the combination is accepted as a flag. If bits six and seven are both 1s, the transmitting station is assumed to be sending an *abort condition* [7].

The *address field* (A) immediately follows the opening flag of a frame and precedes the control field (C). Each station in the network normally has an individual address and a group address. A group address identifies a family of stations. It is used when data messages must be accepted from or destined to more than one user. Normally the address is 8 bits long, providing 256 bit combinations or addresses ($2^8 = 256$). In HDLC (and ADCCP) the address field can be extended in increments of 8 bits. When this is implemented, the least significant bit is used as an extension indicator. When that bit is 0, the following octet is an extension of the address field. The address field is terminated when the least significant bit of an octet is 1. Thus we can see that the address field can be extended indefinitely.

The *control field* (C) immediately follows the address field (A) and precedes the information field (I). The control field conveys commands, responses, and sequence numbers to control the data link. The basic control field is 8 bits long and uses modulo 8 sequence numbering. There are three types of control field: (1) I frame (information frame), (2) S frame (supervisory frame), and (3) U frame (unnumbered frame). The three control field formats are shown in Figure 11.15.

Consider the basic 8-bit format as shown in Figure 11.15. The information flows from left to night. If the frame in Figure 11.14 has a 0 appear as the first bit in the control field, the frame is an I frame (see Figure 11.15a). If the bit is a 1, the frame is an S or a U frame, as shown in Figures 11.15b, c. If that first bit is a 1 followed by a 0, it is an S frame, and if the bit is a 1 followed by a 1, it is a U frame. These bits are called format identifiers.

Turning now to the information (I) frame (Figure 11.15a), its purpose is to carry user data. Bits 2, 3, and 4 of the control field in this case carry the *send* sequence count of transmitted messages (i.e., I frames).

We now digress to describe a *window* of frames. One generally thinks of a receiver on a point-to-point link with one buffer. This works well with stop-and-wait ARQ. Station X sends a message to station Y, which stores the message in a single buffer. On receipt of the entire bit string making up the

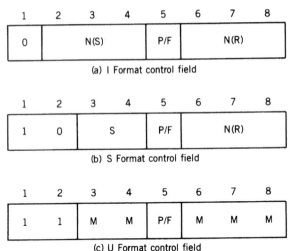

Figure 11.15. The three control field formats of HDLC.

message, the receiver processes the FCS for errors. If the frame is error free, the message or frame is acknowledged. We should note that stop-and-wait ARQ can be slow, tedious, and inefficient.

Suppose now that receiver Y has seven buffers (the number was picked arbitrarily). Thus Y can accept seven frames (or messages) and X is allowed to send seven frames without acknowledgment. To keep track of which frames have been acknowledged, each is labeled with a sequence number 0–7 (modulo 8). Station Y acknowledges a frame by sending the next sequence number expected. For instance, if Y sends a sequence number 3, this acknowledges frame number 2 and is awaiting frame number 3. Such a scheme can be used to acknowledge multiple frames. As an example, Y could receive frames 2, 3, and 4 and withhold all acknowledgments until frame 4 arrives. By sending sequence number 5, it acknowledges the receipt of frames 2, 3, and 4 all at once. Station X maintains a list of sequence numbers that it is allowed to send, and Y maintains a list of sequence numbers it is prepared to receive. These lists are thought of as a *window of frames*.

HDLC allows a maximum window size of 7, or 127 frames. In other words, a maximum number of 7, or 127 unacknowledged frames, can be sent or one less than the modulus 8 or 128. $N(S)$ is the sequence number of the next frame to be transmitted and $N(R)$ is the sequence number of the frame to be received.

Each frame carries a poll/final (P/F) bit. It is bit 5 in each of the three different types of control fields shown in Figure 11.15. The bit serves a function in both command and response frames. In a command frame it is referred to as a poll (P) bit; in a response frame as a final (F) bit. In both cases the bit is sent as a 1.

The P bit is used to solicit a response or sequence of responses from a secondary or balanced station. On a data link only one frame with a P bit set to 1 can be outstanding at any given time. Before a primary or balanced station can issue another frame with a P bit set to 1, it must receive a response frame from a secondary or balanced station with the F bit set to 1. In the NRM mode, the P bit is set to 1 in command frames to solicit response frames from the secondary station. In this mode of operation the secondary station may not transmit until it receives a command frame with the P bit set to 1.

Of course, the F bit is used to acknowledge an incoming P bit. A station may not send a final frame without prior receipt of a poll frame. As can be seen, P and F bits are exchanged on a one-for-one basis. Thus only one P bit can be outstanding at a time. As a result the $N(R)$ count of a frame containing a P or F bit set to 1 can be used to detect sequence errors. This capability is called *check pointing*. It can be used not only to detect sequence errors but to indicate the frame sequence number to begin retransmission when required.

Supervisory (S) frames, shown in Figure 11.15b, are used for flow and error control. Both go-back-n and selective ARQ can be accommodated. There are four types of supervisory or S frames:

1. Receive ready (RR): 1000 P/F $N(R)$
2. Receive not ready (RNR): 1001 P/F $N(R)$
3. Reject (Rej): 1010 P/F $N(R)$
4. Selective reject (SRej): 1011 P/F $N(R)$

The RR frame is used by a station to indicate that it is ready to receive information and acknowledge frames up to and including $N(R) - 1$. Also a primary station may use the RR frame as a command with the poll (P) bit set to 1.

The RNR frame tells a transmitting station that it is not ready to receive additional incoming I frames. It does acknowledge receipt of frames up to and including sequence number $N(R) - 1$. I frames with sequence number $N(R)$ and subsequent frames, if any, are not acknowledged. The Rej frame is used with go-back-n ARQ to request retransmission of I frames with frame sequence number $N(R)$, and $N(R) - 1$ frames and below are acknowledged.

Unnumbered frames are used for a variety of control functions. They do not carry sequence numbers, as the name indicates, and do not alter the flow or sequencing of I frames. Unnumbered frames can be grouped into the following four categories:

1. Mode-setting commands and responses
2. Information transfer commands and responses
3. Recovery commands and responses
4. Miscellaneous commands and responses

The information field follows the control field (Figure 11.14) and precedes the FCS field. The I field is present only in information (I) frames and some unnumbered (U) frames. The I field may contain any number of bits in any code, related to character structure or not. Its length is not specified in the standard (ISO 3309 [7]). Specific system implementations, however, usually place an upper limit on I field size. Some implementations require that the I field contain an integral number of octets.

Frame check sequence (FCS). Each frame includes a frame check sequence (FCS). The FCS immediately follows the I field, or the C field if there is no I field, and precedes the closing flag (F). The FCS field detects errors due to transmission. The FCS field contains 16 bits, which are the result of a mathematical computation on the digital value of all bits excluding the inserted zeros (zero insertion) in the frame and including the address, control, and information fields.

It should be noted that previously we had called the FCS field the BCC or block check count. To most of us, FCS implies the use of CRC for error detection. BCC, to many of us, can have a wider implication. Namely, it may mean only some form of parity check including CRC.

With most HDLC implementations, the FCS field is 16 bits long using the CCITT-recommended CRC (see Chapter 10, Section 5.4.1) (CCITT Rec. V.41 [10]). In some situations that require stringent undetected error rate conditions and/or because of frame length, a 32-bit FCS may be used. This 32-bit CRC is similar to the one used with 802 series LANS; it is described in Chapter 12.

6.3   X.25: A Packet-Switched Network Access Standard

6.3.1   *Introduction to CCITT Rec. X.25.* CCITT Rec. X.25 defines the procedures necessary for a packet mode data terminal to access the services provided by a packet-switched public data network (PDN) [12]. The original CCITT recommendation was approved in 1976 and subsequently has undergone a number of modifications. Figure 11.16 shows the X.25 concept of accessing the PDN. An example of the PDN is a public ISDN, discussed in Chapter 13.

Data terminals defined by X.25 operate in a synchronous full-duplex mode with a data rate of 2400, 4800, 9600, and 14,400 bps; 48, 64, 128, 192, 256, 384, 512, 1024, 1536, and 1920 kbps.

6.3.2   *X.25 Architecture and Its Relationship to OSI.* X.25 spans the lowest three layers of the OSI reference models. Figure 11.17 illustrates the architecture and its relationship to OSI. It can be seen that X.25 is compatible with OSI up to the network layer. In this context there are differences at the network/transport layer boundary. CCITT leans toward the view that the network and transport layer services are identical and that these are provided by X.25 virtual circuits.

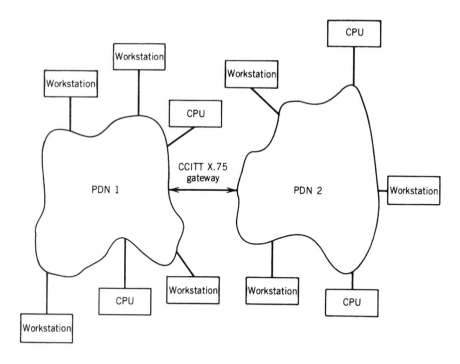

Figure 11.16. X.25 packet communications operates with the public (switched) data network (PDN). CPU = central processing unit or host computer.

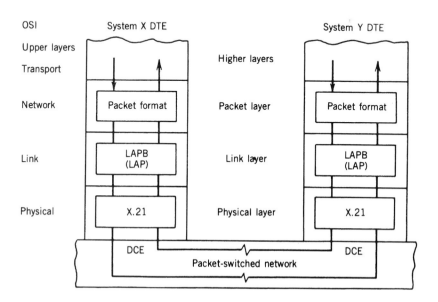

Figure 11.17. X.25's relationship with the OSI reference model.

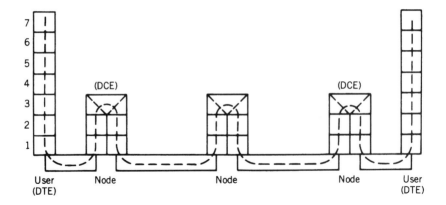

Figure 11.18. X.25 user (DTE) connects through the PDN to a distant user (DTE)

6.3.2.1   User Terminal Relationship to the PDN. CCITT Rec. X.25 calls the user terminal the DTE, and the DCE resides at the related PDN node. The entire recommendation deals with this DTE–DCE interface, not just the physical layer interface. For instance, a node (DCE) may connect to a related user (DTE) with one digital link, which is covered by SLP (single-link procedure) or several links covered by MLP (multilink procedure). Multiple links from a node to a DTE are usually multiplexed on one transmission facility.

The user (DTE) to user (DTE) connectivity through the PDN based on OSI is shown in Figure 11.18. A three-node connection is illustrated in this example. Of course, OSI layers 1–3 are Rec. X.25 specific. Note that the DTE protocol peers for these lower three layers are located in the PDN nodes and not in the distant DTE. The first DTE layer operating end to end is layer 4, the transport layer. Further, the Rec. X.25 protocol operates only at the interface between the DTE and its related PDN node and does not govern internodal network procedures.

6.3.3   The Three Layers of X.25

6.3.3.1   The Physical Layer. The physical layer is layer 1 where the requirements are defined for the functional, mechanical, procedural, and electrical interface between the DTE and DCE. CCITT Rec. X.21 or X.21 bis is the applicable standard for the interface as called out in CCITT Rec. X.25. X.21 bis is similar to EIA-232.

CCITT Rec. X.21 specifies a 15-pin DTE–DCE interface connector (refer to ISO 4903). The electrical characteristics for this interface are the same as CCITT Recs. V.10 and V.11, depending on whether electrically balanced or unbalanced operation is desired [13].

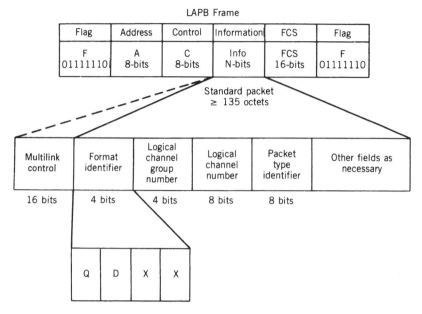

Figure 11.19. Basic X.25 frame structure. Note how the X.25 packet is embedded in the LAPB frame. For the extended LAPB structure, the control field will have 16 bits. There is also an extended modulo-128 structure for the X.25 packet. For a basic data packet, XX = 01. For an extended (modulo-128) data packet, XX = 10. Q = qualifier bit; D = delivery confirmation bit.

6.3.3.2   CCITT Rec. X.25 Link Layer. The Rec. X.25 link layer specifies LAPB (link access protocol B). Earlier versions of Rec. X.25 permitted the use of LAP, which was based on the ISO ARM protocol. LAPB is preferred, but LAP is still permitted. LAPB is fully compatible with HDLC link-layer access protocol, balanced asynchronous class (see Section 6.2.3). The information field in the LAPB frame carries the user data, in this case the layer 3 packet.

LAPB provides several options for link operation. These include modulo-8 or modulo-128 control fields. It also supports MLP or multilink procedures. MLP allows a group of links to be managed as a single transmission facility. It carries out the function of resequencing packets in the proper order at the desired destination. When MLP is implemented, an MLP control field of two octets in length is inserted as the first 16 bits of the information field. It contains a multilink sequence number and four control bits. See Figure 11.19.

6.6.3.3   Datagrams, Virtual Circuits, and Logical Connections. There are three approaches used with Rec. X.25 operation to manage the transfer and routing of packet streams: datagrams, virtual connections (VCs), and permanent virtual connections (PVCs). Datagram service uses optimal routing on a packet-by-packet basis, usually over diverse routes. In the virtual circuit approach, there are two operational modes: virtual connection and permanent virtual connection. These are analogous to a dial-up telephone connection and a leased line connection, respectively. With the virtual connection a logical connection is established before any packets are sent. The packet originator sends a call request to its serving node, which sets up a route in advance to the desired destination. All packets of a particular message traverse this route, and each packet of the message contains a virtual circuit identifier (logical channel number) and the packet data. At any one time each station can have more than one virtual circuit to any other station and can have virtual circuits to more than one station. With virtual circuits routing decisions are made in advance. With the datagram approach ad hoc decisions are made for each packet at each node. There is no call-setup phase with datagrams; there is with virtual connections. Virtual connections are advantageous for high community-of-interest connectivities, datagram for low community-of-interest relations.

Datagram service is more reliable because traffic can be alternately routed around network congestion points. Virtual circuits are fixed-routed for a particular call. Call-setup time at each node is eliminated on a packet basis with the virtual connection technique. Rec. X.25 also allows the possibility of setting up permanent virtual connections and is network assigned. This latter alternative is economically viable only for very high traffic relations; otherwise these permanently assigned logical channels will have long dormant periods.

6.3.4   X.25 Frame Structure: Layer 3, The Packet Layer. The basic data-link layer (LAPB) frame structure is given in Figure 11.19. Its similarity to the HDLC frame structure (Figure 11.14) is apparent. For the X.25 case the packet is embedded in the LAPB information field, as mentioned. When applicable, the other part of the information field contains the MLP, which is appended in front of the X.25 packet and is the first subfield in the information (I) field. Actually the MLP is not part of the actual packet and is governed by the layer 2 LAPB protocol.

6.3.4.1   Structure Common to All Packets.   Table 11.4 shows 17 packet types involved in X.25. Every packet transferred across the X.25 DTE–DCE interface consists of at least three octets.* These three octets contain a general format identifier, a logical channel identifier, and a packet type identifier. Other packets are appended as required. This is shown in Figure 11.19.

Now consider the general format identifier in Figure 11.19. This is a 4-bit sequence. Bit 8 is the qualifier bit only in data packets. In call setup and clearing packets it is the A-bit, and in all other packets it is set to 0. The D-bit, when set to 1, specifies end-to-end delivery confirmation. This confirmation is provided through the packet receive number $[P(R)]$. When set to 01 the XX bits indicate a basic packet (i.e., modulo-8), and when set to 10 they

---

*An octet is an 8-bit sequence. There is a growing tendency to use octet rather than byte when describing such a sequence. It removes ambiguity on the definition of byte.

TABLE 11.4 Packet Type Identifier

Packet Type		Octet 3							
					Bit				
From DCE to DTE	From DTE to DCE	8	7	6	5	4	3	2	1
Call Setup and Cleaning									
Incoming call	Call request	0	0	0	0	1	0	1	1
Call connected	Call accepted	0	0	0	0	1	1	1	1
Clear indication	Clear request	0	0	0	1	0	0	1	1
DCE clear confirmation	DTE clear confirmation	0	0	0	1	0	1	1	1
Data and Interrupt									
DCE data	DTE data	X	X	X	X	X	X	X	0
DCE interrupt	DTE interrupt	0	0	1	0	0	0	1	1
DCE interrupt confirmation	DTE interrupt confirmation	0	0	1	0	0	1	1	1
Flow Control and Reset									
DCE RR (modulo 8)	DTE RR (modulo 8)	X	X	X	0	0	0	0	1
DCE RR (modulo 128)[a]	DTE RR (modulo 128)[a]	0	0	0	0	0	0	0	1
DCE RNR (modulo 8)	DTE RNR (modulo 8)	X	X	X	0	0	1	0	1
DCE RNR (modulo 128)[a]	DTE RNR (modulo 128)[a]	0	0	0	0	0	1	0	1
	DTE REJ (modulo 8)[a]	X	X	X	0	1	0	0	1
	DTE REJ (modulo 128)[a]	0	0	0	0	1	0	0	1
Reset indication	Reset request	0	0	0	1	1	0	1	1
DCE reset confirmation	DTE reset confirmation	0	0	0	1	1	1	1	1
Restart									
Restart indication	Restart request	1	1	1	1	1	0	1	1
DCE restart confirmation	DTE restart confirmation	1	1	1	1	1	1	1	1
Diagnostic									
Diagnostic[a]		1	1	1	1	0	0	0	1
Registration[a]									
	Registration request	1	1	1	1	0	0	1	1
Registration confirmation		1	1	1	1	0	1	1	1

[a] Not necessarily available on every network.
Note: A bit that is indicated as X may be set to either 0 or 1.
Source: ITU-T Rec. X.25, Table 5-2/X.25, page 52 (Ref. 12).

indicate an extended modulo-128 packet. The extension involves sequence number lengths.

Logical channel assignment is shown in Figure 11.20. The logical channel group and logical channel number subfields identify logical channels with the

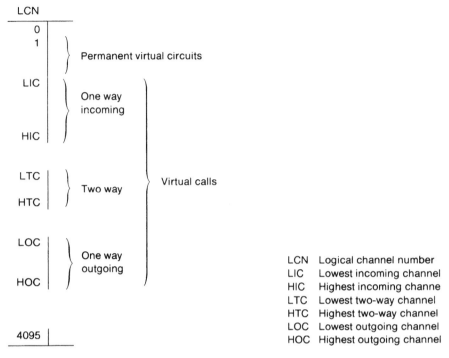

Figure 11.20. Logical channel assignment (Ref. 12, Annex A).

capability of identifying up to 4096 channels (2^{12}). This permits a DTE to establish up to 4095 simultaneous virtual circuits through its DCE to other DTEs. As we mentioned, this is usually done by multiplexing these circuits over a single transmission facility.

Permanent virtual circuits (PVCs) have permanently assigned logical channels, whereas those for virtual calls are assigned channels only for the duration of a call, as shown in Figure 11.20. Channel 0 is reserved for restart and diagnostic functions. To avoid collisions, the DCE starts assigning logical channels at the lowest number end and the DTE from the highest number end. There are one-way and both-way (two-way) circuits. The both-way circuits are reserved for overflow to avoid chances of double seizure (i.e., both ends seize the same circuit).

Octet 3 in Figure 11.19 is the packet type identifier subfield. The packet type and its corresponding coding are shown in Table 11.4. We can see in the table that packet types are identified in associated pairs carrying the same packet identifier (bit sequence). A packet from the calling terminal (DTE) to the network (DCE) is identified by one name. The associated packet delivered by the network to the called terminal (DTE) is referred to by another associated name.

Bits

Octets	8	7	6	5	4	3	2	1
1	General format identifier (Note)				Logical channel group number			
2	Logical channel number							
	Packet type identifier							
3	0	0	0	0	1	0	1	1
4	Address block							
	Facility length							
	Facilities							
	Call user data							

Note: Coded XX01 (modulo 8) or XX10 (modulo 128).

Figure 11.21. Call request and incoming call packet. The general format identifier is coded 0X01 (modulo-8) and 0X10 (modulo-128). From ITU-T Rec. X.25, Figure 5-3/X.25, page 58, ITU-T Organization, Helsinki, March 1993 (Ref. 12).

6.3.4.2 Several Typical Packets

CALL REQUEST AND INCOMING CALL PACKET. This type of packet sets up the call for the virtual circuit. The format of the call request and incoming packet is shown in Figure 11.21. Octets 1–3 have been described in the previous subsection. Octet 4 consists of the address length field indicators for the called and calling DTE addresses. Each address length indicator is binary coded, and bit 1 or 5 is the low-order bit of the indicator. Octet 5 and the following octets consist of the called DTE address, when present, and then the calling DTE address, when present.

The facilities length field (one octet) indicates the length of the facilities field that follows. The facility field is present only when the DTE is using an optional user facility requiring some indication in the call request and incoming call packets. The field must contain an integral number of octets with a maximum length of 109 octets.

Optional user facilities are listed in CCITT Rec. X.2. There are 45 listed. Several examples are listed to give some idea of what is meant by *facilities* in CCITT Rec. X.25:

- Nonstandard default window
- Flow control parameter negotiation
- Throughput class negotiation
- Incoming calls barred
- Outgoing calls barred
- Closed user group (CUG)

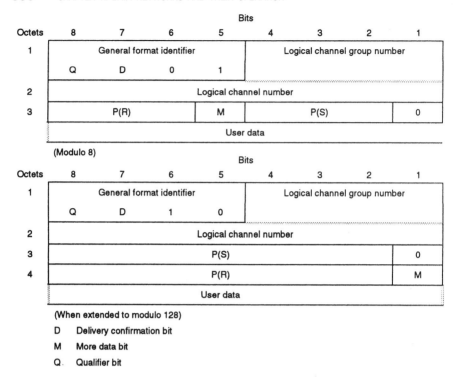

(When extended to modulo 128)
D   Delivery confirmation bit
M   More data bit
Q.   Qualifier bit

Figure 11.22. DTE and DCE packet format. From ITU-T Rec. X.25, Figure 5-7/X.25, page 64, ITU-T Organization, Helsinki, March 1993 (Ref. 12).

- Reverse charging acceptance
- Fast select

DTE AND DCE DATA PACKET. Of the 17 packet types listed in Table 11.4, only one truly carries user information, the DTE and DCE data packet. Figure 11.22 illustrates this packet format.

Octets 1 and 2 have been described. Bits 6, 7, and 8 of octet 3 or bits 2–8 of octet 4, when extended, are used for indicating the packet received sequence number $P(R)$. It is binary coded and bit 6, or bit 2 when extended, is the low-order bit.

In Figure 11.22, M, which is bit 5 in octet 3 or bit 1 in octet 4 when extended, is used for more data (M bit). It is coded 0 for "no more data" and 1 for "more data to follow."

Bits 2, 3, and 4 of octet 3, or bits 2–8 of octet 3 when extended, are used for indicating the packet send sequence number $P(S)$. Bits following octet 3, or octet 4 when extended, contain the user data.

The standard maximum user data field length is 128 octets. CCITT Rec. X.25 (par. 4.3.2) states: "In addition, other maximum user data field lengths may be offered by Administrations from the following list: 16, 32, 64, 256, 512, 1024, 2048, and 4096 octets ... . Negotiation of maximum user data field lengths on a per call basis may be made with the flow parameter negotiation facility."

6.3.5 *Tracing the Life of a Virtual Call.* A call is initiated by a DTE by the transfer to the network of a call request packet. It identifies the logical channel number selected by the originating DTE, the address of the called DTE (destination), and optional facility information and can contain up to 16 octets of user information. The facility and user fields are optional at the discretion of the source DTE. The receipt by the network of the call request packet initiates the call-setup sequence. This same call request packet is delivered to the destination DTE as an incoming call packet. The destination DTE in return sends a "call accepted" packet, and the source DTE receives a "call confirmation" packet. This completes the call-setup phase, and the data transfer can begin.

Data packets (Figure 11.22) carry the user data to be transferred. There may be one or a sequence of packets transferred during a virtual call. It is the M bit that tells the destination that the next packet is a logical continuation of the previous packet(s). Sequence numbers verify correct packet order and are the packet acknowledgment tools.

The last phase in the life of a virtual call is the call clearing (takedown). Either the DTE or the network can clear a virtual call. The "clear" applies only to the logical channel that was used for that call. Three different packet types are involved in the call-clearing phase. The clear request packet is issued by the DTE initiating the clear. The remote DTE receives it as a clear indication packet. Both the DTE and DCE then issue clear confirmation packets to acknowledge receipt of the clear packets.

Flow control is called out by the following packet types: receive ready (RR), receive not ready (RNR), and reject (Rej). Each of these packets is normally three octets long or four octets long using modulo-128 numbering. Sequence numbering also assists flow control.

7 TCP/IP AND RELATED PROTOCOLS

7.1 Background and Scope

The transmission control protocol/internet protocol (TCP/IP) family was developed for the ARPANET (Advanced Research Projects Agency Network; see Figure 11.1). ARPANET was one of the first large advanced packet-switched networks. It was initially designed and operated to interconnect the very large university and industrial defense research community to share research resources. It dates back to 1968 and was well into existence before ISO and CCITT took interest in layered protocols.

The TCP/IP suite of protocols [16–24] has wide acceptance today, especially in the commercial and industrial community worldwide. These protocols are used on both LANs and WANS. They are particularly attractive for

OSI	TCP/IP AND RELATED PROTOCOLS		
Application	File transfer	Electronic mail	Terminal emulation
Presentation	File transfer protocol (FTP)	Simple mail transfer protocol (SMTP)	Telnet protocol
Session			
Transport	Transmission control protocol (TCP)		User datagram protocol (UDP)
Network	Address resolution protocol (ARP)	Internet protocol (IP)	Internet control message protocol (ICMP)
Data link	———— Network interface cards ———— CSMA/CD (Ethernet), token ring, ARCNet, StarLan		
Physical	———— Transmission media ———— Wire pair, fiber optics, coaxial cable, radio		

Figure 11.23. How TCP/IP and associated protocols relate to OSI.

their internetworking capabilities. The internet protocol (IP) [16] competes with CCITT Rec. X.75 protocol [15], but is notably more versatile and has a much wider application.

The architectural model of the IP [16] uses terminology that differs from the OSI reference model.* Figure 11.23 shows the relationship between TCP/IP and related DoD protocols and the OSI reference model. Tracing data traffic from an originating host, which runs an applications program, to another host in another network is shown in Figure 11.24. This may be a LAN-to-WAN-to-LAN connectivity as shown in the figure. It may also be a LAN-to-LAN or it may be a WAN-to-WAN connectivity. The host would enter its own network by means of a network access protocol such as HDLC or an IEEE 802 series protocol (Chapter 12).

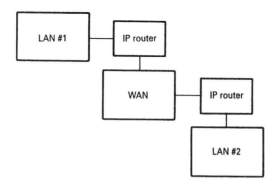

Figure 11.24. Connecting one LAN to another LAN via a WAN with routers equipped with IP.

*IP predates OSI.

A LAN connects via a router (or gateway) to another network. Typically a router (or gateway) is loaded with three protocols. Two of these protocols connect to each of the attached networks (e.g., LAN and WAN), and the third protocol is the IP which provides the network-to-network interface.

Hosts typically are equipped with four protocols. To communicate with routers or gateways, a network access protocol and internet protocol are required. A transport layer protocol assures reliable communication between hosts because end-to-end capability is not provided in either the network access or internet protocols. Hosts also must have application protocols such as E-mail or file transfer protocols (FTPs).

7.2   TCP/IP and Data-Link Layers

TCP/IP is transparent to the type of data-link layer involved, and it is also transparent whether it is operating in a LAN or WAN domain or among them. However, there is document support for Ethernet, IEEE 802 series, ARCNET LANs and X.25 for WANs [18, 19].

Figure 11.25 shows how upper OSI layers are encapsulated with TCP and IP header information and then incorporated into the data-link layer frame.

For the case of IEEE 802 series LAN protocols, advantage is taken of the LLC common to all 802 protocols. The LLC extended header contains the SNAP (sub-network access protocol) such that we have three octets for the LLC header and five octets in the SNAP. The LLC header has its fields fixed as follows (LLC is discussed in Chapter 12):

DSAP = 10101010   (destination service access point)

SSAP = 10101010   (source service access point)

Control = 00000011   [For unnumbered information (UI frame)]

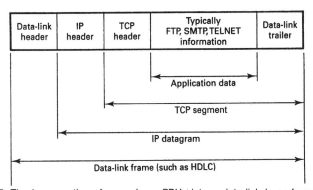

Figure 11.25. The incorporation of upper-layer PDUs into a data-link layer frame showing the relationship with TCP and IP.

TABLE 11.5 EtherType Assignments

Ethernet Decimal	Hex	Description
512	0200	XEROX PUP
513	0201	PUP address translation
1536	0600	XEROX NS IDP
2048	0800	DOD internet protocol (IP)
2049	0801	X.75 internet
2050	0802	NBS internet
2051	0803	ECMA internet
2052	0804	Chaosnet
2053	0805	X.25 level 3
2054	0806	Address resolution protocol (ARP)
2055	0807	XNS compatibility
4096	1000	Berkeley trailer
21000	5208	BBN Simnet
24577	6001	DEC MOP dump/load
24578	6002	DEC MOP remote control
24579	6003	DEC DECnet phase IV
24580	6004	DEC LAT
24582	6005	DEC
24583	6006	DEC
32773	8005	HP probe
32784	8010	Excelan
32821	8035	Reverse ARP
32824	8038	DEC LANBridge
32823	8098	Appletalk

Source: Ref. 17.

The five octets in the SNAP have three assigned for protocol ID or organizational code and two octets for "EtherType." EtherType assignments are shown in Table 11.5. EtherType refers to the general class of LANs based on CSMA/CD. (See Chapter 12 for a discussion of CMSA/CD.)

Figure 11.26 shows the OSI relationships with TCP/IP working with the IEEE 802 LAN protocol group. Figure 11.27 illustrates an IEEE 802 frame incorporating TCP, IP, and LLC (logical link control, Chapter 12).

Often addressing formats are incompatible from one protocol in one network to another protocol in another network. A good example is the mapping of the 32-bit internet address into a 48-bit IEEE 802 address. This problem is resolved with ARP (address resolution protocol). Another interface problem is the limited IP datagram length of 576 octets, where with the 802 series, frames have considerably larger length limits [19].

7.3 The IP Routing Function

In OSI the network layer functions include routing and switching of a datagram through the telecommunications subnetwork. The IP provides this

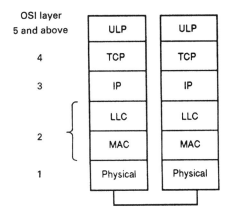

Figure 11.26. How TCP/IP working with IEEE 802 series of LAN protocols relates to OSI.

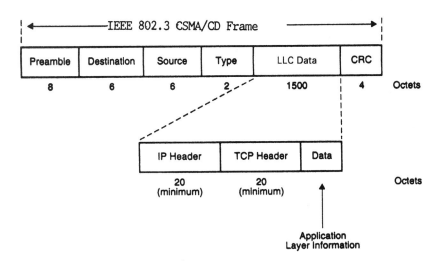

Figure 11.27. A typical IEEE 802 frame showing LLC and TCP/IP functions.

essential function. It forwards the datagram based upon the network address contained within the IP header. Each datagram is independent and has no relationship with other datagrams. There is no guaranteed delivery of the datagram from the standpoint of the internet protocol. However, the next higher layer, the TCP layer, provides for the reliability that the IP lacks. It also carries out segmentation and reassembly functions of a datagram to match frame sizes of data-link layer protocols.

Addresses determine routing, and at the far end, equipment (hardware). Actual routing derives from the IP address, and equipment addresses derive from the data-link layer header (typically the 48-bit Ethernet address) [19].

User data from upper-layer protocols is passed to the IP layer. The IP layer examines the network address (IP address) for a particular datagram and determines if the destination node is on its own local area network or some other network. If it is on the same network, the datagram is forwarded directly to the destination host. If it is on some other network, it is forwarded to the local IP router (gateway). The router, in turn, examines the IP address and forwards the datagram as appropriate. Routing is based on a lookup table residing in each router or gateway.

7.3.1 Detailed IP Operation. The IP provides connectionless service, meaning that there is no call-setup phase prior to exchange of traffic. There are no flow control or error control capabilities incorporated in IP. These are left to the next higher layer, the transmission control protocol (TCP). The IP is transparent to subnetworks connecting at lower layers; thus different types of networks can attach to an IP gateway.

Whereas prior to this discussion we have used the term *segmentation* to mean breaking up a data file into manageable segments, frames, packets, blocks, and so on, the IP specifications refer to this as *fragmentation*. IP messages are called *datagrams*. The minimum datagram length is 576 octets and the maximum length is 65,535 octets. Fragmentation resolves PDU (protocol data unit) sizes of the different networks with which IP carries out an interface function. For example, X.25 packets typically have data fields of 128 octets; Ethernet limits the size of a PDU to 1500 octets, and so forth. Of course, IP does a reassembly of the "fragments" at the opposite end of its circuit [16].

7.3.1.1 Description of the IP Datagram. The IP datagram format is shown in Figure 11.28. The datagram format should be taken in context with Figures 11.26 and 11.27 showing how the IP datagram relates to TCP and the data-link layer. In Figure 11.28, we move from left to right and top down in our description.

The *version* field (4 bits) gives the release number of the IP version for which a particular gateway or router is equipped.

The *header length* (4 bits) is measured in units of 32 bits. A header without certain options, such as QoS, typically has 20 octets. Thus its length is 5 (32 bits or 4 octets per unit; $5 \times 4 = 20$).

The *type of service* (TOS) field (8 bits) is used to identify several QoS parameters provided by IP. The field has eight bits broken down into four active groupings, and the last two bits are reserved. The first is *precedence*, which consists of three bits as follows:

000	routine	001	priority
010	immediate	011	flash
100	flash override	101	CEITIC/ECP
110	internetwork control	111	network control

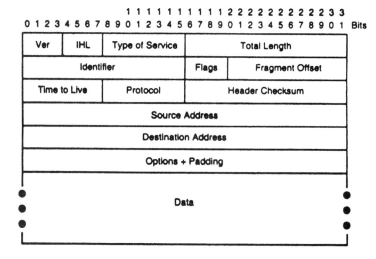

Figure 11.28. The IP datagram format [16].

Delay, 1 bit: 0 = normal, 1 = low
Throughput, 1 bit: 0 = normal, 1 = high
Reliability: 1 bit, 0 = normal, 1 = high

The *total length* field specifies the total length of an IP datagram in question. The unit of measure is the octet, and it includes the length of the header and data fields. The maximum possible length of a datagram is $2^{16} - 1$; the minimum length is 576 octets.

Segmentation (fragmentation) and reassembly are controlled by three fields in the header. These are *identifier* (16 bits), *flags* (3 bits), and *fragment offset* (3 bits). The term *fragment* suggests part of a whole; thus the identifier field identifies a fragment as part of a complete datagram, along with the source address. The flag bits determine if a datagram can be fragmented. When it can be fragmented, one of the flag bits shows whether the fragment is the last fragment of a datagram. The fragmentation offset field gives the relative position of the fragment regarding the original datagram. It is initially set at 0 and then set to the proper number by the fragmenting gateway.

The *time-to-live* (TTL) field's basic purpose is to prevent routing loops. In this we mean a routing that eventually routes back on itself. In telephony, it is sometimes called *ring-around-the-rosy*. It is used to measure the time a datagram has been in the internet. Each internet gateway checks this number and will discard that datagram if the TTL equals zero. There are numerous ways to implement TTL, some being vendor-specific. It is sometimes used for diagnostics by network management features such as SNMP.*

---

*SNMP is discussed in Chapter 18.

The *protocol* field (8 bits) identifies the next-higher-layer protocol the datagram expects at the destination host. Table 11.6, taken from RFC 1060 [17] and Ref. 18, shows the protocol decimal numbering scheme and the corresponding protocol for each number.

The next field is the *header checksum* field (16 bits). This provides error detection on the header.

Source address and *destination address* fields are each 32 bits long. Of course, the source address is the address of the originating host and the destination address is the address of the destination host.

The IP address structure is shown in Figure 11.29. It shows four address formats in which the lengths of the two component fields making up the address field change with each of the different formats. These component fields are the "network address" and the "local address" fields. The first bits of the address field specify the format or "class," there being classes A, B, C, or D. The "local address" is often called "host address."

The *class A* addressing is for very large networks, such as what was ARPANET. The field starts with binary 0, indicating that it is a class A format. In this case the local or host address component field is 24 bits long and has an address capacity of 2^{24}. *Class B* addressing is for medium-sized networks, such as campus networks. The field begins with 10 to indicate that it is a class B format; the network component field is 14 bits long, and the host or local address component field is 16 bits in length. The *class C* format is for small networks with a very large network ID field with an addressing capacity of 2^{24} and a considerably small host ID field of only 8 bits (2^8 addressing capacity). The address field in this case starts with 110. The *class D* format is for multicasting, a form of broadcasting. Its first four bits are the sequence 1110.

7.3.1.2   IP Routing. A gateway (router) needs only the network ID portion of the address to perform its routing function. Each router or gateway has a routing table which consists of: destination network addresses and specified next-hop gateway.

Three types of routing are performed by the routing table:

1. Direct routing to locally attached devices

2. Routing to networks that are reached via one or more gateways

3. Default routing to destination network in case the first types of routing are unsuccessful

Suppose a datagram (or datagrams) is (are) directed to a host which is not in the routing table resident in a particular gateway. Likewise, there is a possibility that the network address for that host is also unknown. These problems may be resolved with the *address resolution protocol* (ARP) [20].

TABLE 11.6   IP Protocol Field Numbering (Assigned Internet Protocol Numbers)

Decimal	Key Word	Protocol
0		Reserved
1	ICMP	Internet control message protocol
2	IGMP	Internet group management protocol
3	GGP	Gateway-to-gateway protocol
4		Unassigned
5	ST	Stream
6	TCP	Transmission control protocol
7	UCL	UCL
8	EGP	Exterior gateway protocol
9	IGP	Interior gateway protocol
10	BBN-MON	BBN-RCC monitoring
11	NVP-II	Network voice protocol
12	PUP	PUP
13	ARGUS	ARGUS
14	EMCON	EMCON
15	XNET	Cross net debugger
16	CHAOS	Chaos
17	UDP	User datagram protocol
18	MUX	Multiplexing
19	DCN-MEAS	DCN measurement subsystems
20	HMP	Host monitoring protocol
21	PRM	Packet radio monitoring
22	XNS-IDP	XEROX NS IDP
23	TRUNK-1	Trunk-1
24	TRUNK-2	Trunk-2
25	LEAF-1	Leaf-1
26	LEAF-2	Leaf-2
27	RDP	Reliable data protocol
28	IRTP	Internet reliable TP
29	ISO-TP4	ISO transport class 4
30	NETBLT	Bulk data transfer
31	MFE-NSP	MFE network services
32	MERIT-INP	MERIT internodal protocol
33	SEP	Sequential exchange
34–60		Unassigned
61		Any host internal protocol
62	CFTP	CFTP
63		Any local network
64	SAT-EXPAK	SATNET and backroom EXPAK
65	MIT-SUBN	MIT subnet support
66	RVD	MIT remote virtual disk
67	IPPC	Internet plur. packet core
68		Any distributed file system
69	SAT-MON	SATNET monitoring
70		Unassigned
71	IPCV	Packet core utility
72–75		Unassigned
76	BRSAT-MON	Backroom SATNET monitoring
77		Unassigned
78	WB-MON	Wideband monitoring
79	WB-EXPAK	Wideband EXPAK
80–254		Unassigned
255		Reserved

Source: RFC 1060 [17] and *Internet Protocol Transition Workbook* [18].

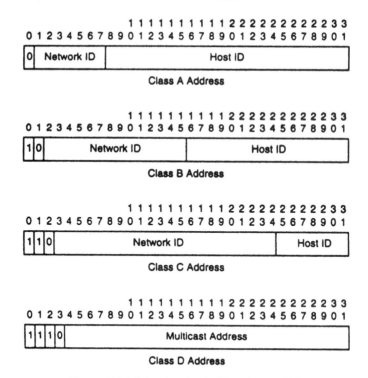

Figure 11.29. Internet protocol address formats [16].

First the ARP searches a mapping table which relates IP addresses with corresponding physical addresses. If the address is found, it returns the correct address to the requester. If it cannot be found, the ARP broadcasts a *request* containing the IP target address in question. If a device recognizes the address, it will reply to the request where it will update its ARP *cache* with that information. The ARP cache contains the mapping tables maintained by the ARP module.

There is also a *reverse address resolution protocol* (RARP) [21]. It works in a fashion similar to that of the ARP, but in reverse order. RARP provides an IP address to a device when the device returns its own hardware address. This is particularly useful when certain devices are booted and only know their own hardware address.

Routing with IP involves a term called *hop*. A hop is defined as a link connecting adjacent nodes (gateways) in a connectivity involving IP. A *hop count* indicates how many gateways (nodes) must be traversed between source and destination.

One part of an IP routing algorithm can be *source routing*. Here an upper-layer protocol (ULP) determines how an IP datagram is to be routed. One option is that the ULP passes a listing of internet addresses to the IP layer. In this case information is provided on the intermediate nodes required for transit of a datagram in question to its final destination.

Each gateway makes its routing decision based on a resident routing list or routing table. If a destination resides in another network, a routing decision is required by the IP gateway to implement a route to that other network. In many cases, multiple hops are involved and each gateway must carry out routing decisions based on its own routing table.

A routing table can be static or dynamic. The table contains IP addressing information for each reachable network and closest gateway for the network, and it is based on the concept of shortest routing, thus routing through the closest gateway.

Involved in IP shortest routing is the *distance metric*, which is a value expressing minimum number of hops between a gateway and a datagram's destination. An IP gateway tries to match the destination network address contained in the header of a datagram with a network address entry contained in its routing table. If no match is found, the gateway discards the datagram and sends an ICMP message back to the datagram source.

7.3.1.3 Internet Control Message Protocol (ICMP). ICMP [22] is used as an adjunct to IP when there is an error in datagram processing. ICMP uses the basic support of IP as if it were a higher-level protocol; however, ICMP is actually an integral part of IP and is implemented by every IP module.

ICMP messages are sent in several situations: for example, when a datagram cannot reach its destination, when a gateway does not have the buffering capacity to forward a datagram, and when the gateway can direct the host to send traffic on a shorter route.

ICMP messages typically report errors in the processing of datagrams. To void the possibility of infinite regress of messages about messages, and so on, no ICMP messages are sent about ICMP messages. Also ICMP messages are only sent about errors in handling fragment zero of fragmented datagrams. (*Note*: Fragment zero has the fragment offset equal to zero.)

MESSAGE FORMATS. ICMP messages are sent using the basic IP header (see Figure 11.28). The first octet of the data portion of the datagram is an ICMP type field. The "data portion" is the last field at the bottom of Figure 11.28. The ICMP type field determines the format of the remaining data. Any field labeled *unused* is reserved for later extensions and is fixed at zero when sent, but receivers should not use these fields (except to include them in the checksum). Unless otherwise noted under individual format descriptions, the values of the internet header fields are as follows:

- Version: 4.
- IHL (internet header length): length in 32-bit words.
- Type of service: 0

- Total length: length of internet header and data in octets.
- Identification, flags, and fragment offset: used in fragmentation, as in basic IP protocol described above.
- Time to live: in seconds; as this field is decremented at each machine in which the datagram is processed, the value in this field should be at least as great as the number of gateways which this datagram will traverse.
- Protocol: ICMP = 1
- Header checksum: the 16-bit one's complement of the one's complement sum of all 16-bit words in the header. For computing the checksum, the checksum field should be zero. The reference RFC (RFC 792) states that this checksum may be replaced in the future.
- Source address: the address of the gateway or host that composes the ICMP message. Unless otherwise noted, this is any of a gateway's addresses.
- Destination address: the address of the gateway or host to which the message should be sent.

There are eight distinct ICMP messages covered in RFC 792:

1. Destination unreachable message
2. Time exceeded message
3. Parameter problem message
4. Source quench message
5. Redirect message
6. Echo or echo reply message
7. Timestamp or timestamp reply message
8. Information request or information reply message

EXAMPLE: DESTINATION UNREACHABLE MESSAGE. The ICMP fields in this case are shown in Figure 11.30.

IP Fields

- Destination address: the source network and address from the original datagram's data.

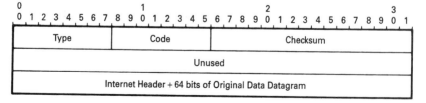

Figure 11.30. A typical ICMP message format; destination unreachable message [22].

ICMP Fields

- Type 3
- Code: 0 = net unreachable
          1 = host unreachable
          2 = protocol unreachable
          3 = port unreachable
          4 = fragmentation needed and DF set
          5 = source route failed.
- Checksum: as above
- Internet header + 64 bits of data datagram

The internet header plus the first 64 bits of the original datagram's data is used by the host to match the message to the appropriate process. If a higher-level protocol uses port numbers, they are assumed to be in the first 64 data bits of the original datagram's data.

DESCRIPTION. If, according to the information in the gateway's routing tables, the network specified in the internet destination field of a datagram is unreachable (e.g., the distance to the network is infinity), the gateway may send a destination unreachable message to the internet source host of the datagram. In addition, in some networks, the gateway may be able to determine if the internet destination host is unreachable. Gateways in these networks may send destination unreachable messages to the source host when the destination host is unreachable.

If, in the destination host, the IP module cannot deliver the datagram because the indicated protocol module or process port is not active, the destination host may send a destination unreachable message to the source host.

Another case is when a datagram must be fragmented to be forwarded by a gateway yet the "Dont Fragment" flag is on. In this case the gateway must discard the datagram and may return a destination unreachable message.

It should be noted that codes 0, 1, 4, and 5 may be received from a gateway; codes 2 and 3 may be received from a host (RFC 792 [22]).

7.4   The Transmission Control Protocol (TCP)

7.4.1   TCP Defined. TCP [23, 24] was designed to provide reliable communication between pairs of processes in logically distinct hosts on networks and sets of interconnected networks. TCP operates successfully in an environment where the loss, damage, duplication or misorder of data, and network congestion can occur. This robustness in spite of unreliable communications media makes TCP well-suited to support commercial, military, and government applications. TCP appears at the transport layer of the protocol hierarchy. Here, TCP provides connection-oriented data transfer

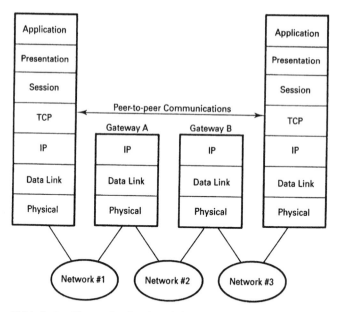

Figure 11.31. Protocol layers showing the relationship of TCP with other layered protocols.

that is reliable, ordered, full duplex, and flow controlled. TCP is designed to support a wide range of upper-layer protocols (ULPs). The ULP can channel continuous streams of data through TCP for delivery to peer ULPs. The TCP breaks the streams into portions which are encapsulated together with appropriate addressing and control information to form a segment—the unit of exchange between TCPs. In turn, the TCP passes the segments to the network layer for transmission through the communication system to the peer TCP.

As shown in Figure 11.31, the layer below the TCP in the protocol hierarchy is commonly the internet protocol (IP) layer. The IP layer provides a way for the TCP to send and receive variable-length segments of information enclosed in internet datagram "envelopes." The internet datagram provides a means for addressing source and destination TCPs in different networks. The IP also deals with fragmentation or reassembly of TCP segments required to achieve transport and delivery through the multiple networks and interconnecting gateways. The IP also carries information on the precedence, security classification, and compartmentation of the TCP segments, so this information can be communicated end-to-end across multiple networks.

7.4.2 TCP Mechanisms. TCP builds its services on top of the network layer's potentially unreliable services with mechanisms such as error detection, positive acknowledgments, sequence numbers, and flow control. These mechanism require certain addressing and control information to be initialized and maintained during data transfer. This collection of information is called a *TCP connection*. The following paragraphs describe the purpose and operation of the major TCP mechanisms.

PAR MECHANISM. TCP uses a positive acknowledgment with retransmission (PAR) mechanism to recover from the loss of a segment by the lower layers. The strategy with PAR is for a sending TCP to retransmit a segment at timed intervals until a positive acknowledgment is returned. The choice of retransmission interval affects efficiency. An interval that is too long reduces data throughput while one that is too short floods the transmission media with superfluous segments. In TCP, the timeout is expected to be dynamically adjusted to approximate the segment round-trip time plus a factor for internal processing; otherwise performance degradation may occur. TCP uses a simple checksum to detect segments damaged in transit. Such segments are discarded without being acknowledged. Hence, damaged segments are treated identically to lost segments and are compensated for by the PAR mechanism. TCP assigns sequence numbers to identify each octet of the data stream. These enable a receiving TCP to detect duplicate and out-of-order segments. Sequence numbers are also used to extend the PAR mechanism by allowing a single acknowledgment to cover many segments' worth of data. Thus, a sending TCP can still send new data although previous data have not been acknowledged.

FLOW CONTROL MECHANISM. TCP's flow control mechanism enables a receiving TCP to govern the amount of data dispatched by a sending TCP. The mechanism is based on a *window* which defines a contiguous interval of acceptable sequence-numbered data. As data are accepted, TCP slides the window upward in the sequence number space. This window is carried in every segment, enabling peer TCPs to maintain up-to-date window information.

MULTIPLEXING MECHANISM. TCP employs a multiplexing mechanism to allow multiple ULPs within a single host and multiple processes in a ULP to use TCP simultaneously. This mechanism associates identifiers, called *ports*, to ULP processes accessing TCP services. A ULP connection is uniquely identified with a *socket*, the concatenation of a port and an internet address. Each connection is uniquely named with a socket pair. This naming scheme allows a single ULP to support connections to multiple remote ULPs. ULPs which provide popular resources are assigned permanent sockets, called *well-known sockets*.

7.4.3 *ULP Synchronization.* When two ULPs wish to communicate, they instruct their TCPs to initialize and synchronize the mechanism information on each to open the connection. However, the potentially unreliable network layer (i.e., IP layer) can complicate the process of synchronization. Delayed or duplicate segments from previous connection attempts might be mistaken for new ones. A handshake procedure with clock-based sequence numbers is used in connection opening to reduce the possibility of such false connections. In the simplest handshake, the TCP pair synchronizes sequence numbers by exchanging three segments, thus the name *three-way handshake*.

7.4.4   ULP Modes.   A ULP can open a connection in one of two modes, passive or active. With a passive open, a ULP instructs its TCP to be *receptive* to connections with other ULPs. With an active open, a ULP instructs its TCP to actually initiate a three-way handshake to connect to another ULP. Usually an active open is targeted to a passive open. This active/passive model supports server-oriented applications where a permanent resource, such as a data base management process, can always be accessed by remote users. However, the three-way handshake also coordinates two simultaneous active opens to open a connection. Over an open connection, the ULP pair can exchange a continuous stream of data in both directions. Normally, TCP groups the data into TCP segments for transmission at its own convenience. However, a ULP can exercise a *push* service to force TCP to package and send data passed up to that point without waiting for additional data. This mechanism is intended to prevent possible deadlock situations where a ULP waits for data internally buffered by TCP. For example, an interactive editor might wait forever for a single input line from a terminal. A push will force data through the TCPs to the awaiting process. A TCP also provides the means for a sending ULP to indicate to a receiving ULP that "urgent" data appear in the upcoming data stream. This urgent mechanism can support, for example, interrupts or breaks. When a data exchange is complete, the connection can be closed by either ULP to free TCP resources for other connections. Connection closing can happen in two ways. The first, called a *graceful close*, is based on the three-way handshake procedure to complete data exchange and coordinate closure between the TCPs. The second, called an *abort*, does not allow coordination and may result in the loss of unacknowledged data.

7.4.5   Scenario.   The following scenario provides a walk-through of a connection opening, data exchange, and connection closing as might occur between the database management process and the user mentioned above. The scenario focuses more on (a) the three-way handshake mechanism in connection with opening and closing and (b) the positive acknowledgment with retransmission mechanism supporting reliable data transfer. Although not pictured, the network layer transfers the information between TCPs. For the purpose of this scenario, the network layer is assumed not to damage, lose, duplicate, or change the order of data unless explicitly noted. The scenario is organized into three parts:

a. A simple connection opening (steps 1–7) (Figure 11.32)

b. Two-way data transfer (steps 8–17) (Figure 11.33)

c. A graceful connection close (steps 18–25) (Figure 11.34)

SCENARIO NOTATION. The following notation is used in the diagrams.

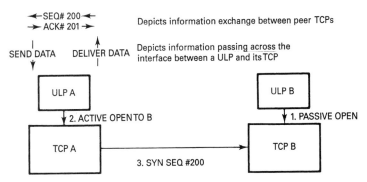

Figure 11.32A. A simple connection opening.

1. ULP B (the DB manager) issues a PASSIVE OPEN to TCP B to prepare for connection attempts from other ULPs in the system.

2. ULP A (the user) issues an ACTIVE OPEN to open a connection to ULP B.

3. TCP A sends a segment to TCP B with an OPEN control flag, called a SYN, carrying the first sequence number (shown as SEQ #200) it will use for data sent to B.

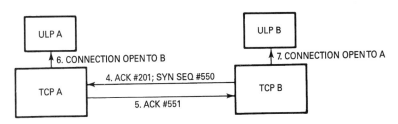

Figure 11.32B. A simple connection opening.

4. TCP B responds to the SYN by sending a positive acknowledgment, or ACK, marked with the next sequence number expected from TCP A. In the same segment, TCP B sends its own SYN with the first sequence number for its data (SEQ #550).

5. TCP A responds to TCP B's SYN with an ACK showing the next sequence number expected from B.

6. TCP A now informs ULP A that a connection is open to ULP B.

7. Upon receiving the ACK, TCP B informs ULP B that a connection has been opened to ULP A.

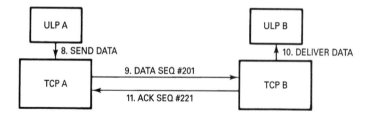

Figure 11.33A. Two-way data transfer.

8. ULP A passes 20 octets of data to TCP A for transfer across the open connection to ULP B.

9. TCP A packages the data in a segment marked with current "A" sequence number.

10. After validating the sequence number, TCP B accepts the data and delivers it to ULP B.

11. TCP 3 acknowledges all 20 octets of data with the ACK set to the sequence number of the next data octet expected.

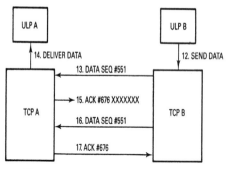

Figure 11.33B. Two-way data transfer.

12. ULP B passes 125 bytes of data to TCP B for transfer to ULP A.

13. TCP B packages the data in a segment marked with the "B" sequence number.

14. TCP A accepts the segment and delivers the data to ULP A.

15. TCP A returns an ACK of the received data marked with the number of the next expected data octet. However, the segment is lost by the network and never arrives at TCP B.

16. TCP B times out waiting for the lost ACK and retransmits the segment. TCP A receives the retransmitted segment, but discards it because the data from the original segment has already been accepted. However, TCP A re-sends the ACK.

17. TCP B gets the second ACK.

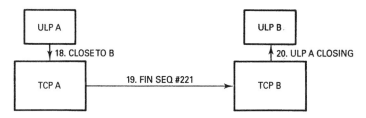

Figure 11.34A. A graceful connection close.

18. ULP A closes its half of the connection by issuing a CLOSE to TCP A.

19. TCP A sends a segment marked with a CLOSE control flag, called a FIN, to inform TCP B that ULP A will send no more data.

20. TCP B gets the FIN and informs ULP B that ULP A is closing.

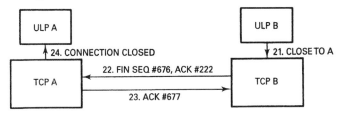

Figure 11.34B. A graceful connection close.

21. ULP B completes its data transfer and closes its half of the connection.

22. TCP B sends an ACK of the first FIN and its own FIN to TCP A to show ULP B's closing.

23. TCP A gets the FIN and the ACK, then responds with an ACK to TCP B.

24. TCP A informs ULP A that the connection is closed.

25. (Not pictured) TCP B receives the ACK from TCP A and informs ULP B that the connection is closed.

(Ref. 24.)

7.4.6 TCP Header Format. The TCP header format is shown in Figure 11.35. It should be noted that TCP works with 32-bit segments.

Source port. The "port" represents the source ULP initiating the exchange. The field is 16 bits long.

Destination port. This is the destination ULP at the other end of the connection. This field is also 16 bits long.

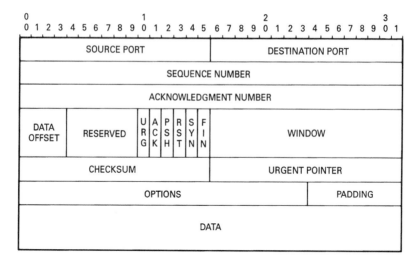

Figure 11.35. TCP header format [23].

Sequence number. Usually, this value represents the sequence number of the first data octet of a segment. However, if an SYN is present, the sequence number is the initial sequence number (ISN) covering the SYN; the first data octet is then numbered SYN + 1. The "SYN" is the *synchronize control flag*. It is the opening segment of a TCP connection. SYNs are exchanged from either end. When a connection is to be closed there is a similar "FIN" sequence exchange.

Acknowledgment Number. If the ACK control bit* is set (bit 2 of the 6-bit control field), this field contains the value of the next sequence number that the sender of the segment is expecting to receive. This field is 32 bits long.

Data Offset. This field indicates the number of 32-bit words in the TCP header. From this value the beginning of the data can be computed. The TCP header is an integral number of 32 bits long. The field size is 4 bits.

Reserved. This is a field of 6 bits set aside for future assignment. It is set to zero.

Control Flags. The field size is six bits covering six items (1 bit per item):

(a) URG: Urgent pointer field significant
(b) ACK: Acknowledgment field significant
(c) PSH: Push function
(d) RST: Reset the connection

---

*ACK: A control bit (acknowledge) occupying no sequence space, which indicates that the acknowledgment field of this segment specifies the next sequence number the sender of this segment is expecting to receive, hence acknowledging receipt of all previous sequence numbers.

(e) SYN: Synchronize sequence numbers

(f) FIN: No more data from sender

Window. The number of data octets beginning with the one indicated in the acknowledgment field which the sender of this segment is willing to accept. The field is two octets in length.

Checksum. The checksum is the 16-bit one's complement of the one's complement sum of all 16 bit words in the header and text. The checksum also covers a 96-bit pseudo-header conceptually prefixed to the TCP header. This pseudo-header contains the source address, the destination address, the protocol, and TCP segment length.

Urgent Pointer. This field indicates the current value of the urgent pointer as a positive offset from the sequence number in this segment. The urgent pointer points to the sequence number of the octet following the urgent data. This field is only to be interpreted in segments with the URG control bit set. The urgent pointer field is two octets long.

Options. This field is variable in size; and, if present, options occupy space at the end of the TCP header and are a multiple of 8 bits in length. All options are included in the checksum. An option may begin on any octet boundary. There are two cases of an option:

(a) Single octet of option-kind

(b) An octet of option-kind, an octet of option length, and the actual option data octets

Options include "end of option list," "no-operation," and "maximum segment size."

Padding. The field size is variable. The padding is used to ensure that the TCP header ends and data begins on a 32-bit boundary. The padding is composed of zeros.

7.4.6.1 TCP Entity State Diagram. Figure 11.36 summarizes TCP operation with a TCP entity state diagram.

Section 7.4 is based on RFC 793 [23] and MIL-STD-1778 [24].

8 IBM SYSTEM NETWORK ARCHITECTURE (SNA)

8.1 Background

The first version of SNA was implemented by IBM in 1974 to bring order out of chaos in resource sharing data circuits, replacing such file access systems as BTAM, RTAM, and TCAM. These were replaced by the virtual tele-communications success method (VTAM).

The traditional SNA is a hierarchy of connected network resources including: host processors such as the IBM 3081, communication controllers

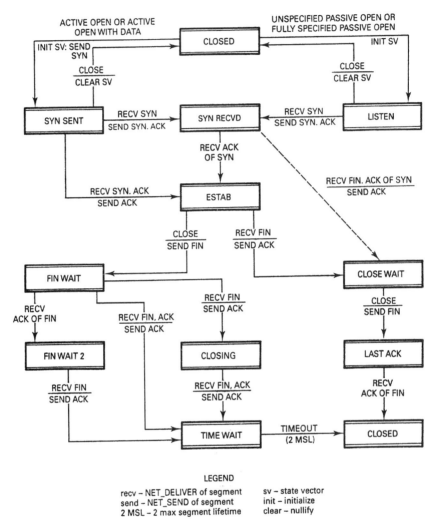

Figure 11.36. TCP entity state summary [24]. Note that this figure is intended only as a summary and does not supersede the formal definitions that precede.

such as the IBM 3725, cluster controllers (typically the IBM 3274), and terminals such as the IBM 3278.

VTAM runs in a host processor where all SNA application subsystems use VTAM as their telecommunications access method. The network control program (NCP) runs a communications controller and provides network management for remote network resources that are attached to the communications controller.

IBM broke with its traditional hierarchy with the introduction of advanced peer-to-peer networking (APPN). APPN allows network nodes to communicate as full peers without going through a mainframe (host). Nodes do not have to know the location of other nodes in order to communicate with them.

SNA continues to go through evolutionary processes. (See Refs. 25 and 26.)

8.2 The SNA Layered Architecture

SNA functions are divided into a hierarchical structure of seven well-defined layers. Each layer in the architecture performs a specific set of functions. Figure 11.37 identifies SNA's seven layers and describes their major functions.

SNA defines formats and protocols between layers that permit equivalent layers (layers at the same level within the hierarchy) to communicate with one another. Similar to OSI, each layer performs services for the next higher layer, requests services from the next lower layer, and communicates with equivalent layers. One example is the encryption of data as required by an end user. The procedure of how the two transmission control layers communicate is outlined in Figure 11.38.

The two transmission layers shown in Figure 11.38 encrypt and decrypt data independently of the functions of any other layer. The transmission control layer in the originating node encrypts the data it receives from the data flow control layer. It then requests that the path control layer route the encrypted data to the destination node. The transmission control layer at the destination node decrypts the data that the path control layer delivered. It then requests that the data flow control layer give the encrypted data to the destination end user.

The interaction of hardware and software components of SNA is shown in Figure 11.39.

The SNA *networking blueprint* is shown in Figure 11.40. The networking blueprint supports the implementation of multiple protocols and also integrates these protocols into a cohesive, modular structure. The networking blueprint defines layers of functions plus a systems management backplane. SNA *advanced peer-to-peer networking* (APPN) is part of the transport layer, and it is one of the protocols that can be used in this layer.

High-performance routing (HPR) is a small but powerful extension to APPN. It enhances data routing performance by decreasing intermediate node processing. HPR increases session reliability via *nondisruptive path switch*. The main components of HPR are *rapid-transport protocol* (RTP) and *automatic network routing* (ANR).

8.3 Advanced Peer-to-Peer Networking (APPN)

APPN ties together diverse platforms, topologies, and applications into a single network. APPN's any-to-any connectivity makes it possible for large

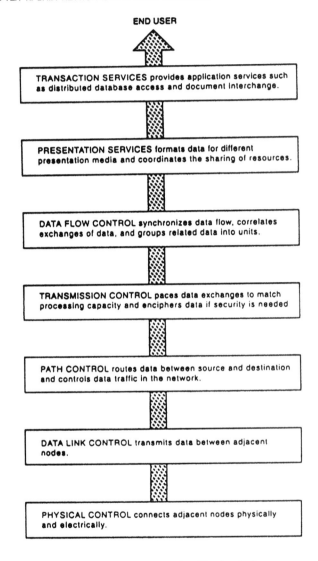

Figure 11.37. SNA layered architecture [25].

and small networks alike to communicate over local and wide area networks, across slow and fast links.

APPN provides two basic functions: (1) keeping track of the location of resources in the network and (2) selecting the best path to route data between resources. APPN nodes dynamically exchange information about each other, eliminating the need for customers having to deal with complicated system and path definitions. APPN nodes limit the information they exchange, enabling more efficient use of network resources.

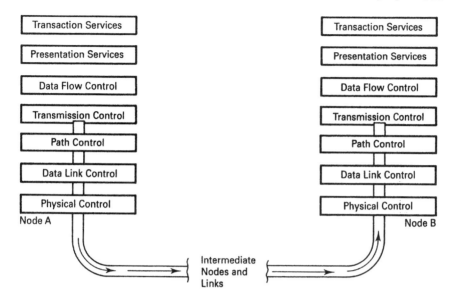

Figure 11.38. Communication between two transmission control layers [25].

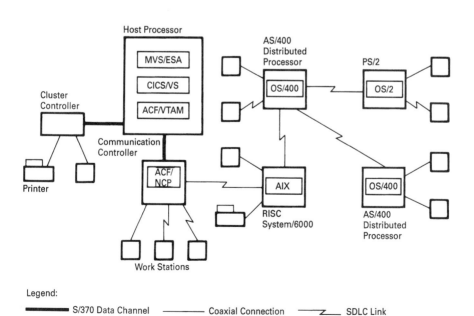

Figure 11.39. Hardware and software components of an SNA network [25].

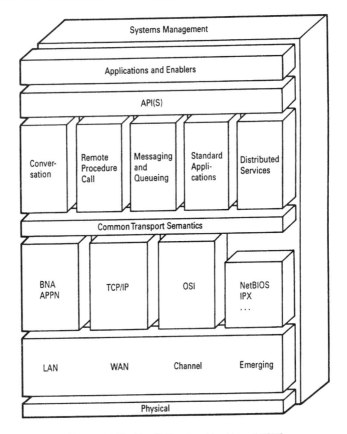

Figure 11.40. The SNA networking blueprint [25].

8.4 Architectural Components of an SNA Network

A data network can be described as a configuration of nodes and links. Nodes are the network component that send data over, and receive data from, the network. A node may be a processor, controller, or workstation. Links are the network components that connect adjacent nodes. Of course, nodes and links work together in transferring data through a network.

A *node* is a set of hardware and associated software components that implement the functions of the seven architectural layers. Although all seven layers are implemented in a given node, nodes can differ based on their architectural components and the sets of functional capabilities they implement. Nodes with different architectural components represent different *node types*. Four types of nodes exist:

Type 5 (T5)

Type 2.0 (T2.0)

Type 4 (T4)

Type 2.1 (T2.1)

Nodes that perform different functions are said to act in different *network roles*. In some cases a given network node can act in different network roles. A T4 node, for example, can perform an interconnection role between nodes at different levels of the subnetwork hierarchy, or between nodes in different subarea networks. The functions performed in these two roles are referred to as *boundary function* and *gateway function*, respectively. T2.1 and T5 nodes can act in several different network roles. Node roles fall into two broad categories: *hierarchical roles* and *peer-oriented roles*.

8.4.1 Hierarchical Roles. Hierarchical roles are those in which certain nodes have a controlling or *mediating* function with respect to the actions of other nodes. SNA hierarchical networks are characterized by nodes of all four types acting in hierarchical roles. Within such networks, nodes are categorized as either *subarea nodes* (SNs) or *peripheral nodes* (PNs). Subarea nodes provide services for and control over peripheral nodes. Networks consisting of subarea and peripheral nodes are referred to as *subarea networks*.

SUBAREA NODES. Type 5 (T5) and type 4 (T4) nodes can act as subarea nodes. T5 subarea nodes provide the SNA functions that control network resources, support transaction programs, support network operators, and provide end-user services. Because these functions are provided by host processors, T5 nodes are also referred to as *host nodes*. T4 subarea nodes provide the SNA functions that route and control the flow of data in a subarea network. Because these functions are provided by communication controllers, T4 nodes are also referred to as *communication controller nodes*.

PERIPHERAL NODES. Type 2.0 (T2.0) and type 2.1 (T2.1) nodes can act as peripheral nodes attached to either T4 or T5 subarea nodes. Peripheral nodes are typically devices such as distributed processors, cluster controllers, or workstations. A T2.1 node differs from a T2.0 node by the T2.1 node's ability to support peer-oriented protocols as well as the hierarchical protocols of a simple T2.0 node. A T2.0 node requires the mediation of a T5 node in order to communicate with another node. Subarea nodes to which peripheral nodes are attached perform a *boundary function* and act as subarea *boundary nodes*.

Although Figure 11.40 represents nodes as particular classes of hardware, there is no architectural association between node type, or node role, and the kind of hardware that implements it. To avoid associating node types and roles with hardware implementations, network architecture diagrams use symbols to represent node types and roles. Figure 11.41 uses symbols to illustrate a subarea network containing the four node types acting as subarea

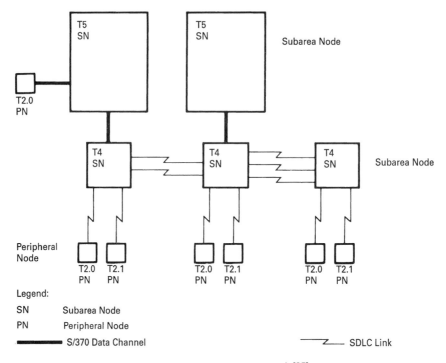

Figure 11.41. A subarea network [25].

and peripheral nodes. The network contains two type 5 subarea nodes, three type 4 subarea nodes, and seven peripheral nodes.

8.4.2 Peer-Oriented Roles. The advanced peer-to-peer networking (APPN) extensions allow greater distribution of network control by enhancing the dynamic capabilities of a node. Nodes with these extensions are referred to as *APPN nodes*, and a network of APPN nodes makes up an *APPN network*. A low-entry networking (LEN) node can also attach to an APPN network. *LEN* offers the lowest-level peer-to-peer networking capability. The LEN allows simple point-to-point connections between adjacent nodes. By adjacent, we mean that the communicating nodes must be connected to one another by a single physical data link or through an SNA subarea network.

An APPN node can dynamically find the location of a partner node, place the location information in directories, compute potential routes to the partner, and select the best route from among those computed. These dynamic capabilities relieve network personnel from having to predefine those locations, directory entries, and routes. APPN nodes can include processors of varying sizes such as the Application System/400, the Enterprise System/9370 (ES/9370) running under Distributed Processing Program Executive/370

(DPPX/370), the Personal System/2 (PS/2) running under Operating System/2 (OS/2), and VTAM running under Multiple Virtual Storage/Enterprise Systems Architecture (MVS/ESA).

Peer-oriented protocols enable nodes to communicate without mediation by a T5 node, giving them increased connection flexibility. APPN defines two possible roles for a node in an APPN network, that of an end node and that of a network node. (Network nodes have additional options that can further distinguish them.)

T2.1 nodes can act as either APPN or LEN nodes. T5 nodes can also act as APPN or LEN nodes, but have additional capability to interconnect sub-area and APPN networks by interchanging protocols between them. In this capacity they are called *interchange nodes*, discussed later. Together with its subordinate T4 nodes, a T5 node can also form a *composite LEN node* or a *composite network node*. As composite nodes they appear as single LEN or network nodes to other LEN or APPN nodes to which they are inter-connected.

If two network nodes do not support the border node option (discussed below) and are located in two separate net-ID subnetworks, CP–CP sessions cannot be established between them. For the two network nodes to com-municate, a LEN connection may be established between the two. This allows the two net-ID subnetworks to communicate, but does not support any APPN function. If NN1 subnet A is to establish a session with NN2 in subnet B, all LUs in subnet A must be predefined to NN2 in subnet B. The network nodes in both subnets are APPN nodes, but since they communicate across net-ID subnetwork boundaries, they are defined to each other via LEN links.

End nodes are located in the periphery of an APPN network. An end node obtains full access to the APPN network through one of the network nodes to which it is directly attached—its *network node server*. There are two kinds of end nodes: APPN end nodes and LEN end nodes. An *APPN end node* supports APPN protocols through explicit interactions with a network node server. Such protocols support dynamic searching for resources and provide resource information for the calculation of routes by network nodes. A *LEN end node* is a LEN node attached to a network node. Although LEN nodes lack the APPN extensions, they are able to be supported in APPN networks using the services provided them by network nodes. In an APPN network, when a LEN node is connected to another LEN node, or to an APPN end node, it is referred to simply as a *LEN node*. When connected to an APPN network node, however, it is referred to as a *LEN end node*.

NETWORK NODES. Together with the links interconnecting them, network nodes form the *intermediate routing network* of an APPN network. Network nodes connect end nodes to the network and provide resource location and route selection services for them. Routes used to interconnect network users

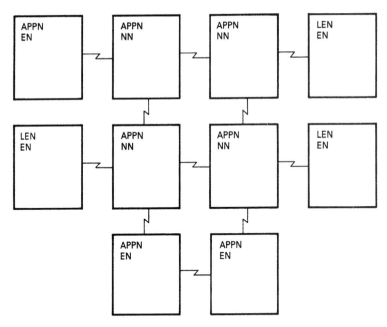

Figure 11.42. An APPN network.

are selected based on network topology information that can change dynamically.

Figure 11.42 illustrates one possible APPN network configuration and contains LEN end nodes as well as APPN nodes.

Network node server is a network node that provides resource location and route selection to the LUs it serves. These LUs can be in the network node itself or in the client end nodes. A network node server uses CP–CP sessions to provide network information for session setup in order to support LUs on served APPN end nodes. In addition, LEN end nodes can also take advantage of the services of the network node server. A LEN end node, unlike an APPN end node, must be predefined by the network operator as a client end node for which the network node acts as server. Any network node can be a network node server for end nodes that are attached to it. The served end nodes are defined as being in that network node server's domain.

Figure 11.43 illustrates an end node connecting to a network node server with a different net ID from its own. The capability allows the end node to dial into different APPN networks. This kind of attachment is called a *nonnative attachment*.

Central directory server (CDS) is a network node that builds and maintains a directory of resources from the network. The purpose of a CDS is to reduce the number of network broadcast searches to a maximum of one per resource. Network nodes and APPN end nodes can register their resources with a CDS, which acts as a focal point for resource location information.

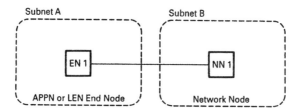

Figure 11.43. An end node attached nonnatively to its network node server.

SUBNETWORKS AND BORDER NODES. A *subnetwork* consists of a group of interconnected nodes, within a larger, composite network, that have some common attribute or characteristic, such as the same network ID, or that share a common topology database or that implement a common protocol. A *net-ID subnetwork* refers to the set of nodes having the same network ID. A *topology subnetwork* consists of all the network nodes exchanging and maintaining the same topology information. A *high-performance routing (HPR) subnetwork* consists of all the interconnected APPN nodes implementing the HPR function.

Within an APPN network it is often desirable to partition the network topology database to reduce its size at each network node and lower the overall topology data interchange traffic. Topology database replication and interchange are confined to a topology subnetwork, with each subnetwork acting independently. The presence of *border nodes* enables the topology subnetworks to be tied together into a larger composite network with the same freedom of LU–LU sessions as if the topology partitioning did not exist. Border nodes are network nodes having additional function and are of two types: *peripheral* and *extended.*

Peripheral border nodes permit APPN networks with different net IDs to interconnect, allowing session setup across net-ID subnetwork boundaries, but in addition allow the partitioning of the topology database among network nodes with the same net ID; this is known as *clustering* and the network partitions are also known as *clusters.* Each cluster is a distinct topology subnetwork.

8.5 Nodes with Both Hierarchical and Peer-Oriented Function

Interchange node is a VTAM product feature merging networks and subarea networks. An interchange node receives network search requests from APPN nodes and transfers them into subarea network searches, without exposing the subarea aspects to the APPN part of the network. At the same time the interchange node can receive search requests from a subarea network and transfer them into APPN requests without exposing the APPN aspects to the subarea network. The interchange node maintains its subarea appearance to other subarea nodes and maintains its APPN appearance to other APPN

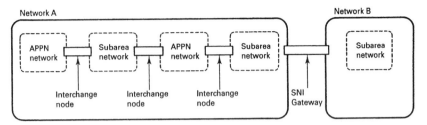

Figure 11.44. Interchange nodes interconnecting APPN and subarea networks [25].

nodes. Figure 11.44 illustrates an intermediate stage of migrating a subarea network to an APPN network. An interchange node permits migration of an existing subarea network to APPN on a node-by-node, link-by-link basis without a need for network-wide coordination. Also, an interchange node permits session establishment across any combination of APPN and subarea networks, including SNA network interconnection (SNI) gateways.

8.6   Links and Transmission Groups

Adjacent nodes in a network are connected to one another by one or more links. A *link* includes both the link stations within the two nodes it connects and the link connection between the nodes.

A *link station* is the hardware or software within a node that enables the node to attach to, and provide control over, a link connection. It exchanges information and controls signals with its partner link station in the adjacent node. Link stations use data-link control protocols to transmit data over a link connection. A *link connection* is the physical medium over which data are transmitted. Examples of transmission media include wire pairs, microwave radio, fiber-optic cables, and satellite circuits. Multiple links between the same two nodes are referred to as *parallel links*.

A *transmission group* (TG) may consist of one or more links between two nodes. A TG comprising two or more parallel links is called a *multilink TG*. Multilink TGs may be defined between two type 4 subarea nodes using SDLC links. In APPN multilink TGs are not defined, so the terms *TG* and *link* are used generally interchangeably in an APPN context. In both subarea networks and APPN networks, multiple (or parallel) TGs may connect two adjacent nodes. Data traffic is distributed dynamically over the links of a multilink TG.

Figure 11.45 shows parallel TGs connecting adjacent T4 nodes via TG2 and TG3, as well as a multilink TG connecting adjacent T4 nodes via TG1. The data traffic flowing between a pair of adjacent T4 nodes is distributed among parallel TGs. If one of the TGs fails, the session is broken; that is, session traffic is not automatically rerouted over the other TG. On the other hand, if at least one link in a multilink TG is still operational, session traffic is not disrupted over the TG in the case of link failure except that throughput

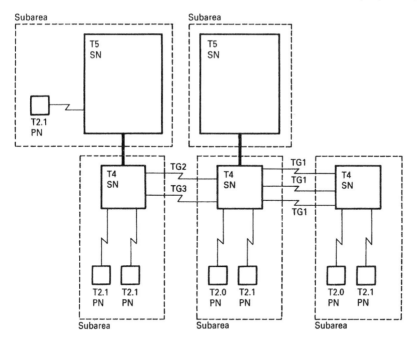

Figure 11.45. Subareas [25].

may suffer. Similarly a link may be restored to operation in a TG without disrupting existing sessions.

8.7   SNA Network Configurations

SNA defines the following network configurations:

- A hierarchical network consisting of subarea nodes and peripheral nodes
- A peer-oriented network consisting of APPN and LEN nodes
- A mixed network that combines one or more hierarchical subnets with one or more peer-oriented subnets

8.7.1   Hierarchical Network Configurations.   The organization of a hierarchical network structure is determined by the way control of network services is maintained. Host nodes containing system services control points (SSCPs) are responsible for overall control of communication in the hierarchical network.

A hierarchical network might include LEN or APPN nodes that are attached as peripheral nodes. These nodes can communicate with each other through a subarea network if the boundary nodes to which they are attached support the basic SSCP-independent LU–LU protocols needed for such peer interactions.

SUBAREAS. A *subarea* consists of one subarea node and the peripheral nodes that are attached to that subarea node. The concept of subarea applies only to subarea networks and composite networks. The network configuration in Figure 11.45 contains five subarea nodes and seven peripheral nodes. Because each subarea node and its attached peripheral nodes constitute a subarea, this configuration contains five subareas.

DOMAINS IN A SUBAREA NETWORK. A *domain* is an area of control. The concept of a domain within a subarea network differs from that within an APPN network. Within a subarea network, a domain is that portion of the network managed by the control point in a T5 subarea node. The control point in a T5 subarea node is called a *system services control point* (SSCP).

When a subarea network has only one T5 node, that node must manage all of the network resources. A subarea network that contains one T5 node is a *single-domain subarea network*. When there are multiple T5 nodes in the network, each T5 node may control a portion of the network resources. A subarea network that contains more than one T5 node is a *multiple-domain subarea network*. In a multiple-domain subarea network, the control of some resources can be shared between SSCPs. Some resources can be shared serially and some concurrently. Figure 11.46 illustrates two domains, A and B, joined by direct-attached T4 nodes to form a multiple-domain subarea network.

8.7.2   Peer-Oriented Network Configurations.   An APPN network constitutes a peer-oriented SNA network. All APPN nodes are considered to be peers and do not rely on other nodes to control communication in the network the way a subarea node controls communication between peripheral nodes. There is, however, a measure of hierarchical control because a network node server provides certain network services to its attached end nodes. The difference is that, in APPN networks, hierarchical control is not determined by product or processor type as it is in subarea networks, where only large host processors contain SSCPs and node types generally reflect product types.

DOMAINS IN AN APPN NETWORK. The domain of a node in an APPN network is that portion of the network served by the control point in the node. The control point in an APPN node is called simply a *control point* (CP). An end node (EN) control point's domain consists solely of its local resources. It is included within the domain of its network node server. A network node control point's domain includes the resources in the network node and in any *client* end nodes (nodes for which the network node is acting as the network node server) attached directly to the network node.

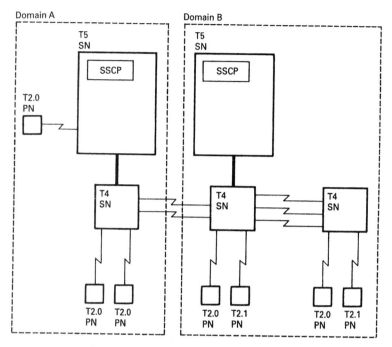

Figure 11.46. Domains in a subarea network [25].

APPN networks are, by definition, multiple-domain networks. Figure 11.47 shows an APPN network containing four network node domains: A, B, C, and D. The domains of the end nodes are included within the domains of their respective node servers.

8.8 The Transport Network and Network Accessible Units

Node and link components exhibit SNA layering through the functions they perform. The lower three architectural layers comprise the *transport network*, and certain components within the upper four layers are referred to as *network accessible units* (NAUs).

8.8.1 The Transport Network. Node and link components within the lower three architectural layers (i.e., the physical control, data-link control, and path control layers) are collectively referred to as the *transport network*. The physical control layer includes both data communication equipment within nodes and the physical link connections between them. Node components within the data-link control layer activate and deactivate links on command from their control points, and they manage link-level data flow. Node components within the path control layer perform routing and congestion control. Together, the distributed components of these three lower

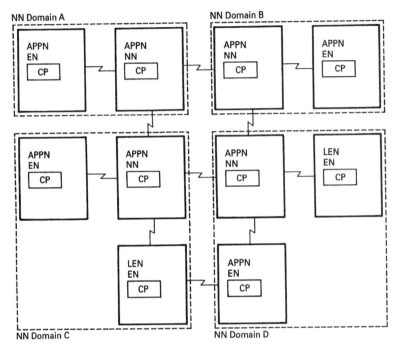

Figure 11.47. Network node domains in an APPN network [25].

layers transport data through the network on behalf of *network accessible units* (NAUs).

8.8.2 Network Accessible Units. Certain components within the upper four architectural layers of a node use transport network services to establish temporary, logical connections with one another called *sessions*. Sessions can be established between two components residing in different nodes or between components within the same node. Components that can establish sessions are referred to as *network accessible units* (NAUs). (*Note*: In earlier literature, NAU was called a "network addressable unit.") NAUs in session with one another are referred to as *session partners*.

A set of physical connections consisting of the links and intermediate nodes between session partners constitutes a *route* for the session. During session-initiation procedures, a route is selected for the session. Once a session is initiated, the session traffic flows over the same route for the duration of the session.

There are three fundamental kinds of network accessible units: physical units, logical units, and control points. *Physical units* (PUs) perform local node functions such as activating and deactivating links to adjacent nodes. PUs exist only in nodes within subarea networks. (*Note*: In APPN networks these functions are performed by control points.) To perform its functions, a

PU must *exchange control data* with its controlling system services control point (SSCP) over an *SSCP–PU* session initiated by the SSCP.

Logical units (LUs) provide network access for end users by helping the end users send and receive data over the network. Nodes in both subarea and APPN networks contain LUs. LUs send and receive control data and *end-user data* over the *LU–LU* sessions established between them.

In a subarea network an LU residing in a peripheral node is classified as either SSCP-dependent or SSCP-independent, depending on the protocols it uses for LU–LU session initiation. An *SSCP-dependent LU*, or simply *dependent LU*, sends a session-initiation request to its controlling SSCP over an *SSCP–LU* session. The LU is dependent on the SSCP to mediate the session initiation with the partner LU and requires an SSCP–LU session for that mediation. The session is activated when the LU receives a session-activation request from the partner LU. Dependent LUs reside in both T2.0 and T2.1 nodes.

An *SSCP-independent LU*, or simply *independent LU*, sends a session-activation request directly to the partner LU. It does not interact with an SSCP to mediate an LU–LU session and does not require an SSCP–LU session. An independent LU may reside in a T2.1 node, but not in a T2.0 node. Independent LU protocols are also referred to as *peer-session protocols*.

In APPN networks, independent LUs send session-activation requests directly to their partners, and all nodes support peer-session protocols. Two LEN nodes directly attached to one another also support peer-session protocols.

Control points (CPs) provide network control functions that include managing the resources in their domains and monitoring and reporting on the status of those resources. The functions of node CPs in subarea networks differ greatly from those in APPN networks.

In subarea networks, SSCPs control the PUs and dependent LUs in their domains by exchanging control data with them over SSCP–PU and SSCP–LU sessions respectively. They may also initiate *SSCP–SSCP sessions* for the control of cross-domain sessions. In APPN networks, a CP does not initiate sessions with LUs in its domain. It does, however, initiate sessions with adjacent CPs, called *CP–CP sessions*, in order to exchange control data for its routing and directory services.

8.8.3 *Interconnecting Session Stages.*
Messages transmitted on a session carry addresses in their headers. The addresses enable *intermediate routing nodes* (the nodes situated along a session route between two partner nodes) to correlate a message with its appropriate session and to route the message accordingly.

As session data traverse a route, not only the endpoints (called *half-sessions*) but also certain components along the route called *session connectors* provide flow control and address translation functions for the session. The parts of a session that are delimited by half-sessions and/or session connectors are called session stages. The three points in an SNA network where session connection occurs are at the boundary function, the gateway function, and the intermediate session routing function.

- The *boundary function* resides in a T4 or T5 node acting as a boundary node. It translates between the network addresses used by a subarea node and the local addresses used by a peripheral node.
- The *gateway function* resides in a T4 node acting as a gateway node. It translates between the network addresses used by one network and the network addresses used by another.
- The *intermediate session routing function* resides in an APPN network node. It translates between the session address used by one session stage and the session address used by another.

8.9   SNA Logical Units (LUs)

All node types can contain logical units (LUs). (However, T4 nodes typically do not contain logical units except for protocol conversion for non-SNA terminals.) The LU supports sessions with control points in type 5 nodes and with LUs in other nodes. A type 6.2 LU using SSCP-independent protocols, however, does not engage in sessions with an SSCP.

End users access SNA networks through logical units. A logical unit manages the exchange of data between end users, acting as an intermediary between the end user and the network. A one-to-one relationship is not required between end users and LUs. The number of end users that can access a network through the same LU is an implementation design option.

Before end users can communicate with one another, their respective LUs must be connected in a session. In some cases multiple, concurrent sessions between the same two logical units are possible. When such sessions are activated, they are called *parallel sessions.*

SNA defines different kinds of logical units as *LU types.* LU types identify sets of SNA functions that support end-user communication. LU–LU sessions can exist only between logical units of the same LU type. For example, an LU type 2 can only communicate with another LU type 2; it cannot communicate with an LU type 3.

The LU types that SNA currently defines, the kind of configuration or application that each type represents, and the hardware or software products that typically use each type of logical unit are listed below.

LU TYPE 1. LU type 1 is for application programs and single- or multiple-device data processing workstations communicating in an interactive, batch data transfer or distributed data processing environment. The data streams used in LU type 1 conform to the SNA character stream or document content architecture (DCA). An example of the use of LU type 1 is an application program running under IMS/VS and communicating with an IBM 8100 Information System at which the workstation operator is correcting a database that the application program maintains.

LU TYPE 2. LU type 2 is for application programs and display workstations communicating in an interactive environment using the SNA 3270 data stream. Type 2 LUs also use the SNA 3270 data stream for file transfer. An example of the use of LU type 2 is an application program running under IMS/VS and communicating with an IBM 3179 display station at which the 3179 operator is creating and sending data to the application program.

LU TYPE 3. LU type 3 is for application programs and printers using the SNA 3270 data stream. An example of the use of LU type 3 is an application program running under CICS/VS and sending data to an IBM 3262 printer attached to an IBM 3174 establishment controller.

LU TYPE 4. LU type 4 is for:

1. Application programs and single- or multiple-device data processing or word processing workstations communicating in interactive, batch data transfer or distributed data processing environments. An example of this use of LU type 4 is an application program running under CICS/VS and communicating with an IBM 6670 information distributor.
2. Peripheral nodes that communicate with each other. An example of this use of LU type 4 is two 6670s communicating with each other.

The data streams used in LU type 4 are the SNA character string (SCS) for data processing environments and Office Information Interchange (OII) Level 2 for word processing environments.

LU TYPE 6.1. LU type 6.1 is for application subsystems communicating in a distributed data processing environment. An example of the use of LU type 6.1 is an application program running under CICS/VS and communicating with an application program running under IMS/VS.

LU TYPE 6.2. LU type 6.2 is for transaction programs communicating in a distributed data processing environment. The LU type 6.2 supports multiple concurrent sessions. The LU 6.2 data stream is either an SNA general data stream (GDS), which is a structured-field data stream, or a user-defined data stream. LU 6.2 can be used for communication between two type 5 nodes, a type 5 node and a type 2.1 node, or two type 2.1 nodes. Examples of the use of LU type 6.2 are:

1. An application program running under CICS/VS communicating with another application program running under CICS/VS
2. An application program in an AS/400 communicating with a personal system/2 (PS/2)

Notes on Acronyms

CICS: Customer information control system. A general-purpose system/370 mainframe-based transaction management system.

IMS: Information management system. An IBM 370/390 host-based database/data communications subsystem that runs in the MVS environment.

MVS: Multiple virtual storage. An IBM primary operating system for the 370/390 IBM mainframes.

8.10   Some SNA Data Formats

SNA defines the following message-unit formats that NAUs, path control elements, and data-link control elements use:

- Network accessible units use basic information units (BIUs)
- Path control elements use path information units (PIUs)
- Data-link control elements use basic link units (BLUs)

8.10.1   *Basic Information Unit.* NAUs use *basic information units* (BIUs) to exchange requests and responses with other NAUs. Figure 11.48 shows the format of a BIU. Basic information units that carry requests contain a request header and a request unit. Basic information units that carry responses consist of (1) both a response header and a response unit or (2) only a response header.

Request Header. Each request that an NAU sends begins with a *request header* (RH). A request header is a 3-byte field that identifies the type of data in the associated request unit. The request header also provides information about the format of the data and specifies protocols for the session. Only NAUs use request header information.

Response Header. Each response that an NAU sends includes a *response header* (RH). Like a request header, a response header is a 3-byte field

Request Header or Response Header	Request Unit or Response Unit

Figure 11.48. Basic information unit (BIU) format.

that identifies the type of data in the associated response unit. A bit called *request/response indicator* (RRI) distinguishes a response header from a request header.

Response Unit. A *response unit* (RU) contains information about the request. Positive responses to command requests generally contain a 1- to 3-byte response unit that identifies the command request. Positive responses to data requests contain response headers, but no response unit. Negative response units are 4–7 bytes long and are always returned with a negative response. Response units are identified as response RUs.

The receiving NAU returns a negative response to the request sender if:

- The sender violates an SNA protocol
- The receiver does not understand the transmission
- An unusual condition, such as a path outage, occurs

The receiving NAU returns a 4- to 7-byte negative response unit to the request sender. The first 4 bytes of the response unit contain sense data explaining why the request is unacceptable. The receiving NAU sends up to three additional bytes that identify the rejected request.

8.10.2 Path Information Unit. The message-unit format used by path control elements is a *path information unit* (PIU) Path control elements form a PIU by adding a transmission header to a basic information unit. Figure 11.49 shows the format of a PIU.

Transmission Header	Request Header or Response Header	Request Unit or Response Unit

Figure 11.49. Path information unit (PIU) format.

Path control uses the *transmission header* (TH) to route message units through the network. The transmission header contains information for the transport network.

SNA defines different header formats and identifies the different formats by a *format identification* (FID) type. Transmission headers vary in length according to their FID type. Path control uses different FID types to route data between different types of nodes.

FID 0. Path control uses this format to route data between adjacent subarea nodes for non-SNA devices. Few networks still use the FID 0;

now a bit is set in the FID 4 transmission header to indicate whether the device is an SNA device or a non-SNA device.

FID 1. Path control uses this format to route data between adjacent subarea nodes if one or both of the subarea nodes do not support explicit and virtual route protocols.

FID 2. Path control uses this format to route data between a subarea boundary node and an adjacent peripheral node, or between adjacent APPN or LEN nodes.

FID 4. Path control uses this format to route data between subarea nodes if the subarea nodes support explicit and virtual route protocols.

FID 5. Path control uses this format to route data over a rapid-transport protocol (RTP) connection. The FID 5 is very similar to the FID 2.

8.10.3 Basic Link Unit. Data-link control uses a message-unit format called a *basic link unit* (BLU) to transmit data across a link. Data-link control forms a BLU by adding a *link header* (LH) and a *link trailer* (LT) to a PIU. Link headers and link trailers contain link control information that manages the transmission of a message unit across a link. Only data-link control elements use link header and link trailer information. (See Section 6.2.3 for a brief discussion of SDLC.) Figure 11.50 shows the format of a BLU.

Link Header	Transmission Header	Request Header or Response Header	Request Unit or Response Unit	Link Trailer

Figure 11.50. Basic link unit (BLU) format.

8.10.4 Network Layer Packet (NLP). A *network layer packet* (NLP) is used to transport data through an HPR network. It is composed of the network layer header (NHDR), transport header (THDR), and the data. The length of a packet must not exceed the maximum packet size of any link over which the packet will flow. The maximum packet size of links is obtained during the route setup protocol.

All of the fields in the NLP are of variable lengths. Figure 11.51 illustrates the format of an NLP.

Figure 11.52 shows the use of SNA data formats.

NHDR	THDR	Data

Figure 11.51. Network layer packet (NLP) format.

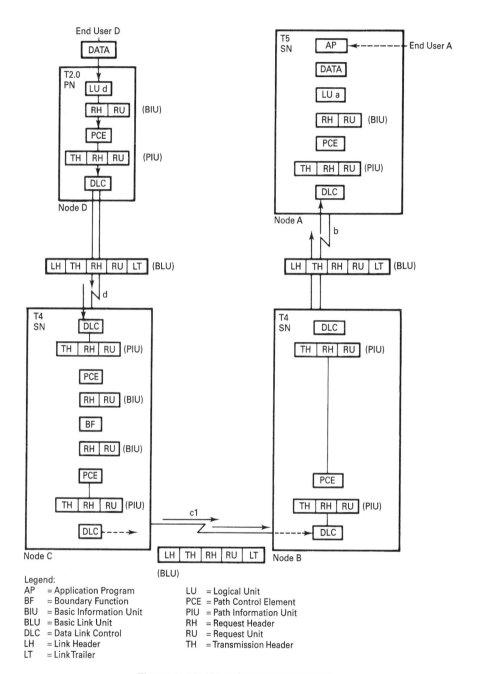

Figure 11.52. Use of SNA data formats [25].

Legend:
AP = Application Program
BF = Boundary Function
BIU = Basic Information Unit
BLU = Basic Link Unit
DLC = Data Link Control
LH = Link Header
LT = Link Trailer

LU = Logical Unit
PCE = Path Control Element
PIU = Path Information Unit
RH = Request Header
RU = Request Unit
TH = Transmission Header

The following paragraphs describe how the message formats shown in Figure 11.52 are used.

Node D

1. End-user D gives end-user data to LU *d*.
2. LU *d* creates a request unit and affixes a request header to it; this forms a basic information unit (BIU). (Multiple BIUs may be needed to accommodate a large message.)
3. LU *d* gives the BIU to a path control element.
4. Path control affixes a transmission header to the BIU; this forms a path information unit (PIU).
5. Path control gives the PIU to a data-link control element.
6. Data-link control affixes both a link header and a link trailer to the PIU; this forms a basic link unit (BLU).
7. Data-link control transmits the BLU over link *d* to node C.

Node C

1. The data-link control element that receives the BLU removes the link header and link trailer and then gives the PIU to a peripheral path control element.
2. This path control element routes the RH and RU (BIU) to the boundary function (BF). The BF then passes the BIU to a subarea path control element, which determines where to route the message unit text. Path control then gives the PIU to another data-link control element.
3. Data-link control affixes both a link header and a link trailer to the PIU.
4. Data-link control transmits the BLU over link *c*1 to node B.

Node B

1. The data-link control element that receives the BLU removes the link header and link trailer and then gives the PIU to path control.
2. Path control uses the information in the transmission header to determine where to route the message unit next. Path control then gives the PIU to another data-link control element.
3. Data-link control affixes both a link header and a link trailer to the PIU.
4. Data-link control transmits the BLU over link *b* to node A.

Node A

1. The data-link control element that receives the BLU removes the link header and link trailer and then gives the PIU to path control.
2. Path control uses information in the transmission header to determine that node A is the destination subarea node. Path control removes the transmission header and then delivers the BIU to the destination NAU.
3. LU *a* removes the request header and gives the RU to end-user A.

The material and figures in this section were extracted from *Systems Network Architecture Technical Overview*, 5th ed., IBM Corp., Research Triangle Park, NC, 1994 (Ref. 25) and are reprinted with permission.

REVIEW QUESTIONS

1. Define *topology* regarding data networks.
2. Give at least four major differences between WANs and LANs.
3. Give six of the seven considerations listed in the text which should be quantified or qualified before data network design begins. We can call them "inputs," if need be.
4. Give the three approaches to data switching. Provide a short description of each.
5. What is the efficiency of start–stop data transmission only from the point of view of useful bits above the physical layer? In this case the stop element is 2 bits long. Assume an ASCII format with one-bit parity.
6. Which is one very prevalent way power utilities make money with their telecommunication network? Carry this thought just one step further!
7. Define a protocol.
8. Describe two early forms of data network access. Let's define a network as some form of interconnectivity with more than two users.
9. Define SAR (segmentation and reassembly). Define encapsulation.
10. In one or two sentences, give the rationale why the OSI 7-layer model was developed.
11. Give the remaining six layers of OSI in proper order starting with the layer just above the physical layer.
12. What are the basic responsibilities of the data-link layer (OSI layer 2)?
13. Both the data-link layer and the network layer have provisions for error control. How can they both do it? Differentiate the responsibilities of each regarding error recovery.

14. What are the four basic types of primitives recommended in OSI for one layer to communicate with an adjoining layer?

15. If we didn't do something about it, what would happen if the bit sequence 01111110 (the flag sequence) appeared in the middle of an HDLC frame?

16. Define the term *transparent*.

17. What are the three generic functions of the control field in the HDLC frame?

18. If an address field is 8 bits long, how many distinct addresses can it indicate?

19. Describe a *window of frames*.

20. X.25 deals primarily with which OSI layer? LAPB, its access protocol, resides in layer 2.

21. LAPB is almost identical to and derives from which data-link layer protocol?

22. Differentiate permanent virtual circuits (PVCs) and virtual calls.

23. X.25 was designed as the basic protocol for a packet network. However, experience has shown that X.25 generally uses permanent virtual circuits. Discuss this with respect to full packet operation.

24. What is the primary purpose of IP? Use the word *interface* in the answer.

25. Even though TCP/IP predates OSI, in what OSI layers would we expect to find IP and TCP?

26. What is the purpose of the *time-to-live field* in IP?

27. What is the purpose of the ARP (address resolution protocol) used in conjunction with IP?

28. How does IP use *source routing*?

29. In what way does ICMP help IP?

30. Give four of the ICMP type messages of the eight listed.

31. Describe the purpose of TCP, especially when it works with IP.

32. Describe the PAR mechanism incorporated with TCP and in support of IP.

33. What is the purpose of the *three-way handshake*?

34. TCP operates on what kind of boundaries or segments (express in bits)?

35. SNA can have a hierarchical structure; it can also have a
_____ _____ _____ _____ structure, which is not hierarchical.

36. In SNA, subarea networks consist of two types of nodes. What are they?

37. Where are end nodes located regarding an APPN network?

38. A central directory server (CDS) is a network node. What is its purpose?

39. Define a *subnetwork* regarding SNA.

40. Define a link and a link station in simple terms, and give the functions of each.

41. Define *domains* in an APPN network. In the definition include the functions of the *control point* and *end node*.

42. What are the three components in SNA's transport network? Relate these components to their respective layers in SNA's layered architecture.

43 Components that can establish *sessions* are called *network addressable units* (NAUs). Just what are sessions?

44. End users access SNA networks via logical units (LUs). What are the functions of a logical unit?

45. Differentiate basic information units (BIUs), path control units (PIUs), and basic link units (BLUs).

REFERENCES

1. *The New IEEE Standard Dictionary of Electrical and Electronics Terms*, 5th ed., IEEE Std 100-1992, IEEE, New York, 1991.
2. W. Stallings, *Handbook of Computer Communications Standards*, Vol. 1, Macmillan, New York, 1987.
3. *Reference Model of Open Systems Interconnection for CCITT Applications*, CCITT Rec. X.200, Fascicle VIII.4, IXth Plenary Assembly, Melbourne, 1988.
4. *Information Processing Systems: Open Systems Interconnection—Basic Reference Model*, ISO 7498, Geneva, 1984.
5. *Open Systems Interconnection Layer Service Definition Conventions*, CCITT Rec. X.210, Fascicle VIII.4, IXth Plenary Assembly, Melbourne, 1988.
6. D. Bertsekas and R. Gallager, *Data Networks*, 2nd ed., Prentice-Hall, Englewood Cliffs, NJ, 1987.
7. *High-Level Data Link Control Procedures—Frame Structure*, ISO 3309, International Standards Organization, Geneva, 1979.
8. *Advanced Data Communication Control Procedures*, X3.66 ANSI, New York, 1979.
9. *High-Level Data Link Control Procedures—Consolidation of Elements of Procedures*, ISO 4335, Geneva, 1980.

10. *Code Independent Error Control Systems*, CCITT Rec. V.41, Fascicle VIII.1, IXth Plenary Assembly, Melbourne, 1988.

11. R. L. Freeman, *Reference Manual for Telecommunications Engineering*, 2nd ed., John Wiley & Sons, New York, 1994.

12. *Interface Between Data Terminal Equipment (DTE) and Data Circuit-Terminating Equipment (DCE) for Terminals Operating in the Packet Mode and Connected to the Public Data Networks by Dedicated Circuit*, ITU-T Rec. X.25, ITU-T Organization, Helsinki, March 1993.

13. *Interface Between Data Terminal Equipment and Data Circuit-Terminating Equipment (DCE) for Synchronous Operation on Public Data Networks*, CCITT Rec. X.21, Geneva, September 1992.

14. *Use on Public Data Networks of Data Terminal Equipment (DTE) Which Is Designed for Interfacing Synchronous V-Series Modems*, CCITT Rec. X.21 bis, Fascicle VIII.2, IXth Plenary Assembly, Melbourne, 1988.

15. *Packet-Switched Signaling System Between Public Networks Providing Data Transmission Services*, ITU-T Rec. X.75, ITU-T Organization, Helsinki, March 1993.

16. *Internet Protocol*, RFC 791, DDN Network Information Center, SRI International, Menlo Park, CA, September 1981.

17. *Assigned Numbers*, RFC 1060, DDN Network Information Center, SRI International, Menlo Park, CA, March 1990.

18. *Internet Protocol Transition Workbook*, SRI International, Menlo Park, CA, March 1982.

19. *A Standard for the Transmission of IP Datagrams over IEEE 802 Networks*, RFC 1042, DDN Network Information Center, SRI International, Menlo Park, CA, February 1988.

20. *An Ethernet Address Resolution Protocol*, RFC 826, DDN Network Information Center, SRI International, Menlo Park, CA, June 1984.

21. *A Reverse Address Resolution Protocol*, RFC 903, DDN Network Information Center, SRI International, Menlo Park, CA, June 1984.

22. *Internet Control Message Protocol*, RFC 792, DDN Network Information Center, SRI International, Menlo Park, CA, September 1981.

23. *Transmission Control Protocol*, RFC 793, DDN Network Information Center, SRI International, Menlo Park, CA, September 1981.

24. Military Standard, *Transmission Control Protocol*, MIL-STD-1778, U.S. Department of Defense, Washington, DC, August 1983.

25. *Systems Network Architecture Technical Overview*, IBM, Research Triangle Park, NC, January 1994.

26. E. R. Coover, *Systems Network Architecture—SNA Networks*, IEEE Computer Society Press, Los Alamitos, CA, 1992.

27. J. M. Rosenberg, *Dictionary of Computers, Data Processing and Telecommunications*, John Wiley & Sons, New York, 1984.

12

LOCAL AREA NETWORKS

1 DEFINITION AND APPLICATIONS

Local area networks (LANs) use a common transmission medium to inter-connect workstations, computers, and/or other related assets over a limited geographical area. Several LAN standards specify capability to serve up to a thousand or more devices on a single LAN. The geographical extension or "local area" may extend no more than several hundred feet (< 100 m) to over 6 mi (> 10 km) or more in other cases. The transmission media providing this connectivity may be wire pair, coaxial cable, or fiber- optic cable. Local area radio schemes are gaining popularity as an extension of personal communication service (PCS). Certain LAN schemes accom-modate other devices as well, such as digital telephones, facsimile, and video equipment. A basic rule on LANs is that only one user at a time may have access to the medium.

Data rates on current LANs vary from 1 Mbps to 100 Mbps. Future LANs on fiber-optic cable hold promise of 1000 Mbps or more. There are two specialized LANs that have transmission rates from 800 to 1600 Mbps. LAN data rates, the number of devices connected to a LAN, the spacing of those devices, and the network extension depend on

- The transmission medium employed
- Transmission technique (i.e., baseband or broadband)
- Network access protocol

Many LANs operate without error correction with bit error rates (BERs) specified in the range of 1×10^{-8} to 1×10^{-12}.

The most common application of a LAN is to interconnect data terminals (workstations) with processing resources, where all the devices reside in a single building or complex of buildings, and usually these resources have a common owner. Cost containment is a driving force toward the implementation of LANs. A LAN permits effective cost sharing of high-value data processing equipment, such as mass storage media, mainframe computers or minicomputers, and high-speed printers. There are other benefits as well. One, of course, is resource sharing. Another is E-mail and similar messaging services leading to a "paperless" environment.

LANs can be extended, up to a certain point, with repeaters or bridges. They can be segmented by means of switching hubs, smart bridges, or routers. Segmenting of a LAN can notably improve performance, especially on a LAN with many users.

The interconnection of LANs in the local area with a high-speed backbone is current practice. LAN interconnection with the outside world such as with distant LANs via a wide area network is becoming prevalent. This is frame relay's principal application. The interface is carried out with a router or gateway.

There are two generic transmission techniques utilized by LANs: baseband and broadband. Baseband transmission can be defined as the direct application of the baseband signal to the transmission medium. Broadband transmission, in this context, is where the baseband signal from the data device is translated in frequency to a particular frequency slot in the RF spectrum. Broadband transmission requires a modem to carry out the translation. Baseband transmission may require some sort of signal conditioning device. With broadband LAN transmission we usually think of simultaneous multiple RF carriers that are separated in the frequency domain. Present broadband technology comes from the cable television (CATV) industry.

Use of the asynchronous transfer mode (ATM) is beginning to find favor in the LAN community and could radically change many of the concepts introduced in this chapter.

2 LAN TOPOLOGIES

There are three types of LAN topology: bus, ring, and star. These are shown in Figure 12.1 along with the *tree network*, which is a subset of the conventional bus topology.

A bus is a stretch of transmission medium from which users tap into, as shown in Figure 12.1A. Originally, the medium was coaxial cable. Today it can also be unshielded twisted pair (UTP) or shielded twisted pair (STP). On higher-speed (e.g., 100 Mbps) LANs, fiber-optics cable is beginning to be used. In fact, one type, fiber distributed data interface (FDDI), is now widely used, particularly as a backbone LAN. FDDI is discussed later in this chapter.

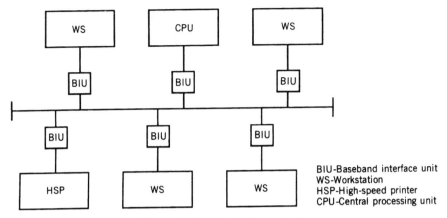

BIU-Baseband interface unit
WS-Workstation
HSP-High-speed printer
CPU-Central processing unit

Figure 12.1A. A typical bus network.

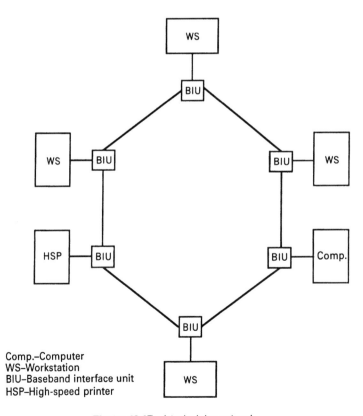

Comp.–Computer
WS–Workstation
BIU–Baseband interface unit
HSP–High-speed printer

Figure 12.1B. A typical ring network.

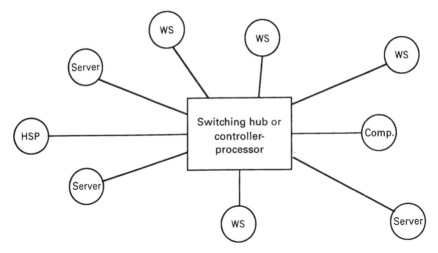

Figure 12.1C. A star network.

A ring is simply a bus that is folded back onto itself. A ring topology is shown in Figure 12.1B. User traffic flows in one direction around the ring. In one approach, which we discuss in this chapter, a second ring is added where the traffic flow is in the opposite direction. Such a dual counter-rotating ring concept improves reliability in case of a failed station or a cut of the ring.

A star network is shown in Figure 12.1C. At the center of the star is a switching device. This could be a switching hub. Users can be paired, two at a time, three at a time, or all at a time, segmented into temporary families of users depending on the configuration of the switch at that moment in time. Such a concept lends itself particularly well to ATM. Each user is connected to the switch on a point-to-point basis.

A tree network is illustrated in Figure 12.1D.

3   THE TWO BROAD CATEGORIES OF LAN TRANSMISSION TECHNIQUES

Baseband and broadband are the two basic transmission techniques employed by LANs, as discussed above.

The baseband technique incorporates single signal transmission of a digital waveform on a transmission medium. Broadband transmission has the capability of transmitting multiple signals simultaneously on a medium, typically coaxial cable. Each signal is assigned a frequency slot in a frequency-division multiplex plan. Broadband technology derives from the CATV (cable television) industry. Baseband lends itself to bus and ring topologies, and broadband lends itself to bus and tree topologies. Broadband systems require

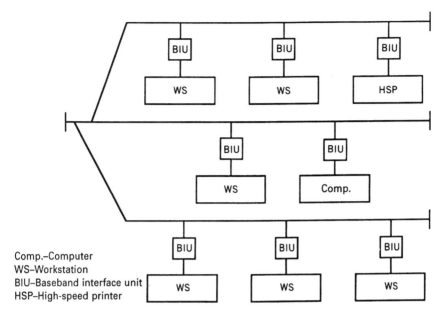

Figure 12.1D. A typical tree configuration.

TABLE 12.1 Comparison of Baseband and Broadband LAN Transmission

Item	Baseband	Broadband
Waveform	Digital: NRZ, RZ, Manchester	RF/FDM
Baseband	Bidirectional	Unidirectional
Topology	Bus or ring	Bus or tree
Access to medium	Tap	Modem or tap
Media	Wire pair, coaxial cable	Coaxial cable
LAN extension	Up to 2 km	Tens of km
Information type	Data only	Data, voice, facsimile, video
Utilization of bandwidth	Single signal occupies entire bandwidth	Multiple simultaneous signals in FDM structure

a modem at each access; baseband systems do not.* Baseband and broadband transmission techniques are compared in Table 12.1.

With both types of LANs (i.e., baseband and broadband) we are dealing with multipoint operation. Two transmission problems arise as a result. The first deals with signal level and signal-to-noise (S/N) ratio, and the second

*The semantics of local area network technology are rather loose. Many baseband systems are not truly baseband. One "baseband" system we discuss uses light; another uses RF. Certainly, if we are truly talking about broad bandwidth, the fiber-optics "baseband" system we describe has an available bandwidth that is extremely broad.

deals with standing waves. Each access on a common medium must have sufficient signal level and S/N such that copied signals have a BER in the range of 1×10^{-8} to 1×10^{-12}. If the medium is fairly long in extension and there are many accesses, the signal level must be high for a transmitting access to reach its most distant destination. The medium is lossy, particularly at the higher bit rates, and each access tap has an insertion loss. This leads to very high signal levels. These may be rich in harmonics and spurious emissions, degrading bit error rate. On the other hand, with insufficient level, the S/N ratio degrades, which will also degrade error performance. A good level balance must be achieved for all users. Every multipoint connectivity must be examined. The number of multipoint connectivities can be expressed by $n(n-1)$, where n is the number of accesses. If, on a particular LAN, 100 accesses are planned, there are 9900 possible connectivities to be analyzed to carry out signal level balance. One way to simplify the job is to segment the network, placing a regenerative repeater (or bridge) at each boundary. This reduces the signal balance job to realizable proportions and ensures that a clean signal of proper level is available at each access tap. For baseband LANs, 50-Ω coaxial cable is favored over the more common 75-ohm cable. The lower-impedance cable is less prone to signal reflections from access taps and provides better protection against low-frequency interference.

The effects of standing waves can be reduced by controlling the spacing between access taps. For example, the Ethernet technical summary [1] recommends spacings no less than 2.5 m for this 10-Mbps system. The technical summary says that by following this placement rule, the chance that objectionable standing waves will result is reduced to a very low (but not zero) probability. Again for Ethernet, up to 100 devices may be placed on a cable segment and the maximum segment length is 500 m. The segments can be connected through regenerative repeaters, and the maximum end-to-end length is 2.5 km [1, 4].

One extremely important consideration for baseband transmission is that only a single thread (transmission line) exists. It can accommodate only one user at time; otherwise there is a high probability of data message collision. Collision is where the electrical signals bearing the traffic of two or more users interfere with one another, corrupting the traffic of each.

3.1 Broadband Transmission Considerations

Broadband transmission permits multiple users to access the medium without collision. Broadband means that we take advantage of the medium's wide bandwidth. This wide bandwidth is broken down into smaller bandwidth segments in an analogous fashion to FDM. Each of these segments is assigned to a family of users. The statement regarding collision is correct if there are no more than two accesses per frequency segment connected on a point-to-point basis, where one access receives while the other transmits.

If, in this case, we assume contention as the access protocol, then as the family increases in number, the chances of collision start to increase. With a little imagination, we can see that, with the proper switching scheme implemented, a user can join any family by simply switching to the proper frequency band of that family. All that is required is a change in modem frequency and possibly modulation waveform. There may also be certain protocol considerations as well.

Unlike their baseband counterparts, broadband systems can be designed to accommodate digital or analog voice, data from kilobit to multimegabit rates, video, and facsimile. Thus broadband systems are versatile. They are also much more expensive than their baseband counterparts and require a higher level of design engineering effort.

As mentioned, much of our present broadband technology derives from cable television technology. Total system bandwidths are on the order of 300–500 MHz. Each access requires a modem to modulate and demodulate the data or other user signal and to translate the modulated frequency to the assigned frequency slot on the cable.

This, then, is RF transmission and, by its very nature, must be one way or unidirectional. Thus a user can only access another user "downstream" from it. If we assume a single medium, usually a 75-Ω coaxial cable, then how does one access another user "upstream"? This is done using a similar approach to that of a two-way or interactive CATV (cable television) system, where two paths are provided on a single coaxial cable. This is accomplished by splitting the cable spectrum into two frequency segments, one segment for one direction and the other for the opposite direction. At a cable terminating point, which some even call a *head end* from CATV terminology, a frequency translator converts and amplifies signals from one direction (frequency segment) into signals for transmission in the opposite direction (frequency segment). Another term for "head end" for broadband LANs is *central retransmission facility* (CRF).

There are two choices of topology for broadband LANs: bus and tree. The head end or CRF is located at some termination point on the bus, and in the case of tree topology the CRF is located at its root, so to speak.

Another approach to achieve dual path operation is to use two cables; one provides the "go" path and the other the "return" path. For single-cable split-band operation, a rather large guard band is left in the center of the cable spectrum to ensure isolation between the two paths. Then we can see that with the provision of two-cable operation the usable bandwidth can be more than doubled. Modem operation is also simpler because, to access a particular net, only one frequency operation is required (i.e., the send and receive frequencies can be the same). With single-cable split-band operation, send and receive frequencies must necessarily be different [2].

Broadband services make up about 5% of the total LAN market. Baseband predominates.

TABLE 12.2 Broadband Channel Allocations for CSMA/CD Services

	Transmit MHz	Receive MHz		
		156.25 Offset		192.25 Offset
Transmit and receive	(1) 35.75–53.75	192.00–210.00	or	228.00–246.00
channel options for	(2) 41.75–59.75	198.00–216.00	or	234.00–252.00
IEEE 802.3b	(3) 47.75–65.75	204.00–222.00	or	240.00–258.00
services (single	(4) 53.75–71.75	210.00–228.00	or	246.00–264.00 (preferred)
cable plant)	(5) 59.75–77.75	216.00–234.00	or	252.00–270.00
	(6) 65.75–83.75	222.00–240.00	or	258.00–276.00
		Transmit and Receive MHz		
Transmit and receive		(1) 36.00–54.00		
channel options for		(2) 42.00–60.00		
IEEE 802.3b		(3) 48.00–66.00		
services (dual		(4) 54.00–72.00		
cable plant)		(5) 60.00–78.00		
		(6) 66.00–84.00 (preferred)		
		(7) 228.00–246.00		
		(8) 234.00–252.00		
		(9) 240.00–258.00		
		(10) 246.00–264.00 (preferred)		
		(11) 252.00–270.00		
		(12) 258.00–276.00		

Source: IEEE Std 802.7, Table 8 (Ref. 5).

Table 12.2 gives broadband channel allocations for CSMA/CD services, and Table 12.3 provides channel allocations for a broadband token bus configuration. Recommended channelization for both single- and dual-cable operation is given.

3.2 Fiber-Optic LANs

Fiber-optic LANs may be considered broadband* in that a class of modem is required to place the digital signal on the fiber. The modem, of course, consists of a light source, detector, and the necessary driver and signal conditioning circuitry. With wavelength-division multiplexing (WDM) we have a true broadband system. At this time, single-wavelength operation prevails.

The type of fiber-optic cable selected for a LAN is a cost trade-off. The extension of a LAN is generally short such that multimode fiber can be used, and the mature short-wavelength technology permits other cost savings. Even plastic fiber may be considered. The losses of the fiber itself will generally be

---

*We will see a contradiction in terminology here when consulting Section 5.6 dealing with FDDI.

TABLE 12.3   Broadband Channel Allocations for Token Ring Services

	Transmit MHz	Receive MHz
Transmit and receive channel options for IEEE 802.4 services (single cable plant)	(1) 59.75–65.75	252.00–258.00
	(2) 65.75–71.75	258.00–264.00
	(3) 71.75–77.75	264.00–270.00
	(4) 77.75–83.75	270.00–276.00
	(5) 83.75–89.75	276.00–282.00
	(6) 89.75–95.75	282.00–288.00
	Transmit and Receive MHz	
Transmit and receive channel options for IEEE 802.4 services (dual cable plant)	(1) 59.75–65.75	
	(2) 65.75–72.75	
	(3) 72.75–77.75	
	(4) 77.75–83.75	
	(5) 83.75–89.75	
	(6) 89.75–95.75	
	(7) 252.00–258.00	
	(8) 258.00–264.00	
	(9) 264.00–270.00	
	(10) 270.00–276.00	
	(11) 276.00–282.00	
	(12) 282.00–288.00	

Note: For 10-Mbps transmission, channels are paired as follows: 1–2, 3–4, 5–6.

Source: IEEE Std 802.7, Table 9 (Ref. 5).

low due to the short-distance operation even at 820-nm operation. A LAN that is 1 km long might display a fiber loss from 2 dB to 5 dB. The major contributor to loss is the taps if a fully passive network is to be implemented.

Fiber lends itself to all three LAN topologies: bus, ring, and star. The use of passive couplers leads to a more reliable system, but the insertion loss of passive couplers is a consideration. With active coupling, the loss of one source or detector can cause the entire network to crash unless bypass switches are used. Figure 12.2 shows passive and active coupler implementations.

4   OVERVIEW OF IEEE/ANSI LAN PROTOCOLS

4.1   General

Many of the widely used LAN protocols have been developed in North America through the offices of the Institute of Electrical and Electronic Engineers (IEEE). The American National Standards Institute (ANSI) has subsequently accepted and incorporated these standards, and they now bear the ANSI imprimatur.

ACTIVE COUPLING

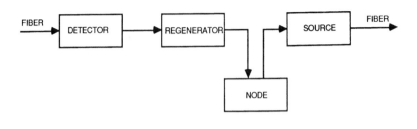

PASSIVE COUPLING

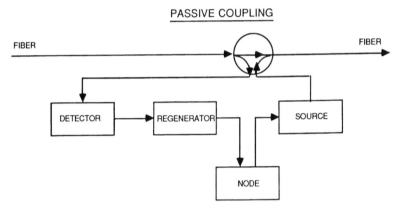

Figure 12.2. Coupling a node to optical fiber cable.

IEEE develops LAN standards in the IEEE 802 committee, which is currently organized into the following subcommittees:

802.1*　High-Level Interface

802.2*　Logical Link Control (LLC)

802.3*　CSMA/CD Networks

802.4*　Token Bus Networks

802.5*　Token Ring Networks

802.6*　Metropolitan Area Networks (MANs)

802.7*　Broadband Technical Advisory Group

802.8　Fiber-Optic Technical Advisory Group

802.9*　ISLAN (Integrated Services Local Area Network) Working Group

802.10*　LAN Security Working Group

802.11*　Wireless LAN Working Group

*Indicates that one or more published standards or draft standards are available.

802.12* Demand Priority Working Group

802.13 —

802.14 Cable TV Working Group

The fiber distributed data interface (FDDI) standard, which is discussed in Section 5.6, has been developed directly by ANSI.

4.2 How LAN Protocols Relate to OSI

LAN protocols utilize only OSI layers 1 and 2, the physical and data-link layers, respectively. The data-link layer is split into two sublayers: medium access control (MAC) and logical link control. These relationships are shown in Figure 12.3.

Stallings [3] presents an interesting and rational argument on the reasoning for limiting the layering to the first two OSI layers. There is no question that the functions of OSI layers 1 and 2 must be incorporated in a LAN architecture. We now ask, Why not layer 3? Layer 3, the network layer, is concerned with routing. There is no routing involved with LANs. There is a direct link involved between any two points. The other functions carried out by OSI layer 3—addressing, sequencing, and flow control—are carried out by layer 2 in LANs. The difference is that layer 2 performs these functions across a single link. OSI layer 3 carries out these functions across a sequence of links required to traverse a network. Of course, there is only one link required to traverse a LAN.

It would seem that layer 3 is required when viewed through an attached device. The reason is that the device sees itself attached to a network

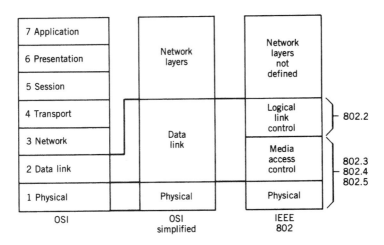

Figure 12.3. LAN 802 architecture related to OSI.

*Indicates that one or more published standards or draft standards are available.

connecting multiple devices. One would think that ensuring delivery of a message to one or more accesses would be a layer 3 function. It was decided that, although the network provides services through layer 3, the characteristics of the network allow these functions to be performed in the first two layers.

As shown In Figure 12.3, the OSI data-link layer is divided into two sublayers: logical link control and medium access control. These sublayers carry out four functions:

1. Provide one or more service access points (SAPs). A SAP is a logical interface between two adjacent layers.

2. Before transmission, assemble data into a frame with address and error-detection fields.

3. On reception, disassemble the frame and perform address recognition and error detection.

4. Manage communications over the link.

The first function and those related to it are performed by the LLC sublayer. The last three functions are handled by the MAC sublayer.

In the following subsections we will describe four common IEEE and ANSI standardized protocols. Logical link control (LLC) is common to all four. They differ in the medium access control (MAC) protocol.

A station on a LAN may have multiple users; oftentimes these are just processes, such as processes on a host computer. These processes may wish to pass traffic to another LAN station which may have more than one "user" in residence. We will find that LLC produces a PDU with its own source and destination address. The source address, in this case, is the address of the originating user. The destination address is the address of a user in residence at a LAN station. Such a user is connected through a service access point (SAP) at the upper boundary of the LLC layer. The resulting LLC PDU is then embedded in the information field of a MAC frame. This is shown in Figure 12.4.

The MAC frame also has source and destination addresses. These direct the traffic to a particular LAN station or stations.

4.3 Logical Link Control (LLC)

The LLC provides services to the upper layers at a LAN station. The upper layers are user defined. The LLC provides two forms of services for its users:

1. Unacknowledged connectionless service.

2. Connection mode service.

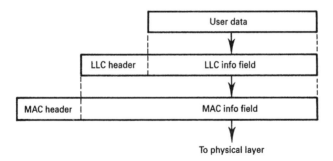

Figure 12.4. A user passes traffic to an LLC where an encapsulation takes place forming an LLC PDU. This traffic is then embedded in a MAC frame. The MAC frame is passed to the physical layer which transmits the traffic on the LAN.

Some brief comments are required to clarify the functions and limitations of each service. With unacknowledged connectionless service a single service access initiates the transmission of a data unit to the LLC, the service provider. From the viewpoint of the LLC, previous and subsequent data units are unrelated to the present unit. There is no guarantee by the service provider of the delivery of the data unit to its intended user, nor is the sender informed if the delivery attempt fails. Furthermore, there is no guarantee of ordered delivery. This type of service supports point-to-point, multipoint, and broadcast modes of operation.

As we might imagine with connection mode service, a logical connection is established between two LLC users. During the data transfer phase of the connection, the service provider at each end of the connection keeps track of the data units transmitted and received. The LLC guarantees that all data will be delivered and that the delivery to the intended user will be ordered (e.g., in the sequence as presented to the source LLC for transmission). When there is a failure to deliver, it is reported to the sender.

IEEE [6] defines the LLC as that part of a data station that supports the LLC functions of one or more logical links. The LLC generates command PDUs (protocol data units) and response PDUs for transmission and interprets received command PDUs and response PDUs. Specific responsibilities assigned to the LLC include:

1. Initiation of control signal interchange.

2. Interpretation of received command PDUs and generation of appropriate response PDUs.

3. Organization of data flow.

4. Actions regarding error-control and error-recovery functions in the LLC sublayer.

LLC is another derivative of HDLC, which was covered in the previous chapter. It is based on the balanced mode of that link-layer protocol with similar formats and functions. This is especially true when operating in the connection mode.

4.3.1 LLC Generic Primitives. The IEEE [7] defines a (service) primitive as *an abstract, implementation-independent interaction between a service user and service provider.* In general, the services of a layer or sublayer are the capabilities which it offers to a user in the next higher layer or sublayer. In order to provide its service, a layer or sublayer builds its functions on the services it requires from the next lower layer or sublayer.

Services are specified by describing the service primitives and parameters that characterize each service. A service may have one or more related primitives that constitute the activity that is related to a particular service. Each service primitive may have zero or more parameters that convey the information required to provide the service.

The generic primitives used in LLC are identical to the generic primitives employed in HDLC (Chapter 11). These are:

REQUEST   The request primitive is passed from the N-user to the N-layer (or sublayer) to request that a service be initiated.

INDICATION The indication primitive is passed from the N-layer (or sublayer) to the N-user to indicate an internal N-layer (or sublayer) event which is significant to the N-user. This event may be logically related to a remote service request, or may be caused by an event internal to the N-layer (or sublayer).

RESPONSE   The response primitive is passed from the N-user to the N-layer (or sublayer) to complete a procedure previously invoked by an indication primitive.

CONFIRM   The confirm primitive is passed from the N-layer (or sublayer) to the N-user to convey the results of one or more associated previous service request(s).

Examples of primitives with parameters comprise a sequence used for a successful connection: DL_CONNECT request, DL_CONNECT indication, DL_CONNECT response, and DL_CONNECT confirm, in that order.

4.3.2 LLC PDU Structure. As shown in Figure 12.4, user data are passed down to the LLC, which appends a header (i.e., encapsulates it). The LLC PDU frame format is shown in Figure 12.5. The header consists of address and control information; the information field contains the user data. The control field is identical with the HDLC control field, shown in Figure 11.15 in Section 6.2.3 of Chapter 11. However, the LLC control field is 2 octets long and there is no provision to extend it to 3 or 4 octets in length as there is in HDLC.

DSAP Address	SSAP Address	Control	Information
8 bits	8 bits	8 or 16 bits	M*8 bits

DSAP Address = destination service access point address field
SSAP Address = source service access point address field
Control = control field (16 bits for formats that include sequence numbering,
       and 8 bits for formats that do not
Information = information field
* = multiplication
M = an integer value equal to or greater than 0. (Upper bound of M is a func-
    tion of the medium access control methodology used.)

Figure 12.5. LLC PDU format [6].

As we mentioned previously, LLC destination address is the user address at an SAP inside the LAN station. It is called the *destination service access point* (DSAP). The SSAP is the *source service access point* and it indicates the message originator inside a particular LAN station. Each has a field of 8 bits as shown in Figure 12.6. However, only the last 7 of those bits are used for actual address. The first bit in the destination address field indicates whether the address is an individual address or a group address (i.e., addressed to more than one SAP). The first bit in the SSAP is the C/R bit which indicates whether a frame is a command frame or a response frame. The control field is briefly described in Section 4.3.4.

4.3.3 *Types and Classes of LLC Operation.* Two types of operation are defined in the referenced standard [6]:

(a) TYPE I operation is where PDUs are exchanged between peer LLCs without the need for the establishment of a data-link connection. In the LLC sublayer these PDUs are not acknowledged, nor is there any error recovery or flow control.

(b) TYPE 2 operation requires a data-link connection to be established between the two peer LLCs prior to the exchange of any information-bearing PDUs. The normal cycle of communication between two TYPE 2 LLCs on a data-link connection consists of the transfer of PDUs containing information from the source LLC to the destination LLC, acknowledged by PDUs in the opposite direction.

With TYPE 2 operation, the control of traffic between a source LLC and destination LLC is by means of a numbering scheme which is cyclic within a modulus of 128 and measured in terms of PDUs. This is the same type of

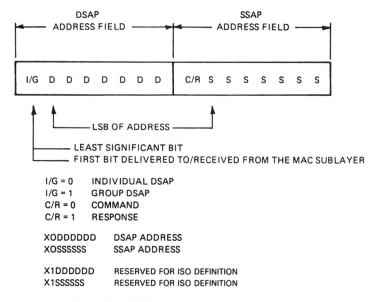

Figure 12.6. DSAP and SSAP address field formats [6].

numbering used with HDLC with its modulus-128 option. See Section 6.2.3.1 in Chapter 11 for a description of its operation.

There are two classes of LLC operation:

(a) Class I LLCs support TYPE 1 operation only. Class I service is applicable to individual, group, global, and null DSAP addressing, as well as to applications requiring no data-link layer acknowledgment or flow control procedures.

(b) Class II LLCs support both TYPE 1 and TYPE 2 operation. In a Class II station, the operation of TYPE 1 procedures and TYPE 2 procedures are completely independent. A Class II LLC is capable of going back and forth between TYPE 1 and TYPE 2 operation on a PDU-to-PDU basis in the same SAP, if necessary.

4.3.4 LLC Control Field and Its Operation. The LLC control field is illustrated in Figure 12.7. It is 16 bits long for formats that include sequence numbering and 8 bits long for formats that do not. The sequence numbering and windows are identical to those used with HDLC and described in Chapter 11, Section 6.2.3.1.

The three formats defined for the control field are used to perform numbered information transfer, numbered supervisory transfer, unnumbered control, and unnumbered information transfer functions. The numbered information transfer and supervisory transfer functions apply only to TYPE

LLC PDU CONTROL FIELD BITS

	1	2	3	4	5	6	7	8	9	10 - 16
INFORMATION TRANSFER COMMAND/RESPONSE (I-FORMAT PDU)	0			N(S)					P/F	N(R)
SUPERVISORY COMMANDS/RESPONSES (S-FORMAT PDUs)	1	0	S	S	X	X	X	X	P/F	N(R)
UNNUMBERED COMMANDS/RESPONSE (U-FORMAT PDUs)	1	1	M	M	P/F	M	M	M		

N(S) = Transmitter send sequence number (Bit 2 = low-order bit)
N(R   = Transmitter receive sequence number (Bit 10 = low-order bit)
S   = Supervisory function bit
M   = Modifier function bit
X   = Reserved and set to zero
P/F   = Poll bit—command LLC PDU transmissions
         Final bit—response LLC PDU transmissions
         (1 = Poll/Final)

Figure 12.7. LLC PDU control field formats [6].

2 operation. The unnumbered control and unnumbered information transfer functions apply to either TYPE 1 or TYPE 2 operation (but not both) depending on the specific function selected.

As shown in Figure 12.7, there are three types of frames as there are in HDLC. These are: I-frames, S-frames, and U-frames, which we describe below.

The I-Frame. The information transfer format (I-format) is used to perform a numbered information transfer in TYPE 2 operation. Except where otherwise specified (e.g., command/response frames UI, TEST, FRMR, and XID*), it is the only LLC PDU that contains an information field. The functions of the sequence numbers N(S) and N(R) and the P/F are independent; that is, each I-format PDU has an N(S) sequence number, an N(R) sequence number which shall or shall not acknowledge additional I-format PDUs at the receiving LLC, and a P/F (poll/final) bit that is set to 1 or 0.

The S-Frame. The supervisory format (S-format) is used to perform data-link supervisory control functions in TYPE 2 operation, such as acknowledging I-format PDUs or requesting retransmission of I-format PDUs and requesting a temporary suspension of transmission of I-format PDUs. The functions of N(R) and P/F are independent; that is, each S-format PDU has an N(R) sequence number that does

---

*These are described in Chapter 11, Section 6.2.3. Remember that the C/R bit in the SSAP defines whether a frame is a command or response.

acknowledge or does not acknowledge additional I-format PDUs at the receiving LLC and a P/F bit that is set to 1 or 0.

The U-Frame. The unnumbered format (U-format) PDUs are used in either TYPE 1 or TYPE 2 operation, depending upon the specific function utilized, to provide additional data-link control functions and to provide unsequenced information transfer. The U-format PDUs contain no sequence numbers, but include a P/F bit which is set to 1 or 0. The P/F bit operates in a manner similar to that in HDLC.

5 LAN ACCESS PROTOCOLS

5.1 Introduction

In this context a protocol includes a means of permitting all users to access a LAN fairly and equitably. Access can be random or controlled. The random access schemes to be discussed include CSMA (carrier sense multiple access) and CSMA/CD, where CD stands for collision detection. The controlled access schemes that are described are token bus and token ring. It should be kept in mind that users accessing the network are unpredictable and the transmission capacity of the LAN should be allocated in a dynamic fashion in response to those needs.

5.2 Background: Contention and Polling

Contention and polling were briefly introduced in Chapter 6. Contention is where there is no discipline at all. On any data communication system where more than two users share a common medium, the risk of collision is always present. Collision occurs where two or more accesses transmit simultaneously on a common medium. The result is that the traffic of both contenders is corrupted. Contention is a form of random access. CSMA and CSMA/CD, which is described in the following subsections, are other forms of random access.

Polling is a form of controlled access. It is commonly used on multipoint configurations. Such a configuration is shown in Figure 11.4B. One station, usually the CPU, is assigned the responsibility of master station. The master station polls each remote station periodically, often sequentially, to determine if there is traffic to be transmitted. When a reply from the remote is in the affirmative, the traffic is then transmitted. Each station on the network has a unique address (e.g., bit sequence). This address is incorporated into the poll. Thus only one indicated remote station will respond to that address. In a similar manner, traffic from the master station to a remote station incorporates the address of that remote station in the header of the message. In such a way that remote station and only that remote station will copy the traffic. This type of polling is called *roll-call polling* by some and *broadcast polling* by others. Because each remote station must respond to the poll (either negative or positive) with that additional overhead, roll-call polling is considered inefficient.

Loop polling is another form of polling. In this case the master station sends a poll request to the first remote station in the loop. If that station has traffic, it is sent to the master station. Once the traffic transmission terminates and all the traffic is sent and acknowledged, that remote station forwards the poll request to the next station in the loop. If the second station has traffic, it is transmitted; if not, the poll request is sent to the third station, and so on, until all remote stations have been polled. The process then starts all over again. The advantages of loop polling are that message transmission and polling operations overlap and no negative responses are transmitted.

Token-passing LAN access methods are a direct outgrowth of polling. Such access schemes can also be called a form of time-division multiple access (TDMA).

5.3 CSMA and CSMA/CD Access Techniques

Carrier sense multiple access (CSMA) is a LAN access technique that some simplistically call "listen before transmit." This "listen before transmit" idea gives insight into the control mechanism. If user 2 is transmitting, user 1 and all others hear that the medium is occupied and refrain from using it. In actuality, when an access with traffic senses that the medium is busy, it backs off for a period of time and tries again. How does one control that period of time? There are three methods of control commonly used. These methods are called *persistence algorithms* and are outlined briefly below:

- **Nonpersistent.** The accessing station backs off a random period of time and then reattempts access.

- **1-persistent.** The station continues to sense the medium until it is idle and then proceeds to send its traffic.

- **p-persistent.** The accessing station continues to sense the medium until it is idle, then transmits with some preassigned probability p. Otherwise it backs off a fixed amount of time, then transmits with a probability p or continues to back off with a probability of $(1 - p)$.

The algorithm selected depends on the desired efficiency of the medium usage and the complexity of the algorithm and resulting impact on firmware and software. With the nonpersistent algorithm collisions are effectively avoided because the two stations attempting to access the medium will back off, most probably with different time intervals. The result is wasted idle time following each transmission. The 1-persistent algorithm is more efficient by allowing one station to transmit immediately after another transmission. However, if more than two stations are competing for access, collision is virtually assured.

The *p*-persistent algorithm lies between the other two and is a compromise, attempting to minimize collisions and idle time.

It should also be noted that with CSMA, after a station transmits a message, it must wait for an acknowledgment from the destination. Here we must take into account the round-trip delay ($2 \times$ propagation time) and the fact that the acknowledging station must also contend for medium access. Another important point is that collisions can occur only when more than one user begins transmitting within the period of propagation time. Thus CSMA is an effective access protocol for packet transmission systems where the packet transmission time is much longer than the propagation time.

The inefficiency of CSMA arises from the fact that collisions are not detected until the transmissions from the two offenders have been completed. With CSMA/CD, which has collision detection, a collision can be recognized early in the transmission period and the transmissions can be aborted. As a result, channel time is saved and overall available channel utilization capability is increased.

CSMA/CD is sometimes called "listen while transmitting." It must be remembered that collisions can occur at any period during channel occupancy, and this includes the total propagation time from source to destination. Even at multimegabit data rates, propagation time is not instantaneous; it remains constant for a particular medium, no matter what the bit rate is. With CSMA the entire channel is wasted. With CSMA/CD one offending station stops transmitting as soon as it detects the second offending station's signal. It can do this because all accesses listen *while* transmitting.

5.3.1 CMSA/CD Description. Carrier sense multiple access with collision detection is defined by ISO/IEC 8802-3 international standard and by the reference ANSI/IEEE Std 802.3 [4, 6]. It is based on the Ethernet approach initiated by Xerox Corporation, Digital Equipment Corporation, and Intel. The IEEE 802.3 version closely resembles Ethernet with changes in packet structure and an expanded set of physical layer options. Figure 12.8 relates CSMA/CD protocol layers to the conventional OSI reference model (see Chapter 11, Section 6.2.1) and identifies acronyms that we use in this description. The bit rates generally encompassed in CSMA/CD are between 1 and 20 Mbps. There are two standards now settling down that cover 100 Mbps on optical fiber. The model used for the discussion below covers the 10-Mbps rate.

The medium is coaxial cable. A user connects to the cable by means of a medium access unit (MAU). This connects through an attachment unit interface (AUI) to the data terminal equipment (DTE). As shown in Figure 12.8, the DTE consists of the physical signaling sublayer (PLS), the medium access control (MAC), and the logical link control (LLC). The PLS is responsible for transferring bits between the MAC and the cable. It uses differential Manchester encoding for the data transfer. With such coding the

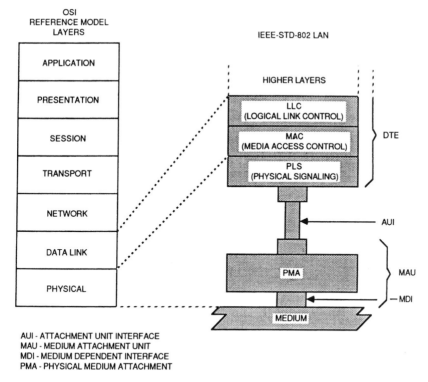

OSI
REFERENCE MODEL
LAYERS

IEEE-STD-802 LAN

AUI - ATTACHMENT UNIT INTERFACE
MAU - MEDIUM ATTACHMENT UNIT
MDI - MEDIUM DEPENDENT INTERFACE
PMA - PHYSICAL MEDIUM ATTACHMENT

Figure 12.8. CSMA/CD LAN relationship to the OSI model. From IEEE/ANSI Std. 802.3 [4]. Courtesy of IEEE, New York.

datum 0 has a transition from high to low at midcell, while the datum 1 has the opposite transition. The line idle is a steady high condition with no transitions.

CSMA/CD LAN systems (or Ethernet) are probably the most widely used type of LANs worldwide. We would say that this is due to their relatively low cost to implement and maintain and to their simplicity. The down side is that their efficiency starts to drop off radically as the number of users increases as well as increased user activity. Thus the frequency of collisions and backoffs increases to a point that throughput can drop to zero. Some users argue that efficiency starts to drop off at around 30% capacity, while others argue that that point is nearer 50%. We will discuss ways to mitigate this problem in our coverage of bridges.

5.3.1.1 *System Operation*

TRANSMISSION WITHOUT CONTENTION. A MAC frame is generated from data from the LLC sublayer. This frame is handed to the transmit media access management component of the MAC sublayer for transmission. To avoid contention with other traffic on the medium, the transmit medium access management monitors the carrier sense signal provided by the physical layer signaling (PLS) component. When the medium is clear, frame transmission is initiated through the PLS interface. When the transmission has been completed without contention (a collision event), the MAC sublayer informs the LLC and awaits the next request for frame transmission.

RECEPTION WITHOUT CONTENTION. At each receiving station, the arrival of a frame is first detected by the PLS, which responds by synchronizing with the incoming preamble and by turning on the carrier sense signal. The PLS passes the received bits up to the MAC sublayer where the leading bits are discarded, up to and including the end of the preamble and start frame delimiter (SFD). In this period the receive media access management component of the MAC sublayer has detected the carrier sense and is waiting for the incoming bits to be delivered. As long as the carrier sense is on, the receive media access management collects bits from the PLS. Once the carrier sense signal has been removed, the frame is truncated at an octet boundary, if required, and then passed to the receive data decapsulation for processing.

It is in receive data decapsulation where the destination address is checked to determine if this frame is destined for this particular LAN station. If it is, the destination address (DA) and source address (SA) and the LLC data unit are passed to the LLC sublayer. It also passes along the appropriate status code indicating that reception is complete or reception too long. It also checks for invalid MAC frames by inspecting the frame check sequence to detect any damage to the frame enroute, as well as by checking for proper octet-boundary alignment of the end of frame.

COLLISION HANDLING. A collision is caused by multiple stations attempting to transmit at the same time, in spite of their attempts to avoid this by deferring. A given station can experience a collision during the initial part of its transmission (the collision window) before its transmitted signal has had time to propagate to all stations on the CSMA/CD medium. Once the collision window has passed, a transmitting station is said to have acquired the medium. Once all stations have noticed that there is a signal on the medium (by way of carrier sense), they defer to it by not transmitting, avoiding any chance of subsequent collision. The time to acquire the medium is thus based on the round-trip propagation time of the physical layer whose elements include the PLS, the physical medium attachment (PMA), and the physical medium itself.

In the event of collision, the transmitting station's physical layer notices a notable increase in standing waves on the medium* and turns on the collision detect (CD) signal. The collision-handling process now starts. First, the transmit media access management enforces the collision by transmitting a bit sequence called jam. This "jam," specified as 32 bits long, ensures that all

---

*This is called "interference" in the ISO/IEC reference standard.

stations involved in the collision are aware that a collision has occurred. After the jam has been sent, the transmit media access management component terminates the transmission and schedules another transmission attempt after a randomly selected time interval. Retransmission is attempted again in the face of repeated collisions. If, on this second attempt, another collision occurs, the transmit media access management attempts to reduce the medium's load by backing off, meaning it voluntarily delays its own retransmissions to reduce the load on the medium. This is accomplished by expanding the interval from which the random retransmission time is selected on each successive transmission attempt. Eventually, either the transmission succeeds or the attempt is abandoned on the assumption that the medium has failed or has become overloaded.

5.3.1.2 MAC Frame Structure. The MAC frame is shown in Figure 12.9. There are eight fields in a frame: preamble, start frame delimiter (SFD), the addresses of the frame's source and destination(s), a length field to indicate the length of the following field containing the LLC data to be transmitted, a field that contains padding if required, and the frame check sequence (FCS) field containing a cyclic redundancy check value to detect errors in received frames. All eight fields are of fixed size except the LLC data and PAD fields, which may contain any integer number of octets between the minimum and maximum values determined by a specific implementation that may be selected.

The minimum and maximum frame size limits refer to that portion of the frame from the destination address field through the frame check sequence field, inclusive. The default maximum frame size is 1518 octets; the minimum size is 64 octets.

The preamble field is 7 octets in length and is used so that the receive PLS can synchronize to the transmitted symbol stream. The SFD is the binary sequence 10101011. It follows the preamble and delimits the start of frame.

There are two address fields: the source address and the destination address. The address field length is an implementation decision. It may be 16 or 48 bits long. In either field length, the first bit specifies whether the address

PREAMBLE (7 OCTETS)	SFD (1 OCTET)	DESTINATION ADDRESS (2 OR 6 OCTETS)	SOURCE ADDRESS (2 OR 6 OCTETS)	LENGTH (2 OCTETS)	LLC DATA	PAD	FCS (4 OCTETS)

LENGTH - GIVES NUMBER OF OCTETS IN DATA FIELD

LLC - LOGICAL LINK CONTROL (OSI LAYER 3 AND ABOVE)

FCS - FRAME CHECK SEQUENCE - A 32-BIT CRC

PAD - ADDS OCTETS TO ACHIEVE MINIMUM FRAME LENGTH WHERE NECESSARY

SFD - START FRAME DELIMITER

Figure 12.9. MAC frame format [4].

is an individual address (bit set to 0) or group address (bit set to 1). In the 48-bit address field, the second bit specifies whether the address is globally administered (bit set to 0) or locally administered (bit set to 1). For broadcast address, the bit is set to 1.

The length field is 2 octets long and indicates the number of LLC data octets in the data field. If the value is less than the minimum required for proper operation of the protocol,* a PAD field (sequence of octets) is appended at the end of the data field and prior to the FCS field. The length field is transmitted and received with the high-order octet first.

The data (LLC data) field contains a sequence of octets that is fully transparent in that any arbitrary sequence of octet values may appear in the data field up to the maximum number specified by the implementation of this standard that is used. The maximum size of the data field supplied by the LLC is determined by the maximum frame size and address size parameters of a particular implementation.

The frame check sequence (FCS) field contains four octets (32 bits) CRC value. This value is computed as a function of the contents of the source address, destination address, length, LLC data, and pad—that is, all fields except the preamble, SFD, and FCS. The encoding is defined by the following generating polynomial:

$$G(x) = x^{32} + x^{26} + x^{23} + x^{22} + x^{16} + x^{12} + x^{11} + x^{10} + x^8 + x^7 + x^5$$
$$+ x^4 + x^2 + x + 1$$

An invalid MAC frame meets at least one of the following conditions:

1. The frame length is inconsistent with the length field.
2. It is not an integral number of octets in length.
3. The bits of the received frame (exclusive of the FCS itself) do not generate a CRC value identical to the one received. An invalid MAC frame is not passed to the LLC.

The minimum frame size is 512 bits for the 10-Mbps data rate [4]. This requires a data field of either 46 or 54 octets, depending on the size of the address field used. The minimum frame size is based on the *slot time*, which for the 10-Mbps data rate is 512 bit times. Slot time is the major parameter controlling the dynamics of collision handling and it is:

• An upper bound on the acquisition time of the medium.
• An upper bound on the length of a frame fragment generated by a collision.
• The scheduling quantum for retransmission.

*Minimum frame length is 64 octets.

To fulfill all three functions, the slot time must be larger than the sum of the physical round-trip propagation time and the MAC sublayer jam time. The propagation time for a 500-m segment of 50-Ω coaxial cable is 2165 ns, assuming that the velocity of propagation of this medium is $0.77 \times 300 \times 10^6$ m/s [4].

5.3.1.3 Transmission Requirements

SYSTEM MODEL. Propagation time is critical for the CSMA/CD access method. The major contributor to propagation time is the coaxial cable and its length. The characteristic impedance of the coaxial cable is $50 \, \Omega \pm 2 \, \Omega$. The attenuation of a 500-m (1640-ft) segment of the cable should not exceed 8.5 dB (17 db/km) measured with a 10-MHz sine wave. The velocity of propagation is $0.77c$.* The referenced maximum propagation times were derived from the physical configuration model described here. The maximum configuration is as follows:

1. A trunk coaxial cable, terminated in its characteristic impedance at each end, constitutes a coax segment. A coax segment may contain a maximum of 500 m of coaxial cable and a maximum of 100 MAUs. The propagation velocity of the coaxial cable is assumed to be $0.77c$ minimum ($c = 300,000$ km/s). The maximum end-to-end propagation delay for a coax segment is 2165 ns.

2. A point-to-point link constitutes a link segment. A link segment may contain a maximum end-to-end propagation delay of 2570 ns and shall terminate in a repeater set at each end. It is not permitted to connect stations to a link segment.

3. Repeater sets are required for segment interconnection. Repeater sets occupy MAU positions on coax segments and count toward the maximum number of MAUs on a coax segment. Repeater sets may be located in any MAU position on a coax segment but shall only be located at the ends of a link segment.

4. The maximum length, between driver and receivers, of an AUI cable is 50 m. The propagation velocity of the AUI cable is assumed to be $0.65c$ minimum. The maximum allowable end-to-end delay for the AUI cable is 257 ns.

5. The maximum transmission path permitted between any two stations is five segments, four repeater sets (including optional AUIs), two MAUs, and two AUIs. Of the five segments, a maximum of three may be coax segments; the remainder are link segments.

The maximum transmission path consists of 5 segments, 4 repeater sets (with AUIs), 2 MAUs, and 2 AUIs as shown in Figure 12.10. If there are two

*$c =$ velocity of light in a vacuum.

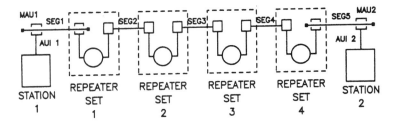

Figure 12.10. Maximum transmission path.

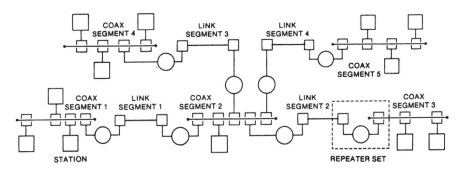

Figure 12.11. An example of a large system with maximum transmission paths.

link segments on the transmission path, there may be a maximum of three coax segments on that path. If there are no link segments on a transmission path, there may be a maximum of three coax segments on that path given current repeater technology. Figure 12.11 shows a large system with maximum transmission paths. It also shows the application of link segments versus coax segments. The bitter ends of coaxial cable segments should be terminated in the coaxial cable characteristic impedance. The coax segments are marked at 2.5-m intervals. MAUs should only be attached at these 2.5-m interval points. This assures nonalignment at fractional wavelength boundaries.

NOTES ON CSMA/CD SYSTEM PARAMETERS

Scheduling of Backoff Attempts. The dynamics of collision handling are largely determined by a single parameter called the *slot time*. This single parameter describes three important aspects of collision handling:

1. It is an upper bound on the acquisition time of the medium.
2. It is an upper bound on the length of a frame fragment generated by a collision, and
3. It is the scheduling quantum for retransmission.

To fulfill all three functions, the slot time must be larger than the sum of the physical layer round-trip time and the media access layer maximum jam time. The slot time is determined by the parameters of implementation. Two examples are given further below.

The scheduling of reattempts is determined by a controlled randomization procedure called *truncated binary exponential backoff*. This defines the delay time before reattempt. The delay is an integer multiple of the slot time. The number of slot times to delay before the nth retransmission attempt is chosen as a uniformly distributed integer r in the range of $0 \leq r < 2^k$, where $k = \min(n, 10)$. Algorithms used to generate the integer r should be designed to minimize correlation between the numbers generated by any two stations at a given time.

Other System Parameters. Two sets of CSMA/CD implementation parameters are given below. The first is for 10BASE5, the 10-Mbps system described herein.

Parameters	Values
slotTime	512 bit times
interFrameGap	9.6 μs
attemptLimit	16
backoffLimit	10
jamSize	32 bits
maxFrameSize	1518 octets
minFrameSize	512 bits (64 octets)
addressSize	48 bits

The second group of parameter values is for 1BASE5, similar to the above but with a 1-Mbps transmission rate.

Parameters	Values
slotTime	512 bit times
interFrameGap	96 μs
attemptLimit	16
backoffLimit	10
jamSize	32 bits
maxFrameSize	1518 octets
minFrameSize	512 bits (64 octets)
addressSize	48 bits

The baseband signal transmitted to the medium by the AUI uses Manchester coding. With Manchester coding there is always a transition in the middle of a bit interval. The first half represents the bit value, and the second half of the bit is the complement of the first half. Thus Manchester coding is rich in transitions for timing extraction and is compatible with the coaxial cable transmission medium [4].

There are a number of different transmission techniques either recommended by Standard 802.3 or derived therefrom. Some of these are shown below:

10BASE5	10BASE2	10BASET
10BROAD36	1BASE5	

The convention used for identifying these techniques uses the first number to indicate the data rate on the medium in megabits per second (Mbps), BASE or BROAD indicates whether it is baseband or broadband transmission, and the final number gives the actual link or segment length in hundreds of meters. 10BASET is a 10-Mbps system on twisted pair.

5.4 Token Bus

A token bus LAN in its simplest version is a length of 75-Ω coaxial cable terminated at each end in its characteristic impedance. Users tap the cable with a coupler and connect to the coupler with a 37.5-Ω stub no longer than 350 mm. Three different transmission regimes are described in IEEE Std 802.4 [8]: phase continuous FSK, phase coherent FSK, and multilevel duobinary AM/PSK. The coaxial cable can be extended by use of regenerative repeaters. In a similar manner tree topologies can be developed. The discussion below covers data rates of 1 Mbps, 5 Mbps, and 10 Mbps.

The 1990 standard [8] also has provision for fiber-optic medium using 62.5-μm/125-μm fiber with data rate options of 5, 10, and 20 Mbps.

With the token bus access technique a short control packet known as the token regulates the right of access to the bus. A station holding the token has the exclusive right to use the network for a specified time period. During this period the station may poll other stations, receive responses, and pass data traffic. When the token holding station has completed operations or when its time period is up, it passes the token to the next station, its successor, in logical sequence.

Figure 12.12 illustrates a typical token bus, showing logical connectivity (dashed line) and that the access method is always sequential in a logical sense. The right to access the medium passes from user to user. It should be noted that physical connectivity has little impact on the order of the logical ring. In fact, stations can respond to a query from a token holder without being part of the *logical* ring. For example, stations H and F can receive data frames but cannot initiate a transmission because they cannot receive the token.

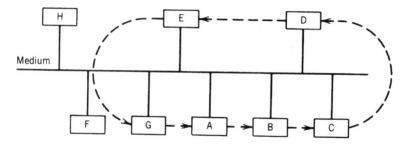

Figure 12.12. Token bus showing logical connectivity (dashed line).

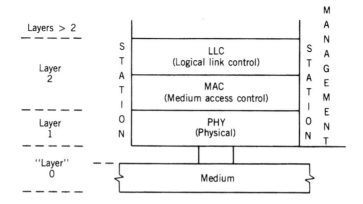

Figure 12.13. IEEE 802.4 model: functional relationship of access control equipment [8].

Token passing ensures equitable access to the network. Such a control also ensures against collision because only the user holding the token can access the medium.

The user access control equipment is shown functionally in Figure 12.13. "Station management," an important part of the control function, is shown on the right. The figure also shows the layer relationship with the OSI model. Also, the similarity with CSMA/CD is apparent. Key, again, is the medium access control.

Specific responsibilities of the MAC include ordered access to the medium, providing a means of admission and deletion of stations on the LAN and the handling of fault recovery. Among the faults handled are:

- Multiple tokens.
- Lost tokens.
- Token pass failure.
- Nonresponsive stations (i.e., a station with an inoperative receiver).
- Duplicate station addresses.

It should be noted that no station takes on an exclusive monitoring and control function.

As with the CSMA/CD protocol, slot time is also an important parameter with the token-passing bus protocol. Slot time in this case is defined as the maximum time any station need wait for an immediate answer from another station. It is measured in octet times and is an integer. Slot time is twice the sum of the propagation delay, station delay, and safety margin. The *response window* equals the slot time. If a station waiting for a response hears a transmission start during the response window, that station does not transmit again until at least the received transmission terminates.

Holding the token gives the right to transmit. The token is passed from station to station in descending numerical order of station address. When a station hears a token frame addressed to itself, it "has the token" and may transmit data frames. After a station has completed transmitting data frames and has completed maintenance functions where necessary, the station passes the token to its successor by sending a *token MAC frame*. It then goes through a procedure to ensure that its successor has the token. For instance, after sending the token MAC frame, the station listens for evidence that the successor has heard the token frame and is active. If the sender hears a valid frame following the token, it can assume that its successor has the token and is transmitting. If the token-sending station does not, it attempts to assess the state of the network.

If the token-sending station hears a noise burst or a frame with an incorrect FCS, it cannot be sure from the source address which station sent the frame. If a noise burst is heard, the token-passing station sets an internal indicator and continues to listen in the *check token pass state* up to four more slot times. If nothing is heard during the four-slot time delay, the station assumes that its successor has the token.

If the token holder does not hear a valid frame after sending the first time, it repeats the token pass procedure once more. If the successor does not transmit after the second attempt, it assumes that the successor has failed. The sender then sends a *who follows frame* with the successor's address in the data field of the frame. All active stations on the LAN then compare the value of the data field of a "who follows frame" with the address of their own predecessor, which is the station that normally would send them the token. The station whose predecessor is the successor of the sending station responds to the "who follows frame" by sending its address in a *set successor frame*. The station holding the token thus establishes a new successor, bridging the failed station out of the logical ring.

Figure 12.14 shows the MAC frame format. The number of octets between the start delimiter (SD) and the end delimiter (ED) should be no greater than 8191. The abort sequence consists of an SD and an ED, each of which is 1 octet long [8].

We define three acronyms dealing with a particular LAN station in question, its predecessor, and its successor in the logical ring of token passing. Again, that logical ring is shown in the example given in Figure 12.12.

WHERE: PREAMBLE - PATTERN FOR SYNC AND SET RECEIVE LEVEL
(1 OR MORE OCTETS)

SD - STARTING DELIMITER (1 OCTET)

FC - FRAME CONTROL (1 OCTET)

DA - DESTINATION ADDRESS(ES) (2 OR 6 OCTETS)

SA - SOURCE ADDRESS (2 OR 6 OCTETS)

DATA - INFORMATION (0 OR MORE OCTETS)

FCS - FRAME CHECK SEQUENCE (4 OCTETS)

ED - ENDING DELIMITER (1 OCTET)

Figure 12.14. The token bus MAC frame format [8].

TS—this station's address

NS—next station's address

PS—previous station's address

A token has seven fields. These are standard preamble, start delimiter (SD), the binary sequence 00001000 for the FC, a DA equal to the value of the station's NS, its SA, an FCS, and an ending delimiter (ED).

An LLC data frame has eight fields: preamble, SD, FC, DA, and SA, then the LLC_data_unit followed by the FCS and an ED.

Four different transmission/modulation schemes are presented in the reference specification [8]. Three of these use 75-Ω coaxial cable as the medium and one uses fiber optics. The bit error rate at the receive physical interface is 1×10^{-8}, with a mean undetected bit error rate of 1×10^{-9}. Extension of the network is accomplished by means of regenerative repeaters. Branched topologies may be implemented by impedance-matched non-directional splitters, which are three-port passive networks that divide the signal incident at one port into two equal parts that are transmitted to the other two ports. Drop cables generally should not exceed 50 m (160 ft).

Phase continuous FSK is one of the several transmission methods given in IEEE Std 802.4 [8]. Successive symbols presented to the physical layer at the MAC interface are encoded, producing a three-PHY-symbol code: H, L, and OFF. These symbols feed a two-tone FSK modulator where the H symbol is the higher frequency tone, the L symbol is the lower frequency tone, and the OFF symbol is no tone. The output signal is then ac-coupled to the coaxial cable.

IEEE Std 802.4 standardizes the line data rate at 1 Mbps with a tolerance of ±0.01% for an originating station and ±0.015% for a repeater station. There are five possible symbols for the FSK implementation: 0, 1, nondata, pad-idle, and silence. Each of these MAC symbols is encoded into a pair of PHY symbols from a different three-symbol H, L, OFF code as follows:

1. Silence (OFF OFF).

2. Pad-idle. Pad-idle symbols are always octets. Each pair of pad-idle symbols is encoded as a sequence of LH, HL.

3. 0 (HL).

4. 1 (LH).

5. Nondata. The MAC layer transmits nondata symbols in pairs, which are encoded as the sequence LL HH.

The start frame delimiter subsequence is *nondata nondata 0 nondata nondata 0*, which is encoded as the sequence LL HH HL LL HH HL. The end-frame delimiter subsequence is *nondata nondata 1 nondata nondata 1*, which is encoded as the sequence LL HH LH HH LH.

The line signal corresponds to an FSK signal with its carrier frequency at 5.00 MHz, varying smoothly between two signaling frequencies of $3.75\ \text{MHz} \pm 80\ \text{kHz}$ and $6.25\ \text{MHz} \pm 80\ \text{kHz}$. When transitioning between two signaling frequencies, the FSK modulator changes its frequency in a continuous and monotonic manner within 100 ns. The output signal level of the modulated carrier frequency is between $+54\ \text{dB}$ and $+60\ \text{dB}$ relative to 1 mV across 37.5 Ω. An S/N ratio of 20 dB at the receiver produces a BER of 1×10^{-8} or better. This is based on an in-band noise floor of $+4\ \text{dB}$ or less relative to 1 mV across 37.5 Ω.

Phase coherent FSK is another transmission method specified in IEEE Std 802.4. The waveform in this case is somewhat different from that of continuous FSK. Again successive MAC symbols presented to the physical layer entity at its MAC interface are applied to an encoder that produces a three-PHY-symbol code: H, L, and OFF. The output is then applied to a two-tone FSK modulator that represents each H symbol as one full cycle of a tone whose period is exactly one-half of the MAC symbol period, each L as one-half cycle of a tone whose full-cycle period is exactly the MAC symbol period, and each OFF as no tone for the same half MAC symbol period. This modulated signal is then ac-coupled to the coaxial cable medium.

The standard data signaling rates for phase coherent FSK systems are 5 Mbps and 10 Mbps. The permitted tolerance for each signaling rate is $\pm0.01\%$ for an originating station and $\pm0.015\%$ for a repeater station. The coding is somewhat different from its continuous FSK counterpart.

For the 5-Mbps data rate, the lower-tone FSK frequency is 5.0 MHz and the higher-tone FSK frequency is 10.0 MHz. For the 10-Mbps data rate the lower-tone FSK frequency is 10.0 MHz and the higher-tone FSK frequency is 20.0 MHz. The output level of the transmitted signal into a 75-Ω resistive load is between $+63\ \text{dBmV}$ and $+66\ \text{dBmV}$, inclusive. The BER at the receiver of 1×10^{-8} will be achieved when a received signal is in the range of $+10\ \text{dBmV}$ to $+66\ \text{dBmV}$ and the maximum noise floor is $-10\ \text{dBmV}$.

Loss budgets are important when engineering any LAN to ensure that each and every LAN station will be provided the desired performance. Based on phase coherent FSK, a loss budget might be worked out as follows.* Both overall length of cable and number of taps should be explicitly combined in the system signal loss budget. Each tap makes a known contribution to the network loss budget, as does each length of the trunk cable. In addition, each tap includes a definite tap loss between the trunk and the drop, which should also be included in the loss budget.

The referenced standard [8] specifies the tap trunk to drop loss value of 20 ± 0.5 dB. Typical values for commercially available two-drop taps conforming to the referenced specification are 0.3-dB insertion loss over a frequency range of 1–30 MHz. An example calculation of the signal loss budget for a simple, unbranched network, including 20 such two-drop taps, might be as follows:

Drop loss at first tap:	20.5 dB
Insertion loss of 18 intervening taps:	5.4 dB
Drop loss at 20th tap:	20.5 dB
Total lumped losses:	46.6 dB

The minimum transmit level is $+63$ dBmV and the minimum receive level is $+10$ dBmV. Thus, the cable loss in this example should not exceed $63 - 10 - 46.4 = 6.6$ dB. Of course, when we specify attenuation for coaxial cable, it always should be specified for the highest frequency of interest. For a 5-Mbps data rate, an RG-11-type cable with an attenuation of 1.5 dB/100 meters at 10 MHz limits the cable length to 440 meters (1430 ft). A longer cable length would be possible if semirigid cable were used.

If fewer taps were used, the cable could be longer. However, the designer should be advised that phase delay distortion may also limit cable length.

The IEEE 802.4 standard also specifies a broadband alternative for token-passing bus using a CATV-like 75-Ω single-cable mid-split configuration. In this configuration, the unidirectional forward and reverse channels are paired to provide bidirectional channels. A frequency offset of 192.25 MHz between forward and reverse channels is used in both the mid-split and high-split configuration. A dual-cable configuration may also be considered.

The following are the three specified data rates, their respective bandwidths, transmitter output power, and receiver noise floors for broadband token bus operation:

Data Rate	Bandwidth	Transmit Power	Receive Noise Floor
1.0 Mbps	1.5 MHz	$+41$ dBmV	-40 dBmV
5.0 Mbps	6.0 MHz	$+47$ dBmV	-34 dBmV
10 Mbps	12.0 MHz	$+50$ dBmV	-31 dBmV

---

*Based on Chapter 13 Appendix of IEEE 802.4 specification of 1990 [8].

The modulation employed is a 3-state duobinary AM/PSK system. A 30-dB S/N is required to comply with BER performance requirements of 1×10^{-8} error rate [8].

5.5   Token Ring

A typical token-passing ring LAN is shown in Figure 12.15. The token ring operation, as specified in IEEE Std 802.5 [9], has the capability of 4-Mbps or 6-Mbps data rate. A ring is formed by physically folding the medium back onto itself. Each LAN station regenerates and repeats each bit and serves as a means of attaching one or more data terminals (e.g., workstations, computers) to the ring for the purpose of communicating with other devices on the network. As a traffic frame passes around the ring, all stations, in turn, copy the traffic. Only those stations included in the destination address field pass that traffic on to the appropriate users that are attached to the station. The traffic frame continues onward back to the originator, who then removes the traffic from the ring. The *pass-back* to the originator acts as a form of acknowledgment that the traffic had at least passed by the destination(s).

The sequential connection of stations removes the need to form a logical ring, as in token bus operation. A reservation scheme is used to accommodate priority traffic. Also, one station acts as a ring monitor to ensure correct network operation. A monitor devolvement scheme to other stations is provided in case a monitor fails or drops off the ring (i.e., shuts down). A station on the ring can become inactive (i.e., close down), and a physical bypass is provided for this purpose.

A station gains the right to transmit frames onto the medium when it detects a token passing on the medium. Any station with traffic to transmit, on detection of the appropriate token, may capture the token by modifying it to a start-of-frame sequence and append the proper fields to transmit the first frame. At the completion of its information transfer and after appropriate

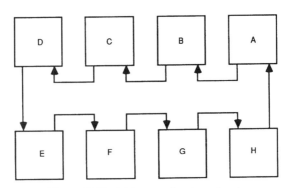

Figure 12.15. A token-passing ring network.

SD	AC	FC	DA	SA	INFO	FCS	ED	FS

|<- SFS ->|<──────── FCS COVERAGE ────────>|<- EFS ->|

WHERE: SFS - START-OF-FRAME SEQUENCE
        SD - STARTING DELIMITER (1 OCTET)
        AC - ACCESS CONTROL (1 OCTET)
        FC - FRAME CONTROL (1 OCTET)
        DA - DESTINATION ADDRESS (2 OR 6 OCTETS)
        SA - SOURCE ADDRESS (2 OR 6 OCTETS)

        INFO - INFORMATION (0 OR MORE OCTETS)*
        FCS - FRAME CHECK SEQUENCE (4 OCTETS)
        EFS - END-OF-FRAME SEQUENCE
        ED - ENDING DELIMITER (1 OCTET)
        FS - FRAME STATUS (1 OCTET)

Figure 12.16. Frame format for token ring based on IEEE Std 802.5 [9].

SD	AC	ED

SD = Starting Delimiter (1 octet)
AC = Access Control (1 octet)
ED = Ending Delimiter (1 octet)

Figure 12.17. Token format for token ring operation [9].

checking for proper operation, the station initiates a new token, which provides other stations with the opportunity to gain access to the ring. Each station has a token-holding timer that controls the maximum period of time a station may occupy the medium before passing the token on.

Figure 12.16 shows the frame format specified in IEEE Std 802.5 [9], and Figure 12.17 shows the token format. In these figures, the leftmost bit is transmitted first. The frame format in Figure 12.16 is used for transmitting both MAC and LLC messages to destination station(s). It may or may not contain the INFO field.

The starting delimiter (SD) consists of the symbol sequence JK0JK000, where J and K are nondata symbols and are described below. Both frames and tokens start with the SD sequence.

The access control (AC) is 1 octet long and contains 8 bits that are formatted PPPTMRRR. The first three bits, PPP, are the priority bits. These are used to indicate the priority of a token and therefore which stations are allowed to use the token. In a system designed for multiple priority, there are eight levels of priority available, where the lowest priority is PPP = 000 and the highest is PPP = 111. The AC field contains the token bit T and the monitor bit M. If T = 0, then the frame is a token (Figure 12.17). The T bit is a 1 on all other frames. The M bit is set by the monitor as part of the procedures for recovering from malfunctions. The bit is transmitted as a 0 in all frames and tokens. The active monitor inspects and modifies this bit. All other stations repeat this bit as received. The three R bits are reservation bits. These bits allow stations with high-priority protocol data units (PDUs) to request that the next token issued be at the requested priority.

The next field in Figure 12.16 is the frame control (FC) field. It is one octet long and defines the type of frame and certain MAC and information frame functions. The first two bits in the FC are designated FF bits, and the last six bits are called ZZZZZZ bits. If FF is 00 the frame contains a MAC PDU, and if FF is 01 the frame contains an LLC PDU.

If the frame-type bits indicate a MAC frame, all stations on the frame interpret and act upon the ZZZZZZ control bits. If the frame-type bits indicate an LLC frame, the ZZZZZZ bits are designated rrrYYY. The rrr bits are reserved and are transmitted as 0s in all the transmitted frames and ignored upon reception. The YYY bits may be used to carry the priority (Pm) of the PDU from the source LLC entity to the target LLC entity or entities.

Each frame (not token) contains a destination address (DA) field and a source address (SA) field. Depending on the token ring LAN system, the address fields may be 16 or 48 bits in length. In either case, the first bit indicates whether the frame is directed to an individual station (0) or to a group of stations (1).

The INFO field carries zero, one, or more octets of user data intended for the MAC, NMT (network management), or LLC. Although there is no maximum length specified for the INFO field, the time required to transmit a frame may be no greater than the token-holding period that has been established for the station. For LLC frames, the format of the information field is not specified in the referenced standard [9]. However, in order to promote interworking among stations, all stations should be capable of receiving frames whose information field is up to and including 133 octets in length.

The frame check sequence (FCS) is a 32-bit sequence based on the standard generator polynomial of degree 32 given in Section 5.3.1. It encompasses the FC, DA, SA, and INFO fields. Its transmission commences with the coefficient of the highest term.

The end delimiter (ED) is one octet in length and is transmitted as the sequence JK1JK1IE. The transmitting station transmits the delimiter as shown. Receiving stations consider the ending delimiter (ED) valid if the first six symbols JK1JK1 are received correctly. The I is the intermediate frame bit and is used to indicate whether a frame transmitted is a singular frame or whether it is a multiple frame transmission. The I bit is set at 0 for the singular frame case. The E bit is the error-detected bit. The E bit is transmitted as 0 by the station that originates the token, abort sequence, or frame. All stations on the ring check tokens and frames for errors such as FCS errors and nondata symbols. The E bit of tokens and frames that are repeated is set to 1 when a frame with an error is detected; otherwise the E bit is repeated as received.

The last field in the frame is the frame status (FS) field. It consists of one octet of the sequence ACrrACrr. The r bits are reserved for future standardization and are transmitted as 0s, and their value is ignored by the receiver. The A bit is the address-recognized bit, and the C bit is the frame-copied bit. These two bits are transmitted as 0 by the frame originator. The A bit is changed to 1 if another station recognizes the destination address as its own

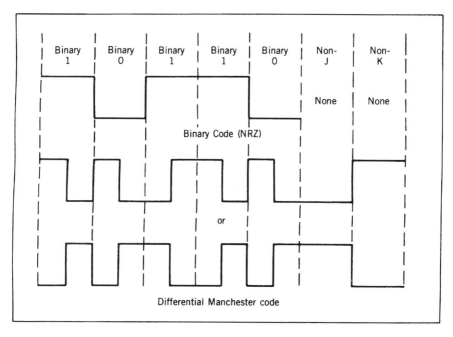

Figure 12.18. Differential Manchester coding format for symbols 1, 0, nondata J, and nondata K (IEEE Std 802.5, Ref. 9).

or relevant group address. If it copies the frame into its buffer, it then sets the C bit to 1. When the frame reaches the originator again, it may differentiate among three conditions:

1. Station nonexistent or nonactive on the ring.
2. Station exists but frame was not copied.
3. Frame copied.

Fill is used when a token holder is transmitting preceding or following frames, tokens, or abort sequences to avoid what would otherwise be an inactive or indeterminate transmitter state. Fill can be either 1s or 0s or any combination thereof and can be any number of bits in length within the constraints of the token-holding timer.

IEEE Std 802.5 describes a true baseband-transmitting waveform using differential Manchester coding. It is characterized by the transmission of two line signal elements per symbol. An example of this coding is shown in Figure 12.18. The figure shows only the data symbols 1 and 0 where a signal element of one polarity is transmitted for one-half the duration of the symbol to be transmitted, followed by the contiguous transmission of a signal element of the opposite polarity for the remainder of the symbol duration. The following advantages accrue from using this type of coding:

- The transmitted signal has no dc component and can be inductively or capacitively coupled.

- The forced midsymbol transition provides inherent timing information on the channel.

The nondata symbols J and K depart from the rule in that a signal element of the same polarity is transmitted for both signal elements of the symbol and therefore there is no midsymbol transition. A J symbol has the same polarity as the preceding symbol. The transmission of nondata symbols occurs in pairs (i.e., JK) to avoid accumulating a dc component.

All stations on the LAN ring are slaved to the active monitor station. They extract timing from the received data by means of a phase-locked loop. *Latency* is the time, expressed in number of bits transmitted, for a signal element to proceed around the entire ring. In order for the token to circulate continuously around the ring when all stations are in the repeat mode, the ring must have a latency of at least the number of bits in the token sequence—that is, 24. Since the latency of the ring varies from one system to another and no a priori knowledge is available, a delay of at least 24 bits should be provided by the active monitor.

IEEE Std 802.5 [9] specifies that the connection of a station to the ring trunk cable be via shielded cable containing two balanced 150-Ω twisted pair. The magnitude of the transmitted signal measured at the medium interface cable when terminated in 150-Ω resistive is specified between 3.0 and 4.5 V peak-to-peak. The amplitude of the positive and negative transmitted signal should be symmetrical within 5%. The trunk cable itself may be twisted pair, coaxial cable, or fiber-optic cable, at the user's discretion.

A LAN station provides an output with an error rate of less than or equal to 1×10^{-9} when the signal-to-noise ratio at the output of the equalizer, specified in paragraph 7.5.2 of the reference document [9], is 22 dB.

As mentioned above, under normal operation there is one station on the ring that is the active monitor. All other stations on the ring are frequency- and phase-locked to this station. They extract timing from the received data by means of a phase-locked loop. The phase-locked loop design is based on the requirement to accommodate a combined total of at least 250 stations and repeaters on the ring [9, 10].

5.6   Fiber Distributed Data Interface

5.6.1   Overview.   A fiber distributed data interface (FDDI) network consists of a set of nodes (e.g., LAN stations) connected by an optical transmission medium (or other medium) into one or more logical rings. A logical ring consists of a set of stations connected as an alternating series of nodes and transmission medium to form a closed loop. This is shown in

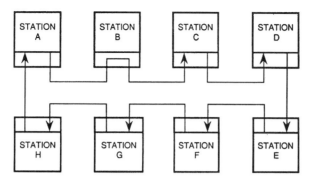

All stations are active except B (illustrated in bypass mode)

Figure 12.19. FDDI token ring; example of logical configuration [12].

Figure 12.19. Information is transmitted as a stream of suitably encoded symbols from one active node to the next. Each active node regenerates and repeats each symbol and serves as a means for attaching one or more devices to the ring for the purpose of communicating with other devices on the ring.

FDDI provides equivalent bandwidth to support a peak data rate of 100 Mbps and a sustained data transfer rate of at least 80 Mbps. With equivalent 4B/5B coding, FDDI line transmission rate is 125 Mbaud (peak). It provides connectivity for many nodes over distances of many kilometers in extent. Certain default parameter values for FDDI (such as timer settings) are calculated on the basis of up to 1000 transmission links or up to 200-km total fiber path length (typically corresponding to 500 nodes and 100 km of dual-fiber cable, respectively). However, the FDDI protocols can support much larger networks by increasing these parameter values.

Two kinds of data service can be provided in a logical ring: packet service and circuit service. With packet service, a given station that holds the token transmits information on the ring as a series of data packets, where each packet circulates from one station to the next. The stations that are addressed copy the packets as they pass. Finally, the station that transmitted the packets effectively removes them from the ring.

In the case of circuit service, some of the logical ring bandwidth is allocated to independent channels. Two or more stations can simultaneously communicate via each channel. The structure of the information stream within each channel is determined by the stations sharing the channel.

Conventional FDDI provides packet service via a token ring. A station gains the right to transmit its information on to the medium when it detects a token passing on the medium. The token is a control signal comprising a unique symbol sequence that circulates on the medium following each series of transmitted packets. Any active station, upon detection of a token, may capture the token by removing it from the ring. The station may then transmit one or more data packets. After transmitting its packets, the station issues a new token, which provides other stations the opportunity to gain access to the ring.

Each station has a token-holding timer (or equivalent means) incorporated, which limits the length of time a station may occupy the medium before passing the token onwards.

FDDI provides multiple levels of priority for independent and dynamic assignment depending on the relative class of service required. The classes of service may be synchronous, which may be typically used for applications such as real-time packet voice; asynchronous, which is typically used for interactive applications; or immediate, which is used for extraordinary applications such as ring recovery. The allocation of ring bandwidth occurs by mutual agreement among users of the ring.

Error detection and recovery mechanisms are provided to restore ring operation in the event that transmission errors or medium transients (e.g., those resulting from station insertion or deletion) cause the access method to deviate from normal operation. Detection and recovery of these cases utilize a recovery function that is distributed among the stations attached to the ring.

One of the more common topologies of FDDI is the counter-rotating ring. Here two classes of station may be defined: dual (attachment) and single (attachment). FDDI trunk rings may be composed only of dual-attachment stations which have two PMD (physical layer medium dependent) entities (and associated PHY entities) to accommodate the dual ring. Concentrators provide additional PMD entities beyond those required for their own attachment to the FDDI network, for the attachment of single-attachment stations which have only one PMD and thus cannot directly attach to the FDDI trunk ring. A dual-attachment station, or one-half of it, may be substituted for a single-attachment station in attaching to a concentrator. The FDDI network consists of all attached stations.

The example of Figure 12.20 shows the concept of multiple physical connections used to create logical rings. As shown in the figure, the logical sequence of MAC connections is stations 1, 3, 5, 8, 9, 10, and 11. Stations 2, 3, 4, and 6 form an FDDI trunk ring. Stations 1, 5, 7, 10, and 11 are attached to the ring by lobes branching out from stations that form it. Stations 8 and 9 are, in turn, attached by lobes branching out from station 7. Stations 2, 4, 6, and 7 are concentrators, serving as the means for attaching multiple stations to the FDDI ring. Concentrators may or may not have MAC entities and station functionality. The concentrator examples of Figure 12.20 do not show any MACs, although their presence is implied by the designation of these concentrators as stations.

5.6.2 FDDI Facilities.

In this section we will discuss symbol sets, protocol data units (PDUs) utilized in FDDI and coding.

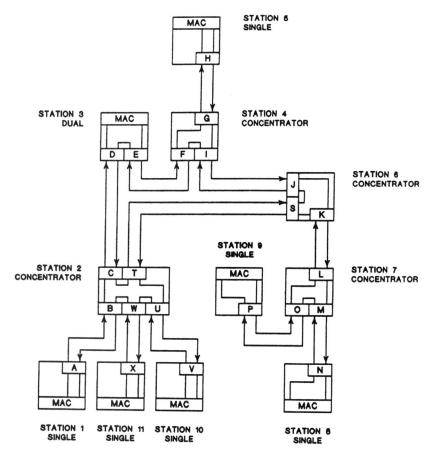

Figure 12.20. FDDI topology example. From Ref. 12.

5.6.2.1 Symbol Set. A symbol is the smallest signaling element used by the FDDI MAC. Symbols can be used to convey three types of information:

1. Line states such as halt (H), quiet (Q), and idle (I).
2. Control sequences such as the starting delimiter (SD), ending delimiter (ED) symbol (T), initial SD symbol (J), and final symbol (K).
3. Data quartets, each representing a group of four ordered data bits.

Peer MAC entities on the ring communicate via a set of fixed-length symbols. These symbols are passed across the MAC-to-PHY interface via defined primitives. The MAC generates PDUs as matched pairs of symbols in accordance with the referenced FDDI standards.

LINE STATE SYMBOLS. These three symbols are reserved for use on the medium between MAC PDUs. The only line state symbol generated by a

MAC is idle. However, the MAC can receive other line state symbols. Detection of any of these symbols within a MAC PDU constitutes an error in the PDU.

Quiet (Q). The quiet symbols indicate the absence of any transitions in the code group.

Halt (H). The halt symbol indicates connection management function (CMT) signaling sequences (in the form of line states). It is also used for filtering line state or code violation symbols from the repeated symbol stream while minimizing the dc component of the NRZI* signal on the transmission medium.

Idle (I). The idle symbol indicates the normal condition of the medium between MAC PDUs. It provides a continuous fill pattern to establish and maintain clock synchronism.

CONTROL SYMBOLS

Starting Delimiter (SD). A starting delimiter (SD) is used to show the starting boundary of a data transmission sequence (e.g., a MAC PDU). The PDU is normally preceded by a preamble of idle symbols, although it may succeed or preempt a previous PDU. The starting delimiter may also succeed or preempt a previous transmission. The SD sequence consists of the uniquely recognizable symbol sequence (JK) that can be recognized independently of previously established symbol boundaries.

Ending Delimiter (ED) Symbol (T). An ending delimiter symbol (T) terminates all MAC PDUs. The T symbol is not necessarily the last symbol in a transmission sequence, since the ED may be followed by one or more control indicator symbols. A sequence of ending delimiter and control indicator symbols are generated by the data-link layer (DLL) as a balanced sequence of symbol pairs (i.e., an even number of R, S, and T symbols). When no control indicators are generated, this sequence is generated as a pair of T symbols.

CONTROL INDICATORS. Control indicators specify logical conditions with a data transmission sequence. They may be independently altered by repeating nodes without altering the normal data in the transmission sequence. A sequence of ending delimiter and control indicator symbols is generated by the MAC as a balanced sequence of symbol pairs (i.e., an even number of R, S, and T symbols). A single ending delimiter symbol followed by an odd number of control indicator symbols is a balanced symbol pair sequence. However, an ending delimiter symbol followed by an even number of control indicator symbols is balanced by adding a final ending delimiter symbol.

*NRZI stands for a nonreturn to zero inverted.

Reset (R). The reset symbol indicates a logical "off" or "false" condition.
Set (S). The set symbol indicates a logical "on" or "true" condition.

DATA QUARTETS (0-F). A data quartet symbol conveys one quartet of binary data within a data transmission sequence. The sixteen data quartet symbols are denoted by the hexadecimal digital values (0-F), and a generic member of the set is denoted by the character "n." A sequence of data quartet symbols is generated as a balanced sequence of symbol pairs (i.e., an even number of n symbols).

5.6.2.2 Coding

CODE BIT. Peer PHY entities communicate via fixed-length code bits. A code bit is the smallest signaling element used by PHY. In the NRZI code, a code bit is represented as a transition (one), or absence of a transition (zero), in the polarity of the signal on the medium.

CODE GROUP. A code group is a consecutive sequence of five code bits. It is used to represent a symbol on the medium. Implicit in the definition of code group is the establishment of code group boundaries. The process of establishing a code group boundary is known as "framing," and the established boundary is known as the "framing boundary." Table 12.4 defines the mapping of symbols to code groups [13].

5.6.3 FDDI Protocol Data Units. Two types of PDUs are used by a MAC: tokens and frames. In the figures that follow, formats are depicted of PDUs in the order of transmission on the medium, with the leftmost symbol transmitted first.

5.6.3.1 Token. Figure 12.22 shows the token format.

```
|        PA        | SD | FC | ED |
```

PA = Preamble (4 or more symbols)
SD = Starting Delimiter (2 symbols)
FC = Frame Control (2 symbols)
ED = Ending Delimiter (2 symbols)

Figure 12.21. The FDDI token format [13].

The token is the means by which the right to transmit MAC SDUs (service data units) (as opposed to the normal process of repeating) is passed from one MAC to another.

5.6.3.2 The FDDI Frame. Figure 12.22 shows the FDDI frame format. The frame format is used for transmitting both MAC recovery information and

TABLE 12.4 FDDI Symbol Coding

Code Group		Symbol	
Decimal	Binary	Name	Assignment

Line State Symbols

00	00000	Q	Quiet
04	00100	H	Halt
31	11111	I	Idle

Starting Delimiter

24	11000	J	First symbol of JK pair
17	10001	K	Second symbol of JK pair

Embedded Delimiter

05	00101	L	Second symbol of IL pair

Data Quartets

			Hexadecimal	**Binary**
30	11110	0	0	0000
09	01001	1	1	0001
20	10100	2	2	0010
21	10101	3	3	0011
10	01010	4	4	0100
11	01011	5	5	0101
14	01110	6	6	0110
15	01111	7	7	0111
18	10010	8	8	1000
19	10011	9	9	1001
22	10110	A	A	1010
23	10111	B	B	1011
26	11010	C	C	1100
27	11011	D	D	1101
28	11100	E	E	1110
29	11101	F	F	1111

Ending Delimiter

13	01101	T	Terminate

Control Indicators

07	00111	R	Reset (logical Zero or Off)
25	11001	S	Set (logical One or On)

Invalid Code Points

01	00001	V or H	These code points shall not be transmitted
02	00010	V or H	because they can generate patterns that
03	00011	V	violate run length or duty cycle
06	00110	V	requirements. Streams of code points 01,
08	01000	V or H	02, 08, and 16 shall be interpreted as Halt
12	01100	V	by the line state detection function.
16	10000	V or H	

(12345) = sequential order of code bit transmission.

Source: ANSI X3.231-1994, Table 1, page 19 (Ref. 12), reprinted with permission.

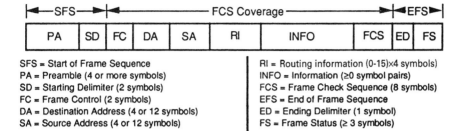

Figure 12.22. The FDDI frame format [13].

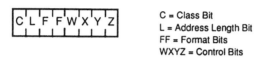

C = Class Bit
L = Address Length Bit
FF = Format Bits
WXYZ = Control Bits

Figure 12.23. Frame control field (FC) [13].

MAC SDUs between peer MAC entities. A frame may or may not have an information field.

FRAME LENGTH. In basic mode the physical layer of FDDI requires limiting the maximum frame length to 9000 symbols, including the four symbols of the preamble. 9000 symbols (quartets) equals 4500 octets.

Preamble (PA). The preamble is transmitted by a PDU originator with a minimum of 4 symbols of the idle pattern when in the hybrid mode,* or 16 symbols of idle pattern in the basic mode.

Starting Delimiter (SD). The starting delimiter was described in Section 5.6.2. A basic mode frame or token starts with a JK symbol pair. No basic mode frame or token is considered valid unless it starts with this explicit sequence.

Frame Control (FC). The frame control field defines the type of frame and associated control functions. This field is shown in Figure 12.23.

FRAME CLASS BIT. The frame class bit indicates the class of service as follows:

C = 0   indicates an asynchronous frame

C = 1   indicates a synchronous frame

*The hybrid mode is another mode of FDDI ring operation involving multiplexing of packet and circuit data and the provision of access to circuit-switched channels.

FRAME ADDRESS BIT LENGTH. The frame address length indicates the length of both MAC addresses (DA and SA) as follows:

L = 0   indicates 16-bit addresses
L = 1   indicates 48-bit addresses

FRAME FORMAT BITS. The FF bit in conjunction with the CL bits and the WXYZ control bits indicate the frame type as follows:

CLFF	WXYZ	to	WXYZ	
0L00	0000			Void frame
1000	0000			Nonrestricted token
1100	0000			Restricted token
0L00	0001	to	1111	Station management frame
1L00	0001	to	1111	MAC frame
CL01	0000	to	1111	LLC frame
CL10	0000	to	1111	Reserved for implementer
CL11	0000	to	1111	Reserved for future standardization

where the W bit is reserved for future standardization in all frames except void frames and tokens. It is transmitted as 0, with the exception of SMT next station addressing frames. It is ignored when received, but is included in the FCS computation for all frames. The referenced specification [13] states that the XYZ bits are reserved for future assignment. These reserved bits are transmitted as 0 and are ignored when received, but are included in the FCS computation.

FF = 00. Frames with FF = 00 are used for management of the FDDI network. They are transmitted with the SA field set to individual_MAC_address of the transmitting MAC and with the RI set to 0. FF = 00 frames are not to be forwarded by bridges.

Control Bits. The control bits are used with associated CLFF bits (see Figure 12.23). All other values (not shown below) and those designated by r are reserved for future assignment.

MAC Beacon Frame (1L00 r010). This frame is transmitted to indicate a serious ring fault and assists in locating persistent faults.

MAC Claim Frame (1LL00 r011). This frame is transmitted during error recovery in the basic mode to determine which MAC creates the token.

MAC Purge Frame (1L00 r100). This frame is used in the hybrid mode during error recovery.

SMT Next Station Addressing Frame (0L00 r111). This frame is transmitted by SMT (station management) to address the next downstream
station that is a member of an addressed group. When the broadcast
address is used, this is the next downstream MAC.

Asynchronous LLC Frame (0L01 WPPP). This frame contains an LLC
PDU using the asynchronous transmission service. The last three
control bits (PPP) indicate the SDU (user) priority, with PPP = 111
being the highest priority and PPP = 000 being the lowest priority.

Synchronous LLC Frame (1L01 rrrr). This frame contains an LLC PDU
using the synchronous transmission service.

Implementer Frame (CL10 rUUU). This frame contains an implementer-
defined MAC SDU. The control bits, UUU, are implementer defined.

DESTINATION AND SOURCE ADDRESSES. The approach here is nearly identical
to the token ring approach in Section 5.5 of this chapter.

Addresses may be either 16 or 48 bits in length. However, all MACs
implement 48-bit address capability. Both 16-bit and 48-bit addresses may be
disabled for specific protocol operations. If an address is disabled or not
implemented, it is equivalent to the null address. A null address is an address
field of all 0s. The group and broadcast address approach is identical to
token ring (Section 5.5). However, specific group addresses have been
assigned for SMT protocols.

The source address rules used in FDDI have some variances with token
ring. There are both 16-bit and 48-bit source addresses. However, bit 1 of the
FDDI source address is the routing indicator (RI). If that bit is 0, there is no
routing information; if it is 1, there is routing information. This routing
information, if present, is contained in a field after the source address (SA)
field. The second bit of the 48-bit source address field is the U/L bit and
indicates whether the address is locally or universally administered.

ROUTING INFORMATION (RI) FIELD. The RI field is used when bridging is
available. When the RI bit of a 48-bit address field is set to 1, it indicates that
the routing information field is included in the frame. The RI field contains 2
to 30 octets (symbol pairs) whose format and meaning are specified by ISO/
IEC 10038 standard on bridging. The length of the RI field in octets is
contained in the first octet of the RI field.

INFORMATION (INFO) FIELD. The INFO field contains zero, one, or more data
symbol pairs whose meaning is determined by the FC field and whose
interpretation is made by the destination entity (e.g., MAC, LLC, or SMT).
The length of the INFO field is variable, but must conform to both the
maximum frame length criteria given above and to the frame validity criteria
defined in the referenced standard [13].

FRAME CHECK SEQUENCE. The frame check sequence generating polynomial is the same as that used for CSMA/CD, token bus, and token ring (Sections 5.3, 5.4, and 5.5, respectively).

ENDING DELIMITER (ED). The symbol T is the ending delimiter of tokens and frames. Ending delimiters and optional control indicators form a balanced symbol sequence (i.e., be transmitted in pairs so as to maintain octet boundaries). This is accomplished by adding a trailing T symbol as required.

FRAME STATUS (FS). The frame status (FS) field consists of an arbitrary-length sequence of control indicator symbols (R and S) that follows an ending delimiter of a frame. It ends if any symbol other than R and S is received. If an expected indicator is not received as R and S, it is reported as not received. A trailing T symbol, if present, is repeated as part of the FS field. The first three control indicators of the frame status field are mandatory, indicating error detected (E), address recognized (A), and frame copied (C). The use of additional control indicators in the frame status field after the C indicator is optional and is implementer-defined. Although the use of optional control indicators in the frame status field is undefined in the ISO/IEC 9314 standard, all conforming FDDI MACs must repeat the entire frame status field.

> *Error-Detected Indicator (E).* The error-detected indicator (E) is transmitted as R by the MAC that originates the frame. All MACs on the ring inspect repeated frames for errors. If an error is detected and the received E indicator is not S, then an error is counted. The E indicator is transmitted as S by a repeating MAC when either an error to be counted is detected, the received E indicator is S, or the E indicator was not received.
>
> *Address-Recognized Indicator (A).* The address-recognized indicator is transmitted as R by the MAC that originates the frame. If another MAC recognizes the destination address as its own individual or group address, it sets the A indicator to S. A MAC does not set the A indicator for a partially filtered group address (i.e., it only sets the A indicator for a precise match). A bridge performing source routing which recognizes a received frame as one it intends to forward by means of the information in the routing information (RI) field sets the A indicator of the repeated frame to S. In all other cases, a repeating MAC transmits this indicator as received.
>
> *Frame-Copied Indicator (C).* The frame-copied indicator is transmitted as R by the MAC that originates the frame. If another MAC has set the A indicator and copied the frame into its receive buffer, it sets the C indicator to S. If a transparent bridge, including optional capability for setting the C indicator, recognizes the destination address as one to be

forwarded, has not set the A indicator, and has copied the frame into its receive buffer for forwarding, it sets the C indicator to S if the A indicator is received as R. If a MAC, including optional capability for clearing the C indicator, has set the A indicator and has not copied the frame into its receive buffer, it transmits the C indicator as R if the received A indicator is R [13].

5.6.4 FDDI Timers. Each MAC maintains three timers to regulate the operation of the ring. These timers are: token-holding timer (THT), valid-transmission timer (TVX), and token-rotation timer (TRT).

The token-holding timer controls how long the MAC may transmit asynchronous frames. The valid-transmission timer is used to recover from transient ring error situations. The token-rotation timer is used to control ring scheduling during normal operation and to detect and recover from serious ring error situations.

5.6.5 FDDI Operation. Access to the physical medium (the ring) is controlled by passing a token around the ring. The token gives the downstream MAC (receiving relative to the MAC passing the token) the opportunity to transmit a frame or a sequence of frames. If a MAC wants to transmit, it strips the token from the ring before the frame control field of the token is repeated. After the token is completely received, the MAC begins transmitting its eligible queued frames. After transmission, the MAC issues a new token for use by a downstream MAC.

MACs that are not transmitting repeat the incoming symbol stream. While repeating the incoming symbol stream, the MAC determines whether frames are intended for this MAC. This is done by matching the DA to its own address or a relevant group address. If a match occurs, the frame is processed by the MAC or sent to SMT or LLC.

FRAME TRANSMISSION. Upon request for service data unit (SDU) transmission, the MAC constructs the PDU or frame from the SDU by placing the SDU in the INFO field of the frame. The SDU remains queued by the requesting entity awaiting the receipt of a token that may be used to transmit it. Upon reception and capture of an appropriate token, the MAC begins transmitting its queued frame(s) in accordance with the rules of token holding. During transmission, the FCS for each frame is generated and appended to the end of the PDU.

After transmission of the frame(s) is completed, the MAC immediately transmits a new token.

FRAME STRIPPING. Each transmitting station is responsible for stripping from the ring the frames that it originated. A MAC strips each frame that it transmits beginning not later than the seventh symbol after the end of the SA field. Normally, this is accomplished by stripping the remainder of each frame whose source address matches the MAC's address from the ring and replacing it with idle symbols.

The process of stripping leaves remnants of frames, consisting at most of PA, SD, FC, DA, and SA and six symbols after the SA field, followed by idle symbols. These remnants exist because the decision to strip a frame is normally based upon recognition of the MAC's address in the SA field, which cannot occur until after the initial part of the frame has already been repeated. These remnants are not recognized as frames because they lack an ending delimiter (ED). The limit of remnant length also prevents remnants from satisfying the minimum frame length criteria. To the level of accuracy required for statistical purposes, they can be distinguished from errored or lost frames because they are always followed by the idle symbol. Remnants are removed from the ring when they encounter a transmitting MAC. Remnants may also be removed by the smoothing function of PHY.

RING SCHEDULING. Transmission of normal PDUs (i.e., PDUs formed from SDUs) on the ring is controlled by a timed token-rotation protocol. This protocol supports two major classes of service:

(a) synchronous—guaranteed bandwidth and response time

(b) asynchronous—dynamic bandwidth sharing

The synchronous class of service is used for those applications whose bandwidth and response limits are predictable in advance, permitting them to be preallocated (via the SMT). The asynchronous class of service is used for those applications whose bandwidth requirements are less critical (e.g., bursty or potentially unlimited) or whose response time requirements are less critical. Asynchronous bandwidth in FDDI is instantaneously allocated from the pool of remaining bandwidth that is unallocated, unused, or both.

5.6.6   Clocking. A local clock is used to synchronize both the internal operation of PHY and its interface to the data-link layer. This clock is derived from a fixed frequency reference. This reference may be created internally within the PHY implementation or supplied to PHY. (A crystal oscillator may be used for this purpose.)

Characteristics of the local clock are as follows:

(a) Nominal symbol time (UI) = 40 ns (1/UI = 25 MHz)

(b) Nominal code bit cell time (UI) = 8 ns (1/UI = 125 MHz)

(c) Frequency accuracy $< \pm 0.005\%$ (± 50 ppm)

(d) Harmonic content (above 125.02 MHz) < -20 dB

(e) Phase jitter (above 20 kHz) $< \pm 8°$ (0.044 UI pp)

(f) Phase jitter (below 20 kHz in hybrid mode) $< \pm 270°$ (1.5 UI pp)

The receive function derives a clock by recovering the timing information from the incoming serial bit stream. This clock is locked in frequency and phase to the transmit clock of the upstream node. The maximum difference between the received bit frequency and the local bit frequency is 0.01% of the nominal frequency. The received frequency can be either slower or faster than the local frequency, resulting in an excess or a deficiency of bits unless some compensation is included. The elasticity buffer function provides this compensation by adding or dropping idle bits in the preamble between DLL PDUs.

The operation of the elasticity buffer function produces variations in the lengths of the preambles between DLL PDUs as they circulate around the logical ring. The cumulative effect on preamble size of PDU propagation through many elasticity buffers can result in excessive preamble erosion and, in hybrid mode, excessive cycle clock jitter. The smoothing function serves to filter out these undesirable effects.

Section 5.6 is based on ANSI X3.231–1994 [12] and ANSI X3.239–1994 [13]. The figures and tables used in Section 5.6 derive directly from these ANSI referenced standards.

5.7 LAN Performance

For this discussion, LANs can be broken down into two categories:

1. CSMA/CD approach with its random access. Unpredictable performance.
2. Token passing. Predictable performance.

CSMA/CD. There are far more CSMA/CD (Ethernet type) LANs in operation than any other type. There are several reasons why this is so. First, CSMA/CD is very economic. Second, it is simple to implement. There is no token monitor or token monitor requirement at each station.

Operational Advantages

A. Once a station is sending packets or frames, it can continue to send them until there are no others to send. There is no time limit as to usage.

B. There is no waiting time once the bus is free. A station can transmit frames or packets as soon as it has an opportunity.

C. Performance for any station is independent from the total number of stations on the bus. Only those stations attempting to transmit at the same time or nearly the same time have any impact on performance.

D. No protocol interchanges are needed. Stations do not need to exchange messages with each other in order to manage access to the bus. This reduces overhead and simplifies station design.

Operational Disadvantages

A. The most serious drawback to CSMA/CD-type LANs is the activity factor versus collision problem. Performance starts to degrade as the load grows. As collisions increase, throughput degrades. It is argued at what point this performance degradation really starts to take hold. One school of thought suggests a point around 25%, whereas another suggests that between 50% and 60% loading is the point where trouble starts [14].

B. There is an inability to incorporate priority traffic for voice or real-time needs. There is no guaranteed access to the bus.

C. With a maximum extension (with repeaters) of 2500 meters, the maximum number of stations that can be accommodated is about 1000. We doubt that 1000 users could be accommodated operationally with any small usage factor at all.

Note: One way this situation can be eased is by segmenting the LAN into user families, where each family has a high community of interest amongst its members. The separation is done physically/operationally using bridges or possibly routers.

TOKEN PASSING. Token-passing schemes have a controlled access. We can readily calculate the performance of a token-passing-type LAN and access can be guaranteed under any load condition.

Operational Advantages

A. Under heavy traffic loading, each station will still receive a fixed level of service. Throughput stabilizes.

B. There is a means of handling traffic with varying priorities.

Operational Disadvantages

A. There is a limitation on the traffic a station can transmit before it must pass the token and wait for its next turn.

B. Token-passing protocols require an exchange of messages—at least the token. This adds to overhead and time consumed in non-revenue-bearing traffic.

C. All stations must be able to act as the token monitor, requiring a replication of function and cost.

D. Any station's performance is a function of the total number of active stations regardless of activity. The more stations, the more each station has to wait for the token.

E. Once a station passes the token, it must wait a minimum time before it can transmit more traffic.

Section 5.7 is based on Ref. 14.

5.8 LAN Internetworking via Spanning Devices

In this section we will briefly describe repeaters, bridges, routers, hubs, and switching hubs.

5.8.1 Repeaters. A repeater is nothing more than a regenerative repeater. It extends a LAN. It does not provide any kind of segmentation of a LAN, except the physical regeneration of the signal. Multiple LANs (with common protocols) can be interconnected with repeaters, in effect making just one large segment. A network using repeaters must avoid multiple paths, as any kind of loop would cause data traffic to circulate indefinitely and could ultimately make the network crash. This concept is shown in Figure 12.24.

The following example shows how a loop can be formed. Suppose two repeaters connect CSMA/CD LAN segments as shown in Figure 12.24. Station #1 initiates an interchange with station #3, both on the same segment (upper in the figure). As data packets or frames are transmitted on the upper

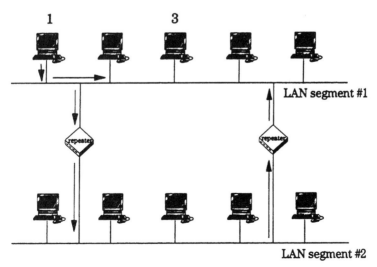

Figure 12.24. Repeaters in multiple paths. Courtesy of Hewlett-Packard Company [15].

segment, each repeater will transmit them unnecessarily to the lower segment. Each repeater will receive the repeated packet on the lower segment and retransmit it once again on to the upper segment. As one can see, any traffic introduced into this network will circulate indefinitely around the loop created by the two repeaters. On larger networks the effects can be devastating, although perhaps less apparent [15].

5.8.2   LAN Bridges.   Whereas repeaters have no intelligence, bridges do. Bridges can connect two LANs at the data-link or MAC protocol level. There are several varieties of bridges depending on the intelligence incorporated.

There is the transparent bridge. It builds a list of nodes it sees transmitting on either side. It isolates traffic. It will not forward traffic that it knows is destined to another station on the same side of the bridge as the sending station. It is able to isolate traffic according to the MAC source and destination address(es) of each individual data frame. MAC-level broadcasts, however, are propagated through the network by the bridges. A bridge can be used for segmenting and extending LAN coverage. Thus it lowers traffic volume for each segment.

There are four varieties of bridges. We covered the transparent bridge. It does not modify any part of a message that it forwards.

The second bridge is the translation bridge. It is used to connect two dissimilar LANS, such as a token ring to CSMA/CD. In order to do this it must modify the MAC-level header and FCS of each frame it forwards in order to make it compatible with the receiving LAN segment. The MAC addresses and the rest of the data frame are unchanged. Translation bridges are far less common than transparent bridges.

The third type of bridge, as shown in Figure 12.25, is the encapsulation bridge. It is also used to connect LANs of dissimilar protocols. But rather than translate MAC header and FCS fields, it simply appends a second MAC layer protocol around the original frame for transport over the intermediate LAN with a different protocol. There is a destination bridge which strips off this additional layer and extracts the original frame for delivery to the destination network segment [15].

The fourth type of bridge is a source routing bridge. It is commonly used in token ring networks. With source routing bridges, each frame carries within it a route identifier (RI) field which specifies the path which that frame is to take through the network. This concept is also shown in Figure 12.25.

Up to this point we have been discussing local bridges. A local bridge spans LANs in the same geographic location. A remote bridge spans LANs in different geographic locations. In this case, an intervening WAN is required. The remote bridge consists of two separate devices that are connected by a wide area network affording transport of data frames between the two. This concept is shown in Figure 12.26.

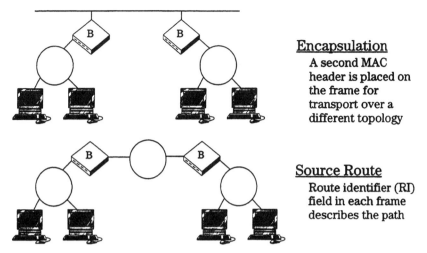

Encapsulation

A second MAC header is placed on the frame for transport over a different topology

Source Route

Route identifier (RI) field in each frame describes the path

Figure 12.25. The concept of bridging. **Top:** Encapsulation bridge. **Bottom:** Source routing bridge. Courtesy of Hewlett-Packard Company [15].

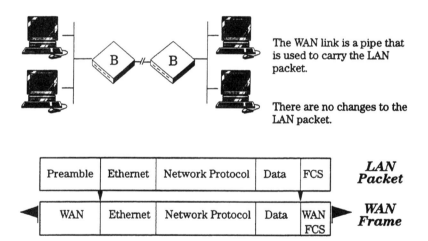

The WAN link is a pipe that is used to carry the LAN packet.

There are no changes to the LAN packet.

Preamble	Ethernet	Network Protocol	Data	FCS	*LAN Packet*
WAN	Ethernet	Network Protocol	Data	WAN FCS	*WAN Frame*

Figure 12.26. The concept of remote bridging. The LAN frame/packet is encapsulated in a WAN frame. Courtesy of Hewlett-Packard Company [15].

As shown in Figure 12.26, the LAN data packet/frame is encapsulated by the remote bridge adding the appropriate WAN header and trailer. The WAN transports the data packet/frame to the distant-end remote bridge which strips the WAN header and trailer, and it delivers the data packet/frame to the far-end LAN.

Remote bridges typically use proprietary protocols such that, in most cases, remote bridges from different vendors do not interoperate.

Bridges are good devices to segment LANS, particularly CSMA/CD LANs. Segmenting breaks a LAN up into user families. It is expected that there is a high community of interest amongst members of a family, but a low community of interest amongst different families. There will be large traffic volumes intra-segment and low traffic volumes inter-segment. It should be pointed out that routers are more efficient at segmenting than bridges [15].

A major limitation of bridges is the inability to balance traffic across two or more redundant routes in a network. The existence of multiple paths in a bridged network can prove to be a bad problem. In such a case, we are again faced with the endless route situation as we were with repeaters. One way to avoid the problem is to use the *spanning tree algorithm*. This algorithm is implemented by having bridges communicate with each other to establish a subset of the actual network topology that is loop-free (often called a *tree*). The idea, of course, is to eliminate duplicate paths connecting one LAN to another, or one segment to another. If there is only one path from one LAN to another, there can be no loop formed [15].

5.8.3 Routers.

Routers carry more intelligence than bridges. Like a bridge, a router forwards data packets/frames. Routers make forwarding decisions based on the destination network layer address. Whereas a bridge worked on the data-link layer, a router operates at the network layer level. Routers commonly connect disparate LANs such as CSMA/CD to token ring and FDDI to CSMA/CD.

Routers are addressable nodes in a network. They carry their own MAC address(es) as well as a network address for each protocol handled. Because routers are addressable, a station desiring the facility of a router must direct its packets/frames to the router in question so that the traffic can be forwarded to the appropriate network. As one would expect, networking software at each station is more complex with a network using routers than one using bridges.

Routers handle only traffic addressed to them. They make decisions about forwarding data packets/frames based on one or several criteria. The decisions may be based on the cost of the link, the number of hops on each path, and the time-to-live.

Routers change packets/frames that pass through them such as MAC source and destination address; they may also modify the network protocol header of each frame (typically decrementing the time-to-live in the case of IP and other protocol fields).

Because routers have more intelligence than bridges, routers will typically have better network management agents installed. This enables them to be remotely configured, to be programmed to pass or not to pass data for security purposes, and to be monitored for performance, particularly error performance. Due to the additional processing performed at routers, they tend to be slower than bridges. Reference 15 suggests that some protocols do not lend themselves to routing, such as DEC LAT, SNA, and NetBIOS among others.

5.8.4 *Hubs and Switching Hubs.* A hub is a multiport device that allows centralization. A hub is usually mounted in a wiring closet or other central location. Signal leads are brought in from workstations/PCs and other data devices, one for each hub port. Physical rings or buses are formed by internally configuring the hub ports. A typical hub may have 8 or 16 ports. Suppose we wished to incorporate 24 devices on our LAN using the hub. We can stack two hubs, one on top of the other (stackables), using one of the hub ports on each for interconnection. In this case we would have a hub with 30 port capacity (32 − 2). Hubs may also have a certain amount of intelligence, such as incorporation of a network management capability. Each hub can incorporate a repeater. However, for CSMA/CD (Ethernet) there is a 4-repeater limit (i.e., a maximum of five 500-m segments). With FDDI and token ring the repeater limit rule does not apply.

There are "intelligent" hubs. These are typically modular, multiprotocol, multimedia, multichannel, fault-tolerant, manageable devices where one can concentrate all the LAN connections into a wiring closet or data center. Since these type hubs are modular (i.e., it has various numbers of slots to install LAN interface boards), it can support CSMA/CD, token ring, FDDI, or ATM (ATM is covered in Chapter 15) simultaneously as well as various transmission media such as twisted pair, fiber cable, and others.

Switching hubs are high-speed interconnecting devices with still more intelligence than the garden variety hub or the intelligent hub. They typically interconnect entire LAN segments and nodes. Full LAN data rate is provided at each port of a switching hub. They are commonly used on CSMA/CD LANs, providing a node with the entire 10-Mbps data rate. Because of a hub's low latency, high data rates and throughputs are achieved.

With a switching hub, nodes are interconnected within the hub itself using its high-speed backplane. As a result, the only place the entire aggregate LAN traffic appears is on that backplane. Traffic between ports on a single card does not even appear on the backplane [15–17].

REVIEW QUESTIONS

1. Define a local area network. Contrast it with a wide area network.

2. Discuss how LANs can make multivendor processing equipment compatible.

3. What are the two basic underlying transmission techniques used for LANs? Compare these using a minimum of eight characteristics.

4. Name the three basic LAN topologies (i.e., network types). Name a fourth type that is a subset of one of the basic topologies.

5. What are the general ranges of bit error rates that can be expected from a well-designed LAN? How is the BER achieved (e.g., by ARQ, channel coding, S/N ratio)?

6. What are the two basic transmission problems that must be faced regarding the medium when designing a LAN?

7. If a LAN has 50 accesses, how many transmission connectivities must be analyzed?

8. Consider conventional broadband LANs. Discuss one- and two-cable operation. Include bandwidth utilization.

9. What is the function of a CRF or head end on a broadband LAN?

10. There are two methods of connecting user facilities to optical fiber LANs. What are they? Discuss the pros and cons of each.

11. What is a LAN access protocol?

12. Compare contention and polling. Describe how each leads to a particular LAN access protocol, such as token passing and CSMA/CD.

13. Why is loop polling more efficient than roll-call polling?

14. Relate the IEEE 802 LAN standards to the seven-layer OSI model. Describe the sublayering involved.

15. What is a SAP, and what is its function?

16. LLC derives from what familiar link-layer protocol?

17. Describe the three basic communication services offered by the LLC. In the description show similarities and differences among the three.

18. Name at least three responsibilities of the LLC.

19. Describe the four fields of the LLC PDU and the function of each field.

20. How are collisions detected on CSMA/CD?

21. When a collision occurs on a CSMA/CD LAN, what happens? This should lead to a discussion of persistence algorithms and their relative efficiencies.

22. Given a 500-m length of coaxial cable with access at each extreme, in what time period can a collision occur after station 1 transmits? Assume a velocity of propagation of 2×10^8 m/s.

23. What is the function of a frame check sequence (FCS)?

24. What is the purpose of the "jam" signal on CSMA/CD?

25. Give three reasons why a CSMA/CD MAC frame may be invalid.

26. In CSMA/CD operation, what is the function of the PAD field? Why is it necessary to have a minimum frame length?

27. Describe differential Manchester coding and its application to baseband LAN transmission. Compare it to conventional NRZ.

28. Define *slot time* regarding CSMA/CD. Why is it an important parameter?

29. How are collisions avoided using a token-passing scheme?

30. The MAC handles fault recovery with the token bus access scheme. Name at least four faults that can be handled by the MAC in this case.

31. Define *slot time* for token bus.

32. How does a LAN user know that a traffic frame is directed to it?

33. Describe how a token holder handles a situation when its successor does not answer a token passed to it.

34. Differentiate between logical connectivity and physical connectivity. On which of the four LAN types (discussed in this chapter) can logical connectivity be used effectively?

35. In the token bus frame format we see the data field. For a "data unit frame," what would be inserted in this field? Be specific.

36. Following the IEEE standards, is token bus truly a baseband system as we have defined it?

37. How does token ring differ from token bus? Name at least three differences.

38. Aside from the token, what are the three types of frames used on token ring?

39. Is token ring, as described, really a baseband system?

40. Define *latency* with respect to the token ring LAN.

41. How is the "I" used in the token ring ending delimiter? It is just one bit.

42. What is the baud rate of FDDI? The bit rate?

43. For both token ring operation and FDDI, name one simple way a transmitting station can assume that its frame has been received by the intended destination station.

44. What is *frame stripping*?

45. Regarding timing, what is the function of the elasticity buffer in FDDI?

46. Compare the performance of token-passing schemes versus CSMA/CD schemes under light traffic loading and under heavy traffic loading conditions.

47. What is the function of LAN repeaters?

48. Four types of bridges were discussed in the text. Name three of them, and with one sentence, describe each of the three.

49. Discuss routers in light of bridges.

50. What is the function of a hub? An "intelligent" hub? A switching hub?

REFERENCES

1. J. F. Schoch, Y. K. Delai, D. D. Redell, and R. C. Crane, "Evolution of the Ethernet Local Computer Network," *Computer* (August 1982).

2. R. N. Dunbar, "Design Considerations for Broadband Coaxial Cable Systems," *IEEE Commun. Mag.*, **24**(6) (June 1986).

3. W. Stallings, *Handbook of Computer-Communications Standards*, Vol. 2, Local Area Networks, Macmillan, New York, 1987.

4. *Carrier Sense Multiple Access with Collision Detection (CSMA/CD) Access Method and Physical Layer Specifications*, ANSI/IEEE Std 802.3, 3rd ed., IEEE, New York, March 1992. (ISO/IEC 8802-3).

5. *Broadband Local Area Networks*, IEEE Std 802.7, IEEE, New York, 1989.

6. *Logical Link Control*, ANSI/IEEE Std 802.3, ISO 8802-2, 1st ed., IEEE, New York, December, 1989.

7. *The New IEEE Standard Dictionary of Electrical and Electronics Terms*, 5th ed., IEEE Std 100-1992, IEEE, New York, 1993.

8. *Token-Passing Bus Access Method and Physical Layer Specifications*, ANSI/IEEE Std 802.4, ISO/IEC 8802-4, 1st ed., IEEE, New York, August 1990.

9. *Token Ring Access Method*, IEEE Std 802.5-1989, IEEE, New York, October 1991.

10. *Recommended Practice for Dual Ring Operation with Wrapback Reconfiguration*, IEEE Std. 802-5c-1991, IEEE, New York, March 1991.

11. *Fiber Data Distributed Interface (FDDI)—Token Ring Physical Layer Medium Dependent (PMD)*, ANSI X3.166-1990, ANSI, New York, 1990.

12. *Fiber Distributed Data Interface (FDDI)—Physical Layer Protocol (PHY-2)*, ANSI X3.231-1994, ANSI, New York, 1994.

13. *Fiber Distributed Data Interface (FDDI): Token Ring Media Access Control—2 (MAC-2)*, ANSI X.239-1994, ANSI, New York, 1994.

14. J. McDonnel, "Internetworking and Advanced Protocols," a seminar, Network Technologies Group, Inc., Boulder, CO, 1985.

15. "Internetwork Troubleshooting Seminar Presentation," Hewlett-Packard Company, Tempe, AZ, January 1995.

16. ChipCom promotional material 1995, ChipCom, Southborough, MA, February 1995.

17. *High Speed Networking—Options and Implications*, ChipCom, Southborough, MA, 1995.

13

INTEGRATED SERVICES DIGITAL NETWORKS

1 BACKGROUND AND GOALS OF ISDN

The analog public switched telecommunications network (PSTN) was based on the 4-kHz voice (VF) channel. It served well in providing speech telephony since the 1880s. In the 19th century the only other service was telegraph, which predated the telephone by some 35 years. The two services evolved separately and distinctly. In some respects they were competitive. Before World War II there was some melding where telegraph and telex were carried as subcarriers on PSTN telephone channels leased by telephone companies or administrations. This might be viewed as the first move toward integrated services. However, it was probably done more for convenience and economy than for any forward thinking regarding integration.

Looking backward over that period, telephony became ubiquitous, with a telephone in every office and in nearly every home. On the other hand, telegraphy evolved into telex (a switched telegraph service), but still took a back seat to telephony. Historically, facsimile was the next service that was integrated rapidly into the telephone network. Facsimile (fax) required a modem to make it compatible with the analog telephone channel. In the office environment, facsimile has almost completely replaced telex for the transmission of record traffic. Then, in the 1950s computer-related data began to emerge, requiring some method of point-to-point relay. This relay facility was carried out again by the ubiquitous telephone network. Once more, a modem was required to integrate the service into the analog telephone network.

By this time the worldwide telephone network was in place and pervasive. Using that network turned out to be a cost-effective method to communicate other information (i.e., other than speech telephony) from point X to point

665

Y. Dial-up telephone connections provided one way of achieving switched service to transport that "other" information on a point-to-point basis, given the transmission limits of the analog VF channel traversing the PSTN.

Digital telephony began to take hold after the discovery, and later development, of the transistor in 1948.* Solid-state circuitry, particularly LSI, made pulse-code modulation (PCM) transmission and later PCM switching cost-effective. The first application of PCM was in the expansion of the trunk cable plant. In the United States and Canada the entire long-distance plant is digital. We expect that the local switching and trunk plant will be all-digital before the year 2000, and the remainder of the world perhaps around 2010.

The present digital network is based on PCM as described in Chapter 8. PCM standards developed along two, or some could argue three, distinct routes. North America and Japan use a basic 24-channel system with in-channel signaling and added-bit framing. Europe and much of the remainder of the world use a 30-channel format with 2 extra channels for separate channel signaling and synchronization/framing. Japan has a distinct higher-level PCM hierarchy.

PCM was designed to serve speech telephony. Even today more than 80% of the traffic on the PSTN is voice telephony. Voice traffic optimization of PCM design is even more evident in North America. In-channel signaling and framing corrupted the basic 64-kbps channel, and thus integrating other services such as computer data required a drop-back to 56 kbps for North American PCM.† Typical of such a drop-back is "switched 56" and AT&T DDS (digital data system).

ISDN has been developed to ease integration of all services, except full motion video. It is based on the 64-kbps channel, variously called DS0 or E0, depending on the standard followed (i.e., European or North American). Whereas 4 kHz was the basic building block of analog telephony, 64 kbps is the basic building block of the digital network and ISDN. The ISDN basic building block is designed to serve, among other services [1]:

- Digital voice
- 64-kbps data, both circuit- and packet-switched
- Telex/teletext
- Facsimile
- Slow-scan video

The goal of ISDN is to provide an integrated facility to incorporate each of the services listed above on a common 64-kbps channel. This chapter provides an overview of that integration, including transmission, switching, and signaling.

---

*The basis of modern digital (PCM) telephony was laid out in 1937. It was not until the development of the IC (integrated circuit) that PCM became feasible.
†64-kbps clear-channel service is available in North America with special conditioning.

As the reader proceeds through this chapter, he/she should be aware that we are dealing with user interfaces into an *existing* digital network (Chapter 10) in which CCITT Signaling System No. 7 (Chapter 16) is operational. In this chapter we are not dealing with the digital network itself, only how it affects the user and how the ISDN user affects the network. We will also discuss the two flavors of ISDN: CCITT and North American.

2 ISDN STRUCTURES

2.1 ISDN User Channels

Here we look from the user into the network. We consider two user classes: residential and commercial. The following are the standard bit rates for user access links:

- B-channel: 64 kbps
- D-channel: 16 or 64 kbps
- H-channels (discussed below)

The B-channel is the basic user channel. It is transparent to bit sequences. It serves all the traffic types listed in Section 1.

In one configuration, called the *basic rate*, the D-channel has a 16-kbps data rate; in another, called the *primary rate*, it is 64 kbps. Its primary use is for signaling. The 16-kbps version, besides signaling, may serve as transport for low-speed data applications, particularly those using X.25 packet data (see Chapter 11).

There are a number of H-channels:

- H_0 channel: 384 kbps
- H_1 channels: 1536 kbps (H_{11}) and 1920 kbps (H_{12})

The H-channel is intended to carry a variety of user information streams. A distinguishing characteristic is that an H-channel does not carry signaling information for circuit switching by the ISDN. User information streams may be carried on a dedicated, alternate (within one call or separate calls), or simultaneous basis, consistent with the H-channel bit rates. The following are examples of user information streams:

- Fast facsimile
- Video, such as video conferencing
- High-speed data
- High-quality audio or sound program channel

- Information streams, each at rates lower than the respective H-channel bit rate (e.g., 64-kbps voice), which have been rate-adapted or multiplexed together

- Packet-switched information

Section 2.1 is based on Ref. 1.

2.2 Basic and Primary User Interfaces

The *basic* rate interface structure is composed of two B-channels and a D-channel referred to as "2B + D." The D-channel at this interface is 16 kbps. The B-channels may be used independently (i.e., two different simultaneous connections). Industry and much of the literature call the basic rate interface the *BRI*.

Appendix I to CCITT Rec. I.412 [1] states that alternatively the basic access may be just one B-channel and a D-channel or just a D-channel.

The *primary* rate interface (PRI) structures are composed of n B-channels and one D-channel, where the D-channel in this case is 64 kbps. There are two primary data rates:

- 1.544 Mbps = 23B + D (from the North American T1 configuration)
- 2.048 Mbps = 30B + D (from the European E1 configuration)

For the user–network access arrangement containing multiple interfaces, it is possible for the D-channel in one structure not only to serve the signaling requirements of its own structure but also to serve another primary rate structure without an activated D-channel. When a D-channel is not activated, the designated time slot may or may not be used to provide an additional B-channel, depending on the situation, such as 24B with 1.544 Mbps.

The primary rate interface H_0-channel structures are composed of H_0 channels with or without a D-channel. When present in the same interface structure the bit rate of the D-channel is 64 kbps.

At the 1544-kbps primary rate interface, the H_0-channel structures are $4H_0$ and $3H_0 + D$. When the D-channel is not provided, signaling for the H_0-channels is provided by the D-channel in another interface.

At the 2048-kbps primary rate interface, the H_0 structure is $5H_0 + D$. In the case of a user–network access arrangement containing multiple interfaces, it is possible for the D-channel in one structure to carry the signaling for H_0-channels in another primary rate interface without a D-channel in use.

The 1536-kbps H_{11}-channel structure is composed of one 1536-kbps H_{11}-channel. Signaling for the H_{11}-channel, if required, is carried on the D-channel of another interface structure within the same user–network access arrangement.

The 1920-kbps H_{12} structure is composed of one 1920-kbps H_{12}-channel and a D-channel. The bit rate of the D-channel is 64 kbps. Signaling for the H_{12}-channel, if required, is carried in this D-channel or the D-channel of another interface structure within the same user–network access arrangement.

3 USER ACCESS AND INTERFACE

3.1 General

The objective of the ISDN designers was to provide a telecommunications service which would be ubiquitous and universal. Whether this ambitious goal is being met can be argued. ISDN transmission rates were to be accommodated on copper wire pairs. Within the next 20 years most of the copper plant will be replaced by fiber optics. Some futurists argue 10 years and others 30 years. Nevertheless, enterprise networks now demand transmission rates in excess of what ISDN has been designed for. The demand has been essentially to provide LAN connectivity over WAN distances. Frame relay (Chapter 9) has removed some of this demand pressure.

Figure 13.1 shows generic ISDN user connectivity to the network. We can select either the basic or primary rate service (e.g., 2B + D, 23B + D, or 30B + D) to connect to the ISDN network. The objectives of any digital interface design, and specifically of ISDN access and interface, are as follows:

1. Electrical and mechanical specification

2. Channel structure and access capabilities

3. User–network protocols

4. Maintenance and operation

5. Performance

6. Services

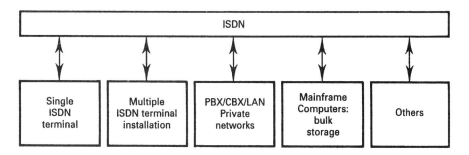

Figure 13.1. ISDN generic users.

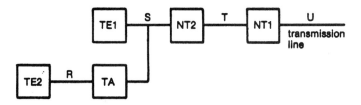

Figure 13.2. ISDN reference model.

ISDN specifications as set out by the ITU-T I. Recommendations and relevant Bellcore/ANSI specifications cover these six items listed above.

Figure 13.2 shows the conventional ISDN reference model. It delineates interface points for the user. In the figure NT1, or network termination 1, provides the physical layer interface; it is essentially equivalent to OSI layer 1. The functions of the physical layer include:

- Transmission facility termination
- Layer 1 maintenance functions and performance monitoring
- Timing
- Power transfer
- Layer 1 multiplexing
- Interface termination, including multidrop termination employing layer 1 contention resolution

Network termination 2 (NT2) can be broadly associated with OSI layers 1, 2, and 3. Among the examples of equipment that provide NT2 functions are user controllers, servers, LANs, and PABXs. Among the NT2 functions are:

- Layers 1, 2, and 3 protocol processing
- Multiplexing (layers 2 and 3)
- Switching
- Concentration
- Interface termination and other layer 1 functions
- Maintenance functions

A distinction must be drawn here between North American and European practice. Historically, telecommunication administrations in Europe have been, in general, national monopolies that are government-controlled. In North America (i.e., United States and Canada) they are private enterprises, often very competitive. Thus in Europe, NT1 is considered as part of the digital network and belongs to the telecommunications administration. The customer ISDN equipment starts at the T interface. In North America, both

NT1 and NT2 belong to the ISDN user, and the U interface defines the network entry point.

It should be noted that there is an overall trend outside of North America to privatize telecommunications such as has happened in the United Kingdom and is scheduled to take place in other countries such as Germany, Mexico, and Venezuela.

TE1 in Figure 13.2 is the terminal equipment and has an interface that complies with ISDN terminal–network interface specifications at the S interface. We will call this equipment *ISDN compatible*. TE1 covers functions broadly belonging to OSI layer 1 and higher OSI layers. Among the equipment are digital telephones, computer workstations, and other devices in the user end-equipment category that are ISDN compatible.

TE2 in Figure 13.2 refers to equipment that does *not* meet the ISDN terminal–network interface at point S. TE2 adapts the equipment to meet the ISDN terminal–network interface. This process is assisted by TA, the terminal adapter.

Reference points T, S, and R are used to identify the interface available at those points. T and S are identical electrically, mechanically, and from the point of view of protocol. Point R relates to the TA interface or, in essence, it is the interface of that nonstandard (i.e., non-ISDN) device. The U-interface is peculiar to the North American version of ISDN.

We will return to user–network interfaces once the stage is set for ISDN protocols looking into the network from the user.

4 ISDN PROTOCOLS AND PROTOCOL ISSUES

When fully implemented, ISDN will provide both circuit and packet switching. For the circuit-switching case, now fairly broadly installed in North America,* the B-channel is fully transparent to the network, permitting the user to utilize any protocol or bit sequence so long as there is end-to-end agreement on the protocol utilized. Of course, the protocol itself should be transparent to bit sequences.

It is the D-channel that carries the circuit-switching control function for its related B-channels. Whether it is the 16-kbps D-channel associated with BRI or the 64-kbps D-channel associated with PRI, it is that channel which transports the signaling information from the user's ISDN terminal from NT to the first serving telephone exchange of the telephone company or administration. Here the D-channel signaling information is converted over to CCITT No. 7 signaling data employing ISUP (ISDN User Part) of SS No. 7. Thus it is the D-channel's responsibility for call establishment (setup), supervision, termination (takedown), and all other functions dealing with network access and signaling control.

Broadly installed means that most of the public carriers can offer the service to their customers.

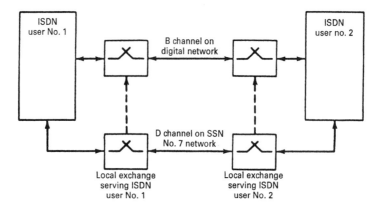

Figure 13.3. Simplified concept of ISDN switching.

The B-channel in the case of circuit switching is serviced by NT1 or NT2 using OSI layer 1 functions only. The D-channel carries out OSI layers 1, 2, and 3 functions such that the B-channel protocol established by a family of ISDN end users will generally make layer 3 null in the B-channel where the networking function is carried out by the associated D-channel.

With packet switching two possibilities emerge. The first basically relies on the B-channel to carry out OSI layers 1, 2, and 3 functions at separate packet-switching facilities (PSFs). The D-channel is used to set up the connection to the local switching exchange at each end of the connection. This type of packet-switched offering provides 64-kbps service. The second method utilizes the D-channel exclusively for lower data rate packet-switched service where the local interface can act as a CCITT (ITU-T) X.25 data communication equipment (DCE) device.

Figure 13.3 is a simplified conceptual diagram of ISDN circuit switching. It shows the B-channel riding on the public digital network and the D-channel, which is used for signaling. Of course, the D-channel is a separate channel. It is converted to a CCITT No. 7 signaling structure and in this form may traverse several signal transfer points (STPs) and may be quasi-associated or disassociated from its companion B-channel(s). Figure 13.4 is a more detailed diagram of the same ISDN circuit-switching concept. The reader should note the following in the figure. (1) Only users at each end have a peer-to-peer relationship available for all seven OSI layers of the B-channel. As the call is routed through the system, there is only layer 1 (physical layer) interaction at each switching node along the call route. (2) The D-channel requires the first three OSI layers for call setup to the local switching center at each end of the circuit. (3) The D-channel signaling data are turned over to CCITT Signaling System No. 7 (SS No. 7) at the near- and far-end local switching centers. (4) SS No. 7 also utilizes the first three OSI layers for circuit establishment, which requires the transfer of control information. In

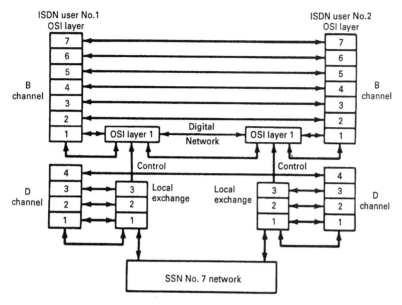

Figure 13.4. Detailed diagram of the ISDN circuit-switching concept.

SS No. 7 terminology, this is called the *message transfer part*. There is a fourth layer called the *user part* in SS No. 7. There are three user parts: telephone user part, data user part, and ISDN user part depending on whether the associated B-channel is in telephone, data, or ISDN service for the user.

Figure 13.5 shows packet service for lower data rates where the D-channel is involved.

5 ISDN NETWORKS

In this context, ISDN networking is seen as a group of access attributes connecting an ISDN user at either end to the local serving exchange (i.e., the local switch). This is shown in Figure 13.6, the basic architectural model of ISDN.

Figure 13.7 shows the ISDN reference configuration of public ISDN connection type. Figure 13.8 illustrates the access connection element model, Figure 13.9 illustrates the national tandem/transit connection element model, and Figure 13.10 shows the private ISDN access connection element.

The national transit network (CCITT terminology) is the public switched digital telephone network. That network, whether DS1–DS4-based or E1–E5-based, provides the two necessary attributes for ISDN compatibility:

1. 64-kbps channelization

2. Separate channel signaling based on CCITT Signaling System No. 7

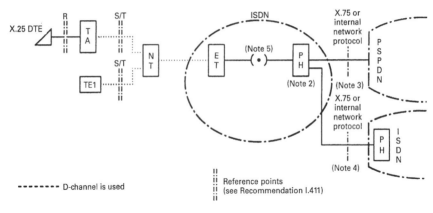

-------- D-channel is used

‖
‖ Reference points
‖ (see Recommendation I.411)
‖

TA   Terminal adaptor
NT   Network termination 2 and/or 1
ET   Exchange termination
TE1 Terminal equipment 1
PH   Packet handling function

NOTES

1 This figure is only an example of many possible configurations and is included as an aid to the text describing the various interface functions.
2 In some implementations, the PH functions logically belonging to the ISDN may reside physically in a node of the PSPDN. The service provided is still the ISDN virtual circuit service.
3 See Recommendation X.325.
4 See Recommendation X.320.
5 This connection is either on demand or semi-permanent, but has no relevance with the user–network procedures. Only internal procedures between the ET and the PH are required.

Figure 13.5. ISDN packet service for lower data rates using the D-channel. From CCITT Rec. X.31, Figure 2/X.31, page 429 (Ref. 2).

Connections from the user at the local connecting exchange interface include:

• Basic service (BRI) 2B + D = 192 kbps (CCITT specified)
                                  160 kbps (North American ISDN 1)
  Both rates include overhead bits.
• Primary service (PRI) 23B + D/30B + D = 1.544/2.048 Mbps

6 ISDN PROTOCOL STRUCTURES

6.1 ISDN and OSI

Figure 13.11 shows the ISDN relationship with OSI. OSI was discussed in Chapter 11. As is seen in the figure, ISDN concerns itself with only the first three OSI layers. OSI layers 4 to 7 are peer-to-peer connections and the end user's responsibility. Remember that the B-channel is concerned with OSI

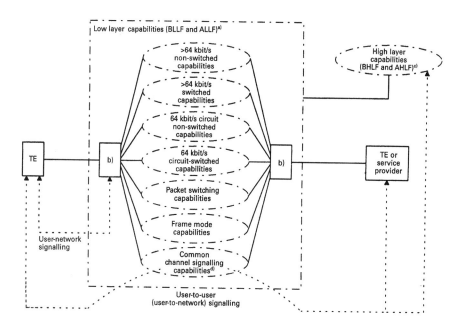

TE   Terminal equipment
BLLF   Basic low layer functions
ALLF   Additional low layer functions
BHLF   Basic high layer functions
AHLF   Additional high layer functions

a) In certain national situations, ALLF may also be implemented outside the ISDN, in special nodes or in certain categories of terminals.
b) The ISDN local functional capabilities correspond to functions provided by a local exchange and possibly including other equipment, such as electronic cross connect equipment, muldexes, etc.
c) These functions may either be implemented within ISDN or be provided by separate networks. Possible applications for basic high layer functions and for additional high layer functions are contained in Recommendation I.210.
d) For signalling between international ISDNs, CCITT Signalling System No. 7 shall be used.

Figure 13.6. The basic architectural model of ISDN. From ITU-T Rec. I.324, Figure 1/I.324, page 3 (Ref. 3).

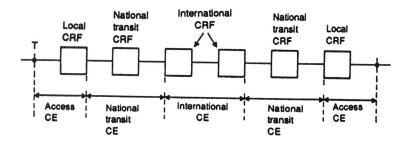

IRP   Internal reference point
CRF   Connection related functions
CE   Connection element

Figure 13.7. Reference configuration of public ISDN connection type. From ITU-T Rec. I.324, Figure 3/I.324, page 8 (Ref. 3).

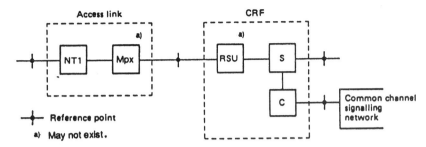

NT1   Network termination 1
S   64 kbit/s circuit switch
C   Signalling handling and exchange control functions
Mpx   (Remote) multiplexer
RSU   Remote switching unit and/or concentrator
CRF   Connection related function

Figure 13.8. Access connection element model. From ITU-T Rec. I.324, Figure 4/I.324, page 9 (Ref. 3).

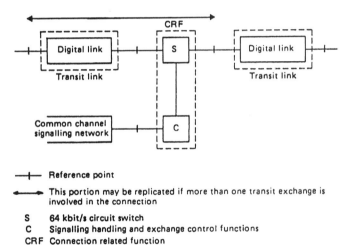

S   64 kbit/s circuit switch
C   Signalling handling and exchange control functions
CRF   Connection related function

Figure 13.9. National tandem/transit element model. From ITU-T Rec. I.324, Figure 5/I:324, page 10 (Ref. 3).

layer 1 only. We showed the one exception; that is, when the B-channel is used for packet service, it will have those first three OSI layers to interface.

The D-channel with its signaling and control functions is the exception to the above statement. The D-channel interfaces with CCITT Signaling System No. 7 at the first serving exchange. D-channels handle three types of information: signaling (s), interactive data (p), and telemetry (t).

The layering of the D-channel has followed the intent of the OSI reference model. The handling of the p and t data can be adapted to the OSI model;

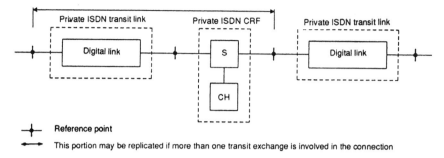

Reference point

This portion may be replicated if more than one transit exchange is involved in the connection

S   64 kbit/s circuit switch
CH   Signalling handling and private ISDN exchange control function
CRF   Connection related function

Figure 13.10. Private ISDN access connection element. From ITU-T Rec. I.324, Figure 8/I.324, page 11 (Ref. 3).

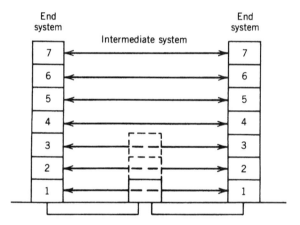

Figure 13.11. A generic communication context showing ISDN's relationship with the seven-layer OSI model. Note that the end system protocol blocks may reside in the subscriber's TE or network exchanges or other ISDN equipment.

the s data, by its very nature, cannot.* Figure 13.12 shows the correspondence between D-channel signaling protocols, SS No. 7 levels, and the OSI seven-layer model.

6.2 Layer 1 Interface, Basic Rate

The S/T interface of the reference model, Figure 13.2 (or layer 1 physical interface), requires a balanced metallic transmission medium (i.e., copper

---

*For CCITT No. 7 signaling system, like any other signaling system, the primary quality-of-service measure is "post dial delay." This is principally the delay in call setup. To reduce the delay time as much as possible, it is incumbent upon system engineers to reduce processing time as much as possible. Thus SS No. 7 truncates OSI to 4 layers, because each additional layer implies more processing time.

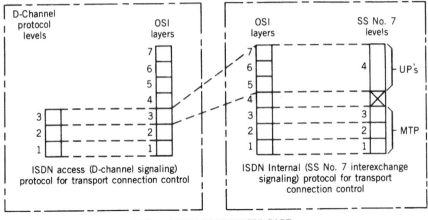

MTP = MESSAGE TRANSFER PART
UP = USER PART

Figure 13.12. Correspondence among the ISDN D-channel, CCITT Signaling System No. 7, and the OSI model. Copyright IEEE, New York, 1985 (Ref. 4).

pair) for each direction of transmission (four-wire) capable of supporting 192 kbps. Again, this is the NT interface of the ISDN reference model.

Layer 1 provides the following services to layer 2 for ISDN operation:

- The transmission capability by means of appropriately encoded bit streams of the B- and D-channels and also any timing and synchronization functions that may be required.

- The signaling capability and the necessary procedure to enable customer terminals and/or network terminating equipment to be deactivated when required and reactivated when required.

- The signaling capability and necessary procedures to allow terminals to gain access to the common resource of the D-channel in an orderly fashion while meeting the performance requirements of the D-channel signaling system.

- The signaling capability and procedures and necessary functions at layer 1 to enable maintenance functions to be performed.

- An indication to higher layers of the status of layer 1.

6.2.1 Primitives between Layer 1 and Other Entities. A *primitive* is data relating to the development or use of software that is employed in developing measures or quantitative descriptions of software. Primitives are directly measurable or countable, or may be given a constant value or condition for a specific measure (IEEE definition [5]).

TABLE 13.1   Primitives Associated with Layer 1

	Specific Name		Parameter		
			Priority	Message	
Generic	Request	Indication	Indicator	Unit	Message Unit Content
L1 ↔ L2					
PH-Data	X[a]	X	X[b]	X	Layer 2 peer-to-peer message
PH-Activate	X	X	—	—	
PH-Deactivate	—	X	—	—	
M ↔ L1					
MPH-Error	—	X	—	X	Type of error or recovery from a previously reported error
MPH-Activate	—	X	—	—	
MPH-Deactivate	X	X	—	—	
MPH-Information	—	X	—	X	Connected/disconnected

[a]PH-Data request implies underlying negotiation between layer 1 and layer 2 for the acceptance of the data.
[b]Priority indication applies only to the request type.
Source: ITU-T Rec. I.430, Table 1/I.430, page 2 (Ref. 6).

In this discussion, primitives represent in an abstract way the logical exchange of information and control between layer 1 and other entities or interfaces.

The primitives to be passed across the boundary between layers 1 and 2 or to the management entity are defined and summarized in Table 13.1. The parameter values associated with these primitives are also summarized in the table. For further information, the reader may consult ITU-T Rec. I.211 [19] which describes the syntax and use of these primitives.

6.2.2   Interface Functions. The S and T functions for the BRI consist of three bit streams that are time-division multiplexed: two 64-kbps B-channels and one 16-kbps D-channel for an aggregate bit rate of 192 kbps. Of this 192 kbps, the $2B+D$ configuration accounts for only 144 kbps. The remaining 48 kbps are overhead bits whose function will be briefly described below.

The functions covered at the interface include bit timing at 192 kbps to enable the TE and NT to recover information from the aggregate bit stream. This octet timing provides 8-kHz octet timing for the NT and TE to recover the time-division multiplexed channels (i.e., $2B+D$ multiplexed). Other functions include D-channel access control, power feeding, deactivation, and activation.

Interchange circuits are required of which there is one in either direction of transmission (i.e., to and from the NT); they are used to transfer digital signals across the interface. All of the functions described above, except for

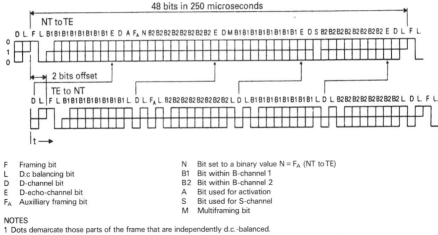

F	Framing bit
L	D.c balancing bit
D	D-channel bit
E	D-echo-channel bit
F_A	Auxilliary framing bit

N	Bit set to a binary value $N = F_A$ (NT to TE)
B1	Bit within B-channel 1
B2	Bit within B-channel 2
A	Bit used for activation
S	Bit used for S-channel
M	Multiframing bit

NOTES

1 Dots demarcate those parts of the frame that are independently d.c.-balanced.

2 The F_a bit in the direction TE to NT is used as a Q bit in every fifth frame if the Q-channel capacility is applied.

3 The nominal 2-bit offset is as seen from the TE. The corresponding offset at the NT may be greater due to delay in the interface cable and varies by configuration.

Figure 13.13. Frame structure at reference points S and T. From ITU-T Rec. I.430, Figure 3/I.430, page 7 (Ref. 6).

power feeding, are carried out by means of a digitally multiplexed signal, described in the next section.

6.2.3 Frame Structure. In both directions the bits are grouped into frames of 48 bits each. The frame structure is identical for all configurations, whether point-to-point or point-to-multipoint. However, the frame structures are different for each direction of transmission. These structures are shown in Figure 13.13 with explanatory notes given in Table 13.2.

TABLE 13.2   Notes on Bit Positions and Groups in Figure 13.13

Bit Position	Group
1 and 2	Framing signal with balance bit
3 to 11	B1-channel (first octet) with balance bit
12 and 13	D-channel bit with balance bit
14 and 15	F_A auxiliary framing bit for Q bit with balance bit
16 to 24	B2-channel (first octet) with balance bit
25 and 26	D-channel bit with balance bit
27 to 35	B1-channel (second octet) with balance bit
36 and 37	D-channel bit with balance bit
38 to 46	B2-channel (second octet) with balance bit
47 and 48	D-channel bit with balance bit

Source: ITU-T Rec. I.430, Table 2/I.430, page 7 (Ref. 6).

Figure 13.14. Pseudoternary line code, example of application.

6.2.4 Line Code. For both directions of transmission, pseudoternary coding is used with 100% pulse width, as shown in Figure 13.14. Coding is performed such that a binary 1 is represented by no line signal, whereas a binary 0 is represented by a positive or negative pulse. The first binary signal following the framing balance bit is the same polarity as the balance bit. Subsequent binary 0s alternate in polarity. A balance bit is a 0 if the number of 0s following the previous balance bit is odd. A balance bit is a binary 1 if the number of 0s following the previous balance bit is even. The balance bits tend to limit the build-up of a DC component on the line.

6.2.5 Timing Considerations. The NT derives its timing from the network clock. A TE synchronizes its bit, octet, and frame timing from the NT, which has derived its timing from the ISDN bit stream being received from the network. The NT uses this derived timing to synchronize its transmitter clock.

6.2.6 BRI Differences in the United States. Bellcore and ANSI prepared ISDN BRI standards fairly well modified from the CCITT I Recommendation counterparts. The various PSTN administrations in the United States are at variance with most other countries. Furthermore, it was Bellcore's intention to produce equipment that was cost effective and marketable and that would easily interface with existing North American telephone plant.

One point, of course, is where the telephone company responsibilities end and customer responsibilities begin. This is called the "U" interface (see Figure 13.2). The tendency toward a two-wire interface rather than a 4-wire interface is another. Line waveform is yet another. Rather than a pseudoternary line waveform, the United States uses 2B1Q. The line bit rate is 160 kbps rather than that recommended by CCITT, namely 192 kbps. More detailed explanation of North American practice follows. The 2B + D + overhead frames differ significantly. The North American version of the BRI is called National ISDN 1. Bellcore uses the generic term DSL for digital subscriber line in its specifying document for National ISDN 1.

DSL BANDWIDTH ALLOCATION AND FRAME STRUCTURE. The DSL, or *digital subscriber line*, modulation rate is 80 kilobauds, equivalent to 160 kbps. The effective 160-kbps signal is divided up into 12 kbps for synchronization words, 144 kbps for 2B + D of customer data, and 4 kbps of DSL overhead.

The synchronization technique used is based on transmission of nine (quaternary) symbols every 1.5 ms, followed by 216 bits of 2B + D data and 6 bits of overhead. The synchronization word provides a robust method of conveying line timing and establishes a 1.5-ms DSL "basic frame" for multiplexing subrate signals. Every eighth synchronization word is inverted (i.e., the 1s become 0s and the 0s become 1s) to provide a boundary for a 12-ms superframe composed of eight basic frames. This 12-ms interval defines an appropriate block of customer data for performance monitoring and permits a more efficient suballocation of the overhead bits among various operational functions.

The 48 overhead bits, sometimes referred to as "maintenance" or *M-bits*, available per superframe are allocated into 24 bits for a 2-kbps *eoc*, 12 bits for a *crc* check covering the 1728 bits of customer data in a superframe, plus 8 particular overhead bits, and one *febe* bit for communicating block errors detected by the *crc* check to the far end. In determining relative positions of the various overhead bit functions, *febe* and *crc* bits were placed away from the start of the superframe to allow time for the *crc* calculation associated with the preceding superframe to be completed.

The *eoc* is the *embedded operations channel* and permits network operations systems to access essential operations functionality in the NT1. There are 12 cyclic redundancy check (*crc*) bits within each superframe, and they cover the 1728 bits of 2B + D customer data plus the eight "M4" bits in the previous superframe. This provides a means of block error check. Two "*crc* bits" are assigned to each of the basic frames 3 through 8. These BRI frame concepts are illustrated in Figures 13.15A and B.

febe stands for "far end block error" bit which is assigned to each superframe. The *febe* bit indicates whether a block error was detected in the 2B + D customer data in a preceding superframe in the opposite direction of transmission. An outgoing *febe* bit with a value of 1 indicates that the *crc* of the preceding incoming superframe matched the value computed from the transmitted data; a value of 0 indicates a mismatch, which, in turn, indicates one or more bit errors.

2B + D CUSTOMER DATA BIT PATTERN. There are 216 2B + D bits placed in each 1.5-ms basic frame, for a customer data rate of 144 kbps. The bit pattern (before conversion to quaternary form and after reconversion to binary form) for the 2B + D data is

$$B_1 B_1 B_1 B_1 B_1 B_1 B_1 B_1 B_2 B_2 B_2 B_2 B_2 B_2 B_2 B_2 DD$$

where B_1 and B_2 are bits from the B_1- and B_2-channels and D is a bit from the D-channel. This 18-bit pattern is repeated 12 times per DSL basic frame.

[8 × 1.5-ms "Basic Frames" → 12-ms Superframe]

		FRAMING	2B + D	Overhead Bits (M1–M6)					
	Quat Positions	1–9	10–117	118s	118m	119s	119m	120s	120m
	Bit Positions	1–18	19–234	235	236	237	238	239	240
Super-frame #	Basic Frame #	Sync Word	2B + D	M1	M2	M3	M4	M5	M6
A	1	ISW	2B + D	eoc_{a1}	eoc_{a2}	eoc_{a3}	act	1	1
	2	SW	2B + D	eoc_{dm}	eoc_{i1}	eoc_{i2}	dea	1	febe
	3	SW	2B + D	eoc_{i3}	eoc_{i4}	eoc_{i5}	1	crc_1	crc_2
	4	SW	2B + D	eoc_{i6}	eoc_{i7}	eoc_{i8}	1	crc_3	crc_4
	5	SW	2B + D	eoc_{a1}	eoc_{a2}	eoc_{a3}	1	crc_5	crc_6
	6	SW	2B + D	eoc_{dm}	eoc_{i1}	eoc_{i2}	1	crc_7	crc_8
	7	SW	2B + D	eoc_{i3}	eoc_{i4}	eoc_{i5}	1	crc_9	crc_{10}
	8	SW	2B + D	eoc_{i6}	eoc_{i7}	eoc_{i8}	1	crc_{11}	crc_{12}
B, C, ...									

NT1-to-LT superframe delay offset by LT-to-NT1 superframe by 60 ± 2 quats (about 0.75 ms).
All bits other than the Sync Word are scrambled.

"1" = reserved bit for future standard; set = 1
( )$_m$, ( )$_s$ "magnitude" bit and "sign" bit for given quat
act = activation bit
crc = cyclic redundancy check: covers 2B + D & M4
dea = deactivation bit

eoc = embedded operations channel
    a = address bit
    dm = data/message indicator
    i = information (data/message)
    febe = far end block error bit

Figure 13.15A. LT to NT1 2B1Q superframe technique and overhead bit assignments. From Bellcore TR-TSY-000397, Issue 1, Figure 3.1a, page 3-9 (Ref. 7).

THE DSL LINE CODE—2B1Q. The average power of a 2B1Q transmitted signal is between +13 and +14 dBm over a frequency band from 0 Hz and 80 kHz, with the nominal peak of the largest pulse being 2.5 volts. The maximum signal power loss at 40 kHz is about 42 dB. As mentioned earlier, the bit rate is 160 kbps and the modulation rate is 80 kbaud.

THE 2B1Q WAVEFORM. It is convenient to express the 2B1Q waveform as +3, +1, −1, −3 because this indicates symmetry about zero, equal spacing between states, and convenient integer magnitudes. The block synchronization word (SW) contains nine quaternary elements repeated every 1.5 ms:

$$+3, +3, -3, -3, -3, +3, -3, +3, +3$$

The 2B1Q waveform is shown below [16].

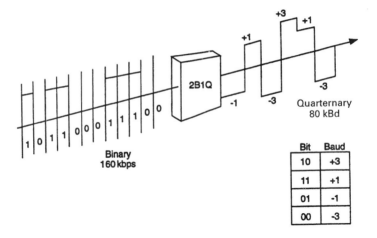

Bit	Baud
10	+3
11	+1
01	-1
00	-3

[8 × 1.5-ms "Basic Frames" → 12-ms Superframe]

		FRAMING	2B + D	Overhead Bits (M1–M6)					
	Quat Positions	1–9	10–117	118s	118m	119s	119m	120s	120m
	Bit Positions	1–18	19–234	235	236	237	238	239	240
Super-frame #	Basic Frame #	Sync Word	2B + D	M1	M2	M3	M4	M5	M6
1	1	ISW	2B + D	eoc_{a1}	eoc_{a2}	eoc_{a3}	act	1	1/nib
	2	SW	2B + D	eoc_{dm}	eoc_{i1}	eoc_{i2}	ps_1	1	febe
	3	SW	2B + D	eoc_{i3}	eoc_{i4}	eoc_{i5}	ps_2	crc_1	crc_2
	4	SW	2B + D	eoc_{i6}	eoc_{i7}	eoc_{i8}	ntm	crc_3	crc_4
	5	SW	2B + D	eoc_{a1}	eoc_{a2}	eoc_{a3}	cso	crc_5	crc_6
	6	SW	2B + D	eoc_{dm}	eoc_{i1}	eoc_{i2}	1	crc_7	crc_8
	7	SW	2B + D	eoc_{i3}	eoc_{i4}	eoc_{i5}	1	crc_9	crc_{10}
	8	SW	2B + D	eoc_{i6}	eoc_{i7}	eoc_{i8}	1	crc_{11}	crc_{12}
2, 3, ...									

NT1-to-LT superframe delay offset by LT-to-NT1 superframe by 60 ± 2 quats (about 0.75 ms). All bits other than the Sync Word are scrambled.

"1" = reserved bit for future standard; set = 1
$( )_m, ( )_s$ "magnitude" bit and "sign" bit for given quat
act = activation bit
crc = cyclic redundancy check: covers 2B + D & M4
cso = cold start only bit
dea = deactivation bit
febe = far end block error bit

eoc = embedded operations channel
a = address bit
dm = data/message indicator
i = information (data/message)
ntm = NT1 in Test Mode bit
nib = network indicator bit from LULT and LUNT to LT
ps_1, ps_2 = power status bits
= 1 from NT1 to LT

Figure 13.15B. NT1 to LT 2B1Q superframe technique and overhead bit assignment. From Bellcore TR-TSY-000397, Issue 1, Figure 3-1b, page 3-10 (Ref. 7).

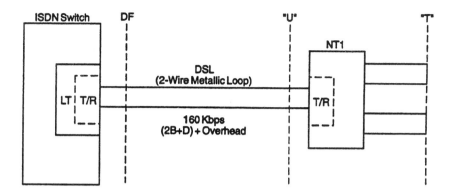

Legend:
DF = Distribution Frame
DSL = Digital Subscriber Line
LT = Line Termination
NT1 = Network Termination 1
"T" = "T" Reference Point (4-wire)
T/R = DSL Transceiver
"U" = "U" Reference Point (2-wire)
B = 64 kbps

Figure 13.16. ISDN basic access rate configuration. From Figure 12-20 (Ref. 16).

LINE CONFIGURATIONS. One basic access is provided by a single DSL interface. This is shown in Figure 13.16. The two-wire interface is shown on the
right-hand side of the figure. Operation on this two-wire line is full-duplex.
To avoid interference between the transmitted and received signals, an echo
canceller with hybrid is used. Echo cancellation involves adaptively forming a
replica of the echo signal arriving at a receiver from its local transmitter and
subtracting it from the signal at the input of the receiver [7].

6.3   Layer 1 Interface, Primary Rate

This interface is applicable for the 1.544- or 2.048-Mbps data rates.

6.3.1   Interface at 1.544 Mbps

6.3.1.1   Bit Rate and Synchronization

NETWORK CONNECTION CHARACTERISTICS. The network delivers (except as
noted below) a signal synchronized from a clock having a minimum accuracy
of 1×10^{-11} (stratum 1; see Chapter 9 for stratum definition). When synchronization by a stratum 1 clock has been interrupted, the signal delivered
by the network to the interface will have a minimum accuracy of 4.6×10^{-6}
(stratum 3).

While in normal operation, the TE1/TA/NT2 transmits a 1.544-Mbps signal having an accuracy equal to that of the received signal by locking the frequency of its transmitter signal to the long-term average of the incoming 1.544-Mbps signal, or by providing equal signal frequency accuracy from another source. ITU-T Rec. I.431 [9] advises against this latter alternative.

RECEIVER BIT STREAM SYNCHRONIZED TO A NETWORK CLOCK

(a) *Receiver Requirements.* Receivers of signals across interface I_a operate with an average transmission rate in the range of 1.544 Mbps $\pm$ 4.6 ppm. However, operation with a received signal transmission rate in the range of 1.544 Mbps $\pm$ 32 ppm is required in any maintenance state controlled by signals/messages passed over the m-bits and by AIS (alarm indication signal). In normal operation the bit stream is synchronized to stratum 1.

(b) *Transmitter Requirements.* The average transmission rate of signals transmitted across interface I_a by the associated equipment is the same as the average transmission rate of the received bit stream.

Note: The I_a and I_b interfaces are located at the input/output port of TE or NT.

TE1/TA OPERATING BEHIND AN NT2 THAT IS NOT SYNCHRONIZED TO A NETWORK CLOCK

(a) *Receiver Requirements.* Receivers of signals across interface I_a operate with a transmission rate in the range of 1.544 Mbps $\pm$ 32 ppm.

(b) *Transmitter Requirements.* The transmitted signal across interface I_a is synchronized to the received bit stream.

SPECIFICATION OF OUTPUT PORTS. The signal specification for output ports is summarized in Table 13.3.

6.3.1.2 Frame Structure. The frame structure is shown in Figure 13.17.

Each time slot consists of consecutive bits, numbered 1 through 8. Each frame is 193 bits long and consists of an F-bit (framing bit) followed by 24 consecutive time slots. The frame repetition rate is 8000 frames per second.

Table 13.4 shows the multiframe structure (called *extended superframe* in the United States*) which is 24 frames long. It takes advantage of the more advanced search algorithms for frame alignment. There are 8000 F-bits (frame alignment bits) transmitted per second (i.e., 8000 frames a second, 1 F-bit per frame). In 24 frames, with these new strategies, only 6 bits are

---

*Note that there are some small differences between the CCITT defined 1.544-Mbps superframe and the ANSI/Bellcore specification.

TABLE 13.3 Digital Interface at 1.544 Mbps

Bit rate:		1544 kbps
Pair(s) in each direction of transmission:		One symmetrical pair
Code:		B8ZS[a]
Test load impedance:		100 ohm resistive
Nominal pulse shape:		See pulse mask[b]
Signal level[b,c]:	Power at 772 kHz	+12 dBm to +19 dBm
	Power at 1544 kHz	At least 25 dB below the power at 772 kHz

[a]B8ZS is modified AMI code in which eight consecutive binary ZEROs are replaced with 000+−0−+ if the preceding pulse was positive (+) and with 000−+0+− if the preceding pulse was negative (−).

[b]The pulse mask and power level requirements apply at the end of a pair having a loss at 772 kHz of 0 to 1.5 dB.

[c]The signal level is the power level measured in a 3-kHz bandwidth at the output port for an all binary ONEs pattern transmitted.

Source: ITU-T Rec. I.431, Table 4/I.431, page 13 (Ref. 9).

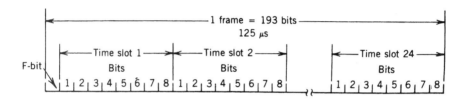

Figure 13.17. Frame structure of 1.544-Mbps interface.

required for frame alignment as shown in Table 13.4. The remaining 18 bits are used as follows. There are 12 m-bits used for control and maintenance. The 6 e-bits are used for CRC6 error checking.

6.3.1.3 Time-Slot Assignment. Time slot 24 is assigned to the D-channel, when this channel is present.

A channel occupies an integer number of time slots and in the same time-slot positions in every frame. A B-channel may be assigned any time slot in the frame, an H_0-channel may be assigned any six slots in a frame in numerical order (not necessarily consecutive), and an H_{11}-channel may be assigned slots 1 to 24. The assignments may vary on a call-by-call basis.

CODES FOR IDLE CHANNELS, IDLE SLOTS, AND INTERFRAME TIME FILL. A pattern including at least 3 binary 1s in an octet is transmitted in every time slot that is not assigned a channel (e.g., time slots awaiting channel assignment on a per-call basis, residual slots on an interface that is not fully provisioned, etc.) and in every time slot of a channel that is not allocated to a call in both directions. Interframe (layer 2) time fill consists of contiguous

TABLE 13.4 Multiframe Structure

Multiframe Frame Number	Multiframe Bit Number	F-bits Assignments		
		FAS	m-bits	CRC
1	1	—	m	—
2	194	—	—	e_1
3	387	—	m	—
4	580	0	—	—
5	773	—	m	—
6	966	—	—	e_2
7	1159	—	m	—
8	1352	0	—	—
9	1545	—	m	—
10	1738	—	—	e_3
11	1931	—	m	—
12	2124	1	—	—
13	2317	—	m	—
14	2510	—	—	e_4
15	2703	—	m	—
16	2896	0	—	—
17	3089	—	m	—
18	3282	—	—	e_5
19	3475	—	m	—
20	3668	1	—	—
21	3861	—	m	—
22	4054	—	—	e_6
23	4247	—	m	—
24	4440	1	—	—

Source: ITU-T Rec. I.431, Table 5/I.431, page 15 (Ref. 9).

HDLC flags transmitted on the D-channel when its layer 2 has no frames to send.

6.3.2 Interface at 2.048 Mbps

6.3.2.1 Frame Structure. There are 8 bits per time slot and 32 time slots per frame, numbered 0 through 31. The number of bits per frame is 256 (i.e., 32×8) and the frame repetition rate is 8000 frames per second. Time slot 0 provides frame alignment and time slot 16 is assigned to the D-channel when that channel is present. A channel occupies an integer number of time slots and the same time-slot position in every frame. A B-channel may be assigned any time slot in the frame and an H_0-channel may be assigned any six time slots, in numerical order, not necessarily consecutive. The assignment of time slots may vary on a call-by-call basis. An H_{12}-channel is assigned time slots 1 to 15 and 17 to 31 in a frame. Time slots 1 to 31 provide bit-sequence-independent transmission. See Chapter 8 for more complete discussion of DS1 and E1.

6.3.2.2 Timing Considerations. The NT derives its timing from the network clock. The TE synchronizes its timing (bit, octet, framing) from the signal received from the NT and synchronizes accordingly the transmitted signal. In an unsynchronized condition—that is, when the access that normally provides network timing is unavailable—the frequency deviation of the free-running clock shall not exceed ± 50 ppm. A TE shall be able to detect and to interpret the input signal within a frequency range of ± 50 ppm.

Any TE which provides more than one interface is declared to be a multiple access TE and is capable of taking the synchronizing clock frequency from its internal clock generator from one or more than one access (or all access links) and synchronize the transmitted signal at each interface accordingly.

6.3.2.3 Codes for Idle Channels and Idle Time Slots, Interframe Fill. A pattern including at least three binary 1s in an octet is transmitted in every time slot that is not assigned to a channel (e.g., time slots awaiting channel assignment on a per-call basis, residual slots on an interface that is not fully provisioned, etc.), and in every time slot of a channel that is not allocated to a call in both directions.

Interframe (layer 2) time fill consists of contiguous HDLC flags (01111110) which are transmitted on the D-channel when its layer 2 has no frames to send.

Frame alignment and CRC procedures can be found in ITU-T Rec. G.706, paragraph 4.

7 OVERVIEW OF THE LAYER 2 INTERFACE: LINK ACCESS PROCEDURE FOR THE D-CHANNEL

The link access procedure (LAP) for the D-channel (LAPD) is used to convey information between layer 3 entities across the ISDN user–network interface using the D-channel.

A *service access point* (SAP) is a point at which the data-link layer provides services to its next higher OSI layer or layer 3. Associated with each data-link layer (OSI layer 2) is one or more data-link connection endpoints (see Figure 13.18). A data-link connection endpoint is identified by a data-link connection endpoint identifier, as seen from layer 3, and by a data-link connection identifier (DLCI), as seen from the data-link layer.

Cooperation between data-link layer entities is governed by a specific protocol to the applicable layer. In order for information to be exchanged

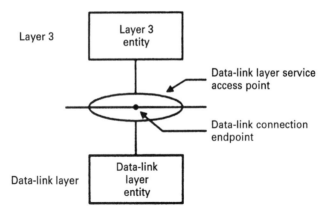

Figure 13.18. Entities, service access points (SAPs), and endpoints. From ITU-T Rec. 920, Figure 2/ Q.920, page 2, Ref. 11.

between two or more layer 3 entities, an association must be established between layer 3 entities in the data-link layer using a data-link layer protocol. This association is provided by the data-link layer between two or more SAPS, as shown in Figure 13.19. Data-link message units are conveyed between data-link layer entities by means of a physical connection. Layer 3 uses *service primitives* to request service from the data-link layer. A similar interaction takes place between layer 2 and layer 1.

Between the data-link layer and its adjacent layers there are four types of service primitives:

1. Request
2. Indication
3. Response
4. Confirm

These functions are shown diagrammatically in Figure 13.20.

The REQUEST primitive is used where a higher layer is requesting service from the next lower layer. The INDICATION primitive is used by a layer providing service to notify the next higher layer of activities related to the REQUEST primitive. The RESPONSE primitive is used by a layer to acknowledge receipt from a lower layer of the INDICATION primitive. The CONFIRM primitive is used by the layer providing the requested service to confirm that the requested activity has been completed.*

Figure 13.21 shows the data-link layer reference model. All data-link layer messages are transmitted in frames delimited by flags, where a flag is a unique

*Remember that LAPD is a direct derivative of HDLC described in Chapter 11.

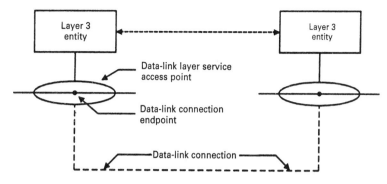

Figure 13.19. Peer-to-Peer relationship. From Figure 3/Q.920, page 3, ITU-T Rec. Q.920, Ref. 11.

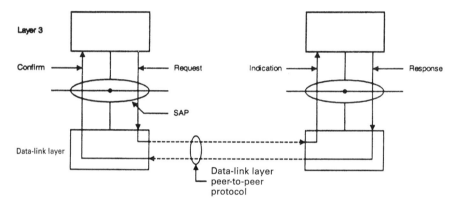

Figure 13.20. Functions of service primitives action sequence. *Note*: The same principle applies for data-link layer–physical layer interactions. From Figure 4/Q.920, page 4, ITU-T Rec. Q.920, Ref. 11.

sequence bit pattern. The frame structure as defined in ITU-T Rec. I.441 (Q.921 [12]) is briefly described in this section.

The LAPD includes functions for:

1. The provision of one or more data-link connections on a D-channel. Discrimination between the data-link connections is by means of a data-link connection identifier (DLCI) contained in each frame.

2. Frame delimiting, alignment, and transparency, allowing recognition of a sequence of bits transmitted over a D-channel as a frame.

3. Sequence control, which maintains the sequential order of frames across a data-link connection.

4. Detection of transmission, format, and operational errors on a data link.

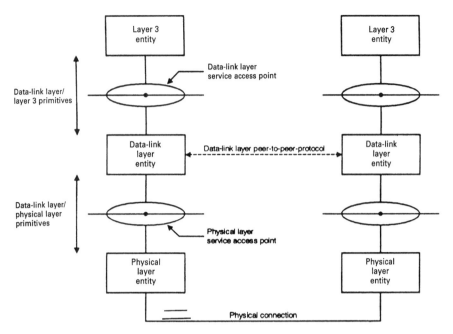

Figure 13.21. Data-link layer reference model. From ITU-T Rec. Q.920, Figure 5/Q.920, page 5 (Ref. 11).

5. Recovery from detected transmission, format, and operational errors. Notification to the management entity of unrecovered errors.

6. Flow control.

There is unacknowledged and acknowledged operation. With unacknowledged operation, information is transmitted in unnumbered information (UI) frames. At the data-link layer the UI frames are unacknowledged. Transmission and format errors may be detected, but no recovery mechanism is defined. Flow control mechanisms are also not defined. With acknowledged operation, layer 3 information is transmitted in frames that are acknowledged at the data-link layer. Error recovery procedures based on retransmission of unacknowledged frames are specified. For errors that cannot be corrected by the data-link layer, a report to the management entity is made. Flow control procedures are also defined.

Unacknowledged operation is applicable for point-to-point and broadcast information transfer. However, acknowledged operation is applicable only for point-to-point information transfer.

There are two forms of acknowledged information that are defined:

- Single-frame operation
- Multiframe operation

For single-frame operation, layer 3 information is sent in sequenced information 0 (SI0) and sequenced information 1 (SI1) frames. No new frame is sent until an acknowledgment has been received for a previously sent frame. This means that only one unacknowledged frame may be outstanding at a time. With multiple-frame operation, layer 3 information is sent in numbered information (I) frames. A number of I frames may be outstanding at the same time. Multiple-frame operation is initiated by a multiple-frame establishment procedure using set asynchronous balanced mode/set asynchronous balanced mode extended (SABM/SABME) command.

7.1 Layer 2 Frame Structure for Peer-to-Peer Communication

There are two frame formats used on layer 2 frames.

1. Format A, for frames where there is no information field.
2. Format B, for frames containing an information field.

These two frame formats are shown in Figure 13.22.

The following discussion briefly describes the frame content (sequences and fields) for the LAPD (layer 2) frame.

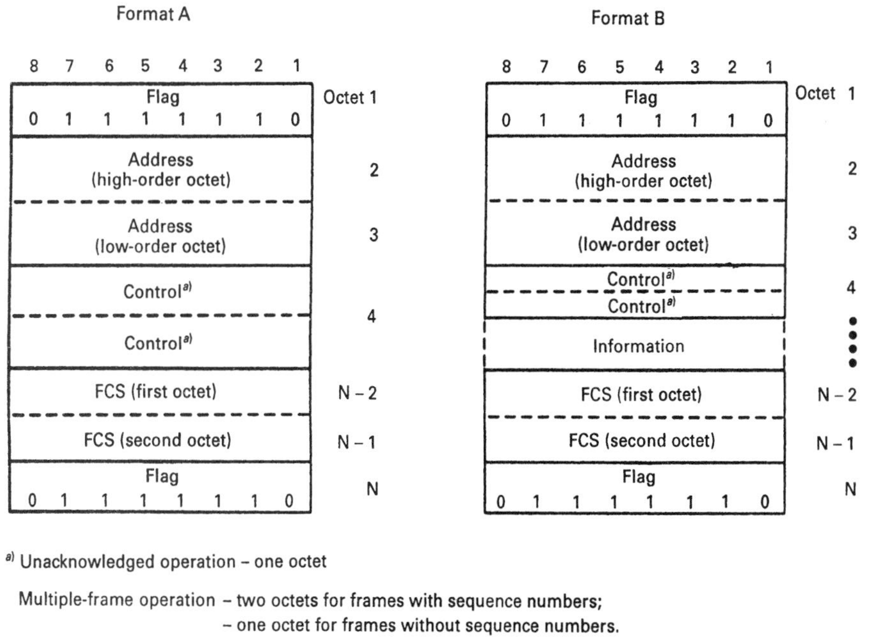

a) Unacknowledged operation – one octet

Multiple-frame operation – two octets for frames with sequence numbers;
– one octet for frames without sequence numbers.

Figure 13.22. Frame formats for layer 2 frames. From ITU-T Rec. Q.921, Figure 1/Q.921, page 20 (Ref. 12).

Flag Sequence. All frames start and end with a flag sequence consisting of one 0 bit followed by six contiguous 1 bits and one 0 bit. These flags are called the *opening* and *closing* flags. If sequential frames are transmitted, the closing flag of one frame is the opening flag of the next frame.

Address Field. As shown in Figure 13.22, the address field consists of two octets and identifies the intended receiver of a command frame and the transmitter of a response frame. A single octet address field is reserved for LAPB operation to allow a single LAPB data-link connection to be multiplexed along with the LAPD data-link connections.

Control Field. The control field consists of one or two octets. It identifies the type of frame, either command or response. It contains sequence numbers where applicable. Three types of control field formats are specified:

1. Numbered information transfer (I format)
2. Supervisory functions (S format)
3. Unnumbered information transfers and control functions (U format)

Information Field. The information field of a frame, when present, follows the control field and precedes the frame check sequence (FCS). The information field contains an integer number of octets:

- For a SAP supporting signaling, the default value is 128 octets.
- For SAPs supporting packet information, the default value is 260 octets.

Frame Check Sequence (FCS) Field. The FCS field is a 16-bit sequence and is the 1's complement of the modulo-2 sum of:

1. The remainder of X raised to the k power:

$$X^{15} + X^{14} \ldots X^1 + 1$$

   divided by the generating polynomial

$$X^{16} + X^{12} + X^5 + 1$$

   where k is the number of bits in the frame existing between but not including the final bit of the opening flag and the first bit of the FCS, excluding bits inserted for transparency, and

2. The remainder by modulo-2 division by the generating polynomial given above of the product of X^{16} by the content of the frame defined above.

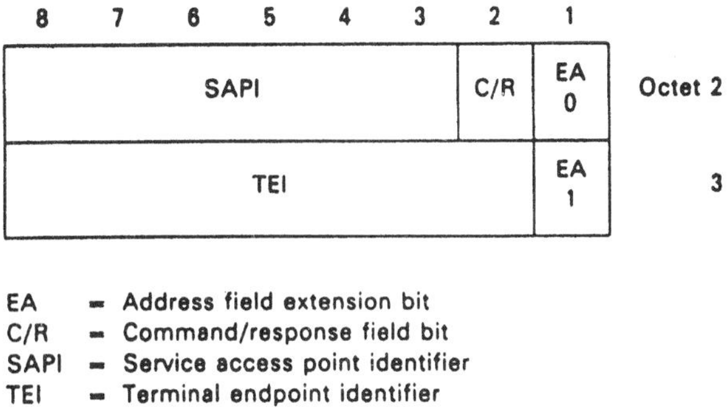

8	7	6	5	4	3	2	1		
		SAPI				C/R	EA 0	Octet 2	
		TEI					EA 1	3	

EA   = Address field extension bit
C/R   = Command/response field bit
SAPI   = Service access point identifier
TEI   = Terminal endpoint identifier

Figure 13.23. LAPD address field format. From CCITT Rec. Q.921, Figure 5/Q.921, page 23 (Ref. 12).

Transparency, mentioned above, ensures that a flag or abort sequence is not initiated within a frame. On the transmit side the data-link layer examines the frame content between the opening and closing flag sequences and inserts a 0 bit after all sequences with five contiguous 1 bits (including the last five bits of the FCS). On the receive side the data-link layer examines the frame contents between the opening and closing flag sequences and discards any 0 bit that directly follows five contiguous 1 bits.

ADDRESS FIELD FORMAT. The address field is shown in Figure 13.23. It contains address field extension bits (EA), command/response indication bit (C/R), a data-link layer service access point identifier (SAPI) subfield, and a terminal endpoint identifier (TEI) subfield.

Address Field Extension Bit (EA). The address field range is extended by reserving the first transmitted bit of the address field to indicate the final octet of the address field. The presence of a 1 in the first bit position of an address field octet signals that it is the final octet of the address field. The double octet address field for LAPD operation has bit 1 of the first octet set to a 0 and bit 1 of the second octet set to a 1.

Command/Response Field Bit (C/R). The C/R bit identifies a frame as either a command or a response. The user side sends commands with the C/R bit set to 0, and it sends responses with the C/R bit set to 1. The network side does the opposite; that is, commands are sent with the C/R bit set to 1, and responses are sent with the C/R bit set to 0.

In keeping with HDLC rules,* commands use the address of the peer data-link entity while responses use the address of their own data-link layer entity. In accordance with these rules, both peer entities on a point-to-point data-link connection use the same data-link connection identifier (DLCI) com-

---

*LAPD, as we know, is a derivative of HDLC.

TABLE 13.5 Allocation of SAPI Values

SAPI Value	Related Layer 3 or Management Entity
0	Call control procedures
1–15	Reserved for future standardization
16	Packet communication conforming to X.25 level 3 procedures
17–31	Reserved for future standardization
63	Layer 2 management procedures
All others	Not available for Q.921 procedures

Source: ITU-T Rec. Q.921, Table 2/Q.921, page 7, Ref. 12.

posed of an SAPI-TEI where SAPI and TEI conform to the definitions contained below.

Service Access Point Identifier (SAPI). The SAPI identifies a point at which data-link services are provided by a data-link layer entity to a layer 3 or management entity. Consequently, the SAPI specifies a data-link layer entity that should process a data-link layer frame. The SAPI allows 64 service access points (SAPs) to be specified, where bit 3 of the address field octet containing the SAPI is the least significant binary digit and bit 8 is the most significant. The SAPI values are allocated as shown in Table 13.5.

Terminal Endpoint Identifier (TEI). The TEI for a point-to-point link connection may be associated with a single terminal equipment (TE). A TE may contain one or more TEIs used for point-to-point data transfer. The TEI for a broadcast data-link connection is associated with all user side data-link layer entities containing the same SAPI. The TEI subfield allows 128 values where bit 2 of the address field octet containing the TEI is the least significant bit and bit 8 is the most significant bit. The following conventions apply for assignment of these values:

1. TEI for broadcast data-link connection. The TEI subfield bit pattern 111 1111 (= 127) is defined as the group TEI. The group TEI is assigned to the broadcast data-link connection associated with the addressed SAP.

2. TEI for point-to-point data-link connection. The remaining TEI values are used for point-to-point data-link connections associated with the addressed SAP. The range of TEI values is as follows:

Values 0–63:   Nonautomatic TEI assignment user equipment
Values 64–126: Automatic TEI assignment user equipment

CONTROL FIELD FORMATS. Control field formats are shown in Table 13.6. The control field identifies the type of frame, either a command or a response.

TABLE 13.6   Control Field Formats

Control Field Bits (Modulo 128)	8	7	6	5	4	3	2	1	
I format	N(S)							0	Octet 4
	N(R)							P	5
S format	X	X	X	X	S	S	0	1	Octet 4
	N(R)							P/F	5
U format	M	M	M	P/F	M	M	1	1	Octet 4

N(S)   Transmitter send sequence number   M   Modifier function bit

N(R)   Transmitter receive sequence number   P/F   Poll bit when issued as a command, final bit
                                                   when issued as a response

S   Supervisory function bit   X   Reserved and set to 0

Source: CCITT Rec. Q.921, Table 4/Q.921, page 28 (Ref. 12).

The control field contains sequence numbers, where applicable. Three types of control field formats are specified: numbered information transfer (I format), supervisory functions (S format), and unnumbered information transfers and control functions (U format).

7.2   LAPD Primitives

The following comments clarify the semantics and usage of primitives. Primitives consist of commands and their respective responses associated with the services requested of a lower layer. The general syntax of a primitive is:

XX-generic name-type: parameters

where XX designates the layer providing the service. For the data-link layer, XX is DL, PH for the physical layer, or MDL for the management entity to the data-link layer interface. Table 13.7 gives the primitives associated with the data-link layer.

8   OVERVIEW OF LAYER 3

The layer 3 protocol, of course, deals with the D-channel and its signaling capabilities. It provides the means to establish, maintain, and terminate network connections across an ISDN between communicating application entities. A more detailed description of the layer 3 protocol may be found in ITU-T Rec. Q.931 [14].

TABLE 13.7   Primitives Associated with the Data-Link Layer

Generic name	Type				Parameters		Message Unit Contents
	Request	Indication	Response	Confirm	Priority Indicator	Message Unit	
L3 ↔ L2							(Note 1)
DL-Establish	X	X	—	X	—	—	
DL-Release	X	X	—	X	—	—	
DL-Data	X	X	—	—	—	X	Layer 3 (peer-to-peer message)
DL-Unit Data	X	X	—	—	—	X	Layer 3 (peer-to-peer message)
M ↔ L2							
MDL-Assign	X	X	—	—	—	X	TEI value, CES (Note 2)
MDL-Remove	X	—	—	—	—	X	TEI value, CES
MDL-Error	—	X	X	—	—	X	Reason for error message
MDL-Unit Data	X	X	—	—	—	X	Management function peer-to-peer message
MDL-XID	X	X	X	X	—	X	Connection management PDU (peer-to-peer XIO frame)
L2 ↔ L1							
PH-Data	X	X	—	—	X	X	Data-link layer peer-to-peer frame
PH-Activate	X	X	—	—	—	—	
PH-Deactivate	—	X	—	—	—	—	
M ↔ L1							
MPH-Activate	—	X	—	—	—	—	
MPH-Deactivate	X	X	—	—	—	—	
MPH-Information	—	X	—	—	—	X	Connected/disconnected

L3 ↔ L2: Layer 3/data-link layer boundary   X Exists
L2 ↔ L1: Data-link layer/physical layer boundary   — Does not exist
M ↔ L2: Management entity/data-link layer boundary
M ↔ L1: Management entity/physical layer boundary

Note 1: Although not shown below, the CES is implicitly associated with each *L3 ↔ L2* primitive, indicating the applicable connection endpoint.
Note 2: TEI value is included only in the MDL-ASSIGN request.

Source· CCITT Rec. Q.921, Table 6/Q.921, page 15 (Ref. 12).

Layer 3 utilizes functions and services provided by its data-link layer, as described in Section 7 under LAPD functions. These necessary layer 2 support functions are listed and briefly described below:

- Establishment of the data-link connection
- Error-protected transmission of data
- Notification of unrecoverable data-link errors
- Release of data-link connections
- Notification of data-link layer failures
- Recovery from certain error conditions
- Indication of data-link layer status

Layer 3 performs two basic categories of functions and services in the establishment of network connections. The first category directly controls the connection establishment. The second category includes those functions relating to the transport of messages in addition to the functions provided by the data-link layer. Among these additional functions are the provision of rerouting of signaling messages on an alternative D-channel (where provided) in the event of D-channel failure. Other possible functions include multiplexing and message segmenting and blocking. The D-channel layer 3 protocol is designed to carry out establishment and control of circuit-switched and packet-switched connections. Also, services involving the use of connections of different types, according to user specifications, may be provided through "multimedia" call control procedures. Functions performed by layer 3 include:

1. The processing of primitives for communicating with the data-link layer.
2. Generation and interpretation of layer 3 messages for peer level communications.
3. Administration of timers and logical entities (e.g., call references) used in call control procedures.
4. Administration of access resources, including B-channels and packet-layer logical channels (e.g., ITU-T X.25).
5. Checking to ensure that services provided are consistent with user requirements, such as compatibility, address, and service indicators.

The following functions may also be performed by layer 3:

1. *Routing and Relaying*. Network connections exist either between users and ISDN exchanges or between users. Network connections may involve intermediate systems which provide relays to other interconnecting subnetworks and which facilitate interworking with other

networks. Routing functions determine an appropriate route between layer 3 addressees.

2. *Network Connection.* This function includes mechanisms for providing network connections making use of data-link connections provided by the data-link layer.

3. *Conveying User Information.* This function may be carried out with or without the establishment of a circuit-switched connection.

4. *Network Connection Multiplexing.* Layer 3 provides multiplexing of call control information for multiple calls onto a single data-link connection.

5. *Segmenting and Reassembly (SAR).* Layer 3 may segment and reassemble layer 3 messages to facilitate their transfer across user–network interface.

6. *Error Detection.* Error detection functions are used to detect procedural errors in the layer 3 protocol. Error detection in layer 3 uses, among other information, error notification from the data-link layer.

7. *Error Recovery.* This includes mechanisms for recovering from detected errors.

8. *Sequencing.* This includes mechanisms for providing sequenced delivery of layer 3 information over a given network connection when requested. Under normal conditions, layer 3 ensures the delivery of information in the sequence it is submitted by the user.

9. *Congestion Control and User Data Flow Control.* Layer 3 may indicate rejection or unsuccessful indication for connection establish requests to control congestion within a network. Typical is the congestion control message to indicate the establishment or termination of flow control on the transmission of User Information messages.

10. *Restart.* This function is used to return channels and interfaces to an idle condition to recover from certain abnormal conditions.

8.1   Layer 3 Specification

The layer 3 specification is contained in ITU-T Rec. Q.930/931 [13, 14]. It includes both circuit-switched and packet-switched operation. There are 23 message types for circuit-mode connection control. These are shown in Table 13.8, and the content elements of each are given in over 50 tables in the specification. One typical table is given in Table 13.9, Setup Message Content. Additional tables are provided for packet switching.

The following are several explanatory notes for the tables found in ITU-T Rec. Q.931 and for Table 13.9. The letters "M" and "O" mean *mandatory* and *optional*, respectively. Letters "n" and "u" refer to network and user, respectively, and give the direction of traffic such as n → u (network to user)

TABLE 13.8   Messages for Circuit-Mode Connection Control

Call establishment messages:
Alerting
Call Proceeding
Connect
Connect Acknowledge
Progress
Setup
Setup Acknowledge

Call information phase messages:
Resume
Resume Acknowledge
Resume Reject
Suspend
Suspend Acknowledge
Suspend Reject
User Information

Call clearing messages:
Disconnect
Release
Release Complete

Miscellaneous messages:
Congestion Control
Facility
Information
Notify
Status
Status Enquiry

Source: CCITT Rec. Q.931, Table 3-1/Q.931, page 10 (Ref. 14).

and u $\rightarrow$ n (user to network). An asterisk (*) in the table means undefined maximum length.

8.1.1 General Message Format and Information Elements Coding
Within this protocol, every message consists of the following parts:

(a) Protocol discriminator

(b) Call reference

(c) Message type

(d) Other information elements, as required

Information elements (a), (b), and (c) are common to all messages and are always present, while information element (d) is specific to each message type. This organization is illustrated in the example shown in Figure 13.24.

TABLE 13.9   Setup Message Content

Information Element	Direction	Type	Length
Protocol discriminator	Both	M	1
Call reference	Both	M	2–*
Message type	Both	M	1
Sending complete	Both	O (Note 1)	1
Repeat indicator	Both	O (Note 2)	1
Bearer capability	Both	M (Note 3)	4–13
Channel identification	Both	M (Note 4)	2–*
Facility	Both	O (Note 5)	2–*
Progress indicator	Both	O (Note 6)	2–4
Network-specific facilities	Both	O (Note 7)	2–*
Display	n → u	O (Note 8)	Note 9
Keypad facility	u → n	O (Notes 10,12)	2–34
Signal	n → u	O (Note 11)	2–3
Switchhook	u → n	O (Note 12)	2–3
Feature activation	u → n	O (Note 12)	2–4
Feature indication	n → u	O (Note 12)	2–5
Calling party number	Both	O (Note 13)	2–*
Calling party subaddress	Both	O (Note 14)	2–23
Called party number	Both	O (Note 15)	2–*
Called party subaddress	Both	O (Note 16)	2–23
Transit network selection	u → n	O (Note 17)	2–*
Low layer compatibility	Both	O (Note 18)	2–16
High layer compatibility	Both	O (Note 19)	2–4
User–user	Both	O (Note 20)	Note 21

Source: CCITT Rec. Q.931, Table 3-16/Q.931, page 30 (Ref. 14).

(continued)

Explanatory Notes to Table 13.9

Note 1: Included if the user or the network optionally indicates that all information necessary for call establishment is included in the Setup message.

Note 2: The Repeat indicator information element is included immediately before the first Bearer capability information element when either the in-call modification procedure or the bearer capability negotiation procedure is used.

Note 3: May be repeated if the bearer capability negotiation procedure is used. For bearer capability negotiation, either two or three Bearer capability information elements may be included in descending order of priority—that is, highest priority first.

Note 4: Mandatory in the network-to-user direction. Included in the user-to-network direction when the user wants to indicate a channel. If not included, its absence is interpreted as "any channel acceptable."

Note 5: May be included for functional operation of supplementary services.

Note 6: Included in the event of interworking or in connection with the provision or in-band information/patterns.

Note 7: Included by the calling user or the network to indicate network-specific facilities information.

Note 8: Included if the network provides information that can be presented to the user.

Note 9: The minimum length is 2 octets; the maximum length is network dependent and is either 34 or 82 octets.

Note 10: Either the Called party number or the Keypad facility information element is included by the user to convey called party number information to the network. The Keypad facility information element may also be included by the user to convey other call establishment information to the network.

Note 11: Included if the network optionally provides additional information describing tones.

Note 12: As a network option, may be used for stimulus operation of supplementary services.

Note 13: May be included by the calling user or the network to identify the calling user.

Note 14: Included in the user-to-network direction when the calling user wants to indicate the calling party subaddress. Included in the network-to-user direction if the calling user included a Calling party subaddress information element in the Setup message.

Note 15: Either the Called party number or the Keypad facility information element is included by the user to convey called party number information to the network. The Called party number information element is included by the network when called party number information is conveyed to the user.

Note 16: Included in the user-to-network direction when the calling user wants to indicate the called party subaddress. Included in the network-to-user direction if the calling user included a Called party subaddress information element in the Setup message.

Note 17: Included by the calling user to select a particular transit network.

Note 18: Included in the user-to-network direction when the calling user wants to pass Low layer compatibility information to the called user. Included in the network-to-user direction if the calling user included a Low layer compatibility information element in the Setup message.

Note 19: Included in the user-to-network direction when the calling user wants to pass High layer compatibility information to the called user. Included in the network-to-user direction if the calling user included a High layer compatibility information element in the Setup message.

Note 20: Included in the user-to-network direction when the calling user wants to pass user information to the called user. Included in the network-to-user direction if the calling user included a user–user information element in the Setup message.

Note 21: The minimum length is 2 octets; the standard default maximum length is 131 octets.

The term "default" implies that the value defined should be used in the absence of any assignment or in the negotiation of alternative values. When a field, such as the call reference value, extends over more than one octet, the order of bit values progressively decreases as the octet number increases. The least significant bit of the field is represented by the lowest-numbered bit of the highest-numbered octet field.

PROTOCOL DISCRIMINATOR. The purpose of the protocol discriminator is to distinguish messages for user–network call control from other messages (to be defined) with ITU-T Rec. Q.931 from those OSI network layer protocol units which are coded to other ITU-T Recommendations and other standards.

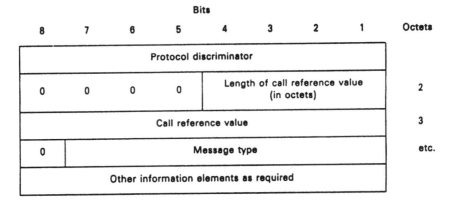

Figure 13.24. General message organization example. From CCITT Rec. Q.931, Figure 4-1/Q.931, page 53 (Ref. 14).

CALL REFERENCE. The purpose of the call reference is to identify the call or facility registration/cancellation request at the local user–network interface to which the particular message applies. The call reference does not have end-to-end significance across ISDNs.

As a minimum, all networks and users must be able to support (1) a call reference value of one octet for a basic user–network interface and (2) a call reference value of two octets for a primary rate interface. The call reference information element includes the call reference value and the call reference flag.

Call reference values are assigned by the originating side of the interface for a call. These values are unique to the originating side within a particular D-channel layer 2 logical link connection. The call reference value is assigned at the beginning of a call and remains fixed for the lifetime of the call (except in the case of call suspension). After a call ends or after a successful suspension, the associated call reference value may be reassigned to a later call. Two identical call reference values on the same D-channel layer 2 logical link connection may be used when each value pertains to a call originated at opposite ends of the link.

The call reference flag can take the values of 0 or 1. The call reference flag is used to identify which end of the layer 2 logical link originated a call reference. The origination side always sets the call reference flag to "0." The destination always sets the call reference flag to "1."

Hence the call reference flag identifies who allocated the call reference value for this call, and the only purpose of the call reference flag is to resolve simultaneous attempts to allocate the same call reference value.

MESSAGE TYPE. The purpose of the message type is to identify the function of the message being sent. For instance, 00000101 indicates a call-setup message.

OTHER INFORMATION ELEMENTS. Forty "other information elements" are listed, such as "sending complete," "congestion level," "call identity," "date/time," and "calling party number."

9 ISDN PACKET-MODE REVIEW

9.1 Introduction

Two main services for packet-switched data transmission are defined for packet-mode terminals connected to the ISDN:

Case A: Access to a PSPDN (PSPDN services) (PSPDN = packet-switched public data network).

Case B: Use of an ISDN virtual circuit service.

9.2 Case A: Configuration When Accessing PSPDN Services

This configuration is shown in Figure 13.25 and refers to the service of Case A, thus implying a transparent handling of packet calls through an ISDN. Only access via the B-channels is possible. In this context, the only support that an ISDN gives to packet calls is a physical 64-kbps circuit-mode

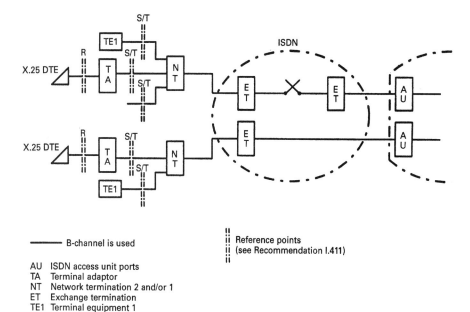

——— B-channel is used

‖ Reference points
‖ (see Recommendation I.411)
‖

AU ISDN access unit ports
TA Terminal adaptor
NT Network termination 2 and/or 1
ET Exchange termination
TE1 Terminal equipment 1

Figure 13.25. Case A, configuration when accessing PSPDN services. From ITU-T Rec. X.31, Figure 2-1/X.31, page 3 (Ref. 2).

semipermanent or demand transparent network connection type between the appropriate PSPDN port and the X.25 DTE + TA or TE1 at the customer premises.

In the case of semipermanent access, the X.25 DTE + TA or TE1 is connected to the corresponding ISDN port at the PSPDN (AU).* The TA, when present, performs only the necessary physical channel rate adaption between the user at the R reference point and the 64-kbps B-channel rate. D-channel layer 3 messages are not used in this case.

In the case of demand access to PSPDN, which is shown in the upper portion of Figure 13.25, the X.25 DTE + TA or TE1 is connected to an ISDN port at the PSPDN (AU). The AU is also able to set up 64-kbps physical channels through the ISDN.

In this type of connection, an originating call will be set up over the B-channel toward the PSPDN port using the ISDN signaling procedure prior to starting X.25 layer 2 and layer 3 functions. This is done by using either hot-line (e.g., direct call) or complete selection methods. Moreover, the TA, when present, performs user rate adaption to 64 kbps. Depending on the data rate adaption technique employed, a complementary function may be needed at the AU of the PSPDN.

In the complete selection case, two separate numbers are used for outgoing access to the PSPDN:

- The ISDN number of the access port of the PSPDN, given in the D-channel layer 3 Setup message (Q.931).

- The address of the called DTE indicated in the X.25 call request packet.

The corresponding service requested in the D-channel layer 3 Setup message is ISDN circuit-mode bearer services.

For calls originated by the PSPDN, the same considerations as above apply. In fact, with reference to Figure 13.25, the ISDN port of the PSPDN includes both rate adaption (if required) and path setting-up functions. When needed, DTE identification may be provided to the PSPDN by using the call establishment signaling protocols in D-channel layer 3 (Q.931). Furthermore, DCE identification may be provided to the DTE, when needed, by using the same protocols.

For the demand access case, X.25 layer 2 and layer 3 operation in the B-channel as well as service definitions are found in ITU-T Rec. X.32. Some PSPDNs may operate the additional DTE identification procedures defined in Rec. X.32 [20] to supplement the ISDN-provided information in Case A.

---

*AU is an ISDN access unit.

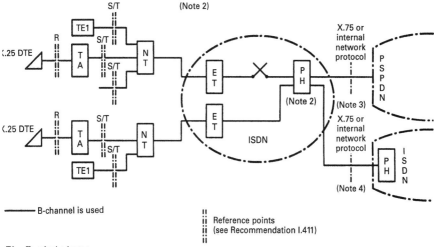

────── B-channel is used

⁞⁞ Reference points
⁞⁞ (see Recommendation I.411)

TA   Terminal adaptor
NT   Network termination 2 and/or 1
ET   Exchange termination
TE1 Terminal equipment 1
PH   Packet handling function

NOTES

1 This figure is only an example of many possible configurations and is included as an aid to the text describing
   the various interface functions.
2 In some implementations, the PH functions logically belonging to the ISDN may reside physically in a node of
   the PSPDN. The service provided is still the ISDN virtual circuit service.
3 See Recommendation X.325.
4 See Recommendation X.320.

Figure 13.26. Case B, configuration for the ISDN virtual circuit service (access via B-channel). From
ITU-T Rec. X.31, Figure 2-2/X.31, page 5 (Ref. 2).

9.3   Case B: Configuration for the ISDN Virtual Circuit Service

This configuration refers to the case where a packet handling (PH) function is
provided within the ISDN. The configuration in Figure 13.26 relates to the
case of X.25 link and packet procedures conveyed through the B-channel. In
this case, the packet call is routed, within an ISDN, to some PH (packet
handling) function where the complete processing of the X-25 call can be
carried out.

The PH function may be accessed in various ways depending on the
related ISDN implementation alternatives. In any case a B-channel connection set up to/from a PH (packet handler) port supports (a) the necessary
processing for B-channel packet calls, (b) standard X.25 functions for layer 2
and layer 3, and (c) possible path setting-up functions for layer 1 and possible
rate adaption.

The configuration shown in Figure 13.5 (Section 4) refers to the case of
X.25 packet layer procedures conveyed through the D-channel. In this case a
number of DTEs can operate simultaneously through a D-channel by using

connection identifier discrimination at ISDN layer 2. The accessed port of PH is still able to support X.25 packet layer procedures.

It should be pointed out that the procedures for accessing a PSDTS* through an ISDN user–network interface over a B- or D-channel are independent of where the service provider chooses to locate packet handling functions, that is,

(a) in a remote exchange or packet-switching module in an ISDN or

(b) in the local exchange.

However, the procedures for packet access through the B-channel or the D-channel are different.

In both cases of B- and D-channel accesses, in the service of Case B, the address of the called DTE is contained in the X.25 call request packet. The establishment of the physical connection from the TA/TE1 to the packet handling function is done on the basis of the requested bearer service (ISDN virtual circuit service), and therefore the user does not provide any addressing information in the layer procedures (D-channel, Q.931).

9.4   Service Aspects

9.4.1   Access to PSPDN Services—Case A

SERVICE CHARACTERISTICS. In this case the ISDN offers a 64-kbps circuit-switched or semipermanent transparent network connection type between the TA/TE1 and the PSPDN port (AU). In the switched access case the AU must be selected by the called address in the D-channel signaling protocol when the TA/TE1 sets up the circuit-switched connection to the AU. In the non-switched access case, layer 3 (D-channel, Q.931) call control messages are not used.

Since the packet-switched service provider is a PSPDN, some DTEs are PSPDN terminals; they are handled by the PSPDN. Other DTEs may access the PSPDN without subscribing to the PSPDN permanently. In the first case, the same services as PSPDN services are maintained, including facilities, quality of service (QoS) characteristics, and DTE–DCE interfaces. In the case where a DTE is not subscribing to the PSPDN, it will be provided with a limited set of PSPDN facilities (see ITU-T Rec. X.32).

Every DTE will be associated with one or more ISDN (E.164 [15]) numbers. In addition, a DTE may be associated with one or more X.121 [17] numbers assigned by the PSPDN(s) associated by the DTE. The method for X.25 packets to convey numbers from the ISDN numbering plan and the relationship with ITU-T Rec. X.121 are described in ITU-T Rec. E.166.

---

*PSDTS = packet-switched data transmission services.

BASIC RULES. Packet data communications, when using a switched B-channel, will be established by separating the establishment phase of the B-channel and the control phase of the X.25 virtual circuits using the X.25 protocol (link layer and packet layer). In general, ISDN has no knowledge of the customer terminal equipment or configuration. The incoming B-channel connection establishment will have to employ the D-channel signaling procedure in ITU-T Rec. Q.931 [14].

9.4.2 Access to the ISDN Virtual Circuit Service—Case B

SERVICE CHARACTERISTICS. The virtual circuit service provided within the ISDN is aligned with what is described in ITU-T X-Series Recommendations (e.g., in terms of facilities, quality of service, etc.).

The service and facilities provided as well as the quality of service characteristics are those of the ISDN. Existing features of the X-Series Recommendations may be enhanced and additional features may also be developed taking into account the new ISDN customer capabilities. A number from the ISDN numbering plan will be associated with one or more TA/TE1. The numbering plan is set out in ITU-T Rec. E.164 [15].

USER ACCESS CAPABILITIES. In this case both B- and D-channels can be used for accessing the ISDN virtual circuit service.

Access Through the B-Channel—Basic Rules. Packet data communications, when using a switched B-channel, is established by separating the establishment phase of the B-channel and the control phase of the virtual circuits using the X.25 protocol (link layer and packet layer).

In general an ISDN has no knowledge of the customer terminal equipment or configuration. In the demand access case the incoming B-channel connection establishment uses the signaling procedures of clause 6 of Q.931 (D-channel layer 3).

Access Through the D-Channel—Basic Rules. The following must be adhered to for TE access to the PSDTS as it is defined in the X-Series Recommendations, particularly X.25.

A single SAPI = 16 LAPD link, as viewed by both the network and the user, must support multiplexing of logical channels at X.25 layer 3. Additionally, because the user may have a multipoint access and because a single TA or TE1 is allowed to operate with more than one TEI (terminal endpoint identifier), the network must support the presence of multiple SAPI = 16 LAPD logical links simultaneously operating at ISDN layer 2. This results in the requirement that the network be able to support simultaneous layer 2 and X.25 layer 3 multiplexing for D-channel packet-mode connections. (*Note:* SAPI = service access point identifier.)

All X.25 packets, including *call request* and *incoming call packets*, must be transported to and from the TE in numbered information frames (I-frames) in a SAPI = 16 LAPD link.

An *incoming call* packet is transmitted to a TE only after the public networks check at least the following:

- Compatibility of user facilities contained in the incoming call packet with the called subscriber profile when present.
- Availability of the X.25 logical channel, either two-way or incoming, on which the incoming call packet is sent.

REVIEW QUESTIONS

1. Name at least three services other than speech telephony that already had been integrated into the PSTN prior to the advent of ISDN.

2. Identify at least two shortcomings of PCM as implemented today regarding its suitability for ISDN. (*Hint*: Think standards.)

3. PCM, as designed, is optimized for speech telephony. Approximately what percentage of traffic on the PSTN is speech telephony?

4. Name at least five communication services that ISDN will support.

5. Distinguish *primary rate* and *basic rate*.

6. Define the B-channel, the D-channel (2 answers here), and the three basic variants of the H-channel.

7. Give at least three applications of the H-channel (different from question 4).

8. 30B + D is the primary rate for E1 configurations. What ever happened to E1's second separate channel?

9. What is the aggregate bit rate of 2B + D? First for CCITT, then for North American ISDN.

10. Distinguish TE1 and TE2 in the standard ISDN network interface. European ISDN would have an NT12 interface. The equipment at this interface would belong to whom? In North America, this interface is split: NT1 and NT2. Discuss the difference.

11. Which OSI layers are involved with the B-channel and D-channel for a voice connection?

12. What is a TA and what purpose does it serve?

13. For ISDN to be a reality, what signaling system has to be implemented in the intervening digital PSTN?

14. Define *primitive* in the context of ISDN protocols.

15. How does an ISDN user (e.g., a TE) derive its timing?

16. Describe *pseudoternary coding* as used as a line signal for ISDN.

17. What is the function of the balancing bit in an ISDN BRI frame?

18. Why must the simulation of a frame alignment signal be prevented?

19. In the United States, a 2B1Q line signal is used where the bit rate is 160 kbps. What is the modulation rate (baud rate) of this signal?

20. North American ISDN practice for the BRI is two-wire for full-duplex operation. What unique feature at the U-interface allows this operation without mutual interference from the outgoing and incoming bit streams on the same wire pair?

21. ISDN PRI operates at 1.544 Mbps $\pm$ 4.6 ppm at the receiver. How can it maintain such an excellent stability?

22. How many B-channels can carry traffic in the normal and conventional North American PRI configuration?

23. The 1920-kbps H_{12}-channel occupies which time slots in an E1 configuration?

24. What are the four LAPD service primitives? Where have we seen them before?

25. Name and describe at least three of the functions carried out by ISDN layer 3.

26. What are the two types of operation which can be carried out by LAPD?

27. What are the three types of LAPD control field formats?

28. What does an SAPI specify?

29. In Case A packet service, what is the responsibility of ISDN besides providing B-channel(s)?

30. What is the packet service protocol recommended for Case A ISDN packet service operation?

REFERENCES

1. *ISDN User–Network Interfaces—Interface Structures and Access Capabilities*, CCITT Rec. I.412, Fascicle III.8, IXth Plenary Assembly, Melbourne, 1988.

2. *Support of Packet Mode Terminal Equipment by an ISDN*, CCITT Rec. X.31, Fascicle VIII.2, IXth Plenary Assembly, Melbourne, 1988.

3. *ISDN Network Architecture*, ITU-T Rec. I.324, ITU Telecommunications Standardization Sector, Geneva, 1991.

4. W. Stallings, ed., *Tutorial: Integrated Services Digital Networks (ISDN)*, IEEE Computer Society Press, Washington, DC, 1985.

5. *The New IEEE Standard Dictionary of Electrical and Electronic Terms*, 5th ed., IEEE Press, New York, January 1993.

6. *Basic User–Network Interface Layer 1 Specification*, ITU-T Rec. I.430, ITU Telecommunications Standardization Sector, Geneva, March 1993.

7. *ISDN Basic Access Transport System Requirements*, Technical Reference TR-TSY-000397, Issue 1, October 1988, Bellcore, Morristown, NJ, 1988.

8. R. L. Freeman, *Telecommunication Transmission Handbook*, 3rd ed., John Wiley & Sons, New York, 1991.

9. *Primary Rate User–Network Interface—Layer 1 Specification*, ITU-T Rec. I.431, ITU Telecommunications Standardization Sector, Geneva, March 1993.

10. *Frame Alignment and Cyclic Redundancy Procedures Relating to Basic Frame Structures Defined in Rec. G.704*, CCITT Rec. G.706, Geneva, 1991.

11. *Digital Subscriber Signalling System No. 1 (DSS1): ISDN User–Network Interface Data Link Layer—General Aspects*, ITU-T Rec. Q.920, ITU Telecommunications Standardization Sector, Geneva, March 1993.

12. *ISDN User–Network Interface—Data Link Layer Specification*, ITU-T Rec. Q.921, ITU Telecommunications Standardization Sector, Geneva, March 1993.

13. *ISDN User–Network Interface: Layer 3—General Aspects*, ITU-T Rec. Q.930, ITU Telecommunications Standardization Sector, Geneva, March 1993.

14. *ISDN User–Network Interface: Layer 3—For Basic Call Control*, ITU-T Rec. Q.931, ITU Telecommunications Standardization Sector, Geneva, March 1993.

15. *Numbering Plan for the ISDN Era*, CCITT Rec. E.164, ITU Geneva, 1991.

16. *BOC Notes on the LEC Networks—1994*, Issue 2, SR-TSV-002275, Bellcore, Piscataway, NJ, April 1994.

17. *International Numbering Plan for Public Data Networks*, CCITT Rec. X.121, Fascicle VIII.3, IXth Plenary Assembly, Melbourne, 1988.

18. R. L. Freeman, *Reference Manual for Telecommunication Engineering*, John Wiley & Sons, New York, 1994.

19. *B-ISDN Service Aspects*, ITU-T Rec. I.211, ITU, Geneva, March 1993.

20. *Interface between Data Terminal Equipment (DTE) and Data Circuit-Terminating Equipment (DCE) for Terminals Operating in the Packet Mode and Accessing a Packet Switched Public Data Network through a Public Switched Telephone Network or an Integrated Services Digital Network or a Circuit Switched Public Data Network*, CCITT Rec. X.32, Fascicle VIII.2, IXth Plenary Assembly, Melbourne, 1988.

14

EMERGING BROADBAND
DATA TECHNOLOGIES

1 INTRODUCTION

In this chapter we discuss three emerging broadband digital technologies: frame relay, distributed queue dual bus (DQDB), and switched multimegabit data service (SMDS). The basic objective of each is to provide a means of local area network (LAN) interconnects over longer distances [i.e., from several miles (km) to virtually around the world].

DQDB is discussed as a way point or backdrop for the section on SMDS. It has some definite circuit length limitations.

SMDS has incorporated portions of the DQDB standard. It is an offering of North American local exchange carriers (LECs). Although it incorporated DQDB, there are no evident distance limitations. SMDS can provide sustained bit rates up to 34 Mbps, which, at least today, can handle most LAN interconnect requirements.

This chapter is also a lead-in to Chapter 15, where we discuss the asynchronous transfer mode (ATM). It is our belief that these three technologies were the forerunners of ATM. Some may argue this point.

2 HOW CAN THE NETWORK BE SPEEDED UP?—FRAME RELAY

2.1 Background and Rationale

It would seem from the terminology that somehow we were making bits travel faster down the pipe. By some means we'd broken the velocity barrier by dramatically increasing the velocity of propagation. Of course, this is eminently not true.

TABLE 14.1   Functional Comparison of X.25 and Frame Relay

Function	X.25 in ISDN (X.31)	Frame Relay
Flag recognition/generation	X	X
Transparency	X	X
FCS checking/generation	X	X
Recognize invalid frames	X	X
Discard incorrect frames	X	X
Address translation	X	X
Fill inter-frame time	X	X
Manage V(S) state variable	X	
Manage V(R) state variable	X	
Buffer packets awaiting acknowledgment	X	
Manager timer T-1	X	
Acknowledge received I-frames	X	
Check received N(S) against V(R)	X	
Generation of rejection message	X	
Respond to poll/final bit	X	
Keep track of number of retransmissions	X	
Act upon reception of rejection message	X	
Respond to receiver not ready (RNR)	X	
Respond to receiver ready (RR)	X	
Multiplexing of logical channels	X	
Management of D bit	X	
Management of M bit	X	
Management of W bit	X	
Management of P(S) packets sent	X	
Management of P(R) packets received	X	
Detection of out-of-sequence packets	X	
Management of network layer RR	X	
Management of network layer RNR	X	

If bandwidth permits,* of course, the bit rate can be increased. That certainly will speed things up. One way to get out of the bandwidth bind is to work around the analog voice channel; use some other means. ISDN was a good step in that direction. Note that 64 kbps gets three or four times as many bits per second than analog voice channel data rates.

Probably the greatest pressure to speed up the network came from LAN users that wished to extend LAN traffic to distant destinations. Ostensibly this traffic, as described in Chapter 12, has local transmission rates from 1 to 100 Mbps. X.25 WAN connectivity was one possible answer. Its packet circuits are robust, but slow and tedious. There must be a better way.

What slows down X.25 service? X.25 is feasible at 64 kbps and even at T1/E1 transmission rates. It is the intensive processing at every node (see Table 14.1) and continual message exchange as to the progress of packets, from

---

*"If bandwidth permits. . . ." There may be other constraints such as group delay, bandwidth coherence, and so forth.

node to originator and from node to destination. On many X.25 connectivities, multiple nodes are involved, slowing service still further. One clue is that X.25 was designed for circuits with poor transmission performance degrading error rates, typically BERs in the range of 1×10^{-4}. Meanwhile, the underlying digital networks in North America have error performance in the range of 1×10^{-7} or better. This begs the question of removing the responsibility of error recovery from the service provider. If errors statistically appear in about 1 in 10 million bits, there is a strong argument for removing error recovery.

In fact, with frame relay the following salient points emerge:

- There is no process for error recovery by the frame relay service provider.
- The service provider does not guarantee delivery nor are there any sort of acknowledgments provided.
- It only uses the first two OSI layers (physical and data-link layer), thus removing layer 3 and its intensive processing requirements.
- Frame overhead is kept to a minimum to minimize processing time and to increase useful throughput.
- There is no control field, no sequence numbering.
- Discarding of frames: Frames are discarded without notifying originator for such reasons as congestion and having encountered an error.
- Operates on a statmultiplex concept.

In sum, the service that the network provides can be speeded up by increasing data rate, eliminating error recovery procedures, and reducing processing time. One source states that a frame relay frame takes some 20 ms to reach the distant end (statistically), where an X.25 packet of similar size takes in excess of 200 ms on terrestrial circuits inside CONUS.

Another advantage of frame relay over a conventional static TDM connection is that it uses virtual connections. Data traffic is often bursty and normally would require much larger bandwidths to support the short data messages and much of the time that bandwidth would remain idle. Virtual connections of frame relay only use the required bandwidth for the period of the burst or usage. This is one reason why frame relay is used so widely to interconnect LANs over a wide area network (WAN). Figure 14.1 shows a typical frame relay network.

2.2 The Genesis of Frame Relay

Frame relay derives from the ISDN D-channel LAPD, or D-channel layer 2. LAPD was discussed in Chapter 13. Its importance has taken on such a magnitude that the ITU-T organization has formulated I.122, *Framework for Frame Mode Bearer Services* [1], and I.233, *Frame Mode Bearer Services* [2].

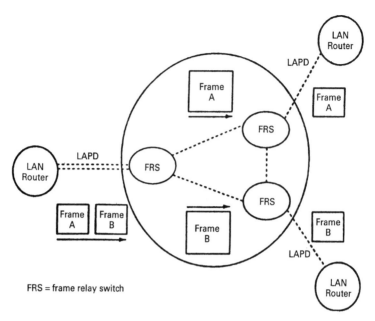

Figure 14.1. A typical frame relay network.

Even the term LAPD, although modified in many cases, continues to be used for frame relay applications.

Frame relay has become an ANSI initiative. There is also the Frame Relay Forum, consisting of manufacturers of frame relay equipment, that many feel is leading this imaginative initiative. So when we discuss frame relay, we must consider what specifications a certain system is designed around:

- ANSI, based on ANSI specifications and their publication dates
- Frame Relay Forum with publication dates
- ITU-T organization and its most current recommendations

There are also equivalent ANSI specifications directly derived from ITU-T recommendations such as ANSI T1.617-1991 [3]. We will see the term *core aspects* of ISDN LAPD or *DL-CORE*. This refers to a reduced subset of LAPD found in Annex A of ITU-T Rec. Q.922 [4]. The basic body of Q.922 presents CCITT/ITU-T specification for frame relay. This derivative is called LAPF rather than LAPD. The material found in ANSI T1.618 [5] is identical for all intents and purposes with Annex A of Q.922.

To properly describe frame relay from our perspective, we will briefly give an overview of the ANSI T1.618-1991 and T1.606-1990 [6]. This will be followed by some fairly well identified variants.

2.3   Introduction to Frame Relay

Frame relay may be considered a cost-effective outgrowth of ISDN, meeting high data rate (e.g., 2 Mbps) and low delay data communications requirements. Frame relay encapsulates data files. These may be considered "packets," although they are called frames. Thus frame relay is compared to CCITT Rec. X.25 packet service. Frame relay was designed for current transmission capabilities of the network with its relatively wider bandwidths* and excellent error performance (e.g., BER better than 1×10^{-7}).

The incisive reader will note the use of the term *bandwidth*. It is used synonymously with bit rate. If we were to admit at first approximation 1 bit per hertz of bandwidth, such use is acceptable. We are mapping frame relay bits into bearer channel bits probably on a one-for-one basis. The bearer channel may be a DS0/E0 64-kbps channel, a 56-kbps channel of a DS1 configuration, or multiple DS0/E0 channels in increments of 64 kbps up to 1.544/2.048 Mbps. We may also map the frame relay bits into a SONET or SDH configuration (Chapter 7). The final bearer channel may require more or less bandwidth than that indicated by the bit rate. This is particularly true for such bearer channels riding on radio systems, and to a lesser extent on a fiber-optic medium or other transmission media. The reader should be aware of certain carelessness of language used in industry publications.

Frame relay works well in the data rate range from 56 kbps up to 1.544/2.048 Mbps. It is being considered for the 45-Mbps DS3 rate for still additional *speed*.

ITU-T's use of the ISDN D-channel for frame relay favors X.25-like switched virtual circuits (SVCs). However, ANSI recognized that the principal application of frame relay was interconnection of LANs, and not to replace X.25. Because of the high data rate of LANs (megabit range), dedicated connections are favored. ANSI thus focused on permanent virtual connections (PVCs). With PVCs, circuits are set up by the network,† not by the endpoints. This notably simplified the signaling protocol. Also, ANSI frame relay does not support voice or video.

As mentioned above, the ANSI frame relay derives from ISDN LAPD core functions. The core functions of the LAPD protocol that are used in frame relay (as defined here) are as follows:

- Frame delimiting, alignment, and transparency provided by the use of HDLC flags and zero bit insertion/extraction.‡
- Frame multiplexing/demultiplexing using the address field.
- Inspection of the frame to ensure that it consists of an integer number of octets prior to zero bit insertion or following zero bit extraction.

---

*We'd rather use the term *greater bit rate capacity*.
†Meaning set up by the switching nodes in the network.
‡LAPD is a derivative of HDLC.

• Inspection of the frame to ensure that it is not too long or too short.

• Detection of (but *not* recovery from) transmission errors.

• Congestion control functions.

In other words, ANSI has selected certain features from the LAPD structure/ protocol, rejected others, and added some new features. For instance, the control field was removed, but certain control functions have been incorporated as single bits in the address field. These are the C/R bit (command/response), DE (discard eligibility), FECN bit (forward explicit congestion notification), and BECN bit (backward explicit congestion notification).

2.4 The Frame Structure

User traffic passed to a FRAD (frame relay access device) is segmented into frames with a maximum length information field or with a default length of 262 octets and a recommended length (ANSI) of at least 1600 octets when the application is LAN interconnectivity. The minimum information field length is one octet.

Figure 14.2 shows the frame relay frame structure. As mentioned before, it uses HDLC flags (01111110) as opening and closing flags. The closing flag may also serve as the opening flag of the next frame; however, receivers must be able to accommodate reception of one or more consecutive flags on a bearer channel.

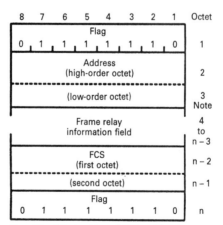

Note: The default address field length is two octets. It may be extended to either three or four octets by bilateral agreement.

Figure 14.2. Frame relay ANSI frame format with a two-octet address. From Ref. 5, reprinted with permission.

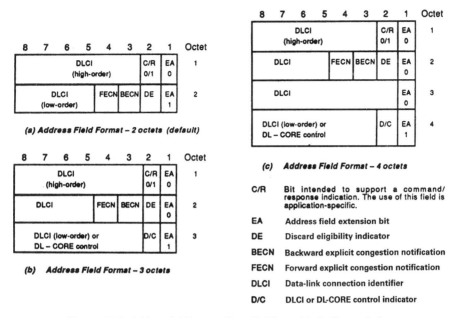

Figure 14.3. Address field formats. From Ref. 5, reprinted with permission.

ADDRESS FIELD. This consists of at least two octets, but may be extended to three or four octets, as shown in Figure 14.3. There is no control field as in HDLC, LAPD, or LAPB.

In its most reduced version, there are just 10 bits allocated to the address field in two octets (the remainder of the bits serve as control functions) supporting up to 1024 logical connections.

It should be noted that the number of addressable logical connections is multiplied because they can be reused at each nodal (switch) interface. That is, an address in the form of a data-link connection identifier (DLCI) has meaning only on one trunk between adjacent nodes. The switch (node) that receives a frame is free to change the DLCI before sending the frame onwards over the next link. Thus, the limit of 1024 DLCIs applies to the link, not the network.

INFORMATION FIELD. This follows the address field and precedes the frame check sequence (FCS). The maximum size of the information field is an implementation parameter, and the default maximum is 262 octets. ANSI chose this default maximum to be compatible with LAPD on the ISDN D-channel which has a two-octet control field and a 260-octet maximum information field. All other maximum values are negotiated between users and networks and between networks. The minimum information field size is one octet. The field must contain an integer number of octets; partial octets are not allowed. A maximum of 1600 octets is encouraged for applications such as LAN interconnects to minimize the need for segmentation and reassembly by user equipment.

TRANSPARENCY. As with HDLC, X.25 (LAPB), and LAPD, the transmitting data-link layer must examine the frame content between opening and closing flags and inserts a 0 bit after all sequences of five contiguous 1s (including the last five bits of the FCS) to ensure a flag or an abort sequence is not simulated within the frame. At the other side of the link, a receiving data-link layer must examine the frame contents between the opening and closing flags and must discard any 0 bit that directly follows five contiguous 1s.

FRAME CHECK SEQUENCE (FCS). This is based on the generator polynomial $X^{16} + X^{12} + X^5 + 1$. The CRC processing includes the content of the frame existing between, but not including, the final bit of the opening flag and the first bit of the FCS, excluding the bits inserted for transparency. The FCS, of course, is a 16-bit sequence. If there are no transmission errors (detected), the FCS at the receiver will have the sequence 00011101 00001111.

ORDER OF BIT TRANSMISSION. The order of bit transmission is that the octets are transmitted in ascending numerical order and that inside an octet, bit 1 is the first bit transmitted.

FIELD MAPPING CONVENTION. When a field is contained within a single octet, the lowest bit number of the field represents the lowest order value. When a field spans more than one octet, the order of bit values progressively decreases as the octet number increases within each octet. The lowest bit number associated with the field represents the lowest order value.

An exception to the preceding field mapping convention is the data-link layer FCS field, which spans two octets. In this particular case, bit 1 of the first octet is the high-order bit and bit 8 of the second octet is the low-order bit.

INVALID FRAMES. An invalid frame is a frame that:

- is not properly bounded by two flags (e.g., a frame abort);
- has fewer than three octets between the address field and the closing flag;
- does not consist of an integral number of octets prior to zero bit insertion or following zero bit extraction;
- contains a frame check sequence error;
- contains a single octet address field;
- contains a data-link connection identifier (DLCI) that is not supported by the receiver.

Invalid frames are discarded without notification to the sender, with no further action.

FRAME ABORT. This consists of seven or more contiguous 1 bits; upon receipt of an abort, the data-link layer ignores the frame currently being received.

2.4.1 *Address Field Discussion.*

Figure 14.3 shows the ANSI-defined address field formats. Included in the field are (a) the address field extension bits, (b) a reserved bit to support a command/response (C/R) indication bit, (c) forward and backward explicit congestion indicator (FECN and BECN) bits, (d) discard eligibility indicator (DE), (e) a data-link connection identification (DLCI) field, and (f) a bit to indicate whether the final octet of a three- or four-octet address field is the low-order part of the DLCI or DL-CORE control information. The minimum and default length of the address field is two octets. However, the address field length may be extended to three or four octets. To support a larger DLCI address range, the three- or four-octet address fields may be supported at the user–network interface or network–network interface based on bilateral agreement.

2.4.2 *Address Field Variables*

ADDRESS FIELD EXTENSION BIT (EA). The address field range is extended by reserving the first transmitted bit of the address field octets to indicate the final octet of the address field. If there is a 0 in this bit position, it indicates that another octet of the address field follows this one. If there is a 1 in the first bit position, it indicates that this octet is the final octet of the address field. As an example, for a two-octet address field, bit one of the first octet is set to 0 and bit one of the second octet is set to 1.

It should be understood that a two-octet address field is specified by ANSI. It is a user's option whether a three- or four-octet field is desired.

COMMAND/RESPONSE BIT (C/R). The C/R bit is not used by the DL-CORE protocol, and the bit is conveyed transparently.

FORWARD EXPLICIT CONGESTION NOTIFICATION (FECN) BIT. This bit may be set by a congested network to notify the user that congestion avoidance procedures should be initiated, where applicable, for traffic in the direction of the frame carrying the FECN indication. This bit is set to 1 to indicate to the receiving end-system that the frames it receives have encountered congested resources. The bit may be used to adjust the rate of destination-controlled transmitters. While setting this bit by the network or user is optional, no network shall ever clear this bit (i.e., set to 0). Networks that do not provide FECN shall pass this bit unchanged.

BACKWARD EXPLICIT CONGESTION NOTIFICATION (BECN). This bit may be set by a congested network to notify the user that congestion avoidance procedures should be initiated, where applicable, for traffic in the opposite direction of the frame carrying the BECN indicator. This bit is set to 1 to indicate to the receiving end-system that the frames it transmits may encounter congested resources. The bit may be used to adjust the rate of source-controlled transmitters.

While setting this bit by the network or user is optional according to the ANSI specification, no network shall ever clear (i.e., set to 0) this bit. Networks that do not provide BECN shall pass this bit unchanged.

DISCARD ELIGIBILITY INDICATOR (DE) BIT. This bit, if used, is set to 1 to indicate a request that a frame should be discarded in preference to other frames in a congestion situation. Setting this bit by the network or user is optional. No network shall ever clear (i.e., set to 0) this bit. Networks that do not provide DE capability shall pass this bit unchanged. Networks are not constrained to only discard frames with DE equal to 1 in the presence of congestion.

DATA-LINK CONNECTION IDENTIFIER (DLCI). This is used to identify the logical connection, multiplexed within the physical channel, with which a frame is associated. All frames carried within a particular physical channel and having the same DLCI value are associated with the same logical connection.

The DLCI is an unstructured field. For two-octet addresses, bit 5 of the second octet is the least significant bit. For three- and four-octet addresses, bit 3 of the last octet is the least significant bit. In all cases, bit 8 of the first octet is the most significant bit.

The structure of the DLCI field may be established by the network at the user–network interface subject to bilateral agreements.

In order to allow for compatibility of call control and layer management between B/H- and D-channels, the following ranges of DLCIs are reserved and preassigned. See Table 14.2. The DLCIs have local significance only.

DLCI ON THE D-CHANNEL. The six most significant bits (bits 8 to 3 of the first octet) correspond to the service access point identifier (SAPI) field in ANSI (standard) T1.602 [9].

The DLCI subfield (bits 8 to 3 of the first octet) values that apply on a D-channel are reserved for specific functions to ensure compatibility with operation of the D-channel that may also use ANSI T1.602 protocols. A two-octet address format for the DLCI is assumed when used on the D-channel. *Note:* For frame relay in the D-channel, only DLCI values in the range 512–991 (SAPI = 31–61) will be assigned.

Table 14.3 gives DLCI values for the D-channel.

TABLE 14.2   DLCI Values for B-Channel and H-Channel Applications

DLCI Values	Function
	Two-Octet Address Format
0	In-channel signaling
1–15	Reserved
16–991	Assigned using frame relay connection procedures[a]
992–1007	Layer 2 management of frame relay bearer service
1008–1022	Reserved
1023	In-channel layer management
	Three-Octet Address Format with D/C = 0
0	In-channel signaling
1–1023	Reserved
1024–63,487	Assigned using frame relay connection procedures[a]
63,488–64,511	Layer 2 management of frame relay bearer service
64,512–65,534	Reserved
65,535	In-channel layer management
	Four-Octet Address Format with D/C = 0
0	In-channel signaling
1–131,071	Reserved
131,072–8,126,463	Assigned using frame relay connection procedures[a]
8,126,464–8,257,535	Layer 2 management of frame relay bearer service
8,257,536–8,388,606	Reserved
8,388,607	In-channel layer management

[a]Some of these values may be assigned to permanent frame relay calls.
Source: Ref. 5, reprinted with permission.

TABLE 14.3   DLCI Values for D-Channel (Two-Octet Address Format)

DLCI Values	Function
512–991	Assigned using frame relay connection procedures

Source: Ref. 5, reprinted with permission.

DLCI OR DL-CORE CONTROL INDICATOR (D/C). The D/C indicates whether the remaining six usable bits of that octet are to be interpreted as the lower DLCI bits or as DL-CORE control bits. This bit is set to 0 to indicate that the octet contains DLCI information. When this bit is set to 1, it indicates that the octet contains DL-CORE control information. The D/C is limited to use in the last octet of the three- or four-octet-type address field. The use of this indication for DL-CORE control is reserved as there have not been any additional control functions defined that need to be carried in the address field. Thus this indicator has been added to provide possible future expansion of the protocol.

2.5 DL-CORE Parameters (As Defined by ANSI)

DLCI VALUE parameter conveys the DLCI agreed to be used between core entities in support of DL-CORE connection. Its syntax and usage are described above.

DL-CORE connection endpoint identifier (CEI) uniquely defines the DL-CORE connection.

Physical connection endpoint identifier (ph-CEI) uniquely identifies a physical connection to be used in support of a DL-CORE connection.

2.6 Procedures

For permanent frame relay bearer connections, information related to the operation of the DL-CORE protocol in support of DL-CORE connection is maintained by DL-CORE management. For demand frame relay bearer connections, layer 3 establishes and releases DL-CORE connections on behalf of the DL-CORE sublayer. Therefore, information related to the operation of the DL-CORE protocol is maintained by coordination of layer 3 management and DL-CORE sublayer management through the operation of the local system environment.

CONNECTION ESTABLISHMENT. When it is necessary to notify the DL-CORE sublayer entity (either because of establishment of a demand frame relay call, because of notification of reestablishment of a permanent frame relay bearer connection, or because of system initialization) that a DL-CORE connection is to be established, the DL-CORE layer management entity signals an MC-ASSIGN request primitive to the DL-CORE sublayer entity.

The DL-CORE sublayer entity establishes the necessary mapping between supporting ph-connection, the core-CEI, and the DLCI. In addition, if it has not already done so, it begins to transmit flags on the physical connection except on the D-channel.

CONNECTION RELEASE. When it is necessary to notify the DL-CORE sublayer entity (either because of release of a demand frame relay call or because of notification of failure of a permanent frame relay bearer connection) that a DL-CORE connection is to be released, the DL-CORE layer management entity signals the MC-REMOVE request primitive to the DL-CORE sublayer entity.

2.7 Traffic and Billing on Frame Relay

Figure 14.4 shows a typical traffic profile on a conventional public telephone network, whereas Figure 14.5 shows a typical profile of bursty traffic over a

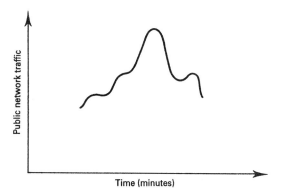

Figure 14.4. Typical traffic profile of a public switched telephone network. Courtesy of Hewlett-Packard Company [7].

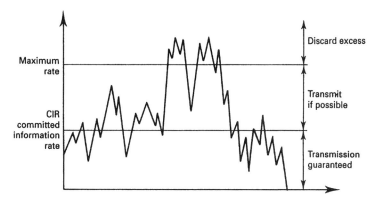

Figure 14.5. Typical bursty traffic of frame relay. Note the traffic levels indicated. Courtesy of Hewlett-Packard Company [7].

frame relay network. Such bursty traffic is typical for a LAN. The primary employment of frame relay is to interconnect LANs at a distance.

Turning to Figure 14.5, the traffic is bursty. With conventional leased data circuits, we have to pay for the bandwidth* whether it is used or not. On the other hand, with frame relay, we only pay for the "time" used. Billing can be handled in one of three ways:

1. CIR (committed information rate) is a data rate subscribed to by a user. This rate may be exceeded for short bursts during the peak period as shown in Figure 14.5.

2. We can pay a flat rate.

3. We can pay per packet (i.e., frame).

*I prefer the use of the expression *bit rate capacity*.

Note that on the right-hand side of Figure 14.5 there is the guaranteed transmission bit rate equivalent to the CIR. Depending on the traffic load and congestion, during short periods a user may exceed the CIR. However, there is a point where the network cannot sustain further increases in traffic without severe congestion resulting. Traffic above such levels is arbitrarily discarded by the network without informing the originator.

2.8   Congestion Control, a Discussion

Congestion in the user plane occurs when traffic arriving at a resource exceeds the network's capacity. It can also occur for other reasons such as equipment failure. Network congestion affects the throughput, delay, and frame loss experienced by the end user.

End users should reduce their offered load in the face of network congestion. Reduction of offered load by an end user may well result in an increase in the effective throughput available to the end user during congestion.

Congestion avoidance procedures, including optional explicit congestion notification, are used at the onset of congestion to minimize its negative effects on the network and its users. Explicit notification is a procedure used for congestion avoidance and is part of the data transfer phase. Users should react to explicit congestion notification (i.e., optional but highly desirable). Users that are not able to act on explicit congestion notification shall have the capability to receive and ignore explicit notification generated by the networks.

Congestion recovery and the associated implicit congestion indication due to frame discard are used to prevent network collapse in the face of severe congestion. Implicit congestion detection involves certain events available to the protocols operating above the core function to detect frame loss (e.g., receipt of a REJECT frame, timer recovery). Upon detection of congestion, the user reduces the offered load to the network. Use of such reduction by users is optional.

2.8.1   Network Response to Congestion. Explicit congestion signals are sent in both forward (towards frame destination) and backward (towards frame source) directions. Forward explicit congestion notification is provided by using the FECN bit in the address field. Backward explicit congestion notification is provided by one of two methods. When timely reverse traffic is available, the BECN bit in an appropriate address field may be used. Otherwise, a single consolidated link layer management message may be generated by the network. The consolidated link layer management (CLLM) message travels on the U-plane physical path. The generation and transport of CLLM by the network are optional.

All networks transport the FECN and BECN bits without resetting.

2.8.2   *User Response to Congestion.* Reaction by the end user to the receipt of explicit congestion notification is rate-based. Annex A to ANSI T1.618-1991 [5] describes user reaction to FECN and BECN.

END-USER EQUIPMENT EMPLOYING DESTINATION-CONTROLLED TRANSMITTERS. End-user reaction to implicit congestion detection or explicit congestion notification (FECN indications), when supported, is based on the values of FECN indications that are received over a period of time. The method is consistent with commonly used destination-controlled protocol suites, such as OSI class 4 transport protocol operated over the OSI connectionless service.

END-USER EQUIPMENT EMPLOYING SOURCE-CONTROLLED TRANSMITTERS. End-user reaction to implicit congestion notification (BECN indication), when supported, is immediate when a BECN indication or a CLLM is received. This method is consistent with implementation as a function of data-link layer elements of procedure commonly used in source-controlled protocols such as CCITT Rec. Q.922 elements of procedure.

2.8.3   *Consolidated Link Layer Management (CLLM) Message.* The CLLM uses XID frames for the transport of functional information. We may remember in the discussion of HDLC (Chapter 11) that an XID frame was an *exchange identification* frame. It was for reporting its station identification. In frame relay, CLLM with its XID frames is used for network management as an alternative for congestion control. CLLM messages originate at network nodes to the frame relay interface usually housed in a router near or incorporated with the user.

As we mentioned, BECN/FECN bits in frames must pass congested nodes in the forward or backward direction. Suppose that for a given user no frames pass in either direction and that the user therefore has no knowledge of network congestion because at that moment the user is not transmitting or receiving frames. Frame relay standards do not permit a network to generate frames with the DLCI of the congested circuit. CLLM covers this contingency. It has DLCI = 1023 reserved.

The use of CLLM is optional. If it is used, it may or may not operate in conjunction with BECN/FECN. The CLLM frame format has one octet for the cause of congestion such as excessive traffic, equipment or facility failure, preemption, or maintenance action.

This same octet indicates whether the cause is expected to be short or long term. Short term is on the order of seconds or minutes, and anything greater is long term. There is also a bit sequence in this octet indicating an unknown cause of congestion and whether short or long term.

CLLM octets 19 and above give the DLCI values that identify logical links that have encountered congestion. This field must accommodate DLCI length such as two-octet, three-octet, and four-octet DLCI fields.

2.8.4   Action of a Congested Node. When a node is congested, it has several alternatives it may use to mitigate or eliminate the problem. It may set the FECN and BECN bits to "1" in the address field and/or use the CLLM message. Of course, the purpose of explicit congestion notification is:

- to inform the "edge" node at the network ingress of congestion so that that edge node can take the appropriate action to reduce the congestion or
- to notify the source that the negotiated throughput has been exceeded or
- to do both.

One of the strengths of the CLLM is that it contains a list of DLCIs that correspond to the congested frame relay bearer connections. These DLCIs indicate not only the sources currently active causing the congestion but also those sources that are not active. The reason for the latter is to prevent those sources that are not active from becoming active and thus causing still further congestion. It may be necessary to send more than one CLLM message if all the affected DLCIs cannot fit into a single frame.

2.9   Policing a Frame Relay Network

2.9.1   Introduction. Frame relay switches may carry out a policing function on accessing users. The result of such action is discarded frames. Similar policing actions show up in SMDS (switched multimegabit data service—Section 3 of this chapter) and in ATM (asynchronous transfer mode) discussed in Chapter 15.

2.9.2   Definitions

Access Rate (AR). The data rate, expressed in bps, of the user access channel (D-, B-, or H-channel). The rate at which users can offer data to the network is bounded by the access rate.

Excess Burst Size (B_e). The maximum amount of uncommitted data (in bits) that the network will attempt to deliver over measurement interval (T). These data may or may not be contiguous (i.e., may appear in one frame or in several frames, possibly with interframe idle flags). B_e is negotiated at call establishment (for demand establishment of communication) or at service subscription time (for permanent establishment of communication). Excess burst data may be marked for discard eligibility (with the DE bit) by the network.

Measurement Interval (T). The time interval over which rates and burst sizes are measured. In general, the duration of T is proportional to the *burstiness* of the traffic. Except as noted below, T is computed as $T = B_c/\text{CIR}$ or $T = B_e/\text{AR}$.

Committed Information Rate (CIR). The rate, expressed in bps, at which the network agrees to transfer information under normal conditions. This rate is measured over the measurement interval T. CIR is negotiated at call establishment or service subscription time. Data marked *discard eligible* (DE) is not accounted for in CIR.

Committed Burst Size (B_c). The maximum amount of data (in bits) that a network agrees to transfer under normal conditions over a measurement interval T. These data may or may not be contiguous (i.e., they may appear in one frame or in several frames, possibly with interframe idle flags). B_c is negotiated at call establishment (for demand establishment of communication) or service subscription time (for permanent establishment of communication).

Fairness. An attempt by the network to maintain the negotiated quality of service for all users operating under normal conditions (i.e., within their CIR and B_c, without discard due to congestion). For example, the network may discard frames offered in excess of CIR and may reject any call attempts that would cause the network resources to be overcommitted.

Offered Load. The bits offered to the network by an end user, to be delivered to the selected destination. The information rate and burst length offered to the network could exceed the negotiated class of service parameters. The offered load consists of the user data portion of frames and therefore excludes flags, FCS, and inserted zeros.

2.9.3 Relationship Among Parameters. Figure 14.6 illustrates the relationship of access rate, excess burst, committed burst, committed information rate, discard eligibility indicator, and measurement interval parameters. The CIR, B_c, and B_e parameters are negotiated at call establishment time for demand establishment of communication or established by subscription for permanent establishment of communication. Access rate is established by subscription for permanent access connections or during demand access connection establishment. Each end user and the network participate in the negotiation of these parameters to agreed-upon values. These negotiated values are then used to determine the measurement interval parameter, T, and when the discard eligibility indicator (if used) is set. These parameters are also used to determine the maximum allowable end-user input levels. The relationship among parameters can be used at any instant of time, T_0, to measure the offered load over the interval $(T_0, T_c + T)$. Similarly, the offered load over any interval $(t - T, t)$ may be measured at any instant of time t, as long as the measurement function retains memory of user activity over the previous interval T. One way of doing this is by use of a "leaky bucket" algorithm described below.

The measurement interval is determined as shown in Table 14.4. The network and the end users may control the operation of the discard eligibility indicator (DE) and the rate enforcer functions by adjusting at call setup the

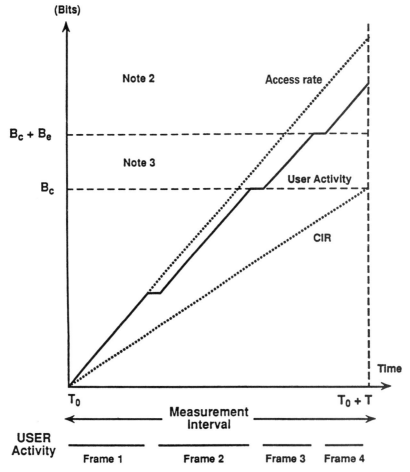

Figure 14.6. Illustration of relationships among parameters. From ANSI T1.606a-1992, Figure 7, page 9 (Ref. 8), reprinted with permission.

TABLE 14.4 Congestion Parameter State

CIR	B_c	B_e	Measurement Interval (T)
>0	>0	>0	$T = B_c/\text{CIR}$
>0	>0	=0	$T = B_c/\text{CIR}$
=0	=0	>0	T is a network-dependent value

Source: Ref. 8, reprinted with permission.

CIR, B_c, and B_e parameters in relation to the access rate. If both of the CIR and B_c parameters are not equal to zero, then $T = B_c/\text{CIR}$. In addition, there are two special conditions:

(a) When CIR = access rate, $B_c = 0$, and $B_e = 0$, both access rates must be equal (i.e., ingress = egress).
(b) When CIR = 0 (B_c must = 0) and $B_e > 0$, then T is a *network-dependent value*.

Notes

1. The ingress and egress access rate do not have to be equal; however, when the ingress access rate is substantially higher than the egress access rate, continuous input of B_e frames at the ingress interface may lead to persistent congestion of the network buffers at the egress interface, and a substantial amount of the input B_e data will be discarded.
2. When a frame, entering the ingress node, consumes the remaining capacity of B_e or B_c, reducing it to zero, the action taken on that frame is network dependent.

Figure 14.6 is a static illustration of the relationship among time, cumulative bits of user data, and rate. In this example, the user sends four frames during the measurement interval $(T_0 + T)$. The *slope* of the line marked "CIR" is B_c/T. Bits are received at the access rate (by the ingress node) of the access channel. Since the sum of the number of bits contained in frames one and two is not greater than B_c, the network does not mark these frames with the discard eligibility indicator (DE). The sum of the number of bits in frames one, two, and three is greater than B_c, but not greater than $B_c + B_e$; therefore frame three is marked discard eligible. Since the sum of the number of bits received by the network in frames one, two, three, and four exceed $B_c + B_e$, frame four is discarded at the ingress node. This figure does not address the case in which the end user sets the DE bit.

Section 2.9 is based on ANSI T1.606a–1992, paragraph 10.2 (Ref. 8)

2.10 Quality of Service Parameters

The quality that frame-relaying service provides is characterized by the values of the following parameters. ANSI adds in Ref. 6 that the specific list of negotiable parameters is for further study.

1. Throughput
2. Transit delay

3. Information integrity

4. Residual error rate

5. Delivered error(ed) frames

6. Delivered duplicated frames

7. Delivered out-of-sequence frames

8. Lost frames

9. Misdelivered frames

10. Switched virtual call establishment delay

11. Switched virtual call clearing delay

12. Switched virtual call establishment failure

13. Premature disconnect

14. Switched virtual call clearing failure

2.10.1 Network Responsibilities. The frames are routed through the network on the basis of an attached label (i.e., the DLCI value in the frame). This label is a logical identifier with local significance. In the virtual call case, the value of the logical identifier and other associated parameters such as layer 1 channel delay, and so on, may be requested and negotiated during call setup. Depending on the value of the parameters, the network may accept or reject the call. In the case of the permanent virtual circuit, the logical identifier and other associated parameters are defined by means of administrative procedures (e.g., at the time of subscription).

The user–network interface structure allows for the establishment of multiple virtual calls or permanent virtual circuits, or both, to many destinations over a single access channel.

Specifically, for each connection, the bearer service:

1. Provides bidirectional transfer of frames.

2. Preserves their order as given at one user–network interface if and when they are delivered at the other end. (*Note:* No sequence numbers are kept by the network. Networks are implemented in such a way that frame order is preserved.)

3. Detects transmission, format, and operational errors such as frames with an unknown label.

4. Transports the user data contents of a frame transparently. Only the frame's address and FCS fields may be modified by network nodes.

5. Does not acknowledge frames.

At the user–network interface, the FRAD (frame relay access device), as a minimum, has the following responsibilities:

1. Frame delimiting, alignment, and transparency provided by the use of HDLC flags and zero bit insertion.

2. Virtual circuit multiplexing/demultiplexing using the address field of the frame.

3. Inspection of the frame to ensure that it consists of an integer number of octets prior to zero bit insertion or following zero bit extraction.

4. Inspection of the frame to ensure it is not too short or too long.

5. Detection of transmission, format, and operational errors.

A frame received by a frame handler may be discarded if the frame:

1. Does not consist of an integer number of octets prior to zero bit insertion or following zero bit extraction.

2. Is too long or too short.

3. Has a frame check sequence (FCS) that is in error.

The network will discard a frame if it:

1. Has a DLCI value that is unknown.

2. Cannot be routed further due to internal network conditions. A frame can be discarded for other reasons, such as exceeding negotiated throughput.

Sections 2.3 through 2.8 were adapted from ANSI T1.618-1991 (Ref. 5). Sections 2.9 and 2.10 were adapted from ANSI T1.606-1990 (Ref. 6) and T1.606a-1992 (Ref. 8).

3 DISTRIBUTED QUEUE DUAL BUS (DQDB)

3.1 Introduction and Purpose

DQDB was prepared for use with metropolitan area networks (MANs). A MAN is similar to a LAN but with considerably greater geographical coverage—a metropolitan area, if you will. In fact, the relevant DQDB specification is covered under the IEEE 802 series (see Chapter 6). Much of the information provided in this section is based on IEEE Std 802.6-1990 [10].

DQDB can be considered a procedure or a protocol. It is not tied to any particular bit rate. It can be implemented on DS1/E1 and DS3, as well as on SONET rates such as 51 and 155 Mbps and higher rates. The DQDB concept was derived from Australia's QPSX Systems which developed a proprietary system that allowed all stations on a dual-path, counterflowing bus to maintain a distributed queue of packets awaiting access to the bus. QPSX was taken up by the IEEE 802.6 committee, and the outcome was the Distributed Queue Dual Bus IEEE 802.6 specification [10].

The DQDB is a subnetwork of a MAN. The purpose of a MAN is to provide integrated services such as data, voice, and video over a large geographic area. The interconnection of DQDB subnetworks could be via routers, bridges, or gateways. (Note the distinct taste of LANs in the wording.)

3.2 An Overview of DQDB

The *dual bus* of DQDB consists of two parallel transmission paths operating in opposite directions. A multiplicity of operational nodes are connected to the dual bus along its route. This concept is shown in Figure 14.7. The buses, denoted A and B in the figure, support communications in opposite directions, allowing full-duplex connectivity between any pair of nodes in the subnetwork. Because both buses are operational at all times, the network has nearly twice the capacity of a single bus. The operation of the two buses in the transfer of data is independent [10].

Figure 14.7 shows a *headend* at the head of each bus. Each headend node generates either DQDB layer management octets or fixed-length slots. The management information octets are used to maintain the operational integrity of the network. The slots are used to carry payload data between/among nodes. Data written into a slot is under the control of an access protocol. All data flow terminates at the end of the bus.

A DQDB network node consists of an attachment unit (AU) and its attachment to both buses. The node access unit performs the DQDB layer functions and attaches to the bus via one read and one write connection. We can roughly associate "read" with a receive function and "write" with a transmit function. The writing of data to the bus is a logical OR of the data from upstream with the data from the access unit. The read connection is

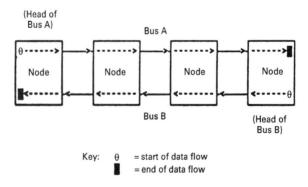

Figure 14.7. Dual bus architecture.

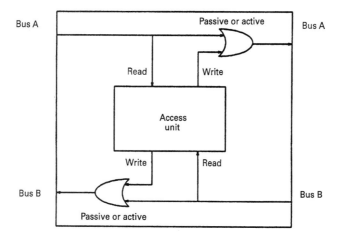

Figure 14.8. An access unit attachment. Note read and write functions on each bus. From Ref. 10, reprinted with permission.

placed logically ahead of the write connection and allows all data to be copied from the bus unaffected by the unit's own writing. An example of an AU connection is shown in Figure 14.8.

There is a single source of slot timing for the subnetwork, which is used to ensure that all nodes of the subnetwork transfer data (i.e., slots and management information octets) at the same average rate. The timing source can be traceable to an external timing source such as the PSTN. Any node or nodes can be designated to provide the external timing source function. If the network is isolated, one node can be designated as the timing source. In such a case, the node must have a clock that has a nominal period of 125 µs.

3.3   Access to the DQDB Network

There are two modes of access control to the dual bus. These are *queued arbitrated* (QA) and *pre-arbitrated* (PA) and they use QA and PA slots for access, respectively. Each slot has an access control field (ACF) and a segment. A segment is further broken down into a header and a payload. Our immediate concern here is the ACF. The QA access is controlled by the distributed queueing protocol and is used to provide nonisochronous services (such as conventional data transfer). The PA access is used in support of isochronous services (typically voice).

3.3.1   DQDB Access Protocol.   The DQDB protocol provides deterministic access (i.e., opposite of random access) to the payload of QA slots on the DQDB bus. The fixed-length payload of a QA slot is called a QA segment. In particular, the DQDB protocol can enable all of the payload bandwidth to be used, and the average slot access delay approximates that of

a perfect scheduler up to high levels of network loading. The protocol operation is very simple. It is based on two control fields in the ACF: the BUSY bit and the REQUEST field. The BUSY bit indicates whether or not a slot is used, whether it contains traffic which is a *segment*; the REQUEST field is used to indicate when a segment has been queued for access. Each node, by counting the number of requests it receives and the unused slots that pass, can determine the number of segments queued (i.e., in line) ahead of it. This counting operation establishes a single queue across the network of segments queued for access to each bus.

Such a queued access protocol lends itself well to establishing levels of priority. A queue can be operated for each level of priority. Segments gain access as soon as capacity becomes available, but priority is always given to segments in higher-level queues.The 802.6-1990 standard [10] specifies three levels of priority.

Distributed queueing is fundamentally different from other LAN MAC protocols. Here there is a form of nodal autonomy in that there is explicit information on the queueing state of the network at each node. Thus, when a node has data to transmit, the node need not first derive information from the network to tell it when it can gain access.

Each node keeps a current record of the number of segments awaiting access. When a node has a segment to transmit, it uses its count to determine its position in the distributed queue. When there are no queued segments, access to the bus is immediate.

THE BASIC ALGORITHM. In Figure 14.9, Bus A is the forward bus and Bus B is the reverse bus when access to Bus A is desired. If we wish access to Bus B, an identical but independent arrangement applies.

The ACF field is key to access. As we mentioned, it contains a BUSY bit and a REQUEST field. This latter contains three bits, one for each priority level. The BUSY bit indicates whether a particular slot is used or idle. The REQUEST bits tell when a QA segment has been placed in queue on the reverse bus.

SINGLE PRIORITY CASE. When an AU has a QA segment for transmission on the forward bus, it will actuate a single REQUEST (REQ) on the reverse bus. This REQ (REQUEST) will be written into the next free REQ bit of the required priority on the reverse bus. Once the REQ bit is written, it will be passed to all upstream AUs, where we define upstream in relation to data flow on the forward bus, in this case Bus A. This REQ bit serves as an indicator to the upstream AUs that an additional QA segment is now queued for access. Each AU is only allowed, at most, one QA segment per priority level, to be queued for access to each bus.

Each AU keeps track of the number of QA segments queued downstream from itself for access to the forward bus by counting the REQ bits as they pass on the reverse bus. This is shown in Figure 14.10A. For each REQ

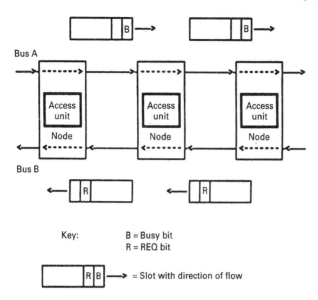

Figure 14.9. Queue formation on Bus A. From Ref. 10, reprinted with permission.

(REQUEST) passing on the reverse bus, the RQ counter is incremented one. For a node that is *not* queued to send, one REQ in the RQ counter is canceled each time an empty slot passes on the forward bus. The node AU does this because the empty slot that passes the AU will be used by one of the downstream queued QA segments. Hence, with these two actions, the RQ counter keeps a record of the number of segments queued downstream.

In addition to issuing the REQ for the reverse bus, an AU with a QA segment to transmit transfers the current value of the RQ counter to another

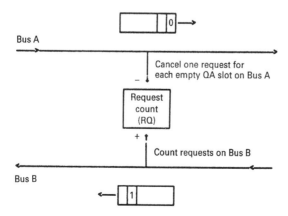

Figure 14.10A. Node not queued to send on Bus A. From Ref. 10, reprinted with permission.

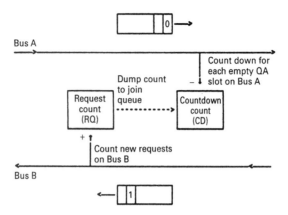

Figure 14.10B. Node queued to send on Bus A. From Ref. 10, reprinted with permission.

counter, the countdown (CD) counter. This is shown in Figure 14.10B, the RQ counter then being reset to zero. This action loads the CD counter with the number of downstream segments queued ahead of it. This, along with the issuing of the REQ for the AU's segment, effectively places the QA segment in the distributed queue. The distributed queue at a given priority approximates a FIFO (first-in, first-out) queue of the QA segments at the heads of the local queues in each node.

To ensure fairness in that the segments registered in the CD counter gain access before the newly queued segments in a given AU, the CD counter is decremented for every empty slot (cell) that passes on the forward bus, shown in Figure 14.10B. A given AU can transmit its QA segment in an empty cell (slot), provided that the CD count is zero. For this single priority description, this is equivalent to claiming the first free slot after the CD count reaches zero, which ensures that no downstream segment that queued after the given segment can access out of order.

Any REQs received from the reverse bus are added to the RQ counter during the time that the AU is waiting for access for its segment. This is shown in Figure 14.10B. Thus the RQ counter still tracks the number of segments newly queued downstream, and the count will be correct for the next QA segment access.

It should be noted that the operation of writing REQ bits and sending QA segments is independent because the control of the access of the QA segment to the forward bus is determined solely by the values of the counters. Access is not inhibited if the value of the CD counter is zero but the REQ bit associated with the QA segment has not yet been written to the reverse bus.

As we can see, a FIFO queue is established for access to the forward bus. This is so because there are two counters in each AU. One counts unsatisfied access requests, and the other counts down before access. The queue formation is also such that, except for bandwidth balancing described later, a slot is never wasted on the network if there is a segment queued for it. This is

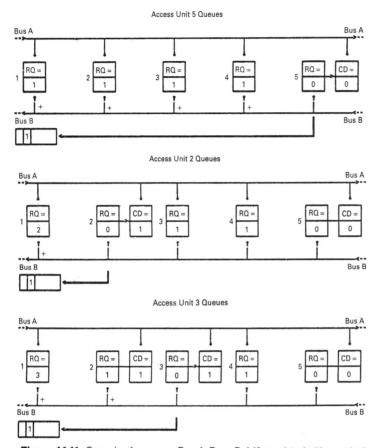

Figure 14.11. Queueing for access Bus A. From Ref. 10, reprinted with permission.

guaranteed since the CD count in AUs with segments queued represents the number of downstream segments ahead in the queue. Since at any point in time (normalized for propagation delay), one segment must have queued first, then at least one AU is guaranteed to have a CD count of zero. It is the most upstream of those that will gain access.

Figure 14.11 is an example of distributed queueing. It shows a network of 5 AUs, 3 of which will queue for access. Let's assume that each request counter, the value of which is indicated by RQ at the AU, and each CD counter, the value of which is indicated by CD at the AU, are all initially zero and that all slots (cells) passing Bus A are used. AUs 5, 2, and 3 queue in that order, use the operations we described above. The counters in the final state are interpreted as follows:

AU1: There are three AUs downstream on Bus A that have QA segments queued for access.

AU2: There is one downstream AU with a QA segment queued for access before AU2, as represented by the one in the CD counter. There is also one downstream AU with a QA segment queued for access after AU2, as represented by the one on the RQ counter.

AU3: There is one downstream AU with a QA segment queued for access before AU3, as represented by the one in the CD counter, and no downstream AU with a QA segment queued after AU3, as represented by the zero in the RQ counter. It should be noted that the REQ from AU2 is not registered at AU3.

AU4: There is one downstream AU with a QA segment queued for access.

AU5: There are no downstream AUs from AU5; hence, both RQ and CD counts are zero.

Now consider Figure 14.12, which shows the AUs gaining access. Assume that no further requests are received on the reverse bus and that empty QA slots pass on Bus A. The AUs then gain access in the order 5, 2, then 3, which is the order of queueing. Note that after AU5 gains access, both AU2 and AU3 have a CD count of zero. Because it is the most upstream, AU2 will gain access first.

PRIORITY DISTRIBUTED QUEUEING. The distributed queueing protocol supports assignment of priority to QA access. There are three priority levels. However, all connectionless data segments are sent at the lowest priority, called priority level 0.

The operation of separate distributed queues for each level of priority is achieved by using a separate REQ bit on the reverse bus for each priority level and separate RQ and CD counters for each level of priority. These counters operate in a similar manner as the single priority case, except that account is taken of REQs at higher priority levels.

That is, for an AU that does not have a QA segment queued at a particular priority level, the RQ counter operating at that level will count REQs at the *same and higher* priority levels. Thus, the RQ counter records all queued segments at equal and higher priorities.

BANDWIDTH BALANCING. Bandwidth balancing involves the occasional skipping the use of QA cells (slots). It may be desirable to implement bandwidth balancing when the bit rate/distance product exceeds one cell duration (53×8 or 424 bits). The following guidelines are based on that product:

2 km at DS3 (44.736 Mbps)

546 meters for STS-3/STM-1 (SONET/SDH) 155.520 Mbps

137 meters for STS-12/STM-4 622.080 Mbps

Ideally for QA access with priority, priority queueing enables selective degradation, under overload conditions, of lower priority services, allowing preferential access to higher priority delay-sensitive services. With bandwidth

Access Units 5, 2, Then 3 Queued

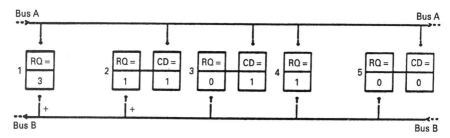

Access Unit 5 Gains Access

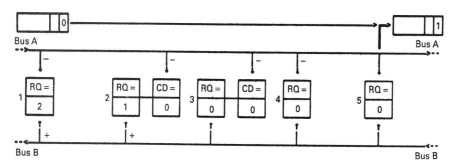

Access Unit 2 Gains Access

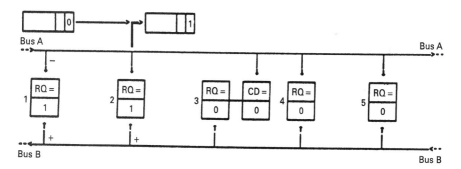

Access Unit 3 Gains Access Next

Figure 14.12. Gaining access (Bus A). From Ref. 10, reprinted with permission.

balancing disabled, the distributed queue mechanism with priority as specified in IEEE Std 802.6 can be relied upon to provide fair sharing of capacity and preferential access if the active stations accessing a given bus span a distance that is less than the equivalent of one 53-octet slot as mentioned above.

Beyond the distances listed above, particularly under overload conditions, the effectiveness of the priority queueing mechanism decreases with increasing number of slots on the bus between contending stations.

Bandwidth balancing may be desirable in a DQDB network when the network physical configuration exceeds the distance–transmission rate products given above. For example, the distance between any two stations exceeds 546 meters on a shared medium whose data rate is 155.52 Mbps.

The bandwidth balancing technique is used to ensure fair sharing of bandwidth between stations operating at a single priority. In this context, *fair* is defined as giving an approximately equal share of bandwidth to all stations attempting to access the medium of transmission. When the physical conditions are as given above and when the offered load to all stations exceeds the bandwidth available on the medium, the use of bandwidth balancing allows all stations to receive an equal share of the bandwidth in the steady-state condition. The performance during the transition to steady state is dependent on the bandwidth balancing modulus (BWB_MOD) and the distance between stations. In the steady state, the medium utilization is less than 100%, being equal to BWB_MOD/(BWB_MOD + 1) × 100% if there is only one active node. The steady-state medium utilization increases with the number of active nodes.

The bandwidth balancing mechanism divides bus bandwidth among the stations and allows some bandwidth to go unused. Take the example of a bus with N stations, if all the following conditions are met:

- No station has any pre-arbitrated traffic.

- Each station always has queued arbitrated (QA) segments waiting to be transmitted on the bus.

- All segments have the same priority.

- The value of the BWB_MOD at each station is M [this means that each station uses a fraction $M/(M + 1)$ of the slots not used by other stations].

- These load conditions persist for a sufficiently long time; then the bandwidth balancing mechanism provides each station with a steady-state average throughput of $1/(N + 1/M)$ per slot time, and the total utilization of the bus in steady state is $N/(N + 1/M)$ segments per slot time.* Note that the total utilization increases as the BWB_MOD increases and as the number of active stations increases. The station throughputs approach their steady-state values gradually. The convergence is faster if the BWB_MOD is smaller.

---

*To understand this formula, recall that the throughput r of a station should equal a fraction $M/(M + 1)$ of the bandwidth not used by the other $N - 1$ stations: $r = [M/(M + 1)] \times [1 - (N - 1) \times r]$ segments per slot time. Solving for r gives the stated result. (10.)

PRE-ARBITRATED ACCESS CONTROL. Pre-arbitrated (PA) slot access is used typically to provide for transfer of isochronous service octets. The access to PA slots (cells) and the use of PA segment payloads differ from that of QA access. The access differs in that PA slots are designated by the node at the head of the bus and that more than one AU may share access to the slot. A PA segment payload consists of a number of octets, each of which can be used by a different AU. Therefore, an AU may write zero, one, or more isochronous service octets into designated positions of a PA segment payload. The AU is notified of the offsets of these octet positions relative to the start of the PA segment payload via the DQDB layer management procedures.

The designation of slots to PA access and the marking of the PA slot and PA segment header is controlled by functions in the node at the head of the bus. In particular, the head of bus function must write the virtual channel identifier (VCI) field into PA slots. The head of bus function must also ensure that the PA slots for each VCI value are provided in a periodic manner on the bus to guarantee that sufficient bandwidth is available for isochronous service users (ISUs).

The access to the PA slots by an AU commences by examining the VCI. For each VCI value that the AU must access, the AU will have a table that indicates which octet offsets within the slot the AU should use for reading and writing. The AU will write isochronous service octets into those write positions and will read from positions the table has marked for reading. The PA slot is ignored if the VCI is not one in use by the AU.

3.4 DQDB Layer Services

A major function of a metropolitan area network (MAN) is to provide LAN connectivity over fairly long distances. The DQDB subnetwork was designed for just that. LAN traffic would derive as a MAC service data unit (MSDU) which is segmented at the source into fixed-length units and the fixed-length units are transferred to the destination, where they are reassembled into the original MSDU.

The segmentation process follows the formation of an initial MAC protocol data unit (IMPDU) by the addition of an IMPDU header, an optional header extension, an optional 32-bit CRC, a common PDU trailer,* and a variable-length PAD field to the MSDU. The PAD field ensures that all of the fields added to the MSDU are 32-bit aligned. The IMPDU is fragmented into fixed-length segmentation units as shown in Figure 14.13, for transfer in QA segments' payloads. There may be padding of the IMPDU with trailing zero octets to ensure complete filling of the last segmentation unit.

---

*A trailer consists of a number of fields attached after the info or data field of a frame, packet, or slot.

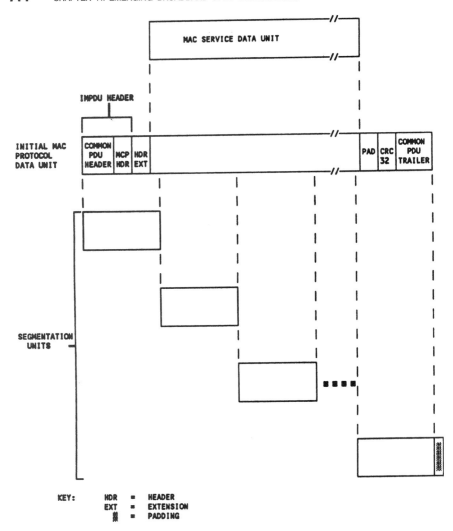

Figure 14.13. Segmentation of an IMPDU.

All segment payloads that support the MAC service (from a LAN) are called derived MAC protocol data units (DMPDUs) and consist of a header field and a trailer field along with the segmentation unit. This is shown in Figure 14.14. The DMPDU header field consists of three subfields. The first is the segment type subfield. The second is the sequence number subfield. The third subfield is the message identifier (MID). The DMPDU trailer consists of two subfields. The first is the payload length subfield and the second is the payload CRC.

The MID is used to provide the logical linking between segmentation units derived from the same IMPDU, and it should be unique on a subnetwork

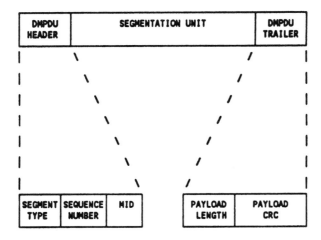

Figure 14.14. Format of a DMPDU. From Ref. 10, reprinted with permission.

while the IMPDU is being transferred. Each AU will have at least one unique MID. The allocation of the MID numbers is controlled by the MID page allocation scheme, which is a distributed method for claiming and keeping MID values that are unique across the whole subnetwork.

The MID identifies all of the DMPDUs derived from a single IMPDU, and it is used in the reassembly of the segmented IMPDU at the destination.

SEGMENTATION AT THE SOURCE. The train of DMPDUs sent as QA segment payloads by the source is shown in Figure 14.15. The first segmentation unit of a multisegmentation unit IMPDU is carried in a beginning of message (BOM) DMPDU. This message is identified by the BOM code in the segment type subfield and signifies the start of a new IMPDU transfer. The MID subfield of this DMPDU carries one of the MIDs obtained by the source and not currently being used for the sending of another IMPDU by the source. The sequence number subfield of the DMPDU carries the initial value of the sequence numbers to be associated with sequential DMPDUs. The segmentation unit of the BOM IMPDU includes the IMPDU header and any header extension plus the first octets of the MSDU, sufficient to fill the segmentation unit.

All subsequent segmentation units of the IMPDU until the last are placed in the payload field of the train of segments following the first segment. The DMPDUs are identified by the COM (continuation of message) code in the segment type subfield. The transfer of the multisegmentation unit IMPDU is completed by sending the last segmentation unit in a DMPDU that contains the EOM (end of message) code in the segment type subfield. The COM and EOM DMPDUs carry the value of sequence number in the BOM DMPDU incremented by one for each successive DMPDU. All COM DMPDUs and

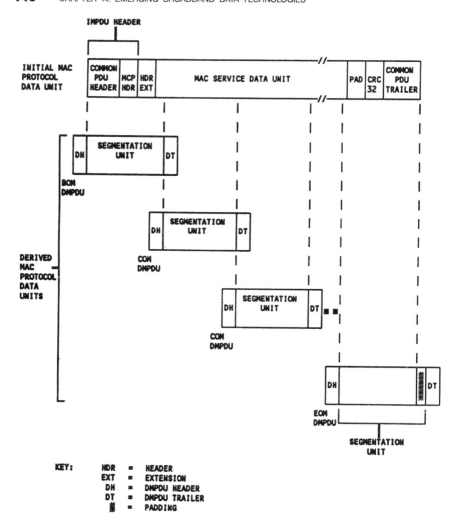

KEY:
HDR	=	HEADER
EXT	=	EXTENSION
DH	=	DMPDU HEADER
DT	=	DMPDU TRAILER
▓	=	PADDING

Figure 14.15. Transfer of an IMPDU; the formation of segmentation units. From Ref. 10, reprinted with permission.

the EOM DMPDU derived from an IMPDU carry the same MID value as the BOM DMPDU.

For the transfer of an IMPDU that only requires a single segmentation unit, the SSM (single segment unit message) code is used in the segment type subfield of the DMPDU. The MID is not used in this case, and thus it is set to the reserved value of zero.

Each DMPDU trailer is constructed by writing into the payload length subfield the number of IMPDU octets used in the segmentation unit. For connectionless MAC service to LLC (i.e., MAC LAN interconnect), this number is always 44 for BOM and COM DMPDUs. The number written into EOM DMPDUs indicates the remaining number of octets in the IMPDU that need to be transferred. This number can be any multiple of 4 in the range of 4 to 44, inclusive. The payload CRC subfield is a CRC computed over all octets of the segment payload, including the DMPDU header, segmentation units, and the DMPDU trailer.

3.5 DQDB Layer Protocol Data Unit (PDU) Formats

A DQDB slot, which we also have been calling a *cell*, has 53 octets. It has a striking similarity to the ATM cell which is described in the next chapter. Those 53 octets consist of an ACF field of one octet, a header of four octets, and a payload of 48 octets. The slot format is shown in Figure 14.16, and the ACF (access control field) is shown in Figure 14.17.

ACF	Segment
(1 octet)	(52 octets)

Figure 14.16. DQDB slot format. Note the term "segment."

BUSY	SL_TYPE	PSR	RESERVED	REQUEST
(1 bit)	(1 bit)	(1 bit)	(2 bits)	(3 bits)

Figure 14.17. The ACF field.

The BUSY bit indicates whether the slot contains information (BUSY = 1) or does not contain information (BUSY = 0). The SL_TYPE bit indicates whether the slot is a QA slot (cell) (SL_TYPE = 0) or a PA slot (cell) (SL_TYPE = 1). Table 14.5 shows the combinations of BUSY and SL_TYPE (bits).

The PSR (previous segment received) bit indicates whether the segment in the previous cell (slot) may be cleared (PSR = 1) or may not be cleared (PSR = 0).

The REQUEST field contains three REQ bits as shown in Figure 14.18. The REQ bits are used in the operation of the three-priority-level distributed queue access mechanism.

TABLE 14.5   Slot Acess Control Field Codings

BUSY	SL_TYPE	Slot State
0	0	Empty QA slot
0	1	Reserved
1	0	Busy QA slot
1	1	PA slot

Each of the REQ_I bits (I = 2, 1, 0) is set to zero by the slot marking function at the head of the bus and may be set to one by the access units along the bus according to the rules of operation for the distributed queue. Each bit is used on Bus x (x = A or B) to request access to a QA slot (cell) on Bus y (y = B or A, respectively). Requests for access to a QA slot are made at one of three priority levels. Priority queue level 2 is the highest priority queue and REQ_2 is the leftmost bit of the REQUEST field. The priority level of the requested queue, REQ_I, decreases with I (I = 2, 1, 0).

REQ_2	REQ_1	REQ_0

Figure 14.18. REQUEST field (3 bits).

3.5.1 The QA (Queued Arbitrated) Cell (Slot).

A QA cell is used to transport a QA segment. All bits of the QA slot are set to 0 by the slot marking function at the head of the bus. The BUSY bit of the ACF set to 0 and the SL_TYPE bit of the ACF set to 0 indicate an empty QA slot.

Access units access QA slots according to the rules of operation of the distributed queue. When an access unit gains access to a QA slot to transfer a QA segment, it marks the QA slot by setting the BUSY bit to 1 and it writes into the QA segment field. Otherwise, the BUSY bit remains unchanged.

QA SEGMENT. A QA segment is the PDU transferred in a QA slot. Each segment contains a header of 4 octets and a payload of 48 octets. This is shown in Figure 14.19.

QA Segment Header	QA Segment Payload
(4 octets)	(48 octets)

Figure 14.19. Format of the QA segment.

The QA segment header contains the fields shown in Figure 14.20. The length of each field is shown in bits. These fields are inserted by the access unit that writes the QA segment into an empty QA slot.

VCI	PAYLOAD TYPE	SEGMENT PRIORITY	HCS
(20 bits)	(2 bits)	(2 bits)	(8 bits)

Figure 14.20. QA segment header fields.

VIRTUAL CHANNEL IDENTIFIER (VCI). The 20-bit VCI provides a means to identify the virtual channel to which the QA segment belongs. There is a single VCI space shared by all services. The VCI value corresponding to all bits being set to zero is not available to refer to an active virtual channel.

DEFAULT CONNECTIONLESS VCI. The VCI value corresponding to all bits being set to one is the default value for the connectionless MAC service provided by the MAC convergence function [see Chapter 12 for a discussion of MAC (medium access control)]. All nodes are able to use the default connectionless VCI for both transmission and reception of QA segments.

PAYLOAD_TYPE. The 2-bit Payload_Type field indicates the nature of the data to be transferred. User data is indicated by the value Payload_Type = 00. All other values are reserved for future use.

SEGMENT PRIORITY. The 2-bit segment priority field is reserved for future use with multiport bridging. The field is set to the 00 value.

SEGMENT HEADER CHECK SEQUENCE (HCS). The 8-bit header check sequence (HCS) field provides for detection of errors and correction of single-bit errors in the QA segment header. The HCS contains an 8-bit cyclic redundancy check (CRC) calculated on the QA segment header field, using the following standard generating polynomial $G(x)$ of degree eight:

$$G(x) = x^8 + x^2 + x + 1$$

The HCS is encoded by the originator of the QA segment header. The contents (treated as a polynomial) of the QA segment header are multiplied by X^8 and then divided (modulo-2) by $G(x)$ to produce a remainder which is inserted in the HCS field.

Error detection using the HCS is mandatory at a receiver, whereas error correction using the HCS is optional.

QA SEGMENT PAYLOAD. The QA segment payload is 48 octets long and its contents are not constrained in any way.

3.5.2 Pre-arbitrated (PA) Slot.
The PA slot is used to transfer isochronous service octets (such as PCM voice time slots).

PA SEGMENT. Each PA segment contains a header of 4 octets and a payload of 48 octets, illustrated in Figure 14.21.

PA Segment Header	PA Segment Payload
(4 octets)	(48 octets)

Figure 14.21. PA segment format.

A PA segment is transported in a PA slot. A PA slot is generated by the slot marking function at the head of the bus with the BUSY bit of the ACF set to 1, the SL_TYPE bit of the ACF set to 1, and all other bits of the ACF set to 0. The BUSY bit and the SL_TYPE bit of the PA slot remain unchanged at all times, but the REQ bits in the ACF may be operated on according to the rules of the distributed queue. The slot marking function at the head of the bus also writes the PA segment header, which is carried in the PA slot. The slot marking function at the head of the bus sets every bit in the PA segment payload to 0.

PA SEGMENT HEADER FIELDS. The PA segment header field is shown in Figure 14.22. These fields are written by the slot marking function at the head of the bus and should remain the same as the slot passes along the bus.

VCI	PAYLOAD TYPE	SEGMENT PRIORITY	HCS
(20 bits)	(2 bits)	(2 bits)	(8 bits)

Figure 14.22. PA segment header fields.

VIRTUAL CHANNEL IDENTIFIER (VCI). The 20-bit VCI provides a means to identify the virtual channel to which the PA segment belongs. There is a single VCI space that is shared by all services. The VCI value corresponding to all bits being set to zero is not available to refer to an active virtual channel. The VCI value with all bits set to one is reserved for connectionless MAC service being supported by QA segments and is not available for use with PA segments.

PAYLOAD_TYPE. The 2-bit Payload_Type field indicates the nature of the data to be transferred. User data is indicated by the value Payload_Type = 00. All other values are reserved for future use. The default value of Payload_Type used in PA segments is 00.

SEGMENT PRIORITY. The 2-bit segment priority field is reserved for future use with multiport bridging. The field is set to the value 00 in the interim.

SEGMENT HEADER CHECK SEQUENCE (HCS). See QA segment HCS above.

PA SEGMENT PAYLOAD. The PA segment payload is 48 octets long. It consists of 48 isochronous service octets.

Isochronous Service Octet. Access to PA segment payloads may be shared among a number of isochronous service users (ISUs). Isochronous service octets are transferred within the PA segment payload of a PA slot (cell) to support service to the ISUs.

Portions of the text of Section 3 have been abstracted from IEEE Std 802.6-1990 (Ref. 10). The figures in this section have been taken from the same reference, and are reprinted with permission.

4 SWITCHED MULTIMEGABIT DATA SERVICE (SMDS)

4.1 Introduction

SMDS provides LAN interconnection over comparatively long distances. In this regard it is like frame relay, but with more capacity, up to 34 Mbps of sustained transmission rate. Figure 14.23 shows this LAN connectivity. Here two (or more) routers are connected to the network using SMDS for the transport mechanism. On the customer premises, those routers would be connected to LANs to which end systems may be attached. Although Figure

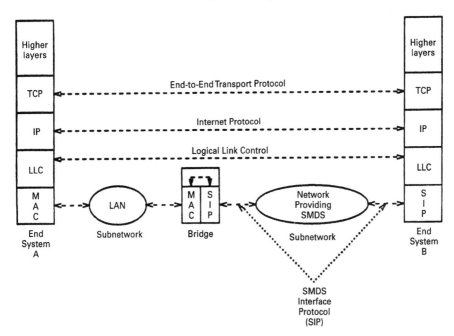

KEY

TCP	= Transmission control protocol
IP	= Internet protocol
LLC	= Logical link control
MAC	= Medium access control (e.g., 802.x, FDDI)
SIP	= SMDS interface protocol
⌐⌐⌐	= Bridging function (e.g., encapsulation) between MAC and SIP

Figure 14.23. Example of use of SMDS for LAN connectivity. From Ref. 12, reprinted with permission.

14.23 shows the application of TCP/IP, an OSI scenario could be used just as well.

4.2   Relationship to DQDB

The SMDS interface protocol (SIP) that operates across a subscriber network interface (SNI) is based on the distributed queue dual bus (DQDB) MAC protocol described above in Section 3 and specified in IEEE Std 802.6 [10]. Operation of the DQDB protocol across the SNI results in an *access DQDB*. The term *access DQDB* distinguishes this from other uses of the DQDB protocol that may occur internally to an *SS* (switching system) and from other DQDB networks that may be present on a subscriber's premises. A switching system (SS) in a network operates as one station on an access DQDB. For the configuration of CPE (customer premises equipment) on the access DQDB, two scenarios are possible:

- A single CPE access arrangement: The access DQDB consists of a simple configuration of two stations, one in the serving telephone network (an SS) and one subscriber-owned station (CPE), as shown in Figure 14.24a.

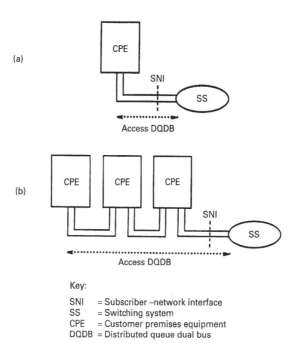

Key:

SNI = Subscriber –network interface
SS   = Switching system
CPE = Customer premises equipment
DQDB = Distributed queue dual bus

Figure 14.24. (a) Single CPE access and (b) multi-CPE access arrangements. From Ref. 12, reprinted with permission.

- A multi-CPE access arrangement: The access DQDB consists of one station in the serving telephone network (an SS) and more than one interconnected subscriber-owned station (CPE) that operate together as a DQDB, as shown in Figure 14.24b.

All CPE units in a multi-CPE access arrangement must belong to the same subscriber. The "shared medium" operation of the DQDB protocol assumes that stations attached to the DQDB network are administered by a single organization. Attaching CPEs from different SMDS subscribers to the same access DQDB would not provide an acceptable degree of security and privacy [12].

4.3   Overview of SMDS

SMDS is a public, packet-switched service that provides for the exchange of variable-length SMDS data units, up to a maximum of 9188 octets of user information per SMDS data unit. In consideration of security and privacy for subscribers, SMDS is offered by means of an access path that is dedicated to an individual subscriber. In addition, the SS validates that the SMDS source address associated with every SMDS data unit is an SMDS address that is legitimately assigned to the SNI from which the SMDS data unit originated.

SMDS uses 10-digit addresses following a prefix of "1," structured according to ITU-T Rec. E.164 and the North American Numbering Plan (NANP) [13]. In addition, SMDS includes a capability for group addressed SMDS data unit transport. Group addressed SMDS data unit transport is a feature analogous to the multicasting feature of LANs. When a CPE sends a group addressed SMDS data unit, the network supporting SMDS delivers copies of the SMDS data unit to a set of SNIs (subscribers) identified by the destination addresses specified by the group address.

SMDS is a multimegabit system. There are two alternatives for the access path between the CPE and the SS providing SMDS:

- a DS3 (T3)-based access path or
- a DS1 (T1)-based access path.

Subscribers to SMDS have a range of traffic requirements, and the capability of CPEs to generate traffic at high-speed rates across the interface to the SS will also vary. Accordingly, SMDS supports different *access classes* across the subscriber's access path. Each access path provides for a different rate of information flow by enforcing limits on the level of sustained information transfer and on the burstiness of information transfer. An access class applies to the information flow from the CPE to the SS only. There is no restriction on the amount of traffic flowing from the SS to the CPE. For DS3 (T3)-based access paths, the access class is selected by the subscriber. For DS1 (T1)-based access paths, a single access class applies.

The SIP (SMDS interface protocol) segments SMDS data units into slots of fixed length. This protocol inherently allows slots belonging to different SMDS data units to be intermingled as they are carried across an SNI. As a result, the SS or a collection of CPE units may concurrently receive and send multiple SMDS data units across an SNI. The slots can also be described as cells of 53 octets, but with only a 44-octet payload.

4.3.1 Individually Addressed SMDS Data Unit Transport. SMDS provides for the transport of individually addressed SMDS data units which can contain up to 8188 octets of user information. Each SMDS data unit is individually addressed and transferred independently of the transfer of any other data unit. The originator of an SMDS data unit specifies both (a) a source address that identifies the SNI from which it was sent and (b) an individual destination address that identifies the SNI of the intended recipient. Both the source address and the destination address, as specified by the originator, are delivered with the SMDS data unit across the recipient's SNI.

4.3.2 Local Communications for Multi-CPE Access Arrangements. As we mentioned, SMDS uses DQDB of the IEEE 802.6 standard as the access protocol, tying CPEs and the local SS together. The DQDB may also serve for communications among a local group of CPEs. Also, some (but not necessarily all) of the local communications among CPEs on the access DQDB will be visible to the SS that serves that SNI. Local communications between CPEs on an access DQDB may include any type of transfer supported by the queued arbitrated part of the DQDB protocol (see Section 3), including connectionless data transfer using the same protocol formats that are used for SMDS data transfer. Therefore, the SS must utilize the destination address of a data unit to distinguish whether the data unit is intended for transfer via SMDS, or intended for local communications and should be ignored by the SS.

4.3.3 Addresses. There are two types of addresses used for SMDS: individual addresses and group addresses. Each individual address is assigned to a unique single SNI. A group address represents a set of individual addresses and thus may identify multiple SNIs, which are the destinations of a group of addressed SMDS data units.

Individual address used for SMDS consists of a 10-digit number following a prefix "1."

A *group address* format consists of a 10-digit number following a prefix of "1." A group address can be distinguished from an individual address by the context of its use.

More than one individual address may be assigned to a single SNI, and any or all of these addresses may be used as valid addresses for communicating with other SNIs. An SS supports the assignment of up to 16 individual addresses to a single SNI.

Address mobility refers to the degree to which an individual address, previously assigned to a particular SNI, can be reassigned to a different SNI belonging to the same subscriber—for example, following a subscriber's relocation.

SOURCE ADDRESS VALIDATION. To ensure that the sender of an SMDS data unit cannot indicate a fraudulent source address, the source address of every SMDS data unit is validated. The SS does not deliver any SMDS data unit whose source address is not one of the set of valid addresses assigned to the SNI from which the data unit was sent.

GROUP ADDRESS COMPOSITION. Group addressed data unit transport is the ability for a CPE to send the same SMDS data unit to several intended recipients. The SMDS service provider is responsible for assigning group addresses and ensuring that each group address identifies uniquely only one set of individual addresses. Since group addresses can be distinguished from individual addresses by the context of their use, any particular 10-digit number could be assigned as an individual address or a group address or both. A group address represents a set of up to 128 individual, and only individual, SMDS addresses. An SS supports the assignment and use of up to 1024 group addresses. A particular individual address can be identified by up to 32 group addresses.

A particular SNI can be identified by up to 48 group addresses. An SNI is identified by a group address if one or more of the individual addresses assigned to the SNI is identified by a group address.

4.3.4 *Access Classes.*

For DS3-based access paths, SMDS provides a range of access classes. Each access class provides for different traffic characteristics (i.e., it prescribes limits on the rate of sustained information transfer from the CPE to the SS and on the burstiness of the information transfer from the CPE to the SS). Keep in mind that access classes are not a means of channelizing or sub-rating the DS3-based access class. When a *burst* of data is transferred across the SNI, by either the SS or the CPE, that data is transferred at the maximum rate achievable using SIP across a DS3-based access path (approximately 34 Mbps after overhead has been removed). A burst of data occurs when one or more SMDS data units are transferred across the SNI without any intervals between data units. However, the duration of the burst is constrained by the access class mechanism. The access class mechanism also constrains the average rate of information transfer. For the duration of a burst, the instantaneous rate of information transfer is 34 Mbps; however, the long-term average rate can be much less.

TABLE 14.6   Choices of SMDS Access Classes

Class No.	SIR (Mbps)
1	4
2	10
3	16
4	25
5	34

Source: Ref. 12, reprinted with permission.

The limits on the rate of information transfer are each defined by a set of parameters that is the basis for *enforcement* by the SS. Access classes apply only to traffic flowing from the CPE to the SS and are determined at subscription time.

There is no access class enforcement for subscribers with DS1-based access paths; the access class in these instances is the maximum effective bandwidth that can be achieved with the SIP operating over a DS1-based access path (approximately 1.17 Mbps after the overhead has been removed).

Table 14.6 shows the available choices for access classes. The column labeled SIR (*sustained information rate*) refers to the rate of transfer of user information that the CPE could sustain over a long period using a particular access class.

Sustained information rate (SIR) is a new concept. There was reference to such a concept on our discussion of frame relay in Section 2 of this chapter. Of course, it is related to the burstiness of LAN traffic. For example, classes 1 through 3 in Table 14.6 allow the CPE to transmit traffic to the SS at sustained rates of 4, 10, and 16 Mbps, respectively. These rates match the following LAN rates: 4 Mbps for IEEE 802.5 (token ring) LAN, 10 Mbps for IEEE 802.3 (CSMA/CD) LAN, and 16 Mbps for a second type of IEEE 802.5 (token ring) LAN.

4.3.5   Multiple Data Units in Transit Concurrently. SMDS data units are transferred across the SNI in slots of fixed length. As we would expect, SMDS data units that are longer than a single slot (cell) are segmented and transferred in multiple slots (cells). The SMDS interface protocol (SIP) inherently allows slots derived from different SMDS data units to be transferred intermingled across the SNI. Thus, it is possible to have multiple SDUs in transit concurrently. This is decided at the time of subscriber subscription. There are two choices in either direction: 1 or up to 16 data units in transit concurrently.

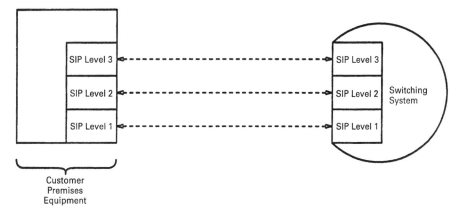

Figure 14.25. The three layers (levels) of the SIP. From Ref. 12, reprinted with permission.

4.4 Subscriber–Network Interface (SNI)

The SNI is defined as the point where the CPE interfaces with the network supporting SMDS. It is at this interface point that the CPE attaches to an access facility that links it to an SS via a dedicated path. That is, only data originating from or destined to a CPE belonging to a particular subscriber is transported across the individual SNI. Thus we see that interfaces are not shared and the data from the CPE of one subscriber cannot be *seen* by the CPE of another subscriber. This dedicated access arrangement is necessary to establish a private and secure network service. Access configurations depicting the SNI are shown in Figure 11.22. Essentially, the SS can be viewed as the head of DQDB bus A and the CPE as head of DQDB bus B.

The SIP defines how the CPE communicates with the network supporting SMDS across the SNI, and thereby accessing the service provided by the network. A *credit manager* mechanism is used to enforce the rates of incoming information (i.e., access classes) across the SNI [12].

4.5 SMDS Interface Protocol (SIP)

The SIP is a connectionless protocol based on the DQDB protocol described in Section 3. There are three protocol layers in the SIP: SIP Level 1, SIP Level 2, and SIP Level 3 whose functions include addressing, framing, error detection, and physical transport. Figure 14.25 shows the relationship of these three protocol levels with the SIP. Bellcore [12] stresses that these layers *do not* correspond to abstract layers in the OSI model.

Figure 14.26 illustrates the functionality of the upper two SIP layers by showing the encapsulation of user information through each level. SIP Level 3 contains the appropriate SMDS addressing information for the data unit passed from the Level 3 user, as well as a means to detect lost L2_PDUs.*

*L2_PDU = level (layer) 2 protocol data unit or PDU.

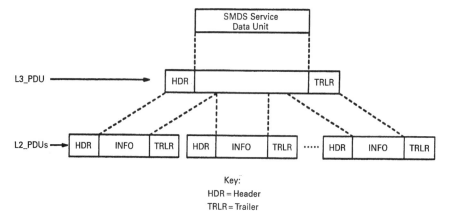

Figure 14.26. Encapsulation of user information by the layers of the SIP. From Ref. 12, reprinted with permission.

The SIP Level 2 functions include bit error detection and framing for a SIP L3_PDU through the use of fixed-length slots. Level 2 also includes segmentation and reassembly (SAR) functions on the variable-length L3_PDU. The SIP Level 1 provides physical layer functions such as bit level transmission across the physical facilities. In the following paragraphs we will discuss the physical layer-independent parts of the SIP.

The following conventions are defined in specifying the SIP.

- A service data unit (SDU) is a parameter containing a unit of data that is passed across an interlayer interface.

- A protocol data unit (PDU) is a unit that is exchanged between peer entities within a particular layer.

- In the specification of PDU formats, fields are depicted in order of their transmission; that is, the leftmost fields are the first to be transmitted. Likewise, the bits comprising a field are listed so that the leftmost, or most significant, bit is the first to be transmitted. When using decimal digits to represent the value of a field, the implication is that the most significant bit of the binary representation is the first to be transmitted.

4.5.1   Level 3 Protocol Data Unit Format. An L3_PDU is equivalent to an *SMDS data unit* described above. The format of the L3_PDU is illustrated in Figure 14.27 with corresponding field lengths. Some of the fields of the format are marked as $X+$. These fields are there to ensure alignment of the SIP format with the DQDB protocol format, and the values so placed in these fields by the user are not processed by the network. In addition, the entire L3_PDU received from the CPE is delivered unchanged by the network with the exception of the BEtag discussed below.

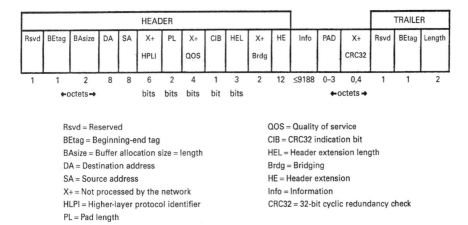

Rsvd = Reserved
BEtag = Beginning-end tag
BAsize = Buffer allocation size = length
DA = Destination address
SA = Source address
X+ = Not processed by the network
HLPI = Higher-layer protocol identifier
PL = Pad length

QOS = Quality of service
CIB = CRC32 indication bit
HEL = Header extension length
Brdg = Bridging
HE = Header extension
Info = Information
CRC32 = 32-bit cyclic redundancy check

Figure 14.27. SIP Level 3 protocol data unit format. From Ref. 12, reprinted with permission..

Refer now to Figure 14.27. The following is a brief description of each field.

- *Reserved.* There are two 1-octet fields, one in the header and one in the trailer of the L3_PDU, that are reserved. These fields are populated with 0s by both the CPE and the SS.
- *BEtag.* There are two 1-octet fields, one in the header and one in the trailer, that contain a *beginning-end tag*. These fields are populated with a binary encoded value ranging from 0 through 255. These two tags are used to form an association between the first and last segments of an L3_PDU.
- *BAsize.* This 2-octet field gives the length, in octets, of that portion of the L3_PDU that extends from the beginning of the *destination address* field up to and including the *CRC-32* field, if present.
- *Destination Address.* This 8-octet field contains two subfields: *Address_Type* and *Address.* The format of this address field is shown in Figure 14.28.
- *The Address_Type Subfield.* This subfield occupies the four most significant bits of the field. The *Address_Type* contains the value 1100 when using an individual 60-bit address and the value 1110 for a 60-bit group address. All other values are not recognized by the SS.

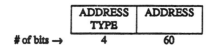

Figure 14.28. Address field.

- *The 60-bit Address Subfield.* This subfield contains the SMDS address for which the L3_PDU is destined. For the initial availability of SMDS, the address uses 10 digits, following the prefix "1." The four most significant bits of the *Address* subfield are populated with the prefix 0001. The 40 bits which follow these four most significant bits contain the BCD* encoded values of the 10 digits. The remaining 16 least significant bits of this field are populated with "1"s. Future addresses used for SMDS may require all 60 bits of this field—for example, international addressing.

- *Source Address.* This 8-octet field also has two subfields: *Address_Type* and *Address*. The format of this field is shown in Figure 14.28. The *Address_Type* subfield occupies the four most significant bits of the field. The *Address_Type* always contains the value 1100 for an individual 60-bit address. All other values are not recognized by the SS. The *Address* subfield has the same format as the *Address* subfield in the *Destination Address* field.

- *Higher-Layer Protocol Identifier.* This 6-bit field is present to ensure alignment of the SIP format with the DQDB protocol format.

- *Pad Length.* This 2-bit field indicates the number of octets in the *PAD* field to make the entire L3_PDU 32-bit aligned. The PAD length may be found from the expression:

$$PAD\ Length = 3 - (\text{number of } information \text{ octets} + 3)(\text{mod } 4)$$

- *Quality of Service.* This 4-bit field is present to ensure alignment of the SIP format with the DQBB protocol format.

- *CRC-32 Indication Bit.* This 1-bit field indicates the presence or absence of the *CRC-32* field. If the CRC-32 has been generated and is present, this bit is set to 1. Otherwise, this bit is set to 0.

- *Header Extension Length.* This 3-bit field indicates the number of 32-bit words in the *Header Extension* field, and it is populated with the value 011 by both the CPE and the SS.

- *Bridging.* This 2-octet field is present to ensure alignment of the SIP format with the DQDB protocol format.

- *Header Extension.* This is a 12-octet field and is shown in Figure 14.29 along with the lengths of different subfields. Octets not used for header

Figure 14.29. Header extension field format.

*BCD = binary coded decimal.

extension elements as defined below are with extra octets, HE PAD, such that the overall *Header Extension* field is 12 octets long.

- *Element Length*. This one-octet field contains the combined lengths of the *Element Length*, *Element Type*, and *Element Value* fields specified in octets for the element.

- *Element Type*. This one-octet field contains a binary encoded value to indicate the type of information found in the *Element Value* field.

- *Element Value*. The content of this variable-length field depends upon the *Element Type* and its intended function.

- *HE PAD*. This field can range from 0 to 9 octets in length. It is sized to make the overall *Header Extension* field 12 octets. If the HE PAD is present (i.e., greater than 0 octets in length), it follows the header extension elements and begins with at least one octet of all zeros.

- *Information*. This is a variable-length field that contains the user's information (or Level 3 SDU) and can be up to 9188 octets in length.

- *PAD*. This field can range from 0 to 3 octets in length and contains zeros. It is sized to make the entire L3_PDU 32-bit aligned for ease of processing, as indicated by the *PAD Length* field.

- *CRC-32*. This 4-octet field may be present or absent, as indicated by the *CRC-32 Indication Bit*. The *CRC-32* performs error detection for the L3_PDU fields from the *Destination Address* through the *CRC-32*. If the *CRC-32* is generated by the CPE, it is performed according to the IEEE 802.6 (DQDB) standard (see Section 3). The SS delivers unchanged the value placed in the *CRC-32* field by the user, if any.

- *Length*. This 2-octet field contains the same length as that found in the *BAsize* field [12].

4.5.2 Level 2 Protocol Data Unit Format. Figure 14.30 illustrates the L2_PDU format. The L2_PDU fields and their purpose are described below.

- *Access Control Field (ACF)*. The *Busy* subfield occupies the most significant bit of the ACF. This one-bit subfield indicates whether the L2_PDU contains information (BUSY = 1) or is empty (BUSY = 0). The

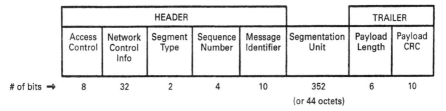

Figure 14.30. SIP Level 2 protocol data unit format. From Ref. 12, reprinted with permission.

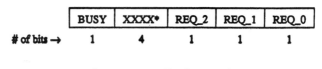

BUSY	XXXX*	REQ_2	REQ_1	REQ_0

\# of bits →   1   4   1   1   1

* not processed by the network

Figure 14.31. Access control field (CPE to SS).

BUSY	0000000

\# of bits →   1   7

Figure 14.32. Access control field (SS to CPE).

remaining 7 bits of the *ACF* may be set differently by the CPE and the SS. Figure 14.31 shows the *ACF* format for L2_PDUs from the CPE to the SS, and Figure 14.32 shows the *ACF* format when sending L2_PDUs from the SS to the CPE. The 4 bits which follow the *Busy* subfield are not interpreted when L2_PDUs are sent from the CPE to the SS. These 4 bits are set to zero for L2_PDUs sent from the SS to the CPE. The *Request* subfield occupies the remaining 3 bits of the *ACF*. Each bit represents a different DQDB priority level and is used for the operation of the three-priority-level distributed queue. The SS sets the request bits to zero when sending L2_PDUs to the CPE. For the single CPE access arrangement, the CPE sets the request bits to zeros when sending L2_PDUs to the SS.

- *Network Control Information.* This 4-octet field contains one of two possible values. For L2_PDUs that contain information, the CPE and the SS populate this field with the value:

111111111111111111000000100010

For empty L2_PDUs, the SS populates this field with zeros.

- *Segment Type.* This 2-bit field indicates how a nonempty L2_PDU is processed by the receiving entity. Table 14.7 shows the meaning of the values of this field. For empty L2_PDUs, the SS populates this field with zeros.

- *Sequence Number.* This 4-bit field is used in reassembling an L3_PDU to verify that all L2_PDUs of an L3_PDU have been received in correct sequence. The value of the *sequence number* is set according to the transmit sequence number counter.

- *Message Identifier (MID).* This 10-bit field contains information used to allow association of segments with an L3_PDU. The *MID* is the same for all segments of a given L3_PDU. Valid MIDs for nonempty

TABLE 14.7   Segment Type Values

Value	Meaning
00	Continuation of message (COM)
01	End of message (EOM)
10	Beginning of message (BOM)
11	Single segment message (SSM)

Source: Ref. 12.

L2_PDUs that are not SSMs (single segment messages) range from 1 to $(2^{10} - 1)$. *MIDs* from 1 to $(2^9 - 1)$ are reserved for L2_PDUs sent from the SS to the CPE. For empty L2_PDUs and for SSMs, the SS populates this field with zeros. The MID *Page Allocation* field is used to allow the CPE to obtain *MIDs* using the page allocation algorithm specified in IEEE Std 802.6 (DQDB) (Section 3).

• *Segmentation Unit*. This 44-octet field is the information portion containing a Level 2 SDU (which is a partial L3_PDU). For empty L2_PDUs, the SS populates this field with zeros.

• *Payload Length*. This 6-bit field indicates how many of the 44 octets of the *segmentation unit* contain actual data. For the BOM (beginning of message) and COM (continuation of message) segments, this field indicates 44 octets. For EOM (end of message) segments, it indicates any value from 28 to 44 that is a multiple of 4 octets. For empty L2_PDUs, the SS populates this field with zeros.

• *Payload CRC*. This 10-bit CRC performs error detection for the *Segment Type, Sequence Number, Message Identifier, Segmentation Unit, Payload Length*, and *Payload CRC fields*, but it does not cover the *Access Control* and *Network Control Information* fields. For nonempty L2_PDUs, the CRC is performed in accordance with the IEEE 802.6 (DQDB) standard. For empty L2_PDUs, the SS populates this field with zeros.

LEVEL 2 PROTOCOL PROCEDURES. When a Level 2 entity has no information to transmit, it sends empty L2_PDUs in which the *Busy* subfield is set to zero.

If a Level 2 entity is passed an SMDS data unit (i.e., an L3_PDU) from the level above, it segments the L3_PDU and it attaches a header and a trailer to each segment to create one or more L2_PDUs for transmission. Each L2_PDU has a 44-octet *Segmentation Unit* field to carry the segmented L3_PDU information. However, the last L2_PDU may only be partially filled since the L3_PDU may not contain an integral number of 44-octet segments. If the last L2_PDU contains less than 44 octets of the Level 3 data unit, it is

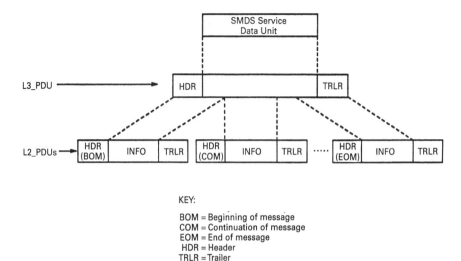

Figure 14.33. Segmentation of an L3_PDU into L2_PDUs. From Ref. 12, reprinted with permission.

padded to 44 octets using zeros. The same applies to an SSM containing a complete L3_PDU that may be less than 44 octets. Figure 14.33 shows the segmentation process. The header of the L2_PDU includes a *Segment_Type* field, a *Sequence Number* field, an *MID* field, *Network Control Information*, and the *Access Control Field (ACF)*. The trailer includes a *Payload Length* field and a *Payload CRC*.

The *Segment_Type field* is used to indicate if an L2_PDU is a single, beginning, continuation, or end of message segment (SSM, BOM, COM, or EOM, respectively). The *MID* field is used to identify the L3_PDU to which an L2_PDU belongs, except in the case where only one L2_PDU is needed to transmit a full L3_PDU. That is, the *MID* is not needed and is set to 0 for L3_PDUs that require an SSM. *MIDs* are only needed for L3_PDUs that require a BOM, possibly one or more COMs, and an EOM. Therefore, in cases where more than one L2_PDU is required to transfer an L3_PDU, the transmitting Level 2 entity assigns an *MID* to each L3_PDU. An L3_PDU's *MID* is unique among L3_PDUs being simultaneously transmitted in a given direction on the access DQDB.

SS Procedures for Sending an L2_PDU. The number of nonzero MIDs that an SS Level 2 entity will use, corresponding to a single SNI, is equal to the maximum number of SMDS data units that may be transferred concurrently from the SS to the CPE, which is set at the time of subscription. The requirement on an SS ensures that CPE reassembly resources (e.g., buffer space) will not be unnecessarily reserved for processing of L3_PDUs that are not completed due to lost EOMs. These reassembly procedures for the CPE are defined in the IEEE 802.6 (DQDB) standard. For each L3_PDU sent by the SS that requires more than one L2_PDU, the SS Level 2 entity maintains a transmit sequence number (TxSN) counter associated with the L3_PDU's MID. Prior to transmitting a BOM or COM, the SS then increments the value of the TxSN counter by one (modulo 16) for use with the next L2_PDU associated with the same MID. The TxSN counter associated with a particular MID is not required to be maintained between L3_PDUs with the same MID, nor is a TxSN counter required to be maintained for the MID value of 0, used for SSM L2_PDUs.

4.6 Access Classes and Enforcement

Across an SNI based on a DS3 (T3) access path, SMDS provides a range of access classes as we described above. The access class that applied to a particular SNI is determined at subscription time. The access class applies to information flowing from the CPE to the SS. Each access class provides for different traffic limits. These limits are each defined by a set of parameters which is the basis for enforcement. Of course, across an SNI based on a DS1 (T1) access path, a single access class applies. There are no enforcement functions for this access class.

4.6.1 Access Class Mechanism. A running balance of the traffic entering the network across each SNI is kept by a credit manager in the SS. The balance, or *credit*, is measured in octets. For every L3_PDU that is sent by the CPE to the SS, the credit manager compares the value *BAsize* 36 with the value of the credit. The value *BAsize* 36 closely approximates the length of the user information field in the L3_PDU. If sufficient credit is available, the L3_PDU is approved for admission into the network and the value of the credit is decremented by the value *BAsize* 36 (or the value of the credit remains unchanged if *BAsize* $\leq$ 36). If there is insufficient credit, the SS will not deliver that L3_PDU and the value of the credit is unchanged.

The information needed to perform the credit check is contained within the first L2_PDU associated with a particular L3_PDU. That first L2_PDU is either a BOM segment or an SSM segment (in which case also the only L2_PDU associated with the L3_PDU). Before the credit manager checks the available credit, the BOM or SSM segment must have passed all the Level 2 protocol procedures (i.e., the BOM or SSM segment is declared to be valid). The L3_PDU destination address, contained in the BOM or SSM segment, must be checked to determine whether that L3_PDU is destined for the network supporting SMDS (i.e., the destination address is formatted properly for SMDS and is not assigned to the originating SNI). The credit manager checks if sufficient credit is available as soon as a valid BOM or SSM segment destined for the network supporting SMDS is received. The access class mechanism may be engaged as soon as a valid BOM segment is received by the SS. It does not require that all segments of the L3_PDU be present.

However, if it is determined after credit has been subtracted that the L3_PDU cannot be successfully transferred to the SS for any reason (for example, a COM segment has a bit error), the credit for the L3_PDU subtracted from the credit balance is *not* returned.

The credit manager also ensures that credit accrues at a constant rate. This means that credit is "earned" at a constant number of octets per unit interval. The rate is governed by the parameters N_{inc} and I_{inc}, which are defined below. However, available credit is capped at some maximum value given by C_{max}.

If traffic entering the network violates the access class mechanism, meaning there is insufficient credit available, access class enforcement occurs. This enforcement involves nondelivery by the SS.

4.6.2 Access Class Parameters. For traffic flowing from the CPE to the SS, a credit manager keeps a running balance of available credit measured in octets. The values for the parameters that define the operation of each credit manager are chosen at the time of subscription and are given by the following:

(a) C_{max} is the maximum amount of credit, measured in octets, that may accrue. This is also the initial value for the credit when operation of the credit manager begins.

(b) I_{inc} is the interval between increments to the credit. I_{inc} is measured by counting L2_PDUs, regardless of whether they are full or empty, as they are received by the SS. For simplicity, only integral values of the I_{inc} are allowed.

(c) N_{inc} is the number of octets that the credit is incremented by whenever I_{inc} L2_PDUs have been counted.

4.6.3 Parameter Values and Relation to Access Classes. Each DS3-based SNI has an access class associated with it. For each access class there is an associated set of parameter values that are used in the execution of the credit manager function.

The value for C_{max}, I_{inc}, and N_{inc} control two aspects of an access class. The C_{max} value is the largest value the credit balance can attain, and its value is related to the allowed burstiness of the information transfer. The values for I_{inc} and N_{inc} control the rate at which credit accumulates, and thus the maximum information transfer rate that can be sustained for long periods. The long-term average can be described by the *sustained information rate (SIR)*, which is the average rate at which user information can be transferred across the SNI.

The SIR is given by the following equation:

$$\text{SIR} = \{8 \text{ bits/octet}\} \times \{12 \text{ L2\_PDUs}/(125 \text{ μs})\} \times \{N_{inc} \text{ octets}/I_{inc} \text{ L2\_PDUs}\}$$

TABLE 14.8 Parameter Values for Access Classes

		Parameters		
Class No.	SIR (Mbps)	N_{inc} (octets)	I_{inc} (L2.PDU counts)	C_{max} (octets)
1	4	26	5	9188
2	10	13	1	9188
3	16	21	1	9188
4	25	32	1	9188
5	34	NA	NA	NA

Source: Ref. 12, reprinted with permission.

Access classes that are defined by different combinations of values for the credit manager parameters C_{max}, I_{inc}, and N_{inc} are given in Table 14.8. Classes 1 through 3 shown in Table 14.8 allow the CPE to sustain transfer of user data at rates of 4 Mbps, 10 Mbps, and 16 Mbps, respectively. The values of C_{max} allow for some degree of burstiness (i.e., the CPE can send data in contiguous L3_PDUs for some period before the credit is exhausted). Class 4 is appropriate for CPEs that transfer data at moderately high sustained rates, but still do not require the full DS3 bandwidth. Class 5 represents the case where there is no access class mechanism operating at all—data is transferred using the maximum access bandwidth achievable with SIP operating across a DS3-based SNI (approximately 34 Mbps).

4.7 SMDS in an Operating Environment

Figure 14.34 shows SMDS in an operational environment. Note the strictly North American scenario.

Portions of Section 4 have been abstracted from *Generic System Requirements in Support of Switched Multi-Megabit Data Service* [12]. The figures in Section 4 derive from the same reference source.

REVIEW QUESTIONS

1. What was the driving force to speed up data network connectivity over WAN distances?

2. X.25 connectivity was one alternative to interconnect LANs over long distances. What were the drawbacks of X.25 for LAN interconnectivity?

3. Frame relay must rely on the excellent performance of the intervening digital network. What performance parameter is so very important here? Quantify it.

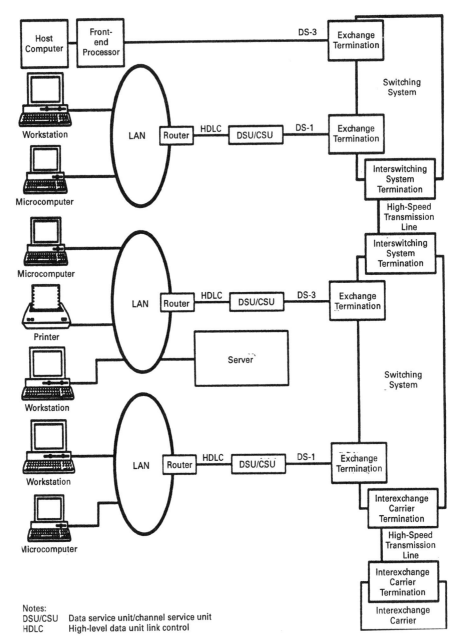

Figure 14.34. SMDS network and switching systems. From Ref. 14, reprinted with permission.

4. Three important factors are mentioned which are necessary to speed up the network. Name them. Can you name a fourth?

5. Frame relay relates to which OSI layers?

6. There are two possible reasons why frame relay discards frame. What are they?

7. How does an end user know that a frame or frames have been lost? How can end-user recipients recoup lost frames?

8. There are three possible sources of standards for frame relay. What are they?

9. Frame relay operation derives from which predecessor?

10. How is the frame relay address field extended? Explain how it works.

11. There are six possible causes for declaring a frame relay frame invalid. Give four of them.

12. How are the FECN and BECN bits used. Distinguish between them. (*Hint*: Use direction of transmission.)

13. Discuss the use of the DE bit. Under what circumstances will the network change the DE bit setting?

14. What are the three possible ways to handle billing on frame relay?

15. Discuss the use of CLLM as an alternative for congestion control.

16. What are the three basic parameters that are negotiated at call establishment or at subscription?

17. If sequence numbers are not used in frame relay, how is sequence maintained at the destination end of a connection?

18. Name four of the five reasons a frame may be discarded by the FRAD and/or switch.

19. What is the length of a DQDB slot (in octets)? How many octets are overhead, and how many are payload?

20. With DQDB, distinguish QA from PA. What type of traffic is QA used for? What type is PA used for?

21. Describe the two control fields in the ACF and their impact on operation.

22. With DQDB, how does a node know when it can get access to the network?

23. What is the function of the *busy* bit?

24. What is the function of the REQUEST field?

25. Bandwidth balancing is involved with _____ _____ _____.

26. The bandwidth balancing equitably divides the bandwidth among user stations and allows some bandwidth to go _____.

27. What was the initial objective of DQDB? (that is, what segment of telecommunications was it originally designed to serve?)

28. With DQDB, traffic blocks are taken from a LAN which are placed in the payload (info field) of an IMPDU/DMPU, where encapsulation takes place. A header and trailer are added. The trailer consists of two important fields. Name them and describe their functions.

29. IMPDU/DMPDUs are segmented into DQDB segments of 53 octets in length. How do we know which segments start, continue, and end a long message block?

30. What is the purpose of the MID?

31. DQDB segments do not use a CRC as we know it. They have a CRC, but what does it cover?

32. What is the principal purpose of SMDS?

33. Describe the SMDS relationship with DQDB.

34. SMDS provides for two different basic access classes based on the transmission rate of the underlying bearer structure. What are they?

35. For the higher data rate access, SMDS provides for different access classes. These are based on what parameters?

36. Discuss peak rate of information transfer and SIR.

37. When is an access class determined with SMDS?

38. How many protocol layers does the SIP have?

39. What is the function of the BEtag in an SMDS data unit?

40. What does the PAD do in the SMDS data unit header? Use some octet values here.

41. What is the length in octets of an L2 segmentation unit (SMDS)?

42. Discuss the operation of the transmit sequence number including its interaction with the MID. This refers to the SMDS L2_PDU.

43. Discuss the SMDS credit manager, especially in regard to BAsize.

44. Would SMDS be attractive for European PTTs?

REFERENCES

1. *Framework for Frame Mode Bearer Services*, ITU-T Rec. I.122, ITU Telecommunications Standardization Sector, Geneva, March 1993.

2. *Frame Mode Bearer Services*, ITU-T Rec. I.233, ITU Telecommunications Standardization Sector, Geneva, 1992.

3. *ISDN Signalling Specification for Frame Relay Bearer Service for Digital Subscriber Signalling System Number 1 (DSS1)*, ANSI T1.617-1991, ANSI, New York, 1991.

4. *ISDN Data Link Layer Service for Frame Mode Bearer Services*, ITU-T Rec. Q.922, ITU Telecommunications Standardization Sector, Geneva, 1992.

5. *Integrated Services Digital Network (ISDN)—Core Aspects of Frame Protocol for Use with Frame Relay Bearer Service*, ANSI T1.618-1991, ANSI, New York, 1991.

6. *ISDN—Architectural Framework and Service Description for Frame Relaying Bearer Service*, ANSI T1.606-1990, ANSI, New York, 1990.

7. *Frame Relay & SMDS*, a seminar, Hewlett-Packard, Burlington, MA, October 1993.

8. *Integrated Services Digital Network (ISDN)—Architectural Framework and Service Description for Frame-Relaying Bearer Service (Congestion Management and Frame Size)*, ANSI T1.606a-1992, ANSI, New York, 1992.

9. *Telecommunications—Integrated Services Digital Network (ISDN)—Data Link Layer Signaling Specification for Application at the User–Network Interface*, ANSI T1.602-1989, ANSI, New York, 1989.

10. *Distributed Queue Dual Bus (DQDB) Subnetwork of a Metropolitan Area Network (MAN)*, IEEE Std 802.6-1990, IEEE Computer Society, IEEE, New York, 1991.

11. W. A. Flanagan, *Frames, Packets and Cells in Broadband Networking*, Telcom Library, New York, 1991.

12. *Generic System Requirements in Support of Switched Multi-Megabit Data Service*, Technical Reference TR-TSV-000772, Issue 1, Bellcore, Morristown, NJ, May 1991.

13. *BOC Notes on the LEC Networks—1990*, SR-TSV-002275, Issue 1, March 1991, Bellcore, Morristown, NJ, March 1991.

14. W. Stallings, *Advances in Local and Metropolitan Area Networks*, IEEE Computer Society Press, Los Alamitos, CA, 1994.

15

THE ASYNCHRONOUS TRANSFER MODE (ATM) AND BROADBAND ISDN

1 WHERE ARE WE GOING?

Frame relay (Chapter 14) was the beginning of the march toward an optimized* format for multimedia transmission (voice, data, video, facsimile).There were new concepts in frame relay. There was a trend toward simplicity where the header was notably shortened. The header was pure overhead, so it was cut back as much as practically possible. The header also implied processing. By reducing the processing, delivery time could be speeded up.

In the effort to speed up delivery, operation was unacknowledged (at least at the frame relay level); there was no operational error correction scheme. It was unnecessary because it was assumed that the underlying transport system had excellent error performance (better than 1×10^{-7}). There was error detection for each frame, and a frame found in error was thrown away. Now that is something that we never did for those of us steeped in old-time data communication. It is assumed that the higher OSI layers would request repeats of the few frames missing (i.e., thrown away).

Frame relay also moved into the flow control arena with the BECN and FECN bits and the CLLM. The method of handling flow control has a lot to do with its effectiveness in this case. It also uses a discard eligibility (DE) bit which set a type of priority to a frame. If the DE bit was set, the frame would be among the first to be discarded in a time of congestion.

DQDB developed by the IEEE 802.6 committee provides a simple and unique access scheme. Even more important, its data transport format is based on the *cell* which the IEEE calls a *slot*. The DQDB slot or cell has a format very similar to the ATM cell which we will discuss at length in this

---

Compromise might be a more appropriate word.

chapter. It even has the same number of octets, 53; 48 of these were payload. This is identical to ATM. Also enter a comparatively new concept of the HCS or header check sequence for detecting errors in the header. The DQDB has no error detection for the body or info portion of the slot or cell. There is a powerful CRC32 in the trailer of the IMPDU (initial MAC protocol data unit), an upper layer of DQDB. It also employs the BOM, COM, and EOM (beginning of message, continuation of message, and end of message, respectively) as well as the message identifier (MID) in the SAR* process. These fields are particularly helpful in message reassembly.

Switched Multimegabit Data Service came on the scene using DQDB as its access protocol. DQDB uses a 5-octet ACF/header and a 48-octet payload with no trailer, which is very similar to the ATM cell. SMDS has a 44-octet payload, a 7-octet header, and a 2-octet trailer, again for a total octet count of 53—in what we will call a cell. Many of the strategies of DQDB followed on down to SMDS. However, SMDS has a credit manager at the switching system (SS) which polices users. This policing concept came from frame relay. It is carried onward into ATM.

2 INTRODUCTION TO ATM

ATM is an outgrowth of the several data transmission format systems discussed above, although some may argue this point. Whereas the formats described above ostensibly were to satisfy the needs of the data world,† ATM (according to some) provides an optimum format or protocol family for data, voice, and image communications, where cells of each can be intermixed as

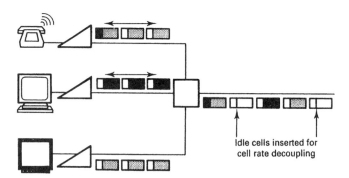

Figure 15.1. ATM links simultaneously carry a mix of voice, data, and image information. Courtesy of Hewlett-Packard Co.

---

*SAR stands for segmentation and reassembly.
†DQDB can also transport voice in its PA (pre-arbitrated) segments.

shown in Figure 15.1. It would really seem to be more of a compromise. Typically, these ATM cells can be transported on SONET, SDH, E1/T1, and other popular digital formats. Cells can also be transported contiguously without an underlying digital network format.

Philosophically, voice and data are worlds apart regarding time sensitivity. Voice cannot wait for long processing and ARQ delays. Most types of data can. So ATM must distinguish the type of service such as constant bit rate (CBR) and variable bit rate (VBR) services. Voice service is typical of constant bit rate or CBR service.

Signaling is another area of major philosophical difference. In data communications, "signaling" is carried out within the header of a data frame (or packet). As a minimum the signaling will have the destination address, and quite often the source address as well. And this signaling information will be repeated over and over again on a long data file that is heavily segmented. On a voice circuit, a connectivity is set up and the destination address, and possibly the source address, is sent just once during call setup. There is also some form of circuit supervision to keep the circuit operational throughout the duration of a telephone call. ATM is a compromise, stealing a little from each of these separate worlds.

Like voice telephony, ATM is fundamentally a connection-oriented telecommunication system. Here we mean that a connection must be established between two stations before data can be transferred between them. An ATM connection specifies the transmission path, allowing ATM cells to self-route through an ATM network. Being connection-oriented also allows ATM to specify a guaranteed quality of service (QoS) for each connection.

By contrast, most LAN protocols are connectionless. This means that LAN nodes simply transmit traffic when they need to, without first establishing a specific connection or route with the destination node.

In that ATM uses a connection-oriented protocol, bandwidth is allocated only when the originating end user requests a connection. This allows ATM to efficiently support a network's aggregate demand by allocating bandwidth on demand based on immediate user need. Indeed it is this concept which lies in the heart of the word *asynchronous*. An analogy would help. New York City is connected to Washington, DC with a pair of railroad tracks for passenger trains headed south and another pair of tracks for passenger trains headed north. On those two pairs of tracks we'd like to accommodate everybody we can when they'd like to ride. The optimum for reaching this goal is to have a continuous train of coupled passenger cars. As the train enters Union Station, it disgorges its passengers and connects around directly for the northward run to Pennsylvania Station. Passenger cars are identical in size, and each have the same number of identical seats.

Of course, at 2 A.M. the train will have very few passengers and many empty seats. Probably from 7 to 9 A.M. the train will be full, no standees allowed, so we'll have to hold potential riders in the waiting room. They'll ride later; those few who try to be standees will be bumped. Others might seek alternate transportation to Washington, DC.

Here we see that the railroad tracks are the transmission medium. Each passenger car is a SONET/SDH frame. The seats in each car are our ATM cells. Each seat can handle a person no bigger than 53 units. Because of critical weight distribution, if a person is not 53 units in size/weight, we'll stick some bricks in the seat to bring the size/weight to 53 units exactly. Those bricks are removed at the destination. All kinds of people ride the train because America is culturally diverse, analogous to the fact that ATM handles all forms of traffic. The empty seats represent idle or unassigned cells. The header information is analogous to the passengers' tickets. Keep in mind that the train can only fill to its maximum capacity of seats. We can imagine the SONET/SDH frame as being full of cells in the payload, some cells busy and some idle/unassigned. At the peak traffic period, all cells will be busy, and some traffic (passengers) may have to be turned away.

We can go even further with this analogy. Both Washington, DC and New York City attract large groups of tourists, and other groups travel to business meetings or conventions. A tour group has a chief tour guide in the lead seat (cell) and an assistant guide in the last seat (cell). There may be so many in the group that they extend into a second car or may just intermingle with other passengers on the train. The tour guide and assistant tour guide keep an exact count of people on the tour. The lead guide wears a badge that says BOM, all tour members wear badges that say COM, and the assistant tour guide wears a badge that says EOM. Each group has a unique MID (message ID). We also see that service is connection-oriented (Washington, DC to New York City).

Asynchronous means that we can keep filling the seats on the train until we reach its maximum capacity. If we look up the word, it means *nonperiodic*, whereas the familiar E1/T1 are periodic (i.e., synchronous). One point that seems to get lost in the literature is that the train has a maximum capacity. Thus the concept "bandwidth on demand" is that we can use the "bandwidth" until we fill to rated capacity. Again the unfortunate use of the word *bandwidth*, because our capacity will be measured in octets, not hertz. For example, SONET's STS-1 has a payload capacity of 87×9 octets (see Chapter 7), not 87×9 Hz.

3 USER–NETWORK INTERFACE (UNI) CONFIGURATION AND ARCHITECTURE

ATM is the underlying packet technology of broadband ISDN (B-ISDN). At times in this section, we will use the terms ATM and B-ISDN interchangeably. Figures 15.2 and 15.3 interrelate the two. Figure 15.2 relates the B-ISDN access reference configuration with ATM user–network interface (UNI). Note the similarities of this figure with Figure 13.2. The only

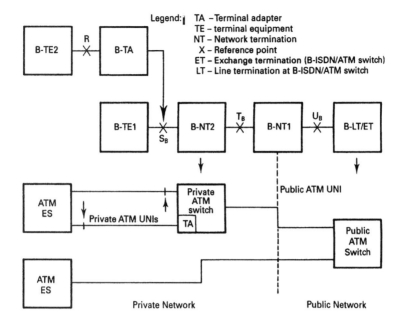

Figure 15.2. ATM reference model and user–network interface configuration. *Source*: Refs. 2–5.

difference is that the block nomenclature has a "B" placed in front to indicate *broadband*. Figure 15.3 is the traditional ITU-T Rec. I.121 [1] B-ISDN protocol reference model showing the extra layer necessary for the several services.

Returning to Figure 15.2 we see that there is an upper part and a lower part. The lower part shows the UNI boundaries. The upper part is the B-ISDN reference configuration with four interface points. The interfaces at reference points U_B, T_B, and S_B are standardized. These interfaces support all B-ISDN services.

There is only one interface per B-NT1 at the U_B and one at the T_B reference points. The physical media is point-to-point (in each case) in the sense that there is only one receiver in front of one transmitter.

One or more interfaces per NT2 are present at the S_B reference point. The interface at the S_B reference point is point-to-point at the physical layer in the sense that there is only one receiver in front of one transmitter and may be point-to-point at other layers.

Consider now the functional groupings in Figure 15.2. B-NT1 includes functions broadly equivalent to OSI layer 1, the physical layer. These functions include:

- Line transmission termination
- Interface handling at T_B and U_B
- OAM functions

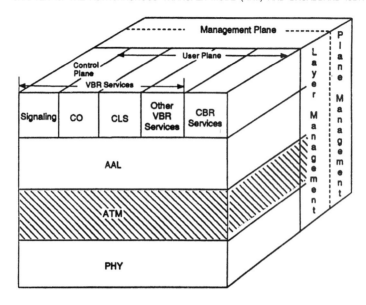

Figure 15.3. B-ISDN protocol reference model. From Ref. 6, reprinted with permission.

The B-NT2 functional group includes functions broadly equivalent to OSI layer 1 and higher OSI layers. The B-NT2 may be concentrated or distributed. In a particular access arrangement, the B-NT2 functions may consist of physical connections. Examples of B-NT2 functions are:

• Adaptation functions for different media and topologies
• Cell delineation
• Concentration; buffering
• Multiplexing and demultiplexing
• OAM functions
• Resource allocation
• Signaling protocol handling

The functional group B-TE (TE stands for terminal equipment) also includes functions of OSI layer 1 and higher OSI layers. Some of these functions are:

• User/user and user/machine dialog and protocol
• Protocol handling for signaling
• Connection handling to other equipment
• Interface termination
• OAM functions

B-TE1 has an interface that complies with the B-ISDN interface. B-TE2, however, has a noncompliant B-ISDN interface. Compliance refers to ITU-T Recs. I.413 and I.432 as well as ANSI T1.624 [3–5].

The terminal adapter (B-TA) converts the B-TE2 interface into a compliant B-ISDN user–network interface.

Four bit rates are specified at the UB, TB, and SB interfaces based on Ref. 4 (ANSI T1.624). These are:

51.840 Mbps (SONET STS-1)

155.520 Mbps (SONET STS-3 and SDH STM-1)

622.080 Mbps (SONET STS-12 and SDH STM-4)

44.736 Mbps (DS3)

These interfaces are discussed subsequently in this chapter.

The following definitions refer to Figure 15.3.

User Plane (in other literature called the U-plane). The user plane provides for the transfer of user application information. It contains physical layer, ATM layer, and multiple ATM adaptation layers required for different service users such as constant bit rate service (CBR) and variable bit rate service (VBR).

Control Plane (in other literature called the C-plane). The control plane protocols deal with call establishment and call release and other connection control functions necessary for providing switched services. The C-plane structure shares the physical and ATM layers with the U-plane as shown in Figure 15.3. It also includes ATM adaptation layer (AAL) procedures and higher-layer signaling protocols.

Management Plane (in other literature called the M-plane). The management plane provides management functions and the capability to exchange information between the U-plane and the C-plane. The M-plane contains two sections: layer management and plane management. The layer management performs layer-specific management functions, while the plane management performs management and coordination functions related to the complete system.

We return to Figure 15.3 and B-ISDN/ATM layering and layer descriptions in Section 6.

4 THE ATM CELL—KEY TO OPERATION

4.1 ATM Cell Structure

The ATM cell consists of 53 octets, 5 of which make up the header and 48 octets are in the payload or "info" portion of the cell. Figure 15.4 shows an

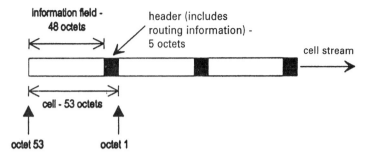

Figure 15.4. An ATM cell stream illustrating the basic makeup of a cell.

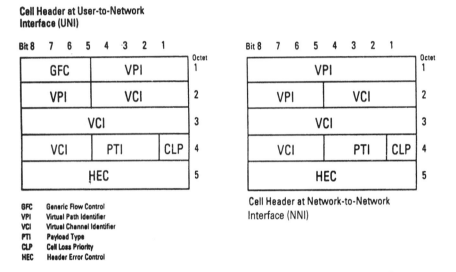

Cell Header at User-to-Network Interface (UNI)

GFC	Generic Flow Control
VPI	Virtual Path Identifier
VCI	Virtual Channel Identifier
PTI	Payload Type
CLP	Cell Loss Priority
HEC	Header Error Control

Cell Header at Network-to-Network Interface (NNI)

Figure 15.5. Basic ATM header structures. (a) UNI cell header structure; (b) NNI header structure.

ATM cell stream delineating the 5-octet header and 48-octet information field of each cell. Figure 15.5 shows the detailed structure of the cell headers at the user–network interface (UNI) (Figure 15.5a) and at the network–node interface (NNI)* (Figure 15.5b).

We digress a moment to discuss why a 53-octet cell was standardized. The cell header contains only 5 octets. It was shortened as much as possible, containing the minimum address and control functions for a working system. It is also non-revenue-bearing overhead. It is the information field that contains the revenue-bearing payload. For efficiency, we'd like the payload to

---

*NNI is variously called network–node interface or network–network interface. It is the interface between two network nodes or switches.

be as long as possible. Yet the ATM designer team was driven to shorten the payload as much as possible. The issue in this case was what is called *packetization delay*. This is the amount of time required to fill a cell at a rate of 64 kbps—that is, the rate to fill the cell with digitized voice samples. According to Ref. 8, the design team was torn between efficiency and packetization delay. One school of thought fought for a 64-octet cell, and another argued for a 32-octet cell size. Thus, the ITU-T opted for a fixed-length 53-octet compromise.

Now let's return to the discussion of the ATM cell and its headers. The left-hand side of Figure 15.5 shows the structure of a UNI header, whereas the right-hand side illustrates the NNI header. The only difference is the presence of the GFC field in the UNI header. The following paragraphs define each header field. By removing the GFC field, the NNI has four additional bits for addressing.

GFC—GENERIC FLOW CONTROL. The GFC field contains 4 bits. When the GFC function is not used, the value of this field is 0000. This field has local significance only and can be used to provide standardized local flow control functions on the customer side. In fact the value encoded in the GFC is not carried end-to-end and will be overwritten by ATM switches (i.e., the NNI interface).

Two modes of operation have been defined for operation of the GFC field. These are *uncontrolled access* and *controlled access*. The "uncontrolled access" mode of operation is used in the early ATM environment. This mode has no impact on the traffic which a host generates. Each host transmits the GFC field set to all zeros (0000). In order to avoid unwanted interactions between this mode and the "controlled access" mode where hosts are expected to modify their transmissions according to the activity of the GFC field, it is required that all CPE (customer premise equipment) and public network equipment monitor the GFC field to ensure that attached equipment is operating in "uncontrolled mode." A count of the number of nonzero GFC fields should be measured for nonoverlapping intervals of $30,000 \pm 10,000$ cell times. If ten or more nonzero values are received within this interval, an error is indicated to layer management [2].

ROUTING FIELD (VPI/VCI). Twenty-four bits are available for routing a cell. There are 8 bits for virtual path identifier (VPI) and 16 bits for virtual channel identifier (VCI). Preassigned combinations of VPI and VCI values are given in Table 15.1. Other preassigned values of VPI and VCI are for further study according to the ITU-T organization. The VCI value of zero is not available for user virtual channel identification. The bits within the VPI and VCI fields used for routing are allocated using the following rules:

• The allocated bits of the VPI field are contiguous.

TABLE 15.1   Combinations of Preassigned VPI, VCI, and CLP Values at the UNI

Use	VPI	VCI	PT	CLP
Meta-signaling	XXXXXXXX	00000000 00000001	0A0	C
(refer to Rec. I.311)	(Note 1)	(Note 5)		
General broadcast signaling	XXXXXXXX	00000000 00000010	0AA	C
(refer to Rec. I.311)	(Note 1)	(Note 5)		
Point-to-point signaling	XXXXXXXX	00000000 00000101	0AA	C
(refer to Rec. I.311)	(Note 1)	(Note 5)		
Segment OAM F4 flow cell	YYYYYYYY	00000000 00000011	0A0	A
(refer to Rec. I.610)	(Note 2)	(Note 4)		
End-to-end OAM F4 flow cell	YYYYYYYY	00000000 00000100	0A0	A
(refer to Rec. I.610)	(Note 2)	(Note 4)		
Segment OAM F5 flow cell	YYYYYYYY	ZZZZZZZZ ZZZZZZZZ	100	A
(refer to Rec. I.610)	(Note 2)	(Note 3)		
End-to-end OAM F5 flow cell	YYYYYYYY	ZZZZZZZZ ZZZZZZZZ	101	A
(refer to Rec. I.610)	(Note 2)	(Note 3)		
Resource management cell	YYYYYYYY	ZZZZZZZZ ZZZZZZZZ	110	A
(refer to Rec. I.371)	(Note 2)	(Note 3)		
Unassigned cell	00000000	00000000 00000000	BBB	0

The GFC field is available for use with all of these combinations.
A   Indicates that the bit may be 0 or 1 and is available for use by the appropriate ATM layer function.
B   Indicates the bit is a "don't care" bit.
C   Indicates the originating signaling entity shall set the CLP bit to 0. The value may be changed by the network.
Notes
1   XXXXXXXX: Any VPI value. For VPI value equal to 0, the specific VCI value specified is reserved for user signaling with the local exchange. For VPI values other than 0, the specified VCI value is reserved for signaling with other signaling entities (e.g., other users or remote networks).
2   YYYYYYYY: Any VPI value.
3   ZZZZZZZZ ZZZZZZZZ: Any VCI value other than 0.
4   Transparency is not guaranteed for the OAM F4 flows in a user-to-user VP.
5   The VCI values are preassigned in every VPC at the UNI. The usage of these values depends on the actual signaling configurations. (See ITU-T Rec. I.311.)
Source: ITU-T Rec. 361, Table 2/I.361, page 3 (Ref. 7).

- The allocated bits of the VPI field are the least significant bits of the VPI field, beginning at bit 5 of octet 2.
- The allocated bits of the VCI field are contiguous.
- The allocated bits of the VCI field are the least significant bits of the VCI field, beginning at bit 5 of octet 4.

PAYLOAD TYPE (PT) FIELD. Three bits are available for PT identification. Table 15.2 gives the payload type identifier coding. The main purpose of the PTI is to discriminate between user cells (i.e., cells carrying user information) and nonuser cells. The first four code groups (000–011) are used to indicate user cells. Within these four, 2 and 3 (010 and 011) are used to indicate

TABLE 15.2 PTI Coding

	PTI Coding	Interpretation
Bits	4 3 2	
	0 0 0	User data cell, congestion not experienced. ATM-user-to-ATM-user indication $= 0$
	0 0 1	User data cell, congestion not experienced. ATM-user-to-ATM-user indication $= 1$
	0 1 0	User data cell, congestion experienced. ATM-user-to-ATM-user indication $= 0$
	0 1 1	User data cell, congestion experienced. ATM-user-to-ATM-user indication $= 1$
	1 0 0	OAM F5 segment associated cell
	1 0 1	OAM F5 end-to-end associated cell
	1 1 0	Resource management cell
	1 1 1	Reserved for future functions

Source: ITU-T Rec. I.361, page 4, para. 2.2.4 (Ref. 7).

congestion has been experienced. The fifth and sixth code groups (100 and 101) are used for VCC level management functions.

Any congested network element, upon receiving a user data cell, may modify the PTI as follows. Cells received with PTI $= 000$ or PTI $= 010$ are transmitted with PTI $= 010$. Cells received with PTI $= 001$ or PTI $= 011$ are transmitted with PTI $= 011$. Noncongested network elements should not change the PTI.

CELL LOSS PRIORITY (CLP) FIELD. Depending on network conditions, cells where the CLP is set (i.e., CLP value is 1) are subject to discard prior to cells where the CLP is not set (i.e., CLP value is 0). The concept here is identical with that of frame relay and the DE (discard eligibility) bit. ATM switches may tag CLP $= 0$ cells detected by the UPC (usage parameter control) to be in violation of the traffic contract by changing the CLP bit from 0 to 1.

HEADER ERROR CONTROL (HEC) FIELD. The HEC is an 8-bit field and it covers the entire cell header. The code used for this function is capable of either single-bit error correction or multiple-bit error detection. Briefly, the transmitting side computes the HEC field value. The receiver has two modes of operation as shown in Figure 15.6. In the default mode there is the capability of single-bit error correction. Each cell header is examined and, if an error is detected, one of two actions takes place. The action taken depends on the state of the receiver. In the *correction mode*, only single-bit errors can be corrected and the receiver switches to the *detection mode*. In the "detection mode," all cells with detected header errors are discarded. When a header is

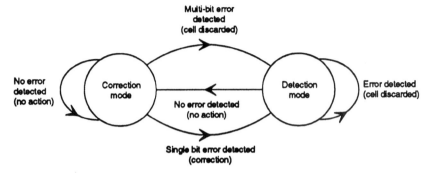

Figure 15.6. HEC: receiver modes of operation. Based on ITU-T Rec. I.432 [3].

examined and found not to be in error, the receiver switches to the "correction mode." The term *no action* in Figure 15.6 means no correction is performed and no cell is discarded.

Figure 15.7 is a flow chart showing the consequence of errors in the ATM cell header. The error protection function provided by the HEC provides for

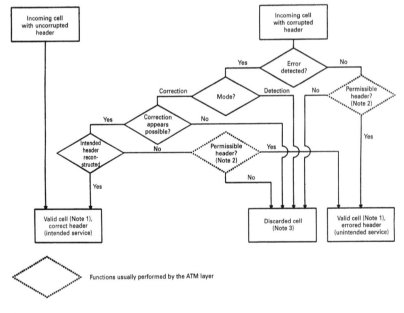

Figure 15.7. Consequences of errors in an ATM cell header. From ITU-T Rec. I.432, Figure 12/I.432, page 18 (Ref. 3).

TABLE 15.3   Header Pattern for Idle Cell Identification

	Octet 1	Octet 2	Octet 3	Octet 4	Octet 5
Header pattern	00000000	00000000	00000000	00000001	HEC = Valid code 01010010

Source: ITU-T Rec. I.432, Table 4/I.432, page 19 (Ref. 3).

both recovery from single-bit errors and a low probability of delivery of cells with errored headers under bursty error conditions. ITU-T Rec. I.432 [3] states that error characteristics of fiber-optic transmission systems appear to be a mix of single-bit errors and relatively large burst errors. Thus, for some transmission systems the error correction capability might not be invoked.

4.2   Idle Cells

Idle cells cause no action at a receiving node except for cell delineation including HEC verification. They are inserted and extracted by the physical layer in order to adapt the cell flow rate at the boundary between the ATM layer and the physical layer to the available payload capacity of the transmission media. This is called *cell rate decoupling*. Idle cells are identified by the standardized pattern for the cell header as shown in Table 15.3. The content of the information field is 01101010 repeated 48 times for an idle cell.

There is some variance in this area between the ITU-T organization documentation and that of the ATM Forum. McDysan and Spohn in Ref. 8 point out the following. ITU-T Rec. I.321 places this function in the TC (transmission convergence) sublayer of the PHY (physical layer) and uses idle cells, whereas the ATM Forum places it in the ATM layer and uses un-assigned cells. This presents a potential low-level incompatibility if different systems use different cell types for cell rate decoupling.

5   CELL DELINEATION AND SCRAMBLING

5.1   Delineation and Scrambling Objectives

Cell delineation allows identification of the cell boundaries. The cell header error control (HEC) field achieves cell delineation. Keep in mind that the ATM signal must be self-supporting in that it has to be transparently transported on every network interface without any constraints from the transmission systems used. Scrambling is used to improve security and robustness of the HEC cell delineation mechanism discussed below. In addition, it helps the randomizing of data in the information field for possible improvement in transmission performance.

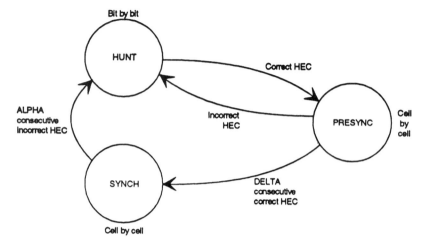

Figure 15.8. Cell delineation state diagram. From ITU-T Rec. I.432, Figure 13/I.432, page 20 (Ref. 3).

Any scrambler specification must not alter the ATM header structure, header error control, and cell delineation algorithm.

5.2 Cell Delineation Algorithm

Cell delineation is performed by using the correlation between the header bits to be protected (32 bits or 4 octets) and the HEC octet which are the relevant control bits (8 bits) introduced in the header using a shortened cyclic code with the generating polynomial $X^8 + X^2 + X + 1$.

Figure 15.8 shows the state diagram of the HEC cell delineation method. A discussion of the figure is given below.

1. In the HUNT state, the delineation process is performed by checking bit by bit for the correct HEC (i.e., syndrome equals zero) for the assumed header field. For the cell-based* physical layer, prior to scrambler synchronization, only the last six bits of the HEC are used for cell delineation checking. For the SDH-based* interface, all 8 bits are used for acquiring cell delineation. Once such an agreement is found, it is assumed that one header has been found, and the method enters the PRESYNCH state. When octet boundaries are available within the receiving physical layer prior to cell delineation as with the SDH-based interface, the cell delineation process may be performed octet by octet.

*Only cell-based and SDH-based interfaces are covered by current ITU-T recommendations. Besides these, we will cover cells riding on other transport means at the end of this chapter.

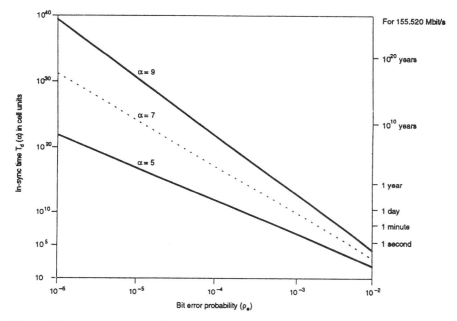

Figure 15.9. In-sync time versus bit error probability. From ITU-T Rec. I.432, Figure B.1/I.432, page 33 (Ref. 3).

2. In the PRESYNCH state, the delineation process is performed by checking cell by cell for the correct HEC. The process repeats until the correct HEC has been confirmed *Delta* times consecutively. If an incorrect HEC is found, the process returns to the HUNT state.

3. In the SYNCH state the cell delineation will be assumed to be lost if an incorrect HEC is obtained *Alpha* times consecutively.

The parameters *Alpha* and *Delta* are chosen to make the cell delineation process as robust and secure as possible while satisfying QoS (quality of service) requirements. Robustness depends on *Alpha* when it is against false misalignments due to bit errors. And robustness depends on *Delta* when it is against false delineation in the resynchronization process.

For the SDH-based physical layer, values of *Alpha* = 7 and *Delta* = 6 are suggested by the ITU-T organization (Rec. I.432 [3]); and for cell-based physical layer, values of *Alpha* = 7 and *Delta* = 8 are suggested. Figures 15.9 and 15.10 give performance information of the cell delineation algorithm in the presence of random bit errors, for various values of *Alpha* and *Delta*.

6 ATM LAYERING AND B-ISDN

The B-ISDN reference model is given in Figure 15.3, and its several planes are described. This section provides brief descriptions of the ATM layers and sublayers.

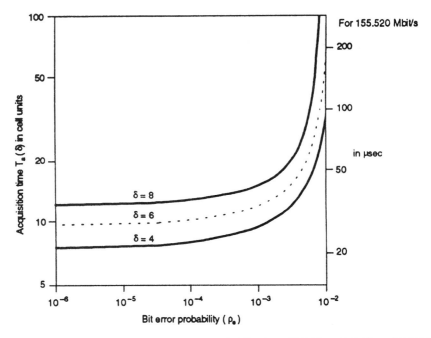

Figure 15.10. Acquisition time versus bit error probability. From ITU-T Rec. I.432, Figure B.2/I.432, page 34 (Ref. 3).

6.1 Functions of Individual ATM/B-ISDN Layers

Figure 15.11 illustrates B-ISDN/ATM layering and sublayering of the protocol reference model. It identifies the functions of the physical layer, the ATM layer and the AAL, and related sublayers.

6.1.1 Physical Layer. The physical layer consists of two sublayers. The physical medium (PM) sublayer includes only physical medium-dependent functions. The transmission convergence (TC) sublayer performs all functions required to transform a flow of cells into a flow of data units (i.e., bits) which can be transmitted and received over a physical medium. The service data unit (SDU) crossing the boundary between the ATM layer and the physical layer is a flow of valid cells. The ATM layer is unique (meaning independent of the underlying physical layer). The data flow inserted in the transmission system payload is physical medium-independent and self-supported. The physical layer merges the ATM cell flow with the appropriate information for cell delineation, according to the cell delineation mechanism described above, and carries the operations and maintenance (OAM) information relating to this cell flow.

The physical medium sublayer provides bit transmission capability including bit transfer and bit alignment as well as line coding and electrical-

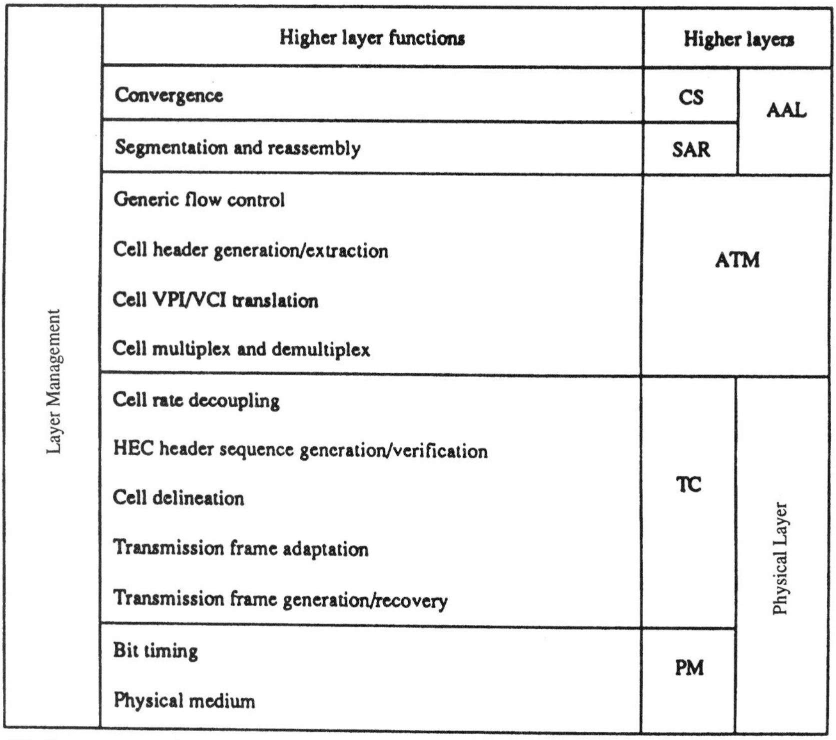

	Higher layer functions	Higher layers	
	Convergence	CS	AAL
	Segmentation and reassembly	SAR	
Layer Management	Generic flow control Cell header generation/extraction Cell VPI/VCI translation Cell multiplex and demultiplex	ATM	
	Cell rate decoupling HEC header sequence generation/verification Cell delineation Transmission frame adaptation Transmission frame generation/recovery	TC	Physical Layer
	Bit timing Physical medium	PM	

CS Convergence sublayer
PM Physical medium
SAR Segmentation and reassembly sublayer
TC Transmission convergence

Figure 15.11. B-ISDN/ATM functional layering.

optical transformation. Of course, the principal function is the generation and reception of waveforms suitable for the medium, the insertion and extraction of bit timing information, and line coding where required. The primitives identified at the border between the PM and TC sublayers are a continuous flow of logical bits or symbols with this associated timing information.

TRANSMISSION CONVERGENCE SUBLAYER FUNCTIONS. Among the important functions of this sublayer is the generation and recovery of transmission frame. Another function is transmission frame adaptation which includes the actions necessary to structure the cell flow according to the payload structure of the transmission frame (transmit direction) and to extract this cell flow out of the transmission frame (receive direction). The transmission frame may be a cell equivalent (i.e., no external envelope is added to the cell flow), an SDH/SONET envelope, an E1/T1 envelope, and so on. In the transmit direction, the HEC sequence is calculated and inserted in the header. In the receive direction, we include cell header verification. Here cell headers are checked for errors and, if possible, header errors are corrected. Cells are discarded where it is determined that headers are errored and are not correctable.

Another transmission convergence function is cell rate decoupling. This involves the insertion and removal of idle cells in order to adapt the rate of valid ATM cells to the payload capacity of the transmission system. In other words, cells must be generated to exactly fill the payload of SDH/SONET, as an example, whether the cells are idle or busy.

Section 12 gives several examples of transporting cells using the convergence sublayer.

6.1.2 The ATM Layer. Table 15.4 shows the ATM layer functions supported at the UNI (U-plane). The ATM layer is completely independent of the physical medium. One important function of this layer is *encapsulation*. This includes cell header generation and extraction. In the transmit direction, the cell header generation function receives a cell information field from a higher layer and generates an appropriate ATM cell header except for the header error control (HEC) sequence. This function can also include the translation from a service access point (SAP) identifier to a VP (virtual path) and VC (virtual circuit) identifier.

In the receive direction, the cell header extraction function removes the ATM cell header and passes the cell information field to a higher layer. As in the transmit direction, this function can also include a translation of a VP and VC identifier into an SAP identifier.

In the case of the NNI (network–node interface) the GFC (generic flow control) is applied at the ATM layer. The flow control information is carried in assigned and unassigned cells. Cells carrying this information are generated in the ATM layer.

In a switch the ATM layer determines where the incoming cells should be forwarded to, resets the corresponding connection identifiers for the next link, and forwards the cell. The ATM layer also handles traffic management

TABLE 15.4 ATM Layer Functions Supported at the UNI

Functions	Parameters
Multiplexing among different ATM connections	VPI/VCI
Cell rate decoupling (unassigned cells)	Preassigned header field values
Cell discrimination based on predefined header field values	Preassigned header field values
Payload type discrimination	PT field
Loss priority indication and selective cell discarding	CLP field, network congestion state
Traffic shaping	Traffic descriptor

Source: Based on Ref. 2 and Ref. 7.

functions and buffers incoming and outgoing cells. It indicates to the next higher layer (the AAL) whether or not there is congestion during transmission. The ATM layer monitors both transmission rates and conformance to the service contract—called *traffic shaping* and *traffic policing*.

CELL DISCRIMINATION BASED ON PREDEFINED HEADER FIELD VALUES. The predefined header field values defined at the UNI are given in Table 15.5A, and those for the NNI are given in Table 15.5B. Of interest are the VPI and VCI values reserved as shown in the tables for six of the seven rows (less invalid pattern).

Meta-signaling cells are used by the meta-signaling protocol for establishing and releasing signaling virtual channel connections (VCCs). For virtual channels allocated permanently (PVC), meta-signaling is not used.

General broadcast signaling cells are used by the ATM network to broadcast signaling information independent of service profiles.

The virtual path connection (VPC) operation flow (F4 flow) is carried via specially designated OAM cells. F4 flow OAM cells have the same VPI value as the user-data cells transported by the VPC but are identified by two unique preassigned virtual channels within this VPC. At the UNI, the virtual channel identified by a VCI value = 3 is used for VP level management

TABLE 15.5A Preassigned Header Field Values at the UNI (Excluding the HEC)

	Value			
Use	Octet 1	Octet 2	Octet 3	Octet 4
Unassigned cell indication	AAAA0000	00000000	00000000	0000BBB0
Meta-signaling (see note 4)	AAAAXXXX	XXXX0000	00000000	00010A0C
General broadcast signaling (see note 4)	AAAAXXXX	XXXX0000	00000000	0100AAC
Point-to-point signaling (see note 4)	AAAAXXXX	XXXX0000	00000000	01010AAC
Invalid pattern	BBBB0000	00000000	00000000	0000BBB1
Segment OAM flow cell (see note 5)	AAAAYYYY	YYYY0000	00000000	00110A0A
End-to-end OAM F4 flow cell (see note 5)	AAAAYYYY	YYYY0000	00000000	01000A0A

Notes

1   A indicates that the bit is available for use by the appropriate ATM layer function.
2   B indicates that the bit is a *don't care* bit.
3   C indicates the originating entity shall set the CLP bit to 0. The value may be changed by the network.
4   These VCI values are reserved for user-to-network signaling with the local exchange, for VPI value, XXXXXXXX, equal to 0, and reserved for user-to-user signaling with other signaling entities (e.g., other users or remote network), for VPI values, XXXXXXXX, other than 0.
5   YYYYYYYY—any VPI value.

Source: ANSI T1.627-1993, Table 1, page 13 (Ref. 6), reprinted with permission.

TABLE 15.5B Preassigned Header Field Values at the NNI (Excluding the HEC)

Use	Value			
	Octet 1	Octet 2	Octet 3	Octet 4
Unassigned cell indication	00000000	00000000	00000000	0000BBB0
Point-to-point signaling	YYYYYYYY	YYYY0000	00000000	01010AAC
Invalid pattern	00000000	00000000	00000000	0000BBB1
Segment OAM F4 flow cell (see note 4)	YYYYYYYY	YYYY0000	00000000	00110A0A
End-to-end OAM F4 flow cell (see note 4)	YYYYYYYY	YYYY0000	00000000	01000A0A

Notes
1   A indicates that the bit is available for use by the appropriate ATM layer function.
2   B indicates that the bit is a *don't care* bit.
3   C indicates the originating entity shall set the CLP bit to 0. The value may be changed by the network.
4   YYYYYYYY—any VPI value.
Source: ANSI T1.627-1993, Table 2, page 13 (Ref. 6), reprinted with permission.

functions between ATM nodes on both sides of the UNI (i.e., single VP link segment) while the virtual channel identified by a VCI value = 4 can be used for VP level end-to-end (user ↔ user) management functions.

What are flows such as "F4 flows"? OAM (operations, administration, and management) flows deal with cells dedicated to fault and performance management of the total system. Consider ATM as a hierarchy of levels, particularly in SDH/SONET, which are the principal bearer formats for ATM. The lowest level where we have F1 flows is the regenerator section (called the *section level* in SONET). This is followed by F2 flows at the digital section level (called the *line level* in SONET). There are the F3 flows for the transmission path (called the *path level* in SONET). ATM adds F4 flows for virtual paths (VPs) and F5 flows for virtual channels (VCs), where multiple VCs are completely contained within a single VP. We discuss VPs and VCs later.

6.1.2.1 ATM Layer Management (M-Plane). Management functions at the UNI require some level of cooperation between customer premises equipment and network equipment. To minimize the coupling required between equipment on both sides of the UNI, the functional requirements have been reduced to a minimal set. The ATM layer management functions supported at the UNI are grouped into the categories under the general heading of fault management.

Fault management contains alarm surveillance and connectivity verification functions. OAM cells are used for exchanging related operation information.

6.1.3 The ATM Adaptation Layer (AAL). The basic purpose of the AAL is to isolate the higher layers from the specific characteristics of the ATM layer by mapping the higher-layer protocol data units (PDUs) into the information field of the ATM cell and vice versa.

6.1.3.1 Sublayering of the AAL. To support services above the AAL, some independent functions are required of the AAL. These functions are organized in two logical sublayers: the convergence sublayer (CS) and the segmentation and reassembly sublayer (SAR). The prime functions of these sublayers are:

- SAR—The segmentation of higher-layer information into a size suitable for the information field of an ATM cell. Reassembly of the contents of ATM cell information fields into higher-layer information.
- CS—Here the prime function is to provide the AAL service at the AAL-SAP (SAP stands for service access point). This sublayer is service-dependent.

6.1.3.2 Service Classification for the AAL. Service classification is based on the following parameters:

- Timing relation between source and destination (this refers to urgency of traffic): required or not required
- Bit rate: constant or variable
- Connection mode: connection-oriented or connectionless

When we combine these parameters, four service classes emerge as shown in Figure 15.12. Examples of services in the classes shown in Figure 15.12 are as follows:

- Class A: constant bit rate such as uncompressed voice or video
- Class B: variable bit rate video and audio, connection-oriented synchronous traffic
- Class C: connection-oriented data transfer, variable bit rate, asynchronous traffic
- Class D: connectionless data transfer, asynchronous traffic such as SMDS

6.1.3.3 AAL Categories or Types. There are five different AAL types or categories. The simplest of these is AAL-0. It just transmits cells down a pipe. That pipe is commonly a fiber-optic link. Ideally we would like the bit rate to

Service Parameters	Class A	Class B	Class C	Class D
Timing Compensation	Required		Not Required	
Bit Rate	Constant		Variable	
Connection Mode	Connection-oriented			Connectionless
AAL Types	AAL1	AAL2	AAL3/4 or AAL5	AAL3/4 or AAL5
Examples	DS1, E1, n × 64-kbps emulation	Packet video, audio	Frame relay X.25	IP, SMDS

Figure 15.12. Service classification for AAL. Based on Refs. 8–10 and 15.

be some multiple of 53×8 or 424 bits. For example, 424 Mbps would handle 1 million cells per second.

6.1.3.3.1 AAL-1. AAL-1 is used to provide transport for synchronous bit streams. Its primary application is to adapt ATM cell transmission to typically E1/T1 and SDH/SONET circuits. Typically, AAL-1 is for voice communications (POTS—plain old telephone service). AAL-1 robs one octet from the payload and adds it to the header, leaving only a 47-octet payload. This octet includes two major fields: sequence number (SN) and sequence number protection (SNP). The principal purpose of these two fields is to check that mis-sequencing of information does not occur by verifying a 3-bit sequence counter. It also allows for regeneration of the original clock timing of the data received at the far end of the link. The SAR-PDU format of AAL-1 is shown in Figure 15.13. The 4-bit sequence number (SN) is broken down into a 1-bit CSI (convergence sublayer indicator) and sequence count. The SNP (sequence number protection) contains a 3-bit CRC and a parity bit. End-to-end synchronization is an important function for the type of traffic carried on AAL-1. With one mode of operation, clock recovery is via a synchronous residual time stamp (SRTS) and common network clock by means of a 4-bit residual time stamp extracted from CSI of cells with odd sequence numbers. The residual time stamp is transmitted over eight cells. It supports DS1 and DS3 and E1 digital streams. Another mode of operation is structured data transfer (SDT). SDT supports an octet-structured nXDS0 service.

Alarm indication in this adaptation layer is via a check of the one's density. When the one's density of the received cell stream becomes significantly different than the density used for the particular PCM line coding scheme in use, it is determined that the system has lost signal and alarm notifications are given.

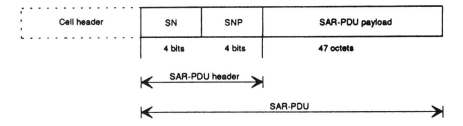

SN Sequence number (4 bits); to detect lost or misinserted cells. A specific value of the sequence number may indicate a special purpose, e.g. the existence of convergence sublayer functions. The exact counting scheme is for further study.

SNP Sequence number protection (4 bits). The SNP field may provide error detection and correction capabilities. The polynomial to be used is for further study.

Figure 15.13. SAR-PDU format for AAL-1. From CCITT Rec. I.363, Figure 1/I.363, page 3 (Ref. 9).

6.1.3.3.2 AAL-2. AAL-2 handles the variable bit rate (VBR) scenario such as MPEG* video. Functions in AAL type 2 include:

(a) Segmentation and reassembly (SAR) of user information

(b) Handling of cell delay variation

(c) Handling of lost and misinserted cells

(d) Source clock frequency recovery at the receiver

(e) Monitoring and handling of AAL-PCI bit errors (PCI stands for protocol control information)

(f) Monitoring of user information field for bit errors and possible corrective action

AAL-2 is still in ITU-T definitive stages. However, an example of an SAR-PDU format for AAL type 2 is given in Figure 15.14.

6.1.3.3.3 aal-3/4. Initially, in ITU-T Rec. I.363, there were two separate AALs, one for connection-oriented variable bit rate data services (AAL-3) and one for connectionless service (AAL-4). As the specifications evolved, the same procedures turned out to be necessary for both of these services, and the specifications were merged to become the AAL-3/4 standard. AAL-3/4 is used for ATM shipping of SMDS, CBDS (Connectionless Broadband Data Services, an ETSI initiative), IP (Internet Protocol), and frame relay.

AAL-3/4 has been designed to take variable-length frames/packets and segment them into cells. The segmentation is done in a way that protects the transmitted data from corruption if cells are lost or mis-sequenced.

*MPEG is a set of video compression schemes. MPEG stands for Motion Picture Experts Group.

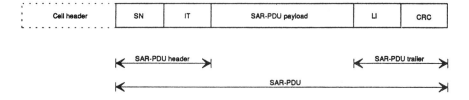

The need of each of the above fields, the position of those fields and their size are for further study.

SN   Sequence number, to detect lost or misinserted cells. A specific value of the sequence number may indicate a specified purpose

IT   Information type, used to indicate beginning of message (BOM), continuation of message (COM), end of message (EOM), timing information and also component of the video or audio signal

LI   Length indicator, check to indicate that the number of octets of the CS-PDU are included in the SAR-PDU payload field

CRC   Cyclic redundancy check code, to correct up to two correlated bit errors

Figure 15.14. Example of an SAR-PDU format for AAL-2. From CCITT Rec. I.363, Figure 2/I.363, page 5 (Ref. 9).

Variable-length packets (up to 64 kbytes) from SMDS/CLNAP (CLNAP stands for connectionless network access protocol) or frame relay frames are padded to an integral word length and encapsulated with a header and a trailer to form what is called the *convergence sublayer PDU* (CS_PDU) and then is segmented into cells. The passing is done to make sure that fields align themselves to 32-bit boundaries, allowing the efficient implementation of the operations in hardware at lower layers. The added header and trailer contain a tag to match the end and the packet length so that the receiving end may allocate the buffer for this packet upon reception of the first fragment. In practice, this buffer allocation size (BAsize) is mostly used as a length check for integrity verification, because the most efficient algorithm is to allocate fixed-size maximum-length buffers. This AAL convergence sublayer protocol data unit (CS_PDU) is shown in Figure 15.15. This example, of course, comes from SMDS, Chapter 14.

Figure 15.16 shows the SAR-PDU format for AAL-3/4 from ITU-T Rec. I.363, and Figure 15.17 shows this same format when used for transmitting SMDS frames.

Turning to Figure 15.17, the segmented portions are 44 octets long (except possibly for the last segment). These portions are then encapsulated with another header (2 octets) and trailer (2 octets) to become a segmentation and reassembly PDU (SAR_PDU) which is inserted into cell payloads. The header at this level with the Segment Type field identifies what kind of a cell it is [i.e., BOM (beginning of message), COM (continuation of message), or EOM (end of message)] so that the individual CS_PDUs can be delineated. The header also includes a sequence number for protection against mis-ordered delivery. It also includes the MID (message identification in SMDS, multiplexing identifier for ATM). The SAR_PDU trailer contains a length indicator to identify how much of the payload is filled. It also has a CRC-10 error check to protect against cell corruption. A complete message contains a

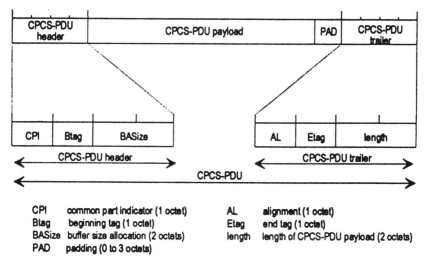

CPI   common part indicator (1 octet)   AL   alignment (1 octet)
Btag   beginning tag (1 octet)   Etag   end tag (1 octet)
BASize   buffer size allocation (2 octets)   length   length of CPCS-PDU payload (2 octets)
PAD   padding (0 to 3 octets)

Figure 15.15. Convergence sublayer protocol data unit (AAL-3/4). *Source*: Refs. 16 and 17.

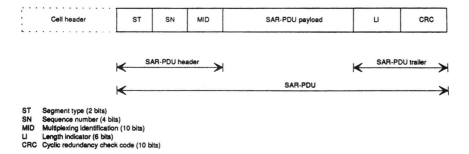

ST   Segment type (2 bits)
SN   Sequence number (4 bits)
MID   Multiplexing identification (10 bits)
LI   Length indicator (6 bits)
CRC   Cyclic redundancy check code (10 bits)

Figure 15.16. SAR-PDU format for AAL-3/4. From ITU-T Rec. I.363, Figure 6/I.363, page 13 (Ref. 9).

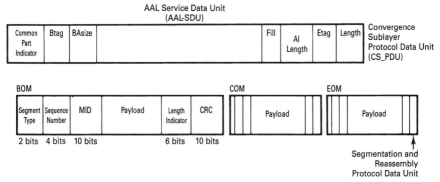

Figure 15.17. AAL-3/4 SAR-PDU as applied to the transmission of SMDS frames. Courtesy of Hewlett-Packard Company [10].

BOM cell, zero or more COM cells, and an EOM cell. If the entire message can fit into one cell, it is called a single segment message (SSM), where the CS_PDU is less than 44 octets long.

AAL-3/4 has several measures to ensure the integrity of the data which has been segmented and transmitted as cells. The contents of the cell are protected by the CRC-10; sequence numbers protect against misordering. Still another measure to ensure against corrupted PDUs being delivered is EOM/ BOM protection. If the EOM of one CPCS_PDU and the BOM of the next are dropped for some reason, the resulting cell stream could be interpreted as a valid PDU. To protect against these kinds of errors, the BEtag numeric values in the CPCS_PDU headers and trailers are compared to ensure that they match (CPCS stands for common part convergence sublayer). Two modes of service are defined for AAL-3/4:

- *Message Mode Service.* This provides for the transport of one or more fixed-size AAL service data units in one or more convergence sublayer protocol data units (CS-PDUs).

- *Streaming Mode Service.* Here the AAL service data unit is passed across the AAL interface in one or more AAL interface data units (IDUs). The transfer of these AAL-IDUs across the AAL interface may occur separated in time, and this service provides the transport of variable-length AAL-SDUs. The streaming mode service includes an abort service by which the discarding of an AAL-SDU partially transferred across the AAL interface can be requested. In other words, in the streaming mode, a single packet is passed to the AAL layer and transmitted in multiple CPCS-PDUs, as and when pieces of the packet are received. Streaming mode may be used in intermediate switches or ATM-to-SMDS routers so they can begin retransmitting a packet being received before the entire packet has arrived. This reduces the latency experienced by the entire packet.

6.1.3.3.4 aal-5. This type of AAL was designed specifically to carry data traffic typically found in today's LANs. AAL-5 evolved after AAL-3/4, which was found to be too complex and inefficient for LAN traffic. Thus AAL-5 got the name "SEAL" for simple and efficient AAL layer. Only a small amount of overhead is added to the CPCS-PDU, and no extra overhead is added when the AAL-5 segments them into SAR-PDUs. There is no AAL level cell multiplexing. In AAL-5 all cells belonging to an AAL-5 CPCS-PDU are sent sequentially.

As shown in Figure 15.18, the CPCS-PDU has only a payload and a trailer. The trailer contains padding, a length field, and a CRC-32 field for error detection. The CPCS-PDUs are padded to become integral multiples of 48, ensuring that there will never be a need to send partially filled cells after segmentation. A bit in the PTI field in the cell headers is used to indicate

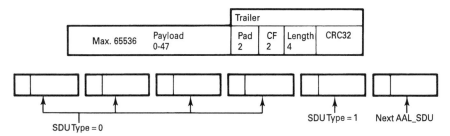

Figure 15.18. AAL type 5. Courtesy of Hewlett-Packard Company [10].

when the last cell of a PDU is transmitted, so that one PDU can be distinguished from the one that follows.

7 SERVICES: CONNECTION-ORIENTED AND CONNECTIONLESS

The issues such as routing decisions and architectures have a major impact on connection-oriented services, where B-ISDN/ATM end nodes have to maintain or get access to lookup tables which translate destination addresses into circuit paths. These circuit path lookup tables which differ at every node must be maintained in a quasi-real-time fashion. This will have to be done by some kind of routing protocol.

One way to resolve this problem is to make it an internal network problem and use a connectionless service as described in ITU-T Rec. I.364 [11]. We must keep in mind that ATM is basically a connection-oriented service. Here we are going to adapt it to provide a connectionless service.

7.1 Functional Architecture

The provision of connectionless data service in the B-ISDN is carried out by means of ATM switches and connectionless service functions (CLSF). ATM switches support the transport of connectionless data units in the B-ISDN between specific functional groups where the CLSF handles the connectionless protocol and provides for the adaptation of the connectionless data units into ATM cells to be transferred in a connection-oriented environment. As shown in Figure 15.19, CLSF functional groups may be located outside the B-ISDN, in a private connectionless network or in a specialized service provider, or inside the B-ISDN.

The ATM switching is performed by the ATM nodes (ATM switch/cross-connect) which are a functional part of the ATM transport network. The CLSF functional group terminates the B-ISDN connectionless protocol and includes functions for the adaptation of the connectionless protocol to the intrinsically connection-oriented ATM layer protocol. These latter functions

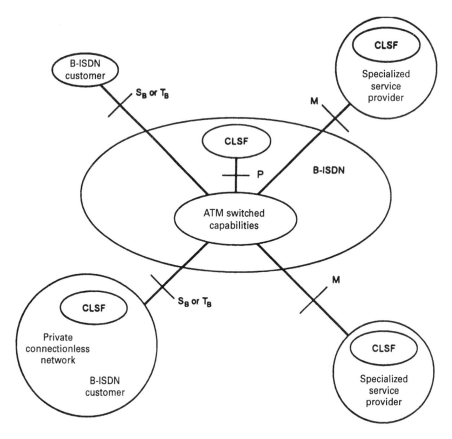

CLSF   Connectionless service functions
P, M, S, T   Reference point

Figure 15.19. Reference configuration for the provision of the CL (connectionless) data service in the B-ISDN. From ITU-T Rec. I.364, Figure 1/I.364, page 2 (Ref. 11).

are performed by the ATM adaptation layer type 3/4 (AAL-3/4), while the CLSF group terminations are carried out by the services layer above the AAL called the CLNAP (connectionless network access protocol). The CL protocol includes functions such as routing, addressing, and QoS (quality of service) selection. In order to perform the routing of CL data units, the CLSF has to interact with the control/management planes of the underlying ATM network.

The general protocol structure for the provision of connectionless (CL) data service is shown in Figure 15-20. Figure 15.21 shows the protocol architecture for supporting connectionless layer service. The CLNAP (connectionless network access protocol) layer uses the type 3/4 AAL unassured service and includes the necessary functionality to provide the connectionless layer service.

Customer
terminal
equipment
CLNAP
AAL 3/4
ATM
Physical
ATM
node
ATM
Physical
ATM switched
capabilities
plus CLSF
CLNAP
AAL 3/4
ATM
Physical
ATM
node
ATM
Physical

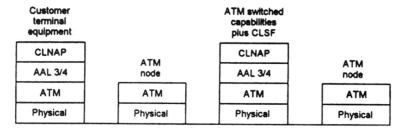

Figure 15.20. General protocol structure for provision of CL data service in B-ISDN.

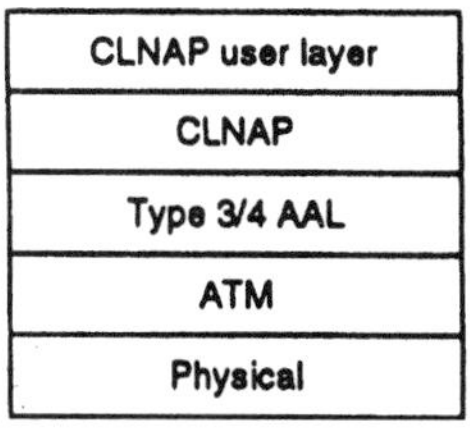

CLNAP user layer
CLNAP
Type 3/4 AAL
ATM
Physical

Figure 15.21. Protocol architecture for supporting connectionless service. CLNAP stands for connectionless network access protocol.

The connectionless service layer provides for transparent transfer of variable-sized data units from a source to one or more destinations in a manner such that lost or corrupted data units are not retransmitted. This transfer is performed using a connectionless technique, including embedding destination and source addresses into each data unit.

7.2 CLNAP Protocol Data Unit (PDU) and Encoding

Figure 15.22 shows the detailed structure of the CLNAP-PDU which contains the following fields:

Destination Address and Source Address. These 8-octet fields each contain a 4-bit *address type* subfield, followed by the 60-bit *address* subfield. The "address-type" subfield indicates whether the "address" subfield contains a publicly administered 60-bit individual address or a publicly administered 60-bit group address. The "address" subfield indicates to which CLNAP-entity(ies) the CLNAP-PDU is destined, and in the case of the source address, it indicates the CLNAP-entity that sourced the CLNAP-PDU. The encoding of the "address-type" and "address" are shown in Figures 15.23A and 15.23B. The address is structured in accordance with CCITT Rec. E.164 [12].

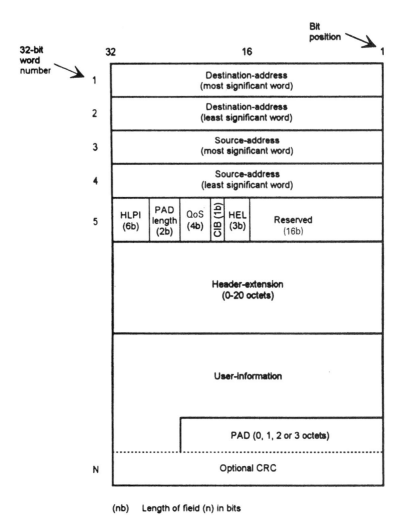

(nb) Length of field (n) in bits

Figure 15.22. Structure of the CLNAP-PDU. From ITU-T Rec. I.364, Figure 5/I.364, page 7 (Ref. 11).

Address type	Meaning
1100	60-bit publicly administered individual address
1110	60-bit publicly administered group address

Figure 15.23A. Destination address field.

Address type	Meaning
1100	60-bit publicly administered individual address

Figure 15.23B. Source address field.

Higher-Layer-Protocol-Identifier (HLPI). This 6-bit field is used to identify the CLNAP user layer entity which the CLNAP-SDU is to be passed to at the destination node. It is transparently carried end-to-end by the network.

PAD Length. This 2-bit field gives the length of the PAD field (0–3 octets). The number of PAD octets is such that the total length of the user-information field and the PAD field together is an integral multiple of four octets (32 bits).

QoS (Quality of Service). 4 bits (assignment under study).

CRC Indication Bit (CIB). This 1-bit field indicates the presence (if $CIB = 1$) or absence (if $CIB = 0$) of a 32-bit CRC field.

Header Extension Length (HEL). This 3-bit field can take on any value from 0 to 5 and indicates the number of 32-bit words in the header extension field.

Reserved. This 16-bit field is reserved for future use. Its default value is 0.

Header Extension. This variable-length field can range from 0 to 20 octets. Its length is indicated by the value of the header extension length field (see above). In the case where the header extension length (HEL) is not equal to zero, all unused octets in the header extension are set to zero. The information carried in the header extension is structured into information entities. An information entity (element) consists (in this order) of element length, element type, and element payload.

Element Length: This is a 1-octet field and contains the combined lengths of the element length, element type, and element payload in octets.

Element Type: This is also a 1-octet field and contains a binary coded value which indicates the type of information found in the element payload field.

Element Payload: This is a variable-length field and contains the information indicated by the element type field.

User Information. This field is variable length up to 9188 octets and is used to carry the CLNAP-SDU.

PAD. This field is 0, 1, 2, or 3 octets in length and is coded as all zeros. Within each CLNAP-PDU the length of this field is selected such that the length of the resulting CLNAP-PDU is aligned on a 32-bit boundary.

CRC. This optional 32-bit field may be present or absent as indicated by the CIB field. The field contains the result of a standard CRC-32 calculation performed over the CLNAP-PDU with the "Reserved" field always treated as if it were coded as all zeros.

Figure 15.24. Relationship between the VC, the VP, and the transmission path.

8   SOME ASPECTS OF A B-ISDN/ATM NETWORK

8.1   ATM Routing and Switching

An ATM transmission path supports virtual paths (VPs), and inside virtual paths are virtual channels (VCs) as shown in Figure 15.24.

As we discussed in Section 4.1, each ATM cell contains a label in its header to explicitly identify the VC to which the cell belongs. This label consists of two parts: a virtual channel identifier (VCI) and a virtual path identifier (VPI).

8.1.1   The Virtual Channel Level.   Virtual channel (VC) is a generic term used to describe a unidirectional communication capability for the transport of ATM cells. A VCI identifies a particular VC link for a given virtual path connection (VPC). A specific value of VCI is assigned each time a VC is switched in the network. A VC link is a unidirectional capability for the transport of ATM cells between two consecutive ATM entities where the VCI value is translated. A VC link is originated or terminated by the assignment or removal of the VCI value.

Routing functions of virtual channels are done at a VC switch/cross-connect.* The routing involves translation of the VCI values of the incoming VC links into the VCI values of the outgoing VC links.

Virtual channel links are concatenated to form a virtual channel connection (VCC). A VCC extends between two VCC endpoints or, in the case of point-to-multipoint arrangements, more than two VCC endpoints. A VCC endpoint is the point where the cell information field is exchanged between the ATM layer and the user of the ATM layer service.

At the VC level, VCCs are provided for the purpose of user–user, user–network, or network–network information transfer. Cell sequence integrity is preserved by the ATM layer for cells belonging to the same VCC.

---

*A VC cross-connect is a network element which connects VC links. It terminates VPCs and translates VCI values and is directed by management plane functions, not by control plane functions.

8.1.2 The Virtual Path Level. Virtual path (VP) is a generic term for a bundle of virtual channel links; all the links in a bundle have the same endpoints.

A VPI identifies a group of VC links, at a given reference point, that share the same VPC. A specific value of VPI is assigned each time a VP is switched in the network. A VP link is a unidirectional capability for the transport of ATM cells between two consecutive ATM entities where the VPI value is translated. A VP link is originated or terminated by the assignment or removal of the VPI value.

Routing functions for VPs are performed at a VP switch/cross-connect. This routing involves translation of the VPI values of the incoming VP links into the VPI values of the outgoing VP links. VP links are concatenated to form a VPC. A VPC extends between two VPC endpoints or, in the case of point-to-multipoint arrangements, there are more than two VPC endpoints. A VPC endpoint is the point where the VCIs are originated, translated, or terminated. At the VP level, VPCs are provided for the purpose of user–user, user–network, and network–network information transfer.

When VPCs are switched, the VPC supporting the incoming VC links are terminated first and a new outgoing VPC is then created. Cell sequence integrity is preserved by the ATM layer for cells belonging to the same VPC. Thus cell sequence integrity is preserved for each VC link within a VPC.

Figure 15.25 is a representation of the VP and VC switching hierarchy where the physical layer is the lowest layer composed of, from bottom up, a regenerator section level, digital section level, and transmission path level. The ATM layer resides just above the physical layer and is composed of VP level, and just above that is the VC level.

Section 8 is based on ITU-T I.311 [13].

9 SIGNALING REQUIREMENTS

9.1 Setup and Release of VCCs

The setup and release of VCCs at the user–network interface (UNI) can be performed in various ways:

- Without using signaling procedures. Circuits are set up at subscription with permanent or semipermanent connections.
- By meta-signaling procedures where a special VCC is used to establish or release a VCC used for signaling. Meta-signaling is a simple protocol used to establish and remove signaling channels. All information inter-changes in meta-signaling are carried out via single cell messages.
- User-to-network signaling procedures such as a signaling VCC to establish or release a VCC used for end-to-end connectivity.

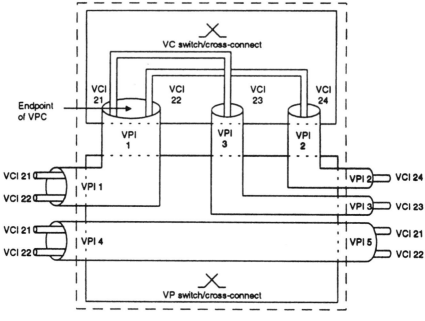

a) Representation of VC and VP switching

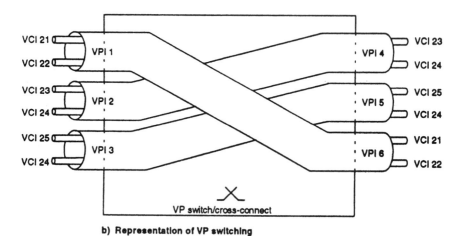

b) Representation of VP switching

Figure 15.25. Representation of the VP and VC switching hierarchy. From ITU-T Rec. I.311, Figure 4/I.311, page 5 (Ref. 13).

- User-to-user signaling procedures such as a signaling VCC to establish or release a VCC within a preestablished VPC between two UNIs.

9.2 Signaling Virtual Channels

9.2.1 Requirements for Signaling Virtual Channels. For a point-to-point signaling configuration, the requirements for signaling virtual channels are as follows:

- One virtual channel connection in each direction is allocated to each signaling entity. The same VPI/VCI value is used in both directions. A standardized VCI value is used for point-to-point signaling virtual channel (SVC).
- In general, a signaling entity can control, by means of associated point-to-point SVCs (signaling virtual channels), user-VCs belonging to any of the virtual paths (VPs) terminated in the same network element.
- As a network option, the user-VCs controlled by a signaling entity can be constrained such that each controlled user-VC is in either upstream or downstream VPs containing the point-to-point SVCs of the signaling entity.

For point-to-multipoint signaling configurations, the requirements for signaling virtual channels are as follows:

(a) *Point-to-Point Signaling Virtual Channels.* For point-to-point signaling, one virtual channel connection in each direction is allocated to each signaling entity. The same VPI/VCI value is used in both directions.

(b) *General Broadcast Signaling Virtual Channel.* The general broadcast signaling virtual channel (GBSVC) may be used for call offering in all cases. In cases where the "point" does not implement service profiles or where "the multipoints" do not support service profile identification, the GBSVC is used for call offering. The specific VCI value for general broadcast signaling is reserved per VP at the UNI. Only when meta-signaling is used in a VP is the GBSVC activated in the VP.

(c) *Selective Broadcast Signaling Virtual Channels.* Instead of the GBSVC, a virtual channel connection for selective broadcast signaling (SBS) can be used for call offering, in cases where a specific service profile is used. No other uses for SBSVCs are foreseen.

9.3 Meta-Signaling

9.3.1 Meta-Signaling Requirements. A meta-signaling channel manages signaling virtual channels only within its own VP pair. In VPI = 0, the meta-signaling virtual channel is always present and has a standardized VCI value.

Meta-signaling VC is activated at VP establishment. The signaling virtual channel (SVC) is assigned and removed when necessary. A specific VCI value for meta-signaling is reserved per VP at the UNI. For a VP with point-to-multipoint signaling configuration, meta-signaling is required and the meta-signaling VC within this VP is activated.

The user negotiates the SVC bandwidth parameter value. The meta-signaling virtual channel (MVSC) bandwidth has a default value. The bandwidth can be changed by mutual agreement between a network operator and user.

9.3.2   Meta-Signaling Functions at the User Access. In order to establish, check, and release the point-to-point and selective broadcast signaling virtual channel connections, meta-signaling procedures are provided. For each direction, meta-signaling is carried out in a permanent virtual channel connection having a standardized VCI value. The channel is called the *meta-signaling virtual channel*. The meta-signaling protocol is terminated in the ATM layer management entity.

The meta-signaling function is required to:

• Manage the allocation of capacity to signaling channels

• Establish, release, and check the status of signaling channels

• Provide a means to associate a signaling endpoint with a service profile if service profiles are supported

• Provide a means to distinguish between simultaneous requests

Meta-signaling should be able to be supported on any VP; however, meta-signaling can only control signaling VCs within its VP.

Section 9 is based on Ref. 11.

10   QUALITY OF SERVICE (QoS)

10.1   ATM Service Quality Review

A basic performance measure for any digital data communication system is bit error rate (BER). Well-designed fiber-optic links will predominate now and into the foreseeable future. We may expect BERs from such links on the order of 1×10^{-10} and with end-to-end performance better than 1×10^{-9} [10].* Thus other performance issues may dominate the scene. These may be called ATM unique QoS items, namely:

*These seem like ambitious goals if end-to-end. They should be considered in light of ITU-T Rec. G.821 and Rec. G.826.

- Cell transfer delay
- Cell delay variation
- Cell loss ratio
- Mean cell transfer delay
- Cell error ratio
- Severely errored cell block ratio
- Cell misinsertion rate

10.2 QoS Parameter Descriptions

10.2.1 Cell Transfer Delay. In addition to the normal delay through network elements and lines, extra delay is added to an ATM network at an ATM switch. The cause of the delay at this point is the statistical asynchronous multiplexing. Using this method, two cells can be directed toward the same output of an ATM switch or cross-connect resulting in output contention.

The result is that one cell or more is held in a buffer until the next available opportunity to continue transmission. We can see that the second cell will suffer additional delay. The delay of a cell will depend upon the amount of traffic within a switch and thus the probability of contention.

The asynchronous path of each ATM cell also contributes to cell delay. Cells can be delayed one or many cell periods, depending on traffic intensity, switch sizing, and the transmission path taken through the network.

10.2.2 Cell Delay Variation (CDV). By definition, ATM traffic is asynchronous, magnifying transmission delay. Delay is also inconsistent across the network. It can be a function of time (i.e., a moment in time), network design/switch design (such as buffer size), and traffic characteristics at that moment of time. The result is cell delay variation (CDV).

CDV can have several deleterious effects. The dispersion effect, or spreading out, of cell inter-arrival times can impact signaling functions or the reassembly of cell user data. Another effect is called *clumping*. This occurs when the inter-arrival times between transmitted cells shorten. One can imagine how this could affect the instantaneous network capacity and how it can impact other services using the network.

There are two performance parameters associated with cell delay variation: 1-point cell delay variation (1-point CDV) and 2-point cell delay variation (2-point CDV).

The 1-point CDV describes variability in the pattern of cell arrival events observed at a single boundary with reference to the negotiated peak rate $1/T$ as defined in ITU-T Rec. I.371 [14]. The 2-point CDV describes variability in the pattern of cell arrival events as observed at the output of a connection portion (MP_2) with reference to the pattern of the corresponding events observed at the input to the connection portion (MP_1).

10.2.3 Cell Loss Ratio. Cell loss may not be uncommon in an ATM network. There are two basic causes of cell loss: error in cell header or network congestion.

Cells with header errors are automatically discarded. This prevents misrouting of errored cells, as well as the possibility of privacy and security breaches.

Switch buffer overflow can also cause cell loss. It is in these buffers that cells are held in prioritized queues. If there is congestion, cells in a queue may be discarded selectively in accordance with their level of priority. Here enters the CLP (cell loss priority) bit discussed in Section 4. Cells with this bit set to 1 are discarded in preference to other, more critical cells. In this way, buffer fill can be reduced to prevent overflow [10].

Cell loss ratio is defined for an ATM connection as:

Lost cells/Total transmitted cells

Lost and transmitted cells counted in severely errored cell blocks should be excluded from the cell population in computing cell loss ratio [2].

10.2.4 Mean Cell Transfer Delay. Mean cell transfer delay is defined as the arithmetic average of a specified number of cell transfer delays.

10.2.5 Cell Error Ratio. Cell error ratio is defined as follows for an ATM connection:

Errored cells/(Successfully transferred cells + Errored cells)

Successfully transferred cells and errored cells contained in cell blocks counted as severely errored cell blocks should be excluded from the population used in calculating cell error ratio.

10.2.6 Severely Errored Cell Block Ratio. The severely errored cell block ratio for an ATM connection is defined as:

Total severely errored cell blocks/Total transmitted cell blocks

A cell block is a sequence of N cells transmitted consecutively on a given connection. A severely errored cell block outcome occurs when more than M errored cells, lost cells, or misinserted cell outcomes are observed in a received cell block.

For practical measurement purposes, a cell block will normally correspond to the number of user information cells transmitted between successive OAM cells. The size of a cell block is to be specified.

10.2.7 Cell Misinsertion Rate. The cell misinsertion rate for an ATM connection is defined as:

$$\text{Misinserted cells/Time interval}$$

This rate may be expressed equivalent as the number of misinserted user information cells per virtual connection.

A misinserted cell is a received cell that has no corresponding transmitted cell on that connection. Cell misinsertion on a particular connection can be caused by an undetected or miscorrected error in the header of a cell originated on a different connection or by an incorrectly programmed translation of VPI or VCI values for cells originated on a different connection [2, 10, 18].

11 TRAFFIC CONTROL AND CONGESTION CONTROL

11.1 Generic Functions

The following functions form a framework for managing and controlling traffic and congestion in ATM networks and are to be used in appropriate combinations from the point of view of ITU-T Rec. I.371 [14].

1. *Network Resource Management (NRM)*. Provision is used to allocate network resources in order to separate traffic flows in accordance with service characteristics.

2. *Connection Admission Control (CAC)*. This is defined as a set of actions taken by the network during the call setup phase or during the call renegotiation phase in order to establish whether a virtual channel (VC) or virtual path (VP) connection request can be accepted or rejected, or whether a request for re-allocation can be accommodated. Routing is part of connection admission control actions.

3. *Feedback Controls*. These are a set of actions taken by the network and by users to regulate the traffic submitted on ATM connections according to the state of network elements.

4. *Usage/Network Parameter Control (UPC/NPC)*. This is a set of actions taken by the network to monitor and control traffic, in terms of traffic offered and validity of the ATM connection, at the user access and network access, respectively. Their main purpose is to protect network resources from malicious as well as unintentional misbehavior, which can affect the QoS of other already established connections, by detecting violations of negotiated parameters and taking appropriate actions.

5. *Priority Control*. The user may generate different priority traffic flows by using the CLP. A congested network element may selectively discard

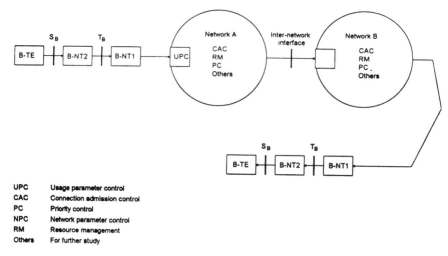

UPC	Usage parameter control
CAC	Connection admission control
PC	Priority control
NPC	Network parameter control
RM	Resource management
Others	For further study

NOTES

1 NPC may apply as well at some intra-network NNIs.

2 The arrows are indicating the direction of the cell flow.

Figure 15.26. Reference configuration for traffic control and congestion control. From ITU-T Rec. I.371, Figure 1/I.371, page 3 (Ref. 14).

cells with low priority, if necessary, to protect as far as possible the network performance for cells with higher priority [14].

Figure 15.26 is a reference configuration for traffic and congestion control.

11.2 Events, Actions, Time Scales, and Response

Figure 15.27 shows the time scales over which various traffic control and congestion control functions can operate. The response time defines how quickly the controls react. For example, cell discarding can react on the order of the insertion time of a cell. Similarly, feedback controls can react on the time scale of round-trip propagation times. Since traffic control and resource management functions are needed at different time scales, no single function is likely to be sufficient.

11.3 Quality of Service, Network Performance, and Cell Loss Priority

QoS at the ATM layer is defined by a set of parameters such as cell delay, cell delay variation sensitivity, cell loss ratio, and so forth.

A user requests a specific ATM layer QoS from the QoS classes which a network provides. This is part of the *traffic contract* at connection establish-

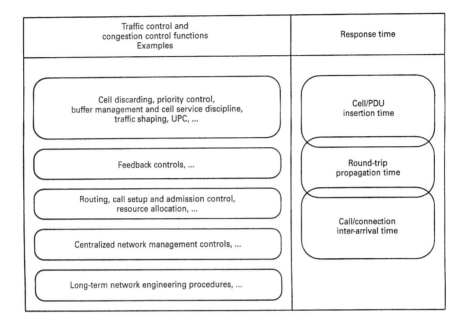

Figure 15.27. Control response times. From ITU-T Rec. I.371, Figure 2/I.371, page 4 (Ref. 14).

ment. It is a commitment for the network to meet the requested QoS as long as the user complies with the traffic contract. If the user violates the traffic contract, the network need not respect the agreed-upon QoS.

A user may request at most two QoS classes for a single ATM connection, which differ with respect to the cell loss ratio objectives. The cell loss priority (CLP) bit in the ATM header allows for two cell loss ratio objectives for a given ATM connection.

Network performance objectives at the ATM SAP (service access point) are intended to capture the network ability to meet the requested ATM layer QoS. It is the role of upper layers, including the AAL, to translate this ATM layer QoS to any specific application requested QoS.

11.4 Traffic Descriptors and Parameters

Traffic parameters describe traffic characteristics of an ATM connection. Traffic parameters are grouped into source traffic descriptors for exchanging information between the user and the network. Connection admission control procedures use source traffic descriptors to allocate resources and derive parameters for the operation of the UPC/NPC.*

We now define the terms *traffic parameters* and *traffic descriptors* [14].

*UPC/NPC stands for user parameter control/network parameter control.

Traffic Parameters. A traffic parameter is a specification of a particular traffic aspect. It may be qualitative or quantitative. Traffic parameters, for example, may describe peak cell rate, average cell rate, burstiness, peak duration, and source type (such as telephone, videophone). Some of these traffic parameters are interdependent, such as burstiness with average and peak cell rate.

Traffic Descriptors. The ATM traffic descriptor is the generic list of traffic parameters which can be used to capture the intrinsic traffic characteristics of an ATM connection. A *source traffic descriptor* is the set of traffic parameters belonging to the ATM traffic descriptor used during the connection setup to capture the intrinsic traffic characteristics of the connection requested by the source.

Connection Traffic Descriptor. This specifies the traffic characteristics of the ATM connection at the public or private UNI. The connection traffic descriptor is the set of traffic parameters in the source traffic descriptor, cell delay variation (CDV) tolerance, and the conformance definition that is used to unambiguously specify the conforming cells of an ATM connection. Connection admission control (CAC) procedures will use the connection traffic descriptor to allocate resources and to derive parameter values for the operation of the UPC. The connection traffic descriptor contains the necessary information for conformance testing of cells of the ATM connection at the UNI.

Any traffic parameter and the CDV tolerance in a connection traffic descriptor should fulfill the following requirements [14]:

- They should be understandable by the user or terminal equipment and conformance testing should be possible as stated in the traffic contract.
- They should be useful in resource allocation schemes, meeting network performance requirements as described in the traffic contract.
- They should be enforceable by the UPC.

11.5   User–Network Traffic Contract

11.5.1   *Operable Conditions.* CAC and UPC/NPC procedures require the knowledge of certain parameters to operate efficiently. For example, they should take into account the source traffic descriptor, the requested QoS, and the CDV tolerance (defined below) in order to decide whether the requested connection can be accepted.

The source traffic descriptor, the requested QoS for any given ATM connection, and the maximum CDV tolerance allocated to the CEQ (customer equipment) define the traffic contract at the T_B reference point (see Figure

15.26). Source traffic descriptors and QoS are declared by the user at connection setup by means of signaling or subscription. Whether the maximum allowable cell delay variation (CDV) tolerance is also negotiated on a subscription or on a per-connection basis is for further study by the ITU-T organization.

The connection admission control (CAC) and usage/network parameter control (UPC/NPC) procedures are operator-specific. Once the connection has been accepted, the value of the CAC and UPC/NPC parameters are set by the network on the basis of the network operator's policy.

ITU-T Rec. I.371 [14] notes that all ATM connections handled by network-connection-related functions (CRF) have to be declared and enforced by the UPC/NPC. ATM layer QoS can only be assured for compliant ATM connections. As an example, individual VCCs inside user end-to-end VPC are neither declared nor enforced at the UPC and hence no ATM layer QoS can be assured for them.

11.5.2 Source Traffic Descriptor, Quality of Service, and Cell Loss Priority.
If a user requests two levels of priority for an ATM connection, as indicated by the CLP bit value, the intrinsic traffic characteristics of both cell flow components have to be characterized in the source traffic descriptor. This is by means of a set of traffic parameters associated with the $CLP = 0$ component and a set of traffic parameters associated with the $CLP = 0 + 1$ component.

As discussed above, the network provides an ATM layer QoS for each of the components ($CLP = 0$ and $CLP = 0 + 1$) of an ATM connection. The traffic contract specifies the particular QoS choice (from those offered by the network operator) for each of the ATM connection components. There may be a limited offering of QoS specifications for the $CLP = 1$ component, according to Ref. 14.

11.5.3 Impact of Cell Delay Variation on UPC/NPC and Resource Allocation.
ATM layer functions such as cell multiplexing may alter the traffic characteristics of ATM connections by introducing cell delay variation as shown in Figure 15.28. When cells from two or more ATM connections are multiplexed, cells of a given ATM connection may be delayed while cells of another ATM connection are being inserted at the output of the multiplexer. Similarly, some cells may be delayed while physical layer overhead or OAM cells are inserted. Therefore, some randomness affects the time interval between reception of ATM cell data requests at the endpoint of an ATM connection to the time that an ATM cell data indication is received at the UPC/NPC. Besides, AAL multiplexing may cause CDV.

The UPC/NPC mechanism will not discard or tag cells in an ATM connection if the source conforms to the source traffic descriptor negotiated at connection establishment. However, if the CDV is not bounded at a point where the UPC/NPC function is performed, it is not possible to design a

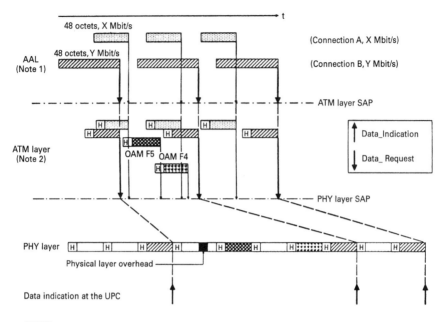

NOTES
1 ATM SDUs are accumulated at the upper layer service bit rate. Besides, CDV may also originate in AAL multiplexing.
2 GFC delay and delay variation are part of the delay and delay variation introduced by the ATM layer.
3 CDV may also be introduced by the network because of random queuing delays which are experienced by each cell in concentrators, switches, and cross-connects.

Figure 15.28. Origins of cell delay variation. From ITU-T Rec. I.371, Figure 3/I.371, page 7 (Ref. 14).

suitable UPC/NPC mechanism and to allocate resources properly. Therefore, ATM specifications (e.g., I.371) require that a maximum allowable value of CDV be standardized edge-to-edge (e.g., between the ATM connection endpoint and T_B, between T_B and an internetwork interface, and between internetwork interfaces. (See Figure 15.28.)

UPC/NPC should accommodate the effect of the maximum CDV allowed on ATM connections within the limit resulting from the accumulated CDV allocated to upstream subnetworks including customer equipment (CEQ). Traffic shaping partially compensates for the effects of CDV on the peak cell rate of the ATM connection. Examples of traffic-shaping mechanisms are respacing cells of individual ATM connections according to their peak cell rate or suitable queue service schemes.

Values of the cell delay variation are network performance issues. It has been found that the definition of a source traffic descriptor and the standardization of a maximum allowable CDV may not be sufficient for a network to allocate resources properly. When allocating resources, the network should take into account the worst-case traffic passing through the UPC/NPC in order to avoid impairments to other ATM connections. The worst-case traffic depends on the specific implementation of the UPC/NPC. The trade-offs between UPC/NPC complexity, worst-case traffic, and optim-ization of network resources are made at the discretion of network operators. The quantity of available network resources and the network performance to be provided for meeting QoS requirements can influence these trade-offs.

11.5.4 Traffic Contract Parameter Specification. Peak cell rate for $CLP = 0 + 1$ is a mandatory traffic parameter to be explicitly or implicitly declared in any source traffic descriptor. In addition to the peak cell rate of an ATM connection, it is mandatory for the user to declare either explicitly or implicitly the cell delay variation tolerance τ within the relevant traffic contract [2].

PEAK CELL RATE (PCR). The following definition applies to ATM connections supporting both CBR and VBR services [14]:

> The peak cell rate in the source traffic descriptor specifies an upper bound on the traffic that can be submitted on an ATM connection. Enforcement of this bound by the UPC/NPC allows the network operator to allocate sufficient resources to ensure that the performance objectives (e.g., for cell loss ratio) can be achieved.

For switched ATM connections, the peak cell rate for $CLP = 0 + 1$ and the QoS class must be explicitly specified for each direction in the connection-establishment SETUP message.

The cell delay variation tolerance must be either explicitly specified at subscription time or implicitly specified.

The sustainable cell rate (SCR) and burst tolerance is an optional traffic parameter set in the source traffic descriptor. If either SCR or burst tolerance is specified, then the other must be specified within the relevant traffic contract.

12 TRANSPORTING ATM CELLS

12.1 In the DS3 Frame

One of the most popular higher-speed digital transmission systems in North America is DS3 operating at a nominal transmission rate of 45 Mbps. It is also being widely implemented for transport of SMDS. The system used to map ATM cells into the DS3 format is the same as that used for SMDS.

DS3 uses the physical layer convergence protocol (PLCP) to map ATM cells into its bit stream. A DS3 PLCP frame is shown in Figure 15.29.

There are 12 cells in a frame. Each cell is preceded by a 2-octet framing pattern (A1, A2), to enable the receiver to synchronize to cells. After the framing pattern there is an indicator consisting of one of 12 fixed bit patterns

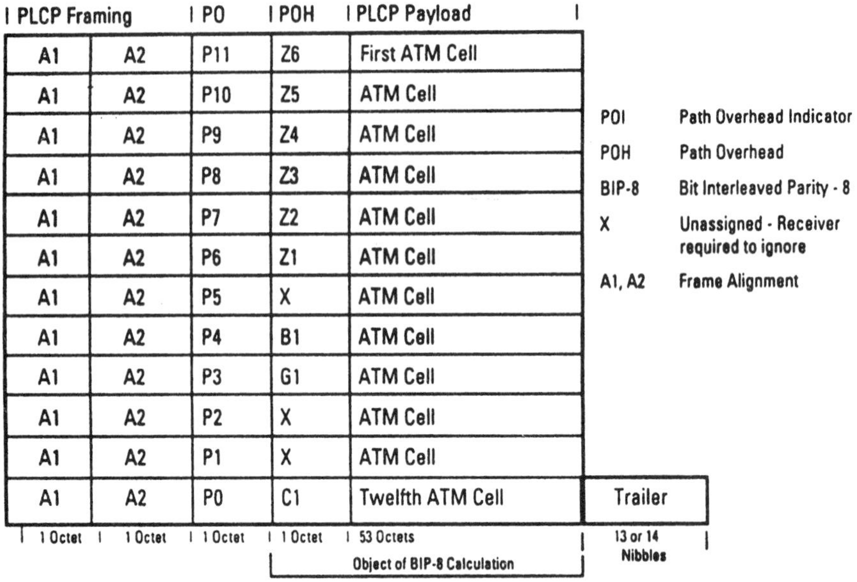

| PLCP Framing | | PO | POH | PLCP Payload | |

A1	A2	P11	Z6	First ATM Cell	
A1	A2	P10	Z5	ATM Cell	
A1	A2	P9	Z4	ATM Cell	
A1	A2	P8	Z3	ATM Cell	
A1	A2	P7	Z2	ATM Cell	
A1	A2	P6	Z1	ATM Cell	
A1	A2	P5	X	ATM Cell	
A1	A2	P4	B1	ATM Cell	
A1	A2	P3	G1	ATM Cell	
A1	A2	P2	X	ATM Cell	
A1	A2	P1	X	ATM Cell	
A1	A2	P0	C1	Twelfth ATM Cell	Trailer

POI — Path Overhead Indicator
POH — Path Overhead
BIP-8 — Bit Interleaved Parity - 8
X — Unassigned - Receiver required to ignore
A1, A2 — Frame Alignment

1 Octet | 1 Octet | 1 Octet | 1 Octet | 53 Octets | 13 or 14 Nibbles

Object of BIP-8 Calculation

Figure 15.29. Format of DS3 PLCP frame. Courtesy of Hewlett-Packard Company [10].

used to identify the cell location within the frame (POI). This is followed by an octet of overhead information used for path management. The entire frame is then padded with either 13 or 14 nibbles (a nibble = 4 bits) of trailer to bring the transmission rate up to the exact DS3 bit rate. The DS3 frame has 125-μs duration.

DS3 has to contend with network slips (added/dropped frames to accommodate synchronization alignment). Thus PLCP is padded with a variable number of stuff (justification) bits to accommodate possible timing slips. The Cl overhead octet indicates the length of padding. The BIP (bit interleaved parity) checks the payload and overhead functions for errors and performance degradation. This performance information is transmitted in the overhead.

12.2 DS1 Mapping

One approach to mapping ATM cells into a DS1 frame is to use a similar procedure as used on DS3 with PLCP. In this case only 10 cells are bundled into a frame, and two of the Z overheads are removed. The padding in the frame is set at 6 octets. The entire frame takes 3 ms to transmit and spans many DS1 ESF (extended superframe) frames. This mapping is shown in Figure 15.30. Note the reference to L2_PDU, taken directly from SMDS (Chapter 14). One must also consider the arithmetic. Each DS1 time slot is 8 bits long, 1 octet. There are then 24 octets in a DS1 frame. This, of course,

1	1	1	1	◄──── 53 Octets ────►
A1	A2	P9	Z4	L2_PDU
A1	A2	P8	Z3	L2_PDU
A1	A2	P7	Z2	L2_PDU
A1	A2	P6	Z1	L2_PDU
A1	A2	P5	F1	L2_PDU
A1	A2	P4	B1	L2_PDU
A1	A2	P3	G1	L2_PDU
A1	A2	P2	M2	L2_PDU
A1	A2	P1	M1	L2_PDU
A1	A2	P0	C1	L2_PDU

OH Byte	Function
A1, A2	Framing Bytes
P9-P0	Path Overhead Identifier Bytes
PLCP Path Overhead Bytes	
Z4-Z1	Growth Bytes
F1	PLCP Path User Channel
B1	BIP-8
G1	PLCP Status
M2-M1	SMDS Control Information
C1	Cycle/Stuff Counter Byte

Trailer = 6 Octets

3 msec

Figure 15.30. DS1 mapping with PLCP. Courtesy of Hewlett-Packard Company [10].

can lead to the second method of carrying ATM cells in DS1, by directly mapping in ATM cells octet for octet (time slot). This is done in groups of 53 octets (1 cell) and would, by necessity, cross DS1 frame boundaries to accommodate an ATM cell.

12.3   E1 Mapping

E1 PCM has a 2.048-Mbps transmission rate. An E1 frame has 256 bits representing 32 channels or time slots, 30 of which carry traffic. Time slots (TS) 0 and 16 are reserved. TS0 is used for synchronization and TS16 for signaling. The E1 frame is shown in Figure 15.31. From bits 9 to 128 and from bits 137 to 256 may be used for ATM cell mapping.

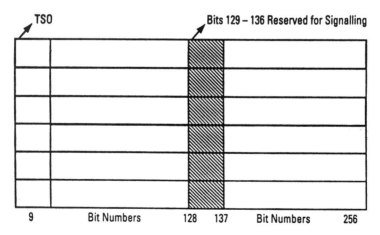

Figure 15.31. Mapping ATM cells directly into E1. Courtesy of Hewlett-Packard Company [10].

ATM cells can also be directly mapped into special E3 and E4 frames. The first has 530 octets available for cells (i.e., 10 cells) and the second has 2160 octets (not evenly divisible).

12.4 Mapping ATM Cells into SDH

12.4.1 AT STM-1 (155.520 Mbps).

SDH is described in Chapter 7. Figure 15.32 shows the mapping procedure. The ATM cell stream is first mapped into the C-4 and then mapped into the VC-4 container along with the VC-4 path overhead. The ATM cell boundaries are aligned with the

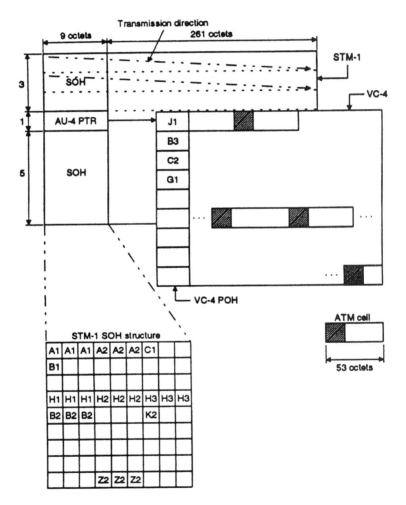

Figure 15.32. The 155.520-Mbps frame structure for SDH-based UNI. From ITU-T Rec. I.432, Figure 8/ I.432, page 13 (Ref. 3).

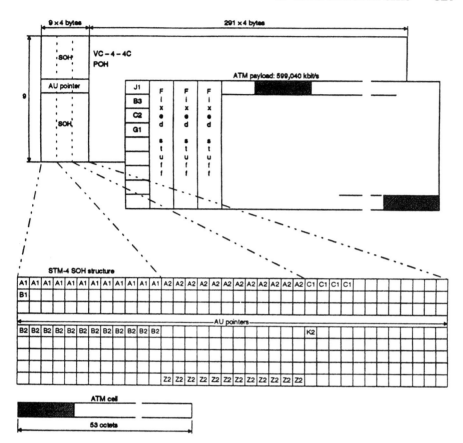

Figure 15.33. The 622.080-Mbps frame structure for SDH-based UNI. From ITU-T Rec. I.432, Figure 10/I.432, page 15 (Ref. 3).

STM-1 octet boundaries. Since the C-4 capacity (2340 octets) is not an integer multiple of the cell length (53 octets), a cell may cross a C-4 boundary.

The AU-4 pointer (octets H1 and H2 in the SOH) is used for finding the first octet of the VC-4.

12.4.2 AT STM-4 (622.080 Mbps). As shown in Figure 15.33, the ATM cell stream is first mapped into C-4-4c and then packed into the VC-4-4c container along with the VC-4-4c path overhead. The ATM cell boundaries are aligned with STM-4 octet boundaries. Since the C-4-4c capacity (9360 octets) is not an integer multiple of the cell length (53 octets), a cell may cross a C-4-4c boundary.

The AU pointers are used for finding the first octet of the VC-4-4c.

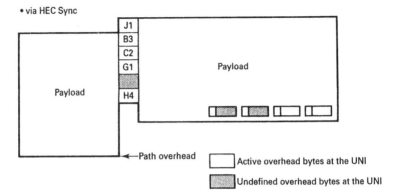

Figure 15.34. Mapping ATM cells into a SONET STS-1 frame. Courtesy of Hewlett-Packard Company [10].

12.5   Mapping ATM Cells into SONET

ATM cells are mapped directly into the SONET payload (49.54 Mbps). As with SDH, the payload in octets is not an integer multiple of the cell length, and thus a cell may cross an STS frame boundary. This mapping concept is shown in Figure 15.34. The H4 pointer can indicate where the cells begin inside an STS frame. Another approach is to identify cell headers, and thus the first cell in the frame.

REVIEW QUESTIONS

1. During the design of ATM, one issue was the length of the cell header. Name and discuss at least two advantages in reducing the size of a frame/block/packet/cell header. Discuss at least one disadvantage.

2. ATM is an "optimum" format for which types of media which are to be carried and switched on the telecommunications network?

3. Relate the media to be transmitted over ATM with the services offered by ATM, namely VBR and CBR.

4. Compare signaling philosophy with standard POTS service and data network service.

5. Discuss *asynchronous* in general, and what it means regarding ATM.

6. Leaving aside Bellcore, there are three ATM standardizing operations bodies. What are they?

7. Give five of the NT-2 functions in B-ISDN/ATM.

8. Give the size, in octets, of the ATM cell. How much is header, and how much is payload? Why must we be careful when we specify this?

9. Describe at least two of the functions of the HEC in the ATM header.

10. Define *packetization delay*. How did packetization delay force ATM designers to shorten cell payload as much as possible?

11. What is probably the most cogent reason for setting the CLP bit?

12. Discuss *cell rate decoupling*.

13. Describe how the cell delineation algorithm works.

14. Layer 6 of ATM is the services layer. Name and briefly describe the functions of the other five layers/sublayers of ATM.

15. What is the purpose of meta-signaling in ATM?

16. Explain what F1 and F2–F4 flows are in ATM.

17. What is the principal purpose of the AAL layer?

18. The service classifications of AAL are based on three parameters. What are the three parameters?

19. What are the five AAL categories? Tell how they fit into classes A, B, C, and D.

20. AAL0 is the simplest of the AALs. What does it do?

21. AAL1 robs one octet of payload and places that octet in the header. This octet has two fields. Name and discuss the two fields.

22. What would be an application of AAL2? Remember what kind of service AAL2 offers, or is restricted to, in this case.

23. Why would we want fields to be aligned to 32-bit boundaries?

24. What is the principal application of AAL5? Why is AAL5 called "SEAL"?

25. Regarding ATM routing and switching, what happens at a VCC endpoint?

26. What does a VPI identify?

27. The setup and release of VCCs can be performed in four ways. Describe each of the four.

28. Meta-signaling channels are provided for what?

29. There are seven ATM-unique QoS items. Name six of them.

30. Describe/define CDV (cell delay transfer) and its deleterious effects.

31. What happens to cells with error(s) found in the header?

32. Define CAC (connection admission control).

33. What is the purpose of user parameter control/network parameter control?

34. Define *traffic parameters* and *traffic descriptors*.

35. There are three items basic to the *traffic contract*. What are they?

36. The *peak cell rate* in the traffic descriptor describes what?

37. Name at least five common digital transmission formats into which ATM cells can be mapped.

38. Name a method of transmitting ATM cells, which we could call a non-format.

39. Consider mapping 53-octet ATM cells into DS1. Describe the two ways of doing this as suggested in the text.

REFERENCES

1. *Broadband Aspects of ISDN*, CCITT Rec. I.121, CCITT Geneva, 1991.

2. *ATM User–Network Interface Specification*, Version 3.0, The ATM Forum. PTR Prentice-Hall, Englewood Cliffs, NJ, 1993.

3. *B-ISDN User–Network Interface—Physical Layer Specification*, ITU-T Rec. I.432, ITU Telecommunication Standardization Sector, Geneva, March 1993.

4. *Broadband ISDN User–Network Interfaces—Rates and Formats Specifications*, ANSI T1.624-1993, American National Standards Institute, New York, 1993.

5. *B-ISDN User–Network Interface*, CCITT Rec. I.413, Geneva, 1991.

6. *Broadband ISDN—ATM Layer Functionality and Specification*, ANSI T1.627-1993, American National Standards Institute, New York, 1993.

7. *B-ISDN ATM Layer Specification*, ITU-T Rec. I.361, ITU Geneva, March 1993.

8. D. E. McDysan and D. L. Spohn, *ATM Theory and Application*, McGraw-Hill, New York, 1995.

9. *B-ISDN ATM Adaptation Layer (AAL) Specification*, CCITT Rec. I.363, CCITT, Geneva, 1991.

10. *Broadband Testing Technologies*, An H-P seminar, Hewlett-Packard, Burlington, MA, October 1993.

11. *Support of Broadband Connectionless Data Service on B-ISDN*, ITU-T Rec. I.364, ITU Telecommunication Standardization Sector, Geneva, March 1993.

12. *Numbering Plan for the ISDN Era*, CCITT Rec. E.164, CCITT Geneva, 1991.

13. *B-ISDN General Network Aspects*, ITU-T Rec. I.311, ITU Telecommunication Standardization Sector, Geneva, March 1993.

14. *Traffic Control and Congestion Control in B-ISDN*, ITU-T Rec. I.371, ITU Telecommunication Standardization Sector, Geneva, March 1993.

15. *B-ISDN ATM Adaptation Layer (AAL) Functional Description*, ITU-T Rec. I.362, ITU Geneva, March 1993.

16. L. C. Cuthbert and J-C Sapanel, "ATM The Broadband Telecommunications Solution," *IEE Telecommunication Series 29*, London, 1993.

17. *Broadband ISDN ATM Adaptation Layer 3/4 Common Part Functions and Specification*, ANSI T1-629-1993, American National Standards Institute, New York, 1993.

18. *B-ISDN ATM Layer Cell Transfer—Performance Parameters*, ANSI T1.511-1994, American National Standards Institute, New York, 1994.

16

CCITT SIGNALING SYSTEM NO. 7

1 INTRODUCTION

CCITT Signaling System No. 7 (SS No. 7) was developed to meet the advanced signaling requirements of the all-digital network based on the 64-kbps channel. It operates in a quite different manner than the signaling discussed in Chapter 4. Nevertheless, it must provide supervision of circuits, address signaling, carry call progress signals, and alerting notification to be eventually passed to the called subscriber. These requirements are no different from those of Chapter 4. The difference is in how it is done. CCITT No. 7 is a data network entirely dedicated to interswitch signaling.*

Simply put, CCITT SS No. 7 is described as an international standardized general-purpose common-channel signaling system that

- Is optimized for operation with digital networks where switches use stored-program control (SPC), such as the DMS-100 and AT&T 4ESS and 5ESS switches discussed in Chapter 9.
- Can meet present and future requirements of information transfer for interprocessor transactions with digital communications networks for call control, remote control, network database access and management, and maintenance signaling.
- Provides a reliable means of information transfer in correct sequence without loss or duplication. (From CCITT, 1980, Yellow Book [2].)

CCITT SS No. 7, in the years since 1980, has become known as the *signaling system for ISDN*. This it is. Without the infrastructure of SS No. 7 embedded in the digital network, there will be no ISDN with ubiquitous

*This would be called *interoffice signaling* in North America.

access. One important point is to be made. CCITT SS No. 7, in itself, is the choice for signaling in the digital PSTN without ISDN. It can and does stand on its own in this capacity.

As we mentioned above, SS No. 7 is a data communication system designed for only one purpose: signaling. It is *not* a general-purpose system. We then must look at CCITT SS No. 7 as (1) a specialized data network and (2) a signaling system [12].

2 OVERVIEW OF SS NO. 7 ARCHITECTURE

The SS No. 7 network model consists of network nodes, termed *signaling points* (SPs), interconnected by point-to-point signaling links, with all the links between two SPs called a *link set*. When the model is applied to a physical network, most commonly there is a one-to-one correspondence between physical nodes and logical entities. But when there is a need (for example, a physical gateway node needs to be a member of more than one network), a physical network node may be logically divided into more than one SP, or a logical SP may be distributed over more than one physical node. These artifices require careful administration to ensure that management procedures within the protocol work correctly.

Messages between two SPs may be routed over a link set directly connecting the two points. This is referred to as the *associated mode* of signaling. Messages may also be routed via one or more intermediate SPs that relay messages at the network layer. This is called *nonassociated mode* of signaling. SS No. 7 supports only a special case of this routing, called *quasi-associated mode*, in which routing is static except for relatively infrequent changes in response to events such as link failures or addition of new SPs. SS No. 7 does not include sufficient procedures to maintain in-sequence delivery of information if routing were to change completely on a packet-by-packet basis.

The function of relaying messages at the network layer* is called the *signaling transfer point* (STP) function. Although this practice results in some confusion, the logical and physical network nodes at which this function is performed are frequently called STPs, even though they may provide other functions as well. An important part in designing an SS No. 7 network is including sufficient equipment redundancy and physical-route diversity so that the stringent availability objectives of the system are met. The design is largely a matter of locating signaling links and SPs with the STP function, so that performance objectives can be met for the projected traffic loads at minimum cost.

---

*Bellcore [3] reports that "purists restrict this further to MTP relaying." (MTP stands for message transfer part.)

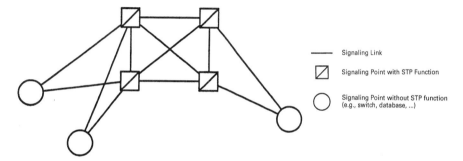

Figure 16.1. Signaling System No. 7 network structure model.

Many telephone companies and administrations deploying SS No. 7 have opted to implement a physical network structure similar to the model illustrated in Figure 16.1 due to the cost of links and STPs. The STP function is concentrated in a relatively small number of nodes that are essentially dedicated to that function. The STPs are paired or *mated*, and pairs of STPs are interconnected with the quad configuration of signaling links as shown in the figure. This has proved to be an extremely reliable backbone network. Other nodes, such as switching systems and service control points (SCPs), are typically homed on one of the mated pairs of STPs with one or more links to each of the mates, depending on traffic volumes (4).

3 SS NO. 7 RELATIONSHIP TO OSI

CCITT SS No. 7 relates to OSI (Chapter 11, Section 6.2.2) up to a certain point. One group believes that SS No. 7 should be fully compatible with the seven layers of OSI. However, the CCITT working groups responsible for the SS No. 7 concept and design were concerned with delay, whether for the data or telephone user of the digital PSN or ISDN. Recall from Chapter 4 that postdial delay is one of the principal measures of performance of a signaling system. To minimize delay, the seven layers of OSI were truncated at layer 4. In fact, CCITT Rec. Q.709 specifies no more than 2.2 s of postdial delay for 95% of calls. To accomplish this, a limit is placed on the number of relay points, called STPs, that can be traversed by a signaling message and by the inherent design of SS No. 7 as a four-layer system. Figure 16.2 relates SS No. 7 protocol layers to OSI.

We should note that SS No. 7 layer 3 signaling network functions include signaling message-handling functions and network management functions. Figure 16.3 shows the general structure of SS No. 7 signaling system.

Schlanger [5] makes the following pertinent observations:

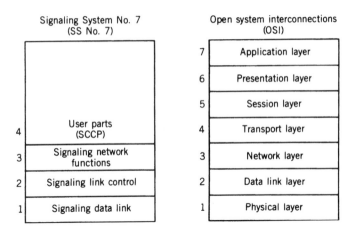

Figure 16.2. How SS No. 7 relates to OSI.

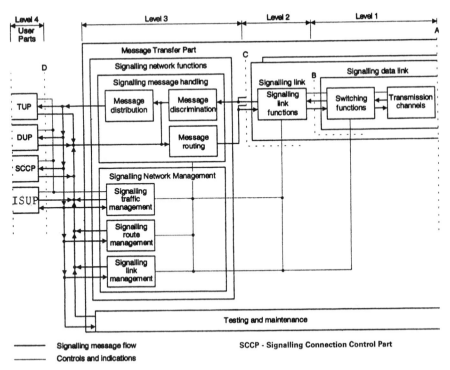

Figure 16.3. General structure of signaling system functions. From ITU-T Rec. Q.701, Figure 6/Q.701, page 8 (Ref. 6).

830

- "Signaling is typically performed to create a communications sub-network for a 'network end user.' As such, some argue that the entire reference model of SS No. 7, as a protocol within the communications subnetwork, should only exist at OSI layer 3 (the network layer) and below."

- "The applications processes within a communications network invoke protocol functionality to communicate with one another in much the same way as 'end users.' Thus, the same seven-layer reference model is felt to apply in this application context."

- "The signaling system protocol is felt to encompass operations, administration, and maintenance (OA & M) activities related to telecommunications. Because craftspeople can be involved in such activities (truly end users) as well as OA & M application processes, the distinction between network layer entities and end users becomes 'fuzzy.'"

There seem to be various efforts to force-fit SS No. 7 into OSI from layer 4 upwards. These efforts have resulted in the sublayering of layer 4 into user parts and SCCP (signaling connection control part).

In Section 4 we briefly describe the basic functions of the four SS No. 7 layers, which are covered in more detail in Sections 5 through 7.

4 SIGNALING SYSTEM STRUCTURE

Figure 16.3, which illustrates the basic structure of SS No. 7, shows two parts of the system: the message transfer part (MTP) and user parts. There are three user parts: telephone user (TUP), data user (DUP), and ISDN user part (ISUP). Figures 16.2 and 16.3 show OSI layers 1, 2, and 3, which make up the MTP. The next paragraphs describe the functions of each of these layers from a system viewpoint.

Layer 1 defines the physical, electrical, and functional characteristics of the signaling data link and the means to access it. In the digital network environment the 64-kbps digital path is the normal basic connectivity. The signaling link may be accessed by means of a switching function that provides the capability of automatic reconfiguration of signaling links.

Layer 2 carries out the signaling link function. It defines the functions and procedures for the transfer of signaling messages over one individual signaling data link. A signaling message is transferred over the signaling link in variable-length signal units. A signal unit consists of transfer control information in addition to the information content of the signaling message. The signaling link functions include:

- Delimitation of a signal unit by means of flags.

- Flag imitation prevention by bit stuffing.
- Error detection by means of check bits included in each signal unit.
- Error control by retransmission and signal unit sequence control by means of explicit sequence numbers in each signal unit and explicit continuous acknowledgments.
- Signaling link failure detection by means of signal unit error monitoring and signaling link recovery by means of special procedures.

Layer 3, signaling network functions, in principle, defines such transport functions and procedures that are common to and independent of individual signaling links. There are two categories of functions in layer 3:

1. *Signaling Message-Handling Functions.* During message transfer, these functions direct the message to the proper signaling link or user part.
2. *Signaling Network Management Functions.* These control real-time routing, control, and network reconfiguration, if required.

Layer 4 is the user part. Each user part defines the functions and procedures peculiar to the particular user, whether telephone, data, or ISDN user part.

The *signal message* is defined by CCITT Rec. Q.701 as an assembly of information, defined at layer 3 or 4, pertaining to a call, management transaction, and so on, which is then transferred as an entity by the message transfer function. Each message contains "service information," including a service indicator identifying the source user part and possibly whether the message relates to international or national application of the user part.

The *signaling information* of the message contains user information, such as data or call control signals, management and maintenance information, and type and format of message. It also includes a "label." The label enables the message to be routed by layer 3 through the signaling network to its destination and directs the message to the desired user part or circuit.

On the signaling link such signaling information is contained in the *message signal units* (MSUs), which also include transfer control functions related to layer 2 functions on the link.

There are a number of terms used in SS No. 7 literature that should be understood before we proceed further.

Signaling Points. Nodes in the network that utilize common-channel signaling.

Signaling Relation (similar to traffic relation). Any two signaling points for which the possibility of communication between their corresponding user parts exist are said to have a signaling relation.

Signaling Links. Signaling links convey signaling messages between two signaling points.

Originating and Destination Points. The originating and destination points are the locations of the source user part function and location of the receiving user part function, respectively.

Signaling Transfer Point (STP). An STP is a point where a message received on one signaling link is transferred to another link.

Message Label. Each message contains a label. In the standard label, the portion that is used for routing is called the *routing label*. The routing label includes:

- Destination and originating points of the message.

- A code used for load sharing, which may be the least significant part of a label component that identifies a user transaction at layer 4.

The standard label assumes that each signaling point in a signaling network is assigned an identification code according to a code plan established for the purpose of labeling.

Message Routing. Message routing is the process of selecting the signaling link to be used for each signaling message. Message routing is based on analysis of the routing label of the message in combination with predetermined routing data at a particular signaling point.

Message Distribution. Message distribution is the process that determines to which user part a message is to be delivered. The choice is made by analysis of the service indicator.

Message Discrimination. Message discrimination is the process that determines, on receipt of a message at a signaling point, whether or not the point is the destination point of that message. This decision is based on analysis of the destination code of the routing label in the message. If the signaling point is the destination, the message is delivered to the message destination function. If not, the message is delivered to the routing function for further transfer on a signaling link.

4.1 Signaling Network Management

4.1.1 Scope. The three signaling network management functional blocks are shown in Figure 16.3. These are signaling traffic management, signaling route management, and signaling link management.

4.1.2 Signaling Traffic Management. The signaling traffic management functions are:

1. To control message routing. This includes modification of message routing to preserve, when required, accessibility of all destination points concerned or to restore normal routing.

2. In conjunction with modifications of message routing, to control the resulting transfer of signaling traffic in a manner that avoids ir-regularities in message flow.

3. Flow control.

Control of message routing is based on analysis of predetermined information about all allowed potential routing possibilities in combination with information, supplied by the signaling link management and signaling route management functions, about the status of the signaling network (i.e., current availability of signaling links and routes).

Changes in the status of the signaling network typically result in modification of current message routing and thus in the transfer of certain portions of the signaling traffic from one link to another. The transfer of signaling traffic is performed in accordance with specific procedures. These procedures are: *changeover, changeback, forced rerouting*, and *controlled rerouting*. The procedures are designed to avoid, as far as circumstances permit, such irregularities in message transfer as loss, mis-sequencing, or multiple delivery of messages.

The changeover and changeback procedures involve communication with other signaling point(s). For example, in the case of changeover from a failing signaling link, the two ends of the failing link exchange information (via an alternative path) that normally enables retrieval of messages that otherwise would have been lost on the failing link.

A signaling network has to have a signaling traffic capacity that is higher than the normal traffic offered. However, in overload conditions (e.g., due to network failures or extremely high traffic peaks) the signaling traffic management function takes flow control actions to minimize the problem. An example is the provision of an indication to the local user functions concerned that the MTP is unable to transport messages to a particular destination in the case of total breakdown of all signaling routes to that destination point. If such a situation occurs at an STP, a corresponding indication is given to the signaling route management function for further dissemination to other signaling points in the network.

4.1.3 Signaling Link Management. Signaling link management controls the locally connected signaling link sets. In the event of changes in the availability of a local link set, it initiates and controls actions with the objective of restoring the normal availability of that link set.

The signaling link management interacts with the signaling link function at level 2 by receipt of indications of the status of signaling links. It also initiates actions, also at level 2, such as initial alignment of an out-of-service link.

The signaling system can be applied in the method of provision of signaling links. Consider that a signaling link probably will consist of a terminal device and data link. It is also possible to employ an arrangement in which any switched connection to the far end may be used in combination with any local signaling terminal device. Here the signaling link management initiates and controls reconfigurations of terminal devices and signaling data links to the extent such reconfigurations are automatic. This implies some sort of switching function at level 1.

4.1.4 *Signaling Route Management.* Signaling route management only relates to the quasi-associated mode of signaling. It transfers information about changes in availability of signaling routes in the signaling network to enable remote signaling points to take appropriate signaling traffic actions. For example, a signaling transfer point may send messages indicating inaccessibility of a particular signaling point via that signaling transfer point, thus enabling other signaling points to stop routing messages to an inoperative route.

5 THE SIGNALING DATA LINK (LAYER 1)

A signaling data link is a bidirectional transmission path for signaling, comprising two data channels operating together in opposite directions at the same data rate. It constitutes the lowest layer (layer 1) in the SS No. 7 functionality hierarchy.

A digital signaling data link is made up of digital transmission channels and digital switches or their terminating equipment, providing an interface to SS No. 7 signaling terminals. The digital transmission channels may be derived from a digital multiplex signal at 1.544, 2.048, or 8.448 Mbps having a frame structure as defined in CCITT Rec. G.704 (see Chapter 8) or from digital multiplex bit streams having a frame structure specified for data circuits in CCITT Recs. X.50, X.51, X.50 bis, and X.51 bis.

The operational signaling data link is exclusively dedicated to the use of SS No. 7 signaling between two signaling points. No other information may be carried by the same channels together with the signaling information.

Equipment such as echo suppressors, digital pads, or A/μ-law converters attached to the transmission link must be disabled in order to ensure full-duplex operation and bit count integrity of the transmitted data stream. In this situation, 64-kbps digital signaling channels are used which are switchable as semipermanent channels in the exchange.

The standard bit rate on a digital bearer is 64 kbps. The minimum signaling bit rate for telephone call control applications is 4.8 kbps. For other applications such as network management, bit rates lower than 4.8 kbps may also be used.

The following is applicable for a digital signaling data link derived from a 2.048-Mbps digital path (i.e., E1). At the input/output interface, the digital multiplex equipment or digital switch block will comply with CCITT Recs.

G.703 for electrical characteristics and G.704 for the functional characteristics—in particular the frame structure. The signaling bit rate is 64 kbps. The standard time slot for signaling is time slot 16. When time slot 16 is not available, any time slot available for 64-kbps user transmission rate may be used. No bit inversion is performed.

For a signaling data link derived from an 8.448-Mbps (E2) digital link, the following applies. At the multiplex input/output interface, there should be compliance with CCITT Recs. 703 for electrical characteristics and G. 704 for functional characteristics—in particular the frame structure. The signaling bit rate is 64 kbps. The standard time slots for use of a signaling data link are time slots 67–70 in descending order of priority. When these time slots are not available, any channel time slot available for 64-kbps user transmission rate may be used. No bit inversion is performed [13].

For North American applications of SS No. 7, *BOC Notes on the LEC Networks—1994* [3] states that data rates from 4.8 to 64 kbps may be used.

6 THE SIGNALING LINK (LAYER 2)

This section deals with the transfer of signaling messages over one signaling link directly connecting two signaling points. Signaling messages delivered by upper hierarchical layers are transferred over the signaling link in variable-length signal units. The signal units include transfer control information for proper operation of the signaling link in addition to the signaling information. The signaling link (layer 2) functions include:

(a) Signaling unit delimitation

(b) Signal unit alignment

(c) Error detection

(d) Error correction

(e) Initial alignment

(f) Signal link error monitoring

(g) Flow control

All these functions are coordinated by the link state control as shown in Figure 16.4.

Signal Unit Delimitation and Alignment. The beginning and end of a signal unit are indicated by a unique 8-bit pattern, called the *flag*. Measures are taken to ensure that the pattern cannot be imitated elsewhere in the unit. Loss of alignment occurs when a bit pattern disallowed by the delimitation procedure (i.e., more than six consecutive 1s) is received, or when a certain

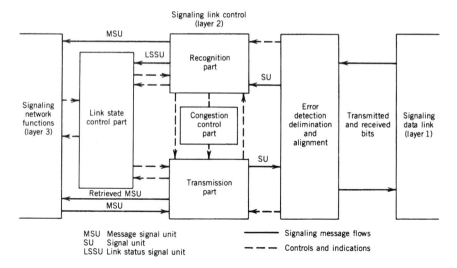

Figure 16.4. Interactions of functional specification blocks for signaling link control. Note: The MSUs, LSSU, and SUs do not include error-control information. From ITU-T Rec. Q.703, Figure 1/Q.703, page 2 (Ref. 7).

maximum length of signal unit is exceeded. Loss of alignment will cause a change in the mode of operation of the signal unit error rate monitor.

Error Detection. The error detection function is performed by means of the 16 check bits provided at the end of each signal unit. The check bits are generated by the transmitting signaling link terminal by operating on the preceding bits of the signal unit following a specified algorithm. At the receiving signaling link terminal, the received check bits are operated on using specified rules which correspond to that algorithm. If consistency is not found between the received check bits and the preceding bits of the signal unit according to the algorithm, then the presence of errors is indicated and the signal unit is discarded.

Error Correction. Two forms of error correction are provided: the *basic method* and the *preventive cyclic retransmission method*.

The basic method applies to (a) signaling links using nonintercontinental terrestrial transmission means and (b) intercontinental signaling links where one-way propagation is less than 15 ms.

The preventive cyclic retransmission method applies to (a) inter-continental signaling links where the one-way delay is equal to or greater than 15 ms and (b) signaling links established via satellite.

In cases where one signaling link with an intercontinental link set is established via satellite, the preventive cyclic retransmission method is used for all signaling links of that set.

The basic method is a noncompelled, positive/negative acknowledgment, retransmission error correction system. A signal unit which has been transmitted is retained at the transmitting signaling link terminal until a positive acknowledgment for that signal unit is received. If a negative acknowledgment is received, then the transmission of new signal units is interrupted and those signal units which have been transmitted but not yet positively acknowledged (starting with that indicated by the negative acknowledgment) will be transmitted once, in the order in which they were first transmitted.

The preventive cyclic retransmission method is a noncompelled, positive acknowledgment, cyclic retransmission forward error correction system. A signal unit which has been transmitted is retained at the transmitting signaling unit terminal until a positive acknowledgment for that signaling unit is received. During the period when there are no new signal units to be transmitted, all signal units which have not been positively acknowledged are retransmitted cyclically.

The forced retransmission procedure is defined to ensure that forward error correction occurs in adverse conditions (e.g., degraded BER and/or high-traffic loading). When a predetermined number of retained, unacknowledged signal units exist, the transmission of new signal units is retransmitted cyclically until the number of unacknowledged signal units is reduced.

Initial Alignment. The initial alignment procedure is used for both first-time initialization (e.g., after "switch-on") and alignment in association with restoration after link failure. The procedure is based on (a) the compelled exchange of status information between the two signaling points concerned and (b) the provision of a proving period. No other signaling link is involved in the initial alignment of any particular link, and the exchange occurs only on the link to be aligned.

Signaling Link Error Monitoring. Two signaling link error rate monitoring functions are provided: one which is employed while a signaling link is in service and which provides one of the criteria for taking the link out of service, and one which is employed while a link is in the proving-in state of the initial alignment procedure. They are called the *signal unit error rate monitor* and the *alignment error rate monitor*, respectively. The characteristics of the signal unit error rate monitor are based on a signal unit error count, incremented and decremented using the "leaky bucket" principle, whereas the alignment error rate monitor is a linear count of signal unit errors. During loss of alignment, the signal unit error rate monitor is incremented in proportion to the period of the loss of alignment.

Link State Control Functions. Link state control is a function of the signaling link which provides directives to the other signaling link functions.

Flow Control. Flow control is initiated when congestion is detected at the receiving end of the signaling link. The congested receiving end of the link notifies the remote transmitting end of the condition by means of an appropriate link status signal and it withholds acknowledgments of all incoming message signal units. When congestion abates, acknowledgments of all incoming signal units are resumed. When congestion exists, the remote transmitting end is periodically notified of this condition. The remote transmitting end will indicate that the link has failed if the congestion continues too long.

6.1 Basic Signal Unit Format

Signaling and other information originating from a user part is transferred over the signaling link by means of signal units. There are three types of signal units used in SS No. 7:

1. Message signal unit (MSU).
2. Link status signal unit (LSSU).
3. Fill-in signal unit (FISU).

These units are differentiated by means of the *length indicator*. MSUs are retransmitted in case of error; LSSUs and FISUs are not. The MSU carries signaling information; the LSSU provides link status information, and the FISU is used during the link idle state—they fill in.

The signaling information field is variable in length and carries the signaling information generated by the user part. All other fields are of fixed length. Figure 16.5 shows the basic formats of the three types of signal units. As shown in the figure, the message transfer control information encompasses eight fixed-length fields in the signal unit that contains information required for error control and message alignment. These eight fields are described below. In the figure we start from right to left, which is the direction of transmission.

The opening *flag* indicates the start of a signal unit. The opening flag of one signal unit is normally the closing flag of the previous signal unit. The flag bit pattern is 01111110. The *forward sequence number* (FSN) is the sequence number of the signal unit in which it is carried. The *backward sequence number* (BSN) is the sequence number of a signal unit being acknowledged. The value of the FSN is obtained by incrementing (modulo 128) the last assigned value by 1. The FSN value uniquely identifies a message signal unit until its delivery is accepted without errors and in correct sequence by the receiving terminal. The FSN of a signal unit other than an MSU assumes the value of the FSN of the last transmitted MSU. The maximum capacity of sequence numbers is 127 message units before reset (modulo 128).

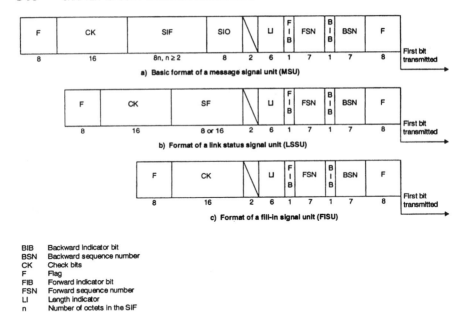

Figure 16.5. Signal unit formats. From ITU-T Rec. Q.703, Figure 3/Q.703, page 5 (Ref. 7).

Positive acknowledgment is accomplished when a receiving terminal acknowledges the acceptance of one or more MSUs by assigning an FSN value of the latest accepted MSU to the BSN of the next signal unit sent in the opposite direction. The BSNs of subsequent signal units retain this value until a further MSU is acknowledged, which will cause a change in the BSN sent. The acknowledgment to an accepted MSU also represents an acknowledgment to all, if any, previously accepted, though not yet acknowledged, MSUs.

Negative acknowledgment is accomplished by inverting the backward indicator bit (BIB) value of the signal unit transmitted. The BIB value is maintained in subsequently sent signal units until a new negative acknowledgment is to be sent. The BSN assumes the value of the FSN of the last accepted signal unit.

As we can now discern, the forward indicator bit (FIB) and the backward indicator bit together with the FSN and BSN are used in the basic error-control method to perform signal unit sequence control and acknowledgment functions.

The *length indicator* (LI) is used to indicate the number of octets following the length indicator octet and preceding the check bits and is a binary number in the range of 0–63. The length indicator differentiates between three types of signal units as follows:

Length indicator = 0   Fill-in signal unit
Length indicator = 1 or 2   Link status signal unit
Length indicator ≥ 2   Message signal unit

The *service information octet* (SIO) is divided into a *service indicator* and a *subservice field*. The service indicator is used to associate signaling information for a particular user part and is present only in MSUs. Each is 4 bits long. For example, a service indicator with a value 0100 relates to the telephone user part, and 0101 relates to the ISDN user part. The subservice field portion of the SIO contains two network indicator bits and two spare bits. The network indicator discriminates between international and national signaling messages. It can also be used to discriminate between two national signaling networks, each having a different routing label structure. This is accomplished when the network indicator is set to 10 or 11.

The *signaling information field* (SIF) consists of an integral number of octets greater than or equal to 2 and less than or equal to 62. In national signaling networks it may consist of up to 272 octets. Of these 272 octets, information blocks of up to 256 octets in length may be accommodated, accompanied by a label and other possible housekeeping information that may, for example, be used by layer 4 to link such information blocks together.

The *link status signal unit* (LSSU) provides link status information between signaling points. The status field can be made up of one or two octets. CCITT Rec. Q.703 shows application of the one-octet field in which the first three bits (from right to left) are used (bits A, B, and C) and the remaining five bits are spare. These first 3-bit values are given in Table 16.1.

Section 6.1 is based on Section 11.1.3 in Ref. 7.

TABLE 16.1   Three-Bit Link Status Indications

	Bits			
C	B	A	Status Indication	Meaning
0	0	0	0	Out of alignment
0	0	1	N	Normal alignment
0	1	0	E	Emergency alignment
0	1	1	OS	Out of service
1	0	0	PO	Processor outage
1	0	1	B	Busy

From para. 11.1.3, ITU-T Rec. Q.703, Ref. 7.

7 SIGNALING NETWORK FUNCTIONS AND MESSAGES (LAYER 3)

7.1 Introduction

In this section we describe the functions and procedures relating to the transfer of messages between signaling points (i.e., signaling network nodes). These nodes are connected by signaling links involving layers 1 and 2 described in Sections 5 and 6. Another important function of layer 3 is to inform the appropriate entities of a fault and, as a consequence, carry out a rerouting of messages through the network. The signaling network functions are broken down into two basic categories:

- Signaling message handling
- Signaling network management

7.2 Signaling Message-Handling Functions

The signaling message-handling function ensures that a signaling message originated by a particular user part at an originating signaling point is delivered to the same user part at the destination point as indicated by the sending user part. Depending on the particular circumstances, the delivery may be made through a signaling link directly interconnecting the originating and destination points or via one or more intermediate signaling transfer points (STPs).

The signaling message-handling functions are based on the label contained in the messages which explicitly identifies the destination and origination points. The label part used for signaling message handling by the MTP is called the *routing label*.

As shown in Figure 16.6, the signaling message handling is divided into the following:

- The *message routing* function, used at each signaling point to determine the outgoing signaling link on which a message is to be sent toward its destination point.

- The *message discrimination* function, used at a signaling point to determine whether or not a received message is destined to that point itself. When the signaling point has the transfer capability, and a message is not destined for it, that message is transferred to the message routing function.

- The *message distribution* function, used at each signaling point to deliver the received messages (destined to the point itself) to the appropriate user part.

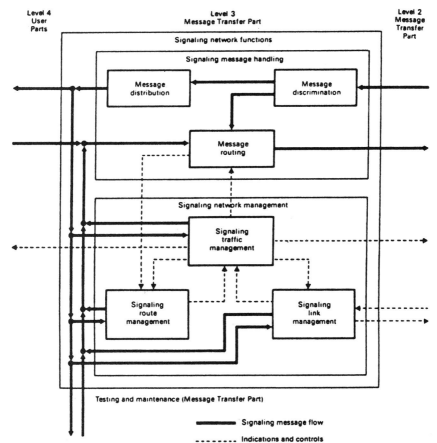

Figure 16.6. Signaling network functions. From CCITT Rec. Q.704, Figure 1/Q.704, page 125 (Ref. 14).

7.2.1 Routing Label. The label contained in a signaling message and used by the relevant user part to identify the particular task to which the message refers (e.g., a telephone circuit) is also used by the message transfer part to route the message toward its destination point.

The part of the message that is used for routing is called the *routing label*, and it contains the information necessary to deliver the message to its destination point. Normally the routing label is common to all services and applications in a given signaling network, national or international. (However, if this is not the case, the particular routing label of a message is determined by means of the service indicator.) The standard routing label should be used in the international signaling network and is applicable in national applications.

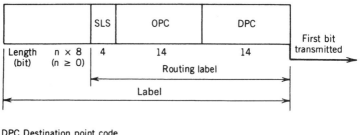

DPC Destination point code
OPC Originating print code
SLS Signaling link selection

Figure 16.7. Routing label structure. Based on Figure 3/Q.704, page 5, CCITT Rec. Q.704 (Ref. 14).

The standard routing label field is 32 bits long and is placed at the beginning of the signaling information field (SIO). Its structure is shown in Figure 16.7.

The *destination point code* (DPC) indicates the destination of the message. The *originating point code* (OPC) indicates the originating point of the message. The coding of these codes is pure binary. Within each field, the least significant bit occupies the first position and is transmitted first.

A unique numbering scheme for the coding of the fields is used for the signaling points of the international network irrespective of the user parts connected to each signaling point. The *signaling link selection* (SLS) field is used, where appropriate, in performing load sharing. This field exists in all types of messages and always in the same position. The only exception to this rule is some message transfer part level 3 messages (e.g., changeover order) for which the message routing function in the signaling point of origin of the message is not dependent on the field. In this particular case the field does not exist as such, but is replaced by other information (e.g., in the case of the changeover order, the identity of the faulty link).

In the case of circuit-related messages of the TUP, the field contains the least significant bits of the circuit identification code [or the bearer identification code in the case of the data user part (DUP)], and these bits are not repeated elsewhere. In the case of all other user parts, the SLS is an independent field. In these cases it follows that the signaling link selection of messages generated by any user part will be used in the load-sharing mechanism. As a consequence, in the case of the user parts which are not specified (e.g., transfer of charging information) but for which there is a requirement to maintain order of transmission of messages, the field is coded with the same value for all messages belonging to the same transaction, sent in a given direction.

In the case of message transfer part level 3 messages, the signaling link selection field exactly corresponds to the signaling link code (SLC) which indicates the signaling link between destination point and originating point to which the message refers.

7.2.2 *Message Routing Function.* Each signaling point will have routing information that allows it to determine the signaling link over which a message has to be sent on the basis of the destination point code and the signaling link selection field, and, in some cases, of the network indicator. Typically, the destination point code is associated with more than one signaling link that may be used to carry the message; the selection of the particular signaling link is made by means of the signaling link selection field, thus effecting load sharing.

Two basic cases of load sharing are defined:

1. Load sharing between links belonging to the same link set
2. Load sharing between links not belonging to the same link set

A load-sharing collection of one or more link sets is called a *combined link set*. The capability to operate in load sharing according to both cases is mandatory for any signaling point in the international network.

In case 1, the traffic flow carried by a link set is shared (on the basis of the signaling link selection field) between different signaling links belonging to the link set. An example of such a case is given by a link set directly interconnecting the originating and destination points in the associated mode of operation. This is shown in Figure 16.8.

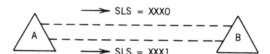

Figure 16.8. Example of load sharing within a link set.

In case 2, traffic relating to a given destination is shared (on the basis of the signaling link selection field) between different signaling links not belonging to the same link set. This situation is shown in Figure 16.9. The load-sharing rule used for a particular signaling relation may or may not apply to all the signaling relations which use one of the signaling links involved. In the example, traffic destined to B is shared between signaling links DE and DF

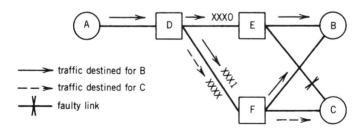

Figure 16.9. Example of load sharing between link sets. Based on Figure 5/Q.704, page 7, CCITT Rec. Q.704 (Ref. 14).

with a given signaling link selection field assignment, whereas that destined to C is sent only on link DF, due to the failure of link EC.

As a result of the message routing function, in normal conditions all messages having the same routing label (e.g., call setup message related to a given circuit) are routed via the same signaling links and STPs [14].

7.3 Signaling Network Management

This basic function deals with maintaining signaling service in view of three possible fault or degraded operation conditions:

- Loss of signaling link
- Loss of signaling point
- Degraded operation due to congestion

As mentioned previously, there are three subsidiary functions used to carry out signaling network management: traffic management, link management, and route management. The signaling traffic management function is used to direct signaling traffic from a link or route to one or more different links or routes or to deload temporarily a link or route in the case of congestion at a signaling point. The following procedures are involved with this function:

- Changeover
- Change back
- Forced rerouting
- Controlled rerouting
- Signaling traffic flow control

The signaling link management function is used to restore failed links, activate idle links (not yet aligned), and deactivate aligned signal links. This function involves the following procedures:

- Signaling link activation
- Link set activation
- Automatic allocation of signaling terminals and signaling data links

The signaling route management function distributes information about signaling network status in order to block or unblock signaling routes. Six procedures for route management are described in CCITT Rec. Q.704 [14].

8 SIGNALING NETWORK STRUCTURE

8.1 Introduction

In this section several aspects in the design of signaling networks are treated. These networks may be national or international. The national and international networks are considered to be structurally independent and, although a particular signaling point may belong to both networks, signaling points are allocated *signaling point codes* according to the rules of each network.

Signaling links are basic components in a signaling network connecting signaling points. The signaling links encompass *level 2* functions which provide for message error control. In addition, provision for maintaining the correct message sequence is provided.

Signaling links connect signaling points at which signaling network functions such as message routing are provided at *level 3* and at which the user functions may be provided at *level 4* if it is also an originating or destination point. A signaling point that *only* transfers messages from one signaling link to another at level 3 serves as a signaling transfer point (STP). The signaling links, STPs, and signaling (originating or destination) points may be combined in many different ways to form a *signaling network*.

8.2 International and National Signaling Networks

The worldwide signaling network is structured into two functionally independent levels: international and national levels. This is shown in Figure 16.10. Such a structure allows a clear division of responsibility for signaling network management and permits numbering plans of signaling points of the international network and the different national networks to be independent of one another.

A signaling point (SP), including an STP, may be assigned to one of three categories:

- National signaling point (NSP) (an STP) which belongs to the national signaling network (e.g., NSP_1) and is identified by a signaling point code (OPC or DPC) according to the national numbering plan for signaling points.

- International signaling point (ISP) (an STP) which belongs to the international signaling network (e.g., ISP_3) and is identified by a signaling point code (OPC or DPC) according to the international numbering plan for signaling points.

- A node that functions both as an international signaling point (STP) and a national signaling point (STP) and therefore belongs to both the international signaling network and a national signaling network and

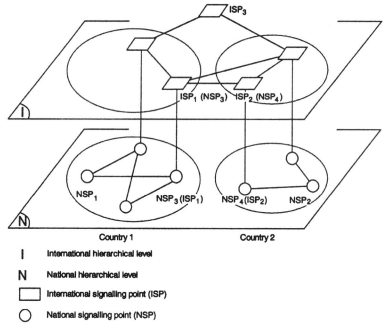

Figure 16.10. International and national signaling networks. From ITU-T Rec. Q.705, Figure 1/Q.705, page 2 (Ref. 8).

accordingly is identified by a specific signaling point code (OPC or DPC)* in each of the signaling networks.

If discrimination between international and national signaling point codes is necessary at a signaling point, the network indicator is used.

9 SIGNALING PERFORMANCE—MESSAGE TRANSFER PART

9.1 Basic Performance Parameters

ITU-T Rec. Q.706 [9] breaks down SS No. 7 performance into three parameter groups:

- Message delay
- Signaling traffic load
- Error rate

*OPC and DPC are discussed in Section 7.

Consider the following parameters and values:

Availability. The unavailability of a signaling route set should not exceed 10 min per year.

Undetected Errors. Not more than 1 in 10^{10} of all signal unit errors will go undetected in the message transfer part.

Lost Messages. Not more than 1 in 10^7 messages will be lost due to failure of the message transfer part.

Messages Out of Sequence. Not more than 1 in 10^{10} messages will be delivered out of sequence to the user part due to failure in the message transfer part. This includes message duplication.

9.2 Traffic Characteristics

Labeling Potential. There are 16,384 identifiable signaling points.

Loading Potential. Loading potential is restricted by the following factors:

- Queuing delay.
- Security requirements (redundancy with changeover).
- Capacity of sequence numbering (127 unacknowledged signal units).
- Signaling channels using bit rates under 64 kbps.

9.3 Transmission Parameters

The message transfer part operates satisfactorily with the following error performance:

- Long-term error rate on the signaling data links of less than 1×10^{-6}.
- Medium-term error rate of less than 1×10^{-4}.

9.4 Signaling Link Delays Over Terrestrial and Satellite Links

Data channel propagation time depends on data rate, the distance between nodes, repeater spacing, and the delays in the repeaters. Data rate (in bps) and repeater delays depend on the type of medium used to transmit the messages. The velocity of propagation of the medium is a most important parameter. Three types of medium are considered in Table 16.2, which gives transmission delays for various call distances. These are wire, fiber-optic cable, and terrestrial radio.

Delay is also a function of the following: processing delays in SPs, number of SPs, number of STPs, number of signaling links, propagation delays of each signaling link, signaling link loading, and message length mix [9].

TABLE 16.2   Calculated Terrestrial Transmission Delays for Various Call Distances

Arc Length (km)	Delay Terrestrial (ms)		
	Wire	Fiber	Radio
500	2.4	2.50	1.7
1,000	4.8	5.0	3.3
2,000	9.6	10.0	16.6
5,000	24.0	25.0	16.5
10,000	48.0	50.0	33.0
15,000	72.0	75.0	49.5
17,737	85.1	88.7	58.5
20,000	96.0	100.0	66.0
25,000	120.0	125.0	82.5

10   NUMBERING PLAN FOR INTERNATIONAL SIGNALING POINT CODES

The numbering plan described in ITU-T Rec. Q.708 [10] has no direct relationship with telephone, data or ISDN numbering. A 14-bit binary code is used for identification of signaling points. An international signaling point code (ISPC) is assigned to each signaling point in the international signaling network. The breakdown of these 14 bits into fields is shown in Figure 16.11. The assignment of signaling network codes is administered by the ITU Telecommunication Standardization Sector (previously CCITT).

All international signaling point codes (ISPCs) consist of three identification subfields as shown in Figure 16.11. The world geographical zone is identified by the NML field consisting of 3 bits. A geographical area or network in a specific zone is identified by the 8-bit field K-D. The 3-bit

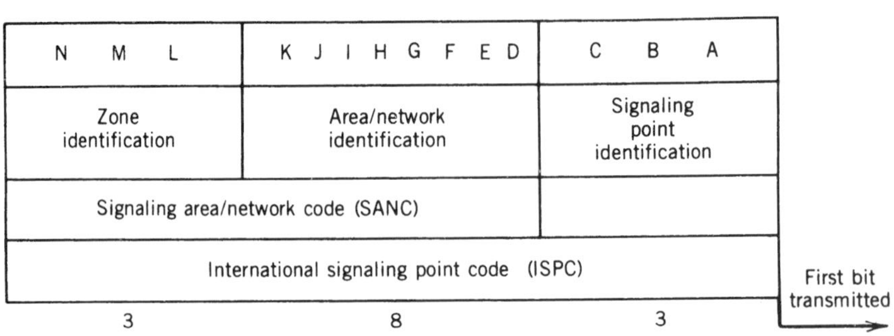

N M L	K J I H G F E D	C B A
Zone identification	Area/network identification	Signaling point identification
Signaling area/network code (SANC)		
International signaling point code (ISPC)		
3	8	3

Figure 16.11. Format for international signaling point code (ISPC). From ITU-T Rec. Q.708, Figure 1/Q.708, page 1 (Ref. 10).

subfield CBA identifies a signaling point in a specific geographical area or network. The combination of the first and second subfields is called a *signaling area/network code* (SANC).

Each country (or geographical area) is assigned at least one SANC. Two of the zone identifications, namely 1 and 0 codes, are reserved for future allocation.

The ISPC system provides for $6 \times 256 \times 8$ (12,288) ISPCs.

If a country or geographical area should require more than 8 international signaling points, one or more additional signaling area/network code(s) would be assigned to it by the ITU-T organization.

A list of SANCs and their corresponding countries can be found in Annex A to ITU-T Rec. Q.708 [10]. The first number of the code identifies the zone. For example, zone 2 is Europe and zone 3 is North America and its environs.

11 HYPOTHETICAL SIGNALING REFERENCE CONNECTIONS

The hypothetical signaling reference connection (HSRC) is composed of signaling points and STPs that are connected in series by signaling data links to produce a signaling connection. The number of signaling points and STPs depends on the size of the network. There are two important parameters we derive from the HSRC: (1) the number of signaling points and STPs in a signaling connection and (2) the signaling message transfer delay, which will directly affect postdial delay.

Definition of Average-Size Country. When the maximum distance between an international switching center and a subscriber who can be reached from it does not exceed 1000 km or, exceptionally, 1500 km, and when the country has less than $n \times 10^7$ subscribers, the country is considered to be average size. A country with a larger distance between an international switching center and a subscriber, or with more than $n \times 10^7$ subscribers, is considered to be of large size. The ITU-T organization states in Ref. 11 that the value of n is for further study.

The number of signaling points in the HSRC (hypothetical signaling reference connection) has been determined by the ITU-T organization by considering the maximum number of links allowed by the Telephone Routing Plan of ITU-T Rec. E.171 (see Chapter 6). Limiting the number of signaling points required reduces signaling delay, because signaling point delay forms the largest component of signaling delay.

The number of STPs in a HRSC is a function of the number of signaling points and the signaling network topology used to connect these signaling points. For an international signaling relationship, no more than two STPs should be used in a signaling relation.

TABLE 16.3   Maximum Overall Signaling Delays

Country Size	Percent of Connections	Delay (ms)[a]; Message Type	
		Simple (e.g., Answer)	Processing Intensive (e.g., IAM)
Large-size to Large-size	50%	1170	1800
	95%	1450	2220
Large-size to Average-size	50%	1170	1800
	95%	1450	2220
Average-size to Average-size	50%	1170	1800
	95%	1470	2240

[a]The values given in the table are mean values.

Source: ITU-T Rec. Q.709, Table 5/Q.709, page 5 (Ref. 11).

TABLE 16.4   Maximum Overall Delay at Signaling Nodes[a,b]

Country Size	Percent of Connections	Delay (ms); Message Type	
		Processing Simple	Processing Intensive
Large-size to Large-size	50%	900	1320
	95%	1270	1900
Large-size to Average-size	50%	900	1320
	95%	1180	1740
Average-size to Average-size	Mean	900	1320
	95%	1200	1760

[a]The maximum signaling nodes delay is the sum of all cross-office involved.
[b]All values are provisional.

Source: ITU-T Rec. Q.709, Table 10/Q.709, page 9 (Ref. 11).

The total unavailability for any signaling route set should not exceed 10 min per year. The overall maximum signaling delays for link-by-link signaling for 50% and 95% of connections are given in Table 16.3 for various combinations of large-size and average-size countries. Average signal point and STP delays at normal loading are assumed. The values given in the table are mean values. These values (in Table 16.3) must be increased by transmission propagation delays.

For end-to-end signaling, 50% and 95% delays are given in Table 16.4 for various combinations of large-size and average-size countries. Average signaling delays at normal loading are assumed.

12 SIGNALING CONNECTION CONTROL PART (SCCP)

12.1 Introduction

The signaling connection control part (SCCP) provides additional functions to the message transfer part (MTP) for both connectionless and connection-oriented network services to transfer circuit-related and non-circuit-related signaling information between switches and specialized centers in telecommunication networks (such as for management and maintenance purposes) via a Signaling System No. 7 network.

Turn now to Figure 16.3 to see where the SCCP appears in a functional block diagram of an SS No. 7 terminal. It is situated above the MTP in level 4 with the user parts. The MTP is transparent and remains unchanged when SCCP services are incorporated in an SS No. 7 terminal. From an OSI perspective, the SCCP carries out the network layer function.

The overall objectives of the SCCP are to provide the means for:

(a) Logical signaling connections within the Signal System No. 7 network.
(b) A transfer capability for network service signaling data units (NSDUs) with or without the use of logical signaling connections.

Functions of the SCCP are also used for the transfer of circuit-related and call-related signaling information of the ISDN user part (ISUP) with or without setup of end-to-end logical signaling connections.

12.2 Services Provided by the SCCP

The overall set of services is grouped into:

• Connection-oriented services
• Connectionless services

Four classes of service are provided by the SCCP protocol, two for connectionless services and two for connection-oriented services. The four classes are:

0 Basic connectionless class
1 Sequenced connectionless class

2 Basic connection-oriented class

3 Flow control connection-oriented class

For connection-oriented services, a distinction has to be made between temporary signaling connections and permanent signaling connections.

Temporary signaling connection establishment is initiated and controlled by the SCCP user. Temporary signaling connections are comparable with dialed telephone connections.

Permanent signaling connections are established and controlled by the local or remote O&M function or by the management function of the node and they are provided for the SCCP user on a semipermanent basis. They can be compared with leased telephone lines.

12.3 Peer-to-Peer Communication

The SCCP protocol facilitates the exchange of information between two peers of the SCCP. The protocol provides the means for:

• Setup of logical signaling connection

• Release of logical signaling connections

• Transfer of data with or without logical signaling connections

A signaling connection is modeled in the abstract by a pair of queues. The protocol elements are objects on the queue added by the origination service user. Each queue represents a flow control function. Figure 16.12 illustrates the SCCP model for connection-oriented service.

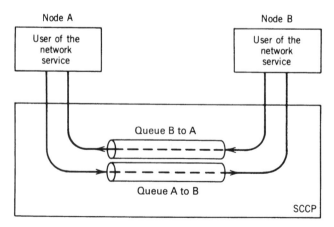

Figure 16.12. Model for internode communication with the SCCP, connection-oriented services. From ITU-T Rec. Q.711, Figure 4/Q.711, page 4 (Ref. 15).

12.4   Primitives and Parameters

Primitives consist of commands and their respective responses associated with the services requested, in this case for the SCCP. Such primitives and parameters are an attempt to conform to OSI, as expressed in CCITT Rec. X.200 series, especially CCITT Rec. X.213. Table 16.5 gives an overview of the primitives to the upper layers and the corresponding parameters for the (temporary) connection-oriented network service. Figure 16.13 shows an overview state transition diagram for the sequence of primitives at a connection endpoint. Here the reader should refer to CCITT Rec. X.213, Network Layer Service Definition of OSI for CCITT application.

TABLE 16.5   Network Service Primitives for Connection-Oriented Services

Primitives		Parameters
Generic Name	Specific Name	
N-CONNECT	Request	Called address
	Indication	Calling address
	Response	Responding address
	Confirmation	Receipt confirmation election
		Expedited data selection
		Quality of service parameter set
		User data
		Connection identification[a]
N-DATA	Request	Confirmation request
	Indication	User data
		Connection identification[a]
N-EXPEDITED DATA	Request	User data
	Indication	Connection identification[a]
N-DATA ACKNOWLEDGE	Request	Connection identification[a]
(for further study)	Indication	
N-DISCONNECT	Request	Originator
	Indication	Reason
		User data
		Responding address
		Connection identification[a]
N-RESET	Request	Originator
	Indication	Reason
	Response	Connection identification[a]
	Confirmation	

[a] In 5.3/X.213, this parameter is implicit. This parameter is for further study.
Source: ITU-T Rec. Q.711, Table 1/Q.711, page 7 (Ref. 15).

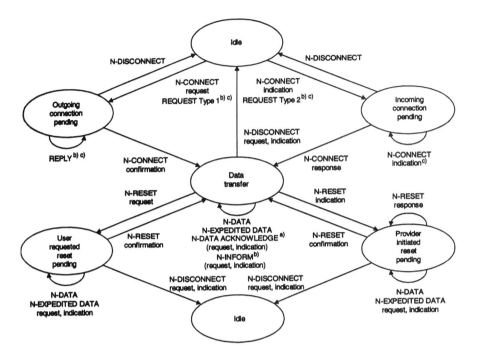

^{a)} The need for this primitive is for further study.
^{b)} This primitive is not in Recommendation X.213.
^{c)} For user part type A only.

Figure 16.13. State transition diagram for sequence of primitives at a connection endpoint (basic functions). From ITU-T Rec. Q.711, Figure 7/Q.711, page 8 (Ref. 15).

12.5 Connection-Oriented Functions: Temporary Signaling Connections

12.5.1 Connection Establishment. A connection setup uses service primitives described in Section 12.4 above. The following are the principal functions used in the connection establishment phase by the SCCP to set up a signaling connection:

- Setup of a signaling connection
- Establishment of the optimum size of NPDUs (network protocol data units)
- Mapping network address onto signalling relations
- Selecting operational functions during data transfer phase (for instance, layer service selection)
- Providing means to distinguish network connections
- Transporting user data (within the request)

12.5.2   Data Transfer Phase. The data transfer phase functions provide the means of a two-way simultaneous transport of messages between two endpoints of a signaling connection. The principal data transport phase functions are listed below. These are used or not used in accordance with the result of the selection function performed in the connection establishment phase. The item with an asterisk requires further study by the ITU-T organization.

- Segmenting/reassembling
- Flow control
- Connection identification
- NSDU delimiting (M-bit)
- Expedited data
- Mis-sequence detection
- Reset
- Receipt confirmation*
- Others

12.5.3   Connection Release Functions. Release functions disconnect the signaling connection regardless of the current phase of the connection. The release may be performed by an upper-layer stimulus or by maintenance of the SCCP itself. The release can start at each end of the connection (symmetric procedure). Of course, the principal function of this phase is disconnection.

12.6   Structure of the SCCP

The basic SCCP structure is shown in Figure 16.14. It consists of four functional blocks as follows [26]:

1. SCCP connection-oriented control. This controls the establishment and release of signaling connections for data transfer on signaling connections.
2. SCCP connectionless control. This provides the connectionless transfer of data units.
3. SCCP management. This functional block provides the capability, in addition to the signal route management and lower control functions of the MTP, to handle the congestion or failure of either the SCCP user or signaling route to the SCCP user.
4. SCCP routing. On receipt of the message from the MTP or from the functions listed above, SCCP routing either forwards the message to the MTP for transfer or passes the message to the functions listed above. A message whose called party address is a local user is passed to functions 1, 2, or 3, whereas one destined for a remote user is forwarded to the MTP for transfer to the distant SCCP.

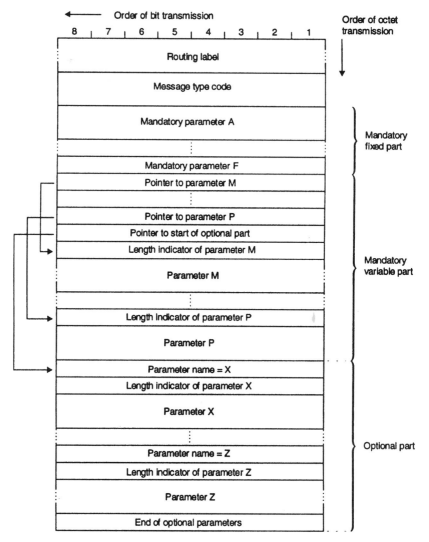

Figure 16.14. General SCCP message format. From ITU-T Rec. Q.713, Figure 2/Q.713, page 3, (Ref. 17).

13 USER PARTS

13.1 Introduction

SS No. 7 user parts, along with the routing label, carry out the basic signaling functions. Turn again to Figure 16.5. There are two fields in the figure we should now discuss: the SIO (service information octet) and the SIF (signaling information field). In the paragraphs that follow, we briefly cover the telephone user part (TUP) and the ISDN user part (ISUP). As shown in

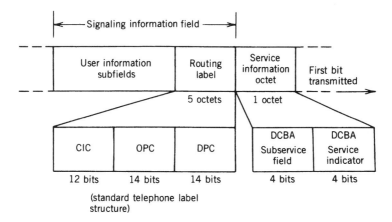

Figure 16.15. Signaling information field (SIF) preceded by the service information octet (SIO). The sequence runs from right to left with the least significant bit transmitted first. DPC, destination point code; OPC, originating point code; CIC, circuit identification code.

Figure 16.15, the user part, OSI layer 4, is contained in the signaling information field to the left of the routing label. ITU-T Rec. Q.723 [21] deals with the sequence of three sectors (fields and subfields of the standard basic message signal unit shown in Figure 16.5).

Turning now to Figure 16.15, we have from right to left, the service information octet (SIO), the routing label, and user information subfields (after the routing label in the SIF.) The SIO is an octet in length made up of two subfields: the service indicator (4 bits) and the subservice field (4 bits). The service indicator, being 4 bits long, has 16 bit combinations with the following meanings (read from right to left):

Bits DCBA	Meaning
0000	Signaling network management message
0001	Signaling network testing and maintenance
0010	Spare
0011	SCCP
0100	Telephone user part
0101	ISDN user part
0110	Data user part (call- and circuit-related message
0111	Data user part (facility registration and cancellation
Remainder (8 sequences)	Spare

It is evident that the SIO directs the signaling message to the proper layer 4 entity, whether SCCP or user part. This is called *message distribution*.

The subservice indicator contains the network bits C and D and two spare bits, A and B. The network indicator is used by signaling message-handling functions determining the relevant version of the user part. If the network indicator is set at 00 or 01, the two spare bits, coded 00, are available for possible future needs. If these two bits are coded 10 or 11, the two spare bits are for national use, such as message priority as an optional flow procedure. The network indicator provides discrimination between international and national usage (bits D and C).

The routing label forms part of every signaling message:

- To select the proper signaling route.
- To identify the particular transaction by the user part (the call) to which the message pertains.

The label format is shown in Figure 16.15. The DPC is the destination point code (14 bits) which indicates the signaling point for which the message is intended. The originating point code (OPC) indicates the source signaling point. The circuit identification code (CIC) indicates the one circuit (speech circuit in the TUP case) among those directly interconnecting the destination and originating points.

For the OPC and DPC, unambiguous identification of signaling points is carried out by means of an allocated code. Separate code plans are used for the international and national networks. The CIC, as shown in the figure, is applicable only to the TUP. CCITT Rec. Q.704 shows a signaling link selection (SLS) field following (to the left) the OPC. The SLS is 4 bits long and is used for load sharing. The ISDN user part address structure is capable of handling E.164 addresses in the calling and called number and is also capable of redirecting address information elements.

13.2   Telephone User Part (TUP)

The core of the signaling information is carried in the SIF (see Figure 16.15). The TUP label was described briefly in Section 7.2.1. Several signal message formats and codes are described in what follows. These follow the label.

One typical message of the TUP is the initial address message (IAM); its format is shown in Figure 16.16. A brief description is given of each subfield, providing further insight of how SS No. 7 operates.

Common to all signaling messages are the subfields H0 and H1. These are the heading codes, each consisting of 4 bits, giving 16 code possibilities in pure binary coding. H0 identifies the specific message group to follow. "Message group" means the type of message. Some samples of message groups are:

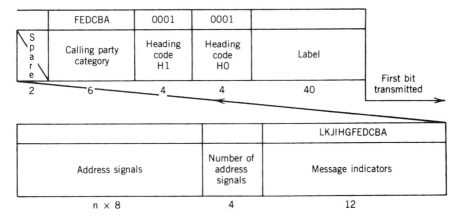

Figure 16.16. Initial address message format. From CCITT Rec. Q.723, Figure 3/Q.723, page 23 (Ref. 21).

Message Group Type	H0 Code
Forward address messages	0001
Forward setup messages	0010
Backward setup messages	0100
Unsuccessful backward setup messages	0101
Call supervision messages	0110
Node-to-node messages	1001

H1 contains a signal code or identifies the format of more complex messages. For instance, there are four types of address message identified by H0 = 0001, and H1 identifies the type of message, such as:

Address Message Type	H0	H1
Initial address message	0001	0001
IAM with additional information	0001	0010
Subsequent address message	0001	0011
Subsequent address message with signal unit	0001	0100

Moving from right to left in Figure 16.16, after H1 we have the calling party subfield consisting of 6 bits. It identifies the language of the operator (Spanish, English, Russian, etc.). An English-speaking operator is coded 000010. It also differentiates the calling subscriber from one with priority, a data call, or a test call. A data call is coded 001100 and a test call is coded 001101. Fifty of the 64 possible code groups are spare.

Continuing to the left in the figure, 2 bits are spare for international allocation. Then there is the message indicator, where the first 2 bits, B and A, give the nature of the address. This is information given in the forward direction indicating whether the associated address or line identity is an international, national (significant), or subscriber number. A subscriber number is coded 00, an international number is coded 11, and a national (significant) number is coded 10.

Bits D and C are the circuit indicator. 00 in this location indicates that there is no satellite circuit in the connection. Remember that the number of space satellite relays in a speech telephone connection is limited to one relay link through a satellite because of propagation delay.

Bits F and E are significant for common-channel signaling systems such as CCIS, CCS No. 6, and SS No. 7. The associated voice channel operates on a separate circuit. Does this selected circuit for the call have continuity? The bit sequence FE is coded:

Bits F and E	Meaning
00	Continuity check not required
01	Continuity check required on this circuit
10	Continuity check performed on previous circuit
11	Spare

Bit G gives echo suppressor information. When coded 0 it indicates that the outgoing half-echo suppressor is not included, and when coded 1 it indicates that the outgoing half-echo suppressor is included. Bit I is the redirected call indicator. Bit J is the all-digital path required indicator. Bit K tells whether any path may be used or whether only SS No. 7 controlled paths may be used. Bit L is spare.

The next subfield has 4 bits and gives the number of address signals contained in the initial address message. The last subfield contains address signals where each digit is coded by a 4-bit group as follows:

Code	Digit	Code	Digit
0000	0	1000	8
0001	1	1001	9
0010	2	1010	Spare
0011	3	1011	Code 11
0100	4	1100	Code 12
0101	5	1101	Spare
0110	6	1110	Spare
0111	7	1111	ST

The most significant address signal is sent first. Subsequent address signals are sent in successive 4-bit fields. As shown in Figure 16.16, the subfield contains *n* octets. A filler code of 0000 is sent to fill out the last octet, if needed. Recall in Chapter 4 that the ST signal is the "end of pulsing" signal and is often used on semiautomatic circuits.

Besides the initial address message, there is the subsequent address message used when all address digits are not contained in the IAM. The subsequent address message is an abbreviated version of the IAM. There is a third type of address message, the initial address message with additional information. This is an extended IAM providing such additional information as network capability, user facility data, additional routing information, called and calling address, and closed under group (CUG). There is also the forward setup message, which is sent after the address messages and contains further information for call setup.

CCITT SS No. 7 is rich with backward information messages. In this group are backward setup request; successful backward setup information message group, which includes charging information; unsuccessful backward setup information message group, which contains information on unsuccessful call setup; call supervision message group; circuit supervision message group; and the node-to-node message group (CCITT Recs. Q.722 and Q.723 [20, 21]).

Label capacity for the telephone user part is given in CCITT Rec. Q.725 [19] as 16,384 signaling points and up to 4096 speech circuits for each signaling point.

13.3 ISDN User Part (ISUP)

13.3.1 Introduction. CCITT Signaling System No. 7 is an integral segment of ISDN. Without SS No. 7 implemented in the PSTN, ISDN does not work. The ISUP encompasses the signaling functions required to provide switched services and user facilities for voice and nonvoice applications in ISDN. We must not forget that ISDN handles voice as well as data and other digital services. The ISUP is also suited for application to dedicated telephone and circuit-switched data networks and in analog and mixed analog and digital networks. In particular the ISUP meets the signaling requirements defined by CCITT for worldwide international semiautomatic and automatic telephone and circuit-switched data traffic [23].

The ISUP is furthermore suitable for national applications. Most signaling procedures, information elements, and message types specified for international use are also required in typical national applications. Moreover, coding space has been reserved in order to allow PSTNs to introduce specific signaling messages and elements of information within the international standardized protocol structure.

Numbering requirements are described in CCITT Rec. E.164 [27].

In the sections that follow we provide a brief description of several key elements of ISUP.

13.3.2 Basic ISUP Signaling Procedures

ADDRESS SIGNALING. In general, the call setup procedure described is standard for both speech and nonspeech connections using en-bloc address signaling for calls between ISDN terminals. Overlap address signaling is also specified.

BASIC PROCEDURES. The basic call control is divided into three phases: call setup, the data/conversation phase, and call cleardown (takedown). Messages on the signaling link are used to establish and terminate different phases of a call. Standard in-band supervisory tones and/or recorded announcements are returned to the caller on appropriate connection types to provide information on call progress. Calls originating from ISDN terminals may be supplied with more detailed call progress information by means of additional messages in the access protocol supported by a range of messages in the network.

Two signaling methods are used with ISUP:

* link-by-link
* end-to-end*

The link-by-link method is primarily used for messages that need to be examined at each intervening exchange. The end-to-end methods are used for messages of endpoint significance. The link-by-link method may also be used for messages of endpoint significance.

ISUP INTERWORKING. In call control interworking between two (ISUP) protocols, the call control provides interworking logic. Peer-to-peer interworking takes place between two exchanges that support different implementations of the same protocol. Interworking is carried out following interpretation of the protocol information received by either exchange.

For this 1992 version of ISUP (ISUP '92), only one ISUP protocol implementation may be present in an exchange because ISUP '92 is backwards compatible with previous versions of ISUP as a result of the following:

* The basic call procedures and supplementary service procedures of ISUP '92 ensure backwards compatibility with the ISUP procedures conforming to the 1988 version (IXth Plenary Assembly, Melbourne, 1988—

---

*See Chapter 4 for conceptual definitions of link-by-link and end-to-end signaling.

Routing label
Circuit identification code
Message type code
Mandatory fixed part
Mandatory variable part
Optional part

Figure 16.17. ISUP message parts.

Blue Books) and those conforming to CCITT Rec. Q.767. No knowledge is required to be stored in the exchange to this effect.

• From ISUP '92 onwards, the forward compatibility is ensured by the guidelines for future protocol enhancements and compatibility procedures as outlined in Section 6 of ITU-T Rec. Q.761 [22].

13.3.3   Formats and Codes of the ISUP. The format of and codes used in the service information octet (SIO) are described in Section 13.1. The service indicator for the ISUP is coded 0101.

The signaling information field of each message signal unit containing an ISUP message consists of an integral number of octets and includes the parts as shown in Figure 16.17.

ROUTING LABEL. The format and codes of the routing label are discussed in Sections 7.2.1 and 13.1 and in ITU-T Rec. Q.704 (Ref. 14, paragraph 2.2). For each individual circuit connection, the same routing label must be used for each message that is transmitted for that connection.

CIRCUIT IDENTIFICATION CODE (CIC). The CIC is an octet long. Bits 1 through 4 are the CIC bits and bits 5 through 8 are spare. The allocation of circuit identification codes to individual circuits is determined by bilateral agreement and/or in accordance with applicable predetermined rules.

For international applications, the four spare bits of the circuit identification field are reserved for CIC expansion, provided that bilateral agreement is obtained before any increase in size is performed. For national applications, the four spare bits can be used as required.

A typical allocation for the 2.048-Mbps digital path is as follows. For circuits derived from a 2.048-Mbps digital path, the circuit identification code contains in the five least significant bits a binary representation of the actual number of the time slot which is assigned to the communication path. The remaining bits in the CIC are used, where necessary, to identify these circuits uniquely among all other circuits of other systems interconnecting an originating and destination point.

MESSAGE TYPE CODE. The message type code consists of a one-octet field and is mandatory for all messages. The message type code uniquely defines the function and format of each ISDN user part message. The allocation is summarized in Table 16.6.

FORMATTING PRINCIPLES. Each message consists of a number of PARAM-ETERS described in ITU-T Rec. Q.763 [24], Section 2. Each parameter has a NAME which is coded as a single octet. There are 59 parameter names listed in Table 5 in Rec. Q.763, plus 10 reserved parameter items. For instance, the first parameter listed is "access delivery information" and is coded 00101110. Between parameters there should be no unused (dummy) octets.

A general format diagram is shown in Figure 16.18.

MANDATORY FIXED PART. Those parameters that are mandatory and of fixed length for a particular message type are contained in the mandatory fixed part. The position, length, and order of the parameters are uniquely defined by the message type, and thus the names of the parameters and the length indicators are not included in the message.

MANDATORY VARIABLE PART. Mandatory parameters of variable length are included in the mandatory variable part. Pointers are used to indicate the beginning of each parameter. Each pointer is encoded as a single octet.

The pointer value (in binary) gives the number of octets between the pointer itself (included) and the first octet (not included) of the parameter associated with that pointer. The pointer value of all zeros is used to indicate that, in the case of optional parameters, no optional parameter is present.

Parameter names are therefore not included in the message. The number of parameters, and thus the number of pointers, is uniquely defined by the message type.

A pointer is also included to indicate the beginning of the optional part. If the message type indicates that no optional part is allowed, then this pointer is not present. If the message type indicates that an optional part is possible, but there is no optional part included in this particular message, then a pointer field containing all zeros is used. The ITU-T organization recommends that all future message types with a mandatory variable part indicate that an optional part is allowed.

All pointers are sent consecutively at the beginning of the mandatory variable part. Each parameter contains the parameter length indicator followed by the contents of the parameters. If there are no mandatory variable parameters, but optional parameters are possible, the start of optional parameters pointer (coded all zeros if no optional parameter is present and coded 00000001 if any optional parameter is present) is included.

TABLE 16.6   Message Type Code Allocation

Message Type	Reference (Table)[a]	Code
Address complete	21	00000110
Answer	22	00001001
Blocking	39	00010011
Blocking acknowledgment	39	00010101
Call progress	23	00101100
Circuit group blocking	40	00011000
Circuit group blocking acknowledgment	40	00011010
Circuit group query @	41	00101010
Circuit group query response @	24	00101011
Circuit group reset	41	00010111
Circuit group reset acknowledgment	25	00101001
Circuit group unblocking	40	00011001
Circuit group unblocking acknowledgment	40	00011011
Charge information @	b	00110001
Confusion	26	00101111
Connect	27	00000111
Continuity	28	00000101
Continuity check request	39	00010001
Facility @	45	00110011
Facility accepted	42	00100000
Facility reject	29	00100001
Facility request	42	00011111
Forward transfer	37	00001000
Identification request	47	00110110
Identification response	48	00110111
Information @	30	00000100
Information request @	31	00000011
Initial address	32	00000001
Loop back acknowledgment @	39	00100100
Network resource management	46	00110010
Overload @	39	00110000
Pass-along @	43	00101000
Release	33	00001100
Release complete	34	00010000
Reset circuit	39	00010010
Resume	38	00001110
Segmentation	49	00111000
Subsequent address	35	00000010
Suspend	38	00001101
Unblocking	39	00010100
Unblocking acknowledgment	39	00010110
Unequipped CIC @	39	00101110
User part available	44	00110101
User part test	44	00110100
User-to-user information	36	00101101
Reserved (used in 1984 version)		00001010
		00001011
		00001111

(*continued*)

TABLE 16.6 *continued*

Message Type	Reference (Table)[a]	Code
Reserved (used in 1984 version) (*continued*)		00100010
		00100011
		00100101
		00100110
Reserved (used in 1988 version)		00011101
		00011100
		00011110
		00100111

[a]Reference to table number in ITU-T Rec. Q.763.
[b]The format of this message is a national matter.

Source: ITU-T Rec. Q.763, Table 4/Q.763, page 7 (Ref. 24).

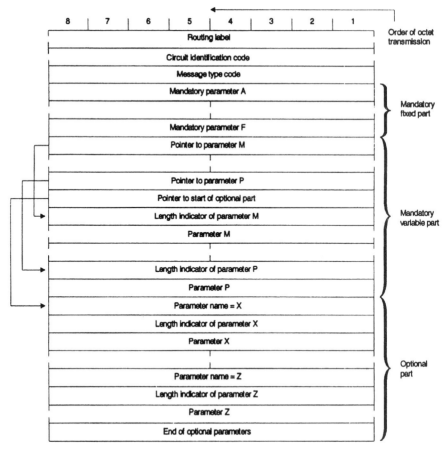

Figure 16.18. ISUP message, general format overview. From ITU-T Rec. Q.763, Figure 3/Q.763, page 5 (Ref. 24).

OPTIONAL PART. The optional part consists of parameters that may or may not occur in any particular message type. Both fixed-length and variable-length parameters may be included. Unless it is explicitly stated to the contrary within the ISUP 1992 Recommendations, an optional parameter cannot occur multiple times within one message. Optional parameters may be transmitted in any order. Each optional parameter includes the parameter name (one octet) and the length indicator (one octet) followed by the parameter contents.

If optional parameters are present and after all optional parameters have been sent, an "end of optional parameters" octet containing all zeros is transmitted. If no optional parameter is present, an "end of optional parameter" octet is not transmitted.

13.3.4   *ISUP Signaling Procedures.* An example of basic call control and signaling procedures is given below for *successful call setup*, forward address signaling, en-bloc operation.

Actions Required at the Originating Exchange

(A) CIRCUIT SELECTION. When the originating gateway exchange has received the complete selection information from the calling party, and has determined that the call is to be routed to another exchange, selection of a suitable, free, interexchange circuit takes place and an initial address message is sent to the succeeding exchange.

Appropriate routing information is either stored at the originating exchange or at a remote database to which a request may be made.

The selection of the route will depend on the called party number, connection type required, and the network signaling capability required. This selection process may be performed at the exchange or with the assistance of the remote database.

In addition, in the case of a subscriber with digital access, the setup message contains bearer capability information which is analyzed by the originating exchange to determine the correct connection type and network signaling capability. The bearer capability information will be mapped into the user service information parameter of the initial address message. The high layer capability information will be mapped into the user teleservice information parameter of the initial address message. The information received from the access interface is used to set the value of the transmission medium requirement parameter.

The connection types allowed are:

- Speech
- 3.1-kHz audio

- 64 kbit/s unrestricted
- 64 kbit/s unrestricted preferred
- 2 × 64 kbit/s unrestricted; multirate connection type
- 384 kbit/s unrestricted; multirate connection type
- 1536 kbit/s unrestricted; multirate connection type
- 1920 kbit/s unrestricted; multirate connection type

The information used to determine the routing of the call by the originating exchange will be included in the initial address message (as transmission medium requirement and forward call indicators) to enable correct routing at intermediate exchanges. The initial address message conveys implicitly the meaning that the indicated circuit has been seized.

(B) ADDRESS INFORMATION SENDING SEQUENCE. The sending sequence of address information on international calls will be the country code followed by the national (significant) number. On national connections, the address information may be the subscriber number or the national (significant) number as required by the administration concerned. For calls to international operator positions (Code 11 and Code 12) refer to Rec. Q.107.

The end-of-pulsing (ST) signal will be used whenever the originating exchange is in a position to know by digit analysis that the final digit has been sent.

(C) INITIAL ADDRESS MESSAGE. The initial address message in principle contains all the information that is required to route the call to the destination exchange and connect the call to the called party.

If the initial address message (IAM) exceeds the 272-octet limit for MTP transfer, it is segmented by the use of the segmentation message (segmentation is discussed in Section 13.3.4.1).

All IAMs include a protocol control indicator and (in the forward call indicator parameter) a transmission medium requirement parameter.

The originating exchange will set the parameters in the protocol control indicator and in the ISDN-user part preference indicator to indicate:

(i) The type of end-to-end method that can be accommodated (see Rec. Q.730)

(ii) The availability of Signaling System No. 7 signaling

(iii) The use of the ISDN-user part

(iv) Network signaling capability required—for example, ISDN-user part required all the way

The ISDN-user part preference indicator is set according to the bearer service, teleservice, and supplementary service(s) requested. The exact setting depends on the service demand conditions and may be different depending on individual cases. In principle, if the service demand requires ISDN-user part to be essential, then the indicator is set to "required," and if the service required is optional but preferred it is set to "preferred." Otherwise it is set to "not required." The indicator is set to either "required" or "preferred," or "not required," according to the most stringent condition required by one or more of the parameters in the initial address message.

The nature of connection indicators is set appropriately based on the characteristics of the selected outgoing circuit.

The transmission medium requirement parameter contains the connection type required information—for example, 3.1-kHz audio.

The propagation delay counter is included according to paragraph 2.6 of Ref. 25.

(D) COMPLETION OF TRANSMISSION PATH. Through-connection of the transmission path will be completed in the backward direction (the transmission path is completed in the forward direction on receipt of a connect or answer message) at the originating exchange immediately after the sending of the initial address message, except in those cases where conditions on the outgoing circuit prevent it.

It is also acceptable that on speech or 3.1-kHz audio calls, through-connection of the transmission path will be completed in both directions immediately after the initial address message has been sent, except in those cases where conditions on the outgoing circuit prevent it.

(E) NETWORK PROTECTION TIMER. When the originating exchange has sent the initial address message, the awaiting address complete timer (T7) is started. If timer (T7) expires, the connection is released and an indication is returned to the calling subscriber.

Actions Required at an Intermediate National Exchange

(A) CIRCUIT SELECTION. An intermediate national exchange, on receipt of an initial address message, will analyze the called party number and the other routing information to determine the routing of the call. If the intermediate national exchange can route the call using the connection type specified in the transmission medium requirement parameter, a free interexchange circuit is seized and an initial address message is sent to the succeeding exchange. Within a network if the intermediate national exchange does not route the call using just the connection type specified in the transmission medium requirement parameter, the exchange may also examine the user service information containing the bearer capability information and/or the user teleservice information containing the high layer capability information, if available, to determine if a suitable route can be selected. In this case if a new connection type is provided, the transmission medium requirement parameter is modified to the new connection type.

(B) PARAMETERS IN THE INITIAL ADDRESS MESSAGE. An intermediate national exchange may modify signaling information received from the preceding exchange according to the capabilities used on the outgoing route. Signaling information that may be changed are nature of connection indicator and propagation delay counter. Other signaling information is passed on transparently—for example, the access transport parameter, user service information, and so on.

The satellite indicator in the nature of connection parameter should be incremented if the selected outgoing circuit is a satellite circuit. Otherwise, the indicator is passed on unchanged.

(C) COMPLETION OF TRANSMISSION PATH. Through-connection of the transmission path in both directions will be completed at an intermediate national exchange immediately after the initial address message has been sent, except in those cases where conditions on the outgoing circuit prevent it.

Actions Required at an Outgoing International Exchange

(A) CIRCUIT SELECTION. On receipt of an initial address message, an outgoing international exchange will analyze the called party number and the other routing information to determine the routing of the call. If the outgoing international exchange can route the call using the connection type specified in the transmission medium requirement parameter, a free interexchange circuit is seized and an initial address message is sent to the succeeding exchange.

If the outgoing international exchange cannot trust that the transmission medium requirement value received from the national network reflects the minimum value of the information transfer susceptance, then the transmission medium requirement value may be modified according to the contents of the information transfer capability and information transfer rate fields of the user service information parameter (if available).

The outgoing international exchange must ensure that the transmission medium requirement parameter is set according to the service requested by the customer (see CCITT Rec. E.172). More specifically this parameter is carried unchanged within the international network.

(B) PARAMETERS IN THE INITIAL ADDRESS MESSAGE. An outgoing international exchange may modify signaling information received from the preceding exchange according to the capabilities used on the outgoing route. Signaling information that may be changed is nature of connection indicator and propagation delay counter; the most significant digits in the called party number may be amended or omitted (country code is removed at the last exchange before the incoming international exchange). Other signaling information is passed on transparently—for example, the access transport parameter, user service information, and so on.

If the outgoing international exchange belongs to a country using μ-law PCM encoding nationally and the transmission medium requirement indicates speech or 3.1-kHz audio, then the user information layer 1 protocol identification field of the user service information parameter must be checked. If it indicates "Rec. G.711 μ-law" this must be changed to "Rec. G.711 A-law" and a μ-law to A-law convertor must be enabled.

The satellite indicator in the nature of connection parameter should be incremented if the selected outgoing circuit is a satellite circuit. Otherwise, the indicator is passed on unchanged.

The outgoing international gateway exchange should include the originating ISC point code parameter in the initial address message. This information is used for statistical purposes—for example, accumulation of the number of incoming calls on an originating international switching center basis.

If a location number parameter is received, the nature of address indicator is checked, and if the nature of address indicator is set to "international number," then the parameter is passed unchanged; otherwise the number is modified to the international number format and the nature of address is set to "international number" before being passed.

The end-of-pulsing (ST) signal will be used whenever the outgoing exchange is in a position to know by digit analysis that the final digit has been sent.

(C) COMPLETION OF TRANSMISSION PATH. Through-connection of the transmission path in both directions will be completed at an outgoing international exchange immediately after the initial address message has been sent, except in those cases where conditions on the outgoing circuit prevent it.

(D) NETWORK PROTECTION TIMER. When an outgoing international exchange has sent the initial address message, the awaiting address complete timer (T7) is started. If timer (T7) expires, the connection is released and an indication is returned to the calling subscriber.

Actions Required at an Intermediate International Exchange

(A) CIRCUIT SELECTION. On receipt of an initial address message, an intermediate international exchange will analyze the called party number and the other routing information to determine the routing of the call. If the intermediate international exchange can route the call using the connection type specified in the transmission medium requirement parameter, a free interexchange circuit is seized and an initial address message is sent to the succeeding exchange.

(B) PARAMETERS IN THE INITIAL ADDRESS MESSAGE. An intermediate international exchange may modify signaling information received from the preceding exchange according to the capabilities used on the outgoing route. Signaling information that may be changed is nature of connection indicator and propagation delay counter; the most significant digits in the called party number may be amended or omitted (country code is removed at the last exchange before the incoming international exchange). Other signaling information is passed on transparently—for example, the access transport parameter, user service information, and so on.

The satellite indicator in the nature of connection parameter should be incremented if the selected outgoing circuit is a satellite circuit. Otherwise, the indicator is passed on unchanged.

(C) COMPLETION OF TRANSMISSION PATH. Through-connection of the transmission path in both directions will be completed at an intermediate international exchange immediately after the initial address message has been sent, except in those cases where conditions on the outgoing circuit prevent it.

When an intermediate international exchange has sent the initial address message, the awaiting address complete timer (T7) is started. If timer (T7) expires, the connection is released and an indication is returned to the calling subscriber.

Actions Required at an Incoming International Exchange

(A) CIRCUIT SELECTION. An incoming international exchange, on receipt of an initial address message, will analyze the called party number and the other routing information to determine the routing of the call. If the incoming international exchange can route the call using the connection type specified in the transmission medium requirement parameter, a free interexchange circuit is seized and an initial address message is sent to the succeeding exchange.

(B) PARAMETERS IN THE INITIAL ADDRESS MESSAGE. An incoming international exchange may modify signaling information received from the preceding exchange according to the capabilities used on the outgoing route. Signaling information that may be changed are nature of connection indicator and propagation delay counter. Other signaling information is passed on transparently—for example, the access transport parameter, user service information, and so on.

The satellite indicator in the nature of connection parameter should be incremented if the selected outgoing circuit is a satellite circuit. Otherwise, the indicator is passed on unchanged.

If the incoming international exchange belongs to a country using μ-law PCM encoding nationally and the transmission medium requirement indicates speech or 3.1-kHz audio, then the user information layer 1 protocol identification field of the user service information parameter must be checked. If it indicates "Rec. G.711 A-law" this must be changed to "Rec. G.711 μ-law" and a μ-law to A-law convertor must be enabled.

The incoming international gateway exchange should delete the originating ISC point code parameter from the initial address message and set up a connection to the national network. This information is used for statistical purposes—for example, accumulation of the number of incoming calls on an originating international switching center basis.

(c) COMPLETION OF TRANSMISSION PATH. Through-connection of the transmission path in both directions will be completed at an incoming international exchange immediately after the initial address message has been sent, except in those cases where conditions on the outgoing circuit prevent it.

(d) NETWORK PROTECTION TIMER. When an incoming international exchange has sent the initial address message, the awaiting address complete timer (T7) is started. If timer (T7) expires, the connection is released and an indication is returned to the calling subscriber.

Actions Required at the Destination Exchange

(A) SELECTION OF CALLED PARTY. Upon receipt of an initial address message, the destination exchange will analyze the called party number to determine to which party the call should be connected. It will also check the called party's line condition and perform various checks to verify whether or not the connection is allowed. These checks will include correspondence of compatibility checks—for example, checks associated with supplementary services.

In this case where the connection is allowed, the destination exchange will set up a connection to the called party. If a continuity check has to be performed on one or more of the circuits involved in a connection, setting up of the connection to the called party must be prevented until the continuity of such circuits has been verified.

(B) SEGMENTED INITIAL ADDRESS MESSAGE. If the initial address message had been segmented by the use of the segmentation message, the remainder of the call setup information is awaited.

13.3.4.1 Segmentation. The simple segmentation procedure uses the segmentation message to convey an additional segment of an overlength message. Any message containing either the optional forward or backward call indicators can be segmented using this method. This procedure provides a mechanism for the transfer of certain messages whose contents are longer than 272 octets but not longer than 544 octets.

The procedure is as follows:

(a) The sending exchange, on detecting that the message to be sent exceeds the 272-octet limit of the message transfer part, can reduce the message length by sending some parameters in a segmentation message sent immediately following the message containing the first segment.

(b) The parameters that may be sent in the second segment using the segmentation message are the user-to-user information, generic digit, generic notification, generic number, and access transport parameters. If the user-to-user information and access transport parameters cannot be carried in the original message and the two together do not fit in the segmentation message, the user-to-user information parameter is discarded.

(c) The sending exchange sets the simple segmentation indicator in the optional forward or backward call indicators to indicate that additional information is available.

(d) When a message is received, at a local exchange, with the simple segmentation indicator set to indicate additional information is available, the exchange starts timer T34 to await the segmentation message. This action may also take place at incoming or outgoing international exchanges if policing of information is required.

(e) When the segmentation message is received, timer T34 is stopped and the call continues.

(f) In case any other message except the ones listed below is received before the segmentation message containing the second segment, the exchange should react as if the second segment is lost; that is, timer T34 is stopped and the call continues. The messages are:

- Continuity
- Blocking
- Blocking acknowledgment
- Circuit group blocking
- Circuit group blocking acknowledgment
- Unblocking
- Unblocking acknowledgment
- Circuit group unblocking
- Circuit group unblocking acknowledgment
- Circuit group query
- Circuit group query response

(g) After expiration of timer T34, the call proceeds and a received segmentation message containing the second segment of a segmented message is discarded.

(h) At an incoming or outgoing international exchange, when following the simple segmentation procedure, it is possible that the exchange has to reassemble an incoming message and subsequently resegment it for onward transmission. In this case it has to be ensured that any unrecognized parameters received in the first, or second, segment are transmitted in the first, or second, segment respectively, when the passing of the parameter is required by the compatibility procedure.

13.3.5 Continuity Check. Because the signaling information in SS No. 7 does not pass over the speech/data channel (i.e., the B-channel), facilities are provided for making a continuity check of the traffic circuit in the circumstances described below.

The application of the continuity check depends on the type of the transmission system used for the circuit. For transmission systems having some inherent fault indication features giving an indication to the switching system in case of fault, a continuity check is not required. However, a per-call continuity check may be needed on fully digital circuits when circuits or bundles of circuits in primary multiplex groups are dropped and inserted en route between switches and alarm indications carried on bits of the primary multiplex frame structure are lost in passing through an intermediate transmission facility that does not relay them transparently. Typical, per-call continuity checks may be needed when the transmission link between switches contains a TDMA satellite system, a digital circuit multiplication system, or a digital access and cross-connection system, where fault indica-tions are lost (see CCITT Rec. Q.33).

When an initial address message is received with a request for a continuity check, a continuity-check loop is connected.

Means should be provided in Signaling System No. 7 to detect circuit identification code misunderstandings between Signaling System No. 7 exchanges.

For exchanges having both analog and digital circuits served by Signaling System No. 7, the continuity check initiated by a continuity-check request message could be used to test for proper alignment of circuit code identities. On those exchanges, reception of a continuity-check request message should always cause a loop to be attached to the circuit.

Alternative methods for detection of circuit identity misunderstandings in exchanges with all-digital circuits may be employed.

The continuity check is not intended to eliminate the need for routine testing of the transmission path.

The continuity check of the circuit will be done, link-by-link, on a per-call basis or by a statistical method prior to the commencement of conversation. Procedures and requirements are specified in 7/Q.724 [20].

The actions to be taken when pilot supervision is used are described in 9/Q.724 [20].

When an initial address message is received with a request for continuity check (either on this circuit or on a previous circuit), timer T8 is started. On receipt of a successful indication of continuity check in a continuity message, timer T8 is stopped. However, if timer T8 expires, the connection is cleared.

If an indication of continuity-check failure is received in a continuity message, timer T27 is started awaiting a continuity recheck request. Also, the connection to the succeeding exchange is cleared. Timer T27 is stopped when the continuity-check request message is received and timer T36 is started awaiting a continuity or release message.

If either timer T27 or timer T36 expires, a reset circuit message is sent to the preceding exchange. On reception of the release complete message, the circuit is set to idle.

Where circumstances require per-call continuity checking for multirate connection type calls, the continuity of the single 64-kbit/s circuit whose circuit identification code is contained in the initial address message shall be checked.

Sections 13.3.4 and 13.3.5 have been extracted from ITU-T Rec. Q.764 [25].

REVIEW QUESTIONS

1. What is the principal rationale for developing and implementing CCITT Signaling System No. 7?

2. Describe SS No. 7's relationship with OSI. Why does it truncate at OSI layer 4?

3. Give the two primary "parts" of SS No. 7. Briefly describe the function of each generic part.

4. With CCITT Signaling System No. 7, OSI layer 4 is subdivided into two sublayers. Name them.

5. Name the three user parts of SS No. 7.

6. What is the normal basic connectivity of SS No. 7? (*Hint*: bit rate.)

7. Layer 2 of SS No. 7 functions as the signaling link. Name five of the seven functions of layer 2. In one sentence describe what layer 2 does.

8. What are the two basic categories of functions of layer 3 of SS No. 7? What is the basic purpose of layer 3? (*Hint*: Think OSI.)

9. What is a signaling relation?

10. How are signaling points defined (identified)?

11. Interrelate message routing, message discrimination, and message distribution.

12. What does a routing label do?

13. What is the purpose of signaling unit delimitation?

14. What is the *basic* method of error correction in SS No. 7?

15. How is flow control initiated when congestion takes place at signaling point(s)?

16. Give the three signaling network management functional blocks.

17. There are three types of signal units used in SS No. 7. What are they? Define the basic function of each.

18. Differentiate forward and backward sequence numbers. When does reset occur for FSN? In essence, what do FSN and BSN accomplish?

19. What is the function of the service indicator field?

20. The routing label is analogous to what in our present telephone system? Name the three basic pieces of information that the routing label provides.

21. Give three of the fault conditions handled by the signaling network management.

22. Define labeling potential.

23. What is the function of an STP? Differentiate between an STP and a signaling point.

24. Why do we wish to limit the number of STPs in a signaling connection on a particular relation?

25. What are the three basic measures of performance of SS No. 7?

26. Give the long-term and medium-term error rates for SS No. 7.

27. What is the biggest contributor to signaling delay in the SS No. 7 network?

28. At what OSI layer does the SCCP reside? What is the basic rationale of implementing SCCP?

29. How can we use a hypothetical signaling reference connection in practice?

30. Regarding user parts, what is the function of the SIO? Of the network indicator?

31. What does the heading code (H0, H1) tell us?

32. What is the purpose of circuit continuity? Explain.

33. Address signals (such as telephone number digits) are sent digit by digit embedded in the last subfield in the SIF. How are they represented? (*Hint*: per digit.)

34. Give two examples of backward information in SS No. 7.

35. What facilitates end-to-end signaling typically for ISUP (i.e., what "part" of SS No. 7)?

36. Give at least three functions of the protocol control indicator, as used, for example, in the IAM.

37. Give the actions taken at an intermediate exchange during call setup using SS No. 7.

38. Give the actions taken by a destination exchange during call setup using SS No. 7.

39. How is an IAM handled that has contents greater than 272 octets?

40. Through what means does SS No. 7 handle A-law/μ-law conversion (i.e., in what message type)?

REFERENCES

1. W. Stallings, *Tutorial, Integrated Services Digital Network (ISDN)*, IEEE Computer Society, Washington, DC, 1985.

2. *Specifications of Signaling System No. 7* (Q.700 series), Fascicle VI. 6, CCITT Recommendations (Yellow Books), VIIth Plenary Assembly, Geneva, 1980.

3. *BOC Notes on the LEC Networks—1994*, Bellcore SR-TSV-002275, Issue 2, April 1994, Bellcore, Piscataway, NJ, 1994.

4. *Introduction to CCITT Signaling System No. 7*, ITU-T Rec. Q.700, Helsinki, March 1993.

5. G. G. Schlanger, "An Overview of Signaling System No. 7," *IEEE J. on Selected Areas in Communications*, **7**(3) (May 1986).

6. *Functional Description of the Message Transfer Part (MTP) of Signaling System No. 7*, ITU-T Rec. Q.701, Helsinki, March 1993.

7. *Signaling System No. 7—Signaling Link*, ITU-T Rec. Q. 703, Helsinki, March 1993.

8. *Signaling System No. 7—Signaling Network Structure*, ITU-T Rec. Q.705, Helsinki, March 1993.

9. *Signaling System No. 7—Message Transfer Part Signaling Performance*, ITU-T Rec. Q.706, Helsinki, March 1993.

10. *Signaling System No. 7—Numbering of International Signaling Point Codes*, ITU-T Rec. Q.708, Helsinki, March 1993.

11. *Signaling System No. 7—Hypothetical Signaling Reference Connection*, ITU-T Rec. Q.709, Helsinki, March 1993.

12. W. C. Roehr, Jr., "Signaling System No. 7," in *Tutorial: Integrated Services Digital Network (ISDN)*, W. Stallings, Ed. IEEE Computer Society, Washington, DC, 1985.

13. R. L. Freeman, *Reference Manual for Telecommunication Engineering*, 2nd ed., John Wiley & Sons, New York, 1994.

14. *Signaling Network, Functions and Messages*, CCITT Rec. Q.704, IXth Plenary Assembly, Melbourne, 1988.

15. *Signaling System No. 7—Functional Description of the Signaling Connection Control Part*, ITU-T Rec. Q. 711, ITU Helsinki, March 1993.

16. *Signaling System No. 7—Definition and Function of SCCP Messages*, ITU-T Rec. Q.712, ITU Helsinki, March 1993.

17. *Signaling System No. 7—SCCP Formats and Codes*, ITU-T Rec. Q.713, ITU Helsinki, March 1993.

18. *Signaling System No. 7—Signaling Connection Control Part (SCCP) Performance*, ITU-T Rec. Q.716, ITU Helsinki, March 1993.

19. *Signaling System No. 7—Signaling Performance in the Telephone Application*, ITU-T Rec. Q.725, ITU Helsinki, March 1993.

20. CCITT Recommendations, Volume VI, Fascicle VI.8, *Specifications of Signaling System No. 7*, Recs. Q.721–Q.766, IXth Plenary Assembly, Melbourne, 1988.

21. *Formats and Codes (Telephone User Part)*, CCITT Rec. Q.723, Fascicle VI.8, IXth Plenary Assembly, Melbourne, 1988.

22. *Functional Description of the ISDN User Part of Signaling System No. 7*, ITU-T Rec. Q.761, Helsinki, March 1993.

23. *General Function of Messages and Signals of the ISDN User Part of Signaling System No. 7*, ITU-T Rec. Q.762, Helsinki, March 1993.

24. *Formats and Codes of the ISDN User Part of Signaling System No. 7*, ITU-T Rec. Q.763, Helsinki, March 1993.

25. *Signaling System No. 7—ISDN User Part Signaling Procedures*, ITU-T Rec. Q.764, Helsinki, March 1993.

26. *Signaling System No. 7—Signal Connection Control Part Procedure*, ITU-T Rec. Q.714, Helsinki, 1993.

27. *Numbering Plan for the ISDN Era*, CCITT Rec. E.164, Geneva, 1991.

17

CELLULAR/MOBILE RADIO AND PCN/PCS

1 INTRODUCTION

1.1 Background

Mobile radio communication dates as far back as Marconi. For a time it was the principal application of radio. This was a ship-to-shore and ship-to-ship communications. The pioneer in this was the Marconi Company of the United Kingdom. It spread to land vehicles and aircraft in the 1920s.

Since 1980, mobile radio communication has taken on a more personal flavor. Cellular radio systems have extended the telephone network to the car and pedestrian. A new and widely used term in our vocabulary is *personal communications*. It it becoming the universal tether. No matter where we go, on land, at sea, and in the air, we can have near instantaneous two-way communications by voice, data, and facsimile. At some time it will encompass video.

Personal radio terminals are becoming smaller. There is the potential of becoming wristwatch size. However, the human interface requires input/output devices that have optimum usefulness. A wristwatch-size keyboard or keypad is rather difficult to operate. A hard-copy printer requires some minimum practical dimensions, and so forth.

1.2 Scope and Objective

This chapter presents an overview of "personal communications." Much of the discussion deals with cellular radio, and it extends this thinking inside of buildings. The coverage most necessarily includes propagation for the several environments, propagation impairments, methods to mitigate these

883

impairments, access techniques, bandwidth limitations, and ways around this problem. It also includes several mobile radio standards and compares a number of existing and planned systems. The chapter objective is to provide an appreciation of mobile/personal communications. Space limitations force us to confine our discussion to what might be loosely called "land mobile systems."

2 SOME BASIC CONCEPTS OF CELLULAR RADIO

Cellular radio systems connect a mobile terminal to another user, usually through the PSTN. The "other user" most commonly is a telephone subscriber of the PSTN. However, the other user may be another mobile terminal. Most of the connectivity is extending POTS to mobile users. Data and facsimile services are in various stages of implementation. Some of the terms used in this section have a strictly North American flavor.

Figure 17.1 shows a conceptual layout of a cellular system. The heart of the system for a specific serving area is the mobile telephone switching office (MTSO). The MTSO is connected by a trunk group to a nearby telephone exchange, providing an interface to, and connectivity with, the PSTN.

The area to be served by a *cellular geographic serving area* (CGSA) is divided into small geographic cells which ideally are hexagonal. Cells are

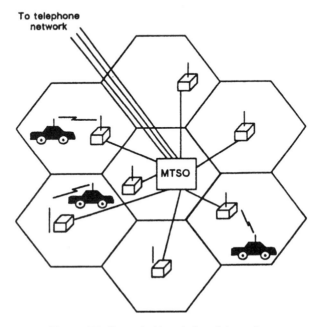

Figure 17.1. Conceptual layout of a cellular system.

initially laid out with centers spaced about 4 to 8 mi (6.4 to 12.8 km) apart. The basic system components are the cell sites, the MTSO, and mobile units. These mobile units may be hand-held or vehicle-mounted terminals.

Each cell has a radio facility housed in a building or shelter. The facility's radio equipment can connect and control any mobile unit within the cell's responsible geographic area. Radio transmitters located at the cell site have a maximum effective radiated power (ERP*) of 100 watts. Combiners are used to connect multiple transmitters to a common antenna on a radio tower, usually between 50 and 300 ft (15 and 92 m) high. Companion receivers use a separate antenna system mounted on the same tower. The receive antennas are often arranged in a space-diversity configuration.

The MTSO provides switching and control functions for a group of cell sites. A method of connectivity is required between the MTSO and the cell site facilities. The MTSO is an electronic switch and carries out a fairly complex group of processing functions to control communications to and from mobile units as they move between cells as well as to make connections with the PSTN. Besides making connectivity with the public network, the MTSO controls cell site activities and mobile actions through command and control data channels. The connectivity between cell sites and the MTSO is often via DS1 on wire pairs or on microwave facilities, the latter being the most common.

A typical cellular mobile unit consists of a control unit, a radio transceiver, and an antenna. The control unit has a telephone handset, a pushbutton keypad to enter commands into the cellular/telephone network, and audio and visual indications for customer alerting and call progress. The transceiver permits full-duplex transmission and reception between a mobile and cell sites. Its ERP is nominally 6 watts. Hand-held terminals combine all functions into one small package that can be easily held in one hand. The ERP of a hand-held is a nominal 0.6 watts.

In North America, cellular communication is assigned a 25-MHz band between 824 and 849 MHz for mobile unit-to-base transmission and a similar band between 869 and 894 MHz for transmission from base to mobile.

The original North American cellular radio system was called *advanced mobile telephone system* (AMPS). The original system description was contained in an entire issue of the Bell System Technical Journal (BSTJ) of January 1979. The present AMPS is based on 30-kHz channel spacing using frequency modulation. The peak deviation is 12 kHz. The cellular bands are each split into two to permit competition. Thus only 12.5 MHz is allocated to one cellular operator for each direction of transmission. With 30-kHz spacing, this yields 416 channels. However, nominally 21 channels are used for control purposes, with the remaining 395 channels available for cellular end users.

---

*Care must be taken with terminology. In this instance, ERP and EIRP are not the same. The reference antenna in this case is the dipole which has a 2.16-dBi gain.

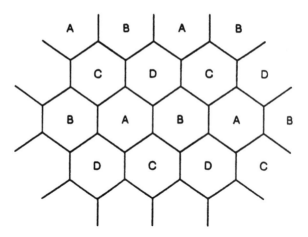

Figure 17.2. Cell separation with four different sets of frequencies.

Common practice with AMPS is to assign 10 to 50 channel frequencies to each cell for mobile traffic. Of course, the number of frequencies used depends on the expected traffic load and the blocking probability. Radiated power from a cell site is kept at a relatively low level with just enough antenna height to cover the cell area. This permits frequency reuse of these same channels in nonadjacent cells in the same CGSA with little co-channel interference. A well-coordinated frequency reuse plan enables tens of thousands of simultaneous calls over a CGSA.

Figure 17.2 illustrates a frequency reuse method. Here four channel frequency groups are assigned in a way that avoids the same frequency set used in adjacent cells. If there were uniform terrain contours, this plan could be applied directly. However, real terrain conditions dictate further geographic separation of cells that use the same frequency set. Reuse plans with 7 or 12 sets of channel frequencies provide more physical separation and are often used depending on the shape of the antenna pattern employed.

With user growth in a particular CGSA, cells may become overloaded. This means that grade of service objectives are not being met due to higher than planned traffic levels during the busy hour (BH). In these cases, congested cells can be subdivided into smaller cells, each with its own base station, as shown in Figure 17.3. These smaller cells use lower transmitter power and antennas with less height, thus permitting greater frequency reuse. These subdivided cells can be split still further for still greater frequency reuse. However, there is a practical limit to cell splitting, often with cells with a 1-mi (1.6-km) radius. Under normal large-cell operation, antennas are usually omnidirectional. When cell splitting is employed, 60° or 120° directional antennas are often used to mitigate interference brought about by increased frequency reuse.

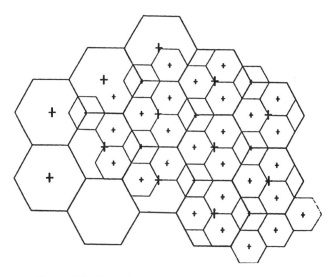

Figure 17.3. Staged growth by cell splitting (subdividing).

Radio system design for cellular operation differs from that used for line-of-sight (LOS) microwave operation. For one thing, mobility enters the picture. Path characteristics are consistently changing. Mobile units experience multipath scattering, reflection, and/or diffraction by obstructions and buildings in the vicinity. There is shadowing, often very severe. The resulting received signal under these conditions varies randomly as the sum of many individual waves with changing amplitude, phase, and direction of arrival. The statistical autocorrelation distance is on the order of one-half wavelength. Space diversity at the base station tends to mitigate these impairments [21].

In Figure 17.1, the MTSO is connected to each of its cell sites by a voice trunk for each of the radio channels at the site. Also two data links (AMPS design) connect the MTSO to each cell site. These data links transmit information for processing calls and for controlling mobile units. In addition to its "traffic" radio equipment, each cell site has installed signaling monitor-ing equipment and a "setup" radio channel to establish calls.

When a mobile unit becomes operational, it automatically selects the setup channel with the highest signal level. It then monitors that setup channel for incoming calls destined for it. When an incoming call is sensed, the mobile terminal in question again samples signal level of all appropriate setup channels so it can respond through the cell site offering the highest signal level, and then it tunes to that channel for response. The responsible MTSO assigns a vacant voice channel to the cell in question, which relays this information via the setup channel to the mobile terminal. The mobile terminal subscriber is then alerted that there is an incoming call. Outgoing calls from mobile terminals are handled in a similar manner.

While a call is in progress, the serving cell site examines the mobile's signal level every few seconds. If the signal level drops below a prescribed level, the system seeks another cell to handle the call. When a more appropriate cell site is found, the MTSO sends a command, relayed by the old cell site, to change frequency for communication with the new cell site. At the same time, the landline subscriber is connected to the new cell site via the MTSO. The periodic monitoring of operating mobile units is known as *locating*, and the act of changing channels is called *handover*. Of course, the functions of locating and handover are to provide subscribers satisfactory service as a mobile unit traverses from cell to cell. When cells are made smaller, handovers are more frequent.

The management and control functions of a cellular system are quite complex. Handover and locating are managed by signaling and supervision techniques which take place on the setup channel. The setup channel uses a 10-kbps data stream which transmits paging, voice channel designation, and overhead messages to mobile units. In turn, the mobile unit returns page responses, origination messages, and order confirmations.

Both digital messages and continuous supervision tones are transmitted on the voice radio channel. The digital messages are sent as discontinuous "blank-and-burst" inband data stream at 10 kbps and include order and handover messages. The mobile unit returns confirmation and messages that contain dialed digits. Continuous positive supervision is provided by an out-of-band 6-kHz tone, which is modulated onto the carrier along with the speech transmission.

Roaming is a term used for a mobile unit that travels such distances that the route covers more than one cellular organization or company. The cellular industry is moving toward technical and tariffing standardization so that a cellular unit can operate anywhere in the United States, Canada, and Mexico [21].

3   RADIO PROPAGATION IN THE MOBILE ENVIRONMENT

3.1   The Propagation Problem

Line-of-site microwave and satellite communications covered in Chapter 7 dealt with fixed systems. Such systems were and are optimized. They are built up and away from obstacles. Sites are selected for best propagation.

Not so with mobile systems. Motion and a third dimension are additional variables. The end-user terminal often is in motion. Or the user is temporarily fixed, but that point can be anywhere within a serving area of interest. Whereas before we dealt with point-to-point, here we deal with point-to-multipoint.

One goal in LOS microwave design was to stretch the distance as much as possible between repeaters by using high towers. In this chapter there are some overriding circumstances where we try to limit coverage extension by reducing tower heights, what we briefly introduced in Section 2. Even more importantly, coverage is area coverage where shadowing is frequently encountered. Examples are valleys, along city streets with high buildings on either side, in verdure such as trees, and inside buildings, to name a few situations. Such an environment is rich with multipath scenarios. Paths can be highly dispersive, as much as 10 µs of delay spread [1]. Due to a user's motion, Doppler shift can be expected.

The radio-frequency bands of interest are UHF, especially around 800 and 900 MHz and 1700 to 2000 MHz.

3.2 Several Propagation Models

We concentrate on cellular operation. There is a fixed station (FS) and mobile stations (MS) moving through the cell. A cell is the area of responsibility of the fixed station, a cell site. It usually is pictured as a hexagon in shape, although its propagation profile is more like a circle with the fixed station in its center. The cell contour is rather ragged due to obstructions in each path radial. Cell radii vary from 1 km (0.6 mi) to 30 km (19 mi) or somewhat more.

3.2.1 Path Loss. We recall the free space loss (FSL) formula discussed in Chapter 7. It simply stated that FSL was a function of the square of the distance and the square of the frequency plus a constant. It is a very useful formula if the strict rules of obstacle clearance are obeyed. Unfortunately in the cellular situation, it is impossible to obey these rules. Then to what extent is this free space loss formula modified by atmospheric effects, the presence of the earth bulge, and the effects of trees, buildings, and hills which exist in, or close to, the transmission path?

3.2.1.1. CCIR Formula#CCIR developed a single path loss formula (CCIR Rec. 370 [12]) for radio and television broadcasting where the frequency term has been factored out. It states that the path loss (L_{dB}):

$$L_{dB} = 40 \log d_m - 20 \log(h_T h_R) \qquad (17.1)$$

Note that the equation is an inverse fourth power law and is fundamental to terrestrial mobile radio. Distance is in meters; h_T is the height of the transmit antenna above plane earth, again in meters; and h_R is the height of the receive antenna above plane earth in meters.

Suppose $d = 1000$ m (1 km) and the product of $h_T \times h_R$ is 10 m^2 (low antennas), then $L = 100$ dB.

Now suppose that $d = 25$ km and $h_T \times h_R = 100$ m^2, where the base station is on high ground.

The loss L is 136 dB.

The CCIR formula only brings in two new, but important, variables: h_T and h_R; it also uses the fourth power rather than the second power.

3.2.1.2. The Amended CCIR Equation#This amended model takes the following into account:

(a) Surface roughness
(b) LOS obstacles
(c) Buildings and trees

The resulting path loss equation is:

$$L_{dB} = 40 \log d - 20 \log(h_T h_R) + \beta \tag{17.2}$$

where β represents the additional losses listed above lumped together.

3.2.1.3 British Urban Path Loss Formula. The following formula was proposed by Allesbrook and Parsons [13]:

$$L_{dB} = 40 \log d_m - 20 \log(h_T h_R) + 20 + f/40 + 0.18L - 0.34H \tag{17.3}$$

where $f =$ frequency in MHz; $L =$ land usage factor, a percentage of the test area covered by buildings of any type (0–100%); and $H =$ terrain height difference between Tx and Rx (i.e., Tx terrain height − Rx terrain height).

Example: Let $d = 2000$ m $f = 900$ MHz
$h_T = 30$ m $L = 50\%$
$h_R = 3.3$ m   and $H = 27$ m

$$h_T h_R = 100 \text{ m}^2$$
$$L_{dB} = 132 - 40 + 20 + 22.5 + 5.4 - 9.18$$
$$= 130.76 \text{ dB}$$

3.2.1.4 The Okumura Model. Okumura et al. [14] carried out a detailed analysis for path predictions around Tokyo for mobile terminals. Hata [15] published an empirical formula based on Okumura's results to predict path loss:

$$L_{dB} = 69.55 + 26.16 \log f - 13.82 \log h_t - A(h_r)$$
$$+ (44.9 - 6.55 \log h_t) \log d \tag{17.4}$$

where f is between 150 and 1500 MHz, h_t is between 30 and 300 m, and d is between 1 and 20 km.

$A(h_r)$ is the correction factor for mobile antenna height and is computed as follows:

For a small or medium-sized city

$$A(h_r) = (1.1 \log f - 0.7)h_r - (1.56 \log f - 0.8) \quad \text{(dB)} \qquad (17.5a)$$

where h_r is between 1 and 10 m.

For a large city

$$A(h_r) = 3.2[\log(11.75h_r)]^2 - 4.97 \text{ dB}$$

$$f \geq 400 \text{ MHz} \qquad (17.5b)$$

Example: Let $f = 900$ MHz
$$h_t = 40 \text{ m}$$
$$h_r = 5 \text{ m}$$
$$d = 10 \text{ km}$$

Calculate $A(h_r)$ for a medium-sized city:

$$A(h_r) = 12.75 - 3.8 = 8.95 \text{ dB}$$
$$L_{dB} = 69.55 + 72.28 - 22.14 - 8.95 + 34.4$$
$$= 145.15 \text{ dB}$$

3.2.2 Building Penetration. For a modern multistory office block at 864 and 1728 MHz, path loss (L_{dB}) includes a value for clutter loss $L(v)$ and is expressed as follows

$$L_{dB} = L(v) + 20 \log d + n_f a_f + n_w a_w \qquad (17.6)$$

where the attenuation in dB of the floors and walls was a_f and a_w, and the number of floors and walls along the line d were n_f and n_w, respectively. The values of $L(v)$ at 864 and 1728 MHz were 32 and 38 dB, with standard deviations of 3 and 4 dB, respectively [1].

Another source [16] provided the following information. At 1650 MHz the floor loss factor was 14 dB, while the wall losses were 3–4 dB for double plasterboard and 7–9 dB for breeze block or brick. The parameter $L(v)$ was 29 dB. When the propagation frequency was 900 MHz, the first floor factor was 12 dB and $L(v)$ was 23 dB. The higher value for $L(v)$ at 1650 MHz was attributed to a reduced antenna aperture at this frequency compared to 900 MHz. For a 100-dB path loss, the base station and mobile terminal distance exceeded 70 m on the same floor, was 30 m for the floor above, and was 20 m for the floor above that, when the propagation frequency was 1650 MHz. The corresponding distances at 900 MHz were 70 m, 55 m, and 30 m. Results will vary from building to building depending on the type of construction of the building, the furniture and equipment it houses, and the number and deployment of the people who populate it.

3.3   Microcell Prediction Model According to Lee

For this section a microcell is defined as a cell with a 1-km or less radius. Such cells are used in heavily urbanized areas where demand for service is high and where large cell coverage would be spotty at best. With this model, LOS conditions are seldom encountered; shadowing is the general rule, as shown in Figure 17.4. The major contribution to loss in such situations is due to the dimensions of intervening buildings.

Lee's model [6] also includes an antenna height–gain function. Reference 6 reports a 9 dB/oct or 30 dB/dec for an antenna height change. This would mean that if we doubled the height of an antenna, 9-dB transmission loss improvement would be achieved.

The Lee model for a microcell breaks the prediction process down into a received signal level (dBm) for the LOS component (P_{LOS}) and then the attenuation due to the building blockage component.

Figure 17.5 provides a microcell scenario which we will use to understand Lee's model. The receive signal level P_r is equal to the receive signal level for LOS conditions (P_{LOS}) minus the blockage loss due to buildings, α_B. B is the blockage distance in feet. From Figure 17.5, $B = a + b + c$, the sum of the distances (in feet) through each building.

$$P_{LOS} = P_t - 77\,\text{dB} - 21.5\log(d/100') + 30\log(h_1/20') \qquad (17.7a)$$

$$\text{for } 100' \leq d < 200'$$

$$= P_t - 83.5\,\text{dB} - 14\log(d/200') + 30\log(h_1/20') \qquad (17.7b)$$

$$\text{for } 200' \leq d < 1000'$$

$$= P_t - 93.3\,\text{dB} - 36.5\log(d/1000') + 30\log(h_1/20') \qquad (17.7c)$$

$$\text{for } 1000' \leq d < 5000'$$

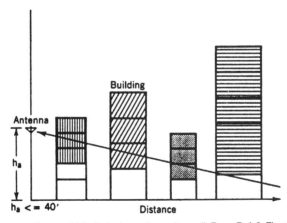

Figure 17.4. A propagation model typical of an urban microcell. From Ref. 6, Figure 2.24. Reprinted with permission.

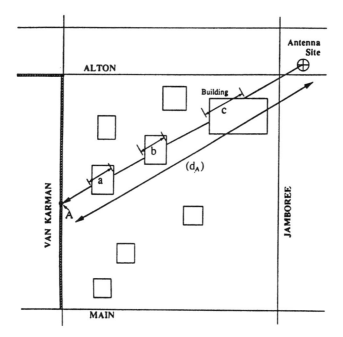

Figure 17.5. Path loss model for a typical microcell in an urban area. Mobile terminal is at location *A*. With antenna site so situated, there is blockage by buildings *a*, *b*, and *c*. Thus $B = a + b + c$. From Ref. 6, Figure 2.26. Reprinted with permission.

$$
\begin{aligned}
\alpha_B &= 0 & 1' \leq B & & (17.8a)\\
\alpha_B &= 1 + 0.5\log(B/10') & 1' \leq B < 25' & & (17.8b)\\
\alpha_B &= 1.2 + 12.5\log(B/25') & 25' \leq B < 600' & & (17.8c)\\
\alpha_B &= 17.95 + 3\log(B/600') & 600' \leq B < 3000' & & (17.8d)\\
\alpha_B &= 20 \text{ dB} & 3000' \leq B &
\end{aligned}
$$

where P_t is the ERP* in dBm, d is the total distance in feet, and h is the antenna height in feet.

Example 1: Mobile terminal is 500 ft from the cell site antenna, which is 30 ft high. There are three buildings in line between the mobile terminal and the cell site antenna with cross section (in line with the ray beam) distance of 50, 100, and 150 ft, respectively. Thus $B = 300$ ft. The ERP $= +30$ dBm (1 watt). First use equation 17.7b:

*ERP = effective radiated power (over a dipole).

$$P_{\text{LOS}} = +30 \text{ dBm} - 83.5 \text{ dB} - 14 \log(500'/200') + 30 \log(30'/20')$$
$$= +30 \text{ dBm} - 83.5 \text{ dB} - 5.57 \text{ dB} + 5.28 \text{ dB}$$
$$= -53.79 \text{ dBm}$$

Use equation 17.8c:

$$\alpha_B = 1.2 + 12.5 \log(300'/25')$$
$$= 1.2 \text{ dB} + 13.5 \text{ dB}$$
$$= 14.7 \text{ dB}$$

The receive signal level, P_r, is

$$P_r = -53.79 \text{ dBm} - 14.7 \text{ dB}$$
$$= -68.47 \text{ dBm}$$

Of course, we must assume that the sum of the gain of the receive antenna and the transmission line loss equals 0 dB so that P_r is the same as isotropic receive level.

Example 2: There are 4000 ft separating the cell site transmit antenna from the receive terminal. The transmit antenna is 40 ft high. There are four buildings causing blockage of 150 ft, 200 ft, 140 ft, and 280 ft. These distances are measured along the ray beam line. Thus $B = 770$ ft. The ERP is +20 dBm. What is the receive signal level (P_r) assuming no gain or loss in the receive antenna system? Use equation 17.7c for P_{LOS}:

$$P_{\text{LOS}} = +20 \text{ dBm} - 93.3 \text{ dB} - 36.5 \log(4000'/1000') + 30 \log(40'/20')$$
$$= -73.3 \text{ dBm} - 21.4 \text{ dB} + 9 \text{ dB}$$
$$= -85.67 \text{ dBm}$$

Now use equation 17.8d to calculate α_B:

$$\alpha_B = 17.95 + 3 \log(770'/600')$$
$$= 18.28 \text{ dB}$$
$$P_r = -85.67 \text{ dBm} - 18.28 \text{ dB}$$
$$= -103.94 \text{ dBm}$$

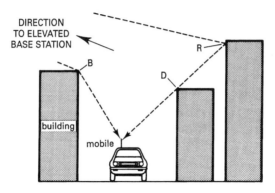

Figure 17.6. Mobile terminal in an urban setting. R, reflection; D, diffraction.

4 IMPAIRMENTS—FADING IN THE MOBILE ENVIRONMENT

4.1 Introduction

Fading in the mobile situation is quite different than in the static LOS microwave setting discussed in Chapter 7. Radio paths are not optimized as in the LOS environment. The mobile terminal may be fixed throughout a telephone call, but it is more apt to be in motion. Even the hand-held terminal may well have micromotion. When a terminal is in motion, the path characteristics are constantly changing.

Multipath propagation is the rule. Consider the simplified multipath pictorial model in Figure 17.6. Commonly multiple rays reach the receive antenna, each with its own delay. The constructive and destructive fading can become quite complex. We must deal with both reflection and diffraction. Energy will arrive at the receive antenna reflected off sides of buildings, streets, lakes, and so on. Energy will also arrive diffracted from knife edges (e.g., building corners) and rounded obstacles (e.g., water tanks, hilltops).

Because the same signal arrives over several paths, each with a different electrical length, the phases of each path will be different, resulting in constructive and destructive amplitude fading. Fades of 20 dB are common, and even 30-dB fades can be expected.

On digital systems, the deleterious effects of multipath fading can be even more severe. Consider a digital bit stream to a mobile terminal with a transmission rate of 1000 bps. Assuming NRZ coding, the bit period would be 1 ms (bit period = 1/bit rate). We find the typical multipath delay spread may be in the order of 10 μs. Thus delayed energy will spill into a subsequent bit (or symbol) for the first 10 μs of the bit period and will have no negative effect on the bit decision. If the bit stream is 64,000 bps, then the bit period is 1/64,000 or 15 μs. Destructive energy from the previous bit (symbol) will spill into the first two-thirds of the bit period, well beyond the mid-bit sampling point. This will cause typical intersymbol interference (ISI), and in this case there is a high probability that there will be a bit error. The bottom line is that the destructive potential of ISI increases as the bit rate increases (i.e., as the bit period decreases).

4.2   Classification of Fading

We consider three types of channels to place bounds on radio system performance. These are:

- Gaussian channel
- Rayleigh channel
- Rician channel

4.2.1   The Gaussian Channel. The Gaussian channel can be considered the ideal channel, and it is only impaired by "additive white Gaussian noise" (AWGN) developed internally by the receiver. We hope to achieve a BER typical of a Gaussian channel when we have done everything we can to mitigate fading and its results. Such measures could be diversity, equalization, FEC coding with interleaving, and so forth. The ideal Gaussian channel is very difficult to achieve in the mobile radio environment.

4.2.2   The Rayleigh Channel. The Rayleigh channel is at the other end of the line, often referred to as a worst-case channel. Remember, in Chapter 7, we treated fading on LOS microwave as Rayleigh fading and it gave us the very-worst-case fading scenario. Figure 17.7 shows a channel approaching Rayleigh fading characteristics. Of course, we are dealing with multipath here. We showed that in the mobile radio scenario, multipath reception commonly had many components. Thus if each multipath component is independent, the probability density function (PDF) of its envelope is Rayleigh.

4.2.3   The Rician Channel. The characteristics of a Rician channel are in between those of a Gaussian channel and those of a Rayleigh channel. The channels can be characterized by a function K (not to be confused with the K-factor in Chapter 7). K is defined as follows:

$$K = \text{power in the dominant path/power in the scattered paths} \qquad (17.9)$$

As cells get smaller, the LOS component becomes more and more dominant. There are many cases, in fact nearly all cases, where there is no full shadowing and where there is an LOS component and scattered components. This is a typical multipath scenario. Turning now to equation 17.9, when $K = 0$, the channel is Rayleigh (i.e., the numerator is 0 and all the received energy derives from scattered paths). When K is infinity, the channel is Gaussian and the denominator is zero. Figure 17.8 gives BER values for some typical

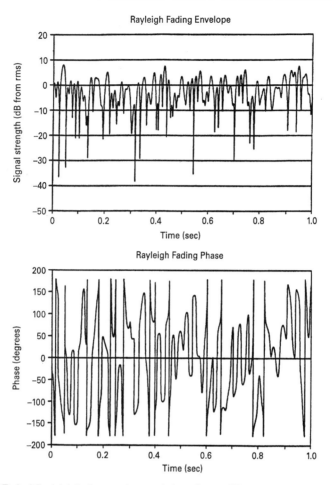

Figure 17.7. Typical Rayleigh fading envelope and phase in a mobile scenario. Vehicle speed about 30 mph; frequency 900 MHz. From Ref. 1, Figure 1.1.

values of K. It shows that those intermediate values of K provide a superior BER than for the Rayleigh channel where $K = 0$. For a microcell mobile scenario, values of K vary from 5 to 30 [1]. Larger cells tend more toward low values of K.

There is also an advantage for Rician fading with higher values of K regarding cochannel interference performance for a desired BER. The smaller the cell, the more fading becomes Rician, approaching the higher values of K.

4.3 Diversity—A Technique to Mitigate the Effects of Fading and Dispersion

4.3.1 Scope. We discuss diversity to reduce the effects of fading and to mitigate dispersion. Diversity was briefly covered in Chapter 7, where we

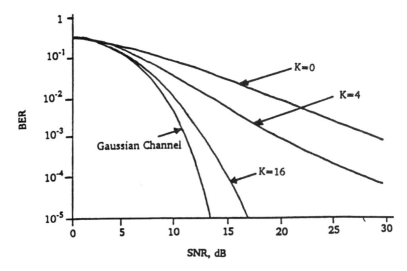

Figure 17.8. BER versus channel SNR for various values of K; noncoherent FSK. From Ref. 1, Figure 1.7.

dealt with LOS microwave radio systems. In that chapter we discussed frequency and space diversity. There is a third diversity scheme called *time diversity*, which can be applied to digital cellular systems.

In principal, such techniques can be employed at the base station and/or at the mobile unit, although different problems have to be solved for each. The basic concept behind diversity is that when two or more radio paths carrying the same information are relatively uncorrelated, when one path is in a fading condition, often the other path is not undergoing a fade. These separate paths can be developed by having two channels, separated in frequency. The two paths can be separated in space. Also, the two paths can be separated in time.

When the two (or more) paths are separated in frequency, we call this *frequency diversity*. However, there must be at least 2% or greater frequency separation for the paths to be comparatively uncorrelated. Because, in the cellular situation, we are so short of spectrum, using frequency diversity (i.e., using a separate frequency with redundant information) is essentially out of the question and will not be discussed further except for its implicit use in CDMA.

4.3.2 Space Diversity. Space diversity is commonly employed at cell sites, and two separate receive antennas are required, separated in either the horizontal or vertical plane. Separation of the two antennas vertically can be impractical for cellular receiving systems. Horizontal separation, however, is quite practical. The space diversity concept is illustrated in Figure 17.9.

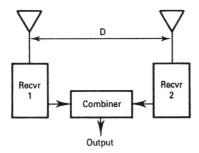

Figure 17.9. The space diversity concept.

One of the most important factors in space diversity design is antenna separation. There is a set of rules for the cell site, and there is another set of rules for the mobile unit.

Space diversity antenna separation, shown as distance D in Figure 17.9, varies not only as a function of the correlation coefficient but also as a function of antenna height, h. The wider the antennas are separated, the lower the correlation coefficient and the more uncorrelated the diversity paths are. Sometimes we find that by lowering the antennas as well as adjusting the distance between the antennas, we can achieve a very low correlation coefficient. However, we might lose some of the height–gain factor advantage.

Lee [6] proposes a new parameter η, where

$$\eta = \text{antenna height/antenna separation} = h/d \qquad (17.10)$$

In Figure 17.10 we relate the correlation coefficient (ρ) with η. α is the orientation of the antenna regarding the incoming signal from the mobile unit. Lee recomends a value of $\rho = 0.7$. Lower values are unnecessary because of the law of diminishing returns. There is much more fading advantage achieved from $\rho = 1.0$ to $\rho = 0.7$ than for $\rho = 0.7$ to $\rho = 0.1$.

Based on $\rho = 0.7$ and $\eta = 11$ from Figure 17.10, we can calculate antenna separation values (for 850-MHz operation). For example, if $h = 50$ ft (15 m), we can calculate d using formula 17.10:

$$d = h/\eta = 50/11 = 4.5 \text{ ft } (1.36 \text{ m})$$

For an antenna 120 ft (36.9 m) high, we find that $d = 120/11 = 10.9$ ft or 3.35 m.

4.3.2.1 Space Diversity on A Mobile Platform. Lee [6] discusses both vertically separated and horizontally separated antennas on a mobile unit. For the vertical and horizontal separation cases, 1.5λ and 0.5λ, respectively, are recommended. At 850 MHz, $\lambda = 35.29$ cm. Then $1.5\lambda = 1.36$ ft or 52.9 cm. For the 0.5λ, the value is 0.45 ft or 17.64 cm.

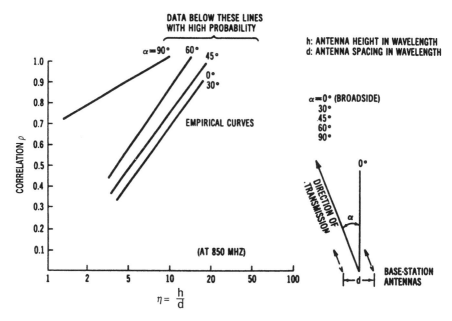

Figure 17.10. Correlation ρ versus the parameter η for two antennas in different orientations. From Ref. 6, reprinted with permission.

Section 4.3.2 is based on Ref. 6.

4.3.3 Frequency Diversity. We pointed out that conventional frequency diversity was not a practical alternative in cellular systems because of the shortage of available bandwidth. However, with CDMA (direct sequence spread spectrum), depending on the frequency spread, many frequency diversity paths are available, and in most CDMA systems we have what is called *implicit diversity*; and multipath can be resolved with the use of a RAKE filter. This is one of the many advantages of CDMA.

4.3.4 Forward Error Correction—A Form of Time Diversity. Forward error correction (FEC) can be used on digital cellular systems not only to improve bit error rate but to reduce the effects of fading. To reduce the effects of fading, an FEC system must incorporate an *interleaver*.

An interleaver pseudorandomly shuffles bits. It first stores a span of bits and shuffles them using a generating polynomial. The span of bits can represent a time period. The rule is that for effective operation against burst errors,* the interleaving span must be greater than the typical fade duration. The deinterleaver used at the receive end is time-synchronized to the interleaver incorporated at the transmit end of the link.

---

*Burst errors are bunched errors due to fading.

4.4 Cellular Radio Path Calculations

Consider the path from the fixed cell site to the mobile platform. There are several mobile receiver parameters that must be considered. The first we derive from EIA/TIA-IS-19 [19]. The minimum SINAD (signal + interference + noise and distortion to interference + noise + distortion ratio) is 12 dB. This SINAD equates to a threshold of −116 dBm or 7 μV/m. This assumes a cellular transceiver with an antenna with a net gain of 1 dBd (dB over a dipole). The gross antenna gain is 2.5 dBd with a 1.5-dB transmission line loss. A 1-dBd gain is equivalent to a 3.16-dBi gain (i.e., 0 dBd = 2.16 dBi). Furthermore, this value equates to an isotropic receive level of −119.16 dBm [19].

One design goal for a cellular system is to more or less maintain a cell boundary at the 39-dBμ contour [20]. 39 dBμ* = −95 dBm (based on a 50-Ω impedance at 850 MHz). Then at this contour, a mobile terminal would have a 24.16-dB fade margin.

If a cellular transmitter had a 10-watt output per channel and an antenna gain of 12 dBi and 2-dB line loss, the EIRP would be +20 dBW or +50 dBm. The maximum path loss to the 39-dBμ contour would be +50 dBm − (−119.16 dBm) or 169 dB.

5 THE CELLULAR RADIO BANDWIDTH DILEMMA

5.1 Background and Objectives

The present cellular radio bandwidth assignment in the 800- and 900-MHz band cannot support the demand for cellular service, especially in urban areas in the United States and Canada. AMPS, widely used in the United States, Canada, and many other nations of the Western Hemisphere, requires 30 kHz per voice channel. This sytem can be called an FDMA (frequency-division multiple access) system, much like the FDMA/DAMA system described in Chapter 7. We remember that the analog voice channel is a nominal 4 kHz, and 30 kHz is seven times that value.

The trend is to convert to digital. As we discussed in Chapters 8 and 9, digital is notorious for being wasteful of bandwidth, when compared to the 4-kHz analog channel. We showed that 8-bit PCM has a 16-times multiplier of the 4-kHz analog channel.

One goal of system designers, therefore, is to reduce the required bandwidth of the digital voice channel without sacrificing too much voice quality. As we will show, they have been quite successful.

The real objective is to increase the ratio of users to unit bandwidth. We will describe two distinctly different methods, each claiming to be more bandwidth conservative than the other. The first method is TDMA (time-division multiple access) and the second is CDMA (code-division multiple access). The former was described in Chapter 7 and the latter was mentioned.

---

*dBμ = dB referenced to 1 μv.

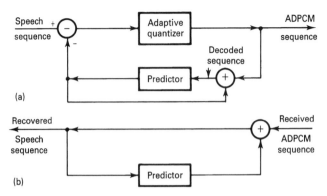

Figure 17.11. An ADPCM codec. (a) Coder, (b) decoder.

5.2 Bit Rate Reduction of the Digital Voice Channel

5.2.1 ADPCM. Adaptive differential PCM (ADPCM) provides nearly equal quality voice but at only 32 kbps. Key to the operation of ADPCM is the predictor, as shown in Figure 17.11. Predictions are based on the decoded sequence (identical to the sequence at the output of the receive decoder in the absence of transmission errors) that consists of the input speech sequence contaminated by quantization noise. With ADPCM the predictors have adaptive coefficients.

5.2.2 Vocoders. Vocoder operation is based on algorithms which attempt to describe the speech production mechanism in terms of a few independent parameters serving as the information-bearing signals. The concept behind vocoder design considers that speech is produced from a *source-filter* arrangement. It is modeled after the human generation of speech. Voiced speech* is the result of exciting the vocal tract (a *filter*) with a series of quasiperiodic glottal pulses generated by the vocal cords which is considered the *source*. As its name implies, a vocoder codes speech. It uses an analysis process based on a speech production model and extracts a set of source-filter parameters which are encoded and transmitted. At the distant-end decoder, the parameters are decoded and are used to control a speech synthesizer based on the speech production model at the transmit end. The synthesized signal at the receiver resembles the original speech signal.

There are two types of vocoders that operate in the frequency domain: the channel and the formant vocoders. The well-known LPC† vocoder is based on a time-domain algorithm. The U.S. Department of Defense uses 2400-bps LPC10 vocoders. They have good intelligibility, but voice quality is considered poor [5].

---

*Unvoiced speech is generated by exciting a filter patterned after the vocal tract with random white noise.
†LPC stands for linear predictive coder.

5.2.3 Sub-Band Coding (SBC).

In sub-band coding the speech signal is filtered into a relatively small number of sub-bands (2 to 16), and each sub-band signal is translated to zero frequency and sampled at the Nyquist rate and adaptively encoded. The number of bits used in the encoding process differs for each sub-band signal, with bits assigned to quantizers in accordance with perceptual criteria. By encoding each sub-band individually, the quantization noise* is confined within its sub-band. Sub-band encoders produce near toll quality speech at 16 kbps. At 9.6 kbps the speech quality is reduced as the "effective" bandwidth of the recovered signal is decreased. Communication quality speech can be achieved at the bit rates between 4.8 and 8 kbps by using sophisticated time-domain coding for the sub-band signals and high-frequency regeneration of "inactive" frequency bands [5].

5.2.4 RELP, a Hybrid Coder.

Many of the residual excited hybrid coding schemes use linear predictive modeling of the synthesis filter. One of the more well-known schemes is the residual excited linear predictive (RELP)-type coder. These coders operate at bit rates between 4.8 and 16 kbps. RELP coders use short-term (and sometimes long-term) linear prediction to derive a difference signal, called "residual," in a feed-forward manner. Macario [5] reports that earlier RELP systems used "baseband" coding and transmitted a low-pass version of the signal. The decoder recovered an approximation of the full band residual signal by employing *high-frequency regeneration* (HFR), which was then used to synthesize speech.

Simple HFR techniques (such as full-wave rectification, spectral folding, and spectral translation) generate considerable distortion in the restored speech signal.

The quality of the recovered speech can be considerably improved for rates below 9.6 kbps if, instead of the "feed-forward" RELP approach, an *analysis by synthesis* (AbS) optimization technique is used to define the excitation signal. This approach leads to the *AbS predictive coder*. In these systems both the "filter" and the "excitation" are defined on a short-term basis using a "closed-loop" optimization process which minimizes a perceptually weighted error measure formed between the input and the decoded output speech signals [5].

5.2.5 CELP Techniques.

One of the most popular voice-coding techniques is the codebook excitation linear predictive or CELP coder. These types of coders provide surprisingly good voice quality at low bit rates. Reference 1 states that the 4.8-kbps CELP coder gives "communication quality" speech, whereas the 8-kbps coder provided near-toll-quality speech.

Schroeder and Atal [17] describe CELP as coders that employ a vocal tract linear predictive (LP)-based model, a codebook-based excitation model, and an error criterion which serves to select an appropriate excitation sequence

---

*Quantization noise or quantization distortion is described in Chapter 8.

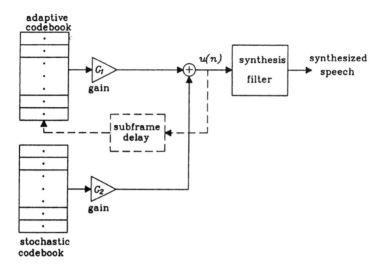

Figure 17.12. Schematic diagram of the CELP synthesis model. From Ref. 1, Figure 3.54, reprinted with permission.

using an AbS optimization process. As a result a CELP system selects that excitation sequence which minimizes a "perceptually" weighted mean square error formed between the input and the "locally" decoded signal.

Reference 5 reports that vocal tract modeling utilizes both a short-term filter, which models the spectral envelope of speech, and a long-term filter, which accounts for pitch periodicity in voiced speech.

Much work has been done on developing codebooks for CELP to reduce the excessive computational load needed to search for the optimum code sequence. These are structured codebooks where the codebook structure enables fast search procedures. There are sparse, ternary, overlapping, and algebraic codebooks.

Figure 17.12 is a schematic diagram of the CELP synthesis model, and some basic ideas on how CELP operates are given below.

After short-term prediction and long-term prediction of the speech signal, redundancies in the speech signal are removed and the residual signal has very little correlation. A Gaussian process with slowly varying power spectrum can be used to represent the residual signal, and the speech waveform is generated by filtering white Gaussian innovation sequences through the time-varying linear long-term and short-term synthesis filters. The optimum innovation sequence from a codebook of random white Gaussian sequence by minimizing the subjectively weighted error between the original and the synthesized speech. An adaptive overlapping codebook carries out the pitch correlation filter function. The address selected from the adaptive codebook and the corresponding gain (i.e., the pitch delay and gain) along with the address selected from the stochastic codebook and the corresponding scaling gain are transmitted to the decoder. The decoder uses the same codebooks to determine the excitation signal at the input of the LPC synthesis filter (assuming no channel errors) to produce the synthesized speech.

The excitation codebook contains L codewords (stochastic vectors) of length N samples. Typically $L = 1024$ and $N = 40$, corresponding to a 5-ms excitation frame. The excitation signal of a speech frame of length N is selected by the exhaustive search of the codebook after scaling the Gaussian vectors by a gain factor β.

Section 5.2.5 is based on Ref. 1.

6 NETWORK ACCESS TECHNIQUES

6.1 Introduction

The objective of a cellular radio operation is to provide a service where mobile subscribers can communicate with any subscriber in the PSTN, where any subscriber in the PSTN can communicate with any mobile subscriber, and where mobile subscribers can communicate amongst themselves via the cellular radio system. In all cases the service is full duplex.

A cellular service company is allotted a radio bandwidth segment to provide this service. Ideally, for full-duplex service, a portion of the bandwidth is assigned for transmission from a cell site to mobile subscriber, and another portion is assigned for mobile user to cell site direction. Our goal here is to select an "access" method to provide this service given our bandwidth constraints.

We will discuss three generic methods of access: frequency-division multiple access (FDMA), time-division multiple access (TDMA), and code-division multiple access (CDMA). It might be useful to the reader to review our discussion of satellite access in Chapter 7 where we describe FDMA and TDMA. However, in this section the concepts are the same, but some of our constraints and operating parameters will be different.

6.2 Frequency-Division Multiple Access (FDMA)

With FDMA, our band of frequencies is divided into segments and each segment is available for one user access at a time. Half of contiguous segments are assigned cell site outbound to mobile users and the other half inbound. A guard band is usually provided between outbound and inbound contiguous channels. This concept is shown in Figure 17.13.

Because of our concern to optimize the number of users per unit bandwidth, the key question is the actual width of one user segment. The North American AMPS system was described in Section 2, where each segment

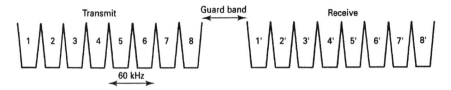

Figure 17.13. A conceptual drawing of FDMA.

width was 30 kHz. The bandwidth of a user segment is greatly determined by the information bandwidth and modulation type. With AMPS the information bandwidth was a single voice channel with a nominal bandwidth of 4 kHz. The modulation was FM and the bandwidth was then determined by Carson's rule (Chapter 7). As we pointed out, AMPS is not exactly spectrum-conservative. On the other hand, it has a lot of the redeeming features that FM provides, such as the noise and interference advantage (FM capture).

Another approach to FDMA would be to convert the voice channel to its digital equivalent using CELP, for example (Section 5.2.5) with a transmission rate of 4.8 kbps. The modulation could be BPSK using a raised cosine filter where the bandwidth could be 1.25% of the bit rate or 6 kHz per voice channel. This alone would increase voice channel capacity five times over AMPS with its 30 kHz per channel. It should be noted that a radio carrier is required for each time slot.

6.3 Time-Division Multiple Access (TDMA)

With TDMA we work in the time domain rather than the frequency domain of FDMA. Each user is assigned a time slot rather than a frequency segment and during the user's turn, the full frequency bandwidth is available for the duration of the user's assigned time slot.

Let's say that there are n users and so the are n time slots. In the case of FDMA, we had n frequency segments and n radio carriers, one for each segment. For the TDMA case, only one carrier is required. Each user gains access to the carrier for $1/n$ of the time and there is generally an ordered sequence of time-slot turns. A TDMA frame can be defined as cycling through n users' turns just once.

A typical TDMA frame structure is shown in Figure 17.14. One must realize that TDMA is only practical with a digital system such as PCM or any of those discussed in Section 5. TDMA is a store and burst system. Incoming user traffic is stored in memory, and when that user's turn comes up, that accumulated traffic is transmitted in a digital burst.

Suppose there are 10 users. Let each user's bit rate be R, then a user's burst must be at least $10R$. Of course, the burst will be greater than $10R$ to accommodate a certain amount of overhead bits as shown in Figure 17.14.

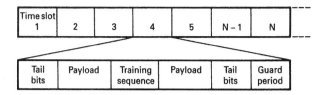

Figure 17.14. A simplified TDMA frame with blow-up of a typical time slot. Tail bits assist in data recovery and act as time-slot marker. Training sequence assists in synchronization and bit timing recovery. Note similarity to GSM frame.

We define *downlink* as outbound from the base station to mobile station(s), and we define *uplink* as a mobile station to base station. Typical frame periods are:

North American IS-54:   40 ms for six time slots
European GSM: 4.615 ms for eight time slots

One problem with TDMA, often not appreciated by the novice, is *delay*. In particular, this is the delay on the uplink. Consider Figure 17.15, where we set up a scenario. A base station receives mobile time slots in a circular pattern and the radius of the circle of responsibility of that base station is 10 km. Let the velocity of a radio wave be 3×10^8 m/s. The time to traverse 1 km is $1000/3 \times 10^8$ or 3.333 µs. Making up an uplink frame is a mobile right on top of the base station with essentially no delay and another mobile right at 10 km with 10×3.33-µs or 33.3-µs delay. A GSM time slot is about 576 µs in duration. The terminal at the 10-km range will have its time slot arriving 33.3 µs late compared to the terminal with no delay. A GSM bit period is about 3.69 µs so that the late arrival eats up roughly 10 bits, and unless something is done, the last bit of the burst will overlap the next burst.

Refer now to Figure 17.16, which illustrates GSM burst structures. Note that the access burst has a guard period of 68.25-bit durations or a "slop" of

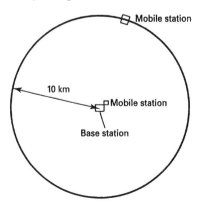

Figure 17.15. A TDMA delay scenario.

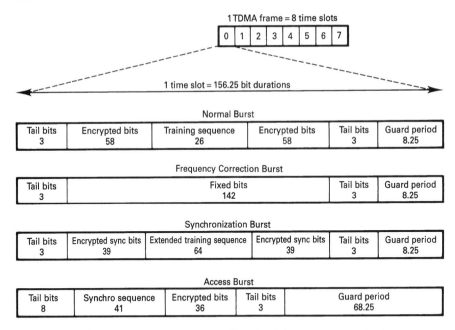

Figure 17.16. GSM frame and burst structures. From Ref. 1, Figure 8.7, reprinted with permission.

3.69 × 68.25 μs which will well accommodate the late arrival of the 10-km mobile terminal of only 33.3 μs.

To provide the same long guard period in the other bursts is a waste of valuable "spectrum."* The GSM system overcomes this problem by using adaptive frame alignment. When the base station detects a 41-bit random access synchronization sequence with a long guard period, it measures the received signal delay relative to the expected signal from a mobile station with zero range. This delay, called the *timing advance*, is transmitted to the mobile station using a 6-bit number. As a result, the mobile station advances its time base over the range of 0 to 63 bits (i.e., in units of 3.69 μs). By this process the TDMA bursts arrive at the base station in their correct time slots and do not overlap with adjacent ones. As a result, the guard period in all other bursts (Figure 17.16) are reduced to 8.25 × 3.69 μs or approximately 30.46 μs, the equivalent of 8.25 bits only. Under normal operations, the base station continuously monitors the signal delay from the mobile station and thus instructs the mobile station to update its time advance parameter. In very large traffic cells there is an option to actively utilize every second time slot only to cope with the larger propagation delays. This is spectrally inefficient, but in large, low-traffic rural cells it is admissible.

Section 6.3 is based on Ref. 1.

*We are equating bit rate, or bit durations, to bandwidth. One could assume 1 bit per hertz as a first-order estimate.

6.3.1 *Some Comments on TDMA Efficiency.* Multichannel FDMA can operate with a power amplifier for every channel, or with a common wideband power amplifier for all channels. With the latter, we are setting up a typical generator of intermodulation products as these carriers mix in a comparatively nonlinear common power amplifier. To reduce the level of IM products, just like in satellite communications discussed in Chapter 7, backoff of power amplifier drive is required. This backoff can be in the order of 3 to 6 dB.

With TDMA (downlink), only one carrier is present on the power amplifier at any moment in time, thus removing most of the causes of IM noise generation. Thus with TDMA, the power amplifier can be operated to full saturation, a distinct advantage. FDMA required some guard bands between frequency segments; there are no guard bands with TDMA. However, as we saw above, a guard time between uplink time slots is required to accommodate the following situations:

- Timing inaccuracies due to clock instabilities
- Delay spread due to propagation*
- Transmission time delay due to propagation distance (see Section 6.3 above)
- Tails of pulsed signals due to transient response

The longer guard times are extended, the more inefficient a TDMA system is.

6.3.2 *Advantages of TDMA.* The introduction of TDMA results in a much improved system signaling operation and cost. Assuming a 25-MHz bandwidth, up to 21.3-times capacity can be achieved with North American TDMA compared to FDMA (AMPS) (see Table II in Ref. 24).

A mobile station can exchange system control signals with the base station without interruption of speech (or data) transmission. This facilitates the introduction of new network and user services. The mobile station can also check the signal level from nearby cells by momentarily switching to a new time slot and radio channel. This enables the mobile station to assist with handover operations and thereby improve the continuity of service in response to motion or signal fading conditions. The availability of signal strength information at both the base and mobile stations, together with suitable algorithms in the station controllers, allows further spectrum efficiency through the use of dynamic channel assignment and power control.

The cost of base stations using TDMA can be reduced by sharing radio equipment with several traffic channels. A reduced number of transceivers leads to a reduction of multiplexer complexity. Outside the major metropolitan areas, the required traffic capacity for a base station may, in many

*Lee [6] reports that a typical urban delay spread is 3 μs.

cases, be served by one or two transceivers. The saving in the number of transceivers results in a significantly reduced overall cost.

A further advantage of TDMA is increased system flexibility. Different voice and nonvoice services may be assigned a number of time slots appropriate to the service. For example, as more efficient speech CODECs are perfected, increased capacity may be achieved by the assignment of a reduced number of time slots for voice traffic. TDMA also facilitates the introduction of digital data and signaling services as well as the possible later introduction of such further capacity improvements as digital speech interpolation (DSI).

Table 17.1 compares threee operational/planned digital TDMA systems.

6.4 Code-Division Multiple Access (CDMA)

CDMA means spread-spectrum multiple access. There are two types of spread spectrum: frequency hop and direct sequence (sometimes called pseudo-noise). In the cellular environment, CDMA means direct sequence spread spectrum [1]. However, the GSM system uses frequency hop, but not as an access technique.

Using spread-spectrum techniques accomplishes just the opposite of what we were trying to accomplish in Chapter 7. Bit packing is used to conserve bandwidth by packing as many bits as possible in 1 Hz of bandwidth. With spread spectrum we do the reverse by spreading the information signal over a very wide bandwidth.

Conventional AM requires about twice the bandwidth of the audio information signal with its two sidebands of information (i.e., $\approx \pm 4$ kHz). On the other hand, depending on its modulation index, frequency modulation could be considered a type of spread spectrum in that it produces a much wider bandwidth than its transmitted information. As with all other spread-spectrum systems, a signal-to-noise advantage is gained with FM, depending on its modulation index. For example, with AMPS, a typical FM system, 30 kHz is required to transmit the nominal 4-kHz voice channel.

If we are spreading a voice channel over a very wide frequency band, it would seem that we are defeating the purpose of frequency conservation. With spread spectrum, with its powerful antijam properties, multiple users can transmit on the same frequency with only some minimal interference one to another. This assumes that each user is employing a different key variable (i.e., in essence, using a different code). At the receiver, the CDMA signals are separated using a correlator that accepts only signal energy from the selected key variable binary sequence (code) used at the transmitter, and then despreads its spectrum. CDMA signals with unmatching codes are not despread and only contribute to the random noise.

CDMA reportedly provides an increase in capacity 15 times that of its analog FM counterpart. It can handle any digital format at the specified input bit rate such as facsimile, data, and paging. In addition, the amount of

TABLE 17.1   Three TDMA Systems Compared

Feature	GSM	North America	Japan

Class of emission
  Traffic channels
  Control channels
	
	271KF7W
	271KF7W
		
		40K0G7WDT
		40K0G1D
			
			tbd[a]
			tbd

Transmit frequency bands (MHz)
  Base stations
 	
  Mobile stations
	
	935–960
	
	890–915
		
		869–894
		
		824–849
			
			810–830
			(1.5 GHz tbd)
			940–960
			(1.5 GHz tbd)

Duplex separation (MHz)	45	45	130 48 (1.5 GHz)
RF carrier spacing (kHz)	200	30	25 interleaved 50
Total number of RF duplex channels	124	832	tbd

Maximum base station erp (W)
  Peak RF carrier
  Traffic channel average
	
	300
	37.5
		
		300
		100
			
			tbd
			tbd

Nominal mobile station transmit power (W): peak and average	20 and 2.5 8 and 1.0 5 and 0.625 2 and 0.25	9 and 3 4.8 and 1.6 1.8 and 0.6 tbd and tbd	tbd

Cell radius (km)
  Min
  Max
	
	0.5
	35
	(up to 120)
		
		0.5
		20
			
			0.5
			20

Access method
  Traffic channels/RF carrier
    Initial
    Design capability
	TDMA
	
	8
	16
		TDMA
		
		3
		6
			TDMA
			
			3
			6

Channel coding	Rate one-half convolutional code with inter- leaving plus error detection	Rate one-half convolutional code	tbd

Control channel structure
  Common control channel
 	
  Associated control channel
  Broadcast control channel
	
	Yes
	
	Fast and slow
	Yes
		
		Shared with
		AMPS
		Fast and slow
		Yes
			
			Yes
			
			Fast and slow
			Yes

(continued)

TABLE 17.1 *continued*

Feature	GSM	North America	Japan
Delay spread equalization capability (µs)	20	60	tbd
Modulation	GMSK[b] (BT = 0.3)	π/4 diff.[c] encoded QPSK (roll off = 0.25)	π/4 diff. encoded QPSK (roll off = 0.5)
Transmission rate (kbit/s)	270.833	48.6	37–42
Traffic channel structure Full-rate speech codec Bit rate (kbit/s)	13.0	8	6.5–9.6
Error protection	9.8 kbit/s FEC +speech processing	5 kbit/s FEC	~3 kbit/s FEC
Coding algorithm Half-rate speech codec	RPE-LTP	CELP	tbd
Initial	tbd	tbd	tbd
Future	Yes	Yes	Yes
Data Initial net rate (kbit/s)	Up to 9.6	2.4, 4.8, 9.6	1.2, 2.4, 4.8
Other rates (kbit/s)	Up to 12	tbd	8 and higher
Handover Mobile assisted	Yes	Yes	Yes
Intersystem capability with existing analog system	No	Between digital and AMPS	No
International roaming capability	Yes >16 countries	Yes	Yes
Design capability for multiple system operators in same area	Yes	Yes	Yes

[a]tbd, to be defined.
[b]GMSK, Gaussian minimum shift keying.
[c]diff., differentially.
Source: CCIR Rep. 1156, Table 1, pages 120–123 (Ref. 24).

transmitter power required to overcome interference is comparatively low when utilizing CDMA. This translates into savings on infrastructure (cell site) equipment, in addition to longer battery life for hand-held terminals. CDMA also provides so-called soft hand-offs from cell site to cell site that make the transition virtually inaudible to the user [25].

Dixon [18] develops from Claude Shannon's classical relationship an interesting formula to calculate the spread bandwidth given the information rate, the signal power, and the noise power:

$$C = W \log_2(1 + S/N) \tag{17.11}$$

where

$$C = \text{capacity of a channel in bits per second}$$
$$W = \text{bandwidth in hertz}$$
$$N = \text{noise power}$$
$$S = \text{signal power}$$

This equation shows the relationship between the ability of a channel to transfer error-free information, compared with the signal-to-noise ratio existing in the channel, and the bandwidth used to transmit the information.

If we let C be the desired system information rate and change the logarithm base to the natural base (e), the result is:

$$C/W = 1.44 \log_e(1 + S/N) \qquad (17.12)$$

and, for an S/N that is very small (e.g., ≤ 0.1) which would be used in an antijam system,* we can say:

$$C/W = 1.44[S/N] \qquad (17.13)$$

From this equation we find that

$$N/S = 1.44W/C \approx W/C \qquad (17.14)$$

and

$$W \approx NC/S \qquad (17.15)$$

This exercise shows that for any given noise-to-signal (N/S) ratio we can have a low information-error rate by increasing the bandwidth used to transfer that information.

If we had a cellular system using a data rate of 4.8 kbps and an N/S of 20 dB (numeric of 100), then the bandwidth for this 4.8-kbps channel would be

$$W = 100 \times 4.8 \times 10^3/1.44$$
$$= 333.333 \text{ kHz}$$

There are two common ways that information can be embedded in the spread-spectrum signal. One way is to add the information to the spectrum-spreading code before the spreading modulation stage. It is assumed that the information to be transmitted is binary because modulo-2 addition is involved in this process. The second method is to modulate the RF carrier with the desired information before spreading the carrier. The modulation is usually PSK or FSK or some other angle modulation scheme [18].

---

*If we are involved in a cellular scenario where one user is transmitting right on the top of others on the same frequency plus adjacent channel interference, we are indeed in an antijam situation.

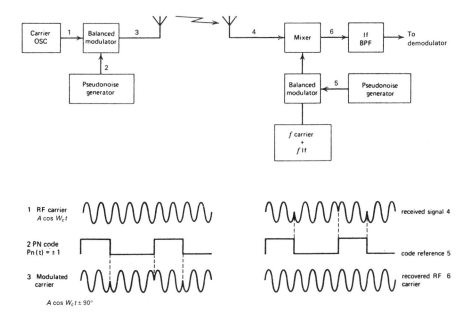

Figure 17.17. A direct sequence spread-spectrum system showing waveforms. From Ref. 18, Figure 2.3, reprinted with permission.

Dixon [18] lists some advantages of spread spectrum:

1. Selective addressing capacity
2. Code-division multiplexing is possible for multiple access
3. Low-density power spectra for signal hiding
4. Message security
5. Interference rejection

Of most importance for the cellular user is the following [18]:

> When codes are properly chosen for low cross correlation, minimum interference occurs between users, and receivers set to use different codes are reached only by transmitters sending the correct code. Thus more than one signal can be unambiguously transmitted at the same frequency and at the same time; selective addressing and code-division multiplexing are implemented by the coded modulation format.

Figure 17.17 shows a direct sequence (pseudo-noise) spread-spectrum system with waveforms.

Processing gain is probably the most commonly used parameter to describe the performance of a spread-spectrum system. It quantifies the signal-to-noise ratio improvement when a spread signal is passed through a "processor." For instance, if a certain processor had an input S/N of 12 dB and an output S/N of 20 dB, the processing gain, then, is 8 dB.

Processing gain is expressed by the following:

$$G_p = \text{spread bandwidth in Hz/information bit rate} \qquad (17.16)$$

More commonly, processing gain is given in a dB value, then

$$G_{p(\text{dB})} = 10\log[\text{spread bandwidth in Hz/information bit rate}] \qquad (17.17)$$

Example: If a certain cellular system voice channel information rate is 9.6 kbps and the RF spread bandwidth is 9.6 MHz, what is the processing gain?

$$\begin{aligned} G_{p(\text{dB})} &= 10\log 9.6 \times 10^6 - 10\log 9600 \\ &= 69.8 - 39.8(\text{dB}) \\ &= 30 \text{ dB} \end{aligned}$$

It has been pointed out in Ref. 1 that the power control problem held back the implementation of CDMA for cellular application. If the standard deviation of the received power from each mobile at the base station is not controlled to an accuracy of approximately ± 1 dB relative to the target receive power, the number of users supported by the system can be significantly reduced. Other problems to be overcome were synchronization and sufficient codes available for a large number of mobile users [1].

Qualcomm, a North American company, has a CDMA design that overcomes these problems and has fielded a cellular system based on CDMA. It operates at the top of the AMPS band using 1.23 MHz for each uplink and downlink. This is the equivalent of 41 AMPS channels (i.e., 30 kHz $\times 41 = 1.23$ MHz) deriving up to 62 CDMA channels (plus one pilot channel and one synchronization channel) or some 50% capacity increase. The Qualcomm system also operates in the 1.7- to 1.8-GHz band [1].

7 FREQUENCY REUSE

Because of the limited bandwidth allocated in the 800-MHz band for cellular radio communications, frequency reuse is crucial for its successful operation. A certain level of interference has to be tolerated. A major source of interference is co-channel interference from a "nearby" cell using the same frequency group as the cell of interest. For the 30-kHz bandwidth AMPS system, Ref. 5 suggests that C/I be at least 18 dB. The primary isolation derives from the distance between the two cells with the same frequency group. In Figure 17.2 there is only one cell diameter for protection.

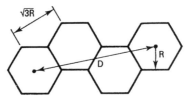

Figure 17.18. Definition of R and D.

Refer to Figure 17.18 for the definition of the parameters R and D. D is the distance between cell centers of repeating frequency groups and R is the "radius" of a cell. We let

$$a = D/R$$

The D/R ratio is a basic frequency reuse planning parameter. If we keep the D/R ratio large enough, co-channel interference can be kept to an acceptable level.

Lee [6] calls a the co-channel reduction factor and relates path loss from the interference source to R^{-4}.

A typical cell in question has six co-channel interferers, one on each side of the hexagon. So there are six equidistant co-channel interference sources. The goal is $C/I \geq 18$ dB or a numeric of 63.1. So

$$C/I = C/\sum I = C/6I = R^{-4}/6D^{-4} = a^4/6 \geq 63.1$$

Then $a = 4.4$. This means that D must be 4.4 times the value of R. If R is 6 mi (9.6 km), then $D = 4.4 \times 6 = 26.4$ mi (42.25 km).

Lee [6] reports that co-channel interference can be reduced by other means such as directional antennas, tilted beam antennas, lowered antenna height, and an appropriately selected site.

If we consider a 26.4-mi path, what is the height of earth curvature at mid-path? From Chapter 7, $h = 0.667(d/2)^2/1.33 = 87.3$ ft (26.9 m). Provided that the cellular base station antennas are kept under 87 ft, the 40-dB/decade rule of Lee holds. Of course, we are trying to keep below line-of-sight conditions.

The total available (one-way) bandwidth is split up into N sets of channel groups. The channels are then allocated to cells, one channel set per cell on a regular pattern, which repeats to fill the number of cells required. As N increases, the distance between channel sets (D) increases, reducing the level of interference. As the number of channel sets (N) increases, the number of channels per cell decreases, reducing the system capacity. Selecting the optimum number of channel sets is a compromise between capacity and quality. Note that only certain values of N lead to regular repeat patterns without gaps. These are $N = 3, 4, 7, 9,$ and 12, and then multiples thereof.

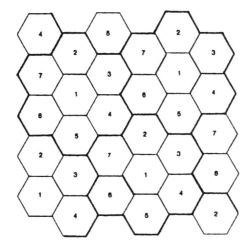

Figure 17.19. A cell layout based on $N = 7$.

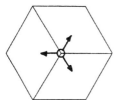 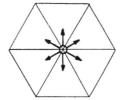

Figure 17.20. Breaking a cell up into three sectors (left) and six sectors (right).

Figure 17.19 shows a repeating 7 pattern for frequency reuse. This means that $N = 7$ or there are 7 different frequency sets for cell assignment.

Cell splitting can take place especially in urban areas in some point in time because the present cell structure cannot support the busy hour traffic load. Cell splitting, in effect, provides more frequency slots for a given area. Macario [5] reports that cells can be split as far down as a 1-km radius.

Co-channel interference tends to increase with cell splitting. Cell sectorization can cut down the interference level. Figure 17.20 shows a three- and six-sector plan. Sectorization breaks a cell into three or six parts, each with a directional antenna. With a standard cell, co-channel interference enters from six directions. A six-sector plan can essentially reduce the interference to just one direction. A separate channel set is allocated to each sector.

The three-sector plan is often used with a seven-cell repeating pattern, resulting in an overall requirement for 21 channel sets. The six-sector plan with its improved co-channel performance and the rejection of secondary interferers allows a four-cell repeat plan (Figure 17.2) to be employed. This results in an overall 24-channel set requirement. Sectorization requires a larger number of channel sets and fewer channels per sector. Outwardly it appears that there is less capacity with this approach; however, the ability to use much smaller cells results in actually a much higher capacity operation [5].

8 PAGING SYSTEMS

8.1 What Are Paging Systems?

Paging is a one-way radio alerting system. The direction of transmission is from a fixed paging transmitter to an individual. It is a simple extension of the PSTN. Certainly paging can be classified as one of the first PCS (personal communication system) operations. The paging receiver is a small box, usually carried on a person's belt. As a minimum, a pager alerts the user that someone wishes to reach him/her by telephone. The person so alerted goes to the nearest telephone and calls a prescribed number. Some pagers have a digital readout which provides the calling number, whereas others give the number and a short message. Most paging systems now operate in the VHF and UHF bands with a 3-kHz bandwidth. Transmitters have 1 to 5 watts of output, and paging receivers have sensitivities in the range of 10 to 100 μV/m [5].

8.2 Radio-Frequency Bands For Pagers

All three ITU regions have some or all of the following frequency bands allocated to mobile services:

26.1–50 MHz	68–88 MHz
146–174 MHz	450–470 MHz
806–960 MHz	

8.3 Radio Propagation into Buildings

Measurement results submitted to the CCIR [22] have indicated that frequencies in the range of 80–460 MHz are suitable for personal radio paging in urban areas with high building densities. It is possible that frequencies in the bands allocated around 900 MHz may also be suitable but that higher frequencies are less suitable. From measurements made in Japan, the following median values of propagation loss suffered by signals in the penetration of buildings (building penetration loss) have been derived. These results are summarized in Table 17.2.

TABLE 17.2   Propagation Loss Suffered by Signals in Penetrating Buildings

Frequency:	150 MHZ	250 MHz	400 MHz	800 MHz
Building penetration loss[a]:	22 dB	18 dB	18 dB	17 dB[b]

[a] The loss is given as the ratio between the median value of the field strengths measured over the lower floors of buildings and the median value of the field strengths measured on the street outside. Similar measurements made in other countries confirm the general trend, but the values of building penetration loss vary about those shown. For instance, measurements made in the United Kingdom indicate that building penetration loss at 160 MHz is about 14 dB and about 12 dB at 460 MHz.

[b] Somewhat less accurate than the other results.

Source: CCIR Rep. 499-5, Table 1, page 59 (Ref. 23).

8.4   Techniques Available for Multiple Transmitter Zones

To cover a service area effectively, it is often necessary to use a number of radio-paging transmitters. When the required coverage area is small, a single RF channel should be used so as to avoid the need for multichannel receivers. In these circumstances, the separate transmitters may operate sequentially or simultaneously. In the latter case, the technique of off-setting carrier frequencies, by an amount appropriate to the coding system employed, is often used. It is also necessary to compensate for the differences in the delay to the modulating signals from the characteristics of the individual landlines to the paging transmitters. One way to do this is to carry out synchronization of the code bits via the radio-paging channel. Of course, information will be required about the bit rates which this synchronization method would permit. It is preferable that the frequency offset of the transmitter carrier frequencies in a binary digital radio-paging system be at least twice the signal fundamental frequency. It is also preferable that delay differences between modulation of the transmitters in a binary digital paging system should be less than a quarter of the duration of a bit if direct FSK, NRZ modulation is employed. For subcarrier systems, the corresponding limit should be less than one-eighth of a cycle of the subcarrier frequency.

8.5   Paging Receivers

Built-in antennas can be designed for 150-MHz operation with reasonable efficiency. A typical radio-paging receiver antenna using a small ferrite rod exhibits a loss factor of about 16 dB relative to a half-wave dipole.

The majority of wide-area paging systems use some form of angle modulation.

Repeated transmission of calls can be used to improve the paging success rate of tone alert pagers. If p is the probability of receiving a single call, then $1 - (1 - p)^n$ is the probability of receiving a call transmitted n times, provided that the calls are uncorrelated. Correlations under Rayleigh fading conditions can be largely removed by spacing the calls more than 1 s apart. Longer delays between subsequent transmission (about 20 s) are required to improve the success rate under shadowing conditions.

Receivers with numeric or alphanumeric message displays can only take advantage of call repetitions if the supplementary messages are used to detect and correct errors.

8.6 System Capacity

The capacity of any paging system is affected by the following:

- Number and characteristics of the radio channels used
- The number of times each channel is reused within the system
- The actual paging location requirements of the individual users
- The peak information (address and message) requirement in a location(s)
- Tolerable paging delay
- Data transmission rate
- Code efficiency
- Method of using the total code capacity throughout the system (this may also affect the system's capabilities for roaming)
- Any inefficiency introduced by battery-saving provisions
- Possible telephone system input restrictions

8.7 Codes and Formats for Paging Systems

The US paging system broadly used across the country is popularly called a *Golay Sequential Code*, referring to the Golay (23, 12) cyclic code with two codewords representing the address. Messages are coded using a BCH (15, 7) code. The code and format provide queueing and numeric/alphanumeric messages flexibility and the ability to operate in a mixed-mode transmission with other formats. The single address capacity of this system is up to 400,000 with noncoded battery saving and up to 4,000,000 with coded preamble.

Japan uses a BCH (31, 16) codeword with a Hamming distance of 7. The format gives approximately 65,000 addresses, 15 groups for battery economy, and a total cycle length of 4185 bits. Each group contains 8 address codewords headed by a 31-bit synchronizing and group-indicating signal.

The UK paging system employs a BCH (31, 21) code plus even parity codeword with a Hamming distance of 6. The code format can handle over 8 million addresses and can be expanded. It can also handle any type of data message such as hexadecimal and CCITT alphabet No. 5. It is designed to share a channel with other codes and to permit mixed simultaneous and sequential multitransmitter operation at the normal 512-bps transmission rate. This code is sometimes referred to as POCSAG and has been adopted as CCIR Radio-Paging Code No. 1 (RPC1).

8.8 Considerations for Selecting Codes and Formats

- Number of subscribers to be served
- Number of addresses assigned to each subscriber
- The calling rate expected including that from any included message facility
- Zoning arrangement
- The data transmission rates possible over the linking network and radio channel(s), taking into account the propagation factors of the radio frequencies to be used
- The type of service: vehicular or personal, urban or rural

Once the data are provided from the above listing of topics, codes may be compared by their characteristics with respect to:

- Code address capacity
- Number of bits per address
- Codeword Hamming distance
- Code efficiency, such as number of information bits compared to the total number of bits per codeword
- Error-detecting capability; error-correcting capability
- Message capability and length
- Battery-saving capability
- Ability to share a channel with other codes
- The capability of meeting the needs of paging systems which vary with respect to size and transmission mode (e.g., simultaneous versus sequential)

Sections 8.2 through 8.8 are based on CCIR Reps. 900-2 and 499-5 (Refs. 22 and 23, respectively).

9 PERSONAL COMMUNICATION SYSTEMS

9.1 Defining Personal Communication

A personal communication system (PCS) is wireless. This simply means that it is radio-based. The user requires no *tether*. The conventional telephone is connected by a wire pair through to the local serving switch. The wire pair is a tether. We can only walk as far with that telephone handset as the "tether" allows.

Both of the systems we have dealt with in the previous sections of this chapter can be classified as PCS. Cellular radio, particularly with the hand-held terminal, gives the user tetherless telephone communication. Paging systems provided the mobile/ambulatory user a means of being alerted that someone wishes to talk to that person on the telephone or can leave that person a short message.

The cordless telephone is certainly another example, which has extremely wide use around the world. By the end of 1994, it was estimated that there were 50 million cordless telephones in use in the United States [26].

New applications are either on the horizon or going through field tests (1995). One that seems to offer great promise in the office environment is the wireless PABX. It will almost eliminate the telecommunication manager's responsibilities with office rearrangements. Another is the wireless LAN (WLAN).

Developments are expected such that PCS can provide not only voice communications, but also facsimile, data, messaging, and possibly video. GSM provides all but video. Cellular digital packet data (CDPD) will permit data services over the cellular system in North America.

Don Cox [26] breaks PCS down into what he calls "high tier" and "low tier." Cellular radio systems are regarded as high-tier PCS, particularly when implemented in the new 1.9-GHz PCS frequency band. Cordless telephones are classified as low tier.

Table 17.3 summarizes some of the more prevalent wireless technologies.

9.2   Narrow-Band Microcell Propagation at PCS Distances

The microcells discussed here have a radial range of < 1 km. One phenomenon is the Fresnel break point which is illustrated in Figure 17.21. Signal level varies with the distance R as A/R^n. For distances greater than 1 km, n is typically between 3.5 and 4. The parameter A describes the effects of environmental features in a highly averaged manner [11].

Typical PCS radio paths can be of an LOS nature, particularly near the fixed transmitter where $n = 2$. Such paths may be down the street from the transmitter. The other types of paths are shadowed paths. One type of shadowed path is found in highly urbanized settings where the signal may be reflected from high-rise buildings. Another is found in more suburban areas where buildings are often just two stories high.

When a signal at 800 MHz is plotted versus R on a logarithmic scale as in Figure 17.21, there are distinctly different slopes before and after the Fresnel break point. We call the break distance (from the transmit antenna) R_B. This is the point for which the Fresnel ellipse about the direct ray just touches the ground. This model is illustrated in Figure 17.22. The distance R_B is approximated by

$$R_B = 4h_1h_2/\lambda$$

For $R < R_B$, n is less than 2, and for $R > R_B$, n approaches 4.

TABLE 17.3 Wireless PCS Technologies

System	High-Power Systems				Low-Power Systems			
	Digital Cellular (High-Tier PCS)				Low-Tier PCS		Digital Cordless	
	IS-54	IS-95 (DS)	GSM	DCS-1800	WACS/PACS	Handi-Phone	DECT	CT-2
Multiple access	TDMA/FDMA	CDMA/FDMA	TDMA/FDMA	TDMA/FDMA	TDMA/FDMA	TDMA/FDMA	TDMA/FDMA	FDMA
Freq. band (MHz) Uplink (MHz) Downlink (MHz)	869–894 824–849 (USA)	869–894 824–849 (USA)	935–960 890–915 (Europe)	1710–1785 1805–1880 (UK)	Emerg. Tech.[a] (USA)	1895–1907 (Japan)	1880–1900 (Europe)	864–868 (Europe and Asia)
RF ch. spacing Downlink (KHz) Uplink (KHz)	30 30	1250 1250	200 200	200 200	300 300	300	1728	100
Modulation	π/4 DQPSK	BPSK/QPSK	GMSK	GMSK	π/4 QPSK	π/4 DQPSK	GFSK	GFSK
Portable txmit Power, max./avg.	600 mW/ 200 mW	600 mW	1 W/ 125 mW	1 W/ 125 mW	200 mW/ 25 mW	80 mW/ 10 mW	250 mW/ 10 mW	10 mW/ 5 mW
Speech coding	VSELP	QCELP	RPE-LTP	RPE-LTP	ADPCM	ADPCM	ADPCM	ADPCM
Speech rate (kb/s)	7.95	8 (var.)	13	13	32/16/8	32	32	32
Speech ch./RF ch.	3	—	8	8	8/16/32	4	12	1
Ch. bit rate (kb/s) Uplink (kb/s) Downlink (kb/s)	48.6 48.6		270.833 270.833	270.833 270.833	384 384	384	1152	72
Ch. coding	1/2 rate conv. 1/3 rate rev.	1/2 rate fwd	1/2 rate conv.	1/2 rate conv.	CRC	CRC	CRC (control)	None
Frame (ms)	40	20	4.615	4.615	2.5	5	10	2

[a]Spectrum is 1.85 to 2.2 GHz allocated by the FCC for emerging technologies; DS is direct sequence.

Source: Ref. 26, Table 1, reprinted with permission of the IEEE.

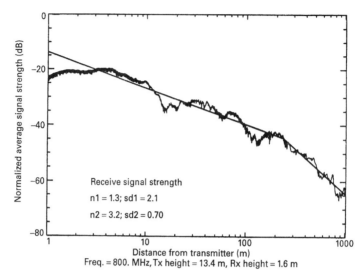

Figure 17.21. Signal variation on a line-of-sight path in a rural environment. Taken from Ref. 11, Figure 3.

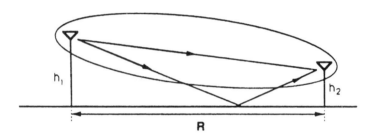

Figure 17.22. Direct and ground reflected rays, and showing the Fresnel ellipse about the direct ray. From Ref. 11, Figure 18.

It was found that on non-LOS paths in an urban environment with low base station antennas and with users at street level, propagation takes place down streets and around corners rather than over buildings. For these non-LOS paths the signal must turn corners by multiple reflections and diffraction at vertical edges of buildings. Field tests reveal that signal level decreases by about 20 dB when turning a corner.

In the case of propagation inside buildings where the transmitter and receiver are on the same floor, the key factor is the clearance height between the average tops of furniture and the ceiling. Bertoni et al. [11] call this clearance W. Here building construction consists of drop ceilings of acoustical material supported by metal frames. That space between the drop ceiling and the floor above contains light fixtures, ventilation ducts, pipes, support beams, and so on. Because the acoustical material has a low dielectric

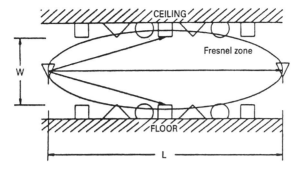

Figure 17.23. Fresnel zone for propagation between transmitter and receiver in clear space between building furnishings and ceiling fixtures. From Ref. 11, Figure 35.

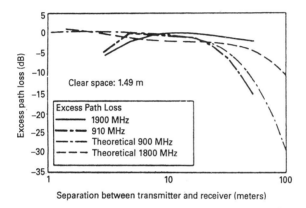

Separation between transmitter and receiver (meters)

Figure 17.24. Measured and calculated excess path loss at 900 and 1800 MHz for a large office building having head-high cubicle partitions, but no floor-to-ceiling partitions. From Ref. 11, Figure 36.

constant, the rays incident on the ceiling penetrate the material and are strongly scattered by the irregular structure, rather than undergoing specular reflection. Floor-mounted building furnishings such as desks, cubicle partitions, filing cabinets, and work benches scatter the rays and prevent them reaching the floor, except in hallways. Thus it is concluded that propagation takes place in the clear space, W.

Figure 17.23 shows a model of a typical floor layout. When both the transmitter and receiver are located in the clear space, path loss can be related to the Fresnel ellipse. If the Fresnel ellipse associated with the path lies entirely in the clear space, the path loss has LOS properties $(1/L^2)$. Now as the separation between the transmitter and receiver increases, the Fresnel ellipse grows in size so that scatters lie within it. This is shown in Figure 17.24. Now the path loss is greater than free space.

Bertoni et al. report on one measurement program where the scatters have been simulated using absorbing screens. It was recognized that path loss will be highly dependent on nearby scattering objects. Figure 17.24 was developed from this program. It was found that the path loss in excess of free space calculated at 900 and 1800 MHz for $W = 1.5$ is plotted in Figure 17.24 as a function of path length L. The figure shows that the excess path loss (over LOS) is small at each frequency out to distances of about 20 and 40 m, respectively, where it increases dramatically.

Propagation between floors of a modern office building can be very complex. If the floors are constructed of reinforced concrete or prefabricated concrete, transmission loss can be 10 dB or more. Floors constructed of concrete poured over steel panels show much greater loss. In this case [11], signals may propagate over other paths involving diffraction rather than transmission through the floors. For instance, signals can exit the building through windows and reenter on higher floors by diffraction mechanisms along the face of the building.

10 MOBILE SATELLITE COMMUNICATIONS

10.1 Background and Scope

In our earlier discussions on cellular mobile radio and PCS, there seemed to be no clear demarcation where one ended and the other began. Cellular hand terminals certainly are used inside of all types of buildings with some fair success. Granted some of the connections are marginal. Often we speak of PCS in the bigger picture of cellular mobile radio. Even CCIR (ITU-R organization) describes FPLMTS* as an integrated system where there is no dividing line between PCS and cellular.

In this section we review satellite services that provide PCS and cellular mobile radio on a worldwide basis. We present a short overview and then discuss Motorola's IRIDIUM system in some detail.

10.2 Overview of Satellite Mobile Services

10.2.1 Existing Systems. INMARSAT [International Maritime Satellite (consortium)] has been providing worldwide full-duplex voice, data, and record traffic service with ships since the mid-1970s. It extended its service to a land-mobile market and to aircraft. At present there are over 25,000 INMARSAT terminals. About 30% of these are land-transportable. INMARSAT satellites are in geostationary earth orbit (GEO).

INMARSAT-M systems provide service to ships and mobile land terminals. By the year 2005, some 600,000 INMARSAT-M terminals are

*FPLMTS stands for Future Public Land Mobile Telecommunication Systems. This is a futuristic concept of an integrated PCS and cellular mobile radio. It includes a satellite adjunct for both PCS and cellular service.

expected to be in operation. The up- and downlinks for mobile terminals are in the 1.5- and 1.6-GHz bands. Services are low-bit-rate voice (5–8 kbps) and data operations. INMARSAT-P is a program specifically directed to the PCS market and is expected to be operational around 1998 [9].

American Mobile Satellite Corporation (AMSC) is also providing voice, data, and facsimile service to the Americas, targeting customers in regions not served by conventional terrestrial cellular systems and terrestrial cellular subscribers which have problems "roaming." These satellites are GEO and provide up- and downlinks in the L-band much like INMARSAT.

10.2.2 Post-1996 Systems. Several satellite communication system companies are launching satellites for low rate services in the VHF band (148- to 149-MHz uplink and 137- to 138-MHz downlink). These systems use satellites in low earth orbit (LEO). All provide two-way message services and do not offer voice service. Some orbits are polar and some are inclined.

TRW expects to have its ODYSSEY system in operation by 1998. The satellites will provide voice, data, facsimile, and paging services worldwide. Serving mobile platforms, uplinks will be in the 1610.5- to 1626.5-MHz band and the companion downlinks will be in the 2483.5- to 2500-MHz band using channelized CDMA access. The orbits will be MEOs (medium earth orbit, about 10,400 km). Twelve satellites will be in three orbital planes.

Loral and Qualcomm have joined forces to offer an LEO system called GLOBALSTAR consisting of a network of 48 satellites in eight orbital planes. GLOBALSTAR specifically is targeting the hand-held terminal market for interconnection with the PSTN. Access will be by channelized CDMA, and services offered are voice, data, facsimile, and position location. Initial operation is expected in 1997 [9].

Constellation Communications of Herndon, Virginia (USA), plans a large LEO system called ARIES which will provide users with voice, data, facsimile, and position location services. These LEO satellites will be in four orbital planes at an average altitude of 1020 km. CDMA access is envisioned using a 16.5-MHz segment of L-band around 1.6 GHz for uplinks and another, similar segment for downlinks around 2.5 GHz. Ten or eleven fixed earth stations are planned for connectivity to the PSTN. These facilities will use standard 30/20-GHz uplinks and downlinks. The system is expected to be operational in 1996.

ELLIPSO is another planned system employing 15 satellites in elliptical inclined orbits in 3 planes and up to 9 satellites in equatorial orbit at maximum altitudes of about 7800 km. L-band connectivity is planned for the mobile user, and C-band connectivity is planned for the feeder uplinks and downlinks. The services offered to customers will be voice, data, facsimile, and paging. Access will be via channelized CDMA, and the first satellites are expected to be operational in 1996.

10.3   System Trends

The low earth orbit offers a number of advantages over the geostationary orbit, and at least one serious disadvantage.

Delay. One-way delay to a GEO satellite is budgeted at 125 ms; one-way up and down is double this value of 250 ms. Round-trip delay is about 0.5 s. Delay to a typical LEO is 2.67 ms and round-trip delay is 4×2.67 ms or about 10.66 ms. Calls to/from mobile users of such systems may be relayed still again by conventional satellite services. Data services do not have to be so restricted on the use of "hand-shakes" and stop-and-wait ARQ as with similar services via a GEO system.

Higher Elevation Angles and "Full Earth Coverage." The GEO orbit provides no coverage above about 80° latitude, and it gives low angle coverage of many of the world's great population centers because of its comparatively high latitude. Typically, cities in Europe and Canada face this dilemma. LEO satellites, depending on orbital plane spacing, can all provide elevation angles $> 40°$. This is particularly attractive in urban areas with tall buildings. Coverage would only be available on the south side of such buildings in the Northern Hemisphere with a clear shot to the horizon. Properly designed LEO systems will not have such drawbacks. Coverage will be available at any orientation.

*Tracking, a Disadvantage of LEOs, MEOs.** At L-band quasi-omni-directional antennas for the mobile user are fairly easy to design and produce. Although such antennas display only modest gain of several decibels, links to LEO satellites can be easily closed with hand-held terminals. However, large feeder, fixed earth terminals will require a good tracking capability as LEOs pass overhead. Hand-off is also required as an LEO disappears over the horizon to another satellite just as it appears over the opposite horizon. The hand-off should be seamless.

The quasi-omnidirectional user terminal antennas will not require tracking, and the hand-off should not be noticeable to the mobile user.

10.4   THE IRIDIUM LOW EARTH ORBIT SATELLITE SYSTEM

10.4.1   Introduction. IRIDIUM is a low earth orbit (LEO) satellite system which will provide cellular/personal communication services worldwide. It is being developed and will be operated by the Motorola Satellite Communications, Inc., a wholly owned subsidiary of Motorola, Inc. Subscribers to this system will use portable and mobile terminals with low

---

*MEO stands for medium earth orbit (i.e., 5000- to 13,000-km orbits).

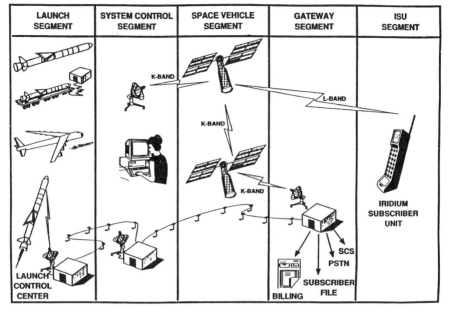

| LAUNCH SEGMENT | SYSTEM CONTROL SEGMENT | SPACE VEHICLE SEGMENT | GATEWAY SEGMENT | ISU SEGMENT |

Figure 17.25. A pictorial overview of the IRIDIUM system.

profile antennas to reach a constellation of 66 satellites. These satellites will be interconnected by crosslinks as they circle the earth in highly inclined polar orbits about 485 statute miles above the Earth. The deployment of the satellites will be in six orbital planes about 31.6° apart. However, planes 1 and 6 will be only 22° apart. The average delay to a satellite is 2.6 ms versus about 125 ms to a geostationary (GEO) satellite.

IRIDIUM offers a wide range of mobile radio services including voice, data, facsimile, and radio position finding. The subscriber communication services can be inerconnected with the PSTN (public-switched telecommunications network). The satellite-subscriber links are at L-band (1.6 GHz); PSTN connectivity is in the 30/20 GHz band and the satellite crosslinks operate in the 23 GHz band. Figure 17.25 gives a pictorial overview of the IRIDIUM system.

10.4.2 Space Segment. The space segment includes the 66 small satellites in low earth orbit (LEO) which are networked together as a switched-digital communication system utilizing cellular techniques to achieve a maximum capability of frequency reuse. The subscriber uplinks and downlinks only occupy the band 1616 to 1626.5 MHz. Each satellite will use up to 48 spot beams to form small cells on the surface of the Earth. These numerous, relatively narrow beams are the consequence of high statellite antenna gains.

The spatial separation of the beams allows increased spectral efficiency by means of time/frequency/spatial reuse over mutiple cells enabling many simultaneous user traffic connectivities over the same frequency channel. The constellation of satellites and its projection of cells on the earth's surface are analogous to a terrestrial cellular telephone system. With conventional cellular operations, a static set of cells serves a large number of mobile/portable users. However, with IRIDIUM, user's motion is relatively slow compared to that of the spacecraft.

Each of the satellites operates crosslinks to support internetting. The crosslinks operate in the 23 GHz band and include both forward and backward looking links to two adjacent satellites in the same orbital plane which are normally at a fixed angle about 2100 nautical miles away. Up to four inter-plane links are also maintained.

Each satellite communicates with earth-based gateways either directly or through other satellites by means of the crosslink network. Initially, there will be two gateways in the United States and about 3 to 20 gateways in other parts of the world. The system can accommodate up to 250 gateway stations. The gateway provides the interface between the IRIDIUM system and the PSTN. Table 17.4 gives a summary of major IRIDIUM satellite characteristics.

10.4.3 Cell Organization and Frequency Reuse

Satellite L-band Antenna Pattern. Each satellite has the capability of projecting 48 L-band spot beams to form a continuous overlapping pattern on the Earth. On a global basis, the entire constellation's beam pattern as projected on the surface of the earth results in approximately 2150 active beams with a frequency reuse of about 180 times. Within the contiguous United States, the system achieves about five times frequency reuse.

Each satellite has 3 multiple-beam phased array antennas. The phased array antennas are located on side panels of the satellite, each of which forms 16 cellular beams. Active transmit–receive (T/R) modules are used to provide power amplification for the transmit function, low noise amplification for the receive function, switch selection between transmit and receive, and digital phase control for active beam steering of both the transmit and receive beams in the phased arrays. The 16-cell pattern of each phased array is repeated for each of the three panels.

Cellular Pattern. IRIDIUM operates with a 7-cell frequency reuse pattern as shown in Figure 17.26. The cells shown as A through G are scanned by the satellite arrays in accordance with the timing pattern and sequence shown in Figure 17.27. During the time slot that the antenna is pointing at a cell, satellite transmissions and receptions from subscriber units (ISUs) may occur during respective transmit and receive intervals. Transmissions may be on the same frequency at the same time in any cells with the same letters (i.e., A, B, C etc.) as shown in Figure 17.26. Each satellite has scanning beam antennas which are programmed to point at the correct cell on the earth at the right

TABLE 17.4 Major IRIDIUM Satellite Characteristics

Orbits (Nominal)	
Number of Operational Satellites	66
Number of Orbital Planes	6
Inclination of Orbital Planes	86.4 degrees
Orbital Period	100 min and 28 sec
Apogee	787 km
Perigee	768 km
Argument of Perigee	90 +/− 10 degrees
Active Service Arc	360 degrees
RAAN	0, 31.6, 63.2, 94.8, 126.3, 157.9 degrees
Earth Coverage	5.9 Million Square (Statute) Miles Per Satellite
Maximum Number of Channels per Satellite	
L-Band Service Links	About 3,840
Intersatellite Links	About 6,000
Gateway/TT & C Links	About 3,000
Frequency Bands[2]	
L-Band Service Links	1616–1626.5 MHz
Intersatellite Links	23.18–23.38 GHz
Gateway/TT & C Links	
Downlinks	19.4–19.6 GHz
Uplinks	29.1–29.3 GHz
Polarization	
L-Band Service Links	Right Hand Circular
Intersatellite Links	Horizontal
Gateway/TT & C Links	
Downlinks	Left Hand Circular
Uplinks	Right Hand Circular
Transmit EIRP[1]	
L-Band Service Links	7.5 to 27.7 dBw
Intersatellite Links	38.4 dBw
Gateway/TT & C Links	13.5 to 23.2 dBw
Final Amplifier Output Power Capability[2]	
L-Band Service Links	0.1 to 3.5 Watts per carrier (Burst)[3]
Intersatellite Links	3.4 Watts per carrier (Burst)
Gateway/TT & C Links	0.1 to 1.0 Watts per channel
Satellite G/T	
L-Band Service Links	−10.6 to −3.1 dBi/K
Intersatellite Links	8.1 dBi/K
Gateway/TT & C	−1.0 dBi/K
Receiving System Noise Temperature	
L-Band Service Links	500 K
Intersatellite	720 K (1188 K with sun)
Gateway/TT & C	1295 K
Gain of Each L-Band Channel	N.A. (Not a transponder)

1. At edge of coverage.
2. Does not include circuit losses.
3. Equipment combined power from phased array antenna

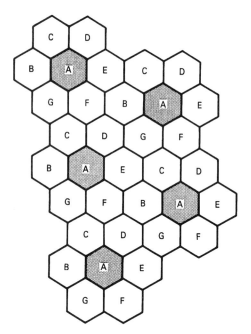

Figure 17.26. Seven cell frequency reuse pattern.

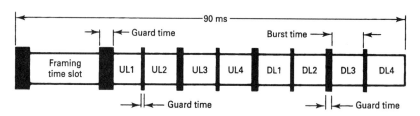

Figure 17.27. TDMA format.

time. Figure 17.28 shows the integration of the 7-cell pattern on a satellite and how it is integrated with satellites whose antenna patterns are contiguous.

Frequency Plans. The frequency plan for L-band operation is shown in Figure 17.29. The total peak capacity for a satellite is 960 voice channels of which 780 are full duplex. As shown in Figure 17.27, each frequency slot has TDMA capability for four full-duplex voice channels. The contiguous United States are covered by approximately 59 beams which yield a capacity of 4720 channels of which 3835 are full-duplex voice channels.

10.4.4 Communications Subsystem. The IRIDIUM communications system provides L-band communications between each satellite and individual subscriber units, K_a-band communications between each spacecraft

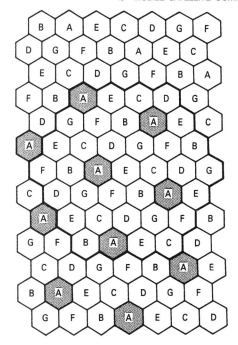

Figure 17.28. Reuse pattern with more than one satellite.

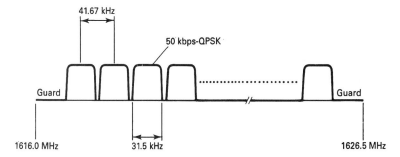

Figure 17.29. L-band uplink/downlink RF frequency plan.

and ground-based facilities, which could be either gateway or system control facilities (TT & C) and K_a-band crosslinks from satellite to satellite. The transfer of Telemetry, Tracking and Control (TT & C) information between the System Control Facilities and each satellite is generally provided via K_a-band communication links, with a dedicated link (via an omni-antenna) as backup.

L-Band Subscriber Terminal Links. The L-band communication subsystem is supported by an antenna complex consisting of three antenna panels which form 48 cellular beams. Of the three beams that can be formed simultaneously (one from each panel), all three can be active during any timeslot, based on projected overlap between cell patterns of adjacent satellites. The L-band communications system is sized to provide, transmit, and receive capability for up to six fully loaded cells.

Each active transmit/receive beam supports up to 80 channels. As shown in Figure 17.29, each FDMA frequency slot is 41.67 kHz wide including guard bands. Each frequency slot supports 4 full-duplex voice channels in a TDMA arrangement shown in Figure 17.27. The resulting usable capacity is 960 channels of which 780 are full-duplex channels. The modulation is QPSK, the coded data rate is 50 kbps per channel, employs $R = 1/2$ convolutional coding, and the frame period is 90 ms.

The L-band communications subsystem is designed to support a bit error rate of 1×10^{-2} end-to-end for voice operation. The improved BER required for data transmission will be supported through the use of processing hardware installed in the subscriber units to apply more robust protocols and coding in order to counter the deep fading experienced with L-band links in the mobile environment.

Gateway Links. The 30/20 GHz gateway links provide full-duplex operation with two ground-based gateways or system support facilities per satellite. Satellite beam center gains for maximum range are 26.9 dBi on the downlink and 30.1 dBi on the uplink. Multiple antennas spaced 34 nautical miles apart provide spatial diversity to avoid sun interference and tend to mitigate rainfall attenuation. Time availabilities on these links are expected in the range of 99.8%.

Each of the two full-duplex gateway links supports 600 simultaneous voice circuits. The frequency plan calls for the allocation of six frequencies each for uplink and downlink operation. The modulation rate in each direction is 12.5 Mbps and the channels are spaced at 15 MHz intervals. Each link provides a BER better than 1×10^{-7}.

10.4.5 Transmission Characteristics.

Digital speech operation on IRIDIUM uses the VSELP compression technique. Conventional PCM requires 64 kbps for a standard voice channel primarily to support speech operation. VSELP* cuts this value to 4.800 kbps or about 15:1 compression.

The modulation and multiple access techniques used in the IRIDIUM system resemble those of a terrestrial cellular system, especially some of the newer schemes such as GSM and the U.S. TDMA TIA IS-54 standard. A

---

*VSELP is vector sum excitation linear predictive (vocoder). This compression scheme of the CELP variety uses a condensed codebook structure to reduce computational burden. It is contained in one chip. A somewhat similar scheme is used on the familiar U.S. DoD STU-III secure telephone.

combined FDMA and TDMA access format is used along with data or vocoded voice and digital modulation techniques (QPSK).

Each subscriber unit operates in a burst mode using a single carrier transmission. The bursts are controlled to occur at the proper time in the TDMA frame. The TDMA format is shown in Figure 17.27. The frame has four user time slots, both uplink and downlink. Each subscriber unit will burst so that its transmission is received by the satellite in the proper time slot, Doppler corrected. The subscriber unit is similar to a conventional cellular hand-held unit in size and shape. It has a quadrifilar helix antenna with nearly hemispherical coverage with a gain of $+1$ dBi. Its peak EIRP is $+6.0$ dBW. The G/T is -23 dB/K. Its required E_b/N_o is 5.8 dB.

Section 10.4 is based on References 27, 28, and 29. Reprinted with permission.

REVIEW QUESTIONS

1. How does a mobile terminal make connectivity to the PSTN in conventional cellular operations?

2. What are typical radiated power levels for a cell site, mobile unit, and hand-held terminal? Reference here is to the North American AMPS system.

3. In North America, what is the assigned bandwidth for a cellular operator outbound? (Of course, the same bandwidth in another part of the spectrum would be assigned inbound.)

4. Cellular radio service has seen explosive growth. What will a cellular operator do when a group of cell sites are overloaded with traffic?

5. Name at least four ways that cellular radio differs from LOS microwave regarding propagation.

6. Describe and discuss *locating* and *handover*.

7. List and discuss actions that go on over the setup channel.

8. What sort of a dispersion value might be encountered on a cellular link? Admittedly, this value approaches a worst-case value. What type of operation will such dispersion affect?

9. Path loss on LOS microwave was a square (2-power) (i.e., $20 \log D$) relationship. What is the relationship for cellular?

10. The amended CCIR equation for transmission loss for cellular radio takes into account three additional factors further adding to the loss. What are they?

11. The British urban path loss formula included the $40 \log d$ factor and a factor to compensate for antenna height–gain. What other factors did it take into account, including a constant?

12. Okumura, a famous researcher on urban path loss for cellular systems, developed a method, applied by HATA, to calculate path loss. In this case, path loss (transmission loss) varied with what factors?

13. What were the basic factors involved in the equation for building penetration loss?

14. For microcells in heavily urbanized areas, what is the major contributor to excess loss (above line of sight)?

15. Explain how dispersion can corrupt a digital signal and show that this corruption, showing up as ISI (intersymbol interference), is a function of the bit rate. Use 10 μs as the value for dispersion.

16. Discuss the three fading models given in the text in view of the function K.

17. Why is space diversity favored over frequency diversity, particularly with cellular operations?

18. Discuss space-diversity antenna separation (1) for a cell site and (2) for a mobile platform. The example should be for the 900-MHz band.

19. Space-diversity antenna separation not only varies as a function of the correlation coefficient but also as a function of _____ _____.

20. Where would we use an interleaver and how does it work?

21. Often cells and cell site transmitter–antenna parameters are built around a signal level contour. What is the value of this contour in dBμ? Also give this dBμ value in dBm based on a 50-Ω impedance at 850 MHz.

22. Convert 1-dBd antenna gain to its equivalent in dBi.

23. Discuss the cellular radio bandwidth dilemma. What is the basic problem anyway?

24. In three sentences, describe how a sub-band coder operates.

25. Describe how a CELP coder works. Down to what bit rate can we expect from a CELP coder without sacrificing too much voice quality/intelligibility?

26. We described three different cellular radio system access techniques. What were they? In two sentences each, describe how they work.

27. TDMA is a leading contender as a digital access technique. Describe how it operates and especially the problem of delay between a nearby user and

another user at cell boundary. Describe one way of mitigating that problem.

28. How many time slots does the GSM TDMA have?

29. Based on North American TDMA (IS 54), and comparing it to AMPS, what is the bandwidth utilization improvement with this type of TDMA?

30. Give at least three advantages of using CDMA as a cellular access technique.

31. What is the processing gain of a direct sequence CDMA system with an information rate of 5 kbps and a spread bandwidth of 5 MHz?

32. Frequency reuse is now vital in cellular radio operations. In large cellular systems, particularly in urban areas, we have to expect a certain level of interference. What value C/I or better is our goal?

33. In a typical hexagonal cell, from how many directions does interference enter? How can we cut down on these interference levels to really just one direction?

34. When we opt for cell splitting to increase system capacity, what would be the minimum practical cell size?

35. Define a paging system? (i.e., What does it do?)

36. Paging system capacity is affected by 10 parameters. List five of them.

37. To improve performance and robustness, almost all paging systems use _____.

38. Explain the knee in a signal level curve (rural environment).

39. For propagation inside buildings, on a single floor, explain the relationship of the "clear space" for path loss calculation and the effects of clutter and obstacles.

40. Discuss advantages and disadvantages of LEO satellite systems compared to GEO systems.

41. Describe how IRIDIUM will cover the entire world. Name three services that IRIDIUM will offer initially.

42. Which is one crucial way IRIDIUM satellites will support internetting?

43. IRIDIUM will use 48 spot beams. Describe their makeup.

44. What sort of cell reuse pattern will IRIDIUM have? (Give number of cells). How many full-duplex channels will there be per cell?

45. Describe the time-slot TDMA format of IRIDIUM. What is the duration of a TDMA frame?

46. What type of modulation and FEC will IRIDIUM use?

47. There is a fairly large guard time between TDMA bursts. Explain why it is comparatively large.

48. IRIDIUM is comparatively robust. What is the worst BER and the average BER for voice communication?

REFERENCES

1. *Mobile Radio Communications*, R. Steele, ed., IEEE Press, New York and Pentech Press, London, 1992.

2. *Land Mobile Communications Engineering*, D. Bodson et al., eds., IEEE Press, New York, 1983.

3. J. D. Parsons and J. G. Gardiner, *Mobile Communication Systems*, Blackie, London and Halsted Press, New York, 1989.

4. J.-P. Linnartz, *Narrowband Land-Mobile Radio Networks*, Artech House, Norwood, MA, 1993.

5. *Personal and Mobile Radio Systems*, R. C. V. Macario, ed., IEE/Peter Peregrinus, London, 1991.

6. W. C. Y. Lee, *Mobile Communications Design Fundamentals*, 2nd ed., John Wiley & Sons, New York, 1993.

7. A. Jagoda and M. de Villepin, *Mobile Communications*, John Wiley & Sons, Chichester, UK, 1991/1992.

8. K. Pahlavan and A. H. Levesque, *Wireless Information Networks*, John Wiley & Sons, New York, 1995.

9. W. W. Wu et al., "Mobile Satellite Communications," *Proc. IEEE*, **82**(9) (September 1994).

10. W. F. Fuhrman and V. Brass, "Performance Aspects of the GSM System," *Proc. IEEE*, **89**(9) (September 1984).

11. H. L. Bertoni et al., "UHF Propagation Prediction for Wireless Personal Communication," *Proc. IEEE*, **89**(9) (September 1994).

12. *VHF and UHF Propagation Curves for the Frequency Range from 30 MHz to 1000 MHz*, CCIR Rec. 370-5, XVIIth Plenary Assembly, Dusseldorf, 1990.

13. K. Allesbrook and J. D. Parsons, "Mobile Radio Propagation in British Cities at Frequencies in the VHF and UHF Bands," *Proc. IEEE*, **124**(2) (1977).

14. Y. Okumura et al., "Field Strength and Its Variability in VHF and UHF Land Mobile Service," *Rev. Electr. Commun. Lab. (Tokyo)*, **16** (1968).

15. M. Hata, "Empirical Formula for Propagation Loss in Land–Mobile Radio Services," *IEEE Trans. VT*, **VT-20** (1980).

16. F. C. Owen and C. D. Pudney, "In-Building Propagation at 900 MHz and 1650 MHz for Digital Cordless Telephones," 6th Int. Conf. on Antennas and Propagation, ICCAP '89, Pt. 2: Propagation, Conf. Pub. No. 301, 1989.

17. M. R. Schroeder and B. S. Atal, "Code-Excited Linear Prediction, High Quality Speech at Low Bit Rates," *IEEE Proc. ICASSP* (1985).

18. R. C. Dixon, *Spread Spectrum Systems with Commercial Applications*, 3rd ed., John Wiley & Sons, New York, 1994.

19. "Recommended Minimum Standards for 800-MHz Cellular Subscriber Units," EIA Interim Standard EIA/IS-19B.

20. "Cellular Radio Systems," a seminar given at the University of Wisconsin–Madison by Andrew H. Lamothe, Consultant, Leesburg, VA, 1993.

21. *Telecommunications Transmission Engineering*, Vol. 2, 3rd ed., Bellcore, Piscataway, NJ, 1992.

22. *Radio-Paging Systems*, CCIR Rep. 900-2, Vol. VIII-1, XVIIth Plenary Assembly, Dusseldorf, 1990.

23. *Radio-Paging Systems*, CCIR Rep. 499-5, Vol. VIII.1, XVIIth Plenary Assembly, Dusseldorf, 1990.

24. *Digital Cellular Public Land Mobile Telecommunication Systems (DCPLMTS)*, CCIR Rep. 1156, Vol. VIII.1, XVIIth Plenary Assembly, Dusseldorf.

25. M. Engelson and J. Hebert, "Effective Characterization of CDMA Signals," Wireless Report, January 1995.

26. D. C. Cox, "Wireless Personal Communications: What Is It?", *IEEE Personal Commun.*, **2**(2) (April 1995).

27. Application of Motorola Satellite Communications, Inc. for IRIDIUM, A Low Earth Orbit Mobile Satellite System," Before the Federal Communications Commission, December 1990.

28. *Minor Amendment* before the Federal Communications Commission (amending Motorola's IRIDIUM System application), August 1992 by Motorola Satellite Communications, Inc.

29. *Minor Amendment* before the Federal Communications Commission (amending Motorola's IRIDIUM System application), November 1994 by Motorola Satellite Communications, Inc.

18

NETWORK MANAGEMENT

1 WHAT IS NETWORK MANAGEMENT?

Effective network management optimizes a telecommunication network's operational capabilities. The key word here is *optimizes*. These are some of the connotations that can be derived:

- It keeps the network operating at peak performance.
- It informs the operator of impending deterioration.
- It provides easy alternative routing and work-arounds when deterioration and/or failure takes place.
- It provides the tools for pinpointing causes of performance deterioration or failure.
- It serves as the frontline command post for network survivability.

There are numerous secondary functions of network management. They are important, but in our opinion they are still secondary. Among these items are:

- It informs in quasi-real time regarding network performance.
- It maintains and enforces network security, such as link encryption and issuance and use of passwords.
- It gathers and files data on network usage.
- It performs a configuration management function.
- It also performs an administrative management function.

2 THE BIGGER PICTURE

Many seem to view network management as a manager of data circuits only. There is a much bigger world out there. Numerous enterprise and government networks serve for the switching and transport of *multimedia communications*. The underlying network will direct (switch) and transport voice, data, and image traffic. Each will have a traffic profile notably differing from the other. Nevertheless, they should be managed as an entity. If for no other reason, it is more cost-effective to treat the whole rather than piecemeal.

There is a tendency in the enterprise scene to separate voice telephony—calling it *telecommunications*—and another world, that of data communications. Perhaps that is why network management seems to often operate on two separate planes. One is data and very sophisticated, and the other is voice, which may have no management facilities at all. This next section treats network management as a whole consisting of its multimedia parts: voice, image, and data, which includes facsimile, telemetry, and CAD/CAM.

3 TRADITIONAL BREAKOUT BY TASKS

There are five tasks traditionally involved with network management:

- Fault management
- Configuration management
- Performance management
- Security management
- Accounting management

3.1 Fault Management

This is a facility that provides information on the status of the network and subnetworks. The "information on the status" should not only display faults (i.e., failures) and their location, but should also provide information on deteriorated performance. One cause of deteriorated performance is congestion. Thus, ideally, we would also like to isolate the cause of the *problem*.

Fault management should include the means to bypass troubled sections of a network or to patch in new equipment for deteriorated or failed equipment.

The complexity of modern telecommunication networks is such that as many network management tasks as possible should be automated. All displays, readouts, and hard-copy records should be referenced to a network time base down to at least 0.1 s. This helps in correlating events, an important troubleshooting tool.

3.2 Configuration Management

Configuration management establishes an inventory of the resources to be managed. It includes resource provisioning (timely deployment of resources to satisfy an expected service demand) and service provisioning (assigning services and features to end users). It identifies, exercises control over, collects data from, and provides data to the network for the purpose of preparing for, initializing, starting, and providing for the operation and termination of services. Configuration management deals with equipment and services, subnetworks, networks, and interfaces. Its functions are closely tied to fault management as we have defined it above.

3.3 Performance Management

Performance management is responsible for monitoring network performance to ensure that it is meeting specified performance. Some literature references [1] add growth management. They then state that the objective of performance and growth management is to ensure that sufficient capacity exists to support end-user communication requirements.

Of course, there is a fine line defining *network capacity*. If too much capacity exists, there will probably be few user complaints, but there is excess capacity. Excess capacity implies wasted resources, thus wasted money. Excess capacity, of course, can accommodate short-term growth. Therefore performance/growth management provides vital information on network utilization. Such data provide the groundwork for future planning.

3.4 Security Management

Security management controls access to and protects both the network and the network management subsystem against intentional or accidental abuse, unauthorized access, and communication loss. It involves link encryption, changes in encryption keys, user authentication, passwords, and unauthorized usage of telecommunication resources.

3.5 Accounting Management

Accounting management processes and records service and utilization records. It generates customer billing reports for services rendered. It identifies costs and establishes charges for the use of services and resources in the network. It may also be a repository for plant-in-place investment for telecommunications plants and provides reports to upper management on return on that investment.

4 SURVIVABILITY—WHERE NETWORK MANAGEMENT REALLY PAYS

The network management center is the frontline command post for the battle for network survivability. We can model numerous catastrophic events affecting a telecommunications network, whether the PSTN or a private/enterprise network. Among such events are fires, earthquakes, floods, hurricanes, terrorism, and public disorder. Telecommunications have brought revolutionary efficiences to the way we do business and is so very necessary even for life itself. Loss of these facilities could destroy a business, even possibly destroy a nation. A properly designed network management system could mitigate losses, even save a network almost in its entirety.

Consider the World Trade Center explosion in New York City. The entire PABX telephone system survived, less the extensions in the immediate area of the bomb, simply because the PABX was mounted on the 6th floor. All offices in the building had communications within the building and with the outside world. Those vital communications were never lost.

This brings in the first rule toward survivability. A network management center is a place of point failure. Here we mean that if the center is lost, probably nearly all means of reconfiguring the network are lost. To avoid such a situation, a second network management center should be installed. This second center should be geographically separate from the principal center. It is advisable that the second center share the network management load and be planned and sized to be able to take the entire load with the loss of the principal center. There should be a communications orderwire between the two centers. Both centers should be provided with no-break power and backup diesel generators.

One simple expedient for survivability is to back up the circuits with an arrangement with the local telephone company. This means that there must be compatibility between the enterprise network and the PSTN as well as one or more points of interface. A major concern is the network clock. One easy solution is to have the enterprise network derive its clock from the digital PSTN.

In the following sections we will describe means to enhance survivability still further.

4.1 Survivability Enhancement—Rapid Troubleshooting

An ideal network management system will advise the operator of one or more events,* indicate where in the network they occurred, and provide other handy troubleshooting data. In fact, the network management system should, in most cases, warn the operator in advance of an impending fault(s). This would be a boon to survivability.

---

*We define an *event* as something out of the ordinary that occurred.

Oftentimes the "ideal" is unachievable or only partially achievable. Human intervention will be required. The troubleshooter should have available a number of units of test equipment, some more automated than others, to aid in pinpointing the cause of an event.

Steve Dauber's article, "Finding Fault" [3], describes four steps to correct network faults. These are:

1. Observing symptoms
2. Developing a hypothesis
3. Testing the hypothesis
4. Forming conclusions

Observing Symptoms. It should be kept in mind that often many symptoms will appear at once, usually a chain reaction. We must be able to spot the real causal culprit or we may spend hours or even days chasing effects, not the root cause.

The reference article suggests these four reminders:

1. Find the range and scope of the symptoms. Does the problem affect all stations (all users)? Does it affect random users or users in a given area? We're looking for an area pattern here.
2. Are there some temporal conditions to the problem? How often does it occur per day, per hour, and so on? Is the problem continuous or intermittent? What is its regularity? Can we set our watches by its occurrence, or is it random in the temporal context?
3. Have there been recent rearrangements, additions to the network, or reconfigurations?
4. Software and hardware have different vintages, called *release dates*. I write this on WordPerfect 6.0, but what was its release date? The question one should ask then would be, Are all items of a certain genre affected in the same way? Different release dates of a workstation may be affected differently, some release dates not at all. The problem may be peculiar to a certain release date. This is particularly true of software.

Before we can move forward in the troubleshooting analysis, we must have firmly at hand the troublesome network's *baseline* performance. What is meant here is that we must have a clear idea of "normal" operation of the network so that we can really qualify and quantify its anomalous operation. For example, what is the expected bit error rate (BER) and is the value related to a time distribution? Can we express error performance in error-free seconds (EFS), errored seconds (ES), and severely errored seconds (SES)? These are defined in CCITT Rec. G.821 [4].

Steven Dauber, in the reference article, lists five network-specific characteristics that the troubleshooter should have familiarity with or data on.

A. *Network Utilization.* What is the average network utilization? How does it vary through the work day? Characteristics of congestion, if any, should be known, as well as where and under what circumstances it might be expected.

B. *Network Applications.* What are the dominant network applications on the network? What version numbers is it running?

C. *Network Protocol Software.* What protocols are running on the network? What are the performance characteristics of the software, and are these characteristics being achieved?

D. *Network Hardware.* Who manufactured the network interface controllers, media attachment units, servers, hubs, and other connection hardware? What versions are they? What are their performance characteristics? Expected? Met?

E. *Internetworking Equipment.* Who manufactured the repeaters, bridges, routers, and gateways on the network? What versions of software and firmware are they running? What are the performance characteristics? What are the characteristics of the interfaced network that are of interest?

Developing a Hypothesis. In this second step, we make a statement as to the cause of the problem. We might say that T1 or E1 frame alignment is lost because of deep fades being experienced on the underlying microwave transport network. Or we might say that excessive frames being dropped on a frame relay network are due to congestion being experienced at node B.

Such statements cannot be made without some strong bases to support the opinion. Here is where the knowledge and experience of the troubleshooter really pays. Certainly there could be other causes of E1 or T1 frame alignment, but if underlying microwave is involved, that would be a most obvious place to look. There could be other reasons for dropping frames in a frame relay system. Errored frames could be one strong reason.

Testing the Hypothesis. We made a statement, and now we must back it up with tests. One test I like is correlation. Are the fades on the microwave correlated with the event occurrences? That test can be done quite easily. If they are correlated, we have some very strong backup that the problem is with the microwave. The frame relay problem may be another matter. First, we could check the FECN and BECN bits to see if there was a change of state passing node B. If there is no change, assuming that flow control is implemented, then congestion may not be the problem. Removing the frame relay from the system and carrying out a bit error rate test (BERT) over some period of time would prove or disprove there is a noise problem.

A network analyzer is certainly an excellent tool in assisting in the localization of faults. Some analyzers have preprogrammed tests which can save the troubleshooter time and effort. Many networks today have some sort of network monitoring equipment incorporated. This equipment may be used in lieu of network analyzers or in conjunction with them. Again we stress the importance of separating cause and effects. Many times network analyzers or network management monitors and testers will only show effects. The root cause may not show at all and must be inferred, or separate tests must be carried out to pinpoint the cause.

We might digress here to talk about what is often called in Spanish *tonterías*. This refers to "silly things." Such tonterías are often brought about by careless installation or careless follow-up repair. Coaxial cable connectors are some of my favorites. Look for intermittents and cold solder joints. A good tool, but not necessarily a foolproof one, is a time-domain reflectometer (TDR). It can spot where a break in a conductor is down to a few feet or less. It can do the same for an intermittent, when in the fault state. In fact, intermittents can prove to be a nightmare to locate. An electrically noisy environment can also be very troublesome.

Forming Conclusions. A conclusion or conclusions are drawn. As we say, "the proof of the pudding is in the eating." The best proof that we were right in our conclusion is to fix the purported fault. Does the problem disappear? If so, our job is done, and the network is returned to its "normal" (baseline) operation.

What conclusions can we draw from this exercise? There are two basic ingredients to elemental network troubleshooting: expertise built on experience of the troubleshooter and the availability of essential test equipment. Troubleshooting time can be reduced (degraded operation or out-of-service time reduction) by having on-line network management equipment. With ideal network management systems, this time can be cut to nearly zero.

5 SYSTEM DEPTH—A NETWORK MANAGEMENT PROBLEM

An isolated LAN is a fairly simple network management problem. There is only a singular transmission medium and, under normal operating conditions, only one user is transmitting information to one or several recipients. It is limited to only two OSI layers. For troubleshooting, often a protocol analyzer will suffice, although much more elaborate network management schemes and equipment are available.

Now connect that LAN to the outside world by means of a bridge or router and network management becomes an interesting challenge. One example from experience was a VAX running DECNET which was a station on a CSMA/CD LAN. The LAN was bridged to a frame relay box which fed a 384-kbps channel with an E1 hierarchy (i.e., 6 E0 channels) via tandemed

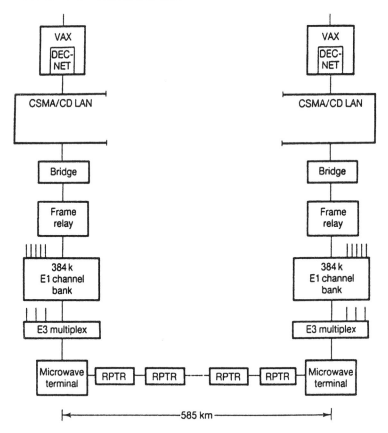

Figure 18.1. A typical multilevel network requiring a network management system. Note multiple convergences.

microwave links to a large facility at the distant end (550 km) with similar characteristics. This connectivity is shown diagrammatically in Figure 18.1. Such is typical of a fairly complex network requiring an overall network management system. To make it even more difficult, portions of the network were leased from the local common carrier, but were soon to be cutover to own ownership.

5.1 Aids in Network Management Provisioning

Modern E1 and T1 digital systems* are provided a means for operational monitoring of performance. The monitoring is done in quasi-real time and while operational (i.e., in traffic).

*This capability is not unique to T1- and E1-type systems. For example, SONET and SDH have even more sophisticated capabilities, often referred to as OA&M (operations, administration, and maintenance). ATM is (will be) rich in such network management aids.

In Chapter 8 we discussed PCM systems and the digital network including T1 (DS1) and E1 hierarchies. Let us quickly review their in-service performance monitoring capabilities.

DS1 or T1 has a frame rate of 8000 frames a second. Each frame is delineated with a framing bit or F-bit. With a modern alignment and synchronization algorithms, to maintain framing repetition of the F-bit 8000 times a second is excessive and unnecessary. Advantage was taken of F-bit redundancy by the development of the extended superframe [(ESF) see Chapter 8 for a discussion]. The ESF consists of 24 consecutive DS1 frames. With 24 frames we expect to have 24 framing bits. Of these, only six bits need to be used for framing, six are used for a cyclic redundancy check (CRC-6) on the frame, and the remaining 12 bits form a 4-kbps datalink for network control and maintenance. It is this channel that can serve as transport for network management information. It can also serve as an ad hoc test link.

Our present concern is in-service monitoring. For instance, we can get real-time error performance with the CRC-6, giving us a measure of errored seconds and severely errored seconds in accordance with CCITT Rec. G.821 [4]. Such monitoring can be done with test equipment such as the HP* 37702A or HP 37741A.

Permanent monitoring can be carried out using typically Newbridge Network monitoring equipment, which can monitor an entire T1 or E1 network for frame alignment loss, errored seconds, and severely errored seconds. One can "look" at each link and examine its performance in 15-min windows for a 24-h period. The Newbridge equipment can also monitor other selected data services such as frame relay, which, in our example above, rides on E1 aggregates. Such equipment can be a most important element in the network management suite.

The E1 digital network hierarchy also provides capability of in-service monitoring and test. We remember from Chapter 8 that E1 has 32 channels or time slots; 30 are used for the payload and 2 channels or time slots serve as support channels. The first of these is channel (or time slot) 0 and the second is channel (time slot) 16. This latter is used for signaling. Time slot (channel) 0 is used for synchronization and framing. Figure 18.2 shows the E1 multiframe structure.

The sequence of bits in the frame alignment (TSO) signal of successive frames is illustrated in Figure 18.2. In frames not containing the frame alignment signal, the first bit is used to transmit the CRC multiframe signal (001011) which defines the start of the sub-multiframe (SMF). Alternate frames contain the frame alignment word (0011011) preceded by one of the CRC-4 bits. The CRC-4 remainder is calculated on all the 2048 bits of the previous sub-multiframe (SMF), and the 4-bit word is sent as C1, C2, C3, C4 of the current SMF. Note that the CRC-4 bits of the previous SMF are set to zero before the calculation is made.

---

*HP stands for Hewlett-Packard, a well-known manufacturer of electronic test equipment.

Sub-multiframe (SMF)	Frame number	Bits 1 to 8 of the frame in timeslot 0							
		1	2	3	4	5	6	7	8
Multiframe I	0	C_1	0	0	1	1	0	1	1
	1	0	1	A	S_{a4}	S_{a5}	S_{a6}	S_{a7}	S_{a8}
	2	C_2	0	0	1	1	0	1	1
	3	0	1	A	S_{a4}	S_{a5}	S_{a6}	S_{a7}	S_{a8}
	4	C_3	0	0	1	1	0	1	1
	5	1	1	A	S_{a4}	S_{a5}	S_{a6}	S_{a7}	S_{a8}
	6	C_4	0	0	1	1	0	1	1
	7	0	1	A	S_{a4}	S_{a5}	S_{a6}	S_{a7}	S_{a8}
II	8	C_1	0	0	1	1	0	1	1
	9	1	1	A	S_{a4}	S_{a5}	S_{a6}	S_{a7}	S_{a8}
	10	C_2	0	0	1	1	0	1	1
	11	1	1	A	S_{a4}	S_{a5}	S_{a6}	S_{a7}	S_{a8}
	12	C_3	0	0	1	1	0	1	1
	13	E	1	A	S_{a4}	S_{a5}	S_{a6}	S_{a7}	S_{a8}
	14	C_4	0	0	1	1	0	1	1
	15	E	1	A	S_{a4}	S_{a5}	S_{a6}	S_{a7}	S_{a8}

Figure 18.2. E1 multiframe, CRC-4 structure. Notes: E = CRC-4 error indication bits; S_{a4} to S_{a8} = spare bits. These bits may be used for a network management (maintenance) link. C_1 to C_4 = CRC-4 bits. A = remote alarm indication. From CCITT Rec. G.704, Table 4b/G.704, page 81 (Ref. 10).

At the receive end, the CRC remainder is recalculated for each SMF and the result is compared with the CRC-4 bits received in the next SMF. If they differ, then the checked SMF is in error. What this is telling us is that a block of 2048 bits had one or more errors. One thousand CRC-4 block error checks are made every second. It should be noted that this in-service error detection scheme does not indicate BER unless one assumes a certain error distribution (random or burst errors), to predict the average errors per block. Rather it provides a block error measurement.

This is very useful for estimating percentage of errored seconds (%ES) which is usually considered the best indication of quality for data transmission—itself a block or frame transmission process. CRC-4 error checking is fairly reliable, with the ability of detecting 94% of errored blocks even under poor BER conditions. (See CCITT Rec. G.706 [5].)

Another powerful feature of E1 channel 0 (when equipped) is the provision of local indication of alarms and errors detected at the far end. When an errored SMF is detected at the far end, one of the E-bits (see Figure 18.2) is changed from a 1 to a 0 in the return path multiframe (TS0). The local end, therefore, has exactly the same block error information as the far-end CRC-4 checker. Counting E-bit changes is equivalent to counting CRC-4 block errors. Thus the local end can monitor the performance of both the go and return paths. This can be carried out by the network equipment itself, or by a test set such as the HP 37722A monitoring the E1 2.048-Mbps data bit stream. In the same way, the A-bits return alarm signals for loss of frame or loss of signal from the remote end.

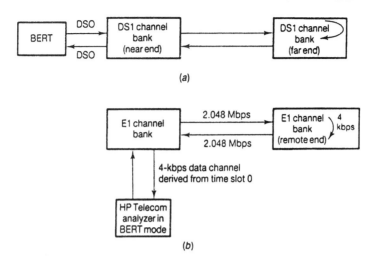

Figure 18.3. (a) Loopback of a DS0 channel with BERT test in place, intrusive or on spare DS0. (b) Loopback of 4-kbps data channel derived from ESF on DS1 or channel 0 on E1.

Loopback testing is a fine old workhorse in our toolbox of digital data troubleshooting aids. There are two approaches for DS1 (T1) and E1 systems: intrusive and nonintrusive. Intrusive, of course, means we interrupt traffic by taking one DS0 or E0 channel out of service, or the entire aggregate. We replace the channel with a pseudo-random binary sequence (PRBS) or other sequence specifically designed to stress the system. Commonly we use conventional bit error rate test (BERT) techniques looping back. This idea of loopback is shown in Figures 18.3a and b.

ESF and channel 0 data channel testing is nonintrusive. It does not interfere with customer traffic. Trouble can also be isolated, whether in the "go" or "return" channel of the loopback. Both intrusive and nonintrusive testing could be automated in the network management suite.

Many frame relay equipments also have forms of in-service monitoring as well as a system of fault alarms. In fact, most complex telecommunication equipment has built-in monitoring and test features. The problem often is that these features are proprietary, whereas our discussion of DS1/E1 systems indicates that they have been standardized by ITU-T and Bellcore recommendations and publications. This is probably the stickiest problem facing network management systems. That is handling, centralizing, and controlling network management features in a multivendor environment.

5.2   Communications Channels for the Network Management System

A network management facility is usually centrally located. It must monitor and control distant communications equipment. It must have some means of communicating with this equipment, which may be widely dispersed geographically. We have seen where DS1 and E1 systems provide a data channel for (OAM) operations and maintenance. Higher levels of the DS1 and E1 hierarchies have special communication channel(s) for OAM. So do SONET and SDH.

The solution for a LAN is comparatively straightforward. The network management facility/LAN protocol analyzer becomes just another active station on the LAN. Network management traffic remains as any other revenue-bearing traffic on the LAN. Of course, the network management traffic should not overpower the LAN with message unit quantity which we might call network management overhead.

WANs vary in their capacity to provide some form of communicating network management information. X.25 provides certain types of frames or messages dedicated for network control and management. However, these frames/messages are specific to X.25 and do not give data on, say, error rate at a particular point in the network. Frame relay provides none with the exception of flow control and the CLLM, which are specific to frame relay. For a true network management system, a separate network management communication channel may have to be provided. It would have to be sandwiched into the physical layer. However, SNMP (described below) was developed to typically use the transport services of TCP/IP. (See Chapter 11 for a discussion of TCP/IP and related protocols.) It is additional overhead, and care must be taken of the percentage of such overhead traffic compared to the percentage of "revenue-bearing" traffic. Some use the terms "in-band" when network management traffic is carried on separate frames on the same medium and "out-of-band" when a separate channel or time slot is used such as with E1/T1.

6 NETWORK MANAGEMENT FROM A PSTN PERSPECTIVE

6.1 Objectives and Functions

The term used by Bellcore [12] for network management is "surveillance and control." The major objectives for network surveillance and control organizations are:

- Maintain a high level of network utilization
- Minimize the effects of network overloads
- Support the BOC's* National Security Emergency Preparedness commitment

*BOC stands for Bell operating company, a local exchange carrier.

There are three important functions that contribute to attaining these objectives:

- Network traffic management (NTM)
- Network service
- Service evaluation

6.2 Network Traffic Management Center

An NTM center provides real-time surveillance and control of message traffic* in local access and transport area (LATA) telephone networks. The goal of an NTM center is to increase call completions and optimize the use of available trunks and switching equipment. Several dedicated operational support systems (OSSs) are employed by an NTM center to achieve this goal by accumulating information on both the flow of traffic and the manual and automatic/dynamic control capabilities provided by the network switching elements. Using this information and call control capability, a network traffic manager can optimize the call-carrying capacity of his/her network. The OSSs also enable the network traffic manager to interact with the network to minimize the adverse effects of traffic overloads and machine and/or facility failures.

6.3 Network Traffic Management Principles

NTM decisions are guided by four principles, which apply regardless of switching technology, network structure, signaling characteristics or routing techniques. All NTM control actions are based on at least one of the following principles:

1. *Keep All Trunks Filled with Messages.*† Since the network is normally trunk-limited, it is important to optimize the ratio of messages to nonmessages on any trunk group. When unusual conditions occur in the network and cause increased short holding-time calls (nonmessages), the number of carried messages decreases because nonmessage traffic is occupying a larger percentage of network capacity.

   NTM controls are designed to reduce nonmessage traffic and allow more call completions. This results in higher customer satisfaction and increased revenues from the network.

2. *Give Priority to Single-Link Connections.* In a network designed to automatically alternate-route calls, the most efficient use of available trunks occurs when traffic loads are at or below normal engineered values. When the engineered traffic load is exceeded, more calls are

---

*Message traffic in this context means telephone traffic.
†Bellcore [12] defines a message as *a call that has a high probability of completion.*

alternate-routed, and therefore they must use more than one link to complete a call. During overload situations, the use of more than one link to complete a call occurs more often, and the possibility of a multilink call blocking other call attempts is greatly increased. Thus in some cases it becomes necessary to limit alternate routing to give first-routed traffic a reasonable chance to complete. Some NTM controls block a portion or all alternate-routed calls to give preference to first-routed traffic.

3. *Use of Available Trunking.* The network is normally engineered to accommodate average business day (ABD) busy-hour calling requirements. Focused overloads (e.g., due to storms, floods, civil disturbances) and holiday calling often result in greatly increased calling and drastic changes from the calling patterns for which the network is engineered. This aberration can also be caused by facility failures and switching system outages. In these cases, some trunk groups are greatly overloaded, while others may be virtually idle. NTM controls can be activated in many of these cases to use the temporary idle capacity in the network. These controls are known as reroutes.

4. *Inhibit Switching Congestion.* A switching system is engineered to handle the expected number of attempts generated over its trunk groups with little or no service degradation. However, large numbers of ineffective attempts that exceed the engineered capacity of the switching system can result in switching congestion. If this switching congestion is not relieved, it not only affects potential messages withing the congested switching system, but it can also cause connected switching systems to become congested. Therefore, NTM controls are available that can remove the ineffective attempts to a congested switching system. These controls will result in inhibiting the switching congestion and preventing its spread to adjacent switching systems.

6.4 Network Traffic Management Functions

The following are several types of overloads for which network traffic management controls can provide complete or partial relief.

1. A *general network overload* is caused by changes in traffic patterns and/ or increased traffic load. These changes may be generated by a reduced business week (heavier calling before and after an extended weekend), holiday traffic, local or seasonal changes such as an increase in tourist traffic, unanticipated growth, and natural or man-made disasters. In cases of general network overload, a large amount of the network capacity may be used to switch calls that have a poor chance of completing. These calls are often regenerated many times by both the calling customers and switching systems before they are completed. This results in contention between the "poor

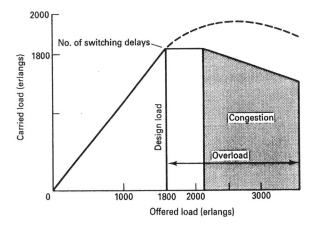

Figure 18.4. Network performance during overload conditions. From Ref. 12, Figure 10-1, courtesy of Bellcore.

completers" and those calls having a better chance of completing. Because of the volume of regeneration to the poor completers, much of the available trunking and switching capacity is used in switching these calls to an overflow condition.

As NTM personnel identify poor completers, appropriate measures are instituted, when necessary, to control congestion and remove some or all of these calls from the network. This is done by using code-blocking, call-gapping, and protective trunk group controls. The control of poor completers can greatly increase the number of messages handled by the network. Figure 18.4 illustrates the typical network performance under a general overload. This figure shows the decrease of completed traffic when the offered load exceeds the engineered capacity and congestion is present and the increased number of calls encountering switching delay.

2. A *focused overload* is generally directed toward a particular location and may result from media stimulation (e.g., news programs, advertisements, call-in contests, telethons) or events that cause mass calling to government or public service agencies, weather bureaus, or public utilities. Without the application of appropriate network controls, the effects of these types of overloads could spread throughout the network. Focused overloads are normally managed using code controls or, if anticipated, trunk-limited or choke networks.

3. A *switching system overload* occurs because each individual switch is engineered to handle a specific load which is known as *engineered capacity*. The engineered capacity is usually less than the total switching capacity. When the load is at or below engineered capacity, the switch handles calls in an efficient and reliable manner. However, when the load increases beyond the engineered capacity, delays can occur internal to the switch. This resulting internal congestion can spread, causing connected systems to wait for start-

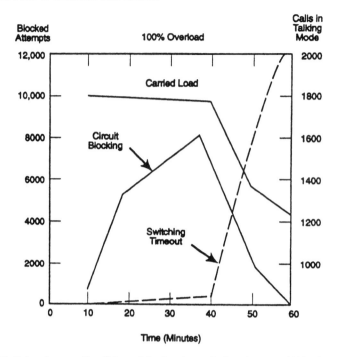

Figure 18.5. Network congestion. This model network was designed to carry 1600 erlangs. From Ref. 12, Figure 10-2, courtesy of Bellcore.

dial indications. This can also cause internal congestion in connected switching systems.

Figure 18.5 shows the buildup of switching delay over time and its effects on the carried load for a model network designed to carry about 1600 erlangs of simultaneous calls. This model was subjected to 100% overload (i.e., twice that of the design load). The left-hand ordinate of the figure indicates the number of attempts lost due to blockage or switching-delay timeouts.

At the onset of overload, also known as *circuit shortage*, the dominant cause for customer blockage is the failure to find an idle circuit. Circuit blockage alone limits the number of extra calls that can be completed but does not cause a significant loss in call-carrying capacity of the network below its maximum. As the overload persists and the network enters a congested state, regeneration-calling pressure changes customer blockage from circuit shortage to switching delays.

Switching delays cause timeout conditions during call setup and occur when switches become severely overloaded. Timeouts are designed into switches to release common-control components after excessively long delay periods and provide the customer with a signal indicating call failure. Switching congestion timeouts with short holding-time attempts on circuit groups replace normal holding-time calls. Switching delays spread rapidly throughout the network.

4. A *trunk-group overload* usually occurs during general or focused overloads and/or atypical busy hours. Some of the overload causes not discussed above are facility outages, inadequate trunk provisioning, and routing errors. The results of trunk-group overload can be essentially the same as those previously discussed for general overloads. However, the adverse effects are usually confined to the particular trunk group or the apex area formed by the trunk group and those groups that alternate-route to the overloaded trunk group. Trunk-group overload problems can often be minimized or handled completely by the use of temporary NTM reroute controls until a more permanent solution can be provided.

6.5   Network Traffic Management Controls

6.5.1   *Circuit-Switched Network Controls.* There are two broad categories of NTM controls:

1. *Protective Controls.*   These controls remove traffic from the network during overload conditions. This traffic is usually removed as close as possible to its origin, thus making more of the network available to other traffic with a higher probability of completion.

2. *Expansive Controls.*   These controls reroute traffic from routes experiencing overflows or failures to other parts of the network that are lightly loaded with traffic because of noncoincident trunk and switching system busy hours.

Implementation of either type of control can be accomplished on a manual or automatic basis. For example, manual controls are activated by network traffic managers, and automatic controls are activated by network components. In some switches, these controls are implemented on a planned control-response basis that is preprogrammed into the switch. In other systems, controls are available on a flexible basis, whereby any control can be assigned to any trunk group on a real-time basis.

The availability of any specific control, its allowable control percentages, and the method of operation can vary with the specific type of switch. In many instances, these network controls can be activated with variable percentages of traffic affected (for example, 25%, 50%, 75%, and 100%) to fine-tune the control to match the magnitude of the problem. Some switches also allow further control selectivity by the use of Hard-to-Reach (HTR) code-determination algorithms or specification of alternate-routed traffic, direct-routed traffic, or combined direct- and alternate-routed traffic-control choices. The most common manual controls are described below:

1. *Cancel controls* consist of two variations (see Figures 18.6 and 18.7). "Cancel From" (CANF) potentially prevents overflow traffic from a

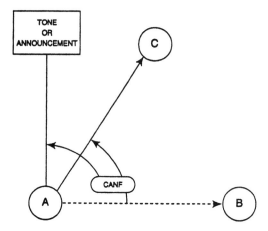

Figure 18.6. Cancel From (CANF). From Ref. 12, Figure 10-4, courtesy of Bellcore.

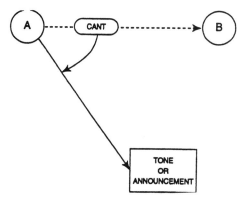

Figure 18.7. Cancel To (CANT). From Ref. 12, Figure 10-5, courtesy of Bellcore.

selected trunk group from advancing to any alternate route. "Cancel To" (CANT) potentially prevents all sources of traffic from accessing a specific route. Some control arrangements permit CANF and CANT to be applied to alternate-routed or direct-routed traffic or both. All cancel controls are implemented on a percentage-of-traffic basis.

2. *Skip route control* directs a percentage of traffic to bypass a specific circuit group and advance to the next route in its normal routing pattern. The control can be adjusted to affect alternate-routed or direct-routed traffic or both. See Figure 18.8.

3. *Code-block control* blocks a percentage of calls routed to a specific destination code. In most cases, a code-block control can also be specified to include the called-station address digits.

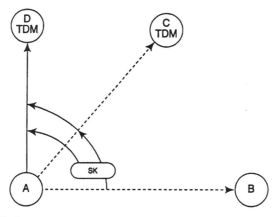

TDM = Tandem

Figure 18.8. Skip (SK). From Ref. 12, Figure 10-6, courtesy of Bellcore.

4. *Call-gapping control*, like code-block control, limits routing to a specific code or station address. Call-gapping is more effective in controlling mass-calling situations than the code-block control. Call-gapping consists of an adjustable timer that stops all calls to a specified code for a time interval selected from 16 different time intervals. After the expiration of the time interval, one call to the specified code or address is allowed access to the network, after which the call-gapping procedure is recycled for another time interval.

5. *Circuit-directionalization control* changes two-way circuits to one-way operation.

6. *Circuit-turndown control* removes one- or two-way circuits from service.

7. *Reroute controls* serve in a variety of ways to redirect traffic from congested or failed routes to other circuit groups not normally included in the route advance chain but that have temporary idle capacity. Reroutes override the normal routing algorithms in a switch. Reroutes can be used on a planned basis, such as on a recurring peak-calling day, or in response to unexpected overloads or failures. "Regular reroute" affects traffic overflowing a trunk group. "Immediate reroute" (IRR) affects traffic before hunting the trunk group for an idle circuit. Reroute controls may redirect traffic to a single or to multiple routes. The multiple option is referred to as a "spray reroute."

6.5.2 Automatic Controls in Modern Digital SPC Switches. Stored program control (SPC) switches may include the following types of automatic controls:

- Selective dynamic overload control (SDOC)
- Selective trunk reservation (STR)
- Dynamic overload control (DOC)
- Trunk reservation
- Selective incoming load control (SILC)

SDOC and STR are considered "selective" protective controls because they can selectively control traffic to HTR points* more severely than other traffic. If the probability of completing through the network is very low and the outgoing trunk groups or connected switches are overloaded, selective protective controls can prevent wasted usage of these overloaded network resources for traffic to HTR points. SDOC responds to switching congestion by dynamically controlling the amount and type offered to an overloaded or failed switch. STR, conversely, responds to trunk congestion in the outgoing trunking field and is triggered on a particular trunk group when less than a certain number of circuits are idle in that group.

SDOC and STR are two-level control systems. The first level indicates less congestion than the second level. The first-level response is typically limited to control of traffic destined for HTR points, whereas the second level applies controls to both HTR points and other traffic, typically alternate-routed traffic. HTR traffic can also be manually enabled.

HTR traffic is automatically detected by the AT&T 4ESS switch based on an analysis of destination-code completion statistics. This analysis is performed on a three- and six-digit basis every 5 min. In the Northern Tele-com DMS-100 and DMS-200 switches, HTR codes can also be manually selected and enabled.

Automatic controls, such as SDOC and STR, are intended to be activated by a switching system within a matter of seconds in response to a switch or trunk-group overload. These controls provide rapid protection for the network and, by their code-selective basis, attempt to restrict traffic that has a low probability of completion. When automatic controls trigger, network traffic managers monitor their operations and adjust system parameters to deal with the particular network condition, whether it is a general overload, a mass call-in, a natural disaster, or a major network-component failure. Among these parameters are call-completion determinations that designate a code "HTR" and control-response options that designate the amount of traffic to be controlled or trunks to be reserved at each triggering level. Since the optimum control response depends on the severity, geographical distribution, and type of overload, maximizing the calls carried by the network requires coordination and combination of automatic and manual control responses.

---

*HTR points are three- or six-digit destination codes to which calls have a very small chance of completing.

7 NETWORK MANAGEMENT SYSTEMS

7.1 What Are Network Management Systems?

Ostensibly a network management system provides an automated means of remotely monitoring a network for:

- Levels of performance (e.g., BER, loss of synchronization, etc.)
- Equipment, module, subassembly, card failures; circuit outages
- Levels of traffic, network usage

The impetus of the systems described below has come from the enterprise network environment, typically from the developer of TCP/IP (U.S. DoD) and later from the OSI world. There are now many proprietary network management systems available to the user. Among these are Hewlett-Packard's Openview, IBM's Netview, and Digital Equipment Corporation's EMA (enterprise management architecture). There has been a distinct trend toward distributed processing in the network management arena.

Such network management systems, especially in the distributed processing environment, require a means to communicate for monitoring and control of the enterprise network. Two network management communication protocols have evolved for this purpose, which we describe below.

7.2 An Introduction to Network Management Protocols

7.2.1 There Are Two Protocols. Two separate communities have been developing network management communication protocols:

- The TCP/IP (ARPANET) community: SNMP
- The ISO/OSI community: CMIP

The most mature and certainly the most implemented by far is SNMP (simple network management protocol). Certain weak points arose in the protocol and a version two has been developed and fielded, called SNMPv.2.

Common management information protocol (CMIP) has been developed for the OSI environment. It is more versatile but requires about five times the memory of SNMP.

7.2.2 An Overview of SNMP. SNMP is probably the dominant method for devices on a network to relay network management information to centralized management consoles which are designed to provide a comprehensive operational view of the network. Having come on-line in about 1990, literally thousands of SNMP systems have been deployed.

There are three components of the SNMP protocol:

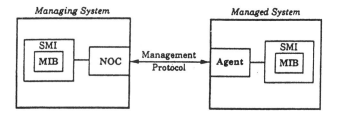

Figure 18.9. SNMP management architecture. SMI, structure of management information; NOC, network operations center; MIB, management information base.

- The management protocol itself
- The MIB (management information base)
- The SMI (structure of management information)

Figure 18.9 shows the classic client–server model. The client runs at the *managing* system. It makes requests and is typically called the *network management system* (NMS) or *network operation center* (NOC). The server is in the *managed* system. It executes requests and is called the *agent*.

SMI. Structure of management information (SMI) defines the general framework within which an MIB can be defined. In other words, SMI is the set of rules which define MIB objects, including generic types used to describe management information. The SNMP SMI uses a subset of Abstract Syntax Notation One (ASN.1) [11] specification language that the International Standards Organization (ISO) developed for communications above the OSI presentation layer. Layer 7, for example, may use ASN.1 standards such as ITU-T Recs. X.400 and X.500. It was designed this way so that SNMP could be aligned with the OSI environment. The SMI organizes MIB objects into an upside-down tree for naming purposes.

MIB. Management information base (MIB) is the set of managed objects or variables that can be managed. Each data element, such as a node table, is modeled as an object and given a unique name and identifier for management purposes. The complete definition of a managed object includes its naming, syntax, definitions, and access method (such as read-only or read-write) which can be used to protect sensitive data and status. By allowing a status of "required" or "optional," the SNMP formulating committee allowed for the possibility that some vendors may not wish to support optional variables. Products are obligated to support required objects if they wish to be compliant with the SNMP standard.

Figure 18.9 also shows an *agent*. One can imagine an agent in every data equipment to be managed as shown in Figure 18.10. Here it shows that the (network) management console manages "agents."

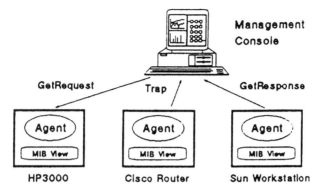

Figure 18.10. The network management console manages agents.

SNMP utilizes an architecture that depends heavily upon communication between one or a small number of managers and a large number of remote agents scattered throughout the network. Agents use the MIB to provide a view of the local data that is available for manipulation by the network management station. In order for a variable, such as the CPU utilization of a remote Sun workstation, to be monitored by the network management station, it must be represented as an MIB object.

The management station sends *get* and *set* requests to remote agents. These agents initiate *traps* to the management station when an unexpected event occurs. In such a configuration, most of the burden for retrieving and analyzing data rests on the management application. Unless data are requested in a proactive way, little information will be shown at the management station. This poll-based approach increases network traffic, especially on the backbone where some users report a 5–10% increase in traffic due to SNMP network management packets (or messages).

SNMP is a connectionless protocol that initially was designed to run over a UDP/IP stack (UDP stands for user datagram protocol). Because of its design, it is traditionally high in overhead, with the ratio of overhead to usable data running about ten bytes to one. A typical SNMP "message" (PDU) embedded in a local network frame is shown in Figure 18.11.

Inside the frame in Figure 18.11 we find an internet protocol (IP; see Chapter 11) datagram which has a header. The header has an IP destination address which directs the datagram to the intended recipient. Following the IP header there is the user datagram protocol (UDP) which identifies the higher-layer protocol process. This is the SNMP message shown in Figure 18.11 as embedded in the UDP. The application here is typically for LANs. If the IP is too long for one frame, it is fragmented (segmented) into one or more additional frames.

Figure 18.12 shows a typical SNMP PDU structure which is valid for all messages but the *trap* format. This structure is embedded as the "message" field of Figure 18.11. The SNMP message itself is divided into two sections:

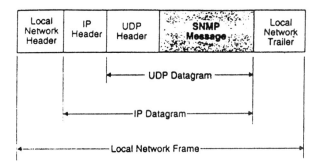

Figure 18.11. An SNMP "message" embedded in a local network frame.

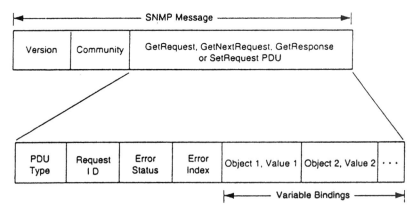

Figure 18.12. An SNMP PDU structure for GetRequest, GetNextRequest, GetResponse, and SetResponse. From Ref. 8, reprinted with permission.

(1) a version identifier plus community name and (2) a PDU. The version identifier and community name are sometimes referred to as the SNMP *authentication header*.

The *Version* field ensures that all parties in the management transaction are using the same version of SNMP protocol. We must remember the origins of SNMP evolved from TCP/IP described in Chapter 11, where we have already seen the use of a "version" field.

Each SNMP message contains a *community name* which is one of the only security mechanisms in SNMP. The agent examines the community name to ensure that it matches one of the authorized community strings loaded in its configuration files or nonvolatile memory. Each SNMP PDU is one of five types (sometimes called *verbs*): *GetRequest*, *GetNextRequest*, *SetRequest*, *GetResponse*, and *Trap*. The trap PDU is shown in Figure 18.13.

The PDU shown in Figure 18.12 has five initial fields. The first field is the *PDU* type. There are five types of PDU as we discussed previously. These are shown in Table 18.1.

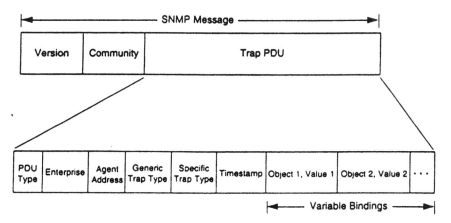

Figure 18.13. SNMP trap PDU format. From Ref. 8, reprinted with permission.

TABLE 18.1 PDU Type Field Values

GetRequest	0
GetNextRequest	1
GetResponse	2
SetRequest	3
Trap	4

TABLE 18.2 SNMP Error Codes

Error Type	Value	Description
noError	0	Success
tooBig	1	Response too large to fit in single datagram
noSuchName	2	Requested object unknown/unavailable
badValue	3	Object cannot be set to specified value
readOnly	4	Object cannot be set
genErr	5	Some other error occurred

The *Request ID* is the second field of the PDU field. It is an INTEGER type field that correlates the manager's request with the agent's response. INTEGER type is a primitive type used in ASN.1 [11].

The *Error Status* field is also an ASN.1 primitive type. It indicates normal operation (noError) or one of five error conditions as shown in Table 18.2.

When an error occurs, the *Error Index* field identifies the entry within the variable bindings list that caused the error. If, for example, a readOnly error occurred, the error index returned would be 4.

A *variable binding* pairs a variable name with its value. A *VarBindList* is a list of such pairings. Note that within the *Variable Binding* field of the SNMP PDU, the word *Object* identifies the variable name (OID* encoding of object type plus the instance) for which a value is being communicated.

A trap is an unsolicited packet sent from an agent to a manager after sensing a prespecified condition such as a cold start, link down, authentication failure, or other such event. Agents always receive SNMP requests on UDP port 161, and network management consoles always receive traps on UDP port 162. This requirement means that multiple applications on the same management station that wish to receive traps must usually pass control of this port to an intermediate software process. This process receives traps and routes them to the appropriate application. Figure 18.13 illustrates the SNMP trap PDU structure. We must appreciate that the trap PDU structure differs from the structure of the other four PDUs shown in Figure 18.12. Like the other PDU structure, the first field, PDU type, will be in this case PDU type = 4. (See Table 18.1.)

The next field is called the *Enterprise* field and identifies the management enterprise under whose registration authority the trap was defined. As an example, the OID prefix {1.3.6.1.4.123} would identify Newbridge Networks Corporation as the enterprise sending the trap. Further identification is provided in the *Agent Address* field, which contains the IP address of the agent. In some circumstances a non-IP transport protocol is used. In this case the value 0.0.0.0 is returned.

After the *Agent Address* field in Figure 18.13, we find the *Generic Trap Type* which provides more specific information on the event being reported. There are seven defined values (enumerated INTEGER types) for this field as shown in Table 18.3.

The next field is the *Time Stamp* field which contains the value of the sysUpTime object. This represents the amount of time elapsed between the last reinitialization of the agent in question and the regeneration of the trap. As shown in Figure 18.12, the last field contains the Variable Bindings.

TABLE 18.3   SNMP Trap Codes

Trap Type	Value
coldStart	0
warmStart	1
linkDown	2
linkUp	3
authenticationFailure	4
egpNeighborLoss	5
enterpriseSpecific	6

*OID stands for object identifier.

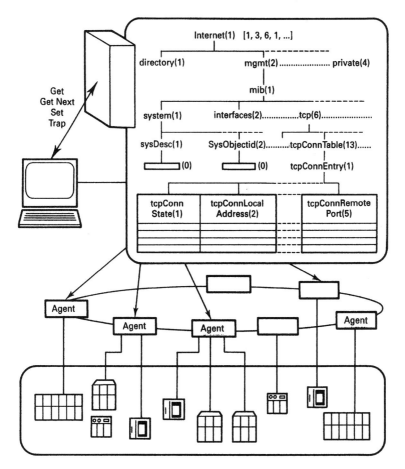

Figure 18.14. An overview of SNMP in place. From Ref. 6, Figure 2-4, courtesy of IEEE Press.

A trap is generated by an agent to alert the manager that a predefined event has occurred. To generate a trap the agent assigns PDU type = 4 and has to fill in Enterprise, Agent Address, Generic Trap, Specific Trap Type, Time Stamp fields, and Variable Bindings list.

Figure 18.14 gives an excellent summary overview of SNMP.

7.3 Remote Monitoring (RMON)

The tendency toward more and more distributed networks has become apparent. The wide area distribution is both geographical and logical. One method of handling this situation is to place remote management devices on the remote segments. These remote management devices are sometimes called *probes*. These probes are the remote sensors of the network management system, providing the centralized management station with the required monitoring data dealing with network operation. There is remote network monitoring (RMON) MIB which standardizes the management information sent to/from these probes.

As an illustration the Ethernet RMON MIB contains nine groups. One of these groups, for example, is titled "alarms." It compares statistical samples with preset thresholds and generates an alarm when a threshold is crossed. Another is "capture" which allow packets to be captured after they pass through a logical channel.

7.4 SNMP Version 2

SNMPv.2 design used the field experience gained by SNMP to sharpen and simplify the mappings to different transports. The management protocol has been separated from the transport environment, encouraging its use over practically any protocol stack.

One of the weaknesses of SNMP is in the unreliable manner of handling trap messages. Of course, management communications are most critical at times when the network is least reliable. A manager's communications with agents is vital. The use of UDP by SNMP means potentially unreliable transport. SNMP leaves the function of recovery from loss to the manager application. The GET-RESPONSE frame confirms respective GET, GET-NEXT, and SET. A network manager can detect the loss of a request when a response does not return. For instance, it can repeat the request. Traps are another matter. Trap messages are generated by the agent and are not confirmed. If a trap is lost, the agent applications would not be aware that there is a problem, nor would the manager for that matter. Since trap messages signal information which is often of great significance, securing their reliable delivery is very important.

Accordingly, SNMPv.2 has pursued an improved mechanism to handle event notifications, namely traps. For example, the trap primitive has been eliminated. It is replaced by an unsolicited GET-RESPONSE frame which is generated by an agent and directed to the *trap manager* (UDP port 162). Now event notifications can be unified as responses to virtual requests by event managers. In SNMPv.2, a special Trap MIB has been added to unify event handling, subscription by managers to receive events, and repetitions to improve reliable delivery.

In the case of SNMP over TCP/IP, we are looking at what one might call in-band communications. In other words we are using the same communications channel for network management as we use for "revenue-bearing" communications. This tends to defeat the purpose of network management. The problem can be somewhat alleviated if the access path to agents is somewhat protected from the entities that they manage.

Unlike SNMP, SNMPv.2 delivers an array of messaging options that enable agents to communicate more efficiently with management stations. Furthermore, SNMPv.2's bulk retrieval mechanism lets management stations obtain reports from agents about a range of variables without issuing repeated requests (i.e., GetBulkRequest). This feature should cut down the level of packet activity dedicated to network management and yet improve network management efficiency.

Another advantage of SNMPv.2 over its predecessor is manager-to-manager communications, which allows a station to act as either manager or agent. This allows SNMPv.2 systems to offer hierarchical management using mid-level managers to offload tasks from the central network management console.

SNMPv.2 does not have to rely on TCP/IP. It can run over a variety of protocol stacks including OSI and Internet Packet Exchange (IPX). It also has notably enhanced security features over its predecessor.

Figure 18.15 shows a generic SNMPv.2-managed configuration.

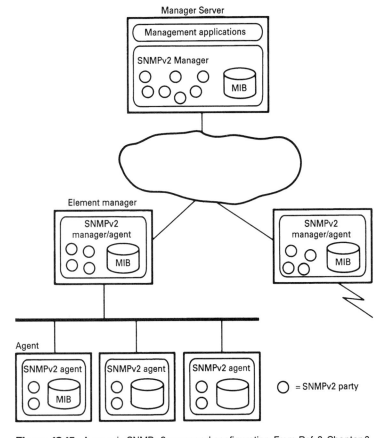

Figure 18.15. A generic SNMPv.2-managed configuration. From Ref. 2, Chapter 3.

7.5   Common Management Information Protocol (CMIP)

SNMP was developed rapidly with an objective to serve as a network management communications standard until CMIP was issued. SNMP met such success that it now has a life of its own. CMIP remains with slow implementation. It has not yet reached the popularity of SNMP.

CMIP is an ISO development and it is designed to operate in the OSI environment. It is considerably more complex than its SNMP counterpart. Figure 18.16 illustrates the OSI management architecture which uses CMIP to access managed information. In Figure 18.16 this managed information is provided by an agent in the LAN hub.

CMIP uses different terminology than SNMP. An agent maintains a management information tree (MIT) as a database; it models platforms and devices using managed objects (MOs). These may represent LANs, ports, and interfaces. CMIP is used by a platform to change, create, retrieve, or delete MOs in the MIT. It can invoke actions or receive event notifications.

CMIP (i.e., OSI management communications) communications are very different than those found in SNMP. These communications are embedded in the OSI application environment and they rely on conventional OSI peer layers for support. They use connection-oriented transport where SNMP uses the datagram (connectionless). In most cases these communications are acknowledged.

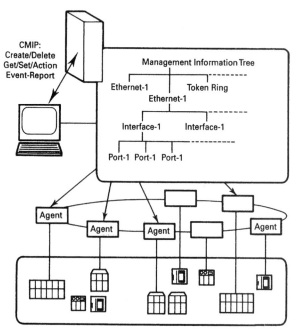

Figure 18.16. A typical overall architecture of an OSI network management system. From Ref. 6, Figure 2-12.

8   NETWORK MANAGEMENT IN ATM

The ITU-T (CCITT) organization and ANSI have left local network management procedures in the M-plane for "further study."

In the interim period until CCITT/ANSI have formulated such standards, SNMP and the ATM UNI management information base (MIB) are required to provide any ATM user device with status and configuration information concerning virtual path and channel connections available at its UNI. (See Chapter 15 for a discussion of ATM and clarification of the acronyms used here.)

The ATM Forum has developed the interim local management interface (ILMI) specification. The ILMI fits into the overall management model for an ATM device as shown in Figure 18.17 as clarified by the following principles and options.

1. Each ATM device supports one or more UNIs.

2. ILMI functions for a UNI provide status, configuration, and control information about link and physical layer parameters.

3. ILMI functions for a UNI also provide for address registration across the UNI.

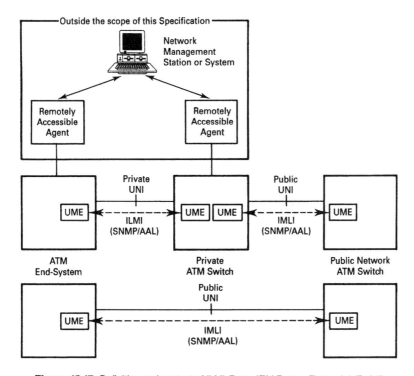

Figure 18.17. Definition and context of ILMI. From ATM Forum, Figure 4-1 (Ref. 7).

4. There is a per-UNI set of managed objects, the UNI ILMI attributes, that is sufficient to support the ILMI functions for each UNI.

5. The UNI ILMI attributes are organized in a standard MIB structure; there is one UNI ILMI MIB structure instance for each UNI.

6. There is one MIB instance per ATM device, which contains one or more UNI ILMI MIB structures. This supports the need for general network management systems to have access to the information in the UNI ILMI MIB structures.

7. For any ATM device, there is a UNI management entity (UME) associated with each UNI that supports the ILMI functions for that UNI, including coordination between the physical and ATM layer management entities associated with that UNI.

8. When two ATM devices are connected across a (*point-to-point*) UNI, there are two UNI management entities (UMEs) associated with the UNI, one UME for each ATM device, and two such UMEs are defined as adjacent UMEs.

9. The ILMI communication takes place between adjacent ATM UMEs.

10. The ILMI communication protocol is an open management protocol (i.e., SNMP/AAL initially).

11. A UNI management entity (UME) can access, via the ILMI communication protocol, the UNI ILMI MIB information associated with its adjacent UME.

12. Separation of the MIB structure from the access methods allows for the use of multiple access methods for management information. For the ILMI function, the access method is an open management protocol (i.e., SNMP/AAL) over a well-known VPI/VCI value. For example, for general network management applications [e.g., from a network management station (NMS) performing generic customer network management (CNM) functions], the access method is also an open management protocol (e.g., SNMP/UDP/IP/AAL) over a specific VPI/VCI value (or a completely separate communications method) allocated to support the general management applications. The peer entity in an ATM device that communicates directly with an NMS is a management agent, not a UME; however, since the management agent can access the MIB instance for the ATM device, it can access all of the UNI ILMI MIB structure instances.

The simple network management protocol (SNMP) without UDP and IP addressing, along with ATM UNI management information base (MIB), was chosen for the ILMI.

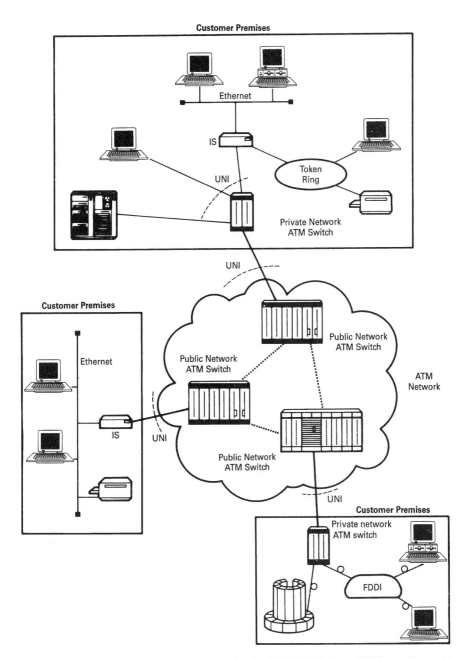

Figure 18.18. Examples of equipment implementing the ATM UNI ILMI. From ATM Forum, Figure 4-2 (Ref. 7).

8.1 Interim Local Management Interface (ILMI) Functions

An ILMI supports bidirectional exchange of management information between UNI management entities (UMEs) related to UNI ATM layer and
physical layer parameters. The communication across the ILMI is protocol
symmetric. In addition, each of the adjacent UMEs supporting ILMI will
contain an agent application and may contain a management application.
Unless stated otherwise for specific portions of the MIB, both of the adjacent
UMEs contain the same management information base (MIB). However,
semantics of some MIB objects may be interpreted differently. As shown in
Figure 18.18, an example list of the equipment that will use the ATM UNI
ILMI is presented below:

- Higher-layer switches such as internet routers, frame relay switches, or
  LAN bridges that transfer their frames within ATM cells and forward
  the cells across an ATM UNI to an ATM switch.
- Workstations and computers with ATM interfaces which send their data
  in ATM cells across an ATM UNI to an ATM switch.
- ATM network switches which send ATM cells across an ATM UNI to
  other ATM devices.

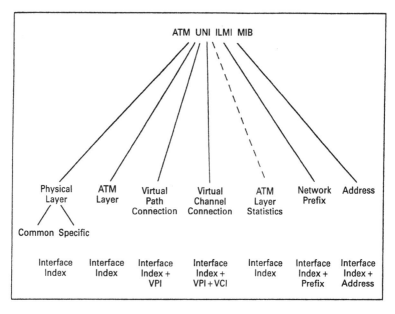

Figure 18.19. ATM UNI ILMI MIB tree structure. From ATM Forum, Figure 4-3 (Ref. 7).

8.2 ILMI Service Interface

The ILMI uses SNMP for monitoring and control operations of ATM management information across the UNI. The ATM UNI management information will be represented in a management information base (MIB). The types of management information that will be available in the ATM UNI MIB are as follows:

- Physical layer
- ATM layer
- ATM layer statistics
- Virtual path (VP) connections
- Virtual channel (VC) connections
- Address registration information

The tree structure of the ATM UNI ILMI MIB is shown in Figure 18.19.

REVIEW QUESTIONS

1. Give at least four major benefits derived by implementing a network management system.

2. Name the five traditional tasks of network management.

3. Discuss fault management and describe some of the capabilities it should incorporate.

4. It has been said that a network management center is the frontline command post for network _____.

5. Give the four steps in "finding (a) fault."

6. Describe how a well-engineered network management system can often cut the time for isolating faults almost to zero.

7. A network management system should be built around on-line monitoring systems that are nonintrusive and that are found in (list such systems).

8. Explain how BERT works.

9. Network management systems require communications, especially to/ from the network management center and remote equipment. Discuss concerns you might have with such systems, particularly if they share traffic-bearing channels.

10. From a PSTN perspective, network management may be called "surveillance and control." What are the two major objectives of surveillance and control?

11. What might nonmessage traffic be?

12. Describe some NTM controls.

13. Give some of the general causes of *general network overload.*

14. What is a *focused overload*? Name two measures to mitigate effects of focused overloads.

15. What are the two broad categories of NTM controls?

16. Describe the two variations of *cancel controls.*

17. List four of the automatic (traffic flow) controls one might encounter in a modern SPC switch.

18. There are two network management communication protocols. Give the name of each and their origin (i.e., what organization was/is responsible for their development).

19. What are the three components of SNMP? Use the acronyms and write out their meaning.

20. What does the MIB do for agents?

21. In SNMP, what does an agent do when an unexpected event occurs?

22. What is the efficiency of SNMP regarding overhead?

23. How does a *probe* fit into the network management operation?

24. The RMON MIB contains nine groups. One of these groups is titled "alarms." How do they work?

25. How does SNMP version 2 improve upon SNMP?

26. CMIP is based on what environment?

27. What is now providing network management in ATM?

28. Name at least five principles and options of ILMI.

29. What types of management information will be available in the ATM UNI MIB? List at least five items.

30. Give a list of example equipment that would use ATM UNI ILMI (at least four items).

REFERENCES

1. G. Held, *Network Management: Techniques, Tools and Systems*, John Wiley & Sons, Chichester, United Kingdom, 1992.

2. W. Stallings, *Network Management*, IEEE Computer Soc. Press, Los Alamitos, CA, 1993.

3. S. M. Dauber, "Finding Fault," *BYTE Magazine* (March 1991).

4. *Error Performance of an International Digital Connection Forming Part of an Integrated Services Digital Network*, CCITT Rec. G.821, Fascicle III.5, IXth Plenary Assembly, Melbourne, 1988.

5. *Frame Alignment and Cyclic Redundancy Check (CRC) Procedures Relating to Basic Frame Structures Defined in Recommendation G.704*, CCITT Rec. G.706, CCITT Geneva, 1991.

6. S. Aidarous and T. Plevyak, *Telecommunications Network Management into the 21st Century*, IEEE Press, Piscataway, NJ, 1993.

7. *ATM User-Network Interface Specification*, Version 3.0, The ATM Forum, PTR Prentice-Hall, Englewood Cliffs, NJ, 1993.

8. M. A. Miller, *Managing Internetworks with SNMP*, M&T Books, New York, 1993.

9. *A Simple Network Management Protocol*, RFC 1157, DDN Network Information Center, SRI International, Menlo Park, CA, May 1990.

10. *Synchronous Frame Structures Used at Primary and Secondary Hierarchical Levels*, CCITT Rec. G.704, Fascicle III.4, IXth Plenary Assembly, Melbourne, 1988.

11. *Information Processing Systems: Open Systems Interconnection—Abstract Syntax Notation One (ASN.1)*, ISO Std 8824, Geneva, 1987.

12. *BOC Notes on the LEC Networks—1994*, Issue 2, Bellcore, Piscataway, NJ, April 1994.

INDEX

An *italic* term denotes extensive coverage of a subject. **Boldface** denotes a definition.

AAL-0, 793–794
AAL-1, 794
AAL-2, 795
AAL-3/4, 798
AAL-5, 798–799
AAL (ATM adaptation layer), 793–799
AAL categories or Types, 793–799
AAL-5 format, 799
AAL multiplexing may cause CDV, 815
AAL-3/4 SAR-PDU, 797
ABD (average business day), 954
A-bits return alarm signals, 950
ABM (asynchronous balanced mode), 543–544
abort condition (HDLC), 545
aborting a frame, 479
abort (TCP), 572
ABS (average busy season), 414
ABSBH (average busy season busy hour), 9
absolute delay, 183
AbS predictive coder, 903
AC (access control) (token ring), 639
access burst, 907
access classes, 755–756
access classes and enforcement, 765–767
access classes (SMDS), 753
access class mechanism, 765, 766
access connection element model, 676
access control field (ACF), 735, 736, 762
access DQDB, 752
access rate (AR), 728
access rate established, 729
access techniques, VSAT, 315–316
access through the B-channel, 709
access through the D-channel—basic rules, 709
access to DQDB network, 735
access to PSPDN services—case A, 708–709
access transport parameter, 874
access unit (AU), DQDB, 735
accounting management, 943
ACD (automatic call distribution), 425
ACF (access control field), 735, 736
ACF (access control field), 761

ACF field, 747
ACH (attempts per circuit hour), 11·
ACK (acknowledgement signal), 472, 473
acknowledgement field significant, 576
acknowledgement number, 576
ACK (TCP), 576
acoustic pressure, 3
acquiring the medium (CSMA/CD), 626
acquisition time vs. BER, 788
AC signaling, 155–159
ac signaling equipment, 156
actions required at an incoming international exchange, 874–875
actions required at an intermediate international exchange, 873–874
actions required at an intermediate national exchange, 871–872
actions required at an outgoing international exchange, 872–873
actions required at originating exchange, 869–871
actions required at the destination exchange, 875–877
active coupling, fiber LAN, 613, 614
active monitor, token ring, 642
activity factor (FDM), 204–206
adapt existing telephone network for data transmission, 489
adaptive logic monitors occurrence of successive 1s, 393
adaptive network management signals, 52
adaptive overlapping codebook, 905
adaptive routing schemes, 229
ADCCP (advanced data communications control protocol), 545
add-drop multiplexing (SONET), 379
additional margin (fiber optics link), 334
address extension bit (EA), 719, 721
address field, 478, 759
address field, frame relay, 719, 721–724
address field extension bit (EA), 695
address field format (LAPD), 695

979

address field formats, frame relay, 719
address field (HDLC), 545
address field (LAPD), 694
address information sending sequence (SS7, ISUP), 870
address (memory location), 421
address message type, 861
address mobility, SMDS, 755
address-recognized indicator (A), 652
address resolution protocol (ARP), 564, 565
address signaling, 151, 159–165, 864
address signals, 862
addressSize, 631
address type subfield, 801
address__type subfield, 759
ad hoc test link, 949
A-digit selector, 115
adjustable equalizer, 362
administrative unit (AU) pointer, 385
administrative unit (AU) pointers, 387
administrative unit-n (AU-n), 384
administrative units in the STM-1 frame, 389
administrative units in the STM-N, 389
ADPCM (adaptive differential PCM), 902
ADPCM codec, 902
advanced mobile telephone system (AMPS), 885, 886, 887
advanced peer-to-pear networking (APPN), 578, 579
advanced peer-to-peer networking (APPN), 579–580
advantages and issues of PCM switching, 404–405
advantages of CSMA/CD, 655–656
advantages of TDMA, 909–910
advantages of token passing, 656
AEN (articulation reference equivalent), 65
AGC (automatic gain control) circuits, 325
agent address, 966
agent(s), 962, 963
aggregate bit stream, 679
AIOD (automatic identified outward dialing), 134
A-law companding, 347, 348
A-law conversion to mu-law, 452–453
ALBO (automatic line buildout), 362
alerting (telephony), 104, 105
algebraic codebook, 904
alignment error rate monitor, 838
all circuits busy, 145
allocation of bits 1 to 8 in TS 0, E1 frame, 365
allocation of SAPI values, 696
allocation of slips, 451–452
ALOHA (an access technique named by the University of Hawaii), 315
Alpha and Delta (ATM synch), 787
alternate (alternative) routing arrangement, 239
alternate mark inversion (AMI), 360
alternate mark inversion (AMI), 485
alternate-routed traffic, 957
alternative access, 426

alternative (alternate) routing, 118, 227
alternative routing, 40–42
alternative routing concept, 41
alternative traffic formula conventions, 20, 31
aluminum as a conductor, 85
AMA (automatic message accounting), 123, 134
AM (administrative module) (5ESS), 417
AM (amplitude modulation), 489
American standard code for information interchange (ASCII), 462, 463, 464, 465, 467
AMI (alternate mark inversion), 341
amplitude distortion, 493
amplitude equalizer, 503
amplitude-frequency response, 182–183, 501
amplitude-frequency response across a voice channel, 493
amplitude modulation, vestigial sideband (VSB), 489, 490
amplitude modulation (AM), double sideband (DSB), 489
AMPS, 901, 905
AMPS (advanced mobile telephone system), 885, 886, 887
AMSC (American Mobile Satellite Corporation), 927
analog applications of fiber optics transmission, 329
analog signal, 339
analog switching in telephony, 99–148
analog voice channel, 485
analysis by synthesis, 903
ANC (all number calling), 139
angle modulation for wide area paging, 919
angle of incidence, 318
angular misalignment, fiber splice, 323
angular spread, large, 324
ANI (automatic number identification), 134
ANR (automatic network routing), 579
ANSI (American National Standards Institute), 613
ANSI frame relay, 717
ANSI/IEEE std 802.3, 624
answered, 165
antenna aperture, 289
antenna height/antenna separation, 899
antenna noise, 305
antenna noise temperature, 306
antijam properties, 910
APD (avalanche photodiode), 326, 327
aperture efficiency, 284
application layer, 542
application of alternate routing, 242
application of EFDAs, 329
application of PCM to interexchange plant in local area, 356
APPN (advanced peer-to-peer networking), 578, 579
APPN end node, 585
APPN multilink TGs, 588
APPN network, 586
APPN nodes, 584

apportionment (G.826), 446–447
apportionment rule (G.826), 446
approaches to PCM switching, 405–412
architectural components of an SNA network, 582–587
architecture of an OSI network management system, 970
area code, 138, 139
areas and exchange relationships, 223
ARIES (a LEO satellite system), 927
ARM (asynchronous response mode), 543
ARP (address resolution protocol), 560
ARPANET (Advanced Research Projects Agency network), 517
ARPANET geographic map, 518
ARP cache, 566
ARQ (automatic repeat request), 317, 472, 473
ARQ delays, 775
ARQ protocol, 527
ARR (automatic rerouting), 231
8-ary PSK, 507, 509
8-ary PSK (phase shift keying), 281, 282
ASCII (American standard code for information interchange), 462, 463, 464, 465, 467
ASN.1 primitive type, 965
ASN.1 standards, 962
assigned internet protocol numbers, 565
assignment and removal of VPI value, 805
associated and disassociated channel signaling, 170–172
associated mode, 828
asynchronous and synchronous transmission, 476–479
asynchronous—ATM, 775, 776
asynchronous balanced mode (ABM), 543–544
asynchronous LLC frame, 651
asynchronous response mode (ARM), 543
asynchronous transfer mode (ATM) and broadband ISDN, 773–822
ATB (all trunks busy), 9, 302
a-t-b (points), 251
ATM adaptation layer (AAL), 793–799
ATM as hierarchy of levels, 792
ATM (asynchronous transfer mode), 713
ATM cell, 779–785
ATM cell mapping, 819
ATM cells in the DS3 frame, 817–818
ATM cell stream, 780
ATM cell structure, 779–785
ATM Forum, 785
ATM four bit rates, 779
ATM header structures, 780
ATM introduction, 774–776
ATM is outgrowth, 774
ATM layer, 785
ATM layer, 790–792
ATM layer functions supported at UNI, 790
ATM layering and B-ISDN, 787–799
ATM layer management, 972
ATM layer management (M-plane), 792
ATM layer QoS, 815

ATM links simultaneously carry a mix of voice, data and image information, 774
atmospheric absorption, 296
ATM reference model, 777
ATM routing and switching, 804
ATM service quality review, 808–809
ATM switch/cross-connect, 799
ATM train analogy, 775
ATM UNI, 780
ATM UNI ILMI implementation example, 973
ATM UNI management, 975
attending-alerting function, 104, 105
attenuation distortion, 182–183, 493
attenuation limit, 71
attenuation per unit length versus wavelength for glass fiber, 325
ATT ESS, 127, 128
AT&T 5ESS, composition of internal 16-bit time slot, 413
AT&T No. 4 ESS, 405
AU (attachment unit), 736, 737
audible tones commonly used in North America, 166
AUG (administrative unit group), 385
AUI (attachment unit interface), 624, 629
AU (=ISDN access unit), 706
AUI uses Manchester coding, 631
A-unit hunter, 115
AU-n pointers, 387, 389
AU-pointers are used, 821
authentication header, 964
autocorrelation distance, 887
automatic controls in modern digital SPC switches, 959–960
automatic equalizer, 503
automatic message accounting (AMA), 123
automatic number identification (ANI), 134
automatic rerouting (ARR), 231–231
automatic service observation, 147
automatic traffic measuring equipment, 145, 146
Autovon, 250
AU writes isochronous service, 743
availability, 12, 446
availability, 110
availability of downstream elements, 227
availability (SS7), 849
available credit, 765
available data transmission options in order of increasing cost, 531
available trunking, use of, 954
average business day (ABD), 954
average busy season busy hour, 9
average delay on all calls, 37
average delay to satellite, 929
average echo tolerance, 254
average holding time, 9
average line bit rate, 432
average percentage of investment in public telephone equipment, 60
average power of a telephone talker, 188
average power per voice channel, 203

average size country, definition, 851
average slot access delay, 735

backbone of a network, 224
backbone route, 48
background block error (BBE), 445
backoffLimit, 631
back-off (TWT operation), 301
backup to fiber optic cable, 267
backward direction (signaling), 164
backward explicit congestion notification (BECN), 721
backward indicator bit (BIB), 840
backward information, 169
backward sequence number (BSN), 839, 840
backward setup request, 863
balance bit, 681
balanced asynchronous class, X.25, 551
balanced configuration, 543
balanced mixer, 194
balanced sequence of symbol pairs, 646
balanced vs. unbalanced receiver, 488
balance return loss, 192
balance return loss, 246, 250
balancing network, 191
balancing network (of a hybrid), 191
band-limited system (noise), 185
bandwidth, unfortunate usage, 776
bandwidth ambiguity—frame relay, 717
bandwidth balancing, 738
bandwidth balancing, 740–742
bandwidth balancing modulus (BWB_MOD), 742
bandwidth bind, 714
bandwidth dilemma, 901–905
bandwidth-distance product, 321
bandwidth limits modulation rate, 501
bandwidth on broadband, 611
bandwidth vs. voice quality, 901
1BASE5, 631
10BASE5, 631
baseband transmission technique, 608, 609
baseband transmitting waveform, true, 641
BASE or BROAD, 632
bases of network configurations, 42–51
base station monitors delay, 908
10BASET on twisted pair, 632
basic and primary user interfaces, 668–669
basic electronic feedback circuit, 391
basic information unit (BIU), 596–597
basic information unit format, 596
basic link unit (BLU), 598
basic link unit (BLU) format, 598
basic rate, ISDN, 667
basic routing fundamentals, 239–243
basic rules (ISDN packet A), 709
basic SDH multiplexing structure, 390
basic signal unit format, 839–841
basics in conventional telephony, 1–57
basic switching functions, 104–106
basic system components, cellular, 885

BAsize 36, 765, 796
BAsize (buffer allocation size), 759
battery, 4
battery (data), 473
battery feed, 62, 429
baud, 483
BBE (background block error), 445
BBE ratio (BEBR), 445
B-bit (X.25), 552
BCC (block check count), 470, 548
BCD code, 464
B-channel, 667
B-channel fully transparent, 671
B-channel packet calls, 707
2B+D, 669
23B + D, 668
30B + D, 668
2B + D customer data pattern, 682
2B + D multiplexed, 679
bearer, 265–266
bearer channel for frame relay, 717
bearer (fiber), 318
BEBR (background block error ratio), 445
BECN (backward explicit congestion notification), 722
BECN indication, 727
beginning-end tag (BEtag), 759
Bell Telephone Manufacturing Company (Antwerp), 357
Bell telephone receiver, 63
bending radii, 322
bent pipe satellite, 294
bent pipe satellite systems, 309
BER, 441
BER: frame relay vs. X.25, 715
BER ad hoc standard for fiber optic links, 330
BER at physical interface, token bus, 635
BER (bit error rate, QoS primary item), 808
BER (bit error rate or ratio), 281
BER (bit error rate) requirements, 269
BER for PCM, signaling threshold, 341
BER for PCM intelligibility, 341
BER for voice, 935
Bernoulli distribution, 16
Bernoulli formula, 17
BER on satellite links, 297
BER range on LANs, 609
BER requirements and objectives, 469
BER specified (PCM), 374
BERT (bit error rate test), 286, 287
BERT (bit error rate test) mode, 951
BER threshold value, 442
BER versus channel SNR for various values of K, 898
BER versus E_b/N_o performance of coherent BPSK/QPSK and 8-ary PSK, 282
BEtag (beginning-end tag), 758, 759
BH (busy hour), 8–9, 225
BH (busy hour) demand, 244
BH (busy hour) peaks, 52
BH (busy hour) service, 205
BH traffic intensity, 40

BH traffic load, 17
bias voltage on APD, 327
BIB (backward indicator bit), 840
binary coding of PCM, 344
binary coding techniques, 461–463
binary convention–removing ambiguity, 460–461
binary digit, 460
binary polar baseband modulation, 494
binary sequence 01111110, 478
binary transmission and the concept of time, 475–485
binomial formula, 31
BIP (bit interleaved parity), 818
Biphase-L, 484, 485
bipolar AMI waveform, 362
bipolar mode in PCM transmission, 360
B-ISDN/ATM functional layering, 789
B-ISDN/ATM network, 804–805
B-ISDN connectionless protocol, 799
B-ISDN protocol reference model, 778
B-ISDN reference configuration, 777
bit, 460
60-bit address subfield, 760
32-bit aligned, 743
7-bit ASCII code, 469
32-bit boundaries, 577, 796
bit error rate (BER), 469
bit error ratio (BER), 469
bit errors accumulate, 397
bit errors accumulate (PCM), 374
bit packing, 281
bit rate and synchronization (PRI), 685–686
bit rate reduction of the digital voice channel, 902–905
bits, bauds and symbols, 483
BITS (building integrated timing supply), 437
BITS clock, 437
3-bit sequence counter, 794
bit-sequence-independent transmission, 689
BITS implementation, 437
bit stuffing, 478, 479
bit stuffing, 545
bit synchronization, 431
8-bit time slot, 351
8-bit to 10-bit mapping, DMS-100, 412
bit transfer and bit alignment, 788
BIU (baseband interface unit), 609
BIU (basic information unit), 596–597
blank-and-burst, 888
block, 470
blockage, lost call and grade of service, 11–12
block check count (BCC), 470
block codes, 471
block diagram of a PCM regenerative repeater, 362
block diagram of a SPADE terminal, 303
blocked, 165
blocked call, 11
block error check, 682, 950

blocking probability, 441
blocking probability can be minimized, 410
BLU (basic link unit), 598
B-NT1 (functional group), 777
B-NT2 functional group, 778
BNZS (binary N-zeros substitution), 360
BOC (Bell operating companies), 952
BOC (Bell operating company), 238
Boltzmann's constant, 185, 279, 308
BOM (beginning of message) DMF-DU, 745, 746
border nodes, 587
BORSCHT, 429
both-way circuits, 52–53
both-way circuits, 106, 554
both-way operation, 52
boundary function, 583, 594
boundary nodes, 583, 589
BPSK/QPSK (binary phase shift keying/quadrature phase shift keying), 281
2B1Q waveform, 681
2B1Q waveform, 683
breaking a cell up into three sectors and six sectors, 917
BRI (basic rate interface), 668, 671
bridged taps removed, 357
bridging (SMDS), 760
BRI differences in the United States, 681–685
Brillouin amplifiers, 328
British urban path loss formula, 890
broadband, 777
broadband alternative for token bus, 637
broadband channel allocations for CSMA/CD services, 612
broadband channel allocations for token ring services, 613
broadband transmission consideration, 610–612
broadband vs. baseband comparison, 609
broadcast address, CSMA/CD, 628
broadcast polling, 622–623
BSBH (busy season busy hour), 9
BSN (backward sequence number), 839, 840
BT (bridged tap), 87
B-TE functional group, 778–779
buffer finite capacity, 367
buffer overflow, 432
buffer overflow/underflow, 367
buffer storage (TDMA), 302
building blockage component, 892
building integrated timing supply (BITS), 437
building penetration, 891
build-out networks removed, 357
bulk billing, 142
bundle of VC links, 805
burst errored seconds, 442
burst errors, 468, 900
burstiness allowed on information transfer, 766
burst of data (SMDS), 755
bursty operation from remote VSAT, 315
bursty traffic of frame relay, 725
busied out, 113
bus network, 607
busy-back, 113

BUSY bit, 736, 748, 750
busy hour, 8–9
busy hour (BH), 886
busy season busy hour, 9
BUSY subfield, 761
busy-test, 118, 120
busy test(ing), 104, 105
BWB_MOD (bandwidth balancing modulus), 742
byte-interleaved multiplexing concept, 380
B3ZS (binary 3 zeros substitution), 360
B8ZS (binary 8 zeros substitution), 360

cable conductors, properties of, 74
cable reel loss variability, 332
CAC and UPC/NPC procedures, 814
CAC (connection admission control), 811
calculated terrestrial transmission delays, 850
calculation of EIRP, 277–278
calculation of isotropic receive level, 278
calculation of receiver noise level, 278–279
calculation of receive signal level (RSL), 278
call-area boundaries, 142
call-attempt dispositions, typical, 39
call-attempt factor, 38, 39
call charging: European vs North American approaches, 134–135
call-completing markers, 123
call concentration, 107
call confirmation packet, 557
call congestion, 19
call control procedures, 230–231
call-gapping, 955
call-gapping, 959
call holding time, 38
call hour (Ch), 10
calling rate, 7
call progress (signaling), 151–152
call-progress tones, 165
call reference flag, 704
call reference values, 704
call request and incoming call packet, 555
call request packet, 557
call setup, end-to-end signaling, 169
call setup for virtual connections, 552
call-store memory (SPC switching), 125, 126
CALT/CHU, 240, 241
CALT (cost of a path on the alternate route), 240
CAMA (centralized automatic message accounting), 134
cancel controls, 957–958
cancel from (CANF), 957
cancel to (CANT), 958
CANF (cancel from), 957, 958
capacity formula of Shannon, 499
carbon microphone, 427
carried traffic, 7
carrier and clock timing recovery pattern, 300
carrier modulation techniques for VSAT, 317
carrier reinsertion frequency offset, 498
carrier sense, 626

carrier-to-noise level in 1 Hz bandwidth, 308
case A: configuration when accessing PSPDN services, 705–706
case B: configuration for ISDN virtual circuit service, 707–710
categories of signaling point (SP), 847
CATV (community antenna television) industry basis of broadband LAN, 608, 611
CATV fiber links, 329
cause of congestion, 727
causes of echo and singing, 245–246
C-bit (DDS), 510
CBR (constant bit rate services), 775
CC (composite clock), 437
CCH (connections per circuit hour), 11
CCIR equation amended, 890
CCIR formula, 889
CCITT alphabet No. 2, 462, 463
CCITT approach (noise in FDM), 207–209
CCITT group, formation of, 196
CCITT modulation plan, 195–200
CCITT mastergroup and supermastergroup, 196
CCITT No. 4 code, 159, 160
CCITT No. 5 code showing variations with R-1 code, 163
CCITT No. 5 signaling code, 163
CCITT pilot frequencies and levels, 206
CCITT Rec. V.110, data over digital network, 510–511
CCITT R-2 signaling system, 158, 159
CCITT signaling system No. 7, 671
CCITT signaling system No. 7, 667
CCITT signaling system No. 7, 827–878
CCITT supergroup, formation of, 196, 198
CCITT supergroup assembly, 200
CCITT synchronization plans, 439–441
CCS (cent call second), 10
CD (collision detect), 626
CD (countdown) counter, 738
CD count of zero, 739
CDMA capacity improvement, 910
CDMA (code-division multiple access), 910, 912–915
CDMA (code division multiple access, implies spread spectrum), 315
CDMA (code division multiple access) diversity, 898
CDO (community dial office), 423
CDPD (cellular digital packet data), 922
CDS (central directory server), 586
CDV be standardized edge-to-edge, 816
CDV (cell delay variation), 809
CDV effects on peak cell rate, 816
CDV tolerance, 814, 815, 817
cell delay variation (CDV), 809
cell delay variation origins, 816
cell delineation and scrambling, 785–787
cell delineation state diagram, 786
cell error ratio, 810
cell layout based on N=7, 917
cell loss priority (CLP) field, 783

cell loss ratio, 810
cell misinsertion rate, 811
cell organization and frequency reuse, 930, 931
16-cell pattern, 931
cell rate decoupling, 785
cells, 884
cell separation with four different sets of
  frequencies, 886
cell sequence integrity, 804
cell splitting, practical limit, 886
cells transported contiguously, 775
cell transfer delay, 809
cellular antijam situation, 913
cellular geographic serving area (CGSA), 884
cellular/mobile radio and PCN/PCS, 883–935
cellular mobile unit, 885
cellular pattern, 931–932
cellular/personal communication services
  worldwide, 929
cellular radio concepts, 884–888
cellular radio frequency band (US), 885
cellular radio influence, 426
**CELP (codebook excitation linear predictive)
  techniques**, 903–905
CELP synthesis model, 904
center of gravity method, 81–82
central directory server (CDS), 586
centralized automatic message accounting
  (CAMA), 134
central office, 47
CEPT30+2 (the same as E1 PCM)(CEPT =
  Conference European Post & Telegraph,
  forerunner of ETSI), 350
CEQ (customer equipment), 814
CGSA (cellular geographic serving area), 884,
  886
change of profile of services, 430
changeover, changeback, 834
changeover order, 844
changing the sense, 474
channel capacity, 499–500
channel counter, 421
channel gate, 355
channelized CDMA access, 927
channel modulators (FDM), 207
channel occupancy, 466
channel time slots, 509
characteristic distortion, 481
characteristic impedance of coaxial cable
  (CSMA/CD), 629
characteristic impedance (subscriber loop), 75,
  76
characteristics of local FDDI clock, 654–655
character parity, 469
check pointing, 547
checksum (TCP), 577
check token pass state, 634
chromatic dispersion, 320, 331
CHU (cost of a trunk on the HU route), 240, 241
CIB (CRC indication bit), 803
C/I (carrier-to-interference) ratio, 915

CIC (circuit identification code), 860
CICS (customer information control system),
  595
CIR (committed information rate), 725, 726, 729
circuit capacity in kbps versus bit rate versus
  time period, 528
circuit-directionalization control, 959
circuit group, 42
circuit identification code (CIC), 844, 860,
  865–866, 877
circuit optimization, 527–530
circuit-path lookup tables, 799
circuit selection (SS7, ISUP), 869–870, 871–872,
  873, 874
circuit shortage, 956
circuit switching (data), 526
circuit-turndown control, 959
circular routings, 220
cladding and core, 318, 319
Class A: constant bit rate, voice and
  uncompressed video, 793
Class B: variable bit rate video, audio,
  connection-oriented synchronous traffic,
  793
class bit, 649
Class C: connection-oriented data transfer,
  VBR, asynchronous traffic, 793
Class D: connectionless data transfer,
  asynchronous traffic such as SMDS, 793
class D format for multicasting (IP), 564
classes of LLC operation, 620
classes of SMDS access classes, 756
CL (connectionless), 800
CL (connectionless) data service reference
  configuration, 800
clearance for Fresnel zone, 273
clear back, 161
clear back, 165
clear confirmation packets, 557
clear forward, 161
clear forward, 165
clear request packet, 557
CLFF bits, 650
CLLM, 952
CLLM (consolidated link layer management),
  726, 727, 728
**CLLM (consolidated link layer management)
  message**, 727
CLLM message, 728
CLNAP (connectionless network access
  protocol), 796, 800
CLNAP-entity, 801
CLNAP-PDU structure, 802
CLNAP protocol data unit (PDU) and encoding,
  801–803
clocking, FDDI, 654–655
clock timing source, 437
closed user group (CUG), 863
closing flag, 694
CLP = 0, 815
CLP = 0 + 1, 815

CLP bit, 813
CLP (cell loss priority), 780, 782, 810
CLP value 0, 783
CLSF (connectionless service functions), 799, 800
clumping, 809
cluster/clustering, 587
clutter loss, 891
CM (communications module) (5ESS), 414, 424
C-message response, 211
C-message weighting, 496
C-message weighting, North America, 210
CMIP (common management information protocol), 961
CMP (communications module processor) (5ESS), 417
C/N calculation example, 310
C/N_o, 308, 309
coaxial feeds, 286
co-channel reduction factor, 916
code bit, 647
code-block control, 958
code-blocking, 955
codec (coder-decoder), 353
code-division multiple access (CDMA), 910, 912–915
code-division multiplexing, 914
code for load coil spacing, 73
code group, FDDI, 647
codes and formats for paging systems, 920
codes for idle channels, idle slots and interframe fill, 687–688, 689
coding, FDDI, 647
coding (data communications), 461–466
coding (PCM), 347–349
coding rate, 311
coefficient of overdispersion, 15
collision handling, 626–627
collision handling dynamics, 628
collision (LAN), 610
collision resolution, 315
collisions and backoff, 625
collisions (LAN), 623, 624
collision window, 626
combinations of preassigned VPI, VCI and CLP values at the UNI, 782
combined station, 543
COM (continuation of message) code, 745
command/response (C/R) bit, 721
command/response field bit (C/R), 695–696
committed burst size, 729
committed information rate (CIR), 725, 726
committed information rate (CIR), 729
common control, 118–122
common control principles, 120–124
common management information protocol (CMIP), 961
common management information protocol (CMIP), 970

communication between two transmission control layers (SNA), 581
communication channels for the network management system, 951–952
communication controller nodes, 583
communications path, AT&T 5ESS, 416–417
communication subsystem (IRIDIUM), 933
community-dial office (CDO), 115
community name, 964
community of interest, 93–94
companding, 346
companding and coding carried out together, 347
companding at syllabic rate, 394
compelled exchange of status information, 838
compelled signaling, 165–168
compelled signaling procedure, 167
complete selection case, 706
completion of transmission path (SS7, ISUP), 871, 872, 873, 874, 875
component degradation of fiber optic system, 333
composite LEN node, 585
composite network node, 585
composite standard deviation (echo loss calculations), 255
compression, a simple graphic representation of, 347
compromise network, 246
computer programs for traffic calculations, 31–37
computer read-write cycle, 127
concentrate light, discretely offered traffic, 215
concentration, 5, 106, 107
concentration and expansion, 107
concentration (telephone switching), 103–104
concentrators and satellites, 132–134
concentrators used with FDDI, 644
concept of bit stuffing (zero insertion)(data), 478
concept of distribution, 107
concept of frame (PCM), 349–353
concept of transition, 480
condensed codebook structure, 935
conditioned channel, 493
condition of marking, 474
condition of spacing, 474
configuration management, 943
configuring data assets, 521
confirm, 618
CONFIRM primitive, 690
congested node action, 728
congestion control and user data flow control, 700
congestion control (frame relay), 726
congestion in telephone network, 144, 145
congestion parameter state, 730
connecting one LAN to another LAN via a WAN with routers equipped with IP, 558
connecting terminals/workstations to a WAN, 522
connection admission control (CAC), 811
connection control, 533
connection establishment, 699

connection establishment, release (frame relay), 724

connection-establishment SETUP message, 817

connection establishment (SS7, SCCP), 856

connection indicators, nature of, 871

connectionless data transfer, 754

connectionless protocol, 963

connectionless service, 536

connectionless service functions (CLSF), 799

connectionless service on B-ISDN, 799

connectionless service (SCCP), 853

connection management, 541

connection memory, 421

connection mode: connection-oriented or connectionless, 793

connection-oriented data transfer (TCP), 569

connection-oriented function: temporary signaling connections, 856–858

connection-oriented network service, 855

connection-oriented services, 799

connection-oriented services (SCCP), 853

connection-oriented transport, 970

connection release functions (SS7, SCCP), 857

connection traffic descriptor, 814

connection types allowed, 869

connectors for fiber cable, 323

consequences of errors in ATM cell header, 784

consolidated link layer management (CLLM), 726, 727, 728

consolidated link layer management (CLLM) message, 727

constant amplitude digital data tones, 497

construction and refractive index properties for step-index and graded index fiber, 321

constructive and destructive fading, 895

container, C-n, 384

contention, 532

contention and polling, 622–623

continuity check, 875

continuity check, 877–878

continuity check (B-channel), 862

continuity re-check request, 878

continuous ARQ, 472

continuous variable-slope delta modulation (CVSD), 392–395

control bits, FDDI, 650–651

control circuits for direct progressive control, 116

control code (switching), 121

control field, 478

control field formats, 696–697

control field formats, 697

control field formats, LLC PDU, 621

control field formats of HDLC, 546

control field (HDLC), 545

control field (LAPD), 694

control field LLC, 619

control flags, 576

control functions, frame relay, 719

control indicators, 646–647

controlled access, 781

controlled performance on a 64 kbps international connection bearer channel, 451

controlled rerouting, 834

control of echo and singing, 246

control of message routing, 834

control packet called a token, 632

control path (5ESS), 417

control plane (ATM), 779

control point (SNA), 590, 593

control response times, 813

control signaling VCs within its VP, 808

control symbols, 646

conventional analog switching in telephony, 99–148

convergence sublayer indicator (CSI), 794

convergence sublayer PDU, 796

convergence sublayer protocol data unit (AAL-3/4), 797

conversation path, 51

convolutional coding, 471

cordless telephone, 922

core and cladding, 318, 319

core aspects of ISDN, 716

cornucopia antenna, 286

corrected reference equivalent, 67, 68

correction mode, 783

correlation coefficient, 899

correlation p versus parameter eta for two antennas in different orientations, 900

COT (central office termination), 375

countdown (CD) counter, 738

counter-rotating ring, 644

counting REQ bits, 736

country code (numbering), 138

coupling a node to a fiber optic cable, 614

coupling efficiency of an LD, 324

coupling loss between two circuits (crosstalk), 186

CP (control point—SNA), 587

CP-CP sessions, 586, 593

CPCS (common part convergence sublayer), 798

CPE (customer premise equipment), 752, 753, 754, 755, 756, 757

CPU (central processing unit), 521, 523, 525, 532

crankback, 231–231

crankback, 235

crankback concept, 231

CRC-6, 364

CRC-16, 470

CRC, CLNAP, 803

CRC-4, CRC-6 remainder, 949

CRC-CCITT, 470

CRC (cyclic redundancy check), 364

CRC (cyclic redundancy check), 470

CRC (cyclic redundancy check), 548

crc (cyclic redundancy check) check, 682

CRC-10 error check, 796

CRC-6 error checking, 687

CRC-32 indication bit, 760
CRC-4 multiframe, 365
CRC-4 multiframe structure, 366
C/R (command/response), 695
C/R (command response) bit, 619, 719, 721
CRC processing, 472
CRC remainder, 950
CRC value, 628
CRE (corrected reference equivalent), 67
CRE (corrected reference equivalent), 68
credit manager, 757
credit manager, 765, 766
credit measured in octets, 766
CREG (concentration range extension with gain), 87
CRF (central retransmission facility), 611
CRF (connection related functions), 675, 815
critical parameters (data transmission), 491–497
crossbar concentrator, 133
crossbar switch, 108
crossbar switch, 113–114
cross correlation, 914
cross-domain sessions, 593
crosslinks interconnection, 929
cross-point matrix, 109, 406, 409
crosstalk, 186
crosstalk, 494
crosstalk coupling loss, 374
crosstalk (PCM), 374–375
crosstalk reduction in PCM cable systems, 356
crow-fly distance tariffs, 220
CSC (common signaling channel), 303
CS (convergence) sublayer, 793
CSI (convergence sublayer indicator), 795
CSMA and CSMA/CD access techniques, 623–632
CSMA (carrier sense multiple access), 622, 623
CSMA/CD (carrier sense multiple access with collision detection), 624, 625
CSMA/CD LAN segments, 657
CSMA/CD relationship to OSI model, 625
CS-PDU (convergence sublayer PDU), 798
CSU (channel service unit), 485
CUG (closed user group), 863
cumulative error in timing, 476
customer satisfaction, 53
customer-to-customer talker echo, 253
CVSD (continuous variable slope delta modulation), 392–395
CVSD encoder/decoder block diagram, 394
cycle clock jitter, 655
cyclic bit interleaving in the tributary numbering order, 371
cyclic distortion, 482
cyclic redundancy check (CRC), 470

DA (destination address), 626
D/A (digital-to-analog) conversion, 260
DA field, 640

DAMA (demand assigned multiple access), 297
DAMA schemes (VSAT), 316
DAMA with slotted ALOHA reservation, 316
data circuit interfaces, physical layer, 486
data communication equipment (DCE), 485
data communications, 459–511
data decapsulation, 626
data frame, block or packet, typical, 529
data frame, generic, 478
data frame starts with unique field, 477
datagram envelopes, 570
datagrams, 562
datagrams, virtual circuits and logical connections, 551–552
data interchange circuits, 486
data interface—the physical layer, 485–489
data-link connection identifier (DLCI), 722
data-link connections, 538
data-link connections between two or more SAPs, 691
data-link layer, 538–539
data-link layer frame, generalized, 539
data-link layer reference model, 692
data link protocols, 479
data loading (FDM), 204
data memory, 420, 421
data message, buildup and breakdown, following OSI model, 537
data modem, 485
data modems, medium rate (CCITT), 506–508
data network operation, 533–557
data networks and their operation, 517–601
data offset, 576
data perishability, 520
data quartets, 648
data quartets (O-F), 647
data sources and destinations, 520
data switching: three approaches, 526–527
data switching methods, summary, 528
data terminal equipment (DTE), 485
data terminals and workstations, 521
data transfer, 541
data transfer phase (SS7, SCCP), 857
data transmission on the digital network, 509–511
data user part (DUP), 844
dBd vs. dBi, 901
dBi (decibels referenced to an isotropic[antenna]), 278
dBmV, 187
dBr, 136, 187
6-dB relationship, 277
dBrnC0 (dB reference noise, C-message weighting, referenced to 0-TLP), 496
dBrnC0 (dB reference noise, C-message weighting, 0) (refering to 0 TLP), 211
39-dBu contour, 901
DCA (document control architecture), 594
D/C (DLCI or DL-CORE indicator), 723–724
DCE (data communication equipment), 485
D-channel, 667
D-channel signaling capabilities, 697

dc loop(data), 473
dc nature of data transmission, 473–474
DDS (digital data system), 509–510
DDS facility compatible with DS1, 509
DDS signals subhierarchy, 510
DE bit, 731
DE (discard eligibility) capability, 722
DE (discard eligibility indicator), 719, 721
default, 703
default connectionless VCI, 749
default maximum frame size, CSMA/CD, 627
defective repeaters, identification of, 360
defining personal communications, 921–922
definition of R and D, 916
definition of time interval error (TIE), 440
degeneration, 110
degradation of timing at a network node,
    maximum permissible, 441
degraded minutes (DM), 448
degree of annoyance (echo), 245
DE indicator, 729
deinterleaver, 472
delay distortion, 492
delay(ed) energy, 289
delay equalizer, 503
delay in TDMA, 907
delay (LEO), 928
delay on SS7 signaling circuits, 849, 850
delay (phase distortion), 183
delay-sensitive services, 740
delineation and scrambling objectives, 785–786
delta encoder, 391
delta encoder/decoder, simplified functional
    block diagram, 392
delta encoding waveform, 392
delta modulation, 391–395
delta-modulation format, 303
demand access case, 706
demand-access pool, 302
demand-assigned multiple access, 297
demand-assignment multiple access (DAMA),
    302–303
DEO (digital end office), 259
derived MAC protocol data unit (DMPDU), 744
design of local area trunks, 88–89
design of long-distance links, 265–334
destination address and source address, CLNAP,
    801
destination address field, CLNAP, 803
destination address (SMDS), 759
destination and source addresses, FDDI, 651
destination-code completion statistics, 960
destination-controlled transmitters, 727
destination point code (DPC), 844
destination point code (DPC), 860
destination port, 575
destination unreachable message, 569
destination unreachable message (ICMP), 568
destructive energy, 895
detailed billing, 134, 142
detection mode, 783

determination of loudness rating, 69
deterministic access, 735
developing a hypothesis, 946
development of a pulse-code modulation
    signal, 341–354
dialog control, 541
dial-pulse signaling, 175
dial-pulse signals, 173
dial-service observation, 146–147
dial-tone delay, 53
dial-tone marker, 123
di-bit coding, 501
DI (data input) (NTI DMS-100), 421
differential delay across a voice channel, 183,
    492
differential Manchester coding, 641
different signals to indicate congestion, 231
digital bandwidth wasteful, 901
digital bearer is 64 kbps, 835
digital communications by satellite, 311
digital conditioning device, 485
digital data system (DDS), 509–510
digital data transmission waveforms, 484
digital data waveforms, 483–485
digital extension to the subscriber, 427–430
digital hierarchy, North American, 368
digital interface at 1.544 Mbps, 687
digital loop carrier (DLC), 375
digital loss, 453
digital modulation of LOS microwave radios,
    281–284
digital network, 425–453
digital network loss plan, 259–260
digital network performance requirements,
    441–453
digital network synchronization, 431–441
digital network synchronization methods, 434
digital radio systems, 280, 281
digital signal, 339
digital speech interpolation (DSI), 910
digital switching and networks, 403–453
digital switching concepts, review, 413–425
digital transmission: advantages and
    disadvantages, 396–397
digital transmission advantages, 339–340
digital transmission network models: CCITT, 431
digital transmission on an analog channel,
    489–511
digital transmission systems, 339–395
digit combiner, 355
dimensioning and efficiency, 39–42
dimensioning of trunks, 93
dimensioning the route, 7
direct and ground-reflected rays, 924
direct-burial optical fiber cable, 322
direct-circuit groups between exchanges, 231
directional antenna, 917
direct or tandem, 216
directory number (DN), 122
direct progressive control, 114
direct progressive control, 115

direct route, 225
direct route, 227
direct sequence spread spectrum, 910
direct-sequence spread-spectrum system
  showing waveforms, 914
direct traffic routing, 220
disadvantages of CSMA/CD, 656
disadvantages of token passing, 656–657
discard eligibility (DE) indicator bit, 722
discard eligible, 731
discarding frames in frame relay, 715
disciplined oscillators, 434
dispersion effect (ATM CDV), 809
dispersion limited, optical fiber, 320
dispersion-limited domain, 330
dispersion limits circuit length (PCM), 372
dispersion loss (switching), 137
dissasociated and associated channel signaling,
  170–172
distance involved in satellite communications,
  295
distance metric (IP), 567
distance to the network is infinity, 569
distorted signals, three typical, 482
distortion, 481–482
distortion defined, 481
distortion measurement equipment, 482
distortion (PCM), 374
distributed queue dual bus (DQDB), 733–751
distributed queueing protocol, 735
distribution (telephone switching), 104
diversity, 289
diversity and hot-standby operation, 289–291
**diversity—a technique to mitigate effects of
  fading**, 897–900
diversity improvement, 291
DLC (digital loop carrier), 87
DLC (digital loop carrier), 375
DLCI, 695
DLCI = 1023, 727
DLCI address range, 721
DLCI (data link connection identifier), 689, 691
DLCI (data link connection identifier), 722
DLCI (data link connection identifier, frame
  relay), 719
DLCI on D-channel, 722
DLCI or DL-CORE indicator (D/C), 723–724
DLCI values for B-channel and H-channel
  applications, 723
DLCI values for D-channel, 723
DL_CONNECT, 618
DL-CORE, 721
DL-CORE management, 724
DL-CORE parameters, 724
DL-CORE protocol, 724
DLL (data link layer, FDDI), 646
DLL PDUs, 655
DLTU (digital line and trunk unit) (5ESS), 416
DM (delta modulation), 403
DMPDU (derived MAC protocol data unit), 744
DMPDU header, 745
DMS-100, 405

DMS-100 with supernode/ENET, overview of,
  417–423
DN (directory number), 122
DNHR (dynamic nonhierarchical routing), 50, 52
DNHR (dynamic non-hierarchical routing),
  234–236
DNHR tandem switch, 235
DN to EN translation, 125
DOC (dynamic overload control), 960
domains in an APPN network, 590–591
domains in a subarea network, 591
domains of a subarea network, 590
dont fragment flag, 569
Doppler shift, 889
double-ended connecting cords, 104
double seizure, 52, 53, 123
downlink and uplink, 907
down-link extra decibels, 307
downward multiplexing, 534
DPC (destination point code), 844, 860
DPPX (distributed processing program
  executive), 584–585
DQDB access, basic algorithm, 736
DQDB access protocol, 735–743
DQDB (distributed queue dual bus), 713
DQDB (distributed queue dual bus), 733–751
DQDB layer protocol data unit (PDU) formats,
  747–751
DQDB layer services, 743–747
DQDB overview, 734–735
DQDB slot, 773
DQDB slot format, 747
drop and insert points, 271
drop cable length, token bus, 635
drop loss at first tap, 637
D/R ratio (D=distance, R=radius), 916
DRT (dynamic routing technique), 242–243
DR-T routing scheme, example of
  nonhierarchical routing, 243
DSAP and SSAP address field formats, 620
DSAP (destination service access point), 559,
  619
DSB (double sideband), 489
DS0 data ports, 437
DS1 ESF, 818
DS0/E0 timeslot, 378
DS1 framing strategy, 363
DSI (digital speech interpolation), 910
**DSL bandwidth allocation and frame structure
  (BRI US)**, 681–682
DSL basic frame, 682
DSL (digital subscriber line), 681
DS-30 links, 417
DS-512 links, 417
DSL line code—2B1Q, 683–684
DS1 mapping (of ATM cells), 818–819
DS1 mapping with PLCP, 819
DS3 PLCP (physical layer convergence
  protocol) frame, 817
DS1 signal format, 352
DS1 system, 342
DS1 system framing, 350

DSU (data service unit), 485
DSX (digital system cross-connect), 442
DTC (digital trunk controller) (NTI DMS-100), 423
DTE and DCE data packet, 556–557
DTE and DCE packet format, 556
DTE (data terminal equipment), 485
DTE-DCE interface, 485, 486, 487
dual-attachment stations, FDDI, 644
dual bus architecture, 734
dump count to join queue, 738
duobinary AM/PSK, 632
DUP (data user part), 831
duplicate and out-of-order segments (TCP), 571
duration of burst (SMDS), 755
dynamic non-hierarchical routing (DNHR), 234–236
dynamic overload control (DOC), 960
dynamic predistortion, 503
dynamic routing, 232–236
dynamic routing algorithm, 243
dynamic routing principles, 223
dynamic routing rules, 235
dynamic routing schemes, 229–230
dynamic routing technique, fundamentals of, 242–243

EA (address extension bit), 719, 721
E1 aggregates, 949
E and M signaling diagram, 155
E and M signaling states, 155
earned credit, 766
earth-based gateways, 930
earth bulge, 272, 273
earth station, 294
earth station engineering, 303–311
earth station receiving system example, 305
EBCDIC (extended binary coded decimal interchange code), 464, 466
EB (errored block), 445
EBHC (equated busy hour call), 10
E bit, token ring, 640
E-bit changes, counting, 950
E-bits, 366
E_b/N_o, 287
E_b/N_o calculation in digital radio systems, 280–281
E_b/N_o subscriber unit, 935
E_b/N_o versus BER for various digital modulation schemes, 283
ECCS (economic hundred call seconds), 240, 241
E1 channel (time slot) assignments, 353
echo, 245–247
echo annoyance, 246, 247, 248
echo canceler, 247
echo cancellation (ISDN), 685
echo characteristics of each segment, 253
echo delay, 247
echo loss (a-b) on established connection, 253
echo path delay, 247, 248
echo path loss, 254

echo paths in a 4-wire circuit, 247
echo (PCM), 375
echo return loss, 246, 254
echo round-trip delay example, 249
echo suppressor, 247, 257, 258
echo tolerance curves, 248, 250
economic hundred call seconds (ECCS), 240, 241
EDD (envelope delay distortion), 184, 492, 501, 502, 503
ED (end delimiter), 634, 635
ED (ending delimiter) (token ring), 639
edge node, inform, 728
E1 digital hierarchy, 369, 371
E2 digital link signaling (paths), 836
E1 European PCM system, 351
EFDA (erbium-doped fiber optic amplifiers), 328–329
effective noise temperature, 279
effective noise temperature of receiving system, 304
effective radiated power (ERP), maximum, 885
effective refractive index reduced, 332
effects of fading, 287
efficiency of antenna, 284, 285
efficiency of CSMA/CD, 625
efficiency of data transmission, 528, 529
efficiency vs. circuit group size, 42
EFS (end-of-frame sequence), 649
EFS (end-of-frame sequence) (token ring), 639
EFS (error free seconds), 442, 945
egress access rate, 731
EIA-422/423, 487
EIA-530, 486–487
EIA-232E, 486
EIA (Electronic Industries Association), 486
EIA-530 interchange circuits, 488
eight-level coding of North American DS1 PCM system, 351
EIRP, 284
EIRP (effective isotropic radiated power) calculation, 277–278
EIRP limitations, 304
EIRP of satellite, 308
EIRP of subscriber unit, 935
EIRP per voice channel, 310
EIRP reduction, 301
elasticity buffer, 655
elastic store buffer, 433
E-lead, 153
electrical coding of binary signals, 483, 484
electroluminescence, 324
electromagnetic compatibility (EMC), 271
electromechanical switches, types of, 108–109
electronic crosspoints, 416
elemental point-to-point data connectivity, 519
element length, payload, type, 803
elements in the calculation of EIRP, 278
element type, element value, 761
elevation angle, 295, 305, 309
eligible queued frames, 653

ELLIPSO, 927
EMA (enterprise management architecture), 961
E1 mapping (of ATM cells), 819
embedded delimiter, 648
EMC (electromagnetic compatibility), 271
emerging broadband technologies, 713–767
emf (electromotive force i.e., voltage)
empty QA slot, 748
E & M signaling, 153–155
E1 multiframe, 950
enablers (SPC switching), 127
enable signal, 409
encapsulation, 533
encapsulation bridge, 658, 659
encapsulation in ATM layer, 790
encapsulation of user information by the layers of the SIP, 758
ending delimiter (ED), 640
ending delimiter (ED), FDDI, 652
ending delimiter (ED) symbol (T), 646
end node attached nonnatively to its network node server, 587
end nodes (SNA), 585
end office, 47, 49
end of pulsing signal, 870
end of pulsing (ST) signal, 873
end-to-end performance better than, 808
end-to-end signaling, 853, 864
end-to-end signaling vs link-by-link, 168–170
end-to-end slip performance, 450
EN (enable) (NTI DMS-100), 42
EN (equipment number), 122
energy concentration at a frequency of half the bit rate, 361
energy per bit per noise spectral density ratio (E_b/N_o), 280, 281
ENET (enhanced network) (NTI), 417
ENET 2X2 switch matrix and time switch, block diagram, 420
enforcement based on set of parameters, 765
enforcement by the SS, 756
engineered capacity, 955
engineered traffic load exceeded, 953
enhancements to E1, 365
entities, service access points (SAPs) and endpoints, 690
envelope delay cumulative, 504
envelope delay distortion, 184
envelope delay distortion (EDD), 492
eoc (embedded operations channel), 682
EO (end office (exchange)), 496
EOM DMPDUs, 745, 746, 747
EOM (end of message) code, 745
E1 path signaling, 835–836
EPL (equivalent peak level), 188
EPO (error performance objective), 447
EPO (error performance objectives), 445–446
equalization, 503–504
equalization for envelope delay, 504
equalization in PCM regenerative repeaters, 361
equated busy hour call (EBHC), 10

equatorial orbit filling, 296
equivalent binary designations: summary of equivalence, 461
equivalent peak level (EPL), 188
equivalent random group, 41
erbium-doped fiber core geometry, 328
erect sideband, 195
Erlang, 10
Erlang and Poisson traffic formulas, 17–37
Erlang B computer program, 31–33
Erlang B formula, 31
Erlang B loss formula, 19
Erlang B tables, 21–26
Erlang C calculates delay, 36
Erlang C computer program, 34–37
Erlang C formula, 31
Erlang C formula, 37
Erlang load, 39
erlangs per subscriber line, 221
ERP, 893, 894
ERP (effective radiated power), 885
error, 468
error control, 534
error correction, 469
error correction, detection (SS No. 7), 837–838
error correction with feedback channel, 472–473
error detected indicator (E), 652
error detection, 469
error detection and error correction, 469–472
error detection using HCS, 749
errored block (EB), 445
errored second (ES), 445
errored seconds, 442
errored seconds (ES), 441
errored SMF, 950
error free seconds (EFS), 442, 945
error index field, 965
error performance, 441–442
error performance, data, 469
error performance—CCITT perspective, 442–447
error performance objectives, 311
error performance objectives (EPO), 445–446
error performance of underlying frame relay network, 717
error rate, 464, 466, 497, 502
error rate for token bus, 642
error rate on a binary transmission system vs. signal-to-rms-noise ratio, 361
error rate on LANs, 605
error rates on PCM systems, 341
error recovery function, FDDI, 644
error recovery procedures, 692
errors, the nature of, 468–469
errors in ATM cell header, 784
errors in data transmission, 464, 466, 467–473
ES (enterprise system), 584
ES (errored seconds), 441
ESF, 951
ESF (extended superframe), 363–365
ESF (extended superframe), 949
ESR (errored second ratio), 445

5ESS (AT&T) is a TST switch, 414
5ESS (AT&T overview), 414–417
5ESS (AT&T) switch, functional block diagram, 416
essentials of traffic engineering, 7–17
establishment and control of circuit-switched and packet-switched connections, 699
E1 system, 342
E1 system, coding segment 4, 349
E1/T1 and SDH/SONET circuits (ATM), 794
Ethernet, 517, 610, 624, 625
EtherType assignments, 560
European E1 framing, 363
European E1 PCM system, 351
even parity, 469
event-dependent routing, 230
event-dependent routing, 234
events, actions, time-scales and response, 812
example downlink budget, 309
example link calculation, fiber optics, 332–334
excess attenuation due to rainfall, 308
excess burst size (Be), 728
excessive accumulating jitter, 448
excess path loss, measured and calculated, 925
exchange code, 139
exchange location, 80–84
exchange location (long-distance network), 220–222
exchange size, 78
exchange sizes, 77
excitation sequence, 903
exhaust, 77
exhaust, 356
EXM-2000 accesses, 424
expansive controls, 957
expected slip rate, 439
explicit notification, 726
extended binary coded decimal interchange code (EBCDIC), 464, 466
extended modulo-128 packet, 553
extended superframe (ESF), 363–365
extended superframe (ESF), 949
extended superframe structure, 364
extended switch module-2000, 424
extension of PSTN, 918
external synchronization, 434
extinction ratio, 333
extinction ratio vs. power penalty, 333

facilities, X.25, 555–556
facilities length field (X.25), 555
factor K, 273
factors affecting numbering, 140–142
fade margin, 287
fades, fading and fade margins, 287–289
fading, 287
fading classification, 896
fading on satellite systems, 307
fairness (frame relay), 729
fair-sharing of capacity, 741
fan-out, 80
fan outs, 221

fan outs, choice of, 222
far-to-near criterion, 48
FAS (frame alignment signal), 364
fault alarms, 951
fault location on PCM span, 357
fault management, 942–943
fault management (ATM), 792
fault recovery handling, token bus, 633
fault tracing of North American PCM cable system, 359
F bits, 369
FCC (Federal Communications Commission [US]), 290
FC (frame control, FDDI), 647
FC (frame control) (token ring), 639
FCS (frame check sequence), 470
FCS (frame check sequence), 548, 627
FDDI classes of station, 644
FDDI facilities, 644–647
FDDI (fiber distributed data interface), 615
FDDI (fiber distributed data interface), 642–654
FDDI frame, 647, 649
FDDI frame format, 649
FDDI line transmission rate
FDDI maximum links and nodes, 643
FDDI operation, 652–654
FDDI peak data rate, 643
FDDI symbol coding, 648
FDDI symbol set, 645–647
FDDI timers, 653
FDDI token format, 647
FDDI token ring, 643
FDDI topology example, 645
FDMA conceptual drawing, 906
FDMA (frequency division multiple access), 297
FDMA (frequency division multiple access), 905–906
FDMA frequency slot, 934
FDMA operation, 298
FDMA/TDMA access format, 935
FDMA with FDM/FM configuration, 311
FDM (frequency division multiplex), 193–208
febe (far end block error), 682
FEC-code the links, 311
FEC (forward-acting error correction), 471–472
FEC (forward error correction), 289, 317, 497
FECN and BECN bits, 726
FECN (forward explicit congestion notification), 721
FECN indications, 727
FEC rate 1/2 convolutional coding, 935
FEC scheme for a channel with burst errors, 472
feedback controls, 811
feedback controls, 812
feedback error correction, 472–473
feedback network, 391
feed bridge resistance, 71
ferrite rod, small, 919
FEXT (far-end crosstalk), 374
F4 flow OAM cells, 791
F1 flows, regenerator section, 792

F2 flows at digital section level, 792
F3 flows for transmission path, 792
F5 flows for virtual channels, 792
F4 flows for virtual paths, 792
FFS—for further study, 445
fiber cable strength, 322
fiber distributed data interface (FDDI), 642–654
fiber-optic communication link, typical, 320
fiber optic LANs, 612–613
fiber optic link design, 329–332
fiber optic medium for token bus, 632
fiber optic repeaters, 330
fiber optics communication links, 317–334
fiber structure, 319
FIB (forward indicator bit), 840
FID (format identification), 597
FID 0 (SNA), 597
field mapping convention (frame relay), 720
FIFO (first-in, first-out), 38
FIFO (first-in, first-out) queue, 738
fill-in signal unit (FISU), 839
fill transmitted by token holder, 641
final (F) bit, 546, 547
final route, 44–45, 47, 224
final routes, 229, 239
final trunk group, 242
finding fault, 945
fine-gauge design techniques, 84–86
fine-motion switches, 108
finite traffic sources, 14
first-choice alternate routes, 241
first-come, first-served discipline, 37
first-offered traffic, 243
first 3 OSI layers required at relay switching
    point, 539
first-routed traffic, 954
FISU (fill-in signal unit), 839
five-level hierarchy, 224
five-unit start-stop stream of bits with a
    1.5-unit stop element, 476
five-unit synchronous bit stream with timing
    error, 476
fixed-length code bits (FDDI), 647
fixed-length payload, 735
fixed-length slots, 734
fixed-loss plan, 249
fixed or dynamic (routing), 228–229
fixed phase alignment, 382
fixed routing scheme, 229
flag as unique bit sequence, 690–691
flag bits (IP), 563
flag pattern, 478, 529
flag sequence, 694
flat channel, 211
flat channel noise measurements, conversion,
    212
flat rate (call charging), 134
flat response, 211
flexibility of TDMA, 910
flow control, 534
flow control, 541
flow control, X.25, 557

flow control connection-oriented class, 854
flow control mechanism (TCP), 571
flow control procedures, 692
flow control (SS7), 839
flow of logical bits or symbols with associated
    timing information, 789
flow parameter negotiation facility, 557
focused overload, 955
forced rerouting, 834
forced retransmission procedure, 838
forecasting, 93
formant vocoder, 902
format for international signaling point code,
    850
format identification (FID), 597
formation of an LLC PDU, 617
formation of DS2 signal from four DS1
    signals, M12 multiplexer, 369
formation of PAM wave and PCM signal, 343
formation of standard CCITT group, 197
format of a DMPDU, 745
format of DS3 PLCP frame, 818
formats and codes for ISUP, 865–869
formatting principles, SS7, 866
fortuitous distortion, 481
forward-acting error correction (FEC), 471–472
forward error correction (FEC), 900
forward explicit congestion notification (FECN),
    721
forward indicator bit (FIB), 840
forward sequence number (FSN), 839, 840
forward signal/reverse signal, 166, 167
four-port balanced transformer, 190
four-stage grid network, 121
four-wire basis, 193
four-wire transmission, 189–190
FPLMTS (future public land mobile
    telecommunication systems), 926
FRAD (frame relay access device), 718, 732
fragmentation, 562
fragmentation (IP), 563
fragment offset (IP), 563
fragment zero of fragmented datagrams, 567
frame abort (frame relay), 721
frame address bit length, 650
frame alignment, 355
frame alignment signal (FAS), 364, 365
frame and superframe, 363
frame check sequence, FDDI, 652
frame check sequence (FCS), 470, 471,
    694
frame check sequence (FCS) (HDLC), 544, 548
frame check sequence (frame relay), 720
frame class bit, 649
frame control (FC), 649
frame control field, FDDI, 649
frame-copied indicator (C), 652
frame delimiting, 691
frame format bits, FDDI, 650
frame format for token ring, 639
frame formats for layer 2 frames, 693
frame handler, 733

frame length, 649
frame mode bearer services (CCITT), 715
frame (PCM), 350
frame period for PCM, 342
frame relay, 713–733
frame relay, 952
frame relay address field formats, 719
frame relay ANSI format, 718
Frame Relay Forum, 716
frame relay introduction, 717–718
frame relay network, 716
frame relay salient points, 715
frame relay vs. X.25, 714
frame repeat, 433
frame size, CSMA/CD, minimum and
 maximum, 627
frame slips, 439
frame status (FS), FDDI, 652
frame stripping, 653–654
frame structure, 686–687
frame structure, frame relay, 718–721
frame structure, X.25, 551
frame structure at S and T (BRI), 680
frame structure (BRI), 680
frame structure (E1), 688–689
frame structure of 1.544 Mbps interface, 687
frame structure of North American DS1 PCM
 system, 352
frame structure (SDH), 387–388
frames with FF, 650
frame synchronization, 432
frame transmission, FDDI, 653
framing, alignment, 353
framing alignment in E1 carried out in
 channel (time slot) 0, 353
framing and synchronization channel, 432
framing bit, 350
framing bit, 432
framing boundary, 647
framing pattern 001011, 432
framing sequence 0011011, 432
free routing structure, 219
free-run accuracy, 436
free running clocks, 367
free space loss, 276–277
free space loss, 309
frequency bands, satellite communication,
 296–297
frequency bands for LOS microwave,
 292
frequency bands for pagers, 918
frequency diversity, 290, 898
frequency diversity, 900
frequency-diversity configuration, 290
frequency division multiple access (FDMA),
 297–298, 905
frequency division multiplex (FDM), 193–208,
 297
frequency hop spread spectrum, 910
**frequency planning and assignment (LOS
 microwave)**, 292–293
frequency plans, IRIDIUM, 933

frequency response, 182–183, 493
frequency reuse, 915–918
frequency reuse method, 886
frequency reuse on global basis, 931
frequency shift keying (FSK), 489, 490
frequency shift modulation, commonly called
 frequency shift keying (FSK), 489, 490
frequency shift modulation (FSK), 501
frequency spectrum above 300 MHz, 318
frequency-translation error, 502
frequency translation errors, 498
Fresnel breakpoint, 922, 924
Fresnel ellipse, 925
Fresnel propagation between transmitter and
 receiver in clear space, 925
Fresnel zone, 273
Fresnel zone clearance, 274
fringe-area considerations, 85
front-to-back ratio, 286
FRS (frame relay switch), 716
FS (frame status) (token ring), 639
FSK (frequency shift keying), 489, 490
FSK NRZ modulation, 919
FSL (free space loss), 284, 308
FSN (forward sequence number), 839, 840
FTP (file transfer protocol), 559
full and limited availability, 111
full availability, 14
full earth coverage (LEO), 928
fully-compelled signaling sequence, 166, 167
fully disassociated channel signaling, 171, 172
functional comparison: frame relay and X.25,
 714
functions of individual ATM/B-ISDN layers,
 788–799
functions of OSI layers, 537–543
functions of service primitives, 691

gaining access bus A, 741
gain stability, 250
gaseous absorption loss, 308, 309
gateway function (SNA), 583, 594
gateway links, 935
gateway's routing tables, 569
gating pulses, 355
Gaussian band-limited channel (GBLC), 499
Gaussian channel, 896
Gaussian channel, 898
Gaussian distribution, 494
Gaussian distribution levels (noise), 184
Gaussian noise, 494, 499, 502
Gaussian vectors, 905
GBLC (Gaussian band-limited channel), 499
GDS (general data stream), 595
general broadcast signaling virtual channel, 807
general-coverage antennas, 297
general format identifier, 552
generating polynomial (LAN), 628
generating polynomial (PCM), 364
generating polynomials, 470
generation and recovery of transmission frame,
 789

generic communication context, ISDN relationship with OSI, 677
generic data frame, 478
generic flow control (GFC), 781
generic trap type, 966
genesis of frame relay, 715–716
geostationary, 294
GET-RESPONSE frame, 968
GFC applied at ATM layer, 790
GFC (generic flow control) field, 780, 781
give priority to single-link messages, 953–954
glare, 173
GLOBALSTAR LEO, 927
GMSK (Gaussian minimum shift keying), 912
go and return paths, 950
go-back-n ARQ, 472, 547
Golay cyclic code, 920
GPS (geographical positioning system), 287
GPS (geographical positioning system) synchronization, 434, 436
GPS receivers, 286
graceful close, 572
graceful connection close (TCP), 574
graded index fiber, 321
grade of service, 6
grade of service, 11, 12
grade of service, 40, 92, 93, 145, 215, 216
grade of service, overall, 225, 226
grading, 12
grading, 111
grading groups, 111
grid network, 523
grid network, two stage, 119
grid networks, 119–121
grid-type switches, 121
gross-motion switches, 108
ground start, 173
ground start interface block diagram, 174
group address, SMDS, 754
group address composition, 755
group addressed SMDS data unit, 753
group delay, 492
group efficiency increases with size, 42
group modulators (FDM), 207
group selector concept, 108
growth management, 943
GSM bit period, 907
GSM burst structures, 907
GSM frame and burst structures, 908
GSM (ground system mobile—a European cellular system), 907
GSM system uses frequency hop, 910
G/T, 304
G/T, receiving system figure of merit, 304–306
G/T subscriber unit, 935
G/T values VSAT, 314
guard period, 908
guard time between uplink and downlink slots, 909
guard times and loss of throughput, 301
GWBASIC, 31

H, L OFF code, 635–636
half-sessions, 593
Hamming distance, 471
Hamming distance, 920
hand-held terminals, 885
handling of lost calls, 13–14
handover, 888
handover operations, 909
handset, telephone, 2–3
handshake procedure with clock-based sequence numbers (TCP), 572
hard-to-reach (HTR), 957
hardware and software components of an SNA network, 581
Hata empirical formula, 890–891
H-channel, 667
H0-channel, 687
H11 channel, 687
H11-channel structure, 668
H0-channel structures, 668
HCS (header check sequence), 749, 750
HDB3 (high density binary 3), 360
HDLC flags, 688, 718
HDLC frame, 544–546
HDLC (high-level data link control), 543–548
HDLC link configurations, 544
HDLC spawns LLC, 617, 618
HDR (header), 758
HDSL (high bit rate digital subscriber line), 429
head end, 611
headend of bus, 734
header checksum field (IP), 564
header error control (HEC) field, 785
header extension, 760, 803
header extension field format, 760
header extension length, 760, 803
header extension optional, 743
header length (IP), 562
header pattern for idle cell identification, 785
head of DQDB bus A, bus B, 757
HEC: receiver modes of operation, 784
HEC cell delineation mechanism, 785
HEC (header error control), 783–785
HEC sequence calculated, 790
HE (header extension) pad, 761
height-gain factor advantage, 899
heterojunction structures, 323–324
hexadecimal representation and BCD code, 463–464
hexagonal shape, 79
HFR (high-frequency regeneration), 903
hierarchical master-slave, 434
hierarchical network configurations, 589–590
hierarchical networks, 45–49
hierarchical network segment, 227
hierarchical network structures, 46–47
hierarchical network with alternative (alternate) routing, 225
hierarchical roles (SNA), 583–584
hierarchical rules, 47
hierarchical tandem switches, 235
high binary error density, 445

higher elevation angles (LEO), 928
higher-layer protocol identifier, 760
higher-layer-protocol-identifier (HLPI), 803
**higher-level multiplex structures internal to a
    digital switch**, 413–414
higher-level nodal points, 49
higher-level PCM multiplex comparison, 371
higher-order PCM multiplex systems, 367–371
higher-order star network, 45
higher-order virtual container, 384
higher-Q tuned circuits, 373
high-frequency regeneration (HFR), 903
high-level data-link control (HDLC), 543–548
high/low current (signaling), 173
highly dispersive paths, 889
high-performance routing (HPR), 579
high-priority PDUs, 639
high-speed backplane in hub, 661
high-usage (HU) trunk group, 239
high-usage route, 45
high-usage trunk group, 47
hits (impulse noise), 187
H loading, 89
HLPI (higher-layer-protocol-identifier), 803
holding time, 7
holding-time distribution, 17
holding time per call, 123
holdover and slip performance, 437–438
holdover stability, 435
holdover stability, 436
hole-electron pair, 326
home on higher-ranking exchange, 48
**homing arrangements and interconnecting
    network in North America**, 49–50
hop and hop count (IP), 566
horizontal antenna separation, 898
host address (IP), 564
host nodes, 583
hot splice, 323
hot-standby operation, 291
how SS7 relates to OSI, 830
HPA (high power amplifier), 289
HPR (high-performance routing), 579
HPR network, 598
HRDL (hypothetical reference digital link), 431
HRP (hypothetical reference path), 446
HRX (hypothetical reference connection), 431
HSP (high speed printer) (LAN), 609
HSRC (hypothetical signaling reference
    connection), 851–853
H12 structure, 669
HTR (hard-to-reach), 957
HTR traffic, 960
hub G/T values, 314
hub of star network, 515
hubs and switching hubs, 661
HU (high-usage) route, 224
HU (high usage [route]), 45, 47, 48
HU (high usage) routes, 216
HU (high usage) trunks, 219, 220
human intervention, 945
HUNT state, 786

HU route, 225, 226
HU trunk group, 239, 240, 241, 242
hybrid-balancing network, 245
hybrid insertion loss, 191
hybrid mode, FDDI, 649
hypothetical reference connection (HRX), 431
hypothetical reference digital link (HRDL), 430
hypothetical reference digital path (HRDP),
    269
hypothetical reference path (HRP), 446
**hypothetical signaling reference connection
    (HSRC)**, 851–853
2600 Hz tone, 157

IAM (initial address message), 860, 861
I bit, token ring, 640
ICMP fields, 569
ICMP (internet control message protocol),
    567–569
ICMP message, 567
ICMP message formats, 567–568
identifier (IP), 563
identifiers (SPC switching), 127
IDLC (integrated digital loop carrier), 260, 429
idle cells, 785
idling signal to maintain synchronization, 480
IDR (intermediate data rate) carriers, 311
IDU (interface data unit), 798
IEC (international electrical commission),
    624
IECs (interexchange carriers), 438
IEEE 802.6-1990, 736
IEEE 802 committee, 614–615
IEEE 802 frame showing LLC and TCP/IP
    functions, 561
IEEE (Institute of Electrical and Electronic
    Engineers), 613
IEEE 802.4 relationship of access control
    equipment and OSI layers, 633
IEEE 802 series LAN protocols, 559
IEEE Std 802.4, 632, 635
IEEE Std. 802.5, 638
I format control field (HDLC), 546
I-frame/I-format, LLC, 621
ILMI ATM UNI MIB tree structure, 974
ILMI functions, 974
ILMI (interim local management interface), 971
ILMI service interface, 975
IM (intensity modulation), 331
IM (intermodulation) product problems, 301
immediate reroute (IRR), 959
IM noise and crosstalk, 202
**impact of cell delay variation on UPC/NPC and
    resource allocation**, 815–816
impairments, basic, 181
impairments—fading in the mobile environment,
    895–901
impairments to voice channel transmission,
    182–189
IMPDU (initial MAC protocol data unit), 743,
    744, 745
impedance match, 193

impedance-matched nondirectional splitters, 635

impedance mismatches, 245

impedance on 2-wire side, 246

impedances of subscriber loop, 75–77

implementer frame, 651

IM products, 185

impulse noise, 187

impulse noise, 496

impulse noise can cause burst errors, 468

impulse noise in switching, 135, 137

impulsive noise (*see* impulse noise), 496

IMS (information management system, SNA), 595

inactive frequency bands, 903

in-band and out-band (out-of-band) signaling systems, 155

in-band noise floor, token bus, 636

in-band SF signaling, 158

in-band signaling, 156

in-band vs out-band (out-of-band) signaling, diagram, 158

inbound/outbound, 315

incidental phase modulation, 498

inclined polar orbits, 929

incoming call packet, 710

incompatibility of standard data rates with digital channel, 509

independent LU, 593

in-dialing, 143–144

indication, 618

INDICATION primitive, 690

individual address, SMDS, 754

individually addressed SMDS data unit transport, 754

infinite and finite traffic sources, 14

infinite regress of messages about messages, 567

information bandwidth, 266

information field, frame relay, 719–720

information field (HDLC), 548

information field (LAPD), 694

information (I) frame (HDLC), 545

information (info) field, 651

ingress node, frame relay, 731

initial address message, 869, 870, 873, 875

initial address message format, 861

initial address message (IAM), 860, 861, 872

initial address message (IAM), 870–871

initial alignment (SS7), 838

initial call setup phase, 231

initial design considerations (data networks), 520–521

initial MAC protocol data unit (IMPDU), 743, 744, 745

initial sequence number (TCP), 576

inlet and outlet, 12

INMARSAT (international marine satellite—) terminals, 926

INMARSAT-M/INMARSAT-P, 927

inquiry/response systems, 530

insertion loss of switch, 135, 137

in-service monitoring, 949

instantaneous rate of information transfer, 755

insufficient credit, 766

in-sync time versus BER, 787

integer number of octets, 719

integrated digital loop carrier (IDLC), 429

integrated services digital networks (ISDN), 665–710

integration of 7-cell pattern on a satellite, 931

intelligent hubs, 661

intelligible crosstalk, 186

INTELSAT (international telecommunication satellite [consortium]), 297, 300, 303, 311

INTELSAT V, VA and VI global beam: voice-channel capacity vs. bandwidth assignments, 298

intensity modulation (IM), 331

interactions of functional blocks for signaling link control, 837

interactive file-transfer traffic (VSAT), 316

interchange equivalent circuit, 487

interchange node, 587

interchange nodes interconnecting APPN and subarea networks, 588

interconnecting session stages, 593–594

interconnecting two AUGs, 390

interconnecting VC-11s, 390

interconnection of STM-1s, 390

interexchange control register, 118

interface at 1.544 Mbps, 685–688

interface at 2.048 Mbps, 688–689

interface functions, 679–680

interference rejection, 914

interframe fill, 689

interFrameGap, 631

interim local management interface (ILMI) specification, 971

interLATA connections, 259

interleaver, 471–472

interleaver pseudorandomly shuffles bits, 900

intermediate routing network, 585

intermediate routing nodes, 593

intermediate session routing function, 594

intermodulation (IM) noise, 185–186

intermodulation noise, 494

intermodulation products with TWT operation, 301

internal bit rate of a switch, 412

internal PCM word, 16 bits (AT&T 5ESS), 413

international and national signaling networks, 847–848

international connection, 218

international gateway exchange, 873

international network, 219–220

international number, 138, 873

international prefix (numbering), 137, 138

international signaling point code (ISPC), 850

International Telecommunications Union (ITU), 63

international telephone routing plan, 219

internet control message protocol (ICMP), 567–569
internet header fields, 567
internet protocol address formats, 566
internetworking equipment, 946
internetwork protocols, 540
interregister address information, 168
interregister signaling, 151
interregister signaling, 159
interswitch signaling, 151
intersymbol interference, 362, 499
intertoll trunk, 257
interworking of telephone exchanges, 7
invalid code points, 648
invalid frames, 720–721
inverted sideband, 195
inverting mark and space signaling sense, 488
IP address (network address), 562
IP address structure, 564
IP datagram, 560
IP datagram description, 562–563
IP datagram format, 563
IP header, 561
IP (internet protocol), 557
IP protocol field numbering, 565
IP router, 562
IP routing algorithm, 566
IP routing function, 560–562
IP shortest routing, 567
IPX (internet packet exchange), 969
IR drop, 3, 427
IRIDIUM, 928–935
IRIDIUM/gateway—interface to PSTN, 930
IRIDIUM pictorial view, 929
IRL (isotropic receive level), 276
IRP (internal reference point), 675
IRR (immediate reroute), 959
ISC (international switching center), 219, 223, 224
ISC point code parameter, 873
ISDN background and goals, 665–666
ISDN basic access rate configuration, 685
ISDN basic architectural model, 675
ISDN BRI standards, 681
ISDN circuit switching concept, 673
ISDN D-channel correspondence to CCITT SS No. 7, 678
ISDN generic users, 669
ISDN (integrated services digital networks), 665–710
ISDN networks, 673–674
ISDN numbering plan, 708
ISDN packet review, 705–710
ISDN packet service for lower data rates in D-channel, 674
ISDN protocols and protocol issues, 671–673
ISDN reference model, 670
ISDN relationship with OSI, 677
ISDN structures, 667–669
ISDN switching, simplified concept, 672
ISDN user channels, 667–668
ISDN user part (ISUP), 841, 853

ISDN user part (ISUP), 863–877
ISDN virtual circuit service—case B, 709–710
ISI (intersymbol interference), 895, 896
ISM events, 447
ISM (in-service (error) measurements), 443
isochronous service octet, 750–751
isochronous services, 735
ISO/IEC 8802-3, 624
ISO (international standards organization), 535, 624
ISO transport protocol, 541
isotropic antenna, 276
isotropic antenna, 278
isotropic receive level calculation, 279
isotropic receive level (IRL), 276, 894
ISPC (international signaling point code), 850
ISP (international signaling point), 847
ISU (isochronous service users), 750
ISUP '92, 864, 865
ISUP interworking, 864
ISUP (ISDN user part), 671, 831, 859
ISUP (ISDN user part), 863–877
ISUP message, 868
ISUP message parts, 865
ISUP signaling procedures, 864, 869–877
ITA#2 code, 462, 463
ITU (International Telecommunications Union), 63
ITU-T Rec. G.826, 443–447
IU (interface unit) of 5ESS, 416
IXC (interexchange carrier), 50, 217, 237, 238

jam bit sequence, 626–627
jamSize, 631
J and K are nondata symbols, 639
jitter, 373–374, 447–449
jitter accumulation, 373
jitter affects error rate, 498
jitter and wander, 431
jitter detector, 450
jitter reducers, 448
jitter reduction on fiber optic repeaters, 330
Johnson noise, 494
junction, 47
junctions, 88
junction(s), 6
junctors, 120
junctors, 412
justification and stuffing, 367
justification rate per tributary, 371

Ka-band communication links, 933
Karlson method, 134
k-bit sequence, 470
8448-kbps digital multiplexing frame structure, 372
keep all trunks filled with messages, 953
keying device, 474
key variable binary sequence, 910
K-factor, 286
K factor (community of interest), 94
K-factor (LOS microwave), 273

K of grazing, 275
kTB, 185

label capacity, 863
labeling potential, 849
lack of frequency synchronization, 433
LAMA (local automatic message accounting), 134
LAN access protocols, 622–661
LAN 802 architecture related to OSI, 615
LAN bridges, 658–660
land mobile systems, 884
LAN geographical extension, 605
LAN interconnect—frame relay, 715
LAN interworking via spanning devices, 657–661
LAN (local area network), 517, 519
LAN MAC protocols, 736
LAN of dissimilar protocols, 658
LAN performance, 655–657
LAN protocols relate to OSI, 615–616
LANs based on CSMA/CD, 560
LAN topologies, 606–609
LAPB (link access protocol B), 551
LAPB (link access protocol B-layer), 549
LAPD address field format, 695
LAPD derives from HDLC, 691
LAPD is derivative of HDLC, 717
LAPD (link access procedure for the D-channel), 689, 691
LAPD primitives, 697
LAPD protocol (frame relay), 717
LAPD protocol used for frame relay, 717
LAPF rather than LAPD in frame relay, 716
LAP (link access procedure), 689
last choice route, 228
last come, first served discipline, 37
LATA (local access and transport area), 238
latency, 642
lateral displacement of fiber axes, 323
layer 3, X.25 frame structure, 552–557
layer 2 frame structure for peer-to-peer communication, 693–697
layer 1 interface, basic rate, 677–685
layer 2 interface: link access procedure for the D-channel (LAPD), 689–697
layer 1 interface, primary rate, 685–689
layer 3 specification, 700–701
layout of a cellular system (diagram), 884
L-band operation, 933
L-band subscriber terminal links, 934–935
L-band uplink/downlink RF plan, 934
LBO (line build out), 8, 909
L-carrier, North American, 198, 200–202
L-carrier and CCITT comparison table, 202
L-carrier mastergroup, 201
LCC (lost calls cleared), 13
LCD (lost calls delayed), 13, 37
LCH concept, 19
LCH (lost calls held), 13
LCM (line concentrating module) (NTI DMS-100), 423, 425
LD (injection laser diode), 323, 324

leading register concept for end-to-end signaling, 168–169
leaky bucket algorithm, 729
leaky bucket principle, 838
LEC (local exchange carrier), 50
LEC (local exchange carrier), 235, 237
LED (light emitting diode), 320, 323, 324, 325
LED power drop off with age, 325
LEN end node, 585, 586
LEN (low entry networking), 584
LEO (low earth orbit), 267
LEO (low earth orbit (satellite)), 927
level, 187–188
level, an important parameter, 182
level 2 protocol data unit format, 761–765
level 2 protocol procedures, 763–764
level-regulating pilot, 206
levels and level variations, 497
levels of performance, 961
levels of traffic, network usage, 961
LFA (loss of frame alignment), 365
LGC (line group controller) (NTI DMS-100), 423
light detectors, 325–327
light emitting diode (LED), 320, 323, 324, 325
light-gathering ability of fiber, 319
light sources, 323–325
LI (length indicator), 840
LI (line interface), 496
limited availability, 13
limited availability, 110
limit of DLCIs, 719
line and trunk switching network with an SPC exchange, 125
linear codec, 353
line bit rate, 681
line build out (LBO), 89
line code (BRI), 681
line code (PCM), 360
line coding techniques, 483, 484, 485
line concentrators, 132
line conditions for R-2 code, 165
line configurations, 685
line finder, 112
line frequency, 195, 196, 197, 198
line losses, 277, 278, 310
line-of-sight conditions, 286
line-of-sight distance, 272
line parameters for ITT/BTM PCM system, 359
line rates for standard SONET interface signals, 377
line selector, 112
line signaling, 159, 160
line signaling, CCITT No. 4, 160
line signaling, R-2, 164
line signaling for CCITT No. 5, 163
line state symbols, 645–646
line unit (LU) (5ESS), 414
line weighting curves for telephone channel noise, 210
link access procedure for the D-channel (LAPD), 689–697

link between transport overhead and SPE, 378
link budget, example, 309
link budget example for fiber optics, 332–335
link budget (LOS microwave), 275–293
link-by-link signaling, 864
link-by-link vs. end-to-end signaling, 168–170
link connection (SNA), 588
link encryption, 943
link gains and losses, LOS microwave, 277
link idle state, 839
link-level data flow, 591
link limitation, 217–218
link margin, 332
links and transmission groups, 588–589
link segment, CSMA/CD, 629, 630
link set, 828
links in tandem, 48
link (SNA), 588
links (switching) examples of, 110
link state control functions, 838
link status signal, 839
link status signal unit (LSSU), 839
link trailer (LT) (SNA), 598
listen before transmit, 623
listen while transmitting, 624
LLC, 617
LLC control field and its operation, 620–622
LLC data, 627, 628
LLC__data__unit, 635
LLC generic primitives, 618
LLC (logical link control), 615, 616, 617
LLC (logical link control) header, 559, 560
LLC PDU structure, 618–619
LLC produces a PDU (protocol data unit), 616
LNA (low noise amplifier), 289, 304, 306
load coils removed, 356
loaded subscriber loops, 76
loading coils, 73
loading formulas (FDM), 203, 204, 205
**loading of multichannel frequency-division-
    multiplex systems**, 202–204
loading of VF trunks, 88–89
loading (outside plant), 73–74
loading potential (SS7), 849
loading with constant amplitude signals,
    202–203
load sharing, 232
load-sharing, 233
load sharing between link sets, example, 845
load sharing within a link set, example, 845
local area, 6
local area, 60
local area network (LAN), 517, 519
local area networks (LANs), 605–661
local clock used to synchronize, FDDI, 654
local code (numbering), 138
**local communications for multi-CPE access
    arrangements**, 754
local connecting exchange interface, 674
local digital host switch, 428
local exchange, 6
local exchange carrier, 50

local flow control functions, 781
local link set availability, 834
local networks, 59–96
local numbering area, 138
local switching exchange concentrates, 103
local transmitter slaved to incoming bit stream,
    686
locating, 888
logarithmic curve, 347
logatom, 64
logical channel assignment, 553, 554
logical channel identifier, 552
logical channel number, 552
logical interface, 616
logical link control (LLC), 616–622
logical OR, 734
logical ring, 632, 633
logical units (LUs), 593
logic gates, 406
logic of routing, 228–230
logic sensing for overload, 393
long distance calls, 7
long-distance links, design of, 265–334
long-distance networks, 215–262
long-distance PCM transmission, 371–375
longitudinal balance, 136
longitudinal redundancy check (LRC), 469
long-term conversational level, 188
long-term error rate (SS7), 849
look-up table, 452
lookup table routing (IP), 562
lookup tables, circuit/path, 799
loopback of a DS0 channel, 951
loop polling, 623
loop resistance, 87
loops, 6
loops (data), 473
loop signaling, basic, 173–174
loop-signaling methods, 174
Loran-C, 436
LOS component, 896
LOS distance, 267
LOS (line-of-sight) microwave, 267
LOS (line-of-sight) microwave radio relay
    system, 268
LOS microwave equipment parameters, 276
LOS microwave frequency bands, 292
LOS microwave repeaters, 292
LOS microwave system block diagram, 270
**loss (a-b) to avoid instability during setup,
    clear-down**, 252–253
loss and resistance per 1000 feet of subscriber
    cable, 72
loss assignment in a network, 88
loss budgets, token bus transmission, 636–637
loss design (fiber optic links), 331–334
loss limit calculation (subscriber loop), 72
loss of frame alignment (LFA), 365
loss plan, CCITT, 251–253
loss plan for evolving digital networks, 259–260
lost call, 11
lost call probability, 49

lost calls, 19
lost calls, 53
lost calls, handling of, 13–14
lost EOMs, 764
lost messages (SS7), 849
loudness loss, 67
loudness loss, 68, 71
loudness rating, 67–71
loudness rating, acceptable, 248
loudness ratings, 69
low community-of-interest connections, 552
low dielectric constant, 925
low earth orbit (LEO) advantages, 928
lower-order virtual container, 384
low frequency ac signaling systems, 155–156
low initial fan out, 221
LPC vocoder, 902
L2_PDU, 761, 762, 763
L3_PDU (level (layer) 3 PDU), 758
L2_PDU (level (layer) 2 protocol data unit or PDU), 757
LRC (longitudinal redundancy check), 469–470
LRD (long route design), 86–87
LSSU (link status signal unit), 839
LT to NT1 2B1Q superframe technique and overhead bit assignments, 683
LU (line unit) (5ESS), 414
LU (logical unit), 587
LU-LU sessions, 593
LU type 1, 594
LU type 2, 595
LU type 3, 595
LU type 4, 595
LU type 6.1, 595
LU type 6.2, 595
LU types, 594

MAC beacon frame, 650
MAC claim frame, 650
MAC frame, 617
MAC frame format, 627
MAC frame structure, 627–629
MAC LAN interconnect, 747
MAC (medium access control), 615
MAC purge frame, 650
MAC service data unit (MSDU), 743
MAC sublayer, 616
MAC sublayer informs LLC, 626
MAC-to-PHY interface, 645
main distribution frame (MDF), 427
make and break on subscriber loop, 153
managed objects (MOs), 970
management information base (MIB), 962
management plane (ATM), 779
Manchester coding, 484, 485, 624, 631, 632
Manchester coding for fiber optic signals, 331
mandatory fixed part, 866
mandatory variable part, 866
MAN (metropolitan area network), 517, 733
manual exchange, 105
map-in-memory, 51

mapping ATM cells directly into E1, 819
mapping ATM cells into a SONET STS-1 frame, 822
mapping ATM cells into SDH, 820–821
mapping ATM cells into SONET, 821–822
mapping between data units in adjacent layers, 537
mapping tables maintained by ARP module, 566
margin to compensate for fading, 287
mark, 460
marker, 120–121
marker functional diagram, 122
marking bias, 482
marking condition, 486
mark space decisions, 471
M-ary channel, 500
master clock in digital switch, 432
master station polls, 622
mated pairs of STPs, 829
material dispersion, 320
material dispersion, 331
mating connectors, 323
MAU attachments, 630
MAU (medium access unit), 624
maximum capability of frequency reuse, 929
maximum overall delay at signaling nodes, 852
maximum overall signaling delays, 852
maximum permissible degradation of timing at a network node, 441
maximum permissible jitter at hierarchical interfaces—E1 hierarchy, 449
maximum permissible output jitter at hierarchical interfaces—DS1 hierarchy, 450
maximum propagation times, CSMA/CD, 629
maximum slip rate is 255 slips per day, 450
maximum transmission path, CSMA/CD, 629, 630
maximum user data field length, X.25, 556
m-bits, 365
M-bits available per superframe, 682
M-bit (X.25), 556, 557
155.520 Mbps frame structure for SDH-based UNI, 820
622.080 Mbps frame structure for SDH-based UNI, 821
MDF (main distribution frame), 375
MDL (management entity to data-link layer interface), 697
mean and variance, 14–15
mean cell transfer delay, 763
mean server-holding time, 38
mean synonymous with average, 15
measured RSL, 287
measurement interval determined, 729
measurement interval (T), 728
measurement of telephone traffic, 9–11
medium- and high-rate modems for a telephone voice channel, 500
medium-term error rate (SS7), 849
memory for control, 406

memory for speech, 406
MEO (medium earth orbit (satellite)), 927
mesh, star and double star configurations, 43
mesh and star network, 217
mesh and star network, 217
mesh connection, 4
mesh connection, 49
message, discrimination, distribution, 833, 842
message format and information elements coding, 701, 703
message group, 860, 861
message-handling time, 468
message identifier (MID), 762–763
message label, 833
message mode service, 798
message organization example, 704
message routing, 833
message routing function, 842
message routing function, 845–846
messages for circuit-mode connection control, 701
message signal unit (MSU), 839
message signal units (MSU), 832
messages out of sequence (SS7), 849
message switching, 526
message traffic, 202
message transfer part (MTP), 673
message type, 704
message type code, 866
message type code allocation, 867
Metaconta L block diagram, 131
Metaconta L (switch), 128–129
metallic trunk signaling, 173–177
meta-signaling, 805
meta-signaling, 807–808
meta-signaling cells are used by, 791
meta-signaling function is required to, 808
meta-signaling virtual channel (MVC), 808
metering pulses and signaling, 151
methods of network synchronization, 433–435
metropolitan area network (MAN), 517
MF (multifrequency) signaling, 157
MF signaling systems, 161, 162, 163
MIB (management information base), 962
MIB object, 963
microcell prediction model according to Lee, 892–895
microwave LOS link gains and losses, 277
microwave LOS system block diagram, 270
MID (message identifier), 744
MID (multiplexing identifier in ATM), 796
MID numbers, 745
MID page allocation, 763
MIL-STD-188-114, 487
minFrameSize, 631
mini-gauge design techniques, 84–86
minimum bending radii, 322
minimum-cost paths, 235
minimum dispersion window, 331
minimum loss (echo), 251
minimum loss in 4-wire circuits, 249

minimum transmit level, token bus, 637
mixed-mode transmission, 920
mixer frequency, 298
mixing (FDM), 194–195
M-lead, 153
MLP control field, 551
MLP (multilink procedure, X.25), 550
MLRD (modified long route design), 87
M12 multiplex frame, 368, 369
mobile satellite communications, 926–935
mobile satellite systems post-1995, 927–928
mobile telephone switching office (MTSO), 884, 885, 887, 888
mobile terminal in an urban setting, 895
mobile units experience multipath, 887
mode dispersion, 320
modem, 489
modem internal timing, 480
modem selection considerations, 501–503
mode-setting commands and responses, 547
modes of operation (HDLC), 543–544
modification of existing plant, switching, 130–132
modulation and multiple access techniques, 935
modulation-demodulation schemes, 489
modulation implementation loss, 281
modulation implementation loss, 282
modulation rate, 501, 502
modulo-2 adders, 471
modulo-2 addition, 913
Molina formula, 19–20
Molina formula, 20
monitor bit M, 639
monitor devolvement scheme, 638
monomode fiber, 320
monomode fiber, 321
MPEG (Motion Picture Experts Group), 795
MSB (most significant bit), 420
MSDU (MAC service data unit), 743
MSU (message signal unit), 832, 841
MTBF (mean time between failures), 324
MTP is transparent, 853
MTP (message transfer part), 831, 834
MTP transfer, 870
MTSO (mobile telephone switching office), 884, 885, 887, 888
mu-law companding, 347, 348
mu-law PCM encoding, 873
multi-CPE access (SMDS), 752
multidrop or multipoint network, 523
multiframe structure, 688
multiframe structure (PCM), 363
multifrequency signaling, 161–165
multifrequency signaling in North America, 161–162
multilevel network, typical, 948
multilink TG, 588
multimedia communications transport, 942
multipath is the most common type of fading, 288
multipath propagation, 895

multipath scenarios, rich in, 889
multipath with mobile units, 887
multiple access of a satellite, 297–298
multiple-beam phased-array antennas, 931
multiple data units in transit concurrently, 756
multiple-domain subarea network, 590
multiples and links, 109–110
multiple transmitter zones (pagers), techniques available for, 919
multiplexing, 193–208, 534
multiplexing mechanism (TCP), 571
multiplexing method directly from container-1 using AU-4, 385
multiplexing method directly from container-3 using AU-3, 386
multipoint of multidrop network, 523
multipoint operation (LAN), 609
multisegment unit IMPDU, 745
multi-stage EDFA designs, 328
multistar or hierarchical network, 522
mutilation of start element, 477
mutual synchronization, 434
MVS (multiple virtual storage), 595
MxN array, 409

NACK (negative acknowledgement), 472
NANP (North American numbering plan), 753
narrow-band microcell propagation at PCS distances, 922, 924–926
narrow-band timing extractor, 373
narrow spectral line width, 324
national and international signaling networks, 848
National ISDN 1, 681
national tandem/transit element model, 676
national transmission plans, 249
nature of connection indicator, 872
NAU (network accessible unit), 591, 592
NCP (network control program)(SNA), 578
NCT (network control and timing) (5ESS), 416, 417, 424
NDF (new data flag), 388
Neal-Wilkinson capacity tables, 241
nearby scattering objects, 926
near-singing, 251
negative impedance repeater, 90
negative justification, 371
negotiated peak rate 1/T, 809
NEP (noise equivalent power), 326
net-ID subnetwork boundaries, 585
Netview, 961
network, 6
network, 7
network accessible units, 592–593
network access techniques, 905–915
network architecture, 228
network blueprint (SNA), 579
network capacity, defining, 943
network clock, 944
network congestion, 956
network connection characteristics, 685–686

network control information, 762
network crash, 657
network defect, 445
network-dependent value (CIR, frame relay), 731
network design, sample, 226
network design and configurations from a Bellcore perspective, 237–243
network design procedures, long-distance, 222–227
network hardware, 946
network indicator, 860
network layer, 539
network layer header (NHDR), 598
network layer packet (NLP), 598, 601
network limits of jitter, 449
network limits of wander, 449
network management, 941–975
network management agents in routers, 660
network management console, 963
network management defined, 941
network management from a PSTN perspective, 952–960
network management functions, 829
network management overhead, 952
network management systems, 961–970
network management tasks, 942–943
network managment in ATM, 971–975
network models for digital transmission, 431
network node clocks, 433
network node domains in an APPN network, 592
network nodes, 585–586
network node server, 587
network node server (SNA), 585, 586
network overload, general, 954–955
network performance during overload conditions, 955
network protection timer, 873
network protection timer, 875
network protection timer (SS7), 871
network protocol software, 946
network resource management (NRM), 811
network response to congestion, 726
network responsibilities, frame relay, 732–733
networks, telephone, 6–7
network service signaling data unit (NSDU)
network slip rate objective, 433
network timing (TDMA), 301
network-to-network interface, 559
network topologies and configurations, 521–525
network topology, 227–227
network traffic management controls, 957–960
network traffic management (NTM), 242, 953
network traffic management principles, 953–954
network utilization, 946
neutral and polar dc transmission systems, 473–474
neutral and polar keying on dc loop, 474
neutral loop, 474
neutral transmission, 473–474

neutral transmission, 484
neutral vs. bipolar bit streams, 341
Newbridge equipment, 949
new loop design techniques for US regional Bell operating companies, 86–88
new token issued, 653
NEXT (near-end crosstalk), 374
NHDR (network layer header), 598
NLP (network layer packet), 598, 601
NMS (network management station), 972
NMT (network management), 640
NNI header, 781
NNI (network-node interface), 780
No. 5 crossbar, 123
node not queued to send on bus A, 737
node queued to send on bus A, 738
node (SNA), 582
nodes with both hierarchical and peer-oriented function, 587–588
node table, 962
node types, 582–583
noise, 184–187, 493–497
noise, additive in a switch, 135, 137
noise accumulation, 339
noise advantage, 210, 211
noise and noise calculations in FDM carrier systems, 206–209
noise bandwidth, 185
noise figure, 279, 280
noise figure of commercial LNA or down-converter, 306
noise figure to noise temperature conversion, 279
noise hits or spikes, 495
noise immunity, 488
noise peaks, 494
noise spectral density, 280
noise stops at the repeater, 361
nominal clock slip rate is 0, 449
nominal slip performance, 451
nonassociated mode, 828
noncoincident BH, 52
noncoincident network busy hour, 235
noncompelled, positive/negative acknowledgement, 838
noncompelled signaling, 166
nondata symbols J and K, 642
non-hierachical network of exchanges, 228, 229
nonlinear envelope delay, 186
nonnative attachment, 586
nonpersistent, 623
non-sequential routing, 230
nonsystematic jitter, 373
nonuniform numbering, 139
nonuniform numbering, 170
normalized frequency departure, 440
normalized frequency (fiber), 319
normal response mode (NRM), 543
North American hierarchical network synchronization, 435
North American high-level (PCM) multiplex,

368–369
North American R-1 code, 162
North American synchronization plan as specified by ANSI/Bellcore, 435–438
NOSFER, 64
notes on CSMA/CD system parameters, 629–632
not useful bits, 530
NPDU (network protocol data unit), 856, 857
N(R) count, 547
NRM (network resource management), 811
NRM (normal response mode), 543
NRZI (non-return to zero inverted), 647
NRZ (non-return to zero) coding, 483, 484, 485
NSDU (network service signaling data unit), 853
NSP (national signaling point), 847
NT1, NT2, 670
NT derives timing, 689
NT interface, 678
NTM center, 953
NTM controls, 954, 957
NTM (network traffic management), 242, 953
NT1 (network termination 1), 670
NT1 to LT 2B1Q superframe technique and overhead bit assignments, 684
NT1 to LT superframe delay offset, 684
null pointer indication, 388
numbering, one basis of switching, 102–103
numbering analysis in routing, 141
numbering and the PABX, 143, 144
numbering concepts for telephony, 137–144
numbering effects on signaling, 170
numbering plan area, 223
numbering plan for international signaling point codes, 850–851
numbering requirements, 863
numbering (telephone), 78
number of pointers, 866
number translation (switching), 117
numerical aperture, 319
Nyquist approach to channel capacity, 499–500
Nyquist bandwidth, 494
Nyquist 2-bit rule, 500
Nyquist rate, 391, 499, 903
Nyquist sampling theorem, 340, 341

OAM (operation, administration, maintenance), 788
OA&M (operations administration and maintenance), 948
objective of routing, 227
observing symptoms, 945
occupation time during busy hour for subscriber line, 77
occupied bandwidth, IDR carriers, 313
OCL (overall connection loss), 254
OC-N level, 377
OC (optical carrier), 377
octal-PSK, 282
octet-boundary alignment, 626
octet of option-kind, 577

47-octet payload, 794
octet slips, 451
odd parity, 469
ODYSSEY, 927
off-contour loss, 308
offered load, 239
offered load (frame relay), 729
offered traffic, 7
offered traffic, 91
off-hook, 3, 61, 112
off-hook, 62
off-hook, on-hook, 113
off-hook and on-hook, 104
off-hook condition, 152
75-ohm coaxial cable, 611
75-ohm coaxial cable, token bus, 632
50-ohm coaxial cable (LAN), 610
ohmic losses generate thermal noise, 305
OID (object identifier), 966
Okumura model, 890–891
OLR (overall loudness loss), 69
O&M (operations and maintenance) functions,
    854
one-element code, 391
one-way and both-way working, 228
one-way circuits, 554
one-way groups, 43
on-hook, off-hook, 3
on-hook, off-hook, 152
on-hook disconnect signal, 153
ONU (optical network unit), 260
OPC (originating point code), 844, 860
open/close (signaling), 173
opening flag, 694, 718
opening flag/closing flag, 839
open systems interconnection (OSI), 534–543
Openview, 961
operable conditions, ATM, 814–815
operational support systems (OSS), 953
operation of a hybrid, 190–192
operation of a hybrid transformer, 191
opm/concentrator, 428
OPM (outside plant module) (NTI DMS-100),
    425
optical fiber amplifiers, 327–329
optical fiber as a transmission medium, 318–320
optical horizon, 268
optimal routing on packet-by-packet basis,
    551
optimized format (ATM), 773
optional part, 869
options to meet throughput requirements, 530–532
ordered delivery, 534
order of bit transmission (frame relay), 720
order of transmission of bytes, 377
order-wire circuit (DAMA), 302
ORE, 88
ORE goal, 66
ORE (overall reference equivalent), 64
originating and destination points, 833
originating and terminating line appearances,
    106

originating call control, 231
originating exchange, 169
originating ISC point code parameter, 875
originating-line appearances, 105, 106
originating point code (OPC), 844
originating point code (OPC), 860
origins and destination, 5
ORM (optical remote module) (5ESS), 424
OSI layers, 947
OSI management architecture, 970
OSI (open systems interconnection), 534–543
OSI reference model, 535
OSS (operational support systems), 953
other loop design techniques, 84–86
outgoing sender, 123
out-of-band, tone-on-when idle, 164
out-of-band 6 kHz tone, 888
out-of-band (out-band) signaling, 157–159
out-of-service time, 947
output contention (ATM), 809
outside plant, 427
outside plant module (OPM), 428
overall connection loss (OCL), 254
overall loudness rating opinion results, 69
overall reference equivalent (ORE), 64
overflow, 10, 52
overflow between circuit groups, 228
overflow count, 11
overflow routes, 91
overflow traffic, 41
overload conditions (SS7), 834
overview of IEEE/ANSI LAN 802 protocols,
    613–622
overview of layer 3, 697–705
overvoltage protection, 427, 429
own-exchange routing, 51

PA access, 735
packetization delay, 781
packet send sequence number P(S), 556
packet service for lower data rates, 673
packet-switched public service (SMDS), 753
packet switching, 526–527
packet type identifier, 552, 553
packet type identifier subfield, 554
packet-type identifier (table), 553
PAD, CLNAP, 803
padding (TCP), 577
padding to ensure complete filling, 743
PAD field (PAD—adding octet fill), 627
PAD length, 803
pad length (SMDS), 760
PAD (SMDS), 761
paging receivers, 919–920
paging systems, 928–921
PAM highway, 355
PAM (pulse amplitude modulation) wave as a
    result of sampling a single sinusoid, 342
PAM stream, 354
p and t data, 676
PA (pre-arbitrated), 735
PA (pre-arbitrated) segment, 749, 750, 751

parabolic antenna gain, 284–286
parabolic antenna with front feed, 285
parabolic dish antennas, 285
parallel links (SNA), 588
parallel sessions, 594
parameters in initial address message, 874
parameters in the initial address message, 872–873, 874–875
parameter values for access classes, 767
parity check, 469
PAR mechanism (TCP), 571
PA segment format, 749
PA segment header fields, 750
PA segment payload, 750
PA slot, 735
PA slots, 742
pass-back to originator, 638
passive coupling, fiber LAN, 614
path analysis, 275–293
path control elements, 597
path engineering (satellite communications), 303–304
path loss, 279
path loss, 889–891
path loss, LOS microwave, 276–277
path loss model for typical microcell in an urban area, 893
path profile, 269–272
path profiling, 272
path/site survey, 286
path terminations, 379
pattern generator transmits digital pattern, 359
payload CRC, 744, 747
payload CRC, 763
payload frame start, 385
payload length, 763
payload mapping, payload demapping, 381
payload pointer, 379
PAYLOAD__TYPE, 750
payload__type, 749
payload type (PT) field, 782–783
PCM 7-bit code, 344
PCM 8-bit code, 344
PCM codec, 303
PCM codewords assigned, 348
PCM frame, 342
PCM framing bit, 949
PCM link simplified block diagram, 358
PCM (pulse code modulation), 340–375
PCM switching, 403–453
PCM switching: advantages and issues, 404–405
PCM system enhancements, 363–366
PCM time slot bit streams, 409
PCM used to expand capacity of wire-pair trunks, 355–357
PCM word, 348
PC (personal computer), 517
PCR (peak cell rate), 817
PCS path of LOS nature, 922
PDH (plesiochronous digital hierarchy—e.g., DS1, DS1C—E1), 376
PDN (public data network), 548, 549

PDU (protocol data unit), 533
PDU size (IP), 562
PDU trailer, 743
PDU type field, 964
PDU type field values, 965
peak busy hour, 9
peak cell rate (PCR), 817
peak distortion, 482
peak factor of many talkers, 205
peak responsivity, 326
peer MAC entities, 645
peer-oriented network configurations, 590–591
peer-oriented protocols, 583, 585
peer-oriented roles, 584–587
peer-oriented roles (SNA), 583
peer-session protocols, 593
peer-to-peer communication, 854
peg count, 10
Pentaconta 1000, 114
Pentaconta 2000, 129
perceptual criteria, 903
perfect receiver at room temperature, 279
perfect scheduler, 736
performance management, 943
performance parameters, basic, 848
performance requirements, digital network, 441–453
per-hop time availability, 288
period of bit, symbol or baud, 483
peripheral border nodes, 587
peripheral nodes, 583–584
permanent virtual circuit (PVC), 554
permanent virtual connection (PVC), 551
persistence algorithms, 623
1-persistent, 623
personal communications, 883
personal communication systems, 921–926
P/F (poll/final) bit, 621, 622
phase coherent FSK, 632, 636
phase continuous FSK, 635
phase discontinuity due network clock, 439
phase distortion, 181
phase distortion, 183–184, 492
phase drift among clocks, 433
phase jitter, 373
phase jitter, 498–499
phase jitter, local clock, 654, 655
phase-locked loop, 480
phase shift, 183, 184
phase-shift keying (PSK), 502
phase shift modulation, commonly called phase shift keying (PSK), 489, 491
phase wander and jitter, 432, 433
philosophical differences in signaling: voice and data, 775
photon counter, 326
photon source, 323
PH (packet handler), 707
physical interface, 532
physical layer, 532
physical layer, 537–538

physical layer, X.25, 550
physical layer—ATM, 788–790
physical units (PUs), 592
PHY symbols, 635
PIDB (peripheral interface data bus) (5ESS), 416
piecewise linear approximation of the mu-law logarithmic curve, 350
pigtail, 324
pilot filters, 106
pilot supervision, 877
pilot tones, 198
pilot tones (FDM), 206
PIN diode response time, 327
PIN (term derives from semiconductor device where intrinsic(I)used between p-n junction of diode), 326
PIU (path information unit), 597–598
planned control-response basis, 957
plant-in-place, 943
PLCP (physical layer convergence protocol), 818
plesiochronous interexchange link, 451
plesiochronous network, 439
plesiochronous operation, 433, 434, 438, 439
PLS interface, 626
PLS (physical signaling sublayer), 624
PMA (physical medium attachment), 625
PMD (physical layer medium dependent), 644
PM (peripheral module) (NTI DMS-100), 423
PM (physical medium) (ATM), 789
POH (path overhead), 382
1-point CDV, 809
2-point CDV, 809
pointers, 866
pointing losses, 308
point of presence (POP), 59, 238
point of termination (POT), 238
points a-t-b, 251
point-to-multipoint signaling configurations, 807
point-to-point network, 522
POI (path overhead identifier), 817
Poisson and Erlang traffic formulas, 17–37
Poisson computer program, 33–34
Poisson distribution, 14
Poisson formula, 20
Poisson formula tables, 27–30
Poisson probability distribution function, 16, 17
polar and neutral electrical waveforms, 475
polarity control, 394
polarity of next digit, 391
polarity reversal (signaling), 173
polarization loss, 308
polar mode, 473–474
polar NRZ, RZ, 484
poll-based approach, 963
poll/final bit, 546
poll frame, 547
polling, 532
polling (access), 525
polling a form of controlled access, 622

polling interval (DAMA), 302
poll interval, 525
poll (P) bit, 546, 547
poor completers, 954, 955
POP (point of presence), 44, 47, 59
ports (TCP), 571
position of AU pointers, 388
positive acknowledgement, 840
positive-going transition, 485
positive justification, 371
positive/zero/negative justification, 371
POS (point of sale), 315
post dial delay, 53, 117, 677, 829
postequalization, 503
POT interface, 238
POT (point of termination), 238
POTS (plain old telephone service), 884
power budget, typical downlink, 308–310
power control in CDMA systems, 915
powering a PCM span, 359
powering fiber repeaters, 330
power limited, optical fiber, 320
power-limited domain, 330
p-persistent, 623
PPM (pulse position modulation), 374
practical applications (PCM), 355–357
practical modem applications, 504–509
practical (PCM) system block diagram, 357–360
practice profile, 274
PRBS (pseudo random binary sequence), 951
preamble, CSMA/CD, 627
preamble (PA), FDDI, 649
pre-arbitrated (PA), 735
pre-arbitrated (PA) access control, 743
pre-arbitrated (PA) slot, 749–751
preassigned, 297
preassigned header field values at NNI, 792
preassigned header field values at UNI, 791
preassigned VPI/VCI values, 781
predictors in ADPCM, 902
preplanned distribution of loadsharing, 233
preplanned routing sequences, 237
presentation layer, 542
PRESYNCH state, 786
preventive cyclic retransmission method, 837
primary center, 47, 224
primary isolation derives from distance, 915
primary rate, ISDN, 667
primary station, 543
primitive, 618, 678
primitives and parameters (SCCP), 855–856
primitives associated with layer 1, 679
primitives associated with the data-link layer, 698
primitives between layer 1 and other entities, 678–679
print cycle start, 475
priority control, 811–812

priority levels, FDDI provides multiple, **644**
PRI (primary rate interface), 674
private ATM switch, 777
private ISDN access connection element, 677
private networks, 2
probability density function, 896
probability-distribution curves, 14–15
probability of bit error in Gaussian noise, 495
probability of blockage, 12
procedures for packet access, 708
proceed to send, 166
processing gain, 914–915
processing satellite, 294
profile of bursty traffic, 725
profile practice, 274
program channel, 340, 342
program store memory (SPC switching), 125, 126
progressive call control (routing), 230–231
progressive control, 114–116
propagation delay counter, 874
propagation in the mobile environment, 888–895
propagation loss suffered by signals penetrating buildings, 919
propagation models, 889–891
propagation model typical of an urban microcell, 892
propagation problem, 888–889
propagation reliability, 291
propagation time, 479
propagation time, CSMA/CD, 629
protective controls, 957
protective trunk group controls, 955
protocol, 532
protocol architecture for supporting connectionless service, 801
protocol data unit (PDU), 533
protocol discriminator, 701
protocol discriminator, 703
protocol field (IP), 564
protocol functions, basic, 533–534
protocol identification field, 873
protocols, 533–548
protocol structure of CL data services, 801
prove-in distance, 356
PRS (primary reference source), 435, 436, 437, 438, 439
pseudorandom binary sequence (PRBS), 951
pseudoternary line code, 681
PSFs (packet switching facilities), 672
PSK modulation, 501
PSK (phase shift keying), 489, 491
psophometric noise weighting, 210
PSPDN (packet switched public data network), 705
PSPDN port, 706
PSR (previous segment received), 747
PSTN connectivity, 929
PSTN network management, 952–960
PSTN (public switched telecommunications network), 665

PSTN (public switch telecommunication network), 2
P tables, 17, 20
PTI coding, 783
PTI field, 798
PTI (payload type identification), 782, 783
PTT (post telephone and telegraph), 383
public ISDN connection reference configuration, 675
pull-in/hold-in, 436
pulse amplitude modulation (PAM) wave, 342
pulse amplitude modulator (PAM), 393
pulse-code modulation system operation, 354–355
pulse distortion, 362
pulse duration, 485
pulse position modulation (PPM), 374
pulse response through a Gaussian band-limited channel (GBLC), 500
pulse-stuffing synchronization, 367
pumping EFDAs, 328
pump power bypass, 328
pump power (fiber amplifiers), 329
push-button tones, codes, 165
push function, 576
push service, 572
PVC favored by ANSI for frame relay, 717
PVC (permanent virtual circuit), 791
PVC (permanent virtual connection), 551
PWAC (present worth of annual charges), 84
pWp (picowatts psophometrically weighted), 211, 212

QA cells (slots), 740
64-QAM, 281
256-QAM, 281
QAM (quadrature amplitude modulation), 281, 508
QAM schemes, 281
QA (queued arbitrated), 735
QA segment, 735, 736, 737, 738
QA segment format, 748
QA segment header, 749
QA segment header fields, 748
QA segment payload, 749
QA segment payloads, 745
QA segments' payloads, 743
QA slot, 735, 747, 748
QoS parameter descriptions, 809–811
QoS (quality of service), CLNAP*PAD length*, 803
QoS quality of service (ATM), 775
QoS (quality of service) characteristics (ISDN), 708
QPSK characteristics and transmission parameters for IDR carriers, 312
QPSK modulation, 935
QPSK operation, 506
QPSK (quaternary phase shift keying), 491, 502
QPSX Australia, 733
quadrature amplitude modulation (QAM), 508
Qualcomm system, 915
qualifier bit, 552
quality of a telephone speech connection, 63–71

quality of service, 53–54
quality of service, 760
quality of service, network performance and cell loss priority, 812–813
quality of service (QoS), 808–811
quality of service (QoS) parameters, frame relay, 731–733
quantization, 342–347
quantization and resulting coding using 16 quantum steps, 345
quantization distortion, 353–354
quantization distortion, reduction of, 346
quantization process, 344
quantizing distortion, 346
quantum efficiency, 326
quantum level inside segment, 349
quarternary phase shift keying (QPSK), 491, 502
quasi-omnidirectional user antennas for LEO operation, 928
queued arbitrated (QA), 735
queued downstream, 738
queue discipline, 37
queue formation on bus A, 737
queueing for access bus A, 739
queue length, 37
queue represents a flow control function, 854
quiet (Q), halt (H), idle (I), 646

R, S and T symbols (FDDI), 646
R-2 address signaling, 164
radio and optical horizon, 268
radio-frequency carrier, 266
radio frequency interference (RFI), 292–293
radio-paging channel, 919
radio position finding, 929
radio propagation into buildings, 918–919
radio transmission, introduction, 266–267
radome, 307
rainfall attenuation, 297
rainfall fades, 307
raised cosine filter, 906
Raman amplifiers, 328
random access, 525
random-access central control, 302
random access (LAN), 622
random errors, 468
randomizing of data (ATM), 785
randomizing polynomial, 471
random noise, 494, 495, 910
random traffic, 16, 19
range extender, 73
range extension with gain, 87
RARP (reverse address resolution protocol), 566
rate enforcer functions, 729
ratio method (exchange location), 82–83
ray beam grazes, 267
Rayleigh channel, 896
Rayleigh channel, 897
Rayleigh distribution, 287
Rayleigh fading, 288, 291

Rayleigh fading on paging channels, mitigation, 919
ray paths for several angles of incidence (fiber optics), 319
RBOC (regional Bell operating company), 217
R-1 code, 161–162
R-2 code, 163–165
RDT (remote digital terminal), 259
read and write function, 735
read function, 734
real-time surveillance, 953
reassembly of content of ATM cell information, 793
receive antenna gain, 278
receive function derives clock, 655
receiver, telephone, 3
receiver bit stream synchronized to network clock, 686
receiver diode sensitivities, 327
receive reference equivalent (RRE), 64
receiver noise figure, LOS microwave, 276
receiver noise floor for broadband token bus, 637
receiver noise level calculation, 278–279
receiver noise performance (VSAT), 314
receiver requirements, 686
receive signal level calculation, 278
receive signal level (RSL), 276, 894
reception without contention, 626
reconstruction integrator, 393
recovery commands and responses, 547
reduction of offered load (frame relay), 726
redundancy and channel efficiency, 466
redundant bit, 509
reference clock, 440
reference equivalent, 53
reference equivalent, 63–67
reference equivalent for subscriber sets in various countries, 66
reference equivalent increase, 247
reference equivalent vs. corrected reference equivalent, 68
reference frequency, 62
reference plane, 305
reference-system noise calculations (CCITT), 209
reflected energy, 275
reflected signal energy (echo), 247
reflection point, 274–275
refractive index (fiber), 318
regeneration, 292
regeneration-calling pressure, 956
regeneration of digital signal, 341
regenerative repeater functional block diagram, 362
regenerative repeater (PCM), 357
regenerative repeaters, number of regarding jitter, 373, 374
regenerative repeaters extends coaxial cable length, 632
regenerative repeaters (LAN), 610

regenerative repeaters (PCM), 361–363
register marker replacement, 124
register progressive control, 114, 115
register progressive control, 117
register translator, 117
regular reroute, 959
REJECT frame, 725
Rej (reject frame), 547
relationship among parameters, frame relay,
      729–730
relationship between VC, VP and transmission
      path, 804
release dates, 945
release guard, 161
RELP (residual excited linear predictive), a
      hybrid coder, 903
remnants of FDDI frames, 654
remote bridging concept, 659
remote concentrator, outside plant module or
      switch, 428
remote monitoring (RMON), 967–968
remote switching capabilities, 423–425
remote switching with AT&T 5ESS, 423–424
remote unit performs SLIC functions, 428
repeaters, LAN, 657–658
repeater sets required, CSMA/CD, 629
repeaters in multiple paths, 657
repeater sites, 269
repeaters (LOS microwave), 269, 270, 271
repeating patterns (frequency reuse),
      917
repeat previous frame, 432
REQ bits, 736, 738, 747
REQ (REQUEST), 736
request, 618
REQUEST field, 736, 748
request header, 596
request ID field, 965
REQUEST primitive, 690
reroute controls, 959
rerouting and reallocation of trunks, 127
reset (R), 647
resident routing list, 567
residual delay, 501
resistance-capacitance balancing network, 190
resistance design (RD), 86
resistance limit, 71–72
resistance limits for several types of exchange,
      82
resistance noise, 494
response, 618
response header (RH), 596
RESPONSE primitive, 690
response unit (RU), 597
response window, token bus, 634
responsivity, 326
restore failed links, 846
return loss, 192, 193
return loss, 245–246
return loss, 250
return loss (switches), 136

reuse pattern with more than one satellite, 934
reuse plans, 886
revenue-bearing payload, 780
reverse battery signaling, 174, 176–177
reversing the sense, 460
RFI (radio frequency interference), 292–293
RG-11 type cable, token bus, 637
Rician channel, 896–897
right-of-way for fiber optic links, 266
right-through routing, 51
ring-around-the-rosy, 232, 563
ring-back, 168
ringing current source, 105
ringing voltage, 427
ring network, 524, 607
ring scheduling, FDDI, 654
RI (routing information), 651
RLCM (remote line concentrating module)
      (NTI DMS-100), 425
RLR (receive loudness rating), 70
RMON MIB, 968
RMON (remote monitoring), 967–968
RNR (receive not ready), X.25, 557
RNR (receive not ready frame), 547
roaming, 888
roll-call polling, 622–623
rotary switch, 108
rough traffic, 16, 17
rounded off values, 344
round-trip delay, 249, 250, 255, 259
round-trip delay (satellite communications), 295
round-trip echo delay and overall connection
      loss (curve), 256
round-trip echo path, 254
round-trip propagation time, 247
route identifier (RI) field, 658
route-plan hierarchy, 221
routers, 660–661
routers for LAN connectivity, 751
routers more efficient than bridges for
      segmenting, 660
routers slower than bridges, 660
route selection, 230
route (SNA), 592
routing and relaying, 699–700
routing diagrams, 94
routing field (ATM), 781–782
routing information (RI) field, 651
routing label, 833, 842, 843
routing label, 842–844
routing label, 859, 860, 865
routing label structure, 841, 844
routing loops, 51
routing methods, 51–52
routing scheme, 228–230
routing scheme given inlet and outlet points
      and traffic intentsities, 215
routing structure, 228
routing table can be static or dynamic (IP), 567
routing techniques, new, 227–228
routing with common channel signaling, 52

RQ counter, 737, 738
RQ (request count), 737
RRD (revised resistance design), 87
RRE (receive reference equivalent), 64
RR (receive ready frame), 547
RSC (remote switching center) (NTI DMS-100), 425
R-2 signaling, 159, 163, 164
RSL (receive signal level), 276, 280, 284
RSM (remote switching module) (5ESS), 424
RSU (remote switching unit), 259, 496
RTP (rapid-transport protocol), 579, 598
RT (remote terminal), 375
rule of bit stuffing, 478
rules of conventional hierarchical networks, 47–49
RU (response unit), 597
RZ (return-to-zero) transmission, 484, 485

SABME (set asynchronous balanced mode [operation] extended), 693
SABM (set asynchronous balanced mode [operation]), 693
SA field, 640
same signal arrives over several paths, 895
sample and hold, 354
sample at presumed center, 475
sampling, quantization and coding, 340
sampling interval, 481
sampling (PCM), 341–342
SANC (signaling area/network code), 851
S and T functions at BRI, 679
SAPI, 709
SAPI = 31–61, 722
SAP identifier, 790
SAPI = 16 LAPD link, 710
SAP inside a LAN station, 619
SAPI (service access point identifier), 695, 696
SAP (service access point), 536, 616
SAR-PDU format for AAL-1, 795
SAR PDU format for AAL-2, 796
SAR-PDU format of AAL-1, 795
SAR-PDUs, 798
SAR process, 774
SAR (segmentation and reassembly), 533
SAR (segmentation and reassembly), 758
SAR—segmenting of higher-layer information, 793
SA (source address), 626
satellite characteristics of IRIDIUM, 931
satellite circuit limitation, 220
satellite communications, 293–317
satellite indicator, 872, 874
satellite mobile services, overview, 926–928
SBC (sub-band coding), 903
S-bit(s), 363
SCCP connectionless control, 857
SCCP connection-oriented control, 857
SCCP management, 857
SCCP message format, 858
SCCP model for internode communication, connection-oriented services, 854

SCCP protocol facilitates, 854
SCCP routing, 857
SCCP (signaling connection control part), 853–858
scheduling of backoff attempts, 630–631
SCPC FM system, 303, 304
SCPC (single channel per carrier), 303, 315
SCP (service control point), 829
scrambling objectives, 785–786
scratch pad memory, 125
SCR cross-point matrix, 125
SCR (silicon-controlled rectifier), 125
SCR (sustainable cell rate), 817
SCS (SNA character string), 595
SDH-based interface, 786
SDH bit rates with SONET equivalents, 383
SDH multiplexing directly from container-1 using AU-3, 386
SDH multiplexing structure, 390
SDH multiplexing structure, generalized, 384
SDH/SONET envelope, 789
SDH standard bit rates, 383
SDH (synchronous digital hierarchy), 375
SDH (synchronous digital hierarchy), 383–390
SDOC (selective dynamic overload control), 960
SDP (severely disturbed period), 445, 447
S/D (signal-to-distortion ratio), 353, 354
S + D (speech plus derived), 505
SD (start delimiter), 634
SDU (service data unit), 536
SDU (service data unit), 653, 654, 758
SEAL (stands for simple and efficient AAL layer, AAL-5), 798
secondary meaning (signaling), 164
secondary station, 543
second-order products, 185
section overhead, 387
section overhead (SOH), 384
security management, 943
15-segment approximation of North American mu-law, 349
13-segment approximation of the A-law curve, E1 PCM, 348
segmentation and reassembly (SAR), 533
segmentation indicator, 876
segmentation message, 876
segmentation of an IMPDU, 744
segmentation of an L3__PDU into L2__PDUs, 764
segmentation process (DQDB), 743
segmentation (SS7, ISUP), 875–876
segmentation unit, 746
segmentation unit, 763
segmentation unit field, 763
segment CSMA/CD LANs into common user families, 660
segment (DQDB), 736
segmented initial address, 875
segment header check sequence (HCS), 749
segmenting and reassembly (SAR), 700
segment priority, 750

segments per slot time, 742
segment type, 762
segment type field, 796
seized, 165
seizure signal, 165
selecting codes and formats for paging, 921
selection of called party, 875
selective addressing, 914
selective ARQ, 472
selective broadcast signaling virtual channels, 807
selective dynamic overload control (SDOC), 960
selective incoming load control (SILC), 960
selective trunk reservation, 230
selective trunk reservation (STR), 960
selectors (SPC switching), 127
semiconductor laser, 324
semipermanent access, 706
sender link control, 123
sense of transmission, 460
sensitivity of paging receiver, 918
separate channel signaling, 171
sequence number, 546
sequence number, 762
sequence number protection (SNP), 794
sequence numbers verify correct packet order, 557
sequence number (TCP), 576
sequential route selection, 230
sequential write, random read, 406
serial stream of time slots, 409
server holding times, 38
server-pool traffic, 37–39
service access point (SAP), 536
service access point (SAP), 689
service aspects, 708–710
service characteristics, ISDN packet B, 709
service classification for AAL, 794
service classification for the AAL, 793
service data unit (SDU), 536
service data unit (SDU), 758
service features, subscriber related (SPC switching), 129
service indication octet (SIO), 841
service indicator, 833, 841, 859, 865
service primitives, 690
service primitives for connection-oriented services, network, 855
services: connection-oriented and connectionless, 799–803
service subscription time, 729
serving area shape, 78–81
serving exchange, 61
serving time, 37
SES criterion, 446
SES event, 447
SES ratio (SESR), 445
SESR (severely errored second ratio), 445
SES (severely errored seconds), 443, 945
SES (severely error second), 445
session layer, 541–542

session partners, 592
sessions (SNA), 592
set asynchronous balanced mode/set asynchronous balanced mode extended (SABM/SABME), 693
set BECN/FECN bits, 728
set (S), 647
set successor frame, 634
setting performance requirements (LOS microwave), 269
setup and release of VCCs, 805, 807
setup channel, 887
setup message content, 702
setup time required for compelled signaling, 167
seven-cell frequency reuse pattern, 932
several packet types, 555–557
severely-errored cell block ratio, 810
severely errored second cell blocks, 810
severely errored second (SES), 445
SFD, 627
SFD (start of frame delimiter), 626
S format control field (HDLC), 546
S frame (HDLC), 545
S-frame/S-format, LLC, 621
SF (single frequency) circuits, 157
shadowing effects, 887
Shannon classic paper, 499
Shannon GBLC equation, 500
Shannon missed intersymbol interference, 499
Shannon's classical relationship, 912
shape of a serving area, 78–81
shaping of a voice channel and its meaning in noise units, 208, 209–211
short holding-time attempts, 956
short holding-time calls, 953
short-term frequency departure, 439
short-term linear prediction, 903
short-wavelength technology, 612
sidetone, 2
sidetone, 3
sidetone on subset, 427
SIF (service information field), 859
SIF (signaling information field), 841
signal distributor (SPC switching), 127
signal element duration (signaling), 165
signaling, 4
signaling, functional breakdown, 152
signaling area/network code (SANC), 851
signaling connection control part (SCCP), 853–858
signaling data link (layer 1), 835–836
signaling delay, 851
signaling for analog networks, 151–177
signaling information, 832
signaling information, means of conveying, 152
signaling information extraction, 355
signaling information field (SIF), 865
signaling information field (SIF) preceded by the service information octet (SIO), 859
signaling in the subscriber loop, 172–173
signaling limits, 63

signaling link code (SLC), 844
signaling link delays over terrestrial and satellite links, 849
signaling link error monitoring, 838
signaling link function, 831, 832
signaling link layer (layer 2), 836–841
signaling link level functions, 836
signaling link management, 834–835
signaling links, 832
signaling link selection field, 845
signaling link selection (SLS), 844
signaling message handling functions, 829
signaling message handling functions, 832
signaling message-handling functions, 842–844
signaling methods, 864
signaling monitoring equipment, 887
signaling network functions, 843
signaling network functions and messages (layer 3), 842–846
signaling network management, 833–835, 846
signaling network management functions, 832
signaling network status, 846
signaling network structure, 847–848
signaling performance—message transfer part, 848–850
signaling point code, 848
signaling point codes, 847
signaling point delay, 851
signaling points, 828, 832
signaling register holding time in seconds, 38
signaling register holding times, 38
signaling relation, 832
signaling requirements, ATM, 805–808
signaling route management, 835
signaling system for ISDN, 827
signaling system functions, general structure of, 830
signaling system No. 7 implementation, 222
signaling traffic management, 833–834
signaling transfer point, 828
signaling transfer points, 842
signaling transfer point (STP), 833
signaling VCC, 791
signaling virtual channels, 807
signaling virtual channel (SVC), 808
signaling with a conventional telephone subset, diagram, 154
signal magnitude, 187
signal message, 832
signals can exit through windows, 926
signal-to-distortion ratio, 347
signal-to-Gaussian-noise ratio on PCM repeated lines, 360–361
signal-to-impulse-noise ratio, 497
signal-to-noise ratio, 188–189
signal-to-noise ratio, 189
signal-to-noise ratio, 279, 502
signal-to-noise ratio improvement, 495
signal-to-noise ratio (LAN), 609
signal-to-quantizing noise ratio, 391
signal-to-quantizing noise ratio for typical military CVSD operation, 395

signal turns corners, 924
signal unit delimitation and alignment, 836–837
signal unit error rate monitor, 838
signal unit formats, 840
signal variation on a LOS path in a rural environment, 924
SILC (selective incoming load control), 960
simple connection opening (TCP), 573
simple network management protocol (SNMP), 952, **961–967**
simple telephone connection, 2–5
SINAD (signal+interference+noise and distortion to interference + noise + distortion ratio), 901
singing, 245, 246–247
single-channel loading (FDM), 203
single CPE access (SMDS), 752
single-domain subarea network, 590
single-ended control, 434
single-frame operation, 693
single-frequency (SF) signaling, 156
single-frequency (SF) signaling, 158
single-frequency signaling circuit, 156
single-mode fiber (same as monomode), 331
single priority case, DQDB, 736–738
SIO (service indication octet), 841
SIO (service information octet), 859
SIP level 1, level 2, level 3, 757
SIP level 3 PDU format, 759
SIP level 2 protocol data unit format, 761
SIP level 3 protocol data unit format, 759
SIP (SMDS interface protocol), 754
SIP three layers, 757
SIR calculation/equation, 766
SIR (sustained information rate), 756, 766
site/path survey, 286
site selection and prepartion of a path profile, 269–275
site selection (LOS microwave), 269–272
size a telephone exchange, 7–8
size of numbering area and number length, 142–143
skip-route control, 958
skip (SK), 959
sky noise, 305, 306
slant range, 295
SLC-96, 174, 428
SLC (signaling link code), 844
SLC-96 (SLC = subscriber loop carrier), 375
SLC (subscriber loop carrier), 424
SLIC block diagram, 429
SLIC (subscriber line interface card), 428, 429
slip, 432–433
slip-free network, 438
slip performance in holdover, 438
slip performance objectives (CCITT), 450–452
slip rate, 433, 439
slips, 373
slips are major impairment, 449
slope of the analog input signal, 392
slope overload, 392
slope overload of delta modulator, 393

slotTime, 631
slot time, 628–629, 630–631
slot time, token bus, 634
slot timing, single source, 735
SLP (single-link procedure, X.25), 550
SLR (send loudness rating), 70
SLS (signaling link selection), 844, 860
SL_TYPE bit, 747
SMDS, application of TCP/IP, 751, 752
SMDS addressing, 753
SMDS addressing, 754–755
SMDS data units, 753, 754
SMDS data unit transport, 754
SMDS for LAN connectivity, example of use, 751
SMDS in an operating environment, 767
SMDS interface protocol (SIP), 757–765
SMDS network and switching systems, 768
SMDS overview, 753–756
SMDS relationship to DQDB, 752–753
SMDS (switched multimegabit data service), 713
SMDS (switched multimegabit data service), 751–767
SMF (sub-multiframe), 949
SMF (sub-multiframes), 365, 366
SMI (structure of management information), 962
smooth, rough and random traffic, 16–17
smooth, rough and random traffic, 18
smooth earth, 267
smoothing function, FDDI, 655
smooth traffic, 16
SMPU (switching module processing unit) (5ESS), 417
SM (switching module) (5ESS), 415, 416, 417, 424
SMT next station addressing frame, 651
SMT next station addressing frames, 650, 651
SMT protocols, 651
SNA data formats, 596–601
SNA layered architecture, 579
SNA layered architecture, 580
SNA logical units (LUs), 594–596
SNA network configurations, 589–591
SNA networking blueprint, 582
SNAP (subnetwork access protocol), 559, 560
SNA (system network architecture)(IBM), 577–601
Snell's law, 318
SNI (SNA network interconnection) gateways, 588
SNI (subscriber-network interface), 752, 754, 755, 756, 757
SNMP, 961–967
SNMP/AAL, 972
SNMP error codes, 965
SNMP in place, overview, 967
SNMP management architecture, 962
SNMP message, 964
SNMP PDU, 963
SNMP PDU structure, 964

SNMP (simple network management protocol), 563, 952
SNMP trap codes, 966
SNMP trap format, 965
SNMPv.2, 961
SNMP version 2 (SNMPv.2), 968–969
S/N of 20 dB produces BER, token ring, 642
SNP (sequence number protection), 794
S/N ratio produces BER, token bus, 636
S/N (signal-to-noise ratio) (LAN), 609, 610, 611
socket (TCP), 571
SOCOTEL, 161
SOCOTEL MF signaling code, 162
soft decision decoding with FEC, 497
soft decisions, 471
soft hand-offs, 912
SOH (section overhead), 384
solar cells, 296
some properties of cable conductors, 74
SONET, 376–383
SONET: section, line and path definitions, 376
SONET and SDH, 375–390
SONET facility hubs, 438
SONET interface (5ESS), 424
SONET rates and formats, 376–383
SONET/SDH frame, 776
SONET (synchronous optical network), 375
source address field, CLNAP, 802
source address (SMDS), 760
source address validation, 755
source-controlled transmitters, 727
source-filter arrangement, 902
source port, 575
source-routing bridge, 658, 659
source routing (IP), 566
sources and sinks, 5–6
source traffic descriptor, 814
source traffic descriptor, 816, 817
source traffic descriptor, quality of service and cell loss priority, 815
space, 460
space diversity, 290
space diversity, 887
space diversity, 898–900
space diversity concept, 899
space diversity configuration, 290
space-division switch, 404
space-segment charge, 297
space segment (IRIDIUM), 929–930
space-stage time slots, 410–411
space switch, 408–409
space switch array, 411
space switch connects time slots in a spatial configuration, 408
space-time-space switch, 411
spacing bias, 482
spacing condition, 486
SPADE (single channel per carrier PCM multiple-access demand assignment equipment), 303
SPADE terminal block diagram, 303
spanning tee algorithm, 660

span (PCM), 357
spatial diversity, 935
spatial separation of beams, 930
SPC exchange, simplified functional block
diagram, 126
SPC (stored program control), 13
SPC (stored program control), 50
SPC (stored program control), 124–132
SPC (stored program control), 959
SPDU (session protocol data unit), 541
SPE assembly/disassembly process, 380–381
SPE assembly process, 381
spectral distribution for LED and LD, 324
spectrum-spreading code, 913
SPE disassembly process, 382
speech loading (FDM), 204
speech path connection, 108
speeding up the network—frame relay, 713–733
SPE (synchronous payload envelope), 378
splices and connectors, 322–323
splicing equipment, 323
split-band operation, 611
splitting, 534
spontaneous emission, 329, 332, 333
spot beam, 308
spot-beam antennas, 297
48 spot beams, 929
spray reroute, 959
spread bandwidth, 915
spread spectrum multiple access, 910
SP (signaling point), 847
spurs or feeders, 271
SREJ (selective reject) ALOHA, 315
SRej (selective reject frame), 547
SRTS (synchronous residual time stamp), 794
SSAP (source service access point), 559, 619
SSB carrier system, 198
SSBSC (single sideband suppressed carrier), 193
SSB (single sideband), 193
SSCP-dependent LU, 593
SSCP-independent LU-LU protocols, 590
SSCP-SSCP sessions, 593
SSCP (system services control point)(SNA),
589, 590
SSM L2__PDUs, 765
SSM (single segment message), 746, 764
SS 7 network structure model, 829
SS No. 7 architecture, 828–829
SS No. 7 relationship to OSI, 829–831
SS procedures for sending an L2__PDU, 764–765
SS (signaling system), 836
SS (switching system—SMDS), 752
stability, 250
stability, 251, 252
stability in the control of singing, 248
staged growth by cell splitting, 887
standard deviation, 15
standard routing label, 843
standard termination impedances, 76
star configuration, 43
star network, 5, 523, 608

star network, higher-order, 45
start and stop elements, 466
start element, 477
starting delimiter (SD), 646
starting delimiter (SD), FDDI, 649
start-space, 477
start-stop ASCII bit sequence with 2-bit stop
element and parity bit, 529
start-stop (asynchronous operation), 476, 477
start-stop synchronization, 479
start-stop transmission, 477
state-dependent routing, 229–230
state-dependent routing example, 232–233
state transition diagram for sequence of
primitives at connection endpoint, 856
station management, token bus, 633
station margin, 307–308
statistical multiplexers, 530
step-by-step or Strowger switch, 111–113
step-by-step switch (drawing), 112
step-by-step switches, 108
step-index fiber, 321
STM-N frame, 385
STM-N frame structure, 387
STM-1 (synchronous transport module), 383
stochastic vectors, 905
stop-and-wait ARQ, 472, 545
stop element, 476
stop element length, 477
stop-mark, 476
store-and-forward switching, 526
**stored program control exchange, European
design**, 128–130
stored program control (SPC), 50, 113
stored program control (SPC), 124–132
STP, 834
STPs are mated, 829
STP (signal or signaling transfer point), 833
STP (signal transfer point), 672
stratum 2 BITS clocks, 438
stratum 1 clock, 436
stratum 3E clock filters, 436
stratum-level specifications, 435
stratum 1 traceability, 439
streaming mode service, 798
Strowger, Almon B., 101
Strowger switch, 108
STR (selective trunk reservation), 960
structure common to all packets, 552–555
structured data transfer (SDT), 794
structure of management information (SMI),
962
structure of the SCCP, 857–858
STS-1 envelope capacity, 378
STS-1 frame structure (SONET), 377
ST signal, 863
STS-Nc SPE, 382
STS-N electrical signals, 377
STS-N frame, 379
STS-N modules, 379
STS path overhead (POH), 379

STS payload pointers, 382
STS (space-time-space) switch or array, 411
STS switches, 405
STS-1 (synchronous transport system 1) frame, 378
stuffed bits, 478
stuffing allowed variance, 433
stuffing and justification, 367
stuffing concept, 367
subarea network, 584
subarea nodes, 583
subareas, 589
subareas, 590
sub-band coding (SBC), 903
subdivision of the HRX for slip allocation, 452
subhierarchy of DDS signals, 510
sublayering of OSI layer 3 to achieve internetworking, 540
sublayering of the AAL, 793
sub-multiframe (SMF), 949
sub-multiframes (SMF), 365, 366
subscriber density, 78
subscriber-line jacks, 104
subscriber loop, 61, 71, 72
subscriber loop design, 61–77
subscriber loop design techniques, 71–77
subscriber loop impedances, 75–77
subscriber loop limits, 62–63
subscriber loop loss, 66
subscriber loop(s), 4, 6
subscriber-network interface (SNI), 757
subscriber number, 138
subscriber signaling, 151
subscriber tones and push-button codes (North America), 165, 166
subscriber unit, 935
subscriber uplinks and downlinks, 929
subsequent address signals, 863
subservice field, 841
subservice indicator, 860
subset (telephone) diagram, 154
sub-STS-1 payloads, 379
successful call setup, 869
summary of limiting conditions: transmission and signaling, 75
Sun workstation, 521
superframe alignment pattern, 363
super rate payloads, 382–383
supervision, 62
supervision function, 106
supervisory information leaves, 153
supervisory scan (SPC switching), 127
supervisory (S) frame (HDLC), 547
supervisory signaling, 152–155
supervisory (signaling), 151
supervisory signaling, subscriber loop, 173
surface roughness, 890
surveillance and control, 952
survey of subscribers for percentage of unsatisfactory calls, 66
survivability, where network management pays, 944–947
survivability and availability enhanced, 525
survivability and troubleshooting, 944
survivability (long-distance network), 216
sustainable cell rate (SCR), 817
sustained information rate (SIR), 756
sustained information transfer, 755
SVC (switched virtual circuit), 717
switch, 5
switch block, 119
switchboard plug/jack, 174
switched multimegabit data service (SMDS), 751–767
switch in both time and space without blocking, 422
switching, analog, 99–148
switching concepts, introductory, 106–108
switching congestion, 960
switching congestion, inhibit, 954
switching congestion timeouts, 956
switching cost per circuit, 217
switching-delay timeouts, 956
switching functions, basic, 104–106
switching hubs, 661
switching in the telephone network, 99–100
switching matrix, 125
switch number capacity, 102
switch-to-switch load, 237
symbol, 483
symbol rate, 281
sync burst, 301
synchronism problem, 476
synchronization methods, 434
synchronization plans, CCITT, 439–441
synchronize control flag, 576
synchronous data systems, 477–479
synchronous digital hierarchy (SDH), 375
synchronous digital hierarchy (SDH), 383–390
synchronous hierarchical rates, 377–379
synchronous LLC frame, 650
synchronous optical network (SONET), 375
synchronous payload envelope (SPE), 378
syncrhonizing code word (E1), 353
SYNC state, 787
syntax used, 542
Syski, 37
systematically added redundancy, 466
systematic code, 471
systematic distortion, 481
systematic jitter, 373
system capacity, paging systems, 920
system control (switching), 114–118
system depth—a network management problem, 947–952
system network architecture (SNA) (IBM), 577–601
system operation, CSMA/CD, 625–627
system parameters for fiber optic links, 330
system services control point (SSCP) (SNA), 589, 590
system test prior to cutover, 286–287

table of mark-space convention, 460, 461
tail bits, 907
tail sluggishness, 327
tails of pulses, 499
TA interface, 671
talk battery, 3
talk battery, 61, 427
talk-down, 157
talker echo tolerance (curve), 248
tandem, direct or high-usage (THD), 91–92
tandem exchange, 42
tandem exchange, 44, 46, 47, 92, 223
tandem routing, 90–93
tariff areas, 220
tariffs (numbering), 142
TCP connection, 571
TCP defined, 569–570
TCP entity state diagram, 577, 578
TCP header format, 575–577
TCP header format, 576
TCP/IP, 952, 961, 964, 968, 969
TCP/IP and related protocols, 557–577
TCP/IP protocols relate to OSI, 558
TCP/IP working with IEEE 802 series LAN
   protocols relates to OSI, 561
TCP layer provides reliability, 561
TCP mechanisms, 570–571
TCP relates to IP with upper layer PDUs into
   data link layer, 559
TCP relationship with other layered protocols,
   570
TCP scenario, 572–575
TCP segments, 570
TCP (transmission control protocol), 557
TC (transmission convergence), 788
TC (transmission convergence) sublayer, 785
TCXO (temperature-compensated crystal
   oscillator), 436
TDMA burst format, 300
TDMA carrier, 317
TDMA efficiency, some comments on, 909
TDMA format, 932
TDMA frame, 300
TDMA frame with time slot blow-up, 907
TDMA improved capacity, 909
TDMA-like slot boundaries, 315
TDMA (time division multiple access), 297,
   906–910
TDMA (time division multiple access) (LAN),
   623
TDM (time division multiplex) bit stream
   (VSAT), 316
TDR (time-domain reflectometer), 947
T1 (DS1) PCM system, 342
TE1, TE2 (TE = terminal equipment), 671
TEI, 709
TEI (terminal endpoint identifier) subfield,
   695, 696
telecommunication networks, 1–2
telegraph channel, 193
telephone connection, 2–5
telephone connection, long-distance (model), 190

telephone density per 100 inhabitants, 221
telephone network, simplified, 64
telephone networks, 6–7
telephone numbers, 77
telephone routing plan, 851
telephone traffic, 9
telephone traffic measurement, 144–146
telephone user part (TUP), 859
telephone user part (TUP), 860–863
telephony, basics of conventional, 1–57
ten DNHR time periods, 237
terminal G/T, 309
terminating jacks, 104, 105
term set, 190
terrain topology, 267
terrestrial station, 294
tertiary center, 224
testing the hypothesis, 946–947
test level point, 187
**TE1/TA operating behind an NT2 that is not
   synchronized to network clock**, 686
TG (transmission group), 588
THDR (transport header), 598
THD (tandem, direct or high-usage), 91–92
theoretical final route, 47
thermal noise, 184–185
thermal noise, 502
thermal noise level, 279
thermal noise (PCM), 374
thermal noise power level, 279
the satellite, 294
third-order products, 185
three and six sector plan, 917
three-bit link status indicators, 841
three layers of the SIP, 757
three layers of X.25, 550–552
three-PHY-symbol code, 635
three-priority-level distributed queue access
   mechanism, 747
three-stage grid network, 120
three-stage interconnection, 222
three-state duobinary AM/PSK, 638
three TDMA systems compared, 911–912
three-way handshake (TCP), 572
threshold current (fiber optics), 333
threshold for given BER, 332
throughput, 468, 530
throughput values, 528
THT (token-holding timer), FDDI,
   652
TIA IS-54, 935
TIE (time interval error), 439
time and space arrays, 412
time availability, 287
time availability, 288
time congestion, 19
time consistent busy hour, 9
time-dependent routing, 229
time-dependent routing example, 233
time dissemination, 435
time dissemination services, 436
time diversity, 898

time division multiple access (TDMA), 297
time division multiple access (TDMA), 906–910
time-division multiple access (TDMA), 300–302
time division multiple access (TDMA) (LAN), 623
time division multiplex, 193
time-division multiplexers, 531
time-division space switch cross-point array showing enabling gates, 408
time-division switch, 404
time-divison switching, 405–412
time-domain reflectometer (TDR), 947
timed token rotation protocol, 654
time fill, 687–688
time/frequency/spatial reuse, 930
time interval error (TIE) and frequency departure, 439–441
time-keeping coincidence, 433
time of dwell, 413
time-out, 170
timeouts, 956
timer T7, 873, 874, 875
timer T34, 876
time scale of round-trip propagation time, 812
time slot 0, 949
time slot 16, 836
time-slot arrangement, 406
time slot 24 assigned to D channel, 687
time slot assignment, 687–688
time-slot counter, 421
time-slot counter or processor, 406
2048-time-slot domain, 414
time-slot idle or busy, 406
time-slot interchange: time switch (T), sequential write, random read, 406
time-slot interchanger, 409, 410
time-slot interchanger switch, 413
time-slot interchange (TSI), 405
time-slot interchanging, 406, 407
time-slot interval, 409
time-slot rate, 414
time slot represents virtual channel, 420
time slot synchronization, 431
time-space-time switch, 409–410
time-space-time (TST) switch, 410
time stamp, 966, 967
time switch, 405–408
time-switch, time-slot interchange (T), random write, sequential read, 407
time synchronization with TDMA, 300
time to live, 568
time-to-live, routers, 660
time-to-live (TTL), 563
time unavailability, 286
timing, 479–481
timing accuracy, 479
timing advance, 908
timing and synchronization requirements, 483
timing considerations (BRI), 681
timing considerations (PRI E1), 689
timing degradations, 373
timing distribution, 438

timing error, 476
timing extracted by phase-locked loop, 642
timing extractor, 362
timing filter bandwidth, 373
timing from bit transitions, 432
timing from network clock, 681
timing jitter, 447–448
timing jitter, maximum permissible at Non-SONET interfaces, 448
timing (network) traceable to stratum 1 clock, 685
timing recovery, 354
timing relation between source and destination, 793
timing signals displaced from optimum position, 448
timing source traceable (DQDB), 735
timing stability and accuracy (data), 475
tip, ring and sleeve, 173, 174
0 TLP, 187, 188
0 TLP (zero test level point), 136
TM (trunk module) (NTI DMS-100), 423
T2.1 nodes (SNA), 585
T4 nodes (SNA), 585
token, a unique signal sequence, FDDI, 643
token bit T, 639
token bus, 632–638
token bus MAC frame format, 635
token bus network extension by regenerative repeaters, 635
token bus shows logical connectivity, 633
token format for token ring operation, 639
token holding station, 632
token holding timer, 641
token MAC frame, 634
token passing, predictable performance, 655
token passing direct outgrowth of polling, 623
token passing ring network, 638
token passing schemes have controlled access, 656
token ring, 638–642
toll-access code, 138
toll area, 6
toll-center placement, 216
toll-connecting trunks, 65
toll exchange, 216, 220
toll traffic matrix, 224
tone data channels, 505
tone-on when idle, 157
tonterias, 947
topographic maps, 272
topology, 517–518
torn-tape systems, 526
total length field (IP), 562
tower height, 275
tower heights, 272–274
TPDU (transport protocol data unit), 541
(T+R)/2, 65
tracing the life of a virtual call, 557
tracking, a disadvantage of LEO, 928
traffic, 5–6

traffic and billing on frame relay, 724–726
traffic calculations by computer programs, 31–37
traffic-carrying DS1s, 438
traffic characteristics, SS7, 849
traffic contract, 812
traffic contract parameter specification, 817
traffic control and congestion control, ATM,
   811–817
traffic control and congestion control, reference
   configuration, 812
traffic density, 9, 10
traffic descriptors, 813, 814
traffic engineering essentials, 7–17
traffic intensity, 7–8
traffic intensity, 9–10
traffic intensity over a typical working day, 8
traffic load and blocking probability, 886
traffic matrix, 223
traffic matrix, sample, 95
traffic measurement, 144–146
traffic parameters and descriptors, 813–814
traffic path, 7
traffic peakiness, 41
traffic peaks, 52
traffic policing, 791
traffic probability distributions, 16, 18
traffic routing in the national network, 227–237
traffic shaping, 791
traffic sources, 17
traffic table: full-mesh connection, 92
traffic to circulate indefinitely, 657
trailer, 743
trailing zero octets, 743
training sequence, 907, 908
transfer of an IMPDU, 746
transit center, 48
transit exchange, 46, 47
transit exchanges, 216
transition, 353
transition region, 486
transition richness, 484
transitions placed in a time reference, 480
transition timing, 480
transit switched traffic, 220
transit traffic, 243
translating of VCI values, 804
translation bridge, 658
translator, 122
transmission, 5
transmission control layers, communications
   between two (SNA), 581
transmission control protocol (TCP), 569–577
transmission convergence sublayer functions,
   789–790
transmission cost per circuit, 217
transmission design to control echo and singing,
   247–248
transmission factors in analog switching, 135–137
transmission factors in long-distance telephony,
   244–260
transmission for telephony, 181–212

transmission group (TG), 588
transmission header (SNA), 597
transmission highway, 265
transmission limit, 71, 72
transmission line losses, 282
transmission loss through reinforced concrete,
   926
transmission medium requirement parameter,
   873
transmission parameters for IDR carriers with
   rate 3/4 coding, 313
transmission parameters (SS7), 849
transmission plan, 88
transmission plans, national, 248, 249
transmission requirements, CSMA/CD, 629–632
transmission without contention, 625–626
transmit and receive intervals, 931
transmit reference equivalent (TRE), 64
transmitter power, 277
transmitter power, LOS microwave, 276
transmitter requirements, 686
transparency (frame relay), 720
transparency (LAPD), 695
transparent, 478
transparent bridge, 652, 658
transponder block diagram of a typical satellite,
   294
transponders in satellite, 297
transporting ATM cells, 817–822
transport layer, 540–541
transport network, 591
transport network and network accessible units
   (NAUs), 591–594
transport overhead, 378, 379
transversal equalizer, 503
trap format, 963
trap manager, 968
tree configuration, 609
tree network, 606
tree topology (broadband LAN), 611
trend away from hierarchical structure, 50–51
TRE (transmit reference equivalent), 64
tributary unit group (TUG), 386
tributary unit-n (TU-n), 385
tributary unit pointer, 385
triggering level, 960
TRI-TAC digital hierarchy, 395
TRI-TAC digital multiplexer, 394, 395
TRLR (trailer), 758
troubleshooting, 944–947
TRT (token rotation timer), 653
truncated binary exponential backoff, 631
trunk congestion, 960
trunk-group overload, 957
trunk groups, 107
trunk-hunting functions, 120
trunk loading capacity based on Erlang B
   formula, full availability, 21–26
trunk loading capacity based on Poisson formula,
   full availability, 27–30
trunk losses with VNL, 258

trunk prefix, 138, 140
trunk reservation, 960
trunks, 6
trunk signaling, 104
trunk-signaling equipment, 153
trunks in tandem, 255
trunk stability, 254
TSG (timing signal generator), 437
TSI (time-slot interchange), 405
TSI (time-slot interchanger), 416
TST architecture of AT&T 5ESS, 416
TST compared to STS, 411–412
TSTS-folded, 405
TSTS-folded architecture, 414
TST switches, 405
TST switch is non-blocking—, 410
TST (time-space-time) switch, 410
T-symbol (FDDI), 646
TTL (time-to-live), 563
TUG-2, 387
TUG (tributary unit group), 386
TU-n pointer, 390
TUP (telephone user part), 831, 844, 859, 860
turnaround time: frame relay vs. X.25, 715
TVX (valid-transmission timer), 653
twist and sway of towers, 289
two administrative units are defined, 385
two broad categories of LAN transmission techniques, 608–613
two-clock problem, 438
two-frequency pulse signaling, 159–161
two-frequency signaling, 157
two-link DNHR with crankback, 236
two-motion switch, 112
two-out-of-five code, 122
two-out-of-six tone frequencies, 163
two-stage grid network, 119
two-stage hierarchy, 221
two-stage interconnection with low initial fan out, 222
two-stage multiplex (SDH), 389
two-way circuit, 52, 53
two-way circuits (both way circuits), 227
two-way data transfer (TCP), 574
two-wire and four-wire switches, 101
two-wire and four-wire transmission, 189–193
two-wire duplex modems, 508
two-wire/four-wire operation, 244
two-wire interface (BRI US), 681
two-wire subscriber lines, 245
two-wire-to-four-wire conversion, 192
TWT (traveling-wave tube), 301
TxSN (transmit sequence number), 765
Type 1, Type 2 operation, LLC, 619
Type 6.2 LU, 594
type of service (IP), 562
types and classes of LLC operation, 619–620
typical levels, 187–188
typical Rayleigh fading envelope and phase in mobile scenario, 897

UC (unit call), 19
UDLC (universal digital loop carrier), 429
UDP/IP stack, 963
UDP port 161/162, 966, 968
UDP use by SNMP, 968
UDP (user datagram protocol), 963
U format control field (HDLC), 546
U frame (HDLC), 545
U-frame/U-format, LLC, 622
U-interface, 670, 681
UI (unit interval) peak-to-peak, 448, 450
ULP channels data through TCP, 570
ULP modes (TCP), 572
ULP synchronization, 571–572
ULP (upper layer protocol), 566
ULP (upper layer protocol(s)), 562
ULP (upper layer protocol) (TCP), 572, 573
UME (UNI management entity), 971, 972, 974
unacknowledged and acknowledged operation, 692
unacknowledged frames, 546
unavailability, 287
unavailability, total for any signaling route, 852
unbalanced configuration, 543
uncontrolled access, 781
underflow condition, 432
undetected errors (SS7), 849
UNI (user-network interface) boundaries, 777
uniform numbering, 139
uniform numbering, 170
uniform-spectrum random noise, 212
uniform system (numbering), 140
UNI header, 781
UNI ILMI attributes, 972
UNI ILMI MIB structures, 972
unintelligible crosstalk, 186
unipolar RZ, 484
unique field, 477
unique field aligns frame, 478
unique word (TDMA), 301
unit call, 10
unit interval (UI), 439, 440
UNI (user-network interface), 777, 778, 779, 780, 971, 972
universal digital loop carrier (UDLC), 429
unknown DLCI value, 733
unnumbered frame, 547
unsystematic redundancy, 466
unwanted variations in zero-crossings, 498
unweighted loss (a-b) on established connections, 253
UPC/NPC complexity, 816
UPC/NPC mechanism, 815
UPC/NPC operation, 813
UPC/NPC (usage/network parameter control), 811
update interval length, 243
updating routing patterns, 229
U-plane physical path, 726

uplink carrier, 299
uplink considerations, 310
uplink/downlink, 296
up-link power levels, 301
upstream vs. downstream (LAN), 611
upward multiplexing, 534
urban delay spread, 909
urgency of data, 527
urgent pointer field significant, 576
urgent pointer (TCP), 577
usable wavelengths for fiber optic transmission, 318
usage/network parameter control (UPC/NPC), 811
usage (telephony, traffic), 10
useful bits, 530
useful data, 468
use of SNA data formats, 599
user access and interface, 669–671
user access capabilities, ISDN packet B, 709
user information, CLNAP, 803
user-network interface (UNI) configuration and architecture, 776–779
user-network signaling, 675
user-network traffic contract, 814–816
user part of SS No. 7, 673
user parts, 858–878
user plane (ATM), 779
user response to congestion, 727
user-to-network signaling procedures, 805
user-VCs controlled by a signaling entity, 807
UTC (universal coordinated time), 434, 436, 439

value of credit decremented, 765
value of sequence number in BOM, 745
variability of offered load, 241
variable automatic equalizer, 503
variable binding pairs, 966
variable bindings, 964
variable length PAD, 743
variable loss plan, 249
variance, 15
variations in traffic flow, 52
VBR scenario in AAL2, 795
VBR (variable bit rate services), 775
VC and VP relationships and transmission path, 804
VC and VP switching, representation of, 806
VCC level management functions, 783
VCC (virtual channel connection), 783
VCI and VPI, 804
VCI value, 750
VCI (virtual channel identifier), 743
VCI (virtual channel identifier, ATM), 781
VC switch/cross-connect, 804
velocity of propagation, 73, 183, 249, 256, 257, 479
velocity of propagation of the medium (SS7), 849
version field (IP), 562
version identifier, 964
vertical antenna separation, 898

vertical redundancy check (VRC), 469
very small aperture (VSAT) networks, 311–317
VFCT (voice frequency carrier telegraph), 504
VF pair usage, 356
2VF supervision, 157
VFTG (voice frequency carrier telegraph), 504
VF trunk design, 88–89
VF trunks, 88
2VF (two frequency) signaling, 156
VF (voice frequency) repeater diagram, 90
VHSIC (very high speed integrated circuit), 405
via net loss factor (VNLF), 256, 257
via net loss (VNL), 253–259
violation concept in PCM line codes, 360
virtual analog switching points, 251
virtual channel identifier, 552
virtual channel identifier (VCI), 749
virtual channel level, 804
virtual circuit service (ISDN), 705
virtual circuits using the X.25 protocol, 709
virtual connection, 527
virtual connections, 551, 552
virtual container-n (VC-n), 384
virtual path connection (VPC), 791
virtual path level, 805
virtual tributaries, 379–380
VLSI (very large scale integration), 405
VNL circuits, 255
VNLF (via net loss factor), 256, 257
VNL trunk losses, 258
VNL (via net loss), 246, 248, 249
VNL (via net loss), 253–259
vocal-tract linear-predictive model, 903
vocoders, 902
voice channel, 181
voice-channel sharing techniques, 531
voiced speech, 902
voice-frequency carrier telegraph (VFCT), 504–506
voice-frequency repeaters, 89–90
voice-grade line, 527
VP and VC identifier, 790
VPC endpoints, 805
VPC operation flow (F4 flow), 791
VPC (virtual path connection), 791
VPI=0, 807
VPI/VCI value, 807, 972
VPI (virtual path identifier), 781
VP switch/cross-connect, 806
VP switching, representation of, 806
VRC (vertical redundancy check), 469
VSAT handicap, 314
VSAT hub, 315
VSAT network characteristics, 314–316
VSAT networks, 525
VSAT network topology, 314
VSAT system operation, 317
VSAT transponder operation, 316–317
VSAT (very small aperture terminal), 267
VSAT (very small aperture terminal) links, 311
VSB (vestigial sideband modulation), 489, 490

VSELP (vector sum excitation linear predictive (vocoder)), 935
VTAM product, 587
VTAM (virtual telecommunications access method), 577, 578
VT frames, four types, 380
VT1.5 packaged in an STS-1 SPE, 381
VU meter, 187-188
VU (volume unit), 203, 204, 205

waiting systems (queueing), 37-39
wander, 373, 448-449
wander and jitter gating issues, 405
WAN (wide area network), 517, 519
waveforms, digital data transmission, 484
waveguide dispersion, 331
waveguide dispersion, 332
WA (write address) (NTI DMS-100), 422
weighted channel, 210
weighted figure for return loss, 254
weighting factor, 212
weighting network, 208
well-known sockets (TCP), 571
white noise, 185, 212, 494
who follows frame, 634
wide area network (WAN), 517, 519
Wilkinson, 41
window of frames, 545-546
window of sequenced-numbered data, 571
window (TCP), 577
wireless PCS technologies, 923

withholds acknowledgements, 839
worst-case BER, 311
worst-case fading, 287, 288
worst quantizing error (distortion), 346

X.75, 549
X.25 architecture and its relationship to OSI, 548-550
X.25 call request packet, 708
X.25 data communication equipment, 672
X.25 functions for layer 2, 707
XID frames, 727
X.25 layer 2 and layer 3 functions, 706
X.25-like switched virtual circuits, 717
X.25 link layer, 551
X.25 packet communications operates with PDN, 549
X.25 packet layer, 552-557
X.25 packet-switched access standard, 548-557
X.25 user (DTE) connects through PDN to a distant user (DTE), 550
X.25 virtual circuits, 548

zero crossings of square wave input, 362
zero-dispersion wavelength, 331
zero dispersion window shifted monomode fiber, 331
zero insertion, 478, 545
zero test-level point, 136
zero transmission-level point, 495
zone beam, 308